Nanostructure Science and Technology

Series editor

David J. Lockwood, FRSC
National Research Council of Canada
Ottawa, Ontario, Canada

More information about this series at http://www.springer.com/series/6331

Kenneth I. Ozoemena • Shaowei Chen
Editors

Nanomaterials for Fuel Cell Catalysis

Springer

Editors
Kenneth I. Ozoemena
Council for Scientific and
Industrial Research
Pretoria, South Africa

Shaowei Chen
Department of Chemistry and Biochemistry
University of California
Santa Cruz, CA, USA

ISSN 1571-5744 ISSN 2197-7976 (electronic)
Nanostructure Science and Technology
ISBN 978-3-030-10412-2 ISBN 978-3-319-29930-3 (eBook)
DOI 10.1007/978-3-319-29930-3

Printed on acid-free paper

This Springer imprint is published by Springer Nature
The registered company is Springer International Publishing AG Switzerland

Preface

Ready availability of sufficient energy resources is critical in virtually every aspect of our life. Whereas fossil fuels have remained our primary energy sources, extensive efforts have been devoted to fuel cell research in the past few decades, which represents a unique technology that will make substantial contributions to our energy needs by converting the chemical energy stored in small (organic) molecule fuels into electricity and, more importantly, exert minimal negative impacts on the environment. In fuel cell electrochemistry, the reactions typically involve the oxidation of fuel molecules at the anode and reduction of oxygen at the cathode. Both reactions require appropriate catalysts such that a sufficiently high current density can be generated for practical applications. Precious metals, in particular, the platinum group metals, have been used extensively as the catalysts of choice. Yet their high prices and limited reserves have severely hampered the widespread commercialization of fuel cell technologies. Therefore, a significant part of recent research efforts has been focused on the development of effective electrocatalysts with reduced or even zero amounts (and hence costs) of precious metals used that exhibit competitive or even improved electrocatalytic performance as compared to state-of-the-art platinum-based catalysts. Toward this end, it is imperative to understand the fundamental mechanisms involved such that an unambiguous structure–activity correlation may be established, from which the activity may then be further enhanced or even optimized. It should be recognized that the reaction mechanisms in fuel cell electrochemistry are rather complicated and not fully understood. Yet advances on the theoretical and experimental fronts have yielded significant insights which offer important guidelines in the design and engineering of fuel cell catalysts.

It is within this context that this book volume is conceived and developed. The chapters are written by some of the leading experts in the field. The key goal is to highlight recent progress in electrocatalysis at both fuel cell anode and cathode, with a focus on the impacts of the design and engineering of electrode catalysts on the catalytic performance. This is a critical first step toward the establishment of a structure–activity correlation. A significant portion is devoted to oxygen reduction reaction, as this has been recognized as a major bottleneck that largely determines the overall fuel cell performance due to its sluggish electron-transfer kinetics and complicated reaction pathways. Relevant research is typically centered around three aspects. The first involves alloying and surface ligand engineering of platinum-based electrocatalysts (the conventional catalysts) through the so-called electronic and geometrical contributions such that the costs may be reduced and concurrently the performance improved. The second entails the development of non-platinum precious metal nanoparticle catalysts. The third is focused on cheap transition-metal oxides or totally metal-free catalysts (e.g., doped carbons). In the last two, while the costs of the catalysts may be substantially reduced as compared to those of the platinum-based counterparts, their performances have mostly remained subpar. Thus, how to further improve their activity is a leading challenge in the field.

Similar issues have been found with the anode reactions. Conventionally, small (organic) molecules, such as hydrogen, methanol, and formic acid, have been used as potential fuels. More recently, C2 molecules such as ethanol and ethylene glycol have also been attracting extensive interest because of their ready availability, low toxicity, and high energy density. Yet the oxidation mechanisms involved are far more complex than those of the C1 counterparts (namely, methanol and formic acid), leading to reduced efficiency in the reaction. Therefore, a major thrust of current research is to unravel the reaction mechanisms involved such that the catalytic performance may be further improved and ultimately optimized.

Whereas these issues represent daunting challenges, one may also choose to accept them as unique opportunities where breakthroughs will help advance fuel cell technologies toward commercial applications. This is no doubt a multidisciplinary endeavor, including materials science, (electro)chemistry, interfacial engineering, and so on. It is our hope that this book will offer a unique glimpse of the state of the art of fuel cell electrocatalysis and therefore may serve as a technical reference for researchers at all levels in the areas of nanoparticle materials and fuel cell technologies.

Santa Cruz, CA, USA Shaowei Chen
Pretoria, South Africa Kenneth I. Ozoemena

Contents

1 **Electrochemistry Fundamentals: Nanomaterials Evaluation and Fuel Cells** 1
Neil V. Rees

2 **Recent Advances in the Use of Shape-Controlled Metal Nanoparticles in Electrocatalysis** 31
Francisco J. Vidal-Iglesias, José Solla-Gullón, and Juan M. Feliu

3 **Pt-Containing Heterogeneous Nanomaterials for Methanol Oxidation and Oxygen Reduction Reactions** 93
Hui Liu, Feng Ye, and Jun Yang

4 **Synthesis and Electrocatalysis of Pt-Pd Bimetallic Nanocrystals for Fuel Cells** 169
Ruizhong Zhang and Wei Chen

5 **Integrated Studies of Au@Pt and Ru@Pt Core-Shell Nanoparticles by *In Situ* Electrochemical NMR, ATR-SEIRAS, and SERS** 225
Dejun Chen, Dianne O. Atienza, and YuYe J. Tong

6 **Recent Development of Platinum-Based Nanocatalysts for Oxygen Reduction Electrocatalysis** 253
David Raciti, Zhen Liu, Miaofang Chi, and Chao Wang

7 **Enhanced Electrocatalytic Activity of Nanoparticle Catalysts in Oxygen Reduction by Interfacial Engineering** 281
Christopher P. Deming, Peiguang Hu, Ke Liu, and Shaowei Chen

8 Primary Oxide Latent Storage and Spillover for Reversible Electrocatalysis in Oxygen and Hydrogen Electrode Reactions 309
Milan M. Jaksic, Angeliki Siokou, Georgios D. Papakonstantinou, and Jelena M. Jaksic

9 Metal-Organic Frameworks as Materials for Fuel Cell Technologies 367
Henrietta W. Langmi, Jianwei Ren, and Nicholas M. Musyoka

10 Sonoelectrochemical Production of Fuel Cell Nanomaterials 409
Bruno G. Pollet and Petros M. Sakkas

11 Direct Ethanol Fuel Cell on Carbon Supported Pt Based Nanocatalysts 435
T.S. Almeida, N.E. Sahin, P. Olivi, T.W. Napporn, G. Tremiliosi-Filho, A.R. de Andrade, and K.B. Kokoh

12 Direct Alcohol Fuel Cells: Nanostructured Materials for the Electrooxidation of Alcohols in Alkaline Media 477
Hamish Andrew Miller, Francesco Vizza, and Alessandro Lavacchi

13 Effects of Catalyst-Support Materials on the Performance of Fuel Cells 517
Paul M. Ejikeme, Katlego Makgopa, and Kenneth I. Ozoemena

14 Applications of Nanomaterials in Microbial Fuel Cells 551
R. Fogel and J.L. Limson

Index 577

Chapter 1
Electrochemistry Fundamentals: Nanomaterials Evaluation and Fuel Cells

Neil V. Rees

1.1 Introduction

The field of electrochemistry has enjoyed a renaissance in the last two decades as its relevance to the ever-expanding realm of nanoscience has been appreciated. As new nanomaterials have been discovered and designed by physical science, many of their properties of interest have been associated with electron transfer, as solid state devices or catalysts, most of which can be probed and investigated via electrochemical techniques. It is in the latter case that we focus this chapter, in particular to fuel cell catalysis. Clearly the aim of a fuel cell is to achieve full oxidation of its fuel to extract the maximum thermodynamic output, and we therefore require efficient catalysts in order to achieve this at low to intermediate temperatures (roughly 333–413 K). Electrochemistry plays two roles in the case of fuel cells therefore: first in characterising and investigating candidate catalysts, and second in understanding and optimising the reduction and oxidation ("redox") processes occurring in the fuel cell itself. This chapter will not attempt to cover the whole of this field: there are many excellent texts on physical electrochemistry [1–4] and a growing number of similarly authoritative works on the fundamental science of fuel cells [5–8], but will provide a brief survey of the key concepts and illustrations from recent literature that those of us working in the field of fuel cell science should be aware.

N.V. Rees (✉)
School of Chemical Engineering, University of Birmingham, Birmingham B15 2TT, UK
e-mail: n.rees@bham.ac.uk

K.I. Ozoemena, S. Chen (eds.), *Nanomaterials for Fuel Cell Catalysis*,
Nanostructure Science and Technology, DOI 10.1007/978-3-319-29930-3_1

1.2 Part I: Ex-Situ Electrochemistry

The aim of ex-situ testing of candidate materials for fuel cells is typically to simplify conditions such that variables can be carefully controlled. In the majority of cases the testing takes the form of solution-phase cyclic voltammetry of candidate electrocatalysts and so the degree of catalytic activity needs to be ascertained. In order to understand how this can be performed correctly, avoiding many common pitfalls, we shall first review the underlying physical electrochemistry of the cyclic voltammetric response.

1.2.1 The Electrode Interface

In general, any interface comprising of two different phases will develop a potential difference due to different electrical potentials of the two phases. This is true also for an electrode placed in a test solution. The existence of a potential difference therefore leads to an electric field gradient within the solution close to the electrode; where there are ions in the solution they migrate under the influence of this field such that an electrical double layer is rapidly established, with the Gouy-Chapman-Stern model most commonly used to describe it (see Fig. 1.1). Clearly as the concentration of ions in solution increases, so the diffuse layer becomes more compressed.

Changes to the electrode interface due to adsorption, etc, can be detected via the flow of non-Faradaic currents. These are current flows required to maintain electroneutrality, and are distinct from Faradaic currents which are associated with electron transfer to/from electroactive species to effect reduction/oxidation.

1.2.2 Mass Transport

Transport of materials through a fluid can occur via diffusion along a concentration gradient, migration along an electric field gradient, natural convection due to thermal gradients, or forced convection (due to stirring, flow, etc. deliberately imposed on the system).

Diffusion is typically described via Fick's Laws

$$j = -D\frac{\partial c}{\partial x} \tag{1.1}$$

$$\frac{\partial c}{\partial t} = D\frac{\partial^2 c}{\partial x^2} \tag{1.2}$$

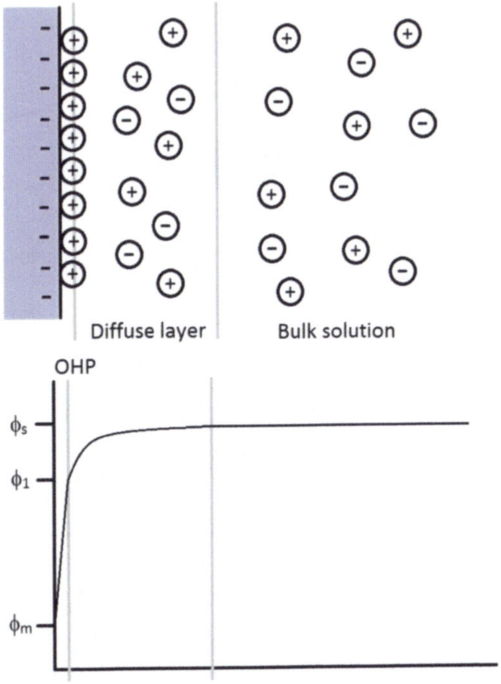

Fig. 1.1 The Gouy-Chapman-Stern model of the electrical double layer, with adsorbed counterions forming the outer Helmholtz plane (OHP), a surfeit of counterions in the diffuse layer and then bulk solution. Accordingly, the potential drops from the electrode potential (ϕ_m) to the solution potential (ϕ_s)

Diffusion to an electrode is also described in terms of the degree of development of the diffusion field around the electrode. As the electrode reaches a potential where electron transfer occurs and so a concentration gradient is established, diffusion is initially linear (or one-dimensional) and the diffusion field gradually develops such that eventually it becomes convergent (see Fig. 1.2).

The time taken for the establishment of convergent diffusion, t_{conv}, can be estimated by the expression [4]

$$t_{conv} \approx \frac{r^2}{D} \tag{1.3}$$

Migration effects are often undesired in ex-situ experiments, and so an excess of inert (or supporting) electrolyte is commonly added to eliminate them. However, in the absence of excess electrolyte (i.e. less than full support), then migration will become significant and needs to be accounted for via the Nernst-Planck and Poisson equations [9]

$$\frac{\partial C}{\partial t} = D\left(\nabla^2 C + \frac{zF}{RT}\left(C\nabla^2\phi + \nabla C\nabla\phi\right)\right) \tag{1.4}$$

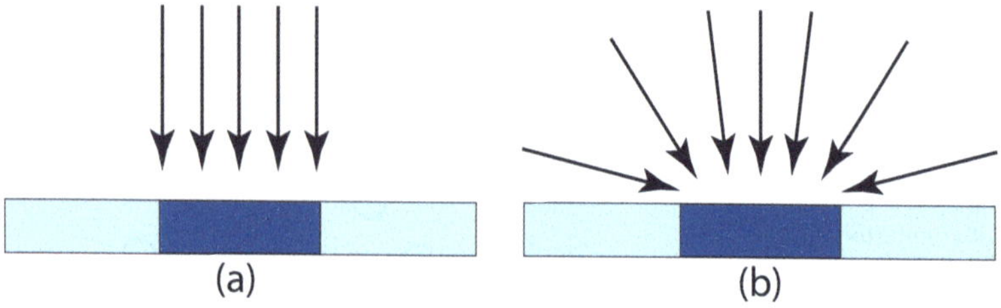

Fig. 1.2 (**a**) Linear and (**b**) convergent diffusion fields to an electrode

$$\nabla^2\phi = -\frac{\rho}{\varepsilon_s\varepsilon_0} \tag{1.5}$$

where z is the species charge, ϕ is the potential, ε_s is the dielectric constant of the solvent medium, ε_0 is the permittivity of free space, and ρ is the local charge density, found from summing all local charges present

$$\rho = F\sum\nolimits_i z_i C_i \tag{1.6}$$

Natural convection effects are notoriously unpredictable and so are generally eliminated by thermostatitng of the electrochemical cell and conducting the experiment over as short a time as practicable, where a timescale of 20–30 s is usually held to be the maximum desired [10].

Forced convection is imposed deliberately on the system in order to increase mass transport. As such the formed of convection is usually chosen such that the hydrodynamics of the system are well-defined [11].

1.2.3 Electrode Kinetics

The simplest case is that of fully reversible behaviour (i.e. where the electron transfer kinetics are effectively infinitely fast on the timescale of the mass transport), where the Nernst equation will hold at all times at the electrode surface [1].

For quasi-reversible and irreversible systems, account must be made of the non-Nernstian condition at the electrode surface and this is most commonly achieved via the Butler-Volmer equations

$$k_f = k_0 e^{\frac{-\alpha F\eta}{RT}} \tag{1.7}$$

$$k_b = k_0 e^{\frac{+\beta F\eta}{RT}} \tag{1.8}$$

where $\alpha + \beta = 1$, and $\eta = E - E^0$.

These are routinely available in commercial simulation packages and are phenomenological insofar as there is limited molecular insight gained from their results. More theoretically informative would be the use of Marcus-Hush theory [12–18], which can be used in a predictive sense as well as modelling experimental data. Here the reductive and oxidative rates constants (k_f and k_b respectively) can be expressed as

$$k_f = \int_{-\infty}^{+\infty} A_{red}(\varepsilon) \frac{exp\left\{-\Delta G^{\neq}_{red}(\varepsilon)/k_B T\right\}}{1 + exp\{-e(\varepsilon - E)/k_B T\}} d\varepsilon \tag{1.9}$$

$$k_b = \int_{-\infty}^{+\infty} A_{ox}(\varepsilon) \frac{exp\left\{-\Delta G^{\neq}_{red}(\varepsilon)/k_B T\right\}}{1 + exp\{+e(\varepsilon - E)/k_B T\}} d\varepsilon \tag{1.10}$$

where e is the electronic charge, k_B the Boltzmann constant, $A_{red/ox}(\varepsilon)$ is the pre-exponential factor that includes the influence of the electronic states on metallic electrode and electroactive species and the density of states of the electrode. If A is assumed to be independent of energy, ε, then the general form for the reductive and oxidative rate constants can be written as

$$k_f = k_0 \left\{ \frac{I_{red}(\eta^*, \lambda^*)}{I_{red}(0, \lambda^*)} \right\} \tag{1.11}$$

$$k_b = k_0 \left\{ \frac{I_{ox}(\eta^*, \lambda^*)}{I_{ox}(0, \lambda^*)} \right\} \tag{1.12}$$

where η^*, λ^* are dimensionless overpotential and reorganisation energy given by $\eta^* = \frac{F}{RT}\left(E - E_f^0\right)$ and $\lambda^* = \frac{F\lambda}{RT}$, respectively. $I(\eta^*, \lambda^*)$ is an integral of the form

$$I_{red}(\eta, \lambda^*) = \int_{-\infty}^{+\infty} \frac{exp\left\{-\Delta G^{\neq}_{red}(\chi)\right\}}{1 + exp\{-\chi\}} d\chi \tag{1.13}$$

$$I_{ox}(\eta, \lambda^*) = \int_{-\infty}^{+\infty} \frac{exp\{-\Delta G^{\neq}_{ox}(\chi)\}}{1 + exp\{+\chi\}} d\chi \tag{1.14}$$

where $\chi = \frac{F}{RT}(\varepsilon - E)$.

It is most commonly encountered in its symmetric representation [19, 20], where

$$\Delta G^{\neq}_{sym,red}(\chi) = \frac{\lambda^*}{4}\left(1 + \frac{\eta^* + \chi}{\lambda^*}\right)^2 \tag{1.15}$$

$$\Delta G^{\neq}_{sym,ox}(\chi) = \frac{\lambda^*}{4}\left(1 - \frac{\eta^* + \chi}{\lambda^*}\right)^2 \tag{1.16}$$

However, this can lead to poor fitting to experimental data in certain circumstances, for example where the electron transfer is extremely irreversible, and so the asymmetric representation is potentially more powerful [19, 20].

$$\Delta G^{\neq}_{asym,red}(\chi) = \frac{\lambda^*}{4}\left(1 + \frac{\eta^* + \chi}{\lambda^*}\right)^2 + \beta^*\left(\frac{\eta^* + \chi}{4\lambda^*}\right)\left\{1 - \left(\frac{\eta^* + \chi}{\lambda^*}\right)^2\right\} + \frac{\beta^{*2}}{16\lambda^*} \tag{1.17}$$

$$\Delta G^{\neq}_{asym,ox}(\chi) = \frac{\lambda^*}{4}\left(1 - \frac{\eta^* + \chi}{\lambda^*}\right)^2 + \beta^*\left(\frac{\eta^* + \chi}{4\lambda^*}\right)\left\{1 - \left(\frac{\eta^* + \chi}{\lambda^*}\right)^2\right\} + \frac{\beta^{*2}}{16\lambda^*} \tag{1.18}$$

1.2.4 *The Cyclic Voltammogram*

The cyclic voltammogram, commonly abbreviated to "CV", shows the current response to a triangular voltage ramp (see Fig. 1.3) and owes its shape to the complex interplay of both electrode kinetics and mass transport effects. As such it can take different forms depending on the exact details of these two factors, often described as transient or steady-state.

1.2.4.1 Transient Cyclic Voltammetry

Figure 1.3a shows the transient form of the CV, which is obtained for linear diffusion systems, most commonly "macro"electrodes (of characteristic dimension >100 μm) or smaller electrodes at very short timescales (i.e. high voltage scan rates) [21]. Here the characteristic shape is due to the interaction of kinetics with mass transport: the current maximum is due to the competing factors of (i) an exponentially increasing rate of electron transfer (see Butler-Volmer equations), and (ii) rapid depletion of the electroactive material in the double layer caused by relatively slow diffusion of fresh material from bulk solution. In a sufficiently large volume of solution, these factors equilibrate at the diffusion limiting current (I_{lim}).

The transient CV is usually characterised in terms of the peak currents (anodic and cathodic, $I_{p,a}$ and $I_{p,c}$) and the separation of the anodic and cathodic peak potentials ($\Delta E_{pp} = E_{p,a} - E_{p,c}$). The reversibility of the redox couple is defined in terms of the peak separation [1] (Fig. 1.4, Table 1.1):

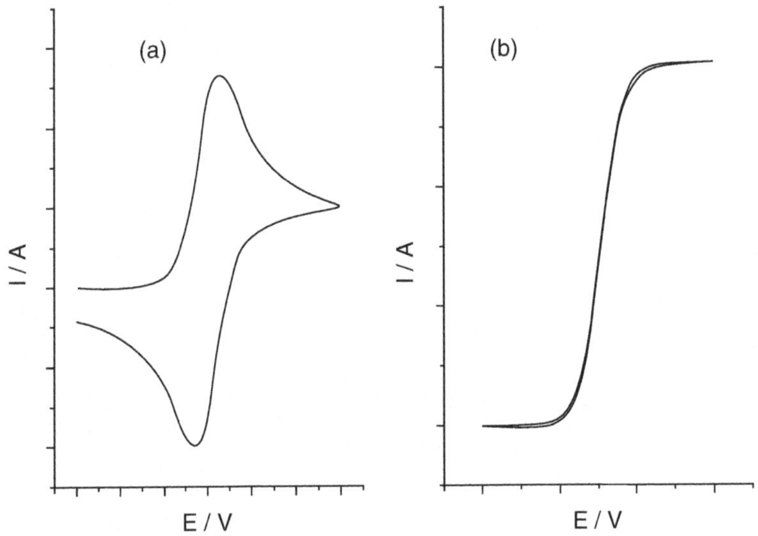

Fig. 1.3 Typical cyclic voltammograms for (**a**) transient and (**b**) steady-state responses, due to linear and convergent diffusion respectively

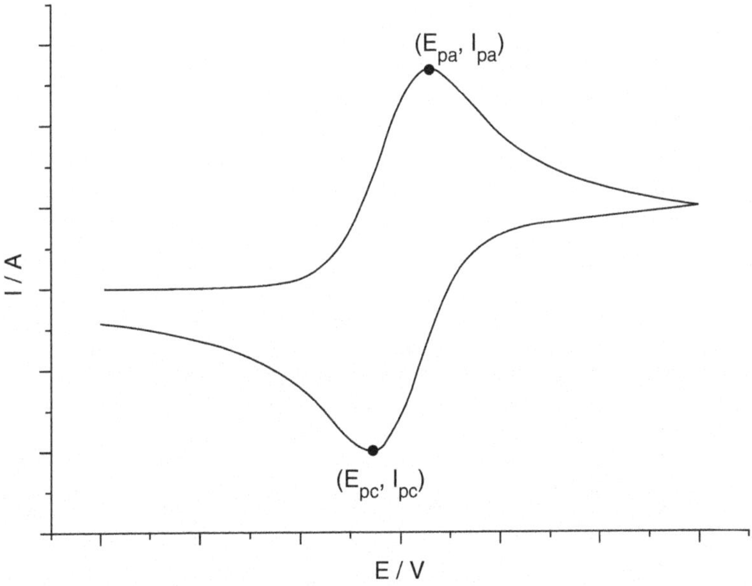

Fig. 1.4 Characteristics of a transient cyclic voltammogram

Table 1.1 Cyclic voltammetric definitions

$\Delta E_{pp} = E_{pa} - E_{pc}$ (mV)	Definition	Other relations
60/n	Reversible	$E^0 = \frac{E_{pa}+E_{pc}}{2}$
$60/n < \Delta E_{pp} < 120/n$	Quasi-reversible	–
>120/n	Irreversible	$E^0 = \frac{E_{pa}+E_{pc}}{2}$

It should be noted here that the reversibility of a redox system is clearly a function of the timescale of the experiment, with a redox couple typically displaying reversible behaviour at slow voltage scan rates, and quasi-reversible or even irreversible behaviour at higher voltage scan rates.

Analytical results for the peak currents are available for the reversible and irreversible cases, due to Randles and Sèvčík [1]:

$$I_{p,rev} = \left(2.69 \times 10^5\right) nAC_{bulk}D^{1/2}\nu^{1/2} \tag{1.19}$$

$$I_{p,irrev} = \left(2.99 \times 10^5\right) n_\alpha^{3/2}\alpha^{3/2}AC_{bulk}D^{1/2}\nu^{1/2} \tag{1.20}$$

where, n is the total number of electrons transferred, n_α is the number of electrons transferred in the rate limiting step, A the geometric electrode area, C_{bulk} the bulk concentration of the electroactive species, D the diffusion coefficient, α the transfer coefficient, and ν the voltage scan rate.

1.2.4.2 Steady-State Voltammetry

In Fig. 1.3b, a steady-state CV is shown: this is most commonly observed for microelectrode voltammetry (for electrodes with characteristic dimension <20 μm), where the diffusion field is fully convergent. The sigmoidal shape is due to the rate of mass transport being sufficiently fast to 'keep up' with the accelerating rate of electron transfer at least until the limiting current (I_{lim}) is reached.

Since the reverse scan retraces the forward scan in the absence of hysteresis effects, the steady-state voltammogram is often recorded as a linear sweep rather than a triangular voltage ramp. The limiting current is given by [1]

$$I_{lim} = 4nFC_{bulk}Dr \tag{1.21}$$

where n, C_{bulk}, and D have their usual meaning, F is the Faraday constant and r the disk radius.

1.2.4.3 The Transition Between Transient and Steady-State

The position of the voltammetric "wave" can be seen to shift to higher overpotentials as the voltammogram shifts from fully transient to fully steady-state, that is, as the mass transport accelerates from linear to convergent

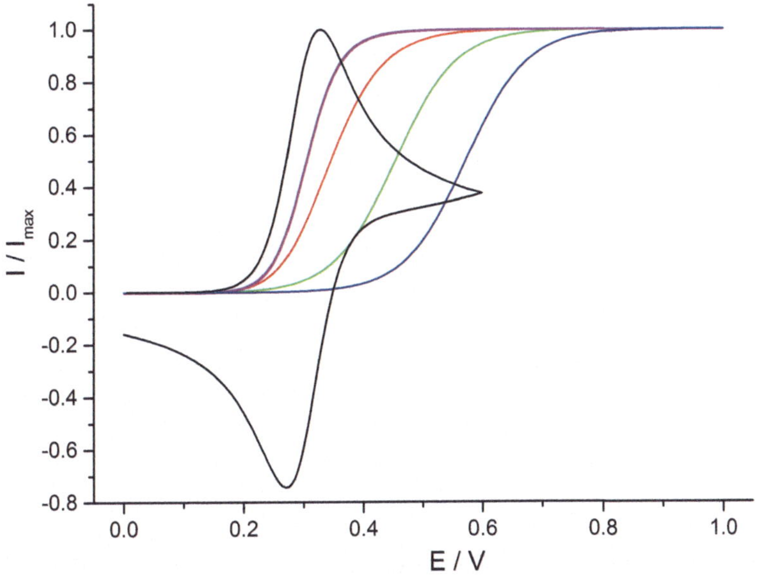

Fig. 1.5 Simulation cyclic voltammograms showing transition from transient to steady state and the subsequent potential shift in the voltammogram. All parameters are the same, except from electrode radius which varies: 1 mm (*black*), 10 μm (*magenta*), 1 μm (*red*), 100 nm (*green*), and 10 nm (*blue*)

(Fig. 1.5). There is no change to the electrode kinetics in this case, the effect is purely one of mass transport. A simple way to understand this is that the higher rates of diffusion sweeps material to and from the electrode more rapidly, hence the electroactive species spends less time in the near vicinity of the electrode and a higher overpotential is required to drive the electron transfer at a fast enough rate for the species to react in that shorter time period.

1.2.4.4 Adsorbed and Thin-Layer Voltammetry

The voltammetry described thus far has been implicitly concerned with a large volume of solution containing a bulk concentration of electroactive species which is not significantly depleted by the electrolysis occurring at the electrode.

Figure 1.6 shows a schematic of both adsorbed and thin-layer systems with their characteristic voltammetric responses. In both cases the current response decays to zero as the concentration of electroactive species is exhausted. Note that in the adsorbed case, the anodic and cathodic peaks are ideally symmetrical about the potential axis due to the absence of mass transport. For thin-layer voltammetry, there is an asymmetry due to the limited mass transport (usually diffusion) occurring in the thin-layer volume.

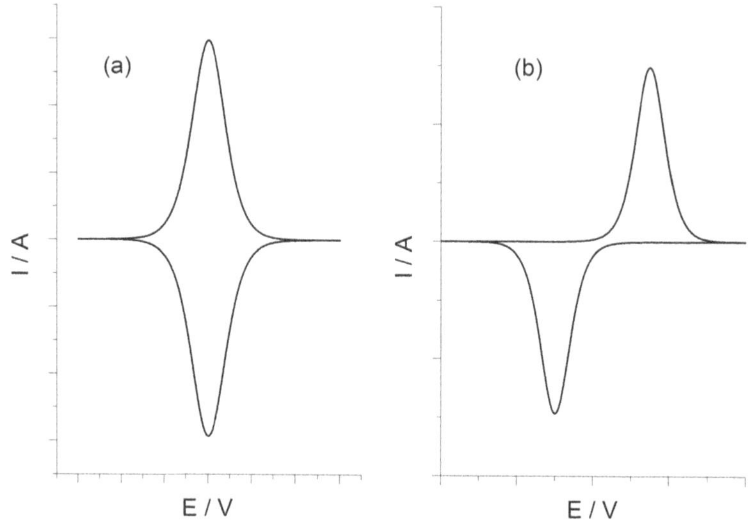

Fig. 1.6 Schematic voltammograms for (**a**) adsorbed species, and (**b**) thin-layer (finite volume) solution

1.2.5 Hydrodynamic Voltammetry

There are a wide range of different hydrodynamic electrodes, where forced convection is applied to achieve a well-defined fluid behaviour [11]. These commonly differ in:

(i) size of electrode—macro vs micro
(ii) geometry—disk, band, tube, ring, ring-disk
(iii) method of convection—flow systems, rotating systems, impinging jet systems
(iv) symmetry—axisymmetric vs non axisymmetric
(v) uniformity of accessibility
(vi) flow regime—laminar vs fully turbulent

Alternative detailed reviews consider this as a topic [11, 22], but here we shall focus on the rotating disk electrode (RDE) and rotating ring-disk electrode (RRDE) as they are commonly used within the fuel cell community (Fig. 1.7).

The RDE (and RRDE) have been in use for many years, having been developed prior to the advent of microelectrodes as a means to shorten the experimental timescale and measure more rapid processes than could be achieved via diffusion-only studies with macroelectrodes [23, 24].

The electrode is mounted axisymmetrically on an insulated rotating shaft, and rotated at sufficient frequency to set up laminar motion in the solution.

In this case the high rates of convective mass transport ensure a steady-state voltammetric response where the limiting current is given by the Levich equation [23]:

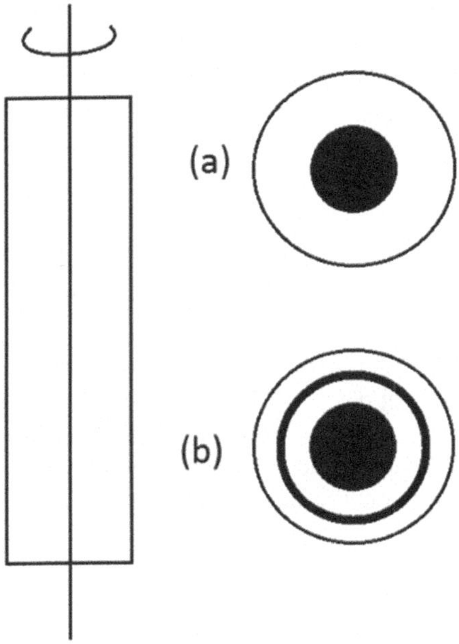

Fig. 1.7 Schematic diagrams of (**a**) rotating disk, and (**b**) rotating ring-disk electrodes

$$I_{lim} = 0.62nFAC_{bulk}D^{2/3}\upsilon^{-1/6}\omega^{1/2} \tag{1.22}$$

where υ is the kinematic viscosity of the solution and ω is the angular speed (rad s^{-1}).

Kinetic parameters can be extracted via Koutecky-Levich analysis, which deconvolutes the total current into kinetic and limiting contributions [25, 26]:

$$I^{-1} = I_{lim}^{-1} + I_k^{-1} \tag{1.23}$$

Plotting $1/I$ vs $\omega^{1/2}$ at a range of potentials yields intercepts of value $1/I_k$. Since I_k is a function of E, given by

$$I_k = nFAC_{bulk}k(E) \tag{1.24}$$

where k(E) is the potential-dependent electrochemical rate constant given by the Butler-Volmer expression

$$k(E) = k_0 exp\left(\frac{-\alpha F\eta}{RT}\right) \tag{1.25}$$

Then a further (Tafel) plot of ln I_k vs η provides α and k_0. The Koutecky-Levich method is widely used for multi-electron systems, most notably oxygen reduction (ORR). However, recent work by Masa et al. suggests that for accurate results it is necessary to take account of the surface roughness of the catalyst deposit, since the ratio of actual to geometric surface area will determine how the apparent rate constant, k_0^{app}, differs from the true rate constant, k_0 [27].

The RRDE is usually used in a generator-collector type experiment which can be illustrated by the (simplified) ORR reaction as follows.

$$O_2(g) + 4H^+(aq) + 4e^- \xrightarrow{k_1} 2H_2O(l)$$

$$O_2(g) + 2H^+(aq) + 2e^- \xrightarrow{k_2} H_2O_2(ads) \xrightarrow{k_3} H_2O_2(aq)$$

$$H_2O_2(ads) + 2H^+(aq) + 2e^- \xrightarrow{k_4} 2H_2O(l)$$

The disk is subjected to a normal voltage scan, in this case to reduce oxygen, whilst the ring potential (E_{ring}) is held at a potential sufficient to oxidise the intermediate (here, hydrogen peroxide). The disk current therefore provides information on the rate of reduction of parent species, whilst the ring current provides information on the amount of intermediate produced.

1.2.6 The Voltammetry of Nanoparticles

The vast majority of catalysts are nanomaterials (assumed spherical) and hence it is important to identify those characteristics that define their voltammetry.

An isolated nanoparticle on a planar substrate behaves as a spherical electrode with a hindered convergent diffusion field (see Fig. 1.8).

For which Bobbert et al. have derived the limiting current expression to be [28]:

$$I_{lim} = 4\pi(ln2)nFC_{bulk}Dr_{np} \tag{1.26}$$

where r_{np} is the radius of the nanoparticle.

Fig. 1.8 Schematic diagram showing convergent diffusion to a (nano)sphere on a plane

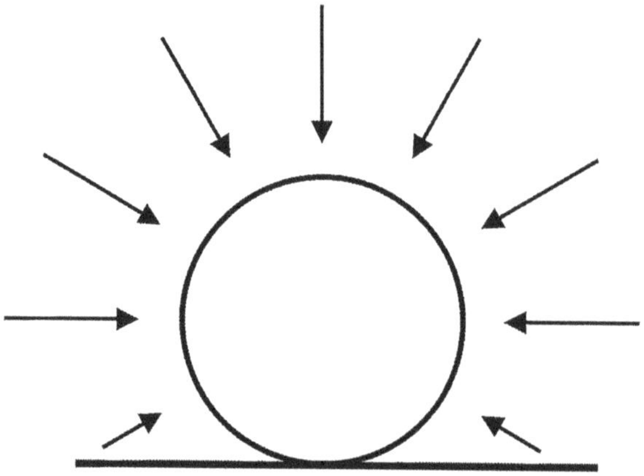

The voltammetry for such as system would appear to be relatively straightforward: the nanoscale size of the 'electrode' (i.e. particle) ensures that all diffusion is convergent on all practical timescales (see Eq. 1.3), and the exceptionally high rates of mass transport associated with this would be expected to cause a shift of the wave to higher overpotentials.

However, it is not common to measure single particle electrochemistry (see later), and instead nanocatalysts are usually studied and used in very large numbers. The catalysts can be deposited directly onto a substrate electrode (usually glassy carbon, GC) or be deposited upon a carbon support material during their fabrication, and the catalyst/support then deposited onto the substrate. In this case the carbon support is usually carbon black (often the commercial Vulcan XC-70), but studies are increasingly using nanotubes, graphenes, etc. There are subtle differences to these two cases, so we consider them in turn.

1.2.6.1 Nanoparticle-Substrate Systems

In this case the nanoparticles are deposited upon the substrate, often via spraying or drop-casting and subsequent evaporation of solvent. The particles are essentially randomly distributed across the surface, and typically are in such numbers that they do not form a monolayer across the entire substrate electrode surface (i.e. coverage $<100\,\%$), shown schematically in Fig. 1.9.

By considering the diffusion to each individual particle, it should be clear that there are two extreme cases (and several intermediate ones) which may occur as a result of the relative spacing of the particles from each other [29].

First, where the particles are very far apart (i.e. extremely low coverage) such that each individual particle develops a convergent diffusion field independent of every other particle. Each particle is diffusionally isolated, and so the voltammetry of the ensemble (or N particles) will appear to be a sigmoidal (steady-state) voltammogram, with limiting current given by

$$I_{lim} = 4N\pi(ln2)nFC_{bulk}Dr_{np} \qquad (1.27)$$

Second, where the particles are very close together (i.e. high coverages) such that each individual particle's diffusion field interferes or overlaps with its neighbour's, then the overall effect is for the ensemble to behave as if the whole area covered is

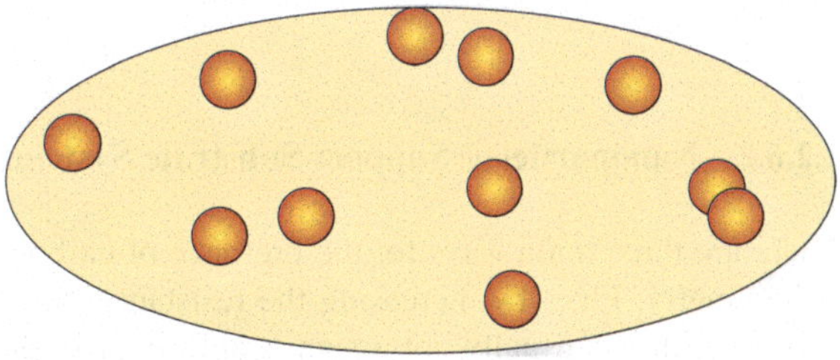

Fig. 1.9 Randomly deposited particles on a substrate electrode

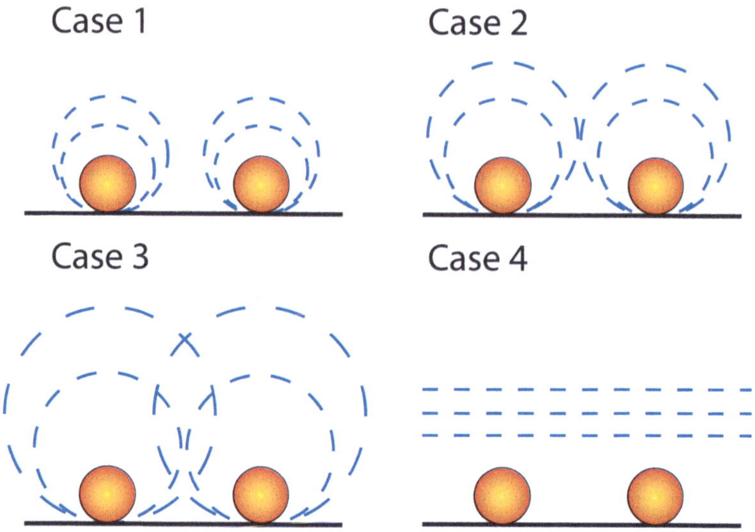

Fig. 1.10 The four cases of diffusion to spheres on a plane (particles on a substrate electrode)

subject to linear diffusion. The voltammetry will therefore appear to be of a transient voltammogram corresponding to an active area equal to that of the geometric area over which the particles are deposited (i.e. the area of the substrate electrode). The usual equations relating to macroelectrode voltammetry (i.e. Randles-Sèvčík) will apply.

These two extremes are often termed Case 1 and Case 4 behaviour respectively, and as they names suggest, two intermediate cases have been identified corresponding to increasing degrees of overlap of diffusion fields associated with neighbouring nanoparticles [19] (Fig. 1.10).

When considering the likely Case in operation for a given system, it is worthwhile considering the diffusion length on the experimental timescale, t [30]:

$$\langle x \rangle \approx \sqrt{2Dt} \tag{1.28}$$

For the reason that even if the particles are small and relatively far apart, if the timescale of the experiment is sufficiently long that $\langle x \rangle$ approaches half of the inter-particle separation, then diffusion field interference will occur and Cases 2–4 will apply.

1.2.6.2 Nanoparticle@Support-Substrate Systems

There are three main ways for the presence of carbon support particles to alter the voltammetry. First, by increasing the resistance between the catalyst particle and substrate: this is usually not a significant effect in the cases of highly conductive carbons (carbon black, nanotubes, etc.), but can become noticeable for less conductive supports such as graphene oxide.

Second, the partial or full occlusion of the catalyst particle. This can occur as part of the catalyst fabrication, as this commonly involves a chemical reduction of the dissolved catalyst precursor salt in the presence of the carbon support in suspension. Any degree of aggregation of the support particles will necessarily occlude catalyst particles. In some cases, it has been reported that during fabrication the catalyst particles may intercalate graphitic particles of carbon support. Alternatively, occlusion may occur during the deposition of the catalyst@support particles onto the substrate through multilayer formation. In these scenarios, the voltammetric changes will be slight, with Cases I–III affected to a decreasing degree and Case IV unaffected. In Case I, the voltammetry will remain steady-state, except the limiting current will reflect the lower apparent radius of the nanoparticles.

Third, the deposition of higher coverages, or strongly aggregated catalyst@support particles will lead to the existence of voids within the catalyst@support deposit. There will be extremely hindered diffusion to these voids from bulk solution and a mixed diffusion regime will be established, where either convergent or linear diffusion occurs to catalyst particles on the surface of the deposit and thin-layer diffusion occurs to catalyst particles within voids in the body of the layer [31–33] (Fig. 1.11).

The resulting voltammetry of such a mixed diffusion regime is therefore some algebraic sum of linear (or convergent is applicable) diffusion and thin-layer diffusion signals. Since the latter, by definition, occurs at a lower overpotential than the former, there is a necessary *shift in overpotential to lower values* than a fully linear diffusion response (Fig. 1.12).

1.2.7 Electrocatalysis

An electrocatalyst increases the rate of an electrochemical reaction by providing a lower energy pathway across the reaction coordinate. Due to the potential dependence of the electrochemical rate constant on overpotential (see Eqs. 1.7 and 1.8)

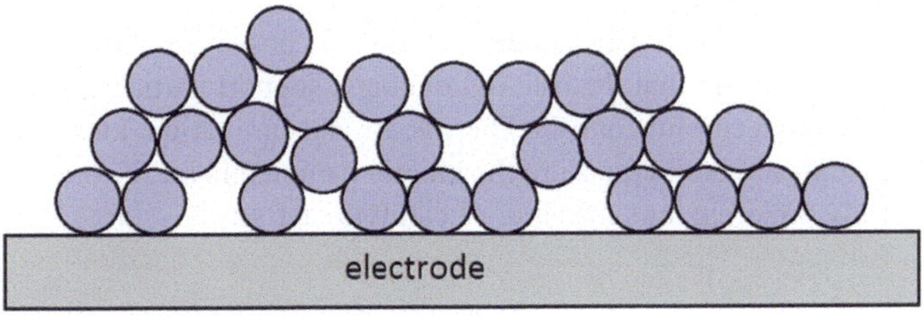

Fig. 1.11 A multi-layer deposit with illustrative voids

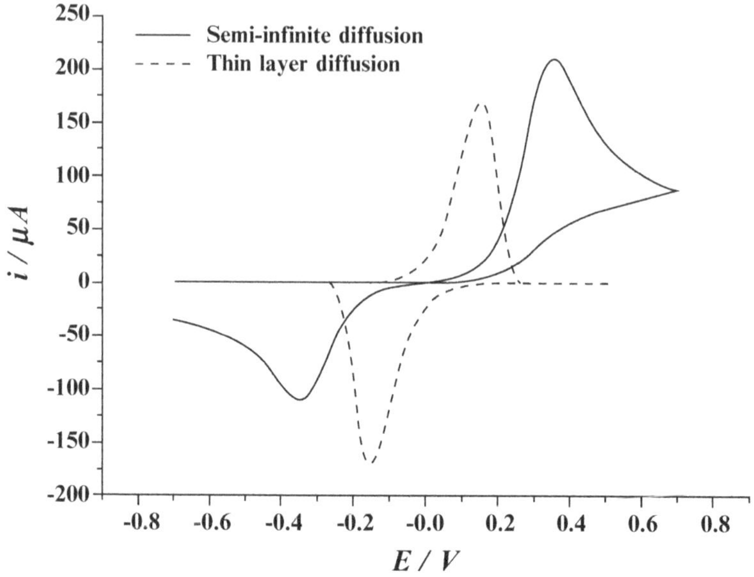

Fig. 1.12 Illustration of how mixed diffusion regime can cause a lowering of overpotential. Reproduced with permission from [34], copyright 2008 Elsevier

this necessarily means that the overpotential required for the electrochemical reaction is decreased and the voltammetric wave is shifted to lower overpotentials. However, the converse is certainly not necessarily true.

There is a widespread naïvety that a candidate catalyst that causes a shift of the voltammetric wave to lower overpotentials must therefore be catalytic. The preceding discussion should make it clear that this conclusion can only be reached if the mass transport to the system has been fully characterised and understood. We have seen that the size of the particle (i.e. electrode) can cause a shift in overpotential through the increasing rates of mass transport via diffusion as particle size decreases. Further, the arrangement of the deposited particles will affect the voltammetric shape (and therefore position on the potential axis) depending on whether Case I–IV behaviour is followed. Finally, and most subtle, the existence of a mixed (linear-thin layer) diffusion regime is itself sufficient to cause a lowering of the observed overpotential.

In any evaluation of nanoparticulate materials as possible catalysts, it is therefore imperative to ensure that the role of coverage, support particles, etc. are taken into account, through careful control experiments and simulation. Ideally, any study of candidate catalysts should measure the kinetic parameters for the reaction, as this is the only truly unambiguous measure of catalytic activity.

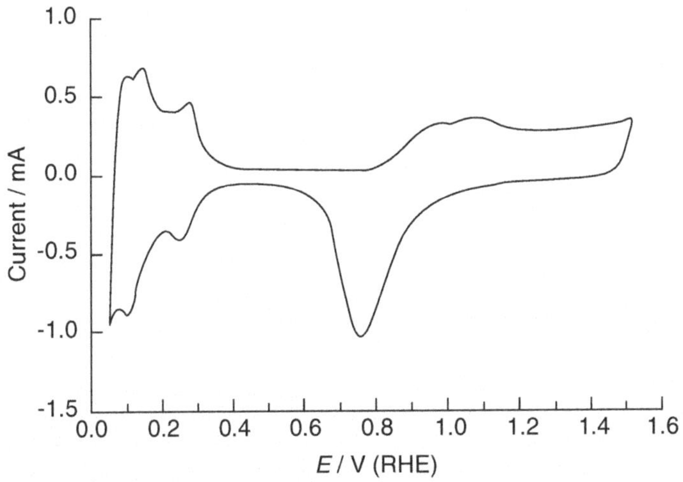

Fig. 1.13 Cyclic voltammogram of Pt film electrode in 0.5 M H_2SO_4 at 50 mV s^{-1}. Reproduced with permission from [35], copyright 2007 Elsevier

1.2.8 Electrochemical Surface Area (ECSA) Measurements

Catalyst preparations are often characterised by their ECSA, usually quoted in $m^2\ g^{-1}$, as a measure of their activity in comparison to metal loading. The most commonly used method for determining the ECSA is to measure the CV of the catalyst in acid solution, determining the charge passed during the adsorption/desorption of protons, and relating this to an average charge per unit area for the metal surface (Fig. 1.13).

This method is particularly used for Pt-based catalysts as the adsorptive charge of 210 μC cm^{-2} is well known from an average of the values for the (1 1 1), (1 1 0), and (1 0 0) surfaces from single crystal studies [36].

However, the measurement of ECSA is often more approximate than acknowledged: the use of surface physisorption as a methodology requires the sample and solution to be scrupulously clean: this is often not achievable for some catalyst testing. The appearance of the resulting CV clearly shows this effect via truncated or poorly resolving adsorption/desorption peaks.

It may therefore be desirable to use additional methods to provide assurance on the ECSA: this can be obligatory for some non-Pt catalysts where hydrogen adsorption is not reliable (e.g. Pd surfaces where H atoms readily absorb). In these cases, CO adsorption is often used, and even the deposition of metals such as Cu [36–38].

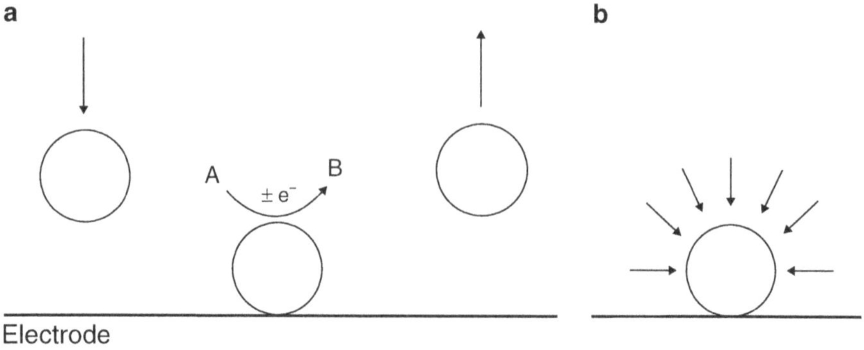

Fig. 1.14 Schematic diagram showing: (**a**) the nanoparticle becoming a nanoelectrode on contact with the substrate electrode. Note the latter is of a different material to the nanoparticle such that the A/B redox couple is inactive at the substrate electrode and only occurs at the nanoparticle whilst it is in electrical contact with the former. (**b**) For the contact period of milliseconds a convergent diffusion field is almost instantly established leading to steady-state currents for the conversion of A to B. Reproduced with permission from [46] copyright 2012, Elsevier

1.2.9 Processes at Single Nanoparticles

There has been considerable recent interest in the ability to measure kinetic processes at individual nanoparticles [39, 40]. This has been pioneered by the groups of Bard and Compton and cover reactions of the nanoparticles themselves (oxidation/reduction) as well as reactions occurring at the particle surface of adsorbed and solution species [41–46].

This is achieved via a conceptually simple, but often experimentally challenging, process of observing the current signal produced when collisions between free nanoparticles and a substrate electrode (usually carbon) which is held at a suitable potential. During the contact phase of the particle-electrode collision, the nanoparticle becomes a spherical electrode on a plane (see Fig. 1.8, Eq. 1.26) and a fully convergent diffusion field is set up within 1 μs of the contact which typically lasts for 1–20 ms at 298 K in aqueous solution for particles of size *ca* 10 nm. It has been shown that kinetic parameters can be extracted from these current signals, as in the study be Kahk et al. into the proton reduction at gold and silver nanoparticles [46] (Fig. 1.14).

In part of this study, Kahk et al. introduced 7 nm radius gold nanoparticles into a solution of 10 mM perchloric acid and 0.6 M sodium perchlorate. A carbon microelectrode was potentiostatted at a range of potentials and reductive current spikes recorded, as illustrated in Fig. 1.15a. These spikes were analysed as a function of potential in Fig. 1.15b, and the data points fitted to a steady-state voltammogram with commercial modelling software (DigiSim™, BASI Inc.) to extract the kinetic parameters $k_0 = (7.0 \pm 5.0) \times 10^{-7}$ cm s^{-1}, and $\alpha = 0.56 \pm 0.05$.

To demonstrate the high rate of mass transport to the nanoparticle in contact with the electrode, Kahk added acetic acid to the reaction, since the rate of dissociation

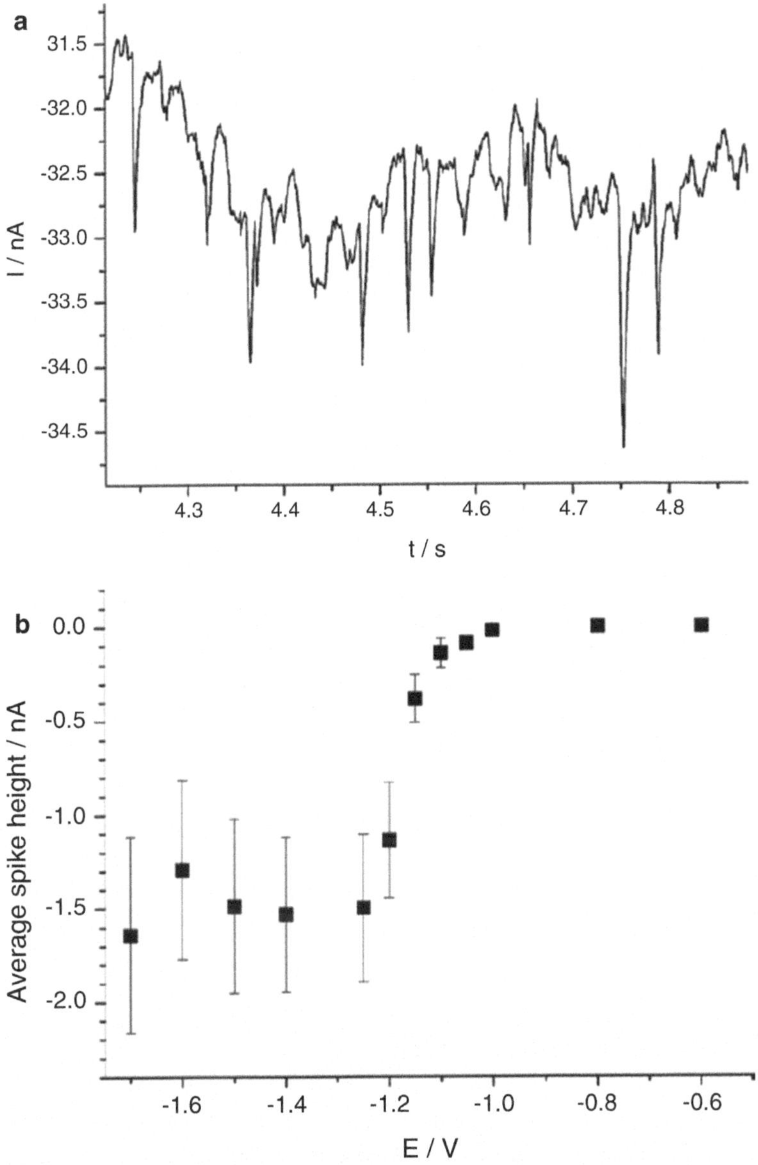

Fig. 1.15 (**a**) AuNP impact spikes for the potential held at −1.40 V, and (**b**) a plot of average spike height vs potential for AuNP impacts. Reproduced with permission from [46] copyright 2012, Elsevier

to release protons is known to be $k_d = 1.3 \times 10^6\ s^{-1}$. Repeat experiments were conducted, and no increase in the apparent proton concentration was detected. Simulation was re-performed to include the dissociation kinetics of acetic acid and identical fits were obtained (Fig. 1.16).

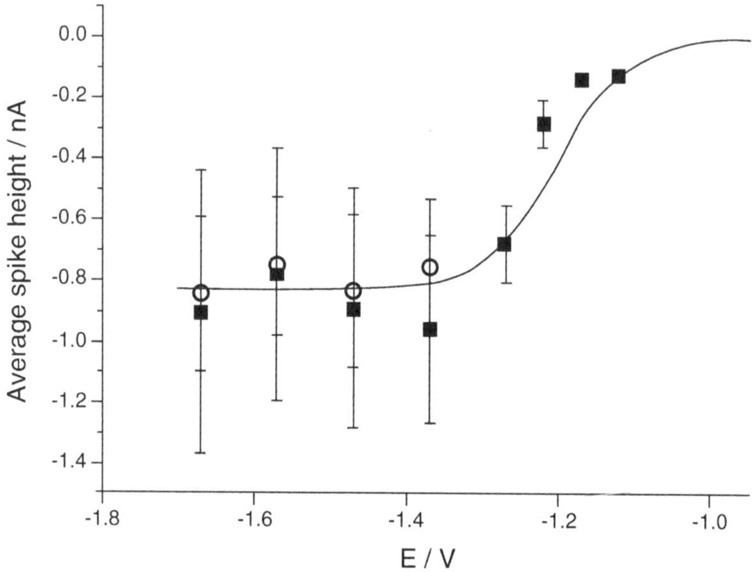

Fig. 1.16 Spike height vs potential plots for 7 nm radius AuNPs in solutions of 10 mM $HClO_4$ and 0.6 M $NaClO_4$ (*circle*) and also including 0.1 M acetic acid (*filled square*). The *solid line* is a computer simulation to include the dissociation of acetic acid and free perchloric acid protons. Reproduced with permission from [46] copyright 2012, Elsevier

1.3 Part II: Examples

In this section we briefly discuss case studies with reference to the foregoing, beginning with screening catalysts for a simple 1-electron hydrogen evolution reaction (HER, or proton reduction) and progressing to a study involving the 4-electron oxygen reduction reaction.

1.3.1 Pt@TiO$_2$ Clusters for HER

Blackmore et al. [47] have recently published a study into fabricating Pt@TiO_2 core shell particles as potential catalysts, and have characterised them via simple cyclic voltammetry. Having made the NPs via a magnetron-sputtering gas condensation cluster beam source, and size selected them via time-of-flight mass selection, the particles were soft landed onto a glassy carbon (GC) target. This formed the active surface of a working electrode, and after TEM imaging, basic cyclic voltammetry was carried out (Fig. 1.17).

In this example, the coverage of the electrode by the nanoparticles was calculated to be $(5 \pm 2)\,\%$. Therefore it is reasonable to assume that no multilayer deposits of particles have occurred, especially in the light of the high resolution

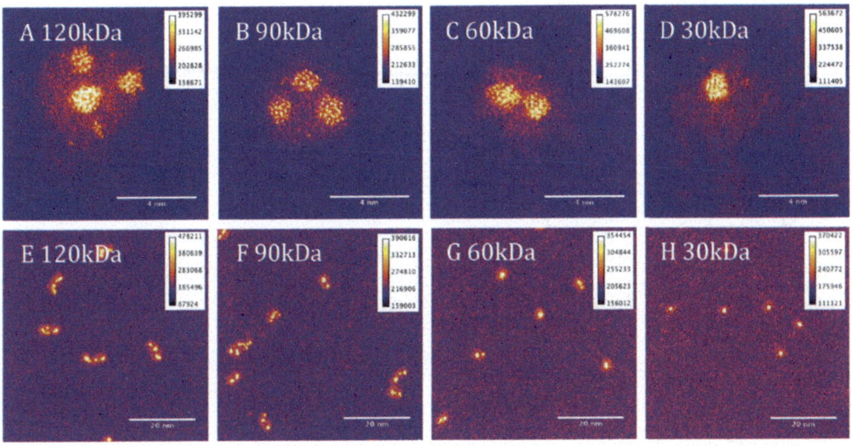

Fig. 1.17 STEM images of Pt-TiO_2 clusters, showing single and multiple Pt cores (*yellow*) in TiO_2 shells (*pink*). Reproduced with permission from [47] copyright 2015, Royal Society of Chemistry

imaging which shows well-dispersed particles. Hence, a simple shift in the voltammetry towards lower overpotentials can be safely interpreted as an increase in catalytic function: this is indeed what is observed in comparison to the control signals obtained from bare GC, and GC modified with TiO_2 particles of a similar size (see Fig. 1.18).

Note that despite the small size of the particles the diffusion length is long enough at this scan rate to give a "transient" shaped CV.

1.3.2 Evaluation of Surfactant-Stabilised Pt NPs for the ORR

In a study of the effects of surfactants as stabilisers in Pt catalyst fabrication Newton et al. [48] compared commercial TKK Pt catalyst with Pt/C catalysts made by the addition of nonylphenolethoxylate (NP9) and tetradecyltrimethylammonium bromide (TTAB) as neutral and cationic surfactants respectively, to the platinum precursor and carbon black. In characterising the three catalysts, they illustrated the difficulty in obtaining reliable ECSA measurements using hydrogen adsorption alone, and compared with CO adsorption measurements (see Fig. 1.19). From this they were able to determine the following ECSA values (all in $m^2\ g^{-1}$): TKK—91, Pt + NP9—22, and Pt + TTAB—1.7.

Due to the fact the catalyst particles were comprised of Pt deposited onto carbon black (nanospheres), it is more appropriate to treat as if multilayer deposits on the electrode are likely. The preferred method therefore is to use a hydrodynamic method, here the RDE, in order to swamp any intra-layer diffusion effects with forced convection to the surface of the deposit (Scheme 1.1).

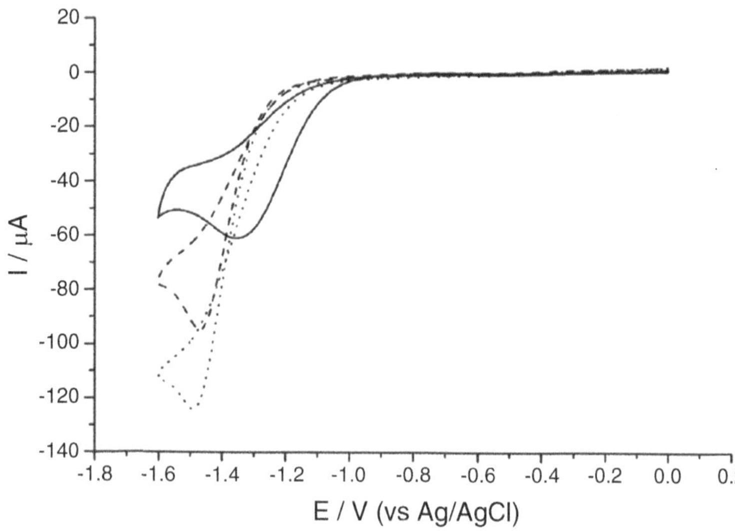

Fig. 1.18 Cyclic voltammograms recorded at 25 mV s^{-1} for a solution of 2 mM $HClO_4$ in 0.1 M $NaClO_4$ at the following surfaces: bare GC (*dotted line*); GC modified with TiO_2 particles (*dashed line*), and GC modified with Pt@TiO_2 particles (*solid line*). Reproduced with permission from [47] copyright 2015, Royal Society of Chemistry

In addition, the use of the RRDE enables quantitative data to be extracted regarding the relative importance of the direct 4e pathway and indirect (2e, 2e) pathway via hydrogen peroxide, for the catalysts under scrutiny (Fig. 1.20).

The number of electrons transferred, n, can be quantified from the disk and ring currents using

$$n = \frac{4I_D}{I_D + (I_R/N)} \tag{1.29}$$

where I_D is the modulus of the total disk current (i.e. the reduction of O_2 and H_2O_2), I_R is the ring current (i.e. oxidation of H_2O_2), and N is the collection efficiency determined to be 0.21 in this case (from prior calibration experiments).

This expression can therefore yield the number of electrons passed on each catalyst—with $n = 4$ the best possible result. This suggests that the Pt + NP9/C catalysts is similar in performance to TKK (Figs. 1.21 and 1.22).

Koutecky-Levich analysis was then performed on the RDE data. Kinetic analysis can be taken further using the model of Hsueh and Chin [49], via plots of I_D/I_R vs $\omega^{-1/2}$ and $I_{DL}/(I_{DL} - I_D)$ vs $\omega^{-1/2}$:

$$\frac{I_D}{I_R} = \frac{1}{N}\left(1 + \frac{2k_1}{k_2}\right) + \left\{\frac{2k_3(1 + k_1/k_2)}{NZ_2}\right\}\omega^{-1/2} \tag{1.30}$$

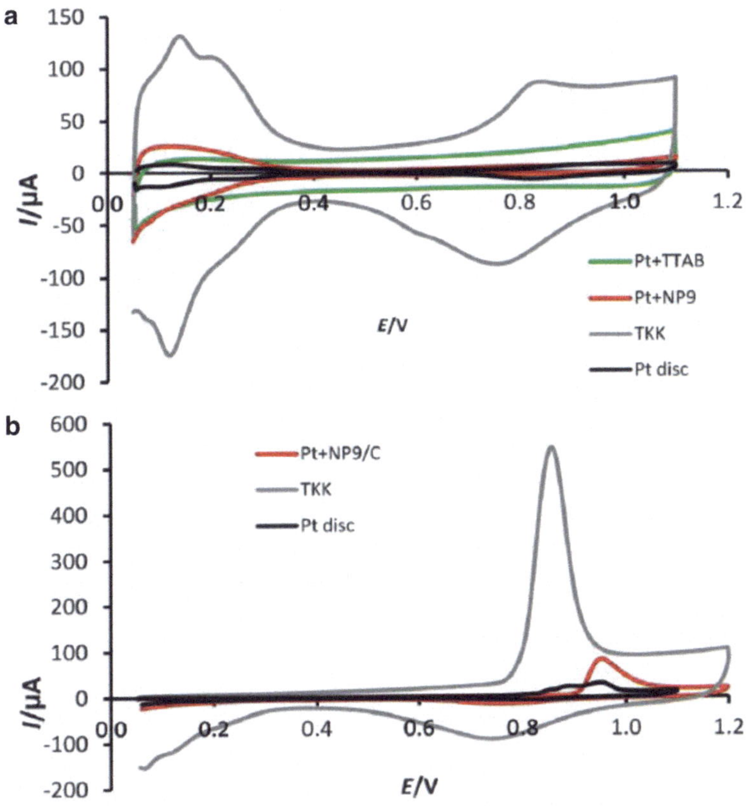

Fig. 1.19 Cyclic voltammograms recorded at 25 mV s^{-1} of Pt/C catalysts of loading 20 μg Pt cm^{-2} on 0.196 cm^2 GC electrodes in N_2-saturated 0.1 M $HClO_4$ at 298 K. Each compared to a 0.196 cm^2 Pt disk electrode. (**a**) Current vs potential, and (**b**) CO-stripping voltammogram. Reproduced with permission from [48] copyright 2014, Royal Society of Chemistry

$$O_2(g) + 4H^+(aq) + 4e^- \xrightarrow{k_1} 2H_2O(l)$$

$$O_2(g) + 2H^+(aq) + 2e^- \xrightarrow{k_2} H_2O_2(ads) \xrightarrow{k_3} H_2O_2(aq)$$

$$H_2O_2(ads) + 2H^+(aq) + 2e^- \xrightarrow{k_4} 2H_2O(l)$$

Scheme 1.1 A simplified ORR mechanism

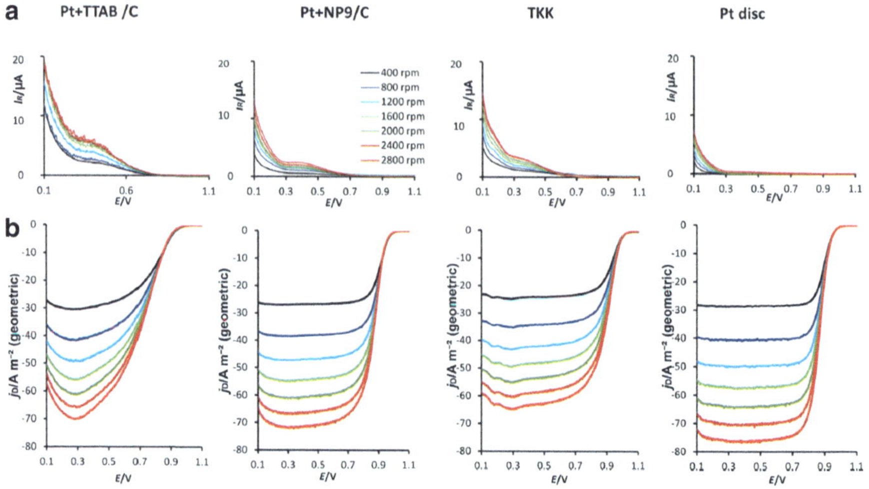

Fig. 1.20 Results for oxygen reduction at an RRDE for Pt + TTAB/C, Pt + NP9/C, TKK Pt catalyst and a Pt disk electrode. (**a**) Ring currents when held at +1.10 V, and (**b**) disk currents at different rotation speeds measured at 298 K in O_2-saturated 0.1 M $HClO_4$ solution on a 0.196 cm^2 substrate electrode. Reproduced with permission from [48] copyright 2014, Royal Society of Chemistry

$$\frac{I_{DL}}{I_{DL} - I_D} = 1 + \frac{(k_1 + k_2)}{Z_1}\omega^{-1/2} \tag{1.31}$$

where I_{DL} is the disk limiting current, $Z_I = 0.62\text{D}(O_2)^{2/3}\nu^{-1/6}$, and $Z_2 = 0.62\text{D}(H_2O_2)^{2/3}\nu^{-1/6}$, and the corresponding plots are (Figs. 1.23 and 1.24):

If the gradients of the plots in Figs. 1.20 and 1.21 are noted as G_1 and G_2 respectively, and the corresponding intercepts C_1 and 1. Then rearranging Eqs. 1.30 and 1.31 enables the rate constants to be evaluated (Fig. 1.25)

$$k_1 = G_2 Z_1 \left(\frac{C_1 N - 1}{C_1 N + 1}\right) \tag{1.32}$$

$$k_2 = \frac{2 Z_1 G_2}{C_1 N + 1} \tag{1.33}$$

$$k_3 = \frac{N Z_2 G_1}{C_1 N + 1} \tag{1.34}$$

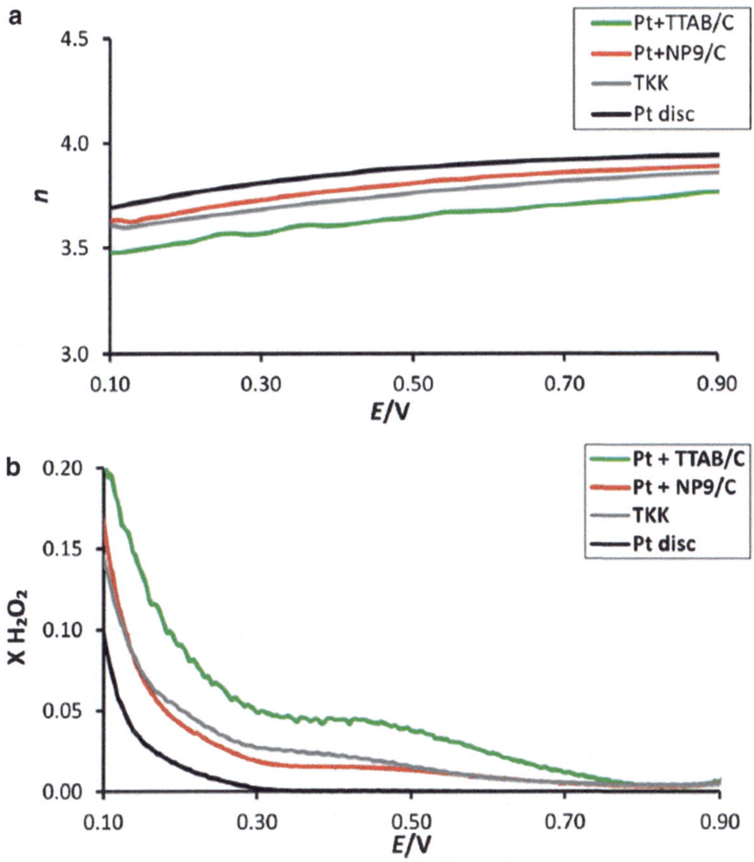

Fig. 1.21 Calculated number of electrons and fraction of hydrogen peroxide produced as a function of potential for each of the three catalysts and a Pt disk electrode. Reproduced with permission from [48] copyright 2014, Royal Society of Chemistry

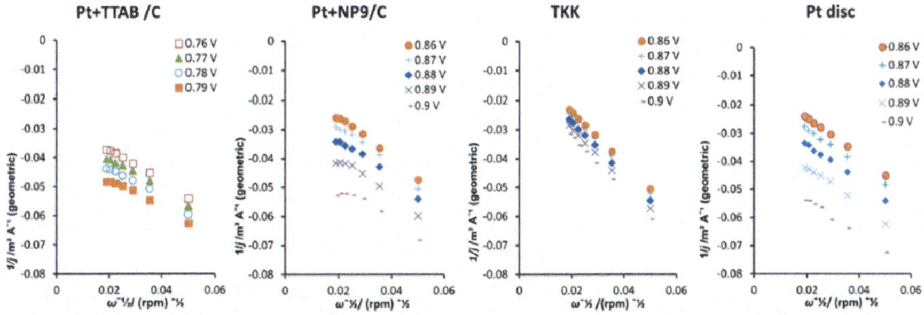

Fig. 1.22 Koutecky-Levich plots for potentials E = 0.86, 0.87, 0.88, 0.89, 0.90 V. Reproduced by permission from [48] from the Royal Society of Chemistry

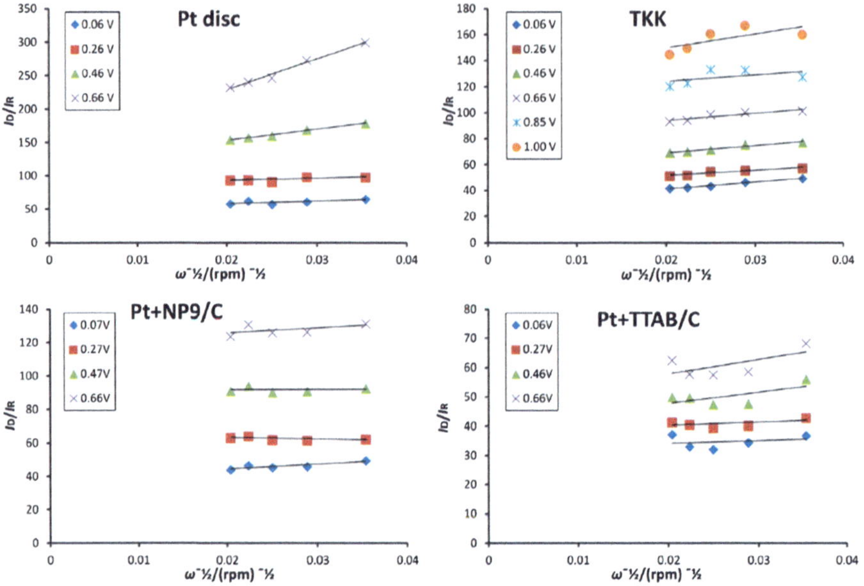

Fig. 1.23 Plots of Eq. 1.30 using the data in Fig. 1.22. Reproduced with permission from [48] copyright 2014, Royal Society of Chemistry

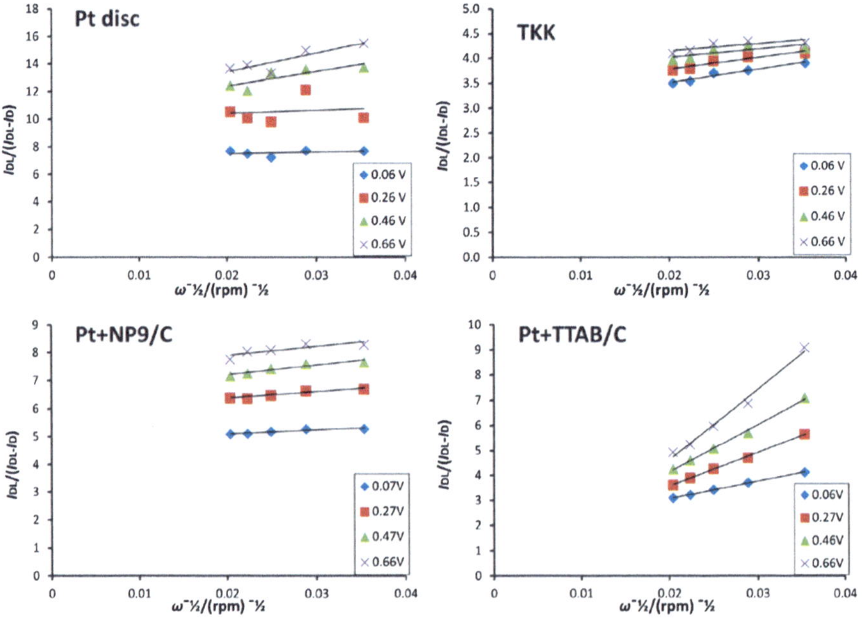

Fig. 1.24 Plots of Eq. 1.31 from the data in Fig. 1.22. Reproduced with permission from [48] copyright 2014, Royal Society of Chemistry

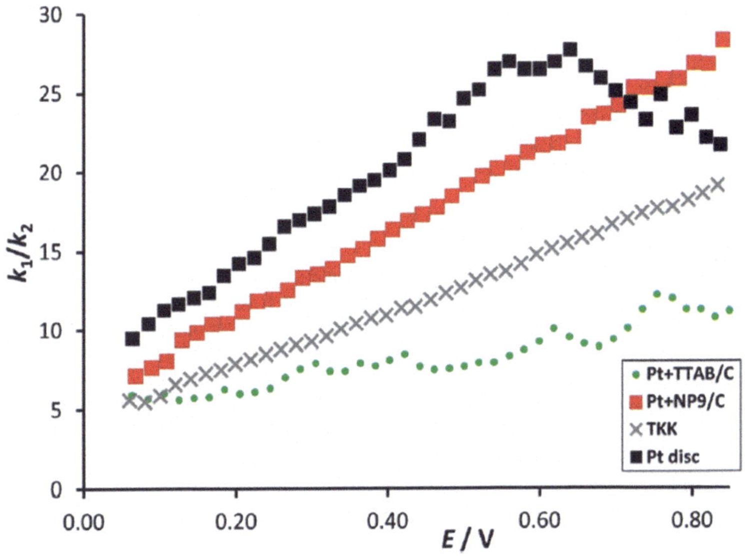

Fig. 1.25 Plot of k_1/k_2 according to above equations. Reproduced with permission from [48] copyright 2014, Royal Society of Chemistry

1.4 Conclusions

The electrochemical evaluation of nanomaterials for use in fuel cells can be straightforward, but it is important to consider the fundamentals of the system before starting ex-situ testing. There are established as well as newly emerging techniques available, but it is vital to fully understand the interplay of kinetics and mass transport to the test material in order to make a reliable judgment on its catalytic activity. Ideally, kinetic parameters should be determined at this stage, enabling true comparison between candidate catalyst materials.

It should then be possible to move on to considering the role of the carbon support, gas diffusion layer, etc in stages so that at each step towards single fuel cell can be optimised and controlled analytically.

Acknowledgement N.V.R. is grateful to Mr. P.H. Robbs for assistance in the preparation of figures for this article.

References

1. Bard AJ, Faulkner LR (2001) Electrochemical methods: from fundamentals to applications, 2nd edn. Wiley, New York
2. Girault HH (2004) Analytical and physical electrochemistry. EPFL Press, Lausanne

3. Gileadi E (2011) Physical electrochemistry: fundamentals, techniques and applications. Wiley, New York
4. Compton RG, Banks CE (2010) Understanding voltammetry, 2nd edn. Imperial College Press, London
5. Santos E, Schmickler W (2011) Catalysis in electrochemistry: from fundamental aspects to strategies for fuel cell development. Wiley, Hoboken
6. Wieckowski A, Koper MTM (2009) Fuel cell catalysis: a surface science approach. Wiley, Hoboken
7. Wieckowski A, Norskov JK (2010) Fuel cell science: theory, fundamentals and biocatalysis. Wiley, Hoboken
8. Barbir F (2011) PEM fuel cells: theory and practice, 2nd edn. Academic, San Diego
9. Dickinson EJF, Limon-Petersen JG, Rees NV, Compton RG (2009) How much supporting electrolyte is required to make a cyclic voltammetry experiment quantitatively "Diffusional"? A theoretical and experimental investigation. J Phys Chem C 113:11157–11171
10. Albery WJ (1975) Electrode kinetics. Oxford University Press, Oxford
11. Brett CMA, Oliveira Brett AMCF (1986) Hydrodynamic electrodes. In: Bamford CH, Tipper CFH, Compton RG (eds) Comprehensive chemical kinetics, vol 26. Elsevier, Amsterdam
12. Marcus RA (1956) On the theory of oxidation-reduction reactions involving electron transfer I. J Chem Phys 24:966
13. Marcus RA (1964) Chemical and electrochemical electron transfer theory. Annu Rev Phys Chem 15:155
14. Hush NS (1958) Adiabatic rate processes at electrodes: energy-charge relationships I. J Chem Phys 28:962
15. Hush NS (1999) Electron transfer in retrospect and prospect 1: Adiabatic electrode processes. J Electroanal Chem 470:170
16. Henstridge MC, Laborda E, Rees NV, Compton RG (2012) Marcus-Hush-Chidsey theory of electron transfer applied to voltammetry: a review. Electrochim Acta 84:12–20
17. Weaver MJ (1987) Redox reactions at metal-solution interfaces. In: Compton RG (ed) Comprehensive chemical kinetics, vol 27. Elsevier, Amsterdam
18. Fletcher S (2010) The theory of electron transfer. J Solid State Electrochem 14:705
19. Laborda A, Henstridge MC, Compton RG (2012) Asymmetric Marcus theory: application to electrode kinetics. J Electroanal Chem 667:48–53
20. Marcus RA (1965) On the theory of electron transfer reactions VI. Unified treatment for homogeneous and electrode reactions. J Chem Phys 43:679
21. Amatore C, Maisonhaute E, Simonneau G (2000) Ultrafast cyclic voltammetry: performing in the few megavolts per second range without ohmic drop. Electrochem Commun 2:81
22. Rees NV, Compton RG (2008) Hydrodynamic microelectrode voltammetry. Russ J Electrochem 44:368–389
23. Levich VG (1962) Physicochemical hydrodynamics. Prentice-Hall, Englewood Cliffs
24. Albery WJ, Hitchman ML (1971) Ring-disc electrodes. Oxford University Press, Oxford
25. Koutecky J, Levich VG (1958) The use of a rotating disk electrode in the studies of electrochemical kinetics and electrolytic processes. Zh Fiz Khim 32:1565–1575
26. Koutecky J, Levich VG (1957) An application of a rotating disk electrode o the studies of kinetic and catalytic processes in electrochemistry. Dokl Akad Nauk SSSR 117:441–444
27. Masa J, Batchelor-McAuley C, Schuhmann W, Compton RG (2014) Koutecky–Levich analysis applied to nanoparticle modified rotating disk electrodes: electrocatalysis or misinterpretation? Nano Res 7:71–78
28. Bobbert PA, Wind MM, Vlieger J (1987) Diffusion to a slowly growing truncated sphere on a substrate. Physica A 141:58
29. Davies TJ, Compton RG (2005) The cyclic and linear sweep voltammetry of regular and random arrays of microdisc electrodes: theory. J Electroanal Chem 58:63–82
30. Einstein A (1956) Investigations of the theory of Brownian movement. Dover, Englewood Cliffs

31. Sims MJ, Rees NV, Dickinson EJF, Compton RG (2010) Effects of thin-layer diffusion in the electrochemical detection of nicotine on basal plane pyrolytic graphite (BPPG) electrodes modified with layers of multi-walled carbon nanotubes (MWCNT-BPPG). Sens Actuators B 144:153–158
32. Henstridge MC, Compton RG (2012) Mass transport to micro- and nanoelectrodes and their arrays: a review. Chem Rec 12:63–71
33. Bae JH, Han J-H, Han D, Chung TD (2013) Effects of adsorption and confinement on nanoporous electrochemistry. Faraday Discuss 164:361–376
34. Streeter I, Wildgoose GG, Shao L, Compton RG (2008) Cyclic voltammetry on electrode surfaces covered with porous layers: an analysis of electron transfer kinetics at single-walled carbon nanotube modified electrodes. Sens Actuators B 133:462–466
35. Kunimatsu K, Senzaki T, Samjeske G, Tsushima M, Osawa M (2007) Hydrogen adsorption and hydrogen evolution reaction on a polycrystalline Pt electrode studied by surface-enhanced infrared absorption spectroscopy. Electrochim Acta 52:5715
36. Wieckowski A (1999) Interfacial electrochemistry: theory, experiment and applications. Marcel Dekker, New York
37. Farias MJS, Herrero E, Feliu JM (2013) Site selectivity for CO adsorption and stripping on stepped and kinked platinum surfaces in alkaline medium. J Phys Chem C 117:2903–2913
38. Fang L-L, Tao Q, Li M-F, Liao L-W, Chen D, Chen Y-X (2010) Determination of the real surface area of palladium electrode. Chin J Chem Phys 23:543
39. Rees NV (2014) Electrochemical insight from nanoparticle collisions with electrodes. A mini-review. Electrochem Commun 43:83
40. Rees NV, Zhou YG, Compton RG (2012) Making contact: charge transfer during particle-electrode collisions. RSC Adv 2:379
41. Bard AJ, Xiao XY (2007) Observing single nanoparticle collisions at an ultramicroelectrode by electrocatalytic amplification. J Am Chem Soc 129:9610
42. Kwon SJ, Fan FRF, Bard AJ (2010) Observing iridium oxide (IROx) single nanoparticle collisions at ultramicroelectrodes. J Am Chem Soc 132:13165
43. Kwon SJ, Zhou H, Fan FRF, Vorobyev V, Zhanf B, Bard AJ (2011) Stochastic electrochemistry with electrocatalytic nanoparticles at inert ultramicroelectrodes—theory and experiments. Phys Chem Chem Phys 13:5394
44. Zhou YG, Rees NV, Compton RG (2011) The electrochemical detection and characterization of silver nanoparticles in aqueous solution. Angew Chem Int Ed 50:4219
45. Rees NV, Zhou YG, Compton RG (2011) The aggregation of silver nanoparticles in aqueous solution investigated via anodic particle coulometry. ChemPhysChem 12:1645
46. Kahk JM, Rees NV, Pillay J, Tshikhudo R, Vilakazi S, Compton RG (2012) Electron transfer kinetics at single nanoparticles. NanoToday 7:174
47. Blackmore CE, Rees NV, Palmer RE (2015) Modular construction of size-selected multiplecore Pt-TiO_2 nanoclusters for electro-catalysis. Phys Chem Chem Phys 17 (42):28005–28009. doi:10.1039/c5cp00285k
48. Newton JE, Preece JA, Rees NV, Horswell SL (2014) Nanoparticle catalysts for proton exchange membrane fuel cells: can surfactant effects be beneficial for electrocatalysis? Phys Chem Chem Phys 16:11435
49. Hsueh KL, Chin DT, Srinivasan S (1983) Electrode kinetics of oxygen reduction: a theoretical and experimental analysis of the rotating ring-disc electrode method. J Electroanal Chem Interfacial Electrochem 153:79–95

Chapter 2
Recent Advances in the Use of Shape-Controlled Metal Nanoparticles in Electrocatalysis

Francisco J. Vidal-Iglesias, José Solla-Gullón, and Juan M. Feliu

2.1 Introduction

The surface structure, that is, the particular arrangement of the atoms at the surface is a very relevant parameter and recognized to strongly determine the electrocatalytic properties of the material under study. In this regard, the extensive and intensive knowledge acquired with the use of metal single crystals is not only of outstanding importance for the establishment of the relationships between reactivity and surface structure but also should be used as a reference for those studies dealing with the understanding and evaluation of the effect of the surface structure on the electrocatalytic properties of metal nanoparticles [1–4]. However, this knowledge should not be limited to foresee the expected results or tendencies but also should be used to learn the methods, protocols and precautions that must be satisfied to properly find such surface structure-reactivity correlations. In the quest of these correlations, the use of shape-controlled metal nanoparticles is of outstanding importance because the shape of a nanoparticle remarkably determines its surface atomic arrangement and coordination. In fact, the shape of a nanoparticle anticipates which surface facets will be present at its surface due to the intrinsic correlation between the shape and the surface structure of a nanoparticle. As described in previous contributions, for fcc metals, it is possible to build a stereographic triangle that correlates the crystalline surface Miller index and the shape of the nanoparticle [5–7]. Thus, for instance, for a perfect Pt octahedron, all the exposed facets would be {1 1 1}, while on a perfect Pt cube these facets would be {1 0 0} domains. However, it is worth noting that these are ideal polyhedral crystals containing perfect surface planes, which is, indeed, an unrealistic situation. Therefore, the real surface of a nanoparticle will contain a determined number of defect,

F.J. Vidal-Iglesias • J. Solla-Gullón • J.M. Feliu (✉)
Instituto de Electroquímica, Universidad de Alicante, Apdo. 99, Alicante E-03080, Spain
e-mail: juan.feliu@ua.es

K.I. Ozoemena, S. Chen (eds.), *Nanomaterials for Fuel Cell Catalysis*,
Nanostructure Science and Technology, DOI 10.1007/978-3-319-29930-3_2

corner, edge, step and kink sites and also ordered surface domains of different dimensions. This complex scenario will particularly determine the resulting electrocatalytic properties of the nanoparticle. Consequently, it is of capital relevance to recall that, from an electrocatalytic point of view, the shape of a nanoparticle is not the key point but its real surface structure. This fact clearly indicates that having tools to characterize in great detail the particular surface structure of a nanoparticle is of paramount importance to understand its intrinsic electrocatalytic activity. This point will be discussed later in this chapter.

To the best of our knowledge, the first examples of shape-controlled metal nanoparticles in electrocatalysis were reported by our group in 2004 [8, 9]. In these contributions, we prepared cubic Pt nanoparticles using the pioneering methodology described in 1996 by El-Sayed and co-workers [10] (this paper is considered to be the first example of synthesis of shape-controlled metal nanoparticles), which consisted in a chemical reduction of K_2PtCl_4 by H_2 in the presence of sodium polyacrylate (NaPA). These cubic Pt nanoparticles were used to electrocatalyze ammonia electrooxidation in alkaline solution and the resulting current density was significantly higher than that found for spherical Pt nanoparticles. It is important to stress that this reaction was selected because of its extreme sensitivity towards the surface structure of single crystal platinum electrodes, taking place almost exclusively on {1 0 0} sites. The enhanced activity was attributed to an increase in the {1 0 0} terrace surface sites as deduced from the voltammetric profile obtained with the cubic Pt nanoparticles in 0.5 M H_2SO_4. This experimental finding clearly illustrated that controlling the surface structure of the nanoparticles was a rational option to optimize their electrocatalytic activity. From these contributions, many other cases have been reported in the literature and most of them appear summarized and described in some relevant reviews about the use of these shaped metal nanoparticles for electrocatalytic applications [11–22].

From the point of view of the synthesis of these nanoparticles, after the aforementioned pioneering contribution by El-Sayed et al. [10], many different approaches have been developed for the preparation of shape-controlled metal nanoparticles. In this chapter, we are not going to go into details about this point and the reader is directed to some of the most relevant reviews dealing with the synthesis of such metal nanoparticles [20, 23–28]. Among many options, the use of colloidal routes is, without any doubt, one of the most, if not the most, conventional approach. This method is essentially based on the chemical reduction of a metallic precursor in the presence of a capping agent. This capping agent is extremely important in the process because it modifies the surface energies of the nanoparticles during their nucleation and growth steps, allowing the thermodynamic limitations to be overcome. As a result, metal nanoparticles with a particular shape can be obtained in a controlled way. Obviously, the nature of the capping agent is not the only parameter determining the resulting shape but also other experimental parameters including temperature, electrolyte, reducing time, nature of the reducing agent, and others, play an important role in defining the particular shape finally obtained.

As previously stated, the presence of a capping agent is of capital importance for the preparation of the shaped metal nanoparticles. However, if these nanoparticles are going to be applied in an electrocatalytic application, the presence, even in residual amounts, of such capping agents at the surface of the nanoparticles represents, from our point of view, one of the most critical aspects and should be evaluated with particular attention. In this sense, it is well-recognized from single-crystal studies that *clean surfaces* are required for a correct understanding of the relationships between surface structure and electrocatalytic surface reactivity. In this sense, the flame annealing protocol developed by Clavilier in 1980 represented the starting point that allowed not only ordered but also clean Pt single crystal surfaces to be obtained and reproduced in a significant number of laboratories worldwide [29]. This historical perspective points out the importance of working with "clean" shaped nanoparticles. Consequently, once the shaped metal nanoparticles are synthesized through appropriate colloidal routes, these must be submitted to specific decontamination procedures in order to completely remove the adsorbed species present at the surface of the nanoparticles. When this step is not performed, the electrocatalytic activity of the nanoparticles is affected in a non-controlled way which does not allow reproducible and comparable results to be obtained. In this chapter, and due to its extreme importance, we will include a particular section in which we will describe in detail some of the specific decontamination protocols described in the literature and applied for shape-controlled metal nanoparticles.

2.2 Surface Cleanness and Surface Site Determination

Once the importance of surface cleanness is understood, one critical question arises, *how should this surface cleanness be evaluated*? In this regard, the detailed knowledge gained with metal single crystals again gives us some interesting options about how to proceed in this aspect. For instance, it is well established that for Pt and Pd electrodes, the so-called hydrogen/anion adsorption-desorption states are an extremely sensitive process for the evaluation of the level of cleanness of the surface of such materials [30, 31]. Thus, when the different hydrogen/anion adsorption-desorption states are well-defined (sharpness) and reversible (symmetry), that is an indisputable proof of the correct cleanness of the surface. In addition, these processes may also provide other relevant aspects such as (i) electroactive surface area of the nanoparticles, the total adsorption or desorption charge being directly proportional to the total amount of exposed surface atoms and (ii) qualitative information about the nature and distribution of the different surface sites present at the surface of the nanoparticles. This latter point is of great interest because it allows the establishment of qualitative correlations between the shape and the surface structure of the nanoparticles to be obtained. In fact, the resulting voltammetric response given by a particular sample in a reference supporting electrolyte can be taken as a fingerprint of its specific surface structure.

Consequently, from the analysis of this electrochemical process it is possible to evaluate not only the level of cleanness of the samples but also their active surface area and surface structure, which are basic parameters to be then correlated with their electrocatalytic properties. Complementary site determination procedures based on irreversibly adsorbed adatoms will be described below.

Similarly, the analysis of the oxide formation/reduction region can be also used to extract information about the surface cleanness and also for the estimation of the active surface area of Pt and Pd and also for Au electrodes. In addition, from the voltammetric profile of the oxide region, it is also possible to evaluate the specific surface structure of the sample. This analysis is based on previous contributions by Hamelin [32, 33] dealing with Au single crystal electrodes and it has been applied to characterize different shape-controlled Au nanoparticles [34, 35]. However, in this case, and due to the high upper potential limit used, the surface stability of these shaped Au nanoparticles could be compromised, thus resulting in a possible surface restructuring. In fact, from the studies of the oxide formation on the gold single crystal electrodes, it was established that gold oxidation/reduction cycles induces the modification of the surface order of the electrodes [36, 37], and after several oxidation/reduction cycles, the initially flat surface of the single crystal electrode becomes rough. Consequently, and despite the changes induced by a single oxidation cycle are smaller than those seen for Pt electrodes [38], this oxidation/reduction treatment in the supporting electrolyte should be avoided.

Another interesting option for Au surfaces was described by El-Deab et al. [39–41], following previous observations by Porter et al. with polyoriented Au electrodes [42, 43]. They were able to estimate the fraction of the surface sites in various gold nanostructures by using the reductive desorption of different thiol compounds such as cysteine, mercaptoacetic acid or cystamine.

With a similar aim of determining the surface structure, other easy-to-do approaches should be also highlighted. Among them, some underpotential deposition (UPD) processes are particularly relevant because these electrochemical reactions are well-known to be very sensitive to the surface structure of the electrodes [44]. In this sense, lead (Pb) UPD on Au [45–51] and copper (Cu) UPD on Pd [52–54] and Pt [55, 56] surfaces have been reported to be very useful for the analysis of their surface structure. In addition, these UPD processes can also be used to calculate the real area of the electrode from the charge involved in the UPD reaction and subsequently normalized with a specific charge density normalization (generally obtained from a polyoriented surface). These different approaches have been very satisfactorily applied for different shape-controlled Au, Pt and Pd nanoparticles. An additional advantage of all these measurements is that they are performed in solution, i.e. the same environment in which the electrochemical reactions will take place. For instance, Fig. 2.1 shows the Pb UPD in alkaline solution of different shaped Au nanoparticles. As it is observed, the different peaks are extremely sensitive towards the shape of the nanoparticle, and therefore towards their surface structure. In particular, the results clearly show a preferential {1 0 0} and {1 1 1} surface structure of the (C) cubic and (D) octahedral Au samples, respectively. In addition, the spherical samples, and independently of their particle

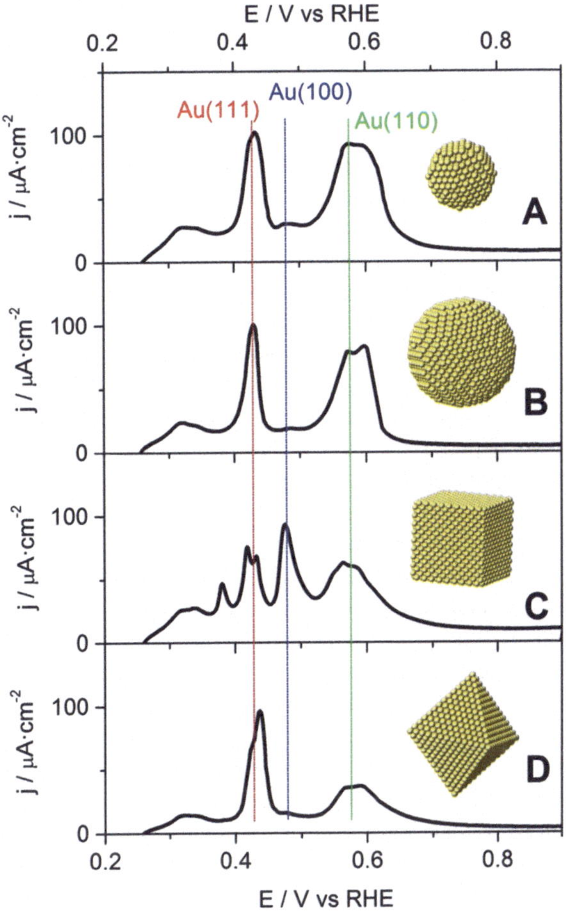

Fig. 2.1 Pb UPD voltammetric profiles of the unsupported gold nanoparticles in Ar-saturated 0.1 M NaOH containing 10^{-3} M Pb $(NO_3)_2$: (**a**) spherical 5 nm (**b**) spherical 30–40 nm (**c**) cubic and (**d**) octahedral nanoparticles. Scan rate: 50 mV s^{-1}

size, show a similar Pb UPD profile, which is characteristic of a polyoriented surface structure.

2.2.1 *Surface Active Sites Quantification*

It is worth noting that these previous approaches are able to provide qualitative information about the surface structure of the clean samples. However, it would be much more relevant to have surface probes to *quantitatively* estimate the different surface sites present at the surface of the nanoparticles. In this regard, to the best of our knowledge, this has been exclusively achieved for shape-controlled Pt nanoparticles using different approaches developed by our group [57]. The methodology is based on a detailed analysis of the redox behaviour of some adatoms (Bi,

Te and Ge) spontaneously adsorbed at the surface of different Pt single crystals, both ideally basal planes and series of vicinal stepped surfaces. Thus, whereas Bi [58, 59] and Te [60] were sensitive to the presence of {1 1 1} terrace domains, Ge [61] was sensitive to the {1 0 0} ordered domains. From the specific response of the adatom for each Pt surface, calibration plots showing the charge density measured under the adatom redox peak and the corresponding ({1 0 0} or {1 1 1}) terrace density were obtained. The quantification of the {1 1 1} and {1 0 0} oriented domains at the surface of the shape-controlled Pt nanoparticles was obtained performing similar irreversibly adsorbed Bi and Ge experiments and then simply using the normalization factors obtained in the calibration plots [57, 59]. This methodology is currently used in different laboratories [62–65].

It is also interesting to note that from the information gained with these Bi and Ge surface probes, we have also reported that other surface structure sensitive reactions could be used as electroanalytical detection tools to monitor the presence of some particular surface sites. Thus for instance, we showed that the desorption-adsorption of hydroquinone-derived adlayers [66] and the ammonia electro-oxidation [67] can be used to selectively quantify the {1 1 1} and {1 0 0} domains at the surface of preferentially oriented Pt nanoparticles, respectively.

Additionally, from the knowledge of the amount of {1 1 1} ordered domains obtained by using Bi or Te, it is also possible to determine the total amount of {1 0 0} and {1 1 0} sites at the surface of the different Pt electrodes (both bulk electrodes and nanoparticles) through pure deconvolution of the voltammograms obtained in the so-called hydrogen region [57]. However, it is important to note some differences between these two different approaches (Ge analysis vs deconvolution). Thus while from deconvolution the total fraction of {1 0 0} surface sites, that is, both step and terrace contributions can be estimated and even separately evaluated, from Ge analysis only those {1 0 0} terrace domains with a terrace width $\geq$4 Pt atoms (each Ge atom requires four platinum atoms to be adsorbed) can be determined. In any case, as expected, for {1 0 0} terrace contributions, both approaches give comparable measurements [57].

Finally, and before moving forward to the next section, it is also important to recall that these previous approaches are convenient for pure metal nanoparticles (Pt, Pd and Au). However, this kind of analysis becomes much more complicated when Pt and/or Pd based alloys or even core-shell nanostructures are used. In this regard, we would like to point out recent contributions by Stamenkovic et al. [68], Shao et al. [69], and Strasser et al. [70] in which different protocols including adsorption in the hydrogen region, CO stripping and Cu UPD charges are used, discussed and compared for different Pt- and Pd-based alloy nanoparticles also including shape-controlled alloy nanoparticles. From these contributions, one may easily realize that a correct determination of the electrochemically active area on these samples is not a trivial task and several reasonable considerations should be taken into account. In addition, the scenario turns almost unexplored if detailed information about the surface structure of these shape-controlled Pt- or Pd-based alloy nanoparticles is needed. These are remaining questions that should be addressed in the forthcoming years and that would allow a comparison of samples prepared with different methodologies to be properly performed.

2.2.2 Surface Cleaning Methodologies

As previously stated, surface cleanness is a critical requirement. If the surface of the nanoparticles is contaminated, it is not possible to reproduce and understand the resulting electrocatalytic activity. Consequently, we have considered of interest to specifically detail those protocols that have been proven to be effective for the cleaning of shape-controlled metal nanoparticles prepared from different colloidal routes, that is, in presence of different capping agents. However, it is worth noting that only those that have unequivocally evaluated the surface cleanness of the samples, using any of the surface cleanness proofs previously described, will be incorporated in this section.

In our first contributions, cubic Pt nanoparticles were prepared in the presence of NaPA [8, 9]. The removal of this NaPA was performed by simple addition of some NaOH pellets to the colloidal suspension. After the addition of NaOH, the colour of the solution changed from golden to black/grey and the Pt nanoparticles began to slowly precipitate. After complete precipitation, the supernatant was discarded and the nanoparticles were collected and transferred to a glass beaker where ultrapure water was added and the nanoparticles redispersed. Once nanoparticles again precipitated (no additional NaOH was required), the sample was again washed with ultrapure water. This water washing was repeated at least two times after which the sample was kept in water for the electrochemical measurements. The cyclic voltammograms obtained in 0.5 M H_2SO_4 showed a very well-defined voltammetric profile in the hydrogen region which clearly indicated the effectiveness of the decontamination procedure. In subsequent contributions, we also demonstrated that this decontamination protocol also resulted effective with cuboctahedral and tetrahedral-octahedral [57, 71–73] Pt nanoparticles prepared using a slightly modified route to that used for the cubic ones but also using NaPA as capping agent. Figure 2.2 shows some representative voltammetric profiles of these clean shape-controlled Pt nanoparticles in different supporting electrolytes (H_2SO_4, $HClO_4$ and NaOH). It is straightforward to notice the different current scales and deduce which test electrolyte is the most indicated to distinguish the surface characteristics of the different samples.

Furthermore, we also reported that this decontamination procedure was appropriate to obtain clean cubic Pd nanoparticles prepared in presence of cetyltrimethylammonium bromide (CTAB) [74–76]. It is worth noting that in all these cases, we always performed a classical CO adsorption stripping step to maximise the surface cleanness of the samples in the electrochemical environment. This CO adsorption-stripping allows the removal of residual impurities on the surface of the nanoparticles without altering their initial surface structure. This decontamination

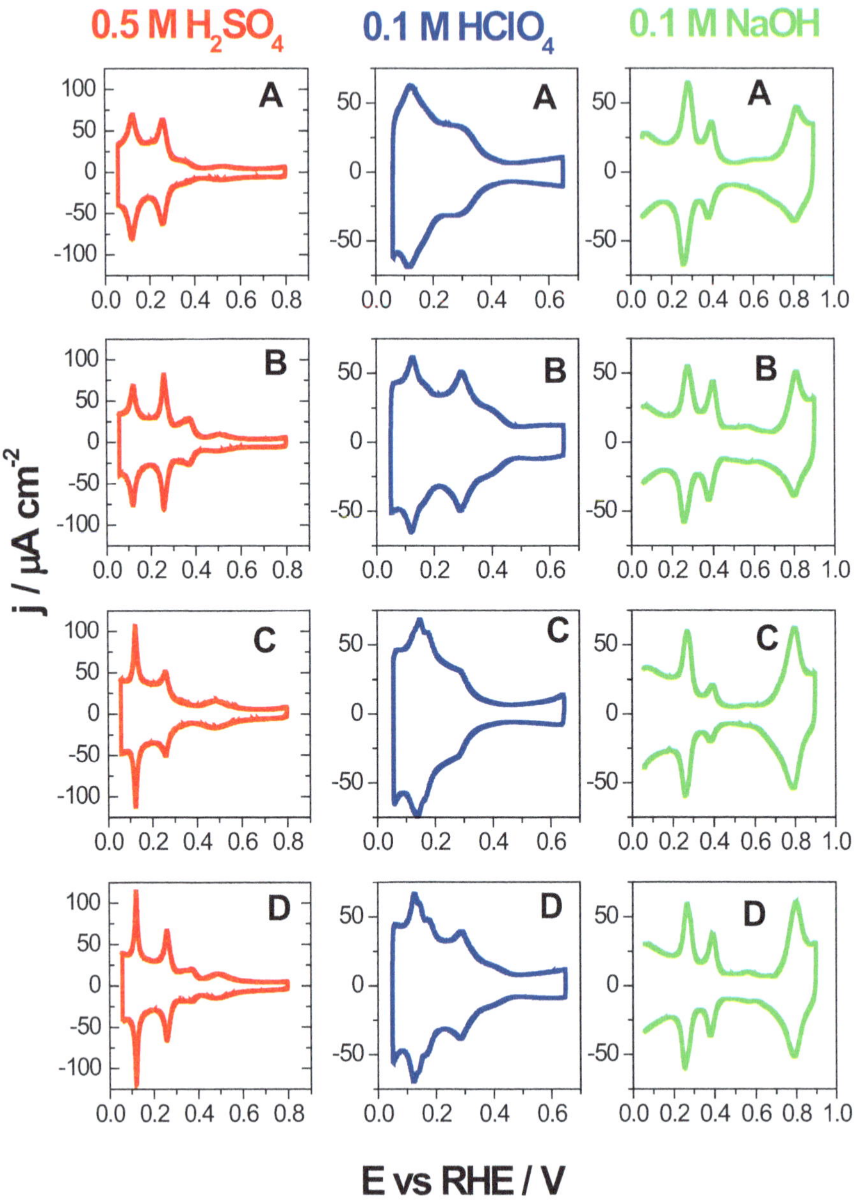

Fig. 2.2 Representative voltammetric profiles of (**a**) quasi-spherical, (**b**) cubic, (**c**) octahedral/tetrahedral and (**d**) cuboctahedral Pt nanoparticles obtained in different supporting electrolytes at 50 mV s^{-1}. Reproduced with permission from [73], © 2012 American Chemical Society

protocol has been adapted in other labs and research groups in other contributions [62, 77, 78].

In this sense, Coutanceau and co-workers [79, 80] also demonstrated that this decontamination procedure was efficient for cubic Pt nanoparticles prepared in the presence of tetradecyltrimethylammonium bromide (TTAB). The voltammetric response of such cubic Pt nanoparticles was very similar to those cubic samples prepared in presence of NaPA [57, 72]. Nevertheless, it is worth noting that this protocol cannot be considered as a universal cleaning method. In fact, Tong and co-workers used this decontamination procedure with shaped Pt nanoparticles (cubic and octahedral/tetrahedral) [81–83] synthesized by one-step polyol-based synthetic procedure, previously developed by Somorjai and coworkers [84], in which polyvinylpyrrolidone (PVP) was used as capping agent. The voltammetric responses of the samples showed insufficient definition and symmetry of the adsorption/desorption peaks which was attributed to the presence of residual PVP blocking the surface sites. Thermogravimetric analyses (TGA) were also performed in order to determine the PVP content of the samples after the NaOH cleaning. The PVP content was determined to be 16 and 11.3 wt% for the octahedral/tetrahedral and cubic samples, respectively thus indicating that a significant amount of PVP still remained at the surface of the nanoparticles. Similarly, Coutanceau and co-workers were also unable to clean similar shaped Pt nanoparticles [79] to those prepared by Tong and co-workers. However, more recently, Tong and co-workers have developed a new protocol to obtain PVP-free cubic and octahedral/tetrahedral Pt nanoparticles using an adapted liquid phase UV photo-oxidation (UVPO) technique [65]. In brief, the PVP-capped Pt nanoparticles were initially centrifuged with acetone to remove the excess ethylene glycol and PVP, until the nanoparticles were clearly separated. The resulting Pt nanoparticles were then dispersed in an ethanol-hexane mixture and again centrifuged. This ethanol-hexane cleaning was repeated at least three times and finally the sample was dispersed in ethanol. Furthermore, and based on previous contributions, a modified procedure of liquid phase UV combined with oxygenated H_2O_2 was applied to remove the residual PVP present at the surface of the nanoparticles. In detail, a small portion of the ethanolic-suspension containing the nanoparticles was dispersed into about 2 mL of H_2O with 200–250 mg NaOH to yield an alkaline environment. On the other hand, a basic and O_2-saturated H_2O_2 solution was prepared by adding 200–250 mg NaOH to a 10 mL H_2O_2 solution which was then bubbled for at least 10 min with O_2 gas. The basic solution containing the nanoparticles was UV irradiated (254 nm) for 1 h while about 1 mL of O_2-saturated H_2O_2 solution was added to the solution every 10 min. After the UV exposure, the solution was finally purified by repetitive centrifugation and precipitation process with ethanol. The voltammetric responses of the shaped Pt nanoparticles in 0.5 M H_2SO_4, Fig. 2.3, were rather similar to those previously reported with similar and clean cubic and octahedral/tetrahedral Pt nanoparticles prepared by our group (see Fig. 2.2). In addition, TGA exhibited no major weight loss due to the desorption of PVP, thus suggesting its absence at the surface of the nanoparticles.

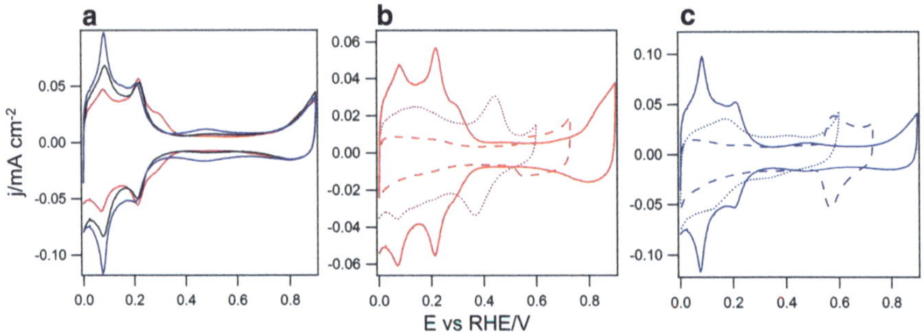

Fig. 2.3 (**a**) Voltammetric profiles of the Pt black (*black*), cubic (*red*) and octahedral/tetrahedral (*blue*) Pt nanoparticles in 0.5 M H_2SO_4. Comparison of CV profiles of pristine (*solid line*) cubic (**b**) and octahedral/tetrahedral (**c**) Pt nanoparticles with CVs after germanium (*dotted line*) and bismuth (*dashed line*) adsorption in 0.5 M H_2SO_4. Reproduced with permission from [65], © 2014 Royal Society of Chemistry

However, this methodology was a modified protocol based on a previous contribution by Koper and co-workers in which PVP-protected Pt nanoparticles were cleaned using a H_2O_2/H_2SO_4 solution [85]. Interestingly, the effectiveness of the methodology was also tested with a Pt(1 1 1) single crystal electrode which demonstrated that the surface was decontaminated without a loss of its initial crystalline surface structure. In addition, they suggested that the decontamination of the surface was in fact a physical removal process of the PVP induced by the oxygen bubbling produced during the decomposition of the H_2O_2. As far as we know, this contribution was the first example of effective cleanness with nanoparticles prepared in the presence of PVP.

More recently, Yang et al. have proposed a new procedure for removing different capping agents commonly used for the synthesis of shape-controlled Pt nanoparticles such as PVP and oleylamine/oleic acid, by a simple electrochemical potential cycling in 0.5 M NaOH [86]. The effectiveness of this protocol was proven by using Pt nanocubes prepared in the presence of such capping agents. The obtained particles were washed by centrifugation with a saturated ethanol solution of NaOH for those nanoparticles prepared in presence of PVP, or in hexane/ethanol mixture for those ones prepared with oleylamine/oleic acid, and finally redispersed in ethanol and hexane, respectively. The different nanoparticles were deposited on a gold electrode and cycled between 0 and 1.0 V (vs RHE) in 0.5 M NaOH with a scan rate of 0.5 V s^{-1}. After 100 cycles, the electrodes were transferred to a second electrochemical cell containing 0.5 M H_2SO_4 where the cleanness was evaluated. The voltammetric response obtained clearly suggested that this electrochemical potential cycling in 0.5 M NaOH solution is an effective way to remove both capping agents, PVP and oleylamine/oleic acid from the surface of the nanoparticles. Unfortunately, this cleaning method has some practical limitations, particularly if large amounts of nanoparticles are needed. In this regard, we have recently developed a new decontamination procedure to obtain a new collection of clean shape-controlled Pt nanoparticles prepared using

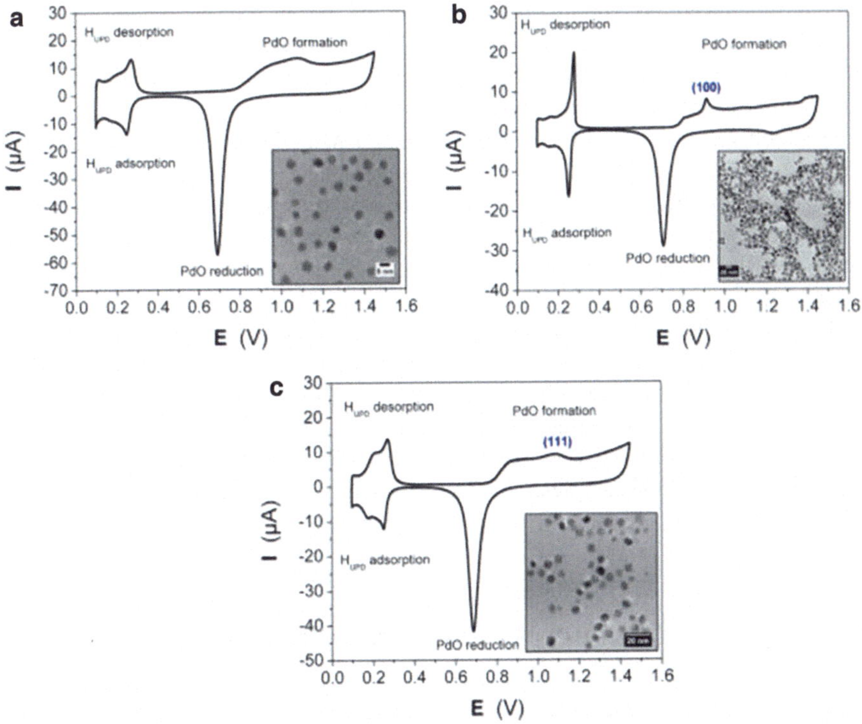

Fig. 2.4 Voltammetric profiles of Pd (**a**) nanospheres, (**b**) nanocubes and (**c**) nanooctahedra in 0.5 M H_2SO_4 at 5 mV s^{-1}. The insets present TEM images of the Pd nanoparticles. Reproduced with permission from [89], © 2014 American Chemical Society

oleylamine/oleic acid as capping material/solvent [87]. The method is not based on electrochemical cleaning but in a chemical one which allows important amounts of Pt nanoparticles to be cleaned.

Coutanceau and co-workers [88, 89] have recently reported that clean shaped controlled Pd nanoparticles prepared in the presence of PVP can be obtained using a NaOH treatment similar to that unsuccessfully used in previous contributions [79, 81–83]. From our point of view, the main differences of this treatment in comparison with the previous ones seem to deal with the required NaOH concentration and the addition of a particular amount of water. In more detail, the resulting colloidal suspension containing the shaped Pd nanoparticles was diluted with H_2O after which NaOH pellets were added to have a 1 M NaOH solution (this NaOH concentration is significantly higher than that previously used). The nanoparticles were collected by precipitation. This protocol was repeated until clean CV voltammetric profiles were obtained, Fig. 2.4.

Neergat and co-workers have recently proposed two new decontamination procedures for the removal of adsorbed stabilisers and capping agents used for the preparation of shape-controlled Pd nanoparticles [90, 91]. The shaped Pd nanoparticles were prepared with PVP also in the presence of other additives such as Br^-, Cl^- and citrate ions. After some preliminary cleaning steps for the removal of the excess of chemicals used during the syntheses, the samples were then treated with solutions containing tert-butylamine [91] or $NaBH_4$ [90] for a particular time, after which were collected by centrifugation, washed with different solvents (ethanol or water) and finally dried. The samples were electrochemically characterized in 0.1 M $HClO_4$ and the response of the treated samples was clearly much better in comparison with the as-prepared ones. This effective surface cleaning also allowed better ORR activities to be recorded as a consequence of the improved surface cleanness.

Another interesting option that has been applied for shape-controlled Au nanoparticles is based on the electrochemical deposition of a film of PbO_2 in alkaline solution during the Pb UPD experiments [45–47, 92]. This methodology was proven to be very effective for the removal of different capping agents (polyethylene glycol dodecyl ether (Brij®30) and CTAB) without altering the initial surface structure of the samples. This fact was demonstrated by testing the protocol with both single crystal and polyoriented Au electrodes.

2.2.3 Surface Cleanness Versus Surface Disordering

As previously stated, surface cleanness is a fundamental requirement to properly correlate the surface structure and reactivity of shape-controlled metal nanoparticles. However, this surface cleaning must be performed without perturbing the intrinsic/initial surface structure of the nanoparticles. To illustrate this situation, we will discuss the use of UV/ozone as decontamination protocol. With the aim of evidencing the importance of an effective surface cleaning, Somorjai and co-workers evaluated the use of UV/ozone and its effects on the catalytic properties of some shape-controlled Pt nanoparticles [93–95]. Interestingly, they observed a remarkably increase in the catalytic activity after the UV/ozone procedure which was justified in terms of a more efficient cleanness of the samples and consequently enhanced catalytic properties. Also, they demonstrated by TEM measurements that both shape and size of the UV/ozone treated nanoparticles remained unmodified in comparison with the untreated samples. It should be stressed that electrocatalysis depends on the real number of clean atoms of the catalyst in contact with the solution, that is, the more clean atoms, the higher the activity. A similar UV/ozone treatment was also used by Kiwi-Minsker to clean PVP-stabilized Pd nanocubes [96]. They also reported improved catalytic activities and, based on high-resolution scanning electron microscopic images, they also stated that the Pd nanocubes were morphologically stable. However, as previously stated, the shape of the nanoparticles is not the key parameter determining the resulting catalytic activity but their surface structure at atomic level and this is important for particular reactions. In fact, our group demonstrated that this UV/ozone treatment strongly perturbed the surface structure of the nanoparticles but, as

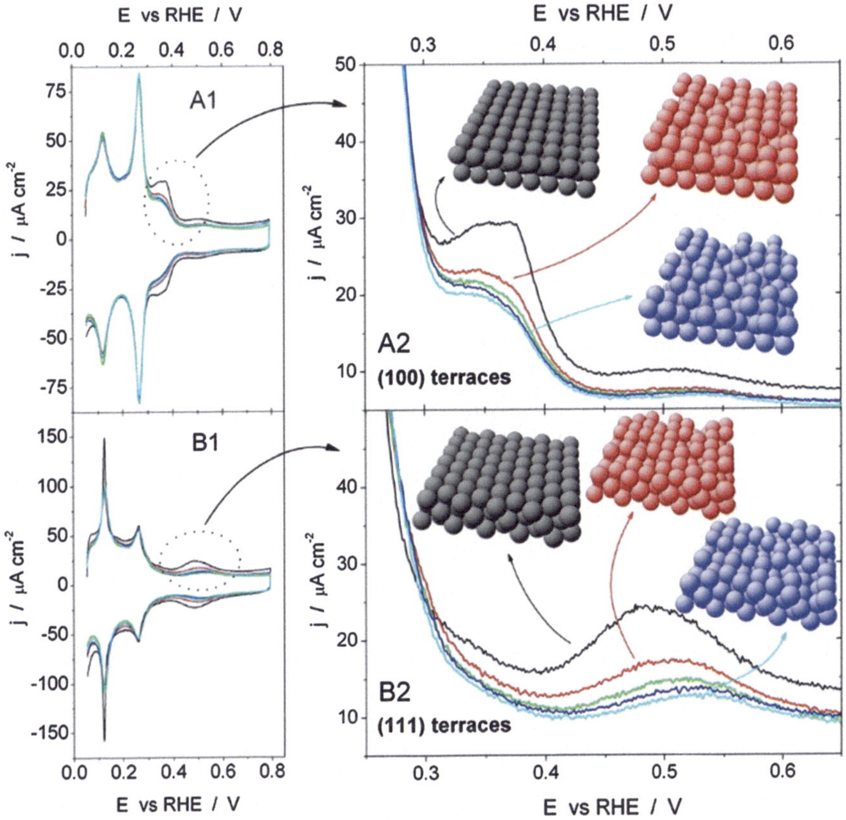

Fig. 2.5 Voltammetric profiles before (*black*) and after 30 (*red*), 60 (*green*), 90 (*blue*), 120 (*cyan*) and 150 (*magenta*) minutes of UV/ozone exposure. (**a**) cubic nanoparticles, (**b**) tetrahedral–octahedral nanoparticles. Test solution: 0.5 M H_2SO_4. Sweep rate 50 mV s^{-1} Reproduced with permission from [97], © 2011 Elsevier

previously shown by Somorjai and co-workers, without altering neither the size nor the shape of the nanoparticles [97]. As a consequence of the perturbation of arrangement of the atoms at the surface of the nanoparticles, in particular those present as terrace domains, the electrocatalytic properties of the samples were drastically modified as illustrated in Fig. 2.5. It should be pointed out that in all cases the surfaces were equally clean, but after the ozone treatment, the surface atoms were less ordered and some particular reactions only take place on ordered domains.

2.2.4 Can the Residual Presence of Capping Agents Have Any Beneficial Effect in Electrocatalysis?

Although this aspect is not going to be analysed in the present chapter, it is important to point out the existence of a number of interesting contributions that claim that a determined presence of certain species, including some capping agents, which are usually regarded as poisoning species because block the active electrocatalytic sites and slow reaction kinetics, may provide beneficial effects in terms of electrocatalytic properties. These effects point out to a third-body role of the surface residues, unfortunately uncontrolled and likely not stable. For those readers interested in this topic we refer to [98–107].

Finally, and before moving forward to the next section, it is also worth noting that shape-controlled metal nanoparticles can also be prepared in the absence of capping agents. In this regard, the electrochemical approach developed by Sun and co-workers is one of the most interesting methodologies [5, 108, 109]. Other interesting approaches are surfactant-free solvothermal synthesis [110–112] or solid-state chemistry methods [113].

2.3 Electrocatalytic Properties of Shape Controlled Metal Nanoparticles

In this section, we will summarize some of the most relevant advances on the use of these shaped metal nanoparticles in electrocatalysis and, in particular, towards reactions of interest for low temperature fuel cells such as formic acid, methanol and ethanol electrooxidations and oxygen reduction. However, only those contributions that, from our point of view, fulfil the particular requirements of surface cleanness and surface structure–electrochemical reactivity correlation will be discussed.

2.3.1 Formic Acid Electrooxidation

Formic acid electrooxidation has been the subject of innumerable contributions, not only because it is a model reaction for a two-electron-transfer reaction, but also for its promising use as fuel in direct formic acid fuel cells (DFAFCs) [3, 114–116]. It is widely accepted that the oxidation of formic acid takes place through a dual path mechanism [117–119]: a direct pathway via an adsorbed active intermediate, the nature and role of which are still under strong discussion at the fundamental level and that readily yield to CO_2, and a second pathway involving the formation of adsorbed CO, which is considered to be a poisoning intermediate, that can be further oxidized to CO_2 at high overpotentials. On Pt surfaces, it is well recognized that both reaction paths are structure sensitive, being the Pt(1 0 0) surface the most active (in the absence of CO poisoning), but also the most poisoned [115]. The remarkable structure sensitivity of this electrochemical process has been also illustrated with shape-controlled Pt nanoparticles. Tian et al. prepared tetrahexahedral (THH) Pt nanocrystals with enhanced electrocatalytic activity towards formic acid electrooxidation using a novel electrochemical approach [108]. The electrocatalytic activity of the THH Pt nanoparticles was always higher than those observed for polycrystalline Pt nanospheres and commercial Pt/C catalyst (E-TEK Co., Ltd), independently of the applied potential range. Interestingly, the best enhancement factors (about four and three times higher than for the nanospheres and the commercial catalyst, respectively) were observed at low potential values of about 0.1 V vs SCE. This contribution was the first synthesis of Pt nanoparticles enclosed with high-index facets and clearly evidenced the great potential of the metal nanoparticles containing high-index facets for some electrocatalytic applications. Some of them will be discussed in this chapter.

Taking into account the previous studies with Pt single crystals in which the Pt (1 0 0) surface was shown to be the most active basal plane, we evaluated the response of different shape-controlled Pt nanoparticles. Thus, spherical, cubic, cubooctahedral and tetrahedral-octahedral Pt nanoparticles were prepared and electrochemically characterized [72]. From the voltammetric profiles of the samples we deduced that the surface structure of these samples was, preferentially polyoriented, (1 0 0), (1 0 0)-(1 1 1) and (1 1 1), respectively, as expected from the stereographic triangle correlation between shape of the nanoparticle and the surface structure of an fcc metal. However, as previously discussed, the results also evidenced that the nanoparticles did not contain a unique type of surface site but that all surface sites were observed, despite having a clear preferential surface structure. In terms of activity towards formic acid electrooxidation, the results obtained perfectly agreed with those obtained with the model Pt surfaces and the different samples showed a similar behaviour to that of their corresponding Pt model surfaces. Thus, for instance, in similarity with a Pt(1 0 0) electrode, those Pt particles having a preferential (1 0 0) orientation showed the best activity in the negative going sweep of the cyclic voltammetric measurements but very low activity during the chronoamperometric experiments as consequence of their fast CO poisoning.

Using similar shape-controlled Pt nanoparticles, we re-evaluated their activity towards formic acid electrooxidation but in this case using pulsed voltammetry [120]. This technique is very useful to independently study both reaction paths and allows the intrinsic activities and the poisoning rates to be determined. Figure 2.6 shows the activities of the (1 0 0) and (1 1 1) preferentially oriented Pt nanoparticles obtained at $t = 0$ (intrinsic activity) corresponding with the activity of the electrode through the active intermediate reaction path in the absence of poison, and that obtained for $t = 1$ s after which the surface poisoning is already observed.

Huang et al. [121] prepared concave Pt nanocrystals having {4 1 1} high-index facets by using amines as capping agents. The electrocatalytic properties of these samples towards formic acid electrooxidation showed that the activity (normalized

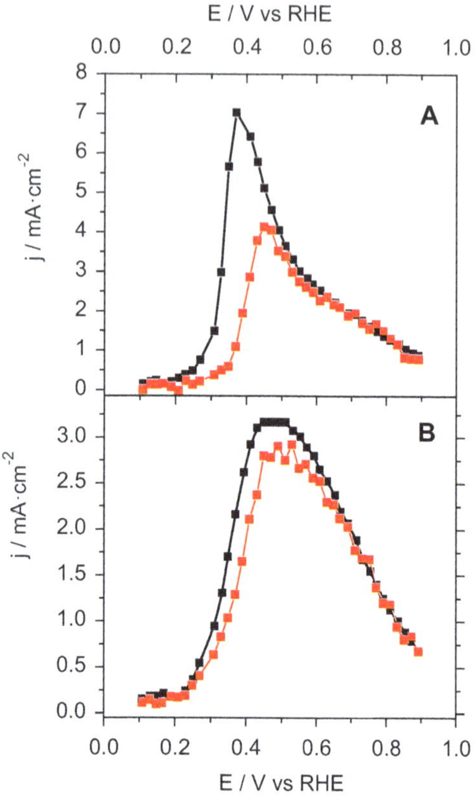

Fig. 2.6 Intrinsic activities (t = 0 s) (*black*) and current densities after t = 1 s (*red*) obtained from pulse voltammetry transients for cubic (**a**) and octahedral/ tetrahedral (**b**) Pt nanoparticles in 0.5 M H_2SO_4 + 0.1 M HCOOH. Reproduced with permission from [120], © 2010 American Chemical Society

to the electrochemically active surface and measured at 0.61 V vs SCE) was two and five times higher than that obtained with a commercial Pt black and Pt/C, respectively. In addition, these concave Pt nanocrystals showed an excellent stability after the electrochemical experiments.

Li et al. [122] reported the preparation of trapezohedral (TPH) Pt nanocrystals enclosed by {5 2 2} high-index facets using a similar electrochemical approach to that developed by Tian et al. [108]. These nanocrystals were tested for different reactions of interest including formic acid electrooxidation. The peak current density (normalized to the electroactive surface area estimated from the so-called hydrogen region) of formic acid electrooxidation at 0.60 V in the positive potential scan was observed to be about three times higher than that obtained with commercial Pt/C samples.

Zhang et al. [123] prepared Pt concave nanocubes with high-index {h k 0} facets using glycine as surface modifier. These nanoparticles showed enhanced specific activities (again normalized to the electroactive surface area estimated from the so-called hydrogen region) towards formic acid electrooxidation in comparison to commercial Pt black and Pt/C catalysts. Unfortunately, the samples were irradiated with a UV lamp (185 and 254 nm) for 12 h in air in order to remove the capping agents. This cleaning methodology was shown to strongly perturb the surface structure of the samples [97].

Xia et al. [124] prepared highly concave Pt nanoframes with high-index facets using an oleylamine-assisted solvothermal method. The resulting Pt nanoframes exhibited a reasonable electrocatalytic activity and an improved electrochemical stability for different reaction of interest including formic acid electrooxidation.

More recently, Korzeniewski et al. reported an interesting surfactant-free solvothermal method for the preparation of <10 nm shape-controlled Pt nanoparticles [111]. The shape of the different Pt nanoparticles (truncated octahedral, cuboctahedral or cubic) was essentiality determined by the water volume fraction (optimal 3 %) in N,N-dimethyformamide (DMF) and the reaction time. Both the voltammetric profiles obtained in 0.5 M H_2SO_4 and the responses towards formic acid electrooxidation were similar to those reported in our previous works [57, 72].

The influence of the shape/surface structure on Pd nanoparticles towards formic acid electrooxidation has been studied more extensively. Interestingly, on Pd surfaces the dehydration step does not take place (CO is not formed) and consequently the reaction directly proceeds through the direct path. Moreover, the onset potential for the oxidation is ca. 200 mV lower than that observed for Pt [125–127]. However, from single crystal studies, it was also shown that the reaction is structure sensitive, being the Pd(1 0 0) the most active surface among the basal planes, both in sulphuric and perchloric acid solutions [128, 129]. As a consequence, the preparation of shape-controlled Pd nanoparticles and particularly cubic nanoparticles in which the (1 0 0) orientation is maximised, has been a subject of great interest. From our point of view, the first relevant contributions were reported by Jin et al. [130] and by Zhang et al. [131]. Jin et al. prepared different Pd nanoparticles including cubes, truncated cubes, cuboctahedra, truncated octahedra, and octahedra. That is, Pd nanoparticles containing different fractions of {1 0 0} and {1 1 1} surface domains. As expected from Pd single crystal studies, the activity systematically improved for those nanoparticles having a higher fraction of {1 0 0} domains, being the cubic sample the most active one, Fig. 2.7. Cu UPD was used in this case for the determination of the electroactive surface area of the nanoparticles.

Similarly, Zhang et al. synthesized rhombic dodecahedral and cubic Pd nanoparticles using a simple one-pot synthesis in the presence of CTAB as capping agent and KI as additive [131]. The voltammetric profiles of these nanoparticles both in H_2SO_4 and in NaOH solutions showed characteristic features corresponding to preferentially oriented {1 1 0} and {1 0 0} Pd nanoparticles, respectively. The Pd nanocubes again exhibited a higher electrocatalytic activity toward formic acid electrooxidation.

Almost in parallel with these contributions, we also studied the electrocatalytic properties of spherical and cubic Pd nanoparticles towards formic acid electrooxidation both in H_2SO_4 and $HClO_4$ solutions [76]. Interestingly, we demonstrated that, to properly compare the intrinsic electrocatalytic properties of both nanoparticles, it was required to minimize the amount of samples during the measurements to avoid diffusional problems of reactants and products that result

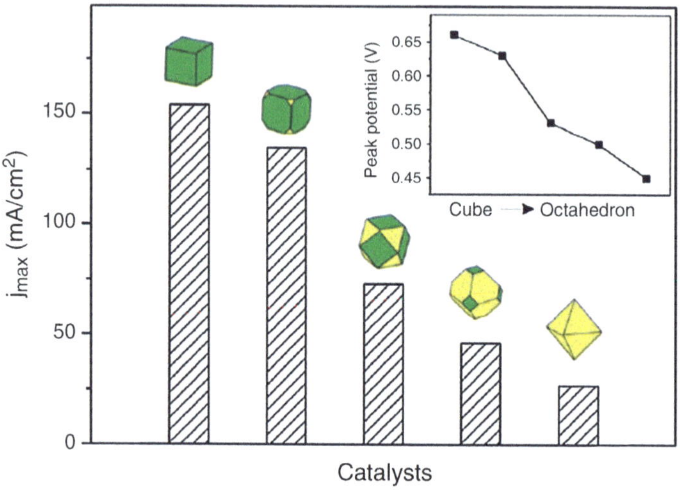

Fig. 2.7 Maximum current densities for formic acid electrooxidation (normalized to the electrochemical surface area) for different shaped Pd nanoparticles in 0.1 M $HClO_4$ + 2 M HCOOH at 10 mV s^{-1}. The inset shows their corresponding peak potentials. Reproduced with permission from [130], © 2012 Royal Society of Chemistry

in a lower apparent activity. Under these controlled conditions, we observed that the activity of the Pd nanocubes was remarkably higher than that found for the spherical nanoparticles. In addition, we reported that the activity found in $HClO_4$ was significantly higher than that obtained in H_2SO_4 in the whole potential range thus pointing out the competitive effect of the adsorbed anions towards formic acid oxidation at these surfaces.

From these initial contributions, new interesting approaches have been reported. Shao et al. prepared different concave Pt nanocrystals (nanocubes, 5-fold twinned nanorods and right bipyramids) [132]. These concave nanocrystals showed enhanced electrocatalytic activity (normalized to electrochemically active surface area estimated from Cu UPD measurements). Kuo et al. studied the effect of lattice strain on the catalytic properties of some shape-controlled Pd nanoparticles [133]. They prepared core–shell AuPd nanocubes and nanooctahedra, and twinned Pd icosahedra that showed lattice strain, detected by X-ray diffraction measurements, due to the intrinsic twinning in the Pd icosahedra and the lattice mismatch between the Au and Pd shells in the AuPd core–shell particles. The electrocatalytic properties of these strained nanoparticles towards formic acid electrooxidation were compared with those obtained with Pd nanocubes and nanooctahedra. The strained nanoparticles showed enhanced activities due to the upshift of the d-band center of Pd as consequence of the lattice expansion.

Xia and co-workers also prepared different shape-controlled Pd nanoparticles including cubes, octahedra, icosahedra and right bipyramids which were used as electrocatalysts towards formic acid electrooxidation [134–136]. Unfortunately, and due to the different experimental conditions used in the electrocatalytic experiments, it is not possible to establish a clear tendency. However, we would like to extract a controversial point from one of these contributions. Shao et al. [136] stated that the use of shape-controlled Pd nanoparticles did not produce any advantage in comparison with conventional Pd nanoparticles. From our point of view, this fact can be explained in terms of an insufficient cleanness of the nanoparticles. In fact, the reported voltammetric profiles are remarkably different to those obtained by Coutanceau et al. [88, 89] using the same synthetic protocol but after a careful surface cleaning of the resulting Pd nanoparticles. This is a clear example of how important the surface cleanness is for electrocatalytic measurements.

Zhang et al. evaluated the electrocatalytic properties of different-shaped and monodispersed (50 nm) palladium nanocrystals, including cubes, octahedra and rhombic dodecahedra, towards formic acid electrooxidation both in $HClO_4$ and H_2SO_4 solutions [137]. The voltammetric responses, in the absence of formic acid, of the different samples showed some of the characteristic features observed with Pd single crystals, which warranties the level of cleanness of the nanoparticles. In both supporting electrolytes, the cubic Pt nanoparticles showed the highest activity as a result of their preferential {1 0 0} surface structure. In addition, the activities reported in $HClO_4$ were higher than those obtained in H_2SO_4 due to the corresponding competitive adsorption of anions. These results were in full agreement with previous observations reported by our group [76].

More recently, Zhang et al. [138] prepared cubic, cuboctahedral and octahedral Pd nanoparticles after HCl oxidative etching, occurring at the corners of cubic Pd nanoparticles. The cubic Pd nanoparticles again showed the highest activity towards formic acid electrooxidation.

Tang et al. have recently reported the synthesis of length tuneable penta-twinned Pd nanorods and their activity toward formic acid electrooxidation in comparison with cubic Pd nanoparticles [139]. Before the electrochemical testing, the different samples were subjected to an electrochemical potential cycling between −1.2 and +0.6 V vs Ag/AgCl at 500 mV s^{-1} and for 100 cycles in a CO-saturated 0.1 M $NaClO_4$ + 1 mM NaOH solution. Despite they claimed that the resulting CVs were characteristic of clean surfaces and similar to those reported in previous contributions, we are not in agreement with this assumption and consider that neither the surface cleanness nor surface structure preservation were demonstrated. These are, from our point of view, critical aspects that must be clearly shown.

It is well-known that the incorporation of a second element both on Pt and Pd surfaces is an interesting approach to improve the electrocatalytic properties of these surfaces towards formic acid electrooxidation [116]. This improvement has been explained in terms of third-body, electronic and bifunctional effects. Third-body effects take mainly place when the surfaces are directly modified by the inclusion of a second element that acts as a mere spectator, and only blocks surface sites, diminishing the activity for an undesired reaction. However, electronic effects occur when the modified surface also changes its electronic properties. In this latter point, the so-called d-band center model [140] is widely accepted and explains the possible modifications of the electronic properties and binding energies of a transition metal surface, after the incorporation of a second metal, and their correlations with the resulting catalytic activity. Finally, bifunctional effects appear when the second metal contributes to the reaction providing a required group at lower potentials, which facilitates the reaction mechanism.

For shape-controlled metal nanoparticles, two main strategies have been employed (i) deposition of different adatoms at the surface of the shaped nanoparticles and (ii) preparation of shape-controlled Pt or Pd alloy based nanoparticles. In the following, we will describe some of the most relevant contributions in these two approaches.

Our group has intensively worked on the use of adatoms to selectively decorate the surface of different shape-controlled Pt nanoparticles towards formic acid electrooxidation. Following previous results obtained with adatom decorated Pt single crystals [115, 141], we have used Bi [142, 143], Pd [144, 145], Sb [146] and more recently Tl [147] as surface modifiers. The reported results showed that improved electrocatalytic activity can be obtained in all cases, although the enhancement factor strongly depends on the surface structure of the substrate and the nature of the adatom. In particular, the systems Pd/Pt_{cubic}, $Bi/Pt_{octahedral}$, $Sb/Pt_{octahedral}$ and Tl/Pt_{cubic} were the most convenient combinations. Nevertheless, it is important to note that the maximum oxidation current values, as well as the optimal potential region, are different in these systems. Thus, for instance, whereas the system Pd/Pt_{cubic} showed a clear shift of the oxidation process towards much lower potential values (maximum currents at 0.3 V) but without a great enhancement in terms of maximum oxidation currents, the system $Bi/Pt_{octahedral}$ provided an important benefit in relation to the maximum oxidation current but at similar potential values (about 0.6 V) to that of the unmodified samples.

In collaboration with the Sun group, we also demonstrated that this strategy also provided interesting results with high-index facet Pt nanoparticles [142]. In this way, THH Pt nanocrystals prepared using the electrochemical approach developed by Tian et al. [108] were decorated with Bi adatoms and their electrocatalytic activity was evaluated for the formic acid electrooxidation. The results obtained again led to an enormous enhanced activity in comparison with the bare THH Pt nanocrystals. In voltammetric experiments an enhancement factor of about 20 (peak current density) was found for the highest Bi coverage ($\theta_{Bi=0.9}$). In addition, chronoamperometric measurements also showed a very significant enhancement that varied from 65 to 1.5 depending on the electrode potential and Bi coverage.

Later, Sun's group published a new contribution in which Au-decorated THH Pt nanocrystals were used towards formic acid electrooxidation [148]. The gold decoration was performed through an initial sub-monolayer Cu UPD which was then exchanged with Au adatoms by simple galvanic replacement. The presence of Au adatoms clearly inhibited the formation of CO trough the dehydration step although, in comparison with the Bi modification, did not provoke an enhanced electrocatalytic activity. This is due to the fact that while in the case of Au decoration, only a third-body effect is occurring, for Bi decoration, where the peak current is more than five times higher than that of Au-modified THH Pt,

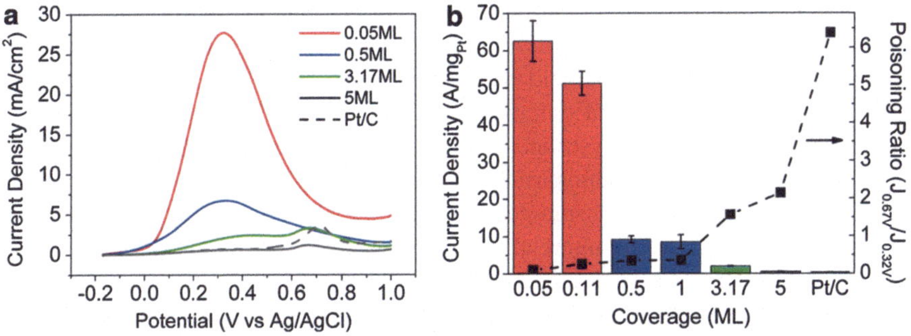

Fig. 2.8 (**a**) Positive scan of the cyclic voltammograms for 0.05 ML, 0.5 ML, 3.17 ML, 5 ML samples, and commercial Pt/C catalyst obtained in 0.5 M HCOOH + 0.1 M H_2SO_4 solution at 50 mV s^{-1}, (**b**) comparison of the mass activities measured at 0.32 V and poisoning ratios (current density at 0.67 V divided by the current density at 0.32 V). Reproduced with permission from [35], © 2013 American Chemical Society

both third-body effect and electronic (ligand) effect are contributing to the enhancement activity.

Yang and Lee reported a very interesting approach dealing with the epitaxially deposition of Pt on gold octahedral nanocrystals enclosed by {1 1 1} facets [35]. The amount/coverage of Pt was finely controlled from fully covered multiple overlayers (about five monolayers) to atomically dispersed submonolayer (0.05 monolayer). Very interestingly, the electrocatalytic activity of Pt modified gold octahedra remarkably increased for decreased Pt coverages. For 0.05 monolayer of Pt a spectacular mass activity of about 63 A $mgPt^{-1}$ was found. This activity was about 120 and 170 times higher than those obtained with 5 ML and with a commercial Pt/C catalyst, Fig. 2.8. Such enormous enhancement was attributed to a third body (lack of ensembles) and bifunctional effects at the Pt-Au sites.

Several shape-controlled Pt and Pd alloy nanoparticles have been tested for the electrooxidation of formic acid. Wang et al. synthesized PdPt alloy nanocubes with tunable compositions [149] and their electrocatalytic activity towards formic acid electrooxidation was compared with those obtained with Pd nanocubes and a commercial Pd black. They observed that the activity of the PdPt nanocube alloys was, independently of their composition, higher than those of the Pd nanocubes and commercial Pd black. In particular, the $Pd_{74.4}Pt_{25.6}$ alloy sample showed the highest formic acid oxidation activity which was about two times higher than that observed for the Pd nanocubes and commercial Pd black.

Fang's group has reported different contributions about the use of shape-controlled Pt alloy nanoparticles as electrocatalysts towards formic acid electrooxidation including Pt_3Fe nanocubes [150], PtCu nanocubes of different atomic composition (Pt_xCu_{100-x} (x = 54–80 at.%)) [151, 152] and Pt-Cu nanooctahedra [153, 154]. In addition, Fang et al. have recently published two interesting papers in which the electrocatalytic activity-enhancement of different shape-controlled Pt-based bimetallic nanocrystals, also including high-index noble metal nanostructures, has been studied [155, 156].

Kang and Murray reported the synthesis and electrocatalytic properties of PtMn nanocubes [157]. The electrocatalytic properties of the PtMn nanocubes (including oxygen reduction and formic acid and methanol oxidation reactions) were evaluated and compared with that obtained with spherical PtMn nanoparticles, with a commercial E-TEK Pt catalysts and with a Pt black. For formic acid oxidation both cubic and spherical PtMn nanoparticles were found to be less active (currents were normalized to surface areas estimated from the charge of the hydrogen adsorption-desorption region) than commercial ETEK Pt/C catalyst, although the cubic PtMn nanoparticles showed higher activity than the spherical ones.

Yu et al. [158] synthesized polyhedral AuPd core-shell structures (60–80 nm) having high-index facets (concave trisoctahedral (TOH) and hexoctahedral (HOH) crystals both with {h k l} facets and THH crystals with {h k 0} facets)). Concave TOH gold nanocrystals were used as seeds onto which Pd layers were heteroepitaxially grown. Their activity for formic acid electrooxidation was also evaluated and compared with that obtained with cubic and octahedral AuPd core-shell nanocrystals enclosed by low-index {1 0 0} and {1 1 1} facets. The maximum current density of formic acid oxidation obtained in the positive going-scan followed the order of octahedra $<$ TOH $<$ HOH $<$ cubes $\cong$ THH.

Zhang et al. [159] obtained HOH AuPd alloy nanoparticles with 48{hkl} facets with a uniform size distribution (~55 nm). Interestingly, these samples showed an activity towards formic acid electrooxidation about five times higher than that obtained with a commercial Pd black catalyst, which was attributed to a synergy effect of the high-index facets and the AuPd alloy composition.

Deng et al. [160] prepared THH PdPt alloy nanocrystals enclosed by $\{10, 3, 0\}$ high-index facets using the electrochemical approach [108]. At the optimal atomic composition ($Pd_{90}Pt_{10}$) the THH nanocrystals exhibited a catalytic activity towards formic acid electrooxidation in perchloric solution that was about three times higher than that on THH Pd nanocrystals, and six times higher than on commercial Pd black catalyst, Fig. 2.9. The enhanced activity was again related to the synergy effect of high-index facets and particular electronic structure of the alloy formed.

Zhang et al. [161] prepared different shape-controlled PtPd bimetallic alloy nanoparticles including PtPd cubes, bars, flowers, concave cubes and dendrites through a one-pot solvothermal synthesis and using different capping agents, surface modifiers and solvents. The electrocatalytic properties of all these PtPd alloy nanoparticles toward formic acid oxidation were higher than those obtained with a commercial Pt black and a Pt/C catalyst. Among the shape-controlled PtPd bimetallic alloy nanoparticles, flowers and dendrites provided the best activities. Nevertheless, the samples were not only irradiated with an UV lamp for 12 h to remove the organic capping agents but also cycled (about 50 cycles) between -0.24 and 1.0 V vs SCE in 0.5 M H_2SO_4 before use. As previously discussed in this chapter, these steps strongly perturb the surface structure at atomic level of the samples.

More recently, Jia et al. [162] described the preparation of novel excavated rhombic dodecahedral (ERD) $PtCu_3$ alloy nanoparticles containing ultrathin nanosheets of high-energy {1 1 0} facets. These ERD $PtCu_3$ nanoparticles showed

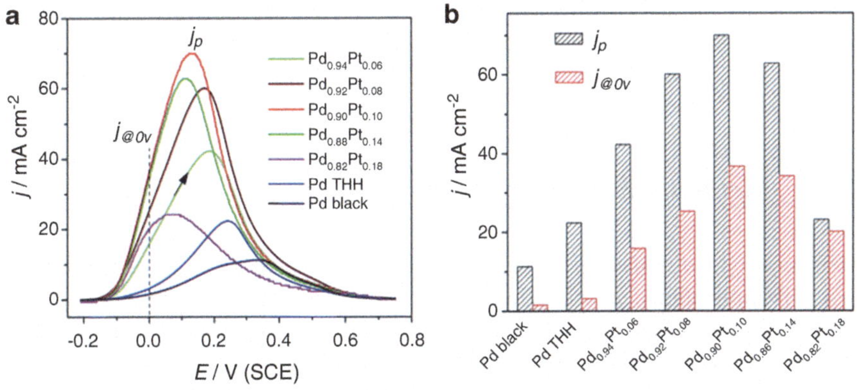

Fig. 2.9 Comparison of electrocatalytic activities of PdPt and Pd THH nanoparticles and commercial Pd black towards formic acid oxidation. (**a**) Current–potential curves recorded at 50 mV s^{-1} in 0.25 M HCOOH + 0.25 M $HClO_4$ (currents normalized to the electroactive surface area of the samples). (**b**) Comparison of oxidation current density at the peak (j_P) and at 0 V ($j_{@0V}$). Reproduced with permission from [160], © 2012 Royal Society of Chemistry

better electrocatalytic activity towards formic acid electrooxidation than a commercial Pt black and octahedral $PtCu_3$ alloy nanoparticles. In addition, CO stripping experiments showed that CO is oxidised at a potential about 50 and 100 mV lower than with the reference materials. As the authors stated, before the electrocatalytic experiments the samples were electrochemically cleaned by continuous potential cycling between −0.2 and +0.9 V vs SCE (the number of cycles is not mentioned). Thus, this potential cycling will induce the formation of a Pt rich surface. As the authors also stated, this Pt enrichment is obviously much more evident after the electrochemical stability test (1000 cycles between −0.2 and 0.8 V vs SCE) although from SEM, TEM and EDX of the ERD $PtCu_3$ nanoparticles, these seem unaltered both in terms of shape and atomic composition. This fact again illustrates the extremely sensitivity of the electrochemical measurements to changes at the surface of the nanoparticles.

Very recently, Sneed et al. [163] reported the preparation of cubic Pd–Ni–Pt core-sandwich-shell nanoparticles and their electrocatalytic properties toward methanol and formic acid electrooxidation. Interestingly, the effect of the Ni layer thickness and the particle size of the core-sandwich-shell on their catalytic activities were also evaluated and compared with that obtained with PdPt nanocubes. Despite the reported activity enhancement, the samples showed an important restructuring, particularly in acidic solution. However, restructuring in acidic conditions suggests a more complex catalytic behaviour from changes in composition.

Finally, Zhang et al. have very recently reported the synthesis of PdCu bimetallic tripods for formic acid oxidation [164]. The PdCu tripods of different atomic composition ($Pd_{87}Cu_{13}$, $Pd_{80}Cu_{20}$ and $Pd_{93}Cu_7$ (prepared after selective etching with HCl)) showed higher specific and mass activities than those found with commercial Pd black. However, in terms of catalytic stability, that sample containing a lower Cu content displayed higher stability during the electrochemical essays.

2.3.2 Methanol Electrooxidation

Methanol oxidation has been extensively studied because of its great interest for fuel cell applications, as using this fuel removes the problem of storing or generating hydrogen found in hydrogen/oxygen fuel cells. Methanol is the simplest alcohol and its electrochemistry is also the easiest [165]. Platinum is the pure metal with the highest activity towards this reaction in spite of its low tolerance for CO, which is an intermediate species and poisons its surface. Due to the sensitivity of this reaction towards the surface structure of platinum [166, 167] and the need to have a second oxophilic metal to minimize the CO poisoning drawback, many research manuscripts have focused during the last 2–3 years on the use of shaped Pt-based bimetallic nanoparticles with the aim of testing them towards this reaction. Other metals have also received attention and some results will be also summarized in this section.

As previously stated, methanol oxidation was found to be sensitive to the Pt surface structure. This dependence was reproduced with shape-controlled (cubic, spherical, octahedral/tetrahedral and truncated octahedral) nanoparticles [72]. For that shape effect comparison, we have to bear in mind that a cube is ideally enclosed by 6 {1 0 0} faces or an octahedron by 8 {1 1 1} faces. Because of the high activity found for Pt(1 0 0) single crystal, platinum nanocubes have been more recently reported to enhance this reaction. Park's group synthesized 4.5 nm cubes by a thermal reduction process in the presence of PVP. Enhanced catalytic activities (2.4 times), in comparison with spherical nanoparticles (no preferential surface structure) were obtained in $HClO_4$ solutions. The authors also concluded that the cubic nanoparticles suffered lower accumulation of residues during the electrooxidation due to a higher ratio in the peak current for the forward and backward scans. It is worth noting that although this ratio is commonly used to gauge CO-tolerance and catalytic activity of Pt-based electrocatalysts, a recent in situ enhanced IR study suggests its inadequacy [168]. Similar sized nanocubes were also prepared by Zheng's group but in this case in the absence of any surfactant or capping agent [169]. They were prepared in benzyl alcohol under a CO atmosphere and 1.2 times higher current densities in comparison with spherical Pt nanoparticles were recorded in H_2SO_4 medium. Capping agents are normally used in the synthesis of shaped nanoparticles but they make the cleaning step a critical one. Concerning the importance of the cleaning step, the same authors reported much lower oxidation current densities for nanoparticles prepared with PVP and oleylamine, which had been only partially cleaned [169]. Tong's group also focused its attention towards the importance of the residual presence of capping agents towards this reaction [98]. In several papers it has been reported the remaining presence of PVP on the nanoparticles [81, 98], which could have an effect on the catalytic activity, and recently the same authors have reported the possibility to remove this capping agent using an adapter liquid phase UV photo-oxidation technique [65]. Interestingly, the authors claimed that the surface-bound residual PVP enhanced the activity for the octahedral/tetrahedral nanoparticles while it was suppressed for cubic nanoparticles. Wei's group was also concerned about the importance of using capping-free nanoparticles because of the great effect these molecules could impose on the catalytic activity of the catalysts [170]. They prepared cubic nanoparticles prepared in the absence of surfactant with an average size of around 3.5 nm. They used $Pt(acac)_2$ which was reduced with ascorbic acid in DMF (N, N-dimethylformamide). The higher activity obtained towards methanol oxidation in KOH medium in comparison with the commercial Pt/C catalysts was ascribed not only to the surface structure (large amount of {1 1 0} and {1 0 0} sites) but also to the clean surface of the catalysts. Sun's group prepared electrochemically THH Pt nanocrystals, bound by well-defined high index crystal planes [171]. The activity was not very high because these planes facilitate, for pure Pt, the dissociative chemisorption of methanol, leading to poisoning by strongly adsorbed CO. Nevertheless, the decoration of these nanoparticles with Ru adatoms exhibited greatly superior catalytic currents and CO_2 yields in the low potential range, when compared with a commercial PtRu alloy nanoparticle catalyst, Fig. 2.10. In addition, the onset potential was shifted 100 mV towards more negative potentials.

Palladium is another pure metal which has been extensively studied as catalysts towards this reaction regardless of its lower activity, due to the generation of CO_{ads}

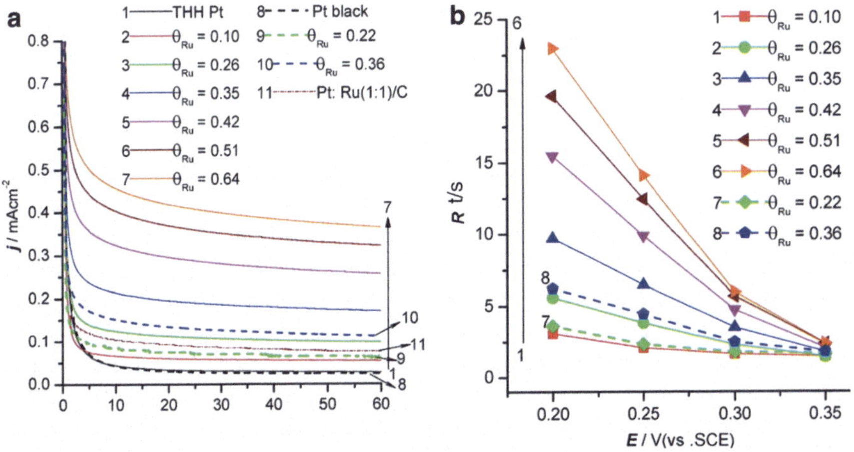

Fig. 2.10 (**a**) Comparison of chronoamperometric current transients recorded at 0.25 V in 1.0 M $CH_3OH + 0.1$ M $HClO_4$ solution for the various Ru-decorated THH Pt nanocrystals (*solid line*), Pt black (*dashed line*) and PtRu/C (*dashed-dotted line*) electrodes. (**b**) Potential dependence of the enhancement factor R (ratio of the chronoamperometric current density to the current density for the relevant clean electrode). Reproduced with permission from [171], © 2012 American Chemical Society

species. This reaction has also proven to be sensitive to the surface structure of palladium, for which the shape has a great influence on the catalytic behavior. Arriaga's group prepared 10 nm cubic Pd nanoparticles in the presence of ascorbic acid, sodium bromide and PVP and observed a 4-fold increase at low potentials in the methanol oxidation in alkaline medium in comparison with commercial Pd [172]. The authors were concerned about the site-blocking due to the PVP. They confirmed by TGA and EDS analyses the difficulty to remove PVP and proposed an electrochemical method, which consisted in cycling the samples in acidic and alkaline media. Although this methodology can be valid to remove the PVP from the surface of the nanoparticles, the high potentials reached during those continuous cycles (oxidation/reduction) modify the surface structure of the nanoparticles, removing a large amount of the inherent {1 0 0} sites of the cubic nanoparticles present at the surface before this cleaning step.

Cubic nanoparticles connected in a chain-like structure were also prepared by Opallo's group and reported 11.5 times higher activity in comparison with the commercial Pd/C catalyst in $HClO_4$ solution [173]. The synthesis was performed by reducing $PdCl_2$ in aqueous solution by 5-hydroxytryptamine and the resulting cubes had an average size of approximately 180 nm. Wang's group also reported the synthesis of cubic Pd nanoparticles dispersed on graphene oxide and observed higher activity and stability than commercial Pd/C catalyst for this reaction in H_2SO_4 medium. The cubes were prepared by reducing Na_2PdCl_4 with ascorbic acid, without the presence of any surfactant or capping agent. In fact the authors mentioned that it is the graphene oxide which acts as protecting and structure-directing agent, ensuring that the nanoparticles have a clean surface, thus allowing a high activity. The same authors published high activities using star-like and concave Pd nanocrystallites [174], and highlighted the importance of the cleaning step even indicating in the title of the manuscript the word "clean" as they do not use any surfactant to obtain these shapes. The synthesis of the nanostars (approx. 250 nm) was performed in aqueous medium adding ascorbic acid (reducing agent) and KBr to a H_2PdCl_4 solution. For the concave nanocrystallites (approx. 50 nm), NaOH was also added before the reducing agent. Those structures showed more than three times higher activity than that of the commercial Pd/C in alkaline medium. Another strategy to have palladium nanoparticles with high activity is to prepare three-dimensional porous Pd nanoflowers with large electrochemically active surface areas. Feng's group prepared this type of structures with a high yield by a simple one-pot wet-chemical method at room temperature using Good's buffers, N-2-hydroxyethylpiperazine-N9-2-ethanesulfonic acid, both as reducing and shape-directing agent [175]. These structures have an average size of 12–68 nm and show, without using any seed or surfactant, a preferential growth along the {1 1 1} directions and exhibited excellent electrocatalytic activity and stability towards methanol oxidation in alkaline medium.

Platinum and palladium are the two pure metals which have attracted most attention towards this reaction and consequently PtPd bimetallic nanoparticles have been also widely studied. It is interesting to highlight that the control over the surface structure/shape of bimetallic systems is not so easy as in the case nanoparticles of only one metal, and only in the last years it has been achieved. Chen and coworkers prepared bimetallic PtPd nanocubes on graphene with a surfactant-free method, in which DMF was used as a bi-functional solvent for the reduction of both metal precursors and graphene oxide [176]. This catalyst, with a mean size of 8 nm, showed a higher catalytic activity in H_2SO_4 and a better tolerance to carbon monoxide poisoning than PtPd nanoparticles deposited on carbon and pure Pt nanoparticles deposited on graphene. This was ascribed to the synergetic effects of Pt and Pd and the enhanced electron transfer by graphene. Yin et al. also prepared PtPd cubes, enriched in {1 0 0} sites, and PtPd nanotetrahedra, enriched in {1 1 1} sites, through a hydrothermal route [177]. The mean particle size was around 8.5 and 4.9 nm for the cubic and the tetrahedral nanoparticles, respectively. In the synthesis procedure some small ions and PVP were used as facet selective ({1 0 0} or {1 1 1}) agents and capping ligand, respectively. Those ions were large amounts of Br^- and tiny amounts of I^- for the {1 0 0} facets and $Na_2C_2O_4$ and formaldehyde for the {1 1 1} facets. This study confirmed the facet-dependent enhanced electrocatalytic activity and durability. The {1 0 0}-facet-enclosed PtPd nanocubes demonstrated the highest activity, whereas the {1 1 1}-facet-enclosed PtPd nanotetrahedra exhibited a better durability in $HClO_4$ medium and also a higher activity than pure platinum nanoparticles.

As in the case of the tetrahedral nanoparticles, octahedral nanoparticles are also ideally enclosed by {1 1 1} facets. Park's group synthesized octahedral PtPd alloy nanoparticles and also reported an improved catalytic activity and stability towards methanol oxidation [178]. The nanoparticles, with a mean size of 8.3 nm, were prepared by means of a polyol process with glycerol as a reducing agent in an aqueous solution. Interestingly, a comparison with pure platinum nanoparticles was made and a stability test was performed on both catalysts, which consisted in holding the catalysts at 0.45 V for 1 h in 0.1 M $HClO_4 + 2$ M CH_3OH, Fig. 2.11. In the case of the Pt/C, the current density at 0.4 V after the stability test seriously decreases (reduction of 78.2 % from the initial value). In contrast, the PtPd alloy

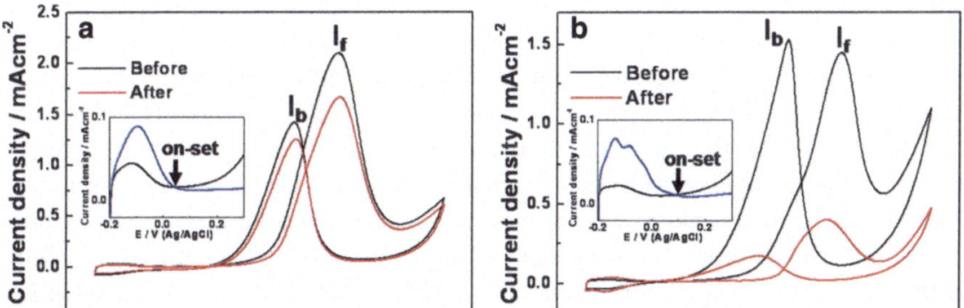

Fig. 2.11 Cyclic voltammograms before and after stability test of (**a**) PtPd alloy/C and (**b**) Pt/C in 0.1 M $HClO_4 + 2$ M CH_3OH at 50 mV s^{-1} and at 25 °C. The insets indicate CVs in 0.1 M $HClO_4$ (*blue line*) and 0.1 M $HClO_4 + 2$ M CH_3OH (*black line*) of the catalysts. Reproduced with permission from [178], © 2011 Royal Society of Chemistry

still had quite high current densities (decrease of 36.5 %) at the end of the test. In addition, the morphology of the catalysts was evaluated before and after the stability test, as this parameter can affect the whole activity. Therefore, shape and size distribution was observed and compared to the initial stage. In the case of the PtPd alloy, the size and morphology of the catalysts still remained after the stability test, thus resulting in much improved catalytic stability. On the other hand, the pure platinum nanoparticles show massive agglomeration and it was observed an increase in the average size from 3.44 to 4.85 nm, together with a wider size distribution after the stability test, thus resulting in a deteriorated catalytic activity.

Yin et al. also prepared multiply twinned PtPd nanoicosahedra and reported them to be highly active electrocatalysts for methanol oxidation [179]. In fact these {1 1 1}-enclosed nanoicosahedra were compared to {1 1 1}-enclosed PtPd nanotetrahedra as well as commercial Pt catalysts. The higher activity of the icosahedra in comparison with the tetrahedra, being both {1 1 1} enriched, was ascribed to the different surface structures between multiply twinned icosahedra and single crystalline tetrahedra, as both of them had practically the same composition and were ideally enclosed by sole {1 1 1} facets. In addition, the larger amount of multiply twinned defects on the icosahedrons, were also likely to play an important role and contribute to the significantly enhanced electrocatalytic activity. The icosahedra had a mean size of 11.2 nm and were prepared under hydrothermal conditions by reducing K_2PtCl_4 and Na_2PdCl_4 with HCHO in the presence of PVP as capping agent and $Na_2C_2O_4$ as a facet-selective agent. In addition, a high concentration of H^+ in solution would hamper the reducing capability of HCHO molecules and thus slow down the reducing rate in synthesis. In this experiment, a proper amount of HCl solution was needed for the preparation of the PtPd nanoicosahedral nanoparticles. Without HCl added, single-crystalline PtPd nanotetrahedra were obtained, but with an excess of HCl, the reducing rate was too slow and only a small amount of nanocrystals with undesired broad size-distributions were obtained.

Concave PtPd nanocubes have been also synthesized and their electrocatalytic activity towards this reaction has been evaluated. Zhan et al. reported how to prepare them in ethylene glycol by reducing H_2PtCl_6 and Na_2PdCl_4 with ascorbic acid and KBr as reducing and capping agents, respectively [180]. PtPd regular nanocubes were initially formed by co-reducing Na_2PdCl_4 and H_2PtCl_6 with ascorbic acid due to the selective chemisorptions of Br^- ions on the {1 0 0} facets. These regular nanocubes were then excavated from the side faces due to bromide-induced galvanic replacement between $[PtCl_6]^{2-}$ ions and Pd atoms. Meanwhile, the overgrowth at corners and edges of the nanocubes arising from the co-reduction of $[PdCl_4]^{2-}$ and $[PtCl_6]^{2-}$ ions eventually led to the PtPd alloy concave nanocubes. It was found that the combination of the galvanic replacement induced by the bromide and the co-reduction was responsible for the formation of these alloy concave nanocubes. By modifying the composition, a great impact on the rate of galvanic replacement was observed and thus the concave content of such alloy. The highest catalytic activity was found for $Pt_{60}Pd_{40}$ concave nanocubes, which showed much improved CO poisoning tolerance, in addition to a 4.6-fold increase in specific activity, in comparison with a commercial Pt catalyst. This enhancement was ascribed to the synergetic effects between Pt and Pd and to the unique surface structure with high-index facets.

Au nanoparticles with different shapes have been also prepared and the structure/shape sensitivity towards this reaction has been confirmed. Zhang et al. published a paper where cubic, rod-like, quasi spherical, concave cubic and trisoctahedral nanoparticles were prepared and their catalytic activity towards methanol oxidation in alkaline medium evaluated [181], being the latter two shapes the most active as a result of the presence of high-index facets on their surface. The different shapes were achieved by reducing $HAuCl_4$ in the presence of trace amounts of $NaBH_4$, ascorbic acid and surfactants. The surfactants used were, depending on the desired shape, CTAB or cetyltrimethylammonium chloride (CTAC).

Cu is another pure metal, the activity of which towards methanol oxidation has been evaluated using shape controlled nanoparticles. Cu is a noble metal with relative low cost and is emerging as a promising alternative to other noble metals such as Pt, Au or Pd. Pesika's group prepared electrochemically platelike Cu crystals enriched in {1 1 1} sites from a copper sulfate solution in the presence of KBr and compared its catalytic activity to that of truncated octahedron-shaped nanoparticles [182]. The former were more active towards this reaction in alkaline medium than the latter. In order to understand this dependency towards the shape/surface structure of the nanoparticles, the authors investigated this reaction with Cu single crystals. They observed that the {1 1 1} and {1 1 0} crystal planes were more active in comparison to the {1 0 0} plane and that was the reason to explain why the {1 1 1} enriched plates were more active than the truncated octahedron-shaped nanoparticles, for which the three crystal facets are exposed almost equally.

Cu has been also used with the aim of reducing the amount of more expensive noble-metal based catalysts, such as Pt or PtPd, by preparing PtCu [183] or PtPdCu [184] catalysts. Qi et al. prepared PtCu ($Pt_{59}Cu_{41}$) concave nanocubes enclosed by high-index {5 1 1} facets by co-reducing $Pt(acac)_2$ and $Cu(acac)_2$ with oleylamine in the presence of CTAB and trioctylphosphine oxide at 180 °C. These concave nanocubes exhibited substantially enhanced electrocatalytic properties for methanol oxidation in $HClO_4$ medium relative to commercial Pt/C in addition to a lower amount of platinum needed [183]. In fact, the concave cubes, with a mean size of 15 nm, showed 3.0 and 4.7 times higher current densities than those of PtCu nanoparticles and commercial Pt/C respectively. Chronoamperometric experiments were performed and concave PtCu cubes had, over the entire time, higher current densities than the other two tested catalysts. The authors ascribed this improvement to the combination of Pt with Cu, which can cause a small compressive strain due to their weak lattice constant, therefore reducing the binding strength of CO on the catalyst. In addition, Cu helps in the oxidation of CO by generating hydroxyl species from water. They also suggested that the surface structure plays an important role in determining the catalytic activity, thus explaining the differences between the PtCu nanoparticles and the PtCu concave nanocubes.

Platinum has been also alloyed with other inexpensive metals, especially 3d transition metals in order not only to decrease costs but also to provide the catalysts with different electronic configurations. These transition metals have the ability to weaken the interaction between Pt and the unwanted PtOH intermediate, thus enhancing the catalytic performance [157, 185]. In fact Fang's group has reviewed the effect of the shape control and the electrocatalytic activity of Pt-based bimetallic nanocrystals, including high-index Pt-M nanostructures [156], in which special attention is paid to the synthesis approach commonly used for their preparation. Hu et al. prepared hollow PtNi nanocatalysts supported on graphene by a galvanic displacement reaction with $PtCl_6^{2-}$ and Ni^{2+} as precursors [186]. First, Ni^{2+} ions were reduced by $NaBH_4$ to Ni nanosheets as well as GO to graphene. Then, the platinum precursor was added and the hollow PtNi nanostructures were formed due to the displacement between the Ni nanosheets and $PtCl_6^{2-}$ ions. The nanoparticles, with a mean size of 30 nm and a pore size of 1 nm showed a high activity and good poison tolerance. The authors ascribed this improvement to the synergistic role of Pt and Ni in the catalysts in comparison to solid PtNi deposited on graphene and to commercial Pt/C nanocatalysts. The Ni hydroxide species, which exist on the surface [187], offer OH species to remove the intermediate CO that blocks active platinum sites. In addition, the open pores in the shells could greatly enhance the surface area and then a better use of the platinum in the catalyst, thus reducing the costs. Furthermore, the mesoporous structured shell provided very large fraction of edges and corner atoms [188, 189], endowing the hollow Pt–Ni–graphene nanocatalysts with good electrocatalytic performance. Graphene also provided a high electrical conductivity and good stability for keeping a stable catalyst structure.

Similar results were obtained by Sun's group with hollow PtNi nanospheres prepared, in this case, by chemical successive reduction method in the presence of PVP [190]. The facet dependent electrocatalytic activity was also confirmed for Pt_3Co nanocrystals [191]. Nanoflowers and nanocubes, enriched in {1 1 1} and {1 0 0} facets, respectively, were prepared and the former showed a higher activity in $HClO_4$ medium towards methanol oxidation than the latter (1.5-fold higher) and Pt_3Co nanoparticles (2.5-fold higher). In the synthesis, oleylamine and oleic acid were used together with benzyl ether and the metal precursors. With a similar synthesis methodology, but using 1-octadecene instead of benzyl ether, PtFe concave nanocubes of 7.6 nm were prepared, showing an enhanced methanol oxidation activity and CO tolerance, in comparison to PtFe nanocubes and commercial Pt catalysts, in $HClO_4$ medium. This enhancement was ascribed to the presence of high-index atomic steps. With the same aim, reduce Pt content while reducing CO poisoning, Zn was also used with Pt to prepare highly active catalysts.

Murray's group prepared PtZn nanocrystals in oleylamine and oleic acid and found the activity of the spherical nanoparticles higher than that of nanocubes [192]. However, upon annealing, the PtZn alloy phase could be transformed into the Pt_3Zn intermetallic phase, which showed better performance than the alloy, together with an excellent poisoning tolerance. The activities recorded in H_2SO_4 medium were very similar to those of commercial platinum while reducing cost and enhancing poisoning tolerance.

2.3.3 Ethanol Electrooxidation

Ethanol electrooxidation [193] is a reaction of great interest as ethanol can be used as fuel in fuel cells or in internal engines. Although methanol has been more extensively studied in the fuel cell field, ethanol is attracting more and more interest because it can be obtained from renewable sources and because its toxicity and corrosive power are lower than those of methanol. An ideal complete oxidation of ethanol to carbon dioxide releases 12 electrons. Nevertheless it is quite difficult to break the C-C bond, and therefore a much lower number of electrons per molecule is obtained. Thus, the main oxidation products in acid media are C_2 molecules, acetaldehyde (two electrons) and acetic acid (four electrons), being the formation of CO_2 quite low at ambient temperature. In acid media, although pure Pt becomes poisoned during the reaction, platinum based catalysts are the most active.

Our group studied the ethanol oxidation by cyclic voltammetry and FTIR techniques using spherical, cubic and octahedral Pt nanoparticles and evaluated the role of the surface structure/shape of the nanoparticles over their catalytic activity. The behavior was that predicted from the results obtained for the single crystal electrodes [194], and the facet geometries of the nanoparticles. Spherical nanoparticles (4 nm) were prepared using a water in oil microemulsion (n-heptane/Brij®30/water), while octahedral and cubic nanoparticles (8–10 nm) were prepared with a colloidal method using NaPA as capping agent. For all the platinum samples, both the incomplete oxidation towards acetaldehyde and acetic acid and the formation of CO was observed, although the ratio of both paths was different. The ethanol oxidation was performed in H_2SO_4 and $HClO_4$ media, Fig. 2.12. Octahedral nanoparticles, enriched in {1 1 1} sites, showed almost exclusively acetic acid formation, therefore showing a low poisoning rate and being very similar the positive and negative sweeps. Cubic nanoparticles are ideally enclosed by 6 {1 0 0} faces, and the Pt(1 0 0) is the most active platinum basal plane at splitting the C-C bond. Thus, cubic nanoparticles showed the largest activity for the cleavage of the C-C bond. Finally, spherical nanoparticles, which show no preferential orientation, possess a high amount of low coordinated atoms on the surface which is also effective to cleave the C-C bond, and CO was also obtained.

Park et al. prepared small Pt cubes (4.5 nm) by a thermal reduction process in the presence of PVP and reported a higher electrocatalytic activity in comparison with spherical nanoparticles (3.5 nm) [195]. They observed current densities 2.4 times higher for the cubic nanoparticles, which also showed a much lower onset potential (0.21 V) in $HClO_4$. In addition to the higher current densities the authors also concluded that the cubic nanoparticles exhibited a lowered poisoning rate in base of the ratio between the current densities in the forward and backward scans, which is much higher in the case of the cubic nanoparticles (0.84 vs 0.57) compared with that of spherical nanoparticles.

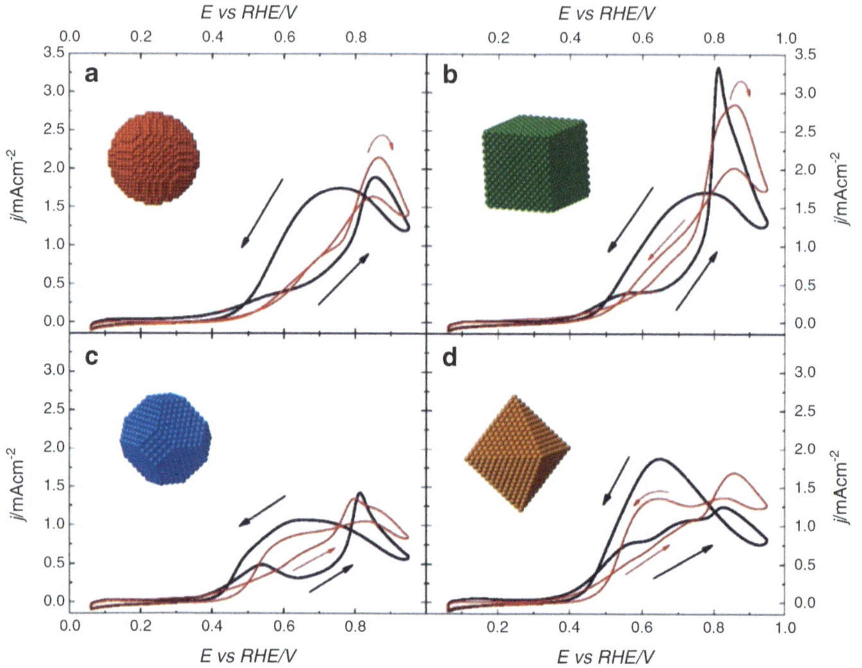

Fig. 2.12 Ethanol oxidation (first cycle) on (**a**) (poly)Pt, (**b**) (1 0 0)Pt, (**c**) (1 0 0)-(1 1 1)Pt and (**d**) (1 1 1)Pt nanoparticles in 0.5 M H_2SO_4 + 0.2 M CH_3CH_2OH (*black line*) and 0.1 M $HClO_4$ + 0.2 M CH_3CH_2OH (*red line*). Sweep rate: 50 mV s^{-1}. Reproduced with permission from [194], © 2013 Royal Society of Chemistry

Sun et al. also studied this reaction using Pt nanoflowers, which were synthesized in deep eutectic solvents and characterized by SEM, TEM, XRD, XPS and electrochemical tests [196]. The size and the shape of these nanoflowers could be controlled and ranged from 45 to 250 nm approximately. Current densities could be up to nearly three times higher in $HClO_4$ medium. These nanoflowers possess sharp single crystal petals and a high density of atomic steps to which the higher electrocatalytic activity was ascribed to.

Probably due to the high activity observed with single crystal surfaces with low Miller indices, several papers have been published with nanoparticles enclosed by high index facets. In the case of platinum Sun et al. published different manuscripts where they studied this reaction with such particles. They observed that concave THH nanocrystals prepared in deep eutectic solvents bounded by {9 1 0} faces were more active than platinum black in $HClO_4$ medium [197]. By modifying the synthesis conditions, the size and shape could be controlled. The electrocatalytic activity was evaluated by cyclic voltammetry and chronoamperometry. The former concluded that the THH were two times more active than Pt black and the latter,

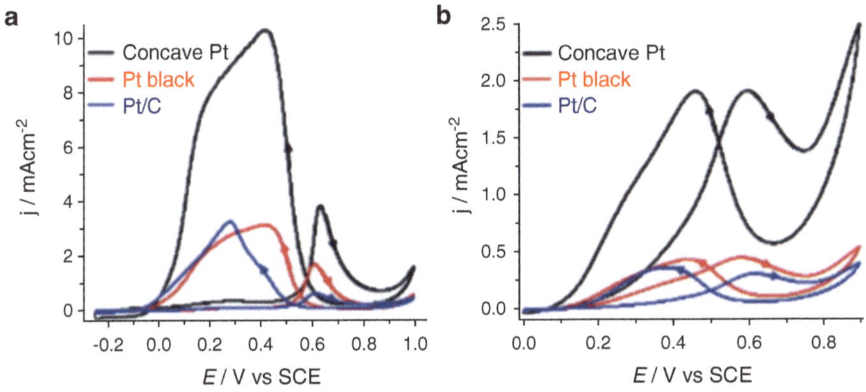

Fig. 2.13 Voltammetric profiles for the electrooxidation of (**a**) formic acid and (**b**) ethanol of concave Pt nanocrystals, commercial Pt black, and Pt/C (E-TEK). The formic acid oxidation was recorded in 0.5 M $H_2SO_4 + 0.25$ M HCOOH solution at a scan rate of 50 mV s^{-1}. The ethanol oxidation was recorded in 0.1 M $HClO_4 + 0.1$ M CH_3CH_2OH solution at a scan rate of 50 mV s^{-1}. Reproduced with permission from [121], © 2011 American Chemical Society

than they were more stable after 2 h at constant potential (0.45 V vs SCE). Also, triambic icosahedral nanoparticles (160 nm, enclosed by {7 7 1} high-index facets) were prepared electrochemically in deep eutectic solvents and showed higher activity (2–4 times) and stability than a commercial Pt black catalyst also in $HClO_4$ medium [198]. The same group reported how Pt nanocubes (10 nm) could be transformed into THH nanocrystals after a short square-wave potential treatment (2 min) in H_2SO_4 with sizes from 6 to 20 nm [199]. With this methodology, low index {1 0 0} facets were transformed into {3 1 0} high-index facets, and the presence of this type of sites was the cause of a remarkable increase in activity towards ethanol oxidation, in comparison with the original Pt nanocubes and commercial platinum nanoparticles. Also, Zheng's group reported an amine-assisted synthesis of concave polyhedral Pt nanoparticles having {4 1 1} high-index faces, which exhibited an enhanced electrocatalytic activity (2.3–5.6 times) over commercial Pt catalysts towards ethanol oxidation reaction in $HClO_4$ [121], Fig. 2.13.

Zhang et al. also reported a higher activity with multipod and concave Pt nanocrystals, mainly exposing {2 1 1} and {4 1 1} high-index facets, respectively [200]. The authors also ascribed the higher activity (in $HClO_4$) in comparison to Pt/C and Pt nanocubes with low-index facets to the high density of surface atomic steps.

Among the other pure metals, Pd and Rh have also been used to prepare shaped nanoparticles, with which ethanol oxidation has been studied. Arriaga et al. prepared cubic Pd nanoparticles with a very high yield by a chemical reduction in aqueous medium employing ascorbic acid, PVP and sodium bromide as reducing agent, surfactant and additive, respectively [172]. The nanoparticles were prepared with (17.5 nm) and without (10.6 nm) Vulcan carbon as support and the PVP was removed electrochemically. Pd nanocubes exhibited a much better performance than commercial Pd nanoparticles towards ethanol oxidation, reporting current densities around three times higher in voltammetric experiments in KOH medium. Pd was also used by Xie et al. to prepare concave nanocubes with high-index {7 3 0} facets and vicinal {h k 0} facets such as {3 1 0} [201]. These were prepared by a simple reduction of an aqueous solution of H_2PdCl_4 with ascorbic acid in the presence of CTAC and CTAB at room temperature. Depending of the ratio CTAC:CTAB, the authors were able to have crystals with different degrees of concavity. Those with the ratio 4:1 were the nanoparticles showing the highest electrocatalytic activity in NaOH medium, showing from 4 to 6.5 times higher catalytic activities than a commercial Pd black catalyst and a good stability.

Shaped pure Rh nanoparticles were also used to study this reaction. Sun's group prepared THH Rh nanocrystals enclosed by {8 3 0} high-index facets through an electrochemical square wave potential method at room temperature in a H_2SO_4 solution [202]. Voltammetric experiments in NaOH gave current densities nearly seven times higher for the THH nanoparticles than for Rh black. Also, for chronoamperometric experiments, after 30 min, these high-index nanoparticles showed higher current densities (1.8 times).

With these metals (Pt, Pd, Rh) and also Au, different bimetallic or trimetallic nanoparticles have been synthesized and their behavior towards ethanol oxidation has been evaluated. Chen et al. reported the synthesis of PtPd alloys with different shapes (spherical, nanoflowers and nanodendrites) deposited in graphene in aqueous solution in the presence of ethanol [203]. For all of them, higher electrocatalytic activities and better tolerance to reaction intermediate poisoning was observed in comparison with Pt nanoflowers supported on graphene, Pd nanoparticles supported on graphene and PtPd nanoparticles supported on carbon black, which was ascribed to the synergetic effects of Pt and Pd and the enhanced electron transfer properties of graphene.

PtRh nanocrystals are also very interesting from a catalytic point of view, because several studies have confirmed that the addition of Rh to Pt can significantly enhance the C-C bond cleavage due to a change of electronic properties [204–206]. In this sense, Sun's group reported the synthesis of PtRh nanocubes with different atomic ratio supported on graphene using the modified polyol method with Br^- for the shape-directing agents and in the absence of any capping agent [207]. The most active catalyst towards ethanol oxidation was that with a Pt:Rh atomic ratio = 9:1, which showed superior specific activity and long-term stability to commercial Pt/C catalyst in $HClO_4$ medium, being current densities for that alloy, twice that for Pt/C for both voltammetric and chronoamperometric experiments. In addition FTIR spectroscopy was used to evaluate the C-C bond cleavage in ethanol. It was found that alloyed Rh can promote the split of this bond, being the catalyst with the ratio Pt:Rh = 1:1 the best for breaking the C-C bond in ethanol. The same group, supported on the advantage of the PtRh nanocrystals and that of nanoparticles with high-index facets for this reaction, also prepared PtRh nanocrystals with high-index facets: {8 3 0}-bound THH and {3 1 1}-bound trapezohedron (TPH) [109]. The nanoparticles were prepared electrochemically and the

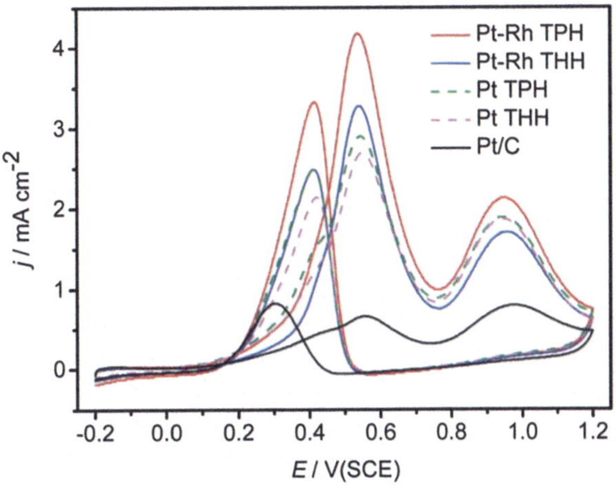

Fig. 2.14 Cyclic voltammograms for ethanol oxidation on Pt–Rh THH, Pt–Rh TPH, Pt THH, Pt TPH and commercial Pt/C catalysts at 50 mV s^{-1} in 0.1 M ethanol + 0.1 M $HClO_4$. Reproduced with permission from [109], © 2013 Royal Society of Chemistry

shapes could be tuned from THH to TPH by altering the electrode potential. They observed excellent electrocatalytic properties and a high ability to break the C-C bond of ethanol in $HClO_4$ medium of both nanoparticles enclosed by high-index facets. The {3 1 1} facets of the trapezohedron nanocrystals have a higher step density than {8 3 0} facets on THH nanoparticles, explaining the highest catalytic activity of the former. That activity was about 6.3 times higher than that of a commercial Pt catalyst and the onset potential was slightly shifted towards more negative potentials, Fig. 2.14.

Au was also used, together with Pd, to synthesize highly active catalysts towards ethanol electrooxidation. AuPd octapodal nanoparticles were prepared by selective etching of {1 0 0} facets by in situ generated Br^- ions in aqueous medium. Their electrocatalytic behavior was compared to that of flower-like AuPd alloy nanoparticles with similar composition and size and that of a commercial Pd/C catalyst [208]. Peak current densities were about two and four times higher, in comparison with those of flower-like and Pd/C catalysts, respectively. Chronoamperometric experiments also revealed a superior electrochemical stability of the octapodal nanoparticles. This enhancement was ascribed to the presence of a number of active sites on their surfaces such as highly active facets, gaps between pods, and some defect sites, which originate from their octapodal structures with rough surface features. Moreover, it was also attributed to the high fraction of {1 0 0} facets in comparison with the other catalysts. By using density functional theory (DFT), Pd(1 0 0) was reported to be the best basal plane for the ethanol dissociation [209]. The same bimetallic system was studied by Zheng's group, which reported the preparation of monodisperse Au–Pd alloy nanoparticles with systematic shape evolution from rhombic dodecahedral (RD) to trisoctahedral (TOH), and hexoctahedral (HOH) structures by varying the concentration of surfactant in the surfactant-mediated synthesis [210]. The catalytic activities toward ethanol electrooxidation were in the order of HOH > RD > TOH, which follows the order of their corresponding surface energies. Finally, Xu's group also prepared

AuPd nanocrystals, but in this case it was Au@Pd concave nanocubes (21 nm) enclosed by high-index facets which were also decorated with Pt on top [211]. The authors normalized their activity to the mass or Pd or Pd + Pt and obtained 7-9 higher current density values than for a Pd black catalyst in KOH medium.

2.3.4 *Oxygen Reduction Reaction*

The oxygen reduction reaction (ORR) is probably the most important electrochemical reaction due to its multiple and relevant applications [212–218]. Consequently, the ORR has been extensively and intensively studied through the years. In the complete reduction of oxygen to water, there are four electrons exchanged, involving the double bond breaking and the formation of four O-H bonds. Thus, this multi-electron transfer process necessarily requires the existence of several steps in its reaction mechanism and therefore involves the possible existence of several adsorption intermediates. However, and although knowledge of the oxygen reduction reaction process has advanced considerably, the exact mechanism remains not fully understood [216, 217, 219, 220].

2.3.4.1 ORR in Shape-Controlled Pt Nanoparticles

Among pure metals, Pt is well-recognized to be the electrode with the highest electrocatalytic activity for the 4 e^- pathway [217, 221]. In addition, from single crystal studies, it is known that the ORR on Pt is a structure-sensitive electrocatalytic reaction. In particular, and considering half-wave potentials, the electrocatalytic activity decreases in the order of Pt(1 1 0) > Pt(1 1 1) > Pt(1 0 0) in perchloric acid solution and of Pt(1 1 0) > Pt(1 0 0) > Pt(1 1 1) in sulphuric acid solution [1, 222]. The different reactivity in both electrolytes points out the important role that the competitive anion adsorption plays on the reaction. Moreover, from studies with stepped Pt single-crystal surfaces, it was clearly observed that these stepped surfaces showed, irrespectively of their step site symmetry, higher ORR catalytic activity than those found with basal low-index crystal surfaces [219, 223–225]. To the best our knowledge, the first use of shape-controlled Pt nanoparticles towards ORR was reported by Inaba et al. [226]. They prepared cubic Pt nanoparticles in presence of NaPA which showed a high activity for ORR in H_2SO_4 solutions. In addition, potential cycling surface structure perturbation led to a decrease on the ORR activity as a consequence of the structural change from the initial {1 0 0} surface orientation to a polycrystalline orientation.

Later, we evaluated the ORR activity of different and clean shape-controlled Pt nanoparticles both in H_2SO_4 and $HClO_4$ solutions by using scanning electrochemical microscopy (SECM) [227]. Under these conditions, we found a good agreement between the reactivity of the nanoparticles and that previously observed with single crystal electrodes. In particular, the activity for the ORR followed the trend

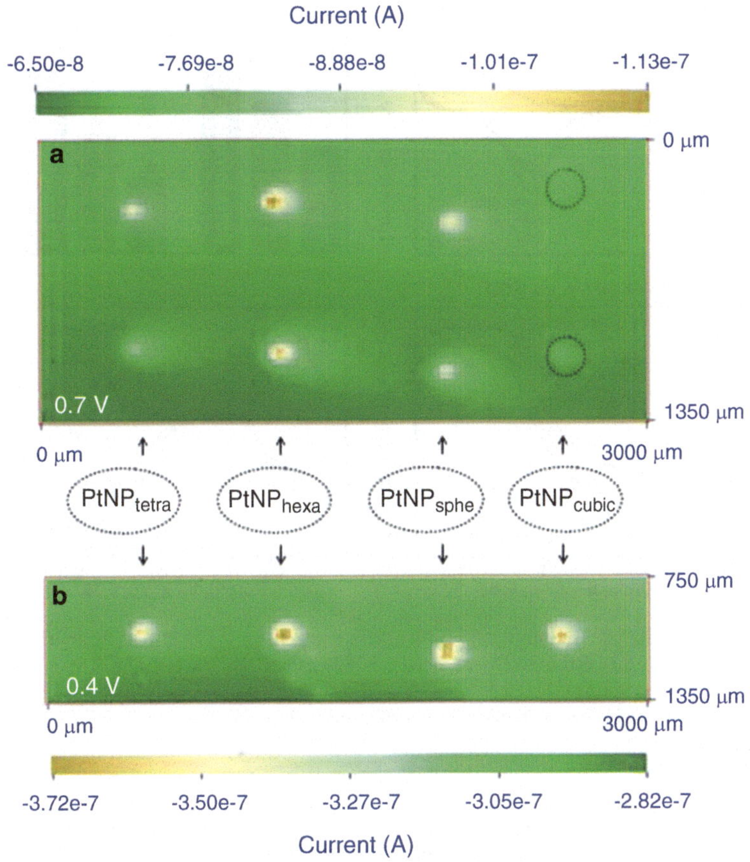

Fig. 2.15 SECM TG/SC images displaying the reduction current collected for ORR in 0.1 M $HClO_4$ solution at a Pt NPs array. This array is formed by spots of four different types of shape-controlled Pt NPs (Tetrahedral (tetra), hexahedral (hexa), spherical (sphe) and cubic (cubic) Pt nanoparticles). Scan rate = 125 μm/s. (**a**) Image of the first and second rows of the array held constant at 0.7 V. (**b**) Image of the second row of the array held constant at 0.4 V. Reproduced with permission from [227], © 2010 American Chemical Society

PtNPhexahedral > PtNPtetrahedral ≈ PtNPspherical > PtNPcubic in 0.1 M $HClO_4$ and the trend PtNPhexahedral > PtNPcubic > PtNPspherical > PtNPtetrahedral in 0.5 M H_2SO_4. Figure 2.15 shows a representative image of the different Pt nanoparticles obtained at two different potentials (0.7 V (kinetic region) and 0.4 V (diffusion region)) in 0.1 M $HClO_4$ solution. At 0.4 V, and due to the fact that the current only depends on the diffusion rate of O_2 toward the surface, which is the same in all cases, the activity of the samples is essentially the same. However, at 0.7 V, the Pt nanoparticles showed different ORR activities, which are determined by their specific surface structure in the electrolyte used.

Later, Yu et al. [228] published the synthesis of concave Pt nanocubes enclosed by high-index facets such as {5 1 0}, {7 2 0}, and {8 3 0}. These Pt concave

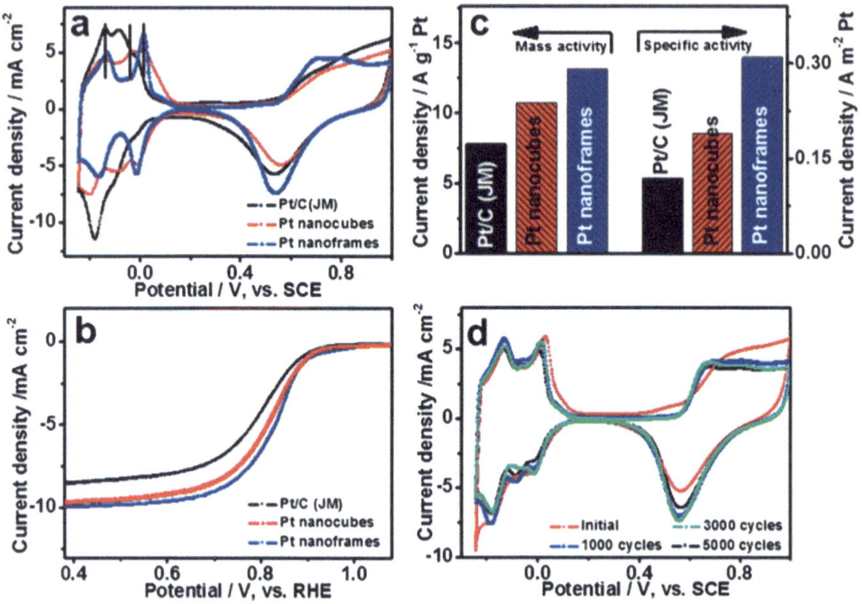

Fig. 2.16 (**a**) Voltammetric profiles of different Pt catalysts in 0.5 M H_2SO_4 at 20 mV s^{-1}. (**b**) Polarization curves for ORR in an O_2 saturated 0.5 M H_2SO_4 solution (1600 rpm at 10 mV s^{-1}). (**c**) Mass and specific activities (at 0.8 V vs. RHE). (**d**) Electrochemical durability test of concave Pt nanoframes in 0.5 M H_2SO_4 at 20 mV s^{-1}. Reproduced with permission from [124], © 2013 Wiley and Sons

nanocubes showed a specific activity towards ORR in $HClO_4$ solutions (normalized to the electroactive surface area) about 2–4 higher than that obtained with Pt nanocubes, cuboctahedral (that is Pt nanoparticles bounded by {1 0 0} and {1 1 1} low-index facets) and commercial Pt/C nanoparticles. These results show the low activity of the Pt cubes and the high activity of the concave Pt nanocubes containing high index surfaces in $HClO_4$ solutions, and agrees with our previous observations [227] and with that expected from Pt single crystal studies [222].

Xia et al. reported an interesting contribution dealing with the preparation of highly concave Pt nanoframes with high-index {7 4 0} facets and their activity towards ORR in H_2SO_4 solution [124]. They found that the concave Pt nanoframes were not only more active (both mass and specific activities, Fig. 2.16b, c) than Pt nanocubes and a commercial Pt/C catalyst (Johnson Matthey, JM 20 wt%) but also much more stable under electrochemical conditions, Fig. 2.16d. Care should be taken, however, with surface cleanness and with the fact that diffusion currents should be the same in all samples for a proper comparison.

Fu et al. [229] again analyzed the activity toward ORR of Pt nanocubes (about 6 nm), prepared using a green hydrothermal method with polyallylamine hydrochloride (PAH) as capping agent and formaldehyde as a reducing agent. The Pt nanocubes displayed an enhanced mass activity and specific activity both in H_2SO_4

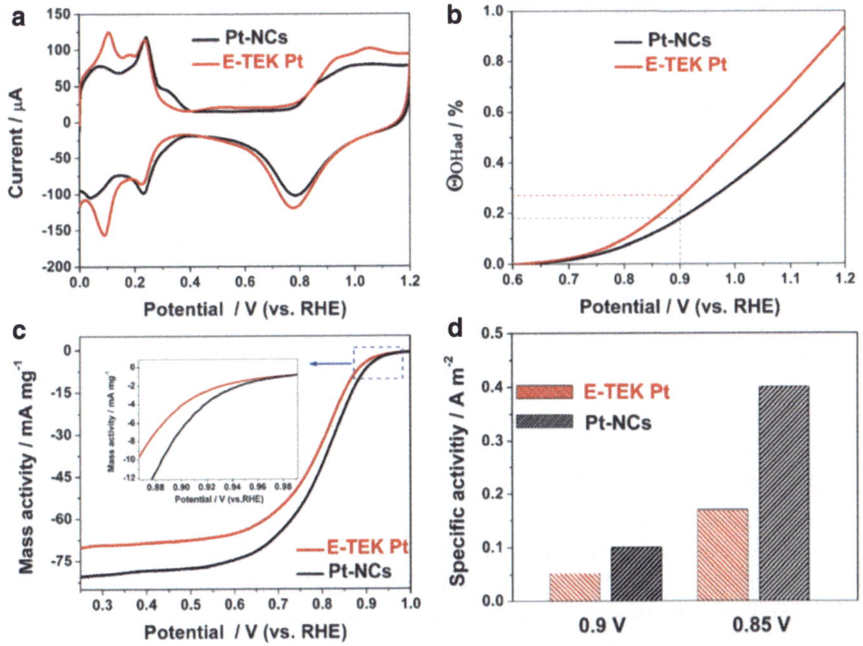

Fig. 2.17 (**a**) CV curves for the Pt nanoparticles and commercial E-TEK Pt black in N_2-saturated 0.5 M H_2SO_4 solutions at 50 mV s^{-1}. (**b**) Hydroxyl surface coverage (Θ_{OH}) for the Pt nanoparticles and E-TEK Pt black. (**c**) ORR polarization curves for the Pt nanoparticles and E-TEK Pt black in O_2-saturated 0.5 M H_2SO_4 solution at 5 mV s^{-1} and rotation rate of 1600 rpm. (**d**) Specific activity of the Pt nanoparticles and commercial E-TEK Pt black at a fixed potential of 0.85 and 0.9 V for the ORR. Reproduced with permission from [229], © 2013 Royal Society of Chemistry

and $HClO_4$ solutions in comparison to a commercial Pt black (E-TEK). Despite this enhancement, it is worth noting that the Pt nanocubes were UV/Ozone treated (185 and 254 nm in air for 4 h) and electrochemically cycled (−0.2 to 1.2 V vs RHE for 50 cycles) in order to remove the capping agent and other possible contaminants. These treatments are known to induce important modifications on their surface structure thus resulting in a more disordered {1 0 0} surface [97]. This effect can be visualized by direct comparison of the characteristic response of the {1 0 0} domains in H_2SO_4 solution of the Pt nanocubes used (Fig. 2.17a) with that obtained in previous contributions, Fig. 2.1 [73]. In addition, as authors stated, from FT-IR experiments they concluded that, owing to its excellent chemical stability, the PAH molecules attached on the surface of the Pt nanocrystals cannot be completely removed by UV/ozone treatment. This strong Pt-N bonding was responsible of the negative shift (about 0.3 eV) of the Pt 4f binding energy of the Pt nanocubes in comparison with the Pt black. Thus, the enhanced electrocatalytic activity of the Pt nanocubes in $HClO_4$ solution was related to this change of the d-band center of Pt that is, to an electronic effect.

Zhou et al. [230] prepared icosahedral Pt nanoparticles (8.8 nm edge length) using the hot injection-assisted GRAILS (gas reducing agent in liquid solution) method [231, 232]. The specific activity (normalized to the ECSA obtained from the charge in the hydrogen region) of the carbon supported Pt icosahedra obtained in $HClO_4$ was about 0.83 mA cm^{-2}Pt which was higher than that found for Pt nanocubes (0.35 mA cm^{-2}Pt) and for a Pt/C reference material (0.2 mA cm^{-2}Pt). This enhancement was attributed to the presence of {1 1 1} domains and twin planes at the surface of the nanoparticles. The stability of the carbon supported Pd icosahedral nanoparticles was evaluated after being cycled in O_2 saturated solutions for 2500 cycles. Unfortunately, this stability was exclusively evaluated (using TEM) in terms of particle shape but not in terms of surface structure.

More recently, Devivaraprasad et al. [63] have reported a detailed electrochemical study of different shape-controlled Pt nanoparticles prepared with NaPA and TTAB as capping agents. The samples were electrochemically characterized using most of the surface probes previously discussed in this chapter, including cyclic voltammetry in different supporting electrolytes (H_2SO_4, $HClO_4$ and NaOH), Cu UPD and Bi and Ge analyses. After this complete electrochemical characterization, the samples were tested towards ORR both the acidic and alkaline electrolytes. They found that the ORR specific activity trends of the shape controlled Pt nanoparticles in the three electrolytes was in good agreement with that expected from single crystals contributions [222]. Thus, whereas the {1 0 0} oriented Pt nanoparticles were more active in H_2SO_4, the {1 1 1} ones displayed better activities both in $HClO_4$ and NaOH. Interestingly, the trends in peroxide formation were also evaluated. Nevertheless, as authors also stated, the analysis of the H_2O_2 formation in this kind of nanomaterials is rather complicated and several aspects should be considered including effect of the particle size, active site density, layer thickness, reaction mechanism and "surface blocking" by adsorbed species.

Interestingly, Tripkovic et al. [233] reported a density functional theory (DFT) study about the effect of the size and shape of Pt nanoparticles on the ORR activity in a non-adsorbing electrolyte. Also, the ORR activity of these shapes at 0.9 V was compared with that of the "equilibrium shape", which was determined by calculating the changes in surface energies with potential for low-index Pt facets. The results showed that the tetrahedral Pt nanoparticles provided the highest activities and the cubic Pt nanoparticles the lowest ones, Fig. 2.18. These findings are in good agreement with previous experimental observations on both shape-controlled Pt nanoparticles and Pt single crystals in non-adsorbing electrolytes such as $HClO_4$.

In this sense, Li et al. [234] have very recently discussed about the activity and stability for ORR in acidic media of size and shape controlled Pt nanoparticles. These aspects were analysed in direct relationship with the knowledge gained from Pt single crystals. They found that under ORR relevant conditions (0.6–1.1 V, 4000 scans, 50 mV s^{-1} and 60 °C), the particle size and shape effects constantly evolved due to the potential-induced modifications. In fact, they stated that for long-term catalytic performance, the shape-controlled activity will disappear and the activity will be determined by that coming from thermodynamically stable round-like nanoparticles that is, the "equilibrium shape" described by Tripkovic et al. [233].

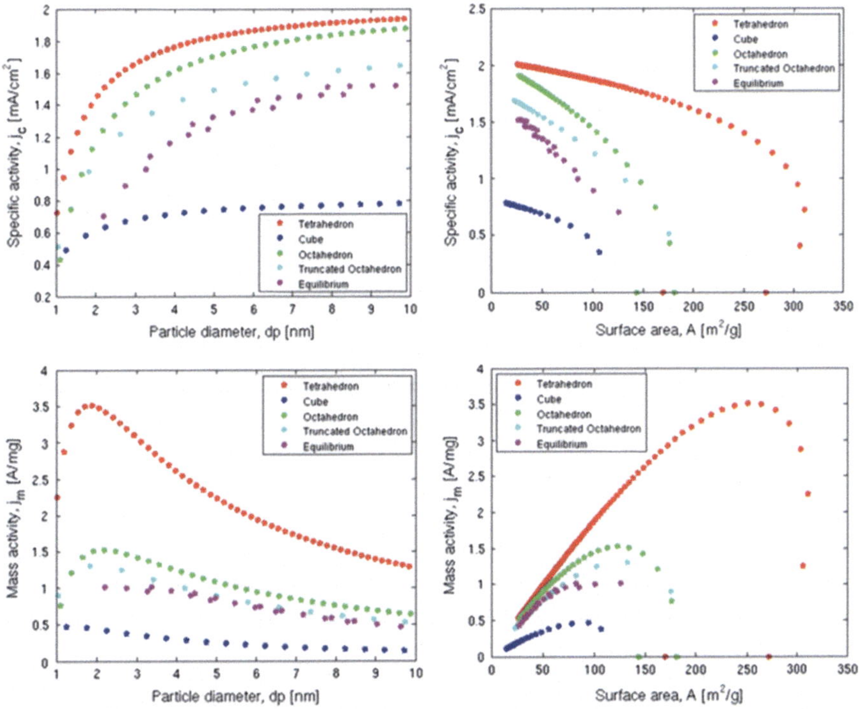

Fig. 2.18 Dependence of the specific activities and mass activities with the particle diameter and specific surface area. Reproduced with permission from [233], © 2014 Springer

2.3.4.2 ORR in Shape-Controlled Pd Nanoparticles

Among the pure metals, Pd is one of the most active catalysts for ORR. From single crystal studies, it was shown that the ORR on Pd is also a structure sensitive reaction. Kondo et al reported that, in $HClO_4$ solution, the ORR activity on Pd single crystals increases in the order of Pd(1 1 0) < Pd(1 1 1) < Pd(1 0 0) (estimated at +0.90 V (RHE)) [235]. In addition, on Pd(S)[n(1 0 0) × (1 1 1)] and Pd(S)[n(1 0 0) × (1 1 0)] single crystals, the activity increased linearly with the increase of the terrace atom density, thus pointing out that the {1 0 0} domains contain the most active surface sites towards ORR [236]. To the best our knowledge, the first example of shape-controlled Pd nanoparticles toward ORR was reported by Xiao et al. [237]. They found that the specific activity of Pd nanorods was about 10 times higher than that of Pd spherical nanoparticles and comparable to that of bulk Pt. This enhancement was attributed, in contrast with the Pd single crystal studies, to the presence of {1 1 0} facets in which, according with their DFT calculations, the interaction between an O adatom and a {1 1 0} site is exceptionally weak.

Almost in parallel, Shao et al. [238] and our group, in collaboration with the Tammeveski group [74], reported the ORR activity of Pd nanocubes, that is nanoparticles containing a preferential {1 0 0} surface structure. Shao et al. showed

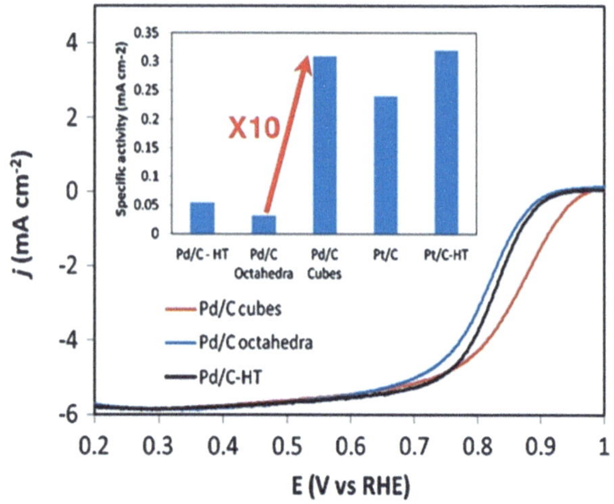

Fig. 2.19 ORR behaviour of Pd cubes and octahedra supported on carbon black and Pd/C annealed at 500 °C for 2 h (Pd/C-HT) in 0.1 M $HClO_4$ at 10 mV s^{-1} and 1600 rpm. Inset: Comparison of specific activities at 0.9 V. Reproduced with permission from [238], © 2011 Royal Society of Chemistry

that the ORR specific activity of Pd nanocubes (about 6–7 nm) in $HClO_4$ was about one order of magnitude higher than that of Pd octahedral (about 5–6 nm edge length) and a commercial cubooctahedral Pd/C catalyst (BASF, about 7 nm (after a thermal treatment)) and comparable to that obtained with the state-of-the-art Pt/C catalysts (TKK, 46.6 wt% about 3 nm), Fig. 2.19. The electroactive surface area of the Pd samples was estimated by integrating the stripping charge of the Cu UPD process, assuming 420, 490, and 460 μC cm^{-2} for full Cu monolayer coverage on cubes, octahedra, and cubooctahedra, respectively.

On the other hand, it was shown [74] that the ORR specific activity (estimated at 0.85 V vs RHE in H_2SO_4 solutions) of clean cubic Pd nanoparticles (about 27 nm) was about three times higher than that obtained with spherical Pd nanoparticles (about 3 nm) and a bulk Pd electrode. In this case, the surface area of the Pd nanoparticles was determined, after oxygen reduction measurements, by charge integration under the oxide reduction peak, assuming the value of 424 μC cm^{-2} as the charge density for the reduction of a monolayer of PdO [239]. In addition, in a forthcoming contribution [75], we also studied the ORR in alkaline solution with similar ~27 nm Pd nanocubes. The ORR specific activity at 0.95 V of the Pd nanocubes was again about three times higher than that obtained with spherical Pt nanoparticles and a bulk Pd electrode.

A similar contribution was published by Lee et al. [240]. They prepared Pd nanocubes with different particle size (27, 48 and 63 nm) and evaluated their ORR activity in alkaline solution. It was found that the ORR specific activity (measured at −0.1 V vs Ag/AgCl) of the 48 nm Pd nanocubes was higher than that obtained with the 27 and 63 nm Pd nanocubes. In addition, the activity of these Pd nanocubes was, independently of their particle size, higher than that obtained with 9 nm Pd nanoparticles. The electrochemical active surface area of the Pd nanocubes was estimated from the integration of the PdO reduction charge.

Shao et al. [136] extended their previous observations about the ORR activity of cubic and octahedral Pd nanoparticles (5–7 nm) in $HClO_4$ [238], H_2SO_4 and alkaline solutions. They observed that, in H_2SO_4, the ORR activity of Pd cubes was again much higher (about 17 times) than that of Pd octahedra. This behaviour was explained as due to the stronger adsorption of (bi)sulfate anion on the {1 1 1} domains present at the surface of the Pd octahedra. Nevertheless, in alkaline solution, the ORR was shown to be almost insensitive to the surface structure of the different nanoparticles, in clear contradiction with previous findings. In this regard, it is worth recalling the important differences in particle size which could change the binding energy of the species involved in the reaction mechanism, in particular, OH_{ad} species. More work would be required to understand these contradictory results.

More recently, Shao [241] and Zhang et al. [242] have published relevant reviews about the recent progress about the electrocatalysis of Pd-based nanomaterials for ORR.

2.3.4.3 ORR in Shape-Controlled Au Nanoparticles

ORR has been also extensively evaluated on Au surfaces. From Au single crystals studies, it is well-established that the ORR is sensitive to the surface structure and particularly to the {1 0 0} bidimensional domains in alkaline solution [214, 243, 244]. In alkaline solution, the ORR on Au(1 0 0) is especially interesting because not only shows an electrocatalytic activity similar to that of Pt but also due to the fact that exchanges four electrons in the potential window between +0.55 and +0.95 V vs RHE. However, the origin of this high catalytic activity is still under discussion. As a practical consequence of this surface structure sensitivity, the ORR in alkaline solution has been explored in different shape-controlled Au nanoparticles. Nevertheless, and taking into account that we have very recently reviewed the Au electrocatalysis for ORR, also including a particular section devoted to the use of shape/surface structure-controlled Au nanoparticles, this aspect will not be included/discussed in this chapter. For those readers interested in this point we refer to [34, 214, 245].

2.3.4.4 ORR in Shape Controlled Pt and Pd Based Alloy Nanoparticles

A great number of contributions have been published on the use of Pt or Pd based alloy nanoparticles to enhance ORR electrocatalysis with two main objectives: (a) improving ORR reactivity and (b) decreasing costs. In this sense, the control of the shape of these Pt- or Pd-based nanoparticles and consequently their surface structure has been also deeply explored [215, 218, 241, 246, 247]. To the best of our knowledge, the first attempt to produce shape controlled PtPd alloy nanoparticles was published in 2009 by Lim et al. [248]. They reported the synthesis of PdPt bimetallic nanodendrites which displayed about 2.5 times higher mass activity than the state-of-the-art Pt/C catalyst. However, the use of these shape controlled Pt alloy nanoparticles is marked by the publication made by Markovic's group [249] in which the specific activity of a single crystal $Pt_3Ni(1\,1\,1)$ surface was shown to be about 18 mA cm^{-2} at 0.9 V, that is, about 90 times higher than that of a commercial Pt/C (about 0.2 mA cm^{-2} Pt) in 0.1 M $HClO_4$ solution. This specific activity is still the highest ORR activity ever measured. This striking enhancement was attributed to a low coverage of hydroxyl species induced by the specific electronic structure of the Pt_3Ni sample. The $Pt_3Ni(1\,1\,1)$ surface was found to form a Pt-skin layer with a peculiar oscillatory near-surface compositional Pt and Ni profile across the 2–4 outermost layers of the {1 1 1} surface. This particular arrangement of Pt and Ni atoms in the top three layers shifts the d-band center relative to the Fermi level of the topmost layer of Pt atoms, thus affecting the adsorption coverage of hydroxyl spectator species and consequently enhancing the ORR activity. Later, Gasteiger and Markovic [250] summarized the most relevant approaches for different oxygen reduction electrocatalysts and, based on the results obtained with $Pt_3Ni(1\,1\,1)$, they estimated that, even with large 30 nm Pt_3Ni octahedral nanoparticles having a Pt skin layer, the mass activity should be about 10 times higher (about 1.6 A mg^{-1} Pt) than the commercial Pt/C. Since then, several contributions have been published in the last years about the development of shape-controlled Pt-alloy nanoparticles. However, and due to its fundamental and practical importance, in the following section we will exclusively focus our attention on the development and ORR activity of shape-controlled Pt-Ni nanomaterials.

The first example of shape-controlled Pt_3Ni nanoparticles was reported by Zhang et al. [251]. They prepared Pt_3Ni nanoctahedra and nanocubes using a wet-chemical approach previously reported [252]. They found that the ORR specific activity of the Pt_3Ni nanoctahedra was about five times higher than that of Pt_3Ni nanocubes in good agreement with what should be expected from the results obtained with $Pt_3Ni(1\,1\,1)$ single crystals, Fig. 2.20. Interestingly, in comparison with a commercial Pt/C electrocatalysts, the specific and mass activities of Pt_3Ni-octahedra/C were about seven and four times higher, respectively, despite the three times larger particle size of the Pt_3Ni nanoctahedra.

Later, the Yang group reported some relevant contributions. They initially showed that the ORR specific activity of truncated octahedral Pt_3Ni nanoparticles was about four times higher than that found for a reference Pt catalyst [253]. In terms of ORR mass activity, a value of 0.53 A mg^{-1} Pt was obtained for the Pt_3Ni sample which was about four times higher than that of the commercial Pt/C catalyst (0.14 A mg^{-1} Pt) and 1.8 times higher than that of carbon-supported octahedral Pt_3Ni catalysts previously used by Zhang et al. (0.3 A mg^{-1} Pt) [251]. Subsequently, they also reported the ORR activity of different truncated octahedral Pt_3Ni nanoparticles in alkaline solution [231, 254]. They found a specific activity of about 0.76 mA cm^{-2}, measured at 0.9 V vs RHE, which was 4.5 times that of a commercial Pt/C catalyst (0.17 mA cm^{-2}). A mass activity of 0.30 A mg^{-1} Pt was obtained which was about twice that of the commercial Pt/C catalyst. Interestingly, they also reported a new approach for the preparation of shape-controlled Pt-M (M: Co, Fe, Ni, Pd) alloy nanoparticles using a gas reducing agent, carbon monoxide in

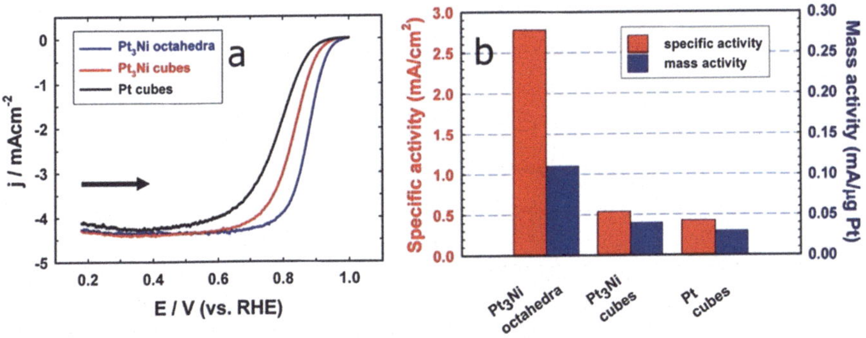

Fig. 2.20 (**a**) ORR behaviour of Pt_3Ni nanoctahedra, Pt_3Ni nanocubes, and Pt nanocubes supported on a rotating GC disk electrode in O_2 saturated 0.1 M $HClO_4$ solution at 20 mV s^{-1} and at 295 K; rotation rate, 900 rpm. (**b**) Comparison of the ORR activities at 0.9 V vs RHE at 295 K. Reproduced with permission from [251], © 2010 American Chemical Society

this case, in liquid solution (GRAILS) method [231]. Using this methodology, they prepared Pt_3Ni nanocubes and nanooctahedra (about 10 nm) which were tested toward ORR in $HClO_4$ solution. The specific activities of the samples were 0.85 mA cm^{-2} Pt for the nanocubes, and 1.26 mA cm^{-2} Pt for the nanooctahedra, which were remarkably higher than those typically obtained with a Pt/C catalyst (~0.20 mA cm^{-2} Pt). In addition, an ORR mass activity of about 0.44 A mg^{-1} Pt was obtained for the octahedral Pt_3Ni catalyst, three times higher than that of the reference Pt/C (0.14 A/mg_{Pt}). Finally they reported an additional improvement by using icosahedral Pt_3Ni nanoparticles (about 13 nm) [255]. The ORR specific and mass activities of the icosahedral samples were about 1.83 mA cm^{-2} Pt and 0.62 A mg^{-1} Pt, respectively. These values were higher than those previously reported with the Pt_3Ni nanooctahedra [231].

Carpenter et al. [112] reported a solvothermal method for the preparation of shape controlled Pt based alloy nanoparticles in which DMF acted as both solvent and reducing agent. They observed a maximum specific activity of about 3 mA cm^{-2} with octahedral PtNi (Pt/Ni ratio about 1) nanoparticles, corresponding with a 15 fold specific activity enhancement in comparison with the state-of-the art Pt/C catalyst. However, in this remarkable improvement, the mass activity was only about five times higher over Pt. Using a similar approach, Cui et al. [110] showed that 9.5 nm octahedral PtNi nanoparticles displayed ORR specific and mass activities of 3.14 mA cm^{-2} Pt and 1.45 A mg^{-1} Pt at 0.9 V vs RHE in $HClO_4$ solutions. These values represented a tenfold specific and mass activity enhancement in comparison with the state-of-art Pt/C catalyst. Such great enhancement was attributed not only to the {1 1 1} surface structure of the octahedral samples but also to the particularly high (about 41 Pt at. %) near-surface Pt-to-Ni atomic surface composition.

Later, Choi et al. [256] increased this previous enhancement using 9 nm PtNi octahedra prepared in presence of oleylamine and oleic acid as capping agents [252], introducing benzyl ether as solvent to decrease the surface coverage of the

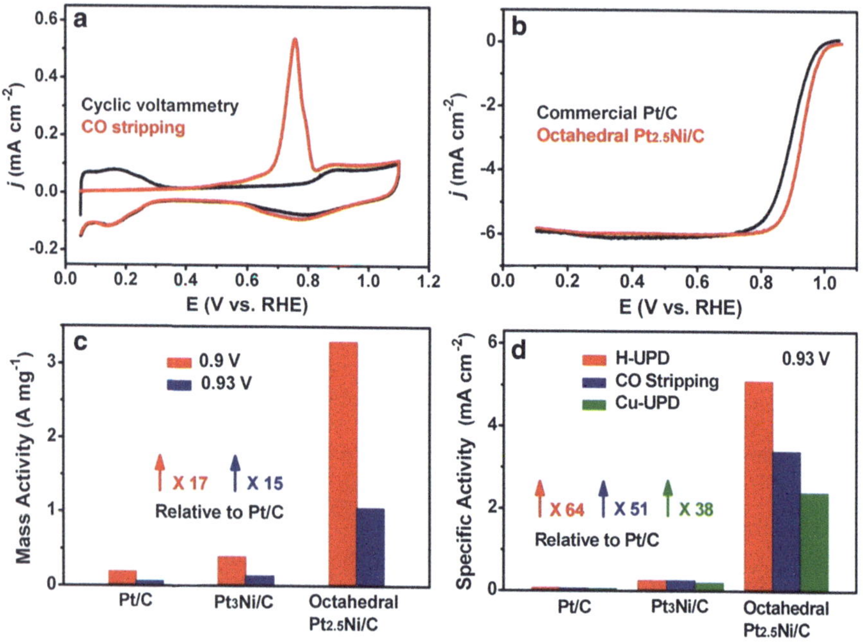

Fig. 2.21 (**a**) CO stripping and subsequent voltammogram of octahedral $Pt_{2.5}Ni/C$ (HAc treatment for 2 h) in a N_2-saturated 0.1 M $HClO_4$ solution at 50 mV s^{-1}. (**b**) ORR behaviour of octahedral $Pt_{2.5}Ni/C$ (HAc treatment for 2 h) and Pt/C in an O_2-saturated 0.1 M $HClO_4$ solution at 10 mV s^{-1}, rotation speed 1600 rpm. (**c**) Mass activities and (**d**) specific activities (calculated from the charges associated with H_{UPD}, CO stripping, and Cu_{UPD}) of Pt/C, Pt_3Ni/C, and octahedral $Pt_{2.5}Ni/C$ at 0.9 and 0.93 V. Reproduced with permission from [256], © 2013 American Chemical Society

surfactant on the PtNi samples, but without altering their shape/surface structure. Additionally, the samples were treated with acetic acid (HAc) at 60 °C in order to maximise the surface cleanness as well as to modify the Pt/Ni near-surface atomic composition. Under optimised conditions, a mass activity of 3.3 mA g^{-1} Pt was obtained at 0.9 V vs RHE, which is about 17 times higher than that of a Pt/C sample and two times higher than that previously reported by Cui et al. [110]. In terms of ORR specific activity, measured at +0.93 V, this optimised octahedral PtNi samples showed an enhancement factor that varied from 64 to 38, as a function of the methodology employed to determine the electroactive surface area (H UPD, CO stripping or Cu UPD), Fig. 2.21. As previously discussed, the determination of the real surface area of this kind of shape-controlled Pt alloy nanoparticles is a critical point that is still under discussion [68–70]. More recently, the ORR activity of these octahedral PtNi nanoparticles has been also evaluated as a function of the particle size (6–12 nm) and the Pt/Ni atomic composition [257]. Octahedral PtNi nanoparticles with a particle size of 9 nm and a Pt/Ni ratio of 2.4 displayed the best ORR specific and mass activities.

Chen et al. [258] have very recently reported an impressive mass activity of 5.7 A mg^{-1} Pt in $HClO_4$ solution at 0.9 V vs RHE, using Pt_3Ni nanoframes. These Pt_3Ni nanoframes were obtained after controlled erosion of rhombic dodecahedral $PtNi_3$ nanoparticles (about 20 nm) in nonpolar solvents. In addition, a subsequent thermal treatment induced the formation of a smooth Pt-skin type structure. Interestingly, not only the ORR activity was improved, but also its durability under ORR operation conditions. The results showed that both the ORR activity and nanoframe structure remained almost identical after the durability test. This stability was attributed to the particular electronic properties of the Pt-skin surface of the nanoframes.

Other interesting octahedral PtNi nanoparticles have been recently prepared and evaluated towards ORR [113, 259–262], but without additional specific and mass activity improvements than those previously discussed. Finally, and despite this remarkable progress on both ORR specific and mass activities, it is important to recall that, from a practical point of view, the stability is a critical issue and has been the subject of some relevant contributions [258, 263–267].

Based on this previous discussion, interesting contributions are being published dealing with the use of shape-controlled core-shell nanoparticles [246, 247, 268–271] and doped shape-controlled nanoparticles for enhanced activity and stability issues [272]. Unfortunately, this aspect will not be covered in this chapter but will represent a step beyond in the understanding of the role of shape/surface structure and atomic composition towards advanced ORR electrocatalysts with improved activity and stability.

2.4 Conclusions

The use of shape-controlled metal nanoparticles in electrocatalysis has produced an important impact of several reactions of interest, both in terms of enhanced activity, selectivity and in some cases stability. Firstly, and due to the fact that most of these materials are prepared in the presence of capping agents through colloidal routes, we have discussed about the importance of surface cleanness and how this cleaning can be achieved by using different approaches. Subsequently, we have summarized some of the most relevant contributions on the use of these shaped clean metal nanoparticles towards reactions of interest for low temperature fuel cells such as formic acid, methanol and, ethanol electrooxidation and oxygen reduction. However it is worth noting that only those contributions fulfilling this surface cleaning requirement have been considered in the present chapter. The results here reviewed clearly point out the outstanding role played by the surface structure of these shaped nanomaterials on their subsequent electrocatalytic properties.

Acknowledgments This work has been financially supported by the MINECO of Spain through projects CTQ2013-44083-P and CTQ2013-48280-C3-3-R and Generalitat Valenciana through project PROMETEOII/2014/013.

References

1. Wieckowski A (1999) Interfacial electrochemistry: theory, experiment, and applications. Marcel Dekker, New York
2. Wieckowski A, Savinova ER, Vayenas CG (2003) Catalysis and electrocatalysis at nanoparticle surfaces. CRC Press, New York
3. Koper MTM (2009) Fuel cell catalysis: a surface science approach. Electrocatalysis and electrochemistry. Wiley, Hoboken
4. Climent V, Feliu JM (2011) Thirty years of platinum single crystal electrochemistry. J Solid State Electrochem 15(7–8):1297–1315. doi:10.1007/s10008-011-1372-1
5. Zhou Z-Y, Tian N, Huang Z-Z, Chen D-J, Sun S-G (2009) Nanoparticle catalysts with high energy surfaces and enhanced activity synthesized by electrochemical method. Faraday Discuss 140:81–92
6. Proussevitch AA, Sahagian DL (2001) Recognition and separation of discrete objects within complex 3D voxelized structures. Comput Geosci 27(4):441–454. doi:10.1016/s0098-3004(00)00141-2
7. Tian N, Zhou ZY, Sun SG (2008) Platinum metal catalysts of high-index surfaces: from single-crystal planes to electrochemically shape-controlled nanoparticles. J Phys Chem C 112 (50):19801–19817
8. Vidal-Iglesias FJ, Solla-Gullón J, Rodríguez P, Herrero E, Montiel V, Feliu JM, Aldaz A (2004) Shape-dependent electrocatalysis: ammonia oxidation on platinum nanoparticles with preferential (100) surfaces. Electrochem Commun 6(10):1080–1084. doi:10.1016/j.elecom.2004.08.010
9. Solla-Gullón J, Vidal-Iglesias FJ, Rodríguez P, Herrero E, Feliu JM, Clavilier J, Aldaz A (2004) In situ surface characterization of preferentially oriented platinum nanoparticles by using electrochemical structure sensitive adsorption reactions. J Phys Chem B 108 (36):13573–13575. doi:10.1021/jp0471453
10. Ahmadi TS, Wang ZL, Green TC, Henglein A, El-Sayed MA (1996) Shape-controlled synthesis of colloidal platinum nanoparticles. Science 272(5270):1924–1926
11. Koper MTM (2011) Structure sensitivity and nanoscale effects in electrocatalysis. Nanoscale 3(5):2054–2073
12. Solla-Gullon J, Vidal-Iglesias FJ, Feliu JM (2011) Shape dependent electrocatalysis. Annu Rep Prog Chem, Sect C 107:263–297
13. Bing Y, Liu H, Zhang L, Ghosh D, Zhang J (2010) Nanostructured Pt-alloy electrocatalysts for PEM fuel cell oxygen reduction reaction. Chem Soc Rev 39(6):2184–2202
14. Chen A, Holt-Hindle P (2010) Platinum-based nanostructured materials: synthesis, properties, and applications. Chem Rev 110(6):3767–3804
15. Peng Z, Yang H (2009) Designer platinum nanoparticles: control of shape, composition in alloy, nanostructure and electrocatalytic property. Nano Today 4(2):143–164
16. Chen J, Lim B, Lee EP, Xia Y (2009) Shape-controlled synthesis of platinum nanocrystals for catalytic and electrocatalytic applications. Nano Today 4(1):81–95
17. Vismadeb Mazumder YL, Sun S (2010) Recent development of active nanoparticle catalysts for fuel cell reactions. Adv Funct Mater 20:1224–1234
18. Wu B, Zheng N (2013) Surface and interface control of noble metal nanocrystals for catalytic and electrocatalytic applications. Nano Today 8(2):168–197. doi:10.1016/j.nantod.2013.02.006
19. Kleijn SEF, Lai SCS, Koper MTM, Unwin PR (2014) Electrochemistry of nanoparticles. Angew Chem Int Ed 53(14):3558–3586
20. You H, Yang S, Ding B, Yang H (2013) Synthesis of colloidal metal and metal alloy nanoparticles for electrochemical energy applications. Chem Soc Rev 42(7):2880–2904
21. Sanchez-Sanchez CM, Solla-Gullon J, Montiel V (2013) Electrocatalysis at nanoparticles. In: Electrochemistry. Nanosystems electrochemistry, vol 11. The Royal Society of Chemistry, London, pp 34–70. doi:10.1039/9781849734820-00034

22. Dai Y, Wang Y, Liu B, Yang Y (2014) Metallic nanocatalysis: an accelerating seamless integration with nanotechnology. Small 11(3):268–289. doi:10.1002/smll.201400847
23. Burda C, Chen X, Narayanan R, El-Sayed MA (2005) Chemistry and properties of nanocrystals of different shapes. Chem Rev 105(4):1025–1102
24. Xia Y, Xiong Y, Lim B, Skrabalak SE (2009) Shape-controlled synthesis of metal nanocrystals: simple chemistry meets complex physics? Angew Chem Int Ed 48(1):60–103
25. Tao AR, Habas S, Yang P (2008) Shape control of colloidal metal nanocrystals. Small 4 (3):310–325
26. Sau TK, Rogach AL (2010) Nonspherical noble metal nanoparticles: colloid-chemical synthesis and morphology control. Adv Mater (Weinheim, Germany) 22(16):1781–1804
27. Leong GJ, Schulze MC, Strand MB, Maloney D, Frisco SL, Dinh HN, Pivovar B, Richards RM (2014) Shape-directed platinum nanoparticle synthesis: nanoscale design of novel catalysts. Appl Organomet Chem 28(1):1–17
28. Gu J, Zhang YW, Tao F (2012) Shape control of bimetallic nanocatalysts through well-designed colloidal chemistry approaches. Chem Soc Rev 41(24):8050–8065
29. Clavilier J, Faure R, Guinet G, Durand R (1980) Preparation of monocrystalline. Pt microelectrodes and electrochemical study of the plane surfaces cut in the direction of the {111} and {110} planes. J Electroanal Chem 107(1):205–209
30. Clavilier J (1999) Flame-annealing and cleaning technique. In: Wieckowski A (ed) Interfacial electrochemistry. Marcel Dekker, New York, pp 231–248
31. Hara M, Linke U, Wandlowski T (2007) Preparation and electrochemical characterization of palladium single crystal electrodes in 0.1 M H_2SO_4 and $HClO_4$ Part I. Low-index phases. Electrochim Acta 52(18):5733–5748
32. Hamelin A (1996) Cyclic voltammetry at gold single-crystal surfaces. 1. Behaviour at low-index faces. J Electroanal Chem 407(1–2):1–11
33. Hamelin A (1996) Cyclic voltammetry at gold single-crystal surfaces. 2. Behaviour of high-index faces. J Electroanal Chem 407(1-2):13–21
34. Ke F-S, Solomon B, Ding Y, Xu G-L, Sun S-G, Wang ZL, Zhou X-D (2014) Enhanced electrocatalytic activity on gold nanocrystals enclosed by high-index facets for oxygen reduction. Nano Energy 7:179–188. doi:10.1016/j.nanoen.2014.01.003
35. Yang S, Lee H (2013) Atomically dispersed platinum on gold nano-octahedra with high catalytic activity on formic acid oxidation. ACS Catal 3(3):437–443. doi:10.1021/cs300809j
36. Nichols RJ, Magnussen OM, Hotlos J, Twomey T, Behm RJ, Kolb DM (1990) An in-situ STM study of potential-induced changes in the surface topography of Au(100) electrodes. J Electroanal Chem 290:21–31
37. Schneeweiss MA, Kolb DM (1997) Oxide formation on Au(111)—an in situ STM study. Solid State Ion 94(1–4):171–179
38. Itaya K, Sugawara S, Sashikata K, Furuya N (1990) Insitu scanning tunneling microscopy of platinum (111) surface with the observation of monatomic steps. J Vac Sci Technol A Vac Surf Films 8(1):515–519
39. El-Deab MS (2009) On the preferential crystallographic orientation of Au nanoparticles: effect of electrodeposition time. Electrochim Acta 54(14):3720–3725
40. El-Deab MS, Sotomura T, Ohsaka T (2005) Size and crystallographic orientation controls of gold nanoparticles electrodeposited on GC electrodes. J Electrochem Soc 152(1):C1–C6
41. El-Deab MS, Arihara K, Ohsaka T (2004) Fabrication of Au(111)-like polycrystalline gold electrodes and their applications to oxygen reduction. J Electrochem Soc 151(6):E213–E218
42. Walczak MM, Alves CA, Lamp BD, Porter MD (1995) Electrochemical and X-ray photoelectron spectroscopic evidence for differences in the binding sites of alkanethiolate monolayers chemisorbed at gold. J Electroanal Chem 396(1–2):103–114
43. Zhong CJ, Zak J, Porter MD (1997) Voltammetric reductive desorption characteristics of alkanethiolate monolayers at single crystal Au(111) and (110) electrode surfaces. J Electroanal Chem 421(1–2):9–13

44. Herrero E, Buller LJ, Abruna HD (2001) Underpotential deposition at single crystal surfaces of Au, Pt, Ag and other materials. Chem Rev 101(7):1897–1930. doi:10.1021/cr9600363
45. Hernández J, Solla-Gullón J, Herrero E, Aldaz A, Feliu JM (2007) Electrochemistry of shape-controlled catalysts: oxygen reduction reaction on cubic gold nanoparticles. J Phys Chem C 111(38):14078–14083. doi:10.1021/jp0749726CCC
46. Hernández J, Solla-Gullón J, Herrero E, Feliu JM, Aldaz A (2009) In situ surface characterization and oxygen reduction reaction on shape-controlled gold nanoparticles. J Nanosci Nanotechnol 9(4):2256–2273. doi:10.1166/jnn.2009.SE38
47. Hernández J, Solla-Gullón J, Herrero E (2004) Gold nanoparticles synthesized in a water-in-oil microemulsion: electrochemical characterization and effect of the surface structure on the oxygen reduction reaction. J Electroanal Chem 574(1):185–196. doi:10.1016/j.jelechem.2003.10.039
48. Hamelin A (1979) Lead adsorption on gold single crystal stepped surfaces. J Electroanal Chem 101(2):285–290
49. Hamelin A, Katayama A (1981) Lead underpotential deposition on gold single-crystal surfaces: the (100) face and its vicinal faces. J Electroanal Chem 117(2):221–232
50. Hamelin A (1984) Underpotential deposition of lead on single crystal faces of gold. Part I. The influence of crystallographic orientation of the substrate. J Electroanal Chem 165 (1–2):167–180
51. Hamelin A, Lipkowski J (1984) Underpotential deposition of lead on gold single crystal faces. Part II. General discussion. J Electroanal Chem 171(1–2):317–330
52. Chierchie T, Mayer C (1988) Voltammetric study of the underpotential deposition of copper on polycrystalline and single crystal palladium surfaces. Electrochim Acta 33(3):341–345
53. Cuesta A, Kibler LA, Kolb DM (1999) A method to prepare single crystal electrodes of reactive metals: application to Pd(hkl). J Electroanal Chem 466(2):165–168
54. Fang LL, Tao Q, Li MF, Liao LW, Chen D, Chen YX (2010) Determination of the real surface area of palladium electrode. Chin J Chem Phys 23(5):543–548
55. Francke R, Climent V, Baltruschat H, Feliu JM (2008) Electrochemical deposition of copper on stepped platinum surfaces in the 01(1)over-bar zone vicinal to the (100) plane. J Electroanal Chem 624(1–2):228–240
56. Danilov AI, Molodkina EB, Rudnev AV, Polukarov YM, Feliu JM (2005) Kinetics of copper deposition on Pt(111) and Au(111) electrodes in solutions of different acidities. Electrochim Acta 50(25–26):5032–5043. doi:10.1016/j.electacta.2005.02.078
57. Solla-Gullón J, Rodríguez P, Herrero E, Aldaz A, Feliu JM (2008) Surface characterization of platinum electrodes. Phys Chem Chem Phys 10(10):1359–1373. doi:10.1039/b709809j
58. Rodríguez P, Solla-Gullón J, Vidal-Iglesias FJ, Herrero E, Aldaz A, Feliu JM (2005) Determination of (111) ordered domains on platinum electrodes by irreversible adsorption of bismuth. Anal Chem 77(16):5317–5323. doi:10.1021/ac050347q
59. Rodríguez P, Herrero E, Solla-Gullón J, Vidal-Iglesias FJ, Aldaz A, Feliu JM (2005) Specific surface reactions for identification of platinum surface domains—surface characterization and electrocatalytic tests. Electrochim Acta 50(21):4308–4317. doi:10.1016/j.electacta.2005.02.087
60. Rodríguez P, Herrero E, Aldaz A, Feliu JM (2006) Tellurium adatoms as an in-situ surface probe of (111) two-dimensional domains at platinum surfaces. Langmuir 22 (25):10329–10337. doi:10.1021/la060981e
61. Rodríguez P, Herrero E, Solla-Gullón J, Vidal-Iglesias FJ, Aldaz A, Feliu JM (2005) Electrochemical characterization of irreversibly adsorbed germanium on platinum stepped surfaces vicinal to Pt(100). Electrochim Acta 50(15):3111–3121. doi:10.1016/j.electacta.2004.10.086
62. Brimaud S, Pronier S, Coutanceau C, Léger JM (2008) New findings on CO electrooxidation at platinum nanoparticle surfaces. Electrochem Commun 10(11):1703–1707

63. Devivaraprasad R, Ramesh R, Naresh N, Kar T, Singh RK, Neergat M (2014) Oxygen reduction reaction and peroxide generation on shape-controlled and polycrystalline platinum nanoparticles in acidic and alkaline electrolytes. Langmuir 30(29):8995–9006
64. Bertin E, Garbarino S, Guay D (2014) Formic acid oxidation on Bi covered Pt electrodeposited thin films: influence of the underlying structure. Electrochim Acta 134:486–495. doi:10.1016/j.electacta.2014.04.111
65. Levendorf AM, Chen D-J, Rom CL, Liu Y, Tong YJ (2014) Electrochemical and in situ ATR-SEIRAS investigations of methanol and CO electro-oxidation on PVP-free cubic and octahedral/tetrahedral Pt nanoparticles. RSC Adv 4(41):21284–21293. doi:10.1039/c4ra00815d
66. Rodriguez-Lopez M, Solla-Gullon J, Herrero E, Tunon P, Feliu JM, Aldaz A, Carrasquillo A (2010) Electrochemical reactivity of aromatic molecules at nanometer-sized surface domains: from Pt(hkl) single crystal electrodes to preferentially oriented platinum nanoparticles. J Am Chem Soc 132(7):2233–2242. doi:10.1021/Ja909082s
67. Martínez-Rodríguez RA, Vidal-Iglesias FJ, Solla-Gullón J, Cabrera CR, Feliu JM (2014) Synthesis and electrocatalytic properties of H_2SO_4-induced (100) Pt nanoparticles prepared in water-in-oil microemulsion. ChemPhysChem 15(10):1997–2001. doi:10.1002/cphc.201400056
68. Van Der Vliet DF, Wang C, Li D, Paulikas AP, Greeley J, Rankin RB, Strmcnik D, Tripkovic D, Markovic NM, Stamenkovic VR (2012) Unique electrochemical adsorption properties of Pt-skin surfaces. Angew Chem Int Ed 51(13):3139–3142
69. Shao M, Odell JH, Choi SI, Xia Y (2013) Electrochemical surface area measurements of platinum- and palladium-based nanoparticles. Electrochem Commun 31:46–48
70. Rudi S, Cui C, Gan L, Strasser P (2014) Comparative study of the electrocatalytically active surface areas (ECSAs) of Pt alloy nanoparticles evaluated by Hupd and CO-stripping voltammetry. Electrocatalysis 5(4):408–418
71. Solla-Gullón J, Vidal-Iglesias FJ, Herrero E, Feliu JM, Aldaz A (2006) CO monolayer oxidation on semi-spherical and preferentially oriented (100) and (111) platinum nanoparticles. Electrochem Commun 8(1):189–194. doi:10.1016/j.elecom.2005.11.008
72. Solla-Gullón J, Vidal-Iglesias FJ, López-Cudero A, Garnier E, Feliu JM, Aldaz A (2008) Shape-dependent electrocatalysis: methanol and formic acid electrooxidation on preferentially oriented Pt nanoparticles. Phys Chem Chem Phys 10(25):3689–3698. doi:10.1039/b802703j
73. Vidal-Iglesias FJ, Aran-Ais RM, Solla-Gullon J, Herrero E, Feliu JM (2012) Electrochemical characterization of shape-controlled Pt nanoparticles in different supporting electrolytes. ACS Catal 2(5):901–910. doi:10.1021/Cs200681x
74. Erikson H, Sarapuu A, Tammeveski K, Solla-Gullón J, Feliu JM (2011) Enhanced electrocatalytic activity of cubic Pd nanoparticles towards the oxygen reduction reaction in acid media. Electrochem Commun 13(7):734–737
75. Erikson H, Sarapuu A, Alexeyeva N, Tammeveski K, Solla-Gullón J, Feliu JM (2012) Electrochemical reduction of oxygen on palladium nanocubes in acid and alkaline solutions. Electrochim Acta 59:329–335. doi:10.1016/j.electacta.2011.10.074
76. Vidal-Iglesias FJ, Aran-Ais RM, Solla-Gullon J, Garnier E, Herrero E, Aldaz A, Feliu JM (2012) Shape-dependent electrocatalysis: formic acid electrooxidation on cubic Pd nanoparticles. Phys Chem Chem Phys 14(29):10258–10265. doi:10.1039/c2cp40992e
77. Brimaud S, Jusys Z, Behm RJ (2014) Shape-selected nanocrystals for in situ spectro-electrochemistry studies on structurally well defined surfaces under controlled electrolyte transport: a combined in situ ATR-FTIR/online DEMS investigation of CO electrooxidation on Pt. Beilstein J Nanotechnol 5:735–746. doi:10.3762/bjnano.5.86
78. Brimaud S, Jusys Z, Behm RJ (2011) Controlled surface structure for in situ ATR-FTIRS studies using preferentially shaped Pt nanocrystals. Electrocatalysis 2(2):69–74. doi:10.1007/s12678-011-0040-7

79. Coutanceau C, Urchaga P, Brimaud S, Baranton S (2012) Colloidal syntheses of shape- and size-controlled Pt nanoparticles for electrocatalysis. Electrocatalysis 3(2):75–87. doi:10.1007/s12678-012-0079-0
80. Urchaga P, Baranton S, Napporn TW, Coutanceau C (2010) Selective syntheses and electrochemical characterization of platinum nanocubes and nanotetrahedrons/octahedrons. Electrocatalysis 1:3–6
81. Susut C, Tong Y (2011) Size-dependent methanol electro-oxidation activity of Pt nanoparticles with different shapes. Electrocatalysis 2(2):75–81. doi:10.1007/s12678-011-0041-6
82. Susut C, Chapman GB, Samjeské G, Osawa M, Tong Y (2008) An unexpected enhancement in methanol electro-oxidation on an ensemble of Pt(111) nanofacets: a case of nanoscale single crystal ensemble electrocatalysis. Phys Chem Chem Phys 10(25):3712–3721
83. Susut C, Nguyen TD, Chapman GB, Tong Y (2008) Shape and size stability of Pt nanoparticles for MeOH electro-oxidation. Electrochim Acta 53(21):6135–6142
84. Song H, Kim F, Connor S, Somorjai GA, Yang P (2005) Pt nanocrystals: shape control and Langmuir-Blodgett monolayer formation. J Phys Chem B 109(1):188–193
85. Monzó J, Koper MTM, Rodriguez P (2012) Removing polyvinylpyrrolidone from catalytic Pt nanoparticles without modification of superficial order. ChemPhysChem 13(3):709–715
86. Yang H, Tang Y, Zou S (2014) Electrochemical removal of surfactants from Pt nanocubes. Electrochem Commun 38:134–137. doi:10.1016/j.elecom.2013.11.019
87. Arán-Ais RM, Vidal-Iglesias FJ, Solla-Gullon J, Herrero E, Feliu JM (2014) Electrochemical characterization of clean shape-controlled Pt nanoparticles prepared in presence of oleylamine/oleic acid. Electroanalysis 27(4):945–956
88. Zalineeva A, Baranton S, Coutanceau C (2013) Bi-modified palladium nanocubes for glycerol electrooxidation. Electrochem Commun 34:335–338
89. Zalineeva A, Baranton S, Coutanceau C, Jerkiewicz G (2015) Electrochemical behavior of unsupported shaped palladium nanoparticles. Langmuir 31(5):1605–1609. doi:10.1021/la5025229
90. Nalajala N, Gooty Saleha WF, Ladewig BP, Neergat M (2014) Sodium borohydride treatment: a simple and effective process for the removal of stabilizer and capping agents from shape-controlled palladium nanoparticles. Chem Commun 50(66):9365–9368
91. Naresh N, Wasim FGS, Ladewig BP, Neergat M (2013) Removal of surfactant and capping agent from Pd nanocubes (Pd-NCs) using tert-butylamine: its effect on electrochemical characteristics. J Mater Chem A 1(30):8553–8559
92. Hernández J, Solla-Gullón J, Herrero E, Aldaz A, Feliu JM (2005) Characterization of the surface structure of gold nanoparticles and nanorods using structure sensitive reactions. J Phys Chem B 109(26):12651–12654. doi:10.1021/jp0521609
93. Park JY, Aliaga C, Renzas JR, Lee H, Somorjai GA (2009) The role of organic capping layers of platinum nanoparticles in catalytic activity of CO oxidation. Catal Lett 129(1-2):1–6
94. Aliaga C, Park JY, Yamada Y, Lee HS, Tsung CK, Yang P, Somorjai GA (2009) Sum frequency generation and catalytic reaction studies of the removal of organic capping agents from Pt nanoparticles by UV-Ozone treatment. J Phys Chem C 113(15):6150–6155
95. Krier JM, Michalak WD, Baker LR, An K, Komvopoulos K, Somorjai GA (2012) Sum frequency generation vibrational spectroscopy of colloidal platinum nanoparticle catalysts: disordering versus removal of organic capping. J Phys Chem C 116(33):17540–17546
96. Crespo-Quesada M, Andanson JM, Yarulin A, Lim B, Xia Y, Kiwi-Minsker L (2011) UV-ozone cleaning of supported poly(vinylpyrrolidone)-stabilized palladium nanocubes: effect of stabilizer removal on morphology and catalytic behavior. Langmuir 27(12):7909–7916
97. Vidal-Iglesias FJ, Solla-Gullón J, Herrero E, Montiel V, Aldaz A, Feliu JM (2011) Evaluating the ozone cleaning treatment in shape-controlled Pt nanoparticles: evidences of atomic surface disordering. Electrochem Commun 13(5):502–505. doi:10.1016/j.elecom.2011.02.033

98. Levendorf A, Sun S-G, Tong Y (2014) In situ FT-IR investigation of methanol and CO electrooxidation on cubic and octahedral/tetrahedral Pt nanoparticles having residual PVP. Electrocatalysis 5(3):248–255. doi:10.1007/s12678-014-0186-1
99. Tong YJ (2012) Unconventional promoters of catalytic activity in electrocatalysis. Chem Soc Rev 41(24):8195–8209
100. Susut C, Chen DJ, Sun SG, Tong YJ (2011) Capping polymer-enhanced electrocatalytic activity on Pt nanoparticles: a combined electrochemical and in situ IR spectroelectrochemical study. Phys Chem Chem Phys 13(16):7467–7474
101. Shen J, Ziaei-Azad H, Semagina N (2014) Is it always necessary to remove a metal nanoparticle stabilizer before catalysis? J Mol Catal A Chem 391(1):36–40
102. Chung YH, Chung DY, Jung N, Sung YE (2013) Tailoring the electronic structure of nanoelectrocatalysts induced by a surface-capping organic molecule for the oxygen reduction reaction. J Phys Chem Lett 4(8):1304–1309
103. Miyabayashi K, Nishihara H, Miyake M (2014) Platinum nanoparticles modified with alkylamine derivatives as an active and stable catalyst for oxygen reduction reaction. Langmuir 30(10):2936–2942
104. Niu Z, Li Y (2013) Removal and utilization of capping agents in nanocatalysis. Chem Mater 26(1):72–83. doi:10.1021/cm4022479
105. Lee H (2014) Utilization of shape-controlled nanoparticles as catalysts with enhanced activity and selectivity. RSC Adv 4(77):41017–41027
106. Newton JE, Preece JA, Rees NV, Horswell SL (2014) Nanoparticle catalysts for proton exchange membrane fuel cells: can surfactant effects be beneficial for electrocatalysis? Phys Chem Chem Phys 16(23):11435–11446
107. Du L, Zhang S, Chen G, Yin G, Du C, Tan Q, Sun Y, Qu Y, Gao Y (2014) Polyelectrolyte assisted synthesis and enhanced oxygen reduction activity of Pt nanocrystals with controllable shape and size. ACS Appl Mater Interfaces 6(16):14043–14049
108. Tian N, Zhou Z-Y, Sun S-G, Ding Y, Wang ZL (2007) Synthesis of tetrahexahedral platinum nanocrystals with high-index facets and high electro-oxidation activity. Science 316 (5825):732–735. doi:10.1126/science.1140484
109. Tian N, Xiao J, Zhou Z-Y, Liu H-X, Deng Y-J, Huang L, Xu B-B, Sun S-G (2013) Pt-group bimetallic nanocrystals with high-index facets as high performance electrocatalysts. Faraday Discuss 162:77–89. doi:10.1039/c3fd20146e
110. Cui C, Gan L, Li HH, Yu SH, Heggen M, Strasser P (2012) Octahedral PtNi nanoparticle catalysts: exceptional oxygen reduction activity by tuning the alloy particle surface composition. Nano Lett 12(11):5885–5889
111. Gumeci C, Marathe A, Behrens RL, Chaudhuri J, Korzeniewski C (2014) Solvothermal synthesis and electrochemical characterization of shape-controlled Pt nanocrystals. J Phys Chem C 118(26):14433–14440
112. Carpenter MK, Moylan TE, Kukreja RS, Atwan MH, Tessema MM (2012) Solvothermal synthesis of platinum alloy nanoparticles for oxygen reduction electrocatalysis. J Am Chem Soc 134(20):8535–8542
113. Zhang C, Hwang SY, Trout A, Peng Z (2014) Solid-state chemistry-enabled scalable production of octahedral Pt-Ni alloy electrocatalyst for oxygen reduction reaction. J Am Chem Soc 136(22):7805–7808
114. Shao M (2013) Electrocatalysis in fuel cells: a non- and low-platinum approach. Springer, London
115. Feliu JM, Herrero E (2003) Formic acid oxidation. In: Vielstich W, Gasteiger H, Lamm A (eds) Handbook of fuel cells, vol 2, Fundamentals, technology and applications. Wiley, Chichester
116. Jiang K, Zhang HX, Zou S, Cai WB (2014) Electrocatalysis of formic acid on palladium and platinum surfaces: from fundamental mechanisms to fuel cell applications. Phys Chem Chem Phys 16(38):20360–20376

117. Capon A, Parsons R (1973) The oxidation of formic acid at noble metal electrodes: I. Review of previous work. J Electroanal Chem 44(1):1–7
118. Capon A, Parsons R (1973) The oxidation of formic acid on noble metal electrodes: II. A comparison of the behaviour of pure electrodes. J Electroanal Chem 44(2):239–254
119. Capon A, Parsons R (1973) The oxidation of formic acid at noble metal electrodes Part III. Intermediates and mechanism on platinum electrodes. J Electroanal Chem 45(2):205–231
120. Grozovski V, Solla-Gullon J, Climent V, Herrero E, Feliu JM (2010) Formic acid oxidation on shape-controlled Pt nanoparticles studied by pulsed voltammetry. J Phys Chem C 114 (32):13802–13812. doi:10.1021/jp104755b
121. Huang X, Zhao Z, Fan J, Tan Y, Zheng N (2011) Amine-assisted synthesis of concave polyhedral platinum nanocrystals having {411} high-index facets. J Am Chem Soc 133 (13):4718–4721
122. Li Y, Jiang Y, Chen M, Liao H, Huang R, Zhou Z, Tian N, Chen S, Sun S (2012) Electrochemically shape-controlled synthesis of trapezohedral platinum nanocrystals with high electrocatalytic activity. Chem Commun 48(76):9531–9533. doi:10.1039/c2cc34322c
123. Zhang Z-C, Hui J-F, Liu Z-C, Zhang X, Zhuang J, Wang X (2012) Glycine-mediated syntheses of Pt concave nanocubes with high-index {hk0} facets and their enhanced electrocatalytic activities. Langmuir 28(42):14845–14848. doi:10.1021/la302973r
124. Xia BY, Wu HB, Wang X, Lou XW (2013) Highly concave platinum nanoframes with high-index facets and enhanced electrocatalytic properties. Angew Chem Int Ed 52 (47):12337–12340. doi:10.1002/anie.201307518
125. Lu GQ, Crown A, Wieckowski A (1999) Formic acid decomposition on polycrystalline platinum and palladized platinum electrodes. J Phys Chem B 103(44):9700–9711
126. Rice C, Ha S, Masel RI, Wieckowski A (2003) Catalysts for direct formic acid fuel cells. J Power Sources 115(2):229–235. doi:10.1016/s0378-7753(03)00026-0
127. Rice C, Ha RI, Masel RI, Waszczuk P, Wieckowski A, Barnard T (2002) Direct formic acid fuel cells. J Power Sources 111(1):83–89
128. Hoshi N, Kida K, Nakamura M, Nakada M, Osada K (2006) Structural effects of electrochemical oxidation of formic acid on single crystal electrodes of palladium. J Phys Chem B 110(25):12480–12484
129. Baldauf M, Kolb DM (1996) Formic acid oxidation on ultrathin Pd films on Au(hkl) and Pt (hkl) electrodes. J Phys Chem 100(27):11375–11381
130. Jin M, Zhang H, Xie Z, Xia Y (2012) Palladium nanocrystals enclosed by {100} and {111} facets in controlled proportions and their catalytic activities for formic acid oxidation. Energy Environ Sci 5(4):6352–6357
131. Zhang H-X, Wang H, Re Y-S, Cai W-B (2012) Palladium nanocrystals bound by {110} or {100} facets: from one pot synthesis to electrochemistry. Chem Commun 48(67):8362–8364. doi:10.1039/c2cc33941b
132. Shao Z, Zhu W, Wang H, Yang Q, Yang S, Liu X, Wang G (2013) Controllable synthesis of concave nanocubes, right bipyramids, and 5-fold twinned nanorods of palladium and their enhanced electrocatalytic performance. J Phys Chem C 117(27):14289–14294
133. Kuo CH, Lamontagne LK, Brodsky CN, Chou LY, Zhuang J, Sneed BT, Sheehan MK, Tsung CK (2013) The effect of lattice strain on the catalytic properties of Pd nanocrystals. Chemsuschem 6(10):1993–2000
134. Xia X, Choi SI, Herron JA, Lu N, Scaranto J, Peng HC, Wang J, Mavrikakis M, Kim MJ, Xia Y (2013) Facile synthesis of palladium right bipyramids and their use as seeds for overgrowth and as catalysts for formic acid oxidation. J Am Chem Soc 135(42):15706–15709
135. LV T, Wang Y, Choi SI, Chi M, Tao J, Pan L, Huang CZ, Zhu Y, Xia Y (2013) Controlled synthesis of nanosized palladium icosahedra and their catalytic activity towards formic-acid oxidation. Chemsuschem 6(10):1923–1930
136. Shao M, Odell J, Humbert M, Yu T, Xia Y (2013) Electrocatalysis on shape-controlled palladium nanocrystals: oxygen reduction reaction and formic acid oxidation. J Phys Chem C 117(8):4172–4180

137. Zhang X, Yin H, Wang J, Chang L, Gao Y, Liu W, Tang Z (2013) Shape-dependent electrocatalytic activity of monodispersed palladium nanocrystals toward formic acid oxidation. Nanoscale 5(18):8392–8397. doi:10.1039/c3nr03100d
138. Zhang J, Feng C, Deng Y, Liu L, Wu Y, Shen B, Zhong C, Hu W (2014) Shape-controlled synthesis of palladium single-crystalline nanoparticles: the effect of HCl oxidative etching and facet-dependent catalytic properties. Chem Mater 26(2):1213–1218. doi:10.1021/cm403591g
139. Tang Y, Edelmann RE, Zou S (2014) Length tunable penta-twinned palladium nanorods: seedless synthesis and electrooxidation of formic acid. Nanoscale 6(11):5630–5633. doi:10.1039/c4nr00299g
140. Nørskov JK, Abild-Pedersen F, Studt F, Bligaard T (2011) Density functional theory in surface chemistry and catalysis. Proc Natl Acad Sci 108(3):937–943. doi:10.1073/pnas.1006652108
141. Boronat-Gonzalez A, Herrero E, Feliu JM (2014) Fundamental aspects of HCOOH oxidation at platinum single crystal surfaces with basal orientations and modified by irreversibly adsorbed adatoms. J Solid State Electrochem 18(5):1181–1193. doi:10.1007/s10008-013-2209-x
142. Chen QS, Zhou ZY, Vidal-Iglesias FJ, Solla-Gullon J, Feliu JM, Sun SG (2011) Significantly enhancing catalytic activity of tetrahexahedral Pt nanocrystals by Bi adatom decoration. J Am Chem Soc 133(33):12930–12933. doi:10.1021/ja2042029
143. López-Cudero A, Vidal-Iglesias FJ, Solla-Gullón J, Herrero E, Aldaz A, Feliu JM (2009) Formic acid electrooxidation on Bi-modified polyoriented and preferential (111) Pt nanoparticles. Phys Chem Chem Phys 11(2):416–424. doi:10.1039/b814072c
144. Vidal-Iglesias FJ, Solla-Gullon J, Herrero E, Aldaz A, Feliu JM (2010) Pd adatom decorated (100) preferentially oriented Pt nanoparticles for formic acid electrooxidation. Angew Chem Int Ed 49(39):6998–7001. doi:10.1002/anie.201002501
145. Vidal-Iglesias FJ, Lopez-Cudero A, Solla-Gullon J, Aldaz A, Feliu JM (2012) Pd-modified shape-controlled Pt nanoparticles towards formic acid electrooxidation. Electrocatalysis 3(3–4):313–323. doi:10.1007/s12678-012-0094-1
146. Vidal-Iglesias FJ, López-Cudero A, Solla-Gullón J, Feliu JM (2013) Towards more active and stable electrocatalysts for formic acid electrooxidation: antimony-decorated octahedral platinum nanoparticles. Angew Chem Int Ed 52(3):964–967. doi:10.1002/anie.201207517
147. Buso-Rogero C, Perales-Rondon JV, Farias MJS, Vidal-Iglesias FJ, Solla-Gullon J, Herrero E, Feliu JM (2014) Formic acid electrooxidation on thallium-decorated shape-controlled platinum nanoparticles: an improvement in electrocatalytic activity. Phys Chem Chem Phys 16(27):13616–13624. doi:10.1039/c4cp00304g
148. Liu HX, Tian N, Brandon MP, Pei J, Huangfu ZC, Zhan C, Zhou ZY, Hardacre C, Lin WF, Sun SG (2012) Enhancing the activity and tuning the mechanism of formic acid oxidation at tetrahexahedral Pt nanocrystals by Au decoration. Phys Chem Chem Phys 14(47):16415–16423
149. Yuan Q, Zhou Z, Zhuang J, Wang X (2010) Pd-Pt random alloy nanocubes with tunable compositions and their enhanced electrocatalytic activities. Chem Commun 46(9):1491–1493. doi:10.1039/b922792j
150. Zhang J, Yang H, Yang K, Fang J, Zou S, Luo Z, Wang H, Bae IT, Jung DY (2010) Monodisperse Pt3Fe nanocubes: synthesis, characterization, self-assembly, and electrocatalytic activity. Adv Funct Mater 20(21):3727–3733
151. Yang H, Dai L, Xu D, Fang J, Zou S (2010) Electrooxidation of methanol and formic acid on PtCu nanoparticles. Electrochim Acta 55(27):8000–8004. doi:10.1016/j.electacta.2010.03.026
152. Xu D, Bliznakov S, Liu Z, Fang J, Dimitrov N (2010) Composition-dependent electrocatalytic activity of Pt-Cu nanocube catalysts for formic acid oxidation. Angew Chem Int Ed 49(7):1282–1285. doi:10.1002/anie.200905248

153. Zhang J, Yang H, Martens B, Luo Z, Xu D, Wang Y, Zou S, Fang J (2012) Pt-Cu nanoctahedra: synthesis and comparative study with nanocubes on their electrochemical catalytic performance. Chem Sci 3(11):3302–3306
154. Bromberg L, Fayette M, Martens B, Luo ZP, Wang Y, Xu D, Zhang J, Fang J, Dimitrov N (2013) Catalytic performance comparison of shape-dependent nanocrystals and oriented ultrathin films of Pt4Cu alloy in the formic acid oxidation process. Electrocatalysis 4 (1):24–36
155. Quan Z, Wang Y, Fang J (2013) High-index faceted noble metal nanocrystals. Acc Chem Res 46(2):191–202
156. Porter NS, Wu H, Quan Z, Fang J (2013) Shape-control and electrocatalytic activity-enhancement of Pt-based bimetallic nanocrystals. Acc Chem Res 46(8):1867–1877. doi:10.1021/ar3002238
157. Kang Y, Murray CB (2010) Synthesis and electrocatalytic properties of cubic Mn–Pt nanocrystals (nanocubes). J Am Chem Soc 132(22):7568–7569. doi:10.1021/ja100705j
158. Yu Y, Zhang Q, Liu B, Lee JY (2010) Synthesis of nanocrystals with variable high-index Pd facets through the controlled heteroepitaxial growth of trisoctahedral Au templates. J Am Chem Soc 132(51):18258–18265
159. Zhang L, Zhang J, Kuang Q, Xie S, Jiang Z, Xie Z, Zheng L (2011) Cu2 + -assisted synthesis of hexoctahedral Au-Pd alloy nanocrystals with high-index facets. J Am Chem Soc 133 (43):17114–17117
160. Deng YJ, Tian N, Zhou ZY, Huang R, Liu ZL, Xiao J, Sun SG (2012) Alloy tetrahexahedral Pd-Pt catalysts: enhancing significantly the catalytic activity by synergy effect of high-index facets and electronic structure. Chem Sci 3(4):1157–1161
161. Zhang ZC, Hui JF, Guo ZG, Yu QY, Xu B, Zhang X, Liu ZC, Xu CM, Gao JS, Wang X (2012) Solvothermal synthesis of Pt-Pd alloys with selective shapes and their enhanced electrocatalytic activities. Nanoscale 4(8):2633–2639
162. Jia Y, Jiang Y, Zhang J, Zhang L, Chen Q, Xie Z, Zheng L (2014) Unique excavated rhombic dodecahedral PtCu3 alloy nanocrystals constructed with ultrathin nanosheets of high-energy {110} facets. J Am Chem Soc 136(10):3748–3751. doi:10.1021/ja413209q
163. Sneed BT, Young AP, Jalalpoor D, Golden MC, Mao S, Jiang Y, Wang Y, Tsung C-K (2014) Shaped Pd–Ni–Pt core-sandwich-shell nanoparticles: influence of Ni sandwich layers on catalytic electrooxidations. ACS Nano 8(7):7239–7250. doi:10.1021/nn502259g
164. Zhang L, Choi S-I, Tao J, Peng H-C, Xie S, Zhu Y, Xie Z, Xia Y (2014) Pd-Cu bimetallic tripods: a mechanistic understanding of the synthesis and their enhanced electrocatalytic activity for formic acid oxidation. Adv Funct Mater 24(47):7520–7529. doi:10.1002/adfm.201402350
165. Li M, Adzic R (2013) Low-platinum-content electrocatalysts for methanol and ethanol electrooxidation. In: Shao M (ed) Electrocatalysis in fuel cells, vol 9, Lecture notes in energy. Springer, London, pp 1–25. doi:10.1007/978-1-4471-4911-8_1
166. Xia XH, Iwasita T, Ge F, Vielstich W (1996) Structural effects and reactivity in methanol oxidation on polycrystal line and single crystal platinum. Electrochim Acta 41(5):711–718
167. Lamy C, Leger JM, Clavilier J, Parsons R (1983) Structural effects in electrocatalysis—a comparative-study of the oxidation of CO, HCOOH and CH_3OH on single-crystal Pt electrodes. J Electroanal Chem 150(1-2):71–77
168. Hofstead-Duffy AM, Chen D-J, Sun S-G, Tong YJ (2012) Origin of the current peak of negative scan in the cyclic voltammetry of methanol electro-oxidation on Pt-based electrocatalysts: a revisit to the current ratio criterion. J Mater Chem 22(11):5205–5208. doi:10.1039/c2jm15426a
169. Chen G, Tan Y, Wu B, Fu G, Zheng N (2012) Carbon monoxide-controlled synthesis of surface-clean Pt nanocubes with high electrocatalytic activity. Chem Commun 48 (22):2758–2760. doi:10.1039/c2cc17984a

170. Chen J, Mao J, Zhao J, Ren M, Wei M (2014) Surfactant-free platinum nanocubes with greatly enhanced activity towards methanol/ethanol electrooxidation. RSC Adv 4 (55):28832–28835. doi:10.1039/c4ra03446e
171. Liu H-X, Tian N, Brandon MP, Zhou Z-Y, Lin J-L, Hardacre C, Lin W-F, Sun S-G (2012) Tetrahexahedral Pt nanocrystal catalysts decorated with Ru adatoms and their enhanced activity in methanol electrooxidation. ACS Catal 2(5):708–715. doi:10.1021/cs200686a
172. Arjona N, Guerra-Balcázar M, Ortiz-Frade L, Osorio-Monreal G, Álvarez-Contreras L, Ledesma-García J, Arriaga LG (2013) Electrocatalytic activity of well-defined and homogeneous cubic-shaped Pd nanoparticles. J Mater Chem A 1(48):15524–15529
173. Kannan P, Maiyalagan T, Opallo M (2013) One-pot synthesis of chain-like palladium nanocubes and their enhanced electrocatalytic activity for fuel-cell applications. Nano Energy 2(5):677–687. doi:10.1016/j.nanoen.2013.08.001
174. Qin Y-L, Zhang X-B, Wang J, Wang L-M (2012) Rapid and shape-controlled synthesis of "clean" star-like and concave Pd nanocrystallites and their high performance toward methanol oxidation. J Mater Chem 22(30):14861–14863. doi:10.1039/c2jm32682e
175. Wang A-J, Li F-F, Zheng J-N, Xi H-X, Meng Z-Y, Feng J-J (2013) Green synthesis of porous flower-like palladium with high electrocatalytic activity towards methanol oxidation. RSC Adv 3(26):10355–10362. doi:10.1039/c3ra40556g
176. Chen X, Cai Z, Chen X, Oyama M (2014) Synthesis of bimetallic PtPd nanocubes on graphene with N,N-dimethylformamide and their direct use for methanol electrocatalytic oxidation. Carbon 66:387–394. doi:10.1016/j.carbon.2013.09.014
177. Yin AX, Min XQ, Zhang YW, Yan CH (2011) Shape-selective synthesis and facet-dependent enhanced electrocatalytic activity and durability of monodisperse sub-10 nm Pt-Pd tetrahedrons and cubes. J Am Chem Soc 133(11):3816–3819
178. Lee Y-W, Ko AR, Han S-B, Kim H-S, Park K-W (2011) Synthesis of octahedral Pt-Pd alloy nanoparticles for improved catalytic activity and stability in methanol electrooxidation. Phys Chem Chem Phys 13(13):5569–5572. doi:10.1039/c0cp02167a
179. Yin A-X, Min X-Q, Zhu W, Wu H-S, Zhang Y-W, Yan C-H (2012) Multiply twinned Pt-Pd nanoicosahedrons as highly active electrocatalysts for methanol oxidation. Chem Commun 48(4):543–545. doi:10.1039/c1cc16482a
180. Zhan F, Bian T, Zhao W, Zhang H, Jin M, Yang D (2014) Facile synthesis of Pd-Pt alloy concave nanocubes with high-index facets as electrocatalysts for methanol oxidation. CrystEngComm 16(12):2411–2416. doi:10.1039/c3ce42362j
181. Zhang J, Xi C, Feng C, Xia H, Wang D, Tao X (2014) High yield seedless synthesis of high-quality gold nanocrystals with various shapes. Langmuir 30(9):2480–2489. doi:10.1021/la404602h
182. Venkatasubramanian R, He J, Johnson MW, Stern I, Kim DH, Pesika NS (2013) Additive-mediated electrochemical synthesis of platelike copper crystals for methanol electrooxidation. Langmuir 29(43):13135–13139. doi:10.1021/la4027078
183. Qi Y, Bian T, Choi S-I, Jiang Y, Jin C, Fu M, Zhang H, Yang D (2014) Kinetically controlled synthesis of Pt-Cu alloy concave nanocubes with high-index facets for methanol electro-oxidation. Chem Commun 50(5):560–562. doi:10.1039/c3cc48061e
184. Yin A-X, Min X-Q, Zhu W, Liu W-C, Zhang Y-W, Yan C-H (2012) Pt-Cu and Pt-Pd-Cu concave nanocubes with high-index facets and superior electrocatalytic activity. Chem Eur J 18(3):777–782. doi:10.1002/chem.201102632
185. Choi S-I, Choi R, Han SW, Park JT (2011) Shape-controlled synthesis of Pt3Co nanocrystals with high electrocatalytic activity toward oxygen reduction. Chem Eur J 17 (44):12280–12284. doi:10.1002/chem.201101138
186. Hu Y, Wu P, Zhang H, Cai C (2012) Synthesis of graphene-supported hollow Pt–Ni nanocatalysts for highly active electrocatalysis toward the methanol oxidation reaction. Electrochim Acta 85:314–321. doi:10.1016/j.electacta.2012.08.080
187. Hu Y, Wu P, Yin Y, Zhang H, Cai C (2012) Effects of structure, composition, and carbon support properties on the electrocatalytic activity of Pt-Ni-graphene nanocatalysts for the

methanol oxidation. Appl Catal Environ 111–112:208–217. doi:10.1016/j.apcatb.2011.10.001
188. Wang L, Nemoto Y, Yamauchi Y (2011) Direct synthesis of spatially-controlled Pt-on-Pd bimetallic nanodendrites with superior electrocatalytic activity. J Am Chem Soc 133 (25):9674–9677. doi:10.1021/ja202655j
189. Wang L, Yamauchi Y (2011) Synthesis of mesoporous pt nanoparticles with uniform particle size from aqueous surfactant solutions toward highly active electrocatalysts. Chem Eur J 17 (32):8810–8815. doi:10.1002/chem.201100386
190. Zhou X-W, Zhang R-H, Zhou Z-Y, Sun S-G (2011) Preparation of PtNi hollow nanospheres for the electrocatalytic oxidation of methanol. J Power Sources 196(14):5844–5848. doi:10.1016/j.jpowsour.2011.02.088
191. Luo X, Liu Y, Zhang H, Yang B (2012) Shape-selective synthesis and facet-dependent electrocatalytic activity of CoPt3 nanocrystals. CrystEngComm 14(10):3359–3362. doi:10.1039/c2ce25088h
192. Kang Y, Pyo JB, Ye X, Gordon TR, Murray CB (2012) Synthesis, shape control, and methanol electro-oxidation properties of Pt-Zn alloy and Pt 3Zn intermetallic nanocrystals. ACS Nano 6(6):5642–5647
193. Koper MTM, Lai SCS, Herrero E (2009) Mechanisms of the oxidation of carbon monoxide and small organic molecules at metal electrodes. In: Koper MTM (ed) Fuel cell catalysis. A surface science approach. Wiley, Hoboken, pp 159–208
194. Buso-Rogero C, Grozovski V, Vidal-Iglesias FJ, Solla-Gullon J, Herrero E, Feliu JM (2013) Surface structure and anion effects in the oxidation of ethanol on platinum nanoparticles. J Mater Chem A 1(24):7068–7076. doi:10.1039/c3ta10996h
195. Lee YW, Han SB, Kim DY, Park KW (2011) Monodispersed platinum nanocubes for enhanced electrocatalytic properties in alcohol electrooxidation. Chem Commun 47 (22):6296–6298
196. Wei L, Fan YJ, Wang HH, Tian N, Zhou ZY, Sun SG (2012) Electrochemically shape-controlled synthesis in deep eutectic solvents of Pt nanoflowers with enhanced activity for ethanol oxidation. Electrochim Acta 76:468–474
197. Wei L, Fan YJ, Tian N, Zhou ZY, Zhao XQ, Mao BW, Sun SG (2012) Electrochemically shape-controlled synthesis in deep eutectic solvents—a new route to prepare Pt nanocrystals enclosed by high-index facets with high catalytic activity. J Phys Chem C 116(2):2040–2044
198. Wei L, Zhou ZY, Chen SP, Xu CD, Su D, Schuster ME, Sun SG (2013) Electrochemically shape-controlled synthesis in deep eutectic solvents: triambic icosahedral platinum nanocrystals with high-index facets and their enhanced catalytic activity. Chem Commun 49(95):11152–11154
199. Zhou ZY, Shang SJ, Tian N, Wu BH, Zheng NF, Xu BB, Chen C, Wang HH, Xiang DM, Sun SG (2012) Shape transformation from Pt nanocubes to tetrahexahedra with size near 10 nm. Electrochem Commun 22(1):61–64
200. Zhang L, Chen D, Jiang Z, Zhang J, Xie S, Kuang Q, Xie Z, Zheng L (2012) Facile syntheses and enhanced electrocatalytic activities of Pt nanocrystals with {hkk} high-index surfaces. Nano Res 5(3):181–189
201. Zhang J, Zhang L, Xie S, Kuang Q, Han X, Xie Z, Zheng L (2011) Synthesis of concave palladium nanocubes with high-index surfaces and high electrocatalytic activities. Chem Eur J 17(36):9915–9919
202. Yu N-F, Tian N, Zhou Z-Y, Huang L, Xiao J, Wen Y-H, Sun S-G (2014) Electrochemical synthesis of tetrahexahedral rhodium nanocrystals with extraordinarily high surface energy and high electrocatalytic activity. Angew Chem Int Ed 53(20):5097–5101. doi:10.1002/anie.201310597
203. Chen X, Cai Z, Chen X, Oyama M (2014) Green synthesis of graphene-PtPd alloy nanoparticles with high electrocatalytic performance for ethanol oxidation. J Mater Chem 2 (2):315–320. doi:10.1039/c3ta13155f

204. de Souza JPI, Queiroz SL, Bergamaski K, Gonzalez ER, Nart FC (2002) Electro-oxidation of ethanol on Pt, Rh, and PtRh electrodes. A study using DEMS and in-situ FTIR techniques. J Phys Chem B 106(38):9825–9830. doi:10.1021/jp014645c
205. Lima FHB, Gonzalez ER (2008) Ethanol electro-oxidation on carbon-supported Pt-Ru, Pt-Rh and Pt-Ru-Rh nanoparticles. Electrochim Acta 53(6):2963–2971
206. Yuan Q, Zhou Z, Zhuang J, Wang X (2010) Seed displacement, epitaxial synthesis of Rh/Pt bimetallic ultrathin nanowires for highly selective oxidizing ethanol to CO_2. Chem Mater 22 (7):2395–2402
207. Rao L, Jiang Y-X, Zhang B-W, Cai Y-R, Sun S-G (2014) High activity of cubic PtRh alloys supported on graphene towards ethanol electrooxidation. Phys Chem Chem Phys 16 (27):13662–13671. doi:10.1039/c3cp55059a
208. Hong JW, Lee YW, Kim M, Kang SW, Han SW (2011) One-pot synthesis and electrocatalytic activity of octapodal Au-Pd nanoparticles. Chem Commun 47 (9):2553–2555. doi:10.1039/c0cc04856a
209. Wang ED, Xu JB, Zhao TS (2010) Density functional theory studies of the structure sensitivity of ethanol oxidation on palladium surfaces. J Phys Chem C 114 (23):10489–10497. doi:10.1021/jp101244t
210. Zhang J, Hou C, Huang H, Zhang L, Jiang Z, Chen G, Jia Y, Kuang Q, Xie Z, Zheng L (2013) Surfactant-concentration-dependent shape evolution of Au–Pd alloy nanocrystals from rhombic dodecahedron to trisoctahedron and hexoctahedron. Small 9(4):538–544. doi:10.1002/smll.201202013
211. Zhang G-R, Wu J, Xu B-Q (2012) Syntheses of sub-30 nm Au@Pd concave nanocubes and Pt-on-(Au@Pd) trimetallic nanostructures as highly efficient catalysts for ethanol oxidation. J Phys Chem C 116(39):20839–20847. doi:10.1021/jp304570c
212. Xing W, Yin G, Zhang J (2014) Rotating electrode methods and oxygen reduction electrocatalysts. Elsevier Science, Amsterdam
213. Katsounaros I, Cherevko S, Zeradjanin AR, Mayrhofer KJJ (2014) Oxygen electrochemistry as a cornerstone for sustainable energy conversion. Angew Chem Int Ed 53(1):102–121
214. Vidal-Iglesias F, Solla-Gullón J, Herrero E, Feliu J (2013) Au electrocatalysis for oxygen reduction. In: Shao M (ed) Electrocatalysis in fuel cells, vol 9, Lecture notes in energy. Springer, London, pp 483–512. doi:10.1007/978-1-4471-4911-8_16
215. Guo S, Zhang S, Sun S (2013) Tuning nanoparticle catalysis for the oxygen reduction reaction. Angew Chem Int Ed 52(33):8526–8544
216. Lee J, Jeong B, Ocon JD (2013) Oxygen electrocatalysis in chemical energy conversion and storage technologies. Curr Appl Phys 13(2):309–321
217. Gewirth AA, Thorum MS (2010) Electroreduction of dioxygen for fuel-cell applications: materials and challenges. Inorg Chem 49(8):3557–3566. doi:10.1021/ic9022486
218. Wu J, Yang H (2013) Platinum-based oxygen reduction electrocatalysts. Acc Chem Res 46 (8):1848–1857. doi:10.1021/ar300359w
219. Gomez-Marin AM, Rizo R, Feliu JM (2014) Oxygen reduction reaction at Pt single crystals: a critical overview. Catal Sci Technol 4(6):1685–1698. doi:10.1039/C3cy01049j
220. Gomez-Marin AM, Rizo R, Feliu JM (2013) Some reflections on the understanding of the oxygen reduction reaction at Pt(111). Beilstein J Nanotechnol 4:956–967. doi:10.3762/Bjnano.4.108
221. Adzic RR (1998) Recent advances in the kinetics of oxygen reduction. In: Lipkowski J, Ross PN (eds) Electrocatalysis. Wiley-VCH, New York, pp 197–242
222. Markovic NM, Gasteiger HA, Ross PN (1995) Oxygen reduction on platinum low-index single-crystal surfaces in sulfuric-acid-solution—rotating ring-Pt(Hkl) disk studies. J Phys Chem 99(11):3411–3415
223. Kuzume A, Herrero E, Feliu JM (2007) Oxygen reduction on stepped platinum surfaces in acidic media. J Electroanal Chem 599(2):333–343. doi:10.1016/j.jelechem.2006.05.006
224. Bandarenka AS, Hansen HA, Rossmeisl J, Stephens IEL (2014) Elucidating the activity of stepped Pt single crystals for oxygen reduction. Phys Chem Chem Phys 16(27):13625–13629

225. Maciá MD, Campina JM, Herrero E, Feliu JM (2004) On the kinetics of oxygen reduction on platinum stepped surfaces in acidic media. J Electroanal Chem 564(1-2):141–150. doi:10.1016/j.jelechem.2003.09.035
226. Inaba M, Ando M, Hatanaka A, Nomoto A, Matsuzawa K, Tasaka A, Kinumoto T, Iriyama Y, Ogumi Z (2006) Controlled growth and shape formation of platinum nanoparticles and their electrochemical properties. Electrochim Acta 52(4):1632–1638
227. Sanchez-Sanchez CM, Solla-Gullon J, Vidal-Iglesias FJ, Aldaz A, Montiel V, Herrero E (2010) Imaging structure sensitive catalysis on different shape-controlled platinum nanoparticles. J Am Chem Soc 132(16):5622–5624. doi:10.1021/ja100922h
228. Yu T, Kim DY, Zhang H, Xia Y (2011) Platinum concave nanocubes with high-index facets and their enhanced activity for oxygen reduction reaction. Angew Chem Int Ed 50 (12):2773–2777
229. Fu G, Wu K, Jiang X, Tao L, Chen Y, Lin J, Zhou Y, Wei S, Tang Y, Lu T, Xia X (2013) Polyallylamine-directed green synthesis of platinum nanocubes. Shape and electronic effect codependent enhanced electrocatalytic activity. Phys Chem Chem Phys 15(11):3793–3802
230. Zhou W, Wu J, Yang H (2013) Highly uniform platinum icosahedra made by hot injection-assisted GRAILS method. Nano Lett 13(6):2870–2874
231. Wu J, Gross A, Yang H (2011) Shape and composition-controlled platinum alloy nanocrystals using carbon monoxide as reducing agent. Nano Lett 11(2):798–802. doi:10.1021/nl104094p
232. Kang Y, Ye X, Murray CB (2010) Size- and shape-selective synthesis of metal nanocrystals and nanowires using CO as a reducing agent. Angew Chem Int Ed 49(35):6156–6159
233. Tripković V, Cerri I, Bligaard T, Rossmeisl J (2014) The influence of particle shape and size on the activity of platinum nanoparticles for oxygen reduction reaction: a density functional theory study. Catal Lett 144(3):380–388
234. Li D, Wang C, Strmcnik DS, Tripkovic DV, Sun X, Kang Y, Chi M, Snyder JD, van der Vliet D, Tsai Y, Stamenkovic VR, Sun S, Markovic NM (2014) Functional links between Pt single crystal morphology and nanoparticles with different size and shape: the oxygen reduction reaction case. Energy Environ Sci 7:4061–4069. doi:10.1039/c4ee01564a
235. Kondo S, Nakamura M, Maki N, Hoshi N (2009) Active sites for the oxygen reduction reaction on the low and high index planes of palladium. J Phys Chem C 113 (29):12625–12628
236. Hitotsuyanagi A, Kondo S, Nakamura M, Hoshi N (2011) Structural effects on the oxygen reduction reaction on n(1 1 1)-(1 0 0) series of Pd. J Electroanal Chem 657(1–2):123–127
237. Xiao L, Zhuang L, Liu Y, Lu J, Abruña HD (2008) Activating Pd by morphology tailoring for oxygen reduction. J Am Chem Soc 131(2):602–608. doi:10.1021/ja8063765
238. Shao M, Yu T, Odell JH, Jin M, Xia Y (2011) Structural dependence of oxygen reduction reaction on palladium nanocrystals. Chem Commun 47(23):6566–6568
239. Grdeń M, Łukaszewski M, Jerkiewicz G, Czerwiński A (2008) Electrochemical behaviour of palladium electrode: oxidation, electrodissolution and ionic adsorption. Electrochim Acta 53 (26):7583–7598
240. Lee C-L, Chiou H-P, Liu C-R (2012) Palladium nanocubes enclosed by (100) planes as electrocatalyst for alkaline oxygen electroreduction. Int J Hydrogen Energy 37 (5):3993–3997. doi:10.1016/j.ijhydene.2011.11.118
241. Shao M (2013) Palladium-based electrocatalysts for oxygen reduction reaction. In: Shao M (ed) Electrocatalysis in fuel cells, vol 9, Lecture notes in energy. Springer, London, pp 513–531. doi:10.1007/978-1-4471-4911-8_17
242. Zhang H, Jin M, Xiong Y, Lim B, Xia Y (2012) Shape-controlled synthesis of Pd nanocrystals and their catalytic applications. Acc Chem Res 46(8):1783–1794. doi:10.1021/ar300209w
243. Rodriguez P, Koper MTM (2014) Electrocatalysis on gold. Phys Chem Chem Phys 16 (27):13583–13594. doi:10.1039/c4cp00394b

244. Mei D, He ZD, Zheng YL, Jiang DC, Chen Y-X (2014) Mechanistic and kinetic implications on the ORR on a Au(100) electrode: pH, temperature and H-D kinetic isotope effects. Phys Chem Chem Phys 16(27):13762–13773. doi:10.1039/c4cp00257a
245. Erikson H, Sarapuu A, Tammeveski K, Solla-Gullón J, Feliu JM (2014) Shape-dependent electrocatalysis: oxygen reduction on carbon-supported gold nanoparticles. ChemElectroChem 1(8):1338–1347. doi:10.1002/celc.201402013
246. Gan L, Cui C, Rudi S, Strasser P (2014) Core-shell and nanoporous particle architectures and their effect on the activity and stability of Pt ORR electrocatalysts. Top Catal 57 (1–4):236–244
247. Oezaslan M, Hasché F, Strasser P (2013) Pt-based core-shell catalyst architectures for oxygen fuel cell electrodes. J Phys Chem Lett 4(19):3273–3291
248. Lim B, Jiang M, Camargo PHC, Cho EC, Tao J, Lu X, Zhu Y, Xia Y (2009) Pd-Pt bimetallic nanodendrites with high activity for oxygen reduction. Science 324(5932):1302–1305
249. Stamenkovic VR, Fowler B, Mun BS, Wang GF, Ross PN, Lucas CA, Markovic NM (2007) Improved oxygen reduction activity on $Pt_3Ni(111)$ via increased surface site availability. Science 315(5811):493–497
250. Gasteiger HA, Markovic NM (2009) Just a dream or future reality? Science 324(5923):48–49
251. Zhang J, Yang H, Fang J, Zou S (2010) Synthesis and oxygen reduction activity of shape-controlled Pt_3Ni nanopolyhedra. Nano Lett 10(2):638–644
252. Zhang J, Fang J (2009) A general strategy for preparation of Pt 3d-transition metal (Co, Fe, Ni) nanocubes. J Am Chem Soc 131(51):18543–18547
253. Wu J, Zhang J, Peng Z, Yang S, Wagner FT, Yang H (2010) Truncated octahedral Pt_3Ni oxygen reduction reaction electrocatalysts. J Am Chem Soc 132(14):4984–4985
254. Wu J, Yang H (2011) Synthesis and electrocatalytic oxygen reduction properties of truncated octahedral Pt_3Ni nanoparticles. Nano Res 4(1):72–82
255. Wu J, Qi L, You H, Gross A, Li J, Yang H (2012) Icosahedral platinum alloy nanocrystals with enhanced electrocatalytic activities. J Am Chem Soc 134(29):11880–11883
256. Choi SI, Xie S, Shao M, Odell JH, Lu N, Peng HC, Protsailo L, Guerrero S, Park J, Xia X, Wang J, Kim MJ, Xia Y (2013) Synthesis and characterization of 9 nm Pt-Ni octahedra with a record high activity of 3.3 A/mgPt for the oxygen reduction reaction. Nano Lett 13 (7):3420–3425
257. Sang-Il Choi SX, Shao M, Lu N, Guerrero S, Odell JH, Park J, Wang J, Kim MJ, Xia Y (2014) Controlling the size and composition of nanosized Pt–Ni octahedra to optimize their catalytic activities toward the oxygen reduction reaction. Chemsuschem 7(5):1476–1483
258. Chen C, Kang Y, Huo Z, Zhu Z, Huang W, Xin HL, Snyder JD, Li D, Herron JA, Mavrikakis M, Chi M, More KL, Li Y, Markovic NM, Somorjai GA, Yang P, Stamenkovic VR (2014) Highly crystalline multimetallic nanoframes with three-dimensional electrocatalytic surfaces. Science 343(6177):1339–1343. doi:10.1126/science.1249061
259. Sakamoto R, Omichi K, Furuta T, Ichikawa M (2014) Effect of high oxygen reduction reaction activity of octahedral PtNi nanoparticle electrocatalysts on proton exchange membrane fuel cell performance. J Power Sources 269:117–123. doi:10.1016/j.jpowsour.2014.07.011
260. Chou S-W, Lai Y-R, Yang YY, Tang C-Y, Hayashi M, Chen H-C, Chen H-L, Chou P-T (2014) Uniform size and composition tuning of PtNi octahedra for systematic studies of oxygen reduction reactions. J Catal 309:343–350. doi:10.1016/j.jcat.2013.09.008
261. Xu X, Zhang X, Sun H, Yang Y, Dai X, Gao J, Li X, Zhang P, Wang H-H, Yu N-F, Sun S-G (2014) Synthesis of Pt–Ni alloy nanocrystals with high-index facets and enhanced electrocatalytic properties. Angew Chem Int Ed 126:12730–12735. doi:10.1002/anie.201406497
262. Huang X, Zhao Z, Chen Y, Zhu E, Li M, Duan X, Huang Y (2014) A rational design of carbon-supported dispersive Pt-based octahedra as efficient oxygen reduction reaction catalysts. Energy Environ Sci 7(9):2957–2962. doi:10.1039/c4ee01082e
263. Wu J, Yang H (2012) Study of the durability of faceted Pt_3Ni oxygen–reduction electrocatalysts. ChemCatChem 4(10):1572–1577. doi:10.1002/cctc.201200242

264. Cui C, Gan L, Heggen M, Rudi S, Strasser P (2013) Compositional segregation in shaped Pt alloy nanoparticles and their structural behaviour during electrocatalysis. Nat Mater 12(8):765–771. doi:10.1038/nmat3668
265. Ahmadi M, Behafarid F, Cui C, Strasser P, Cuenya BR (2013) Long-range segregation phenomena in shape-selected bimetallic nanoparticles: chemical state effects. ACS Nano 7(10):9195–9204. doi:10.1021/nn403793a
266. Tuaev X, Rudi S, Petkov V, Hoell A, Strasser P (2013) In situ study of atomic structure transformations of Pt–Ni nanoparticle catalysts during electrochemical potential cycling. ACS Nano 7(7):5666–5674. doi:10.1021/nn402406k
267. Cui C, Ahmadi M, Behafarid F, Gan L, Neumann M, Heggen M, Cuenya BR, Strasser P (2013) Shape-selected bimetallic nanoparticle electrocatalysts: evolution of their atomic-scale structure, chemical composition, and electrochemical reactivity under various chemical environments. Faraday Discuss 162:91–112. doi:10.1039/c3fd20159g
268. Wu J, Shi M, Yin X, Yang H (2013) Enhanced stability of (111)-surface-dominant core–shell nanoparticle catalysts towards the oxygen reduction reaction. Chemsuschem 6 (10):1888–1892. doi:10.1002/cssc.201300388
269. Choi S-I, Shao M, Lu N, Ruditskiy A, Peng H-C, Park J, Guerrero S, Wang J, Kim MJ, Xia Y (2014) Synthesis and characterization of Pd@Pt–Ni core–shell octahedra with high activity toward oxygen reduction. ACS Nano 8(10):10363–10371. doi:10.1021/nn5036894
270. Xie S, Choi S-I, Lu N, Roling LT, Herron JA, Zhang L, Park J, Wang J, Kim MJ, Xie Z, Mavrikakis M, Xia Y (2014) Atomic layer-by-layer deposition of Pt on Pd nanocubes for catalysts with enhanced activity and durability toward oxygen reduction. Nano Lett 14 (6):3570–3576. doi:10.1021/nl501205j
271. Shao M, He G, Peles A, Odell JH, Zeng J, Su D, Tao J, Yu T, Zhu Y, Xia Y (2013) Manipulating the oxygen reduction activity of platinum shells with shape-controlled palladium nanocrystal cores. Chem Commun 49(79):9030–9032. doi:10.1039/c3cc43276a
272. Li Y, Quan F, Chen L, Zhang W, Yu H, Chen C (2014) Synthesis of Fe-doped octahedral Pt_3Ni nanocrystals with high electro-catalytic activity and stability towards oxygen reduction reaction. RSC Adv 4(4):1895–1899. doi:10.1039/c3ra46299d

Chapter 3
Pt-Containing Heterogeneous Nanomaterials for Methanol Oxidation and Oxygen Reduction Reactions

Hui Liu, Feng Ye, and Jun Yang

3.1 Introduction

The strong growing interest in using direct methanol fuel cells (DMFCs) as portable and mobile power sources is rooted in their desirable features of relatively small environmental footprint, compact system design, and higher volumetric energy densities compared with existing technologies [1, 2]. DMFC relies on the oxidation of methanol on a catalyst layer to form carbon dioxide (CO_2). Water (H_2O) is consumed at the anode and is produced at the cathode. Hydrogen ions are transported across the proton exchange membrane to the cathode where they react with oxygen (O_2) to produce water. Electrons are transported through an external circuit from anode to cathode, providing power to external devices. The overall reaction occurred in a DMFC can be described as $2CH_3OH + 3O_2 \rightarrow 4H_2O + 2CO_2$.

It has been well recognized that the success of a DMFC technology depends largely on the electrocatalysts. Currently, platinum (Pt)-based nanomaterials are the most effective electrocatalysts to facilitate both the anodic (methanol oxidation reaction, MOR) and cathodic (oxygen reduction reduction, ORR) reactions in DMFCs [3–6]. However, they are susceptible to CO poisoning in the MOR and their ORR activity is still limiting the fuel cell performance [7–11]. Other than the classical approaches of increasing the Pt catalytic performance through particle size reduction, particle shape control, and alloying with oxophilic metals [5, 6, 12–14], there is increasing interest in combining structural engineering and the synergistic effects of an adjuvant metal to enhance the Pt catalytic properties [15–17].

H. Liu • F. Ye • J. Yang (✉)
State Key Laboratory of Multiphase Complex Systems, Institute of Process Engineering, Chinese Academy of Sciences, Beijing 100190, China
e-mail: jyang@ipe.ac.cn

K.I. Ozoemena, S. Chen (eds.), *Nanomaterials for Fuel Cell Catalysis*, Nanostructure Science and Technology, DOI 10.1007/978-3-319-29930-3_3

Although a wide variety of recipes have been developed to synthesize, assemble and package nanomaterials/nanostructures into forms more amenable to applications, controllable synthesis of nanomaterials with heterogeneous structure, e.g. core-shell, hollow interiors, cage-bell structures, stellated/dendritic morpholoies, dimeric, or composite construction, is much more complicated [14, 18–24]. The synthesis of heterogeneous nanomaterials can be carried out in the solid, gaseous, or solution state. For example, multi-metallic compounds can be traditionally obtained using metallurgical techniques which melts two or more kinds of bulk metals under proper conditions [25]. This solid state method requires high-temperature heating and annealing for long periods of time, and it is difficult to obtain nanocrystalline compounds with high surface areas or desired structures which are urgently needed in various applied fields such as energy, environment, and catalysis [26]. In contrast, the solution-based method is more powerful and versatile and has been the preferred method by many researchers lately toward the synthesis of heterogeneous nanomaterials [14, 15, 24]. In a solution-based synthetic system, the nucleation and growth process of the nanomaterials can be easily controlled by adjusting the reaction parameters including the concentration of reactants, the mole ratio between precursors and surfactants, and the reaction temperature and time. The synthesis, property, and application of Pt-based nanostructured materials have been recently reviewed by Chen and Holt-Hindle [14]. Therefore, in the following, after a brief update of the literature, we prefer to devote this chapter to summarize the Pt-containing nanomaterials with heterogeneous structures prepared *via* solution-based approaches, their characterization and potential applications as electrocatalysts for MOR and ORR so as to provide the readers a systematic and coherent picture of the field. Most of these works have only been carried out in the last several years, particularly by the authors in different laboratories.

3.2 Pt-Based Heterogeneous Nanomaterials with Core-Shell Constructions

Bi- and multi-metallic nanoparticles with a core–shell construction are a type of conventional heterogeneous nanomaterials. However, they still have attracted tremendous research interest in past few years [27–36]. The complex electron interaction between/among the two or more electron-rich elements, the lattice strain created in these core–shell particles, as well as the heterometallic bonding interactions, modify the surface electronic properties of the nanoparticles [37]. Therefore, core–shell nanoparticles often exhibit improved catalytic properties compared to their alloyed counterparts or to mixtures of monometallic nanoparticles [38, 39].

The seed-mediated growth reaction based on the successive reduction of two or more metal precursor salts is one of the commonest methods to prepare core-shell structured nanoparticles. Turkevich and Kim in the 1970s had tried to obtain

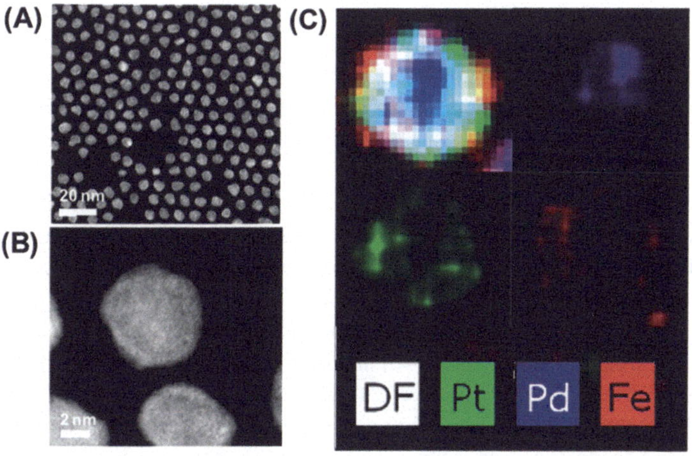

Fig. 3.1 (**a**) HAADF-STEM, (**b**) high-resolution HAADF-STEM, and (**c**) elemental mapping images of the 5 nm/1 nm Pd/FePt NPs. The experiments were carried out on an aberration-corrected JEOL 2200FS microscope. HAADF-STEM images were acquired with a convergence angle of 27 mrad and an inner collection angle of 100 mrad. EDS analysis was carried out with an electron beam size of ~2 Å. Reproduced with permission from [27] Copyright American Chemical Society

gold-layered Pd nanoparticles by this method [40]. The deposition of one metal on preformed nanoparticles of another metal is convenient, provided that the deposition can be carried out selectively on the core particles.

Among the latest investigations into the synthesis and electrocatalytic properties of Pt-containing multi-metallic core-shell nanoparticles, the work from Sun group at Brown University is especially relevant. As a typical example, in 2010, Sun and co-workers reported a unique approach for synthesizing core-shellstructured Pd-FePt NPs with a 5 nm Pd core and a FePt shell whose thickness is tunable from 1 to 3 nm [27]. They first prepared 5 nm Pd seed particles by reducing the Pd $(acac)_2$ in oleylamine and borane *t*-butylamine at 75 °C, and then they coated the Pd seeds with FePt shell in a mixture consisting of oleylamine, oleic acid and 1-octadecene. the core-shell structure of the Pd-FePt nanoparticles was confirmed by transmission electron microscopy (TEM) and aberration-corrected high-angle annular dark-field scanning TEM (HAADF-STEM), as shown in Fig. 3.1. The authors demonstrated that the ORR catalysis of the core-shell Pd-FePt nanoparticles is dependent on the thickness of the FePt shell and that the thin (1 nm or less) FePt shell is both active and durable for ORR in 0.1 M $HClO_4$ solution. The authors proposed that the much-enhanced ORR stability of the 5–1 nm Pd-FePt nanoparticles may arise from the thin FePt coating, as a thin metallic shell can offer the desired thermodynamic stability for a metallic core/shell structure [41].

Subsequently, the Sun group replaced the Pd core with Au, and prepared multimetallic $Au\text{-}FePt_3$ nanoparticles with core-shell construction [42]. They found that the multimetallic core-shell $Au\text{-}FePt_3$ nanoparticles possess both the high catalytic activity of Pt-bimetallic alloys and the superior durability of the tailored morphology and composition profile, with mass-activity enhancement of more than one order of magnitude over Pt catalysts. Experimental observations suggest that the Au core could enhance the ORR catalytic performance of $FePt_3$ alloy shell through altering the surface segregation and hindering place-exchange mechanisms. The obtained fundamental insight into the synergy among these materials could enable the researchers to resolve, define, and utilize the exact role of each constituent in a ternary system, which has guided the synthesis of tailored nanoparticles possessing favorable coordination of surface active sites, distribution of elements and amount of Pt in the system.

Pt nanoparticles with controlled morphologies, such as tetrahedral [43], cubic [12, 44, 45], truncated octahedral [46] and high-index tetra hexahedrons [47], have been demonstrated as highly active catalysts for various reactions. In particular, icosahedral nanoparticles with a high density of compressed twins on the surfaces are expected to be the most active catalysts [48]. The excellent activity of the multiply twinned particles (MTPs) of Pt with (1 1 1) facet of icosahedral morphology with a high density of twins and corners on their surface was also verified computationally [49]. However, unlike palladium (Pd) [48], silver (Ag) [50] and gold (Au) [51], MTPs of Pt were rarely formed [24].

In recent, Yang and co-workers reported for the first time the synthesis of monodispersed core-shell AgPd-Pt MTPs with icosahedral morphology, and investigated their application in catalyzing ORR at room and moderate temperatures [52]. In this method, oleylamine was used as solvent, reducing agent and stabilizer for the formation of nanoparticles. The Ag component of the AgPd alloy inner core was crucial for the construction of the multiply twinned structure of the core-shell nanoparticles, while the Pd component was employed to reduce the tensile strain effect of the Ag on the deposited Pt layers. The core-shell AgPd-Pt MTPs exhibited superior catalytic activity toward ORR, as compared to the commercial Pt nanoparticles. The enhancement was attributed to the higher surface fraction of atoms on the (1 1 1) facets of icosahedral Pt MTPs.

In the first step, multiply twinned Ag nanoparticles with an average diameter of ca. 7 nm and an icosahedral morphology were prepared in oleylamine at elevated temperature. These Ag MTPs were then used as seeds for the formation of core-shell nanoparticles. To prepare core-shell AgPd-Pt MTPs, the $PdCl_2$ and $Pt(acac)_2$ metal precursors were immediately added to Ag colloidal solution in oleylamine, which was kept at 250 °C for 1 h under a nitrogen flow and magnetic stirring. Upon the addition of Pd(II) and Pt(II) salts, the replacement reaction between Ag nanoparticles and Pd(II) ions occurred immediately to form AgPd alloy nanoparticles, analogous to the formation of AgAu or AgPd alloy nanoparticles by replacement reaction between Ag nanoparticles and Au or Pd metal ions in aqueous solution [53]. The alloying process was realized by the rapid interdiffusion of metal atoms as a result of the reduced dimension of the silver templates, the elevated temperature, and the large number of interfacial vacancy defects generated by the replacement reaction. The subsequent reduction of Pt(II) precursors resulted in the homogeneous deposition of Pt on the preformed AgPd alloy nanoparticles.

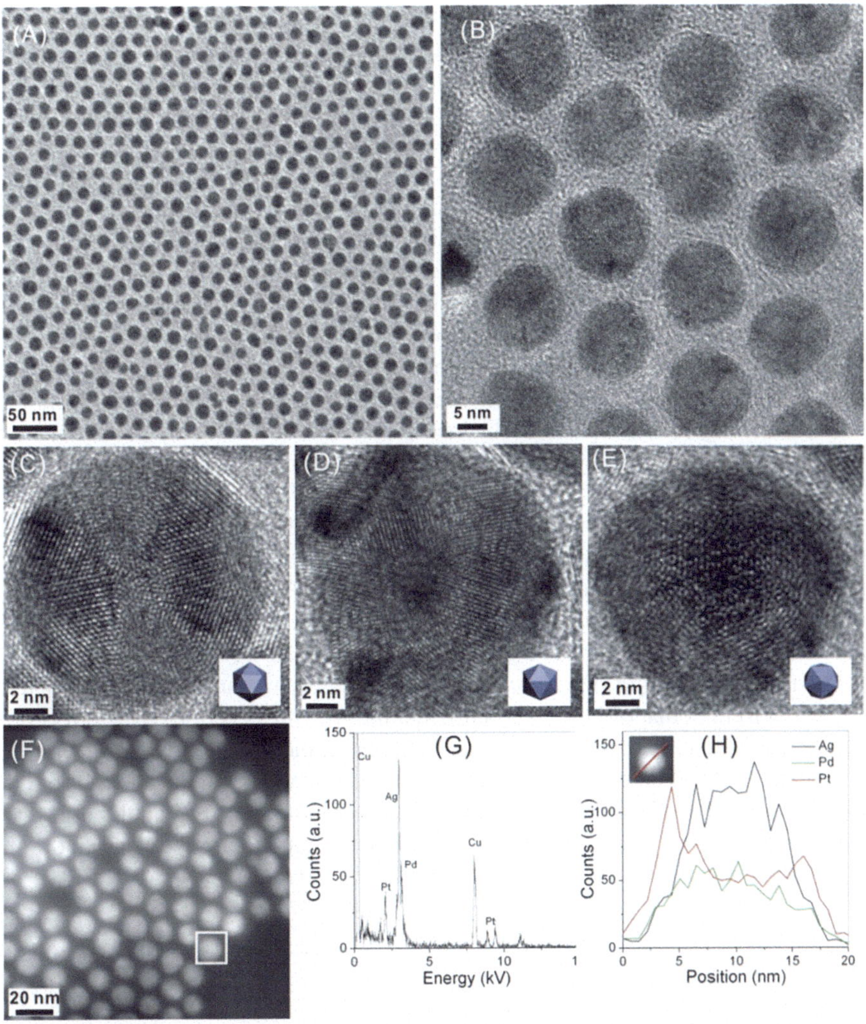

Fig. 3.2 (**a**) TEM and (**b**) HRTEM images of icosahedral core-shell AgPd@Pt nanoparticles. HRTEM images of AgPd@Pt MTP along a (**c**) two-fold, (**d**) three-fold, and (**e**) five-fold symmetry axis. (**f**) HAADF-STEM image. (**g**) EDX spectrum of a single AgPd@Pt MTP boxed in (**f**). (**h**) Ag, Pd and Pt elemental profiles along the *red line* across the MTP (inset of **h**). Reproduced with permission from [52] Copyright American Chemical Society

The as-prepared colloidal solution was then cooled down to room temperature, and the nanoparticles were precipitated, washed with methanol, and re-dispersed in a non-polar organic solvent (e.g. toluene, hexane or chloroform). A typical TEM image of the core-shell AgPd-Pt nanoparticles was shown in Fig. 3.2a. The nanoparticles are nearly monodispersed, with an average diameter of ca. 15 nm. The narrow size distribution of nanoparticles results in the formation of two-dimensional close-packed hexagonal arrays on the TEM grid. The core-shell AgPd-Pt nanoparticles are predominantly multiply twinned with an icosahedral morphology, as illustrated by the high-resolution TEM (HRTEM) image (Fig. 3.2b). They were found in different orientations (along a two-fold symmetry axis in Fig. 3.2c, along a three-fold symmetry axis in Fig. 3.2d, and along a five-fold symmetry axis in Fig. 3.2e) [51, 54, 55]. The energy-dispersive X-ray (EDX) spectroscopy analysis (Fig. 3.2g) of an arbitrary single particle boxed in the HAADF-STEM image of Fig. 3.2f confirms that the particle is indeed composed of Ag, Pd and Pt. Since the lattice mismatch between the AgPd alloy core and the Pt shell was only ca. 1.1 % [52], the core-shell nanoparticles were difficult to determine accurately using standard TEM. However, the formation of core-shell AgPd-Pt nanoparticles could be confirmed by the elemental profiles of these particles. As shown in Fig. 3.2h, the Pt signal is noted across the entire particle (ca. 15 nm), whereas the Ag and Pd signals is obtained only across the core (ca. 10 nm).

The addition of Pd(II) or Pt(II) salt precursor alone to icosahedral Ag seeds would result in AgPd alloy nanoparticles of ca. 10 nm or core-shell Ag-Pt nanoparticles of ca. 12 nm, respectively, which are also predominantly multiply twinned with an icosahedral morphology [52]. The XRD patterns of Ag, AgPd alloy, core-shell Ag-Pt, core-shell AgPd-Pt, and pure Pt nanoparticles prepared in oleylamine in the absence of any seeds were presented in Fig. 3.3. As compared to pure Pt nanoparticles, a shift to a lower angle is observed for the (1 1 1) diffraction peak of core-shell Ag-Pt (Fig. 3.3c). This is because Ag has larger lattice parameter (0.4090 nm) than that of Pt (0.3923 nm), and the Ag cores exert a tensile effect on the Pt atoms deposited on their surface. In contrast, the (1 1 1) diffraction peak of core-shell AgPd-Pt is very close to that of pure Pt, indicating that the presence of Pd in alloy AgPd cores could reduce the tensile effect induced by Ag on the Pt layer. Based on the calculation from Fig. 3.3b, the lattice parameter for the (1 1 1) facet of AgPd alloy was 0.3967 nm, very similar to that of pure Pt.

Fig. 3.3 XRD patterns of (**a**) Ag, (**b**) alloy Ag/Pd, (**c**) core-shell Ag-Pt, (**d**) core-shell Ag/Pd-Pt, and (**e**) Pt nanoparticles. Reproduced with permission from [52] Copyright American Chemical Society

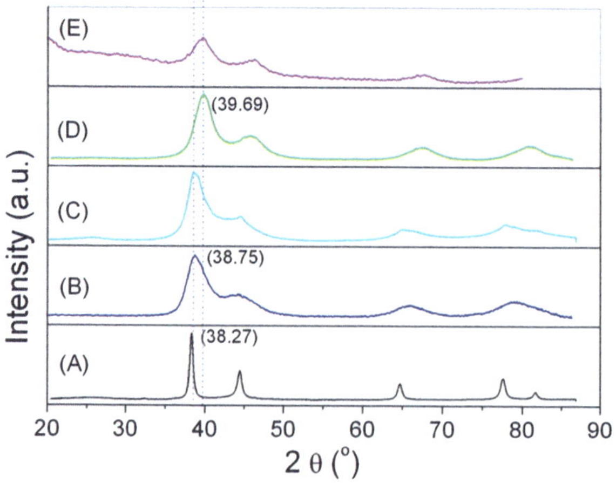

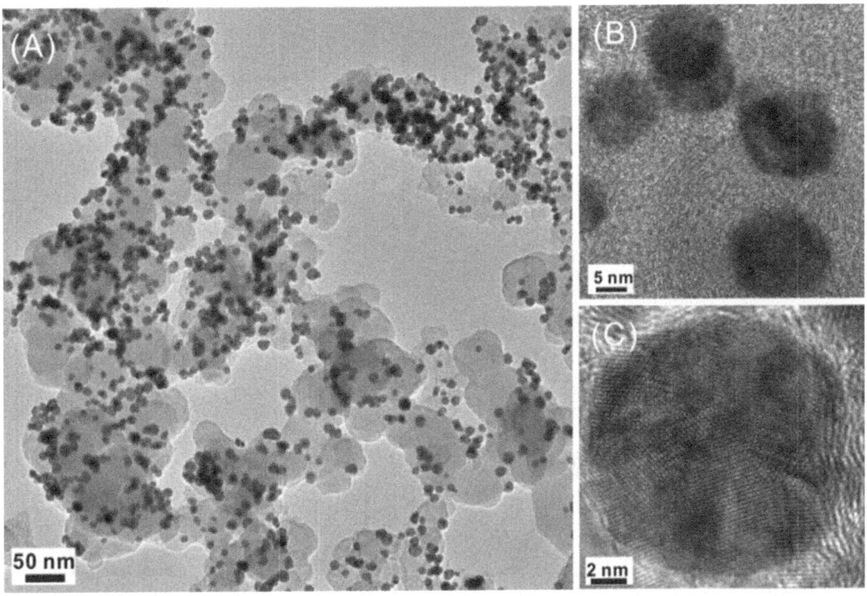

Fig. 3.4 (**a**) TEM and (**b**) HRTEM images of Ag/Pd-Pt/C catalysts. (**c**) HRTEM image of a single Ag/Pd-Pt icosahedral nanoparticle supported on XC-72 carbon. Reproduced with permission from [52] Copyright American Chemical Society

For the electrochemical characterizations, the core-shell AgPd-Pt nanoparticles were loaded on Vulcan XC-72 carbon supports. TEM and HRTEM images of the AgPd-Pt/C nanoparticles show that the core-shell AgPd-Pt nanoparticles are well dispersed on XC-72, maintaining their particle size and the multiply twinned structure with icosahedral morphology (Fig. 3.4).

The catalytic activity of the core-shell AgPd-Pt MTPs with icosahedral morphology towards ORR was evaluated by sweep voltammetry in O_2-saturated $HClO_4$, and benchmarked against that of the commercial Pt/C catalysts with an average diameter of ca. 3 nm. Cyclic voltammograms of commercial Pt/C and core-shell AgPd-Pt/C in argon-purged 0.1 M $HClO_4$ at room temperature were used to obtain the electrochemically active surface areas (ECSAs) from the hydrogen adsorption/desorption regions (0.0–0.3 V vs. RHE in Fig. 3.5a). In the cathodic scan, the oxide (OH_{ads}) stripping peak (0.8 V) for core-shell AgPd-Pt/C is higher than that for commercial Pt/C (0.78 V). The positive shift of ca. 0.02 V in the oxide stripping peak for core-shell AgPd-Pt/C suggests a weaker binding of the OH_{ads} species on the surface of this catalyst [56]. This could be attributed to a weaker interaction with the adsorbed species on the dominant (1 1 1) facets [57]. Figure 3.5b shows the ORR polarization curves of these two catalysts in oxygen-saturated 0.1 M $HClO_4$ at room temperature and 60 °C. As indicated, at room temperature, the core-shell AgPd-Pt/C MTPs exhibit a more positive half-wave potential

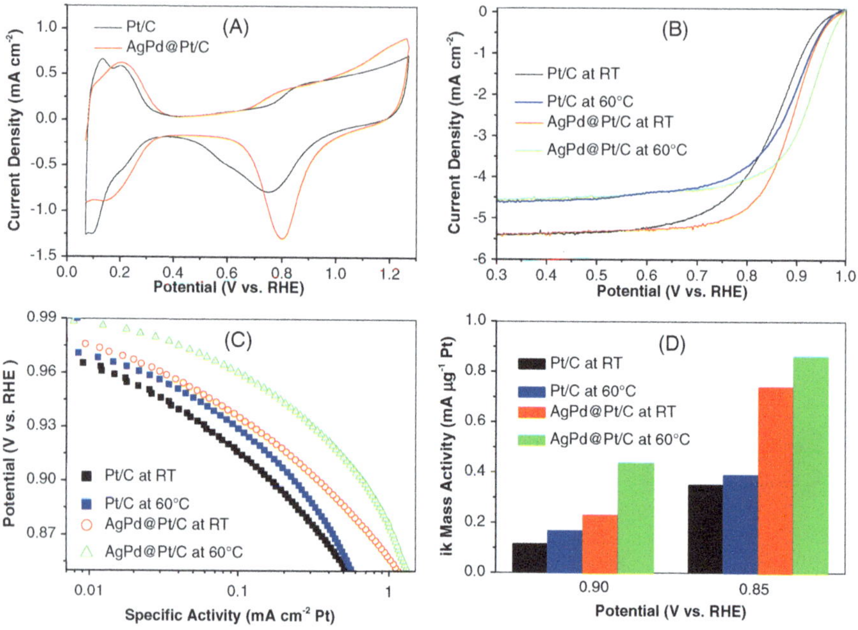

Fig. 3.5 (**a**) Cyclic voltammograms of Pt/C and Ag/Pd-Pt/C in argon-purged 0.1 M $HClO_4$; sweep rate = 20 mV/s, room temperature. (**b**) Linear sweep voltammograms in oxygen-saturated 0.1 M $HClO_4$ showing the positive-going scans; sweep rate = 20 mV/s, room temperature or 60 °C, 1600 rpm. (**c**) Tafel plot for ORR at high potential, normalized by Pt ECSA. (**d**) Kinetic mass activities of ORR at 0.85 and 0.90 V RHE, normalized by Pt mass. Reproduced with permission from [52] Copyright American Chemical Society

(0.88 V) than that of commercial Pt/C catalysts (0.86 V), indicating that they have a higher ORR activity.

By normalizing with the Pt surface area, the specific activities of core-shell AgPd-Pt/C and commercial Pt/C were obtained (Fig. 3.5c). At room temperature and 0.9 V vs. RHE, a specific activity of 0.36 mA cm^{-2} Pt was attained for core-shell AgPd-Pt/C, which is >2 times higher than that for commercial Pt/C (0.16 mA cm^{-2} Pt). Kinetic mass activity calculated from Koutecký-Levìch equation (Fig. 3.5d) also confirms that the core-shell AgPd-Pt/C (0.23 mA μg^{-1} Pt) is superior to commercial Pt/C (0.11 mA μg^{-1} Pt). In addition, regardless of whether Pt surface area or Pt mass is used as the basis for normalizing the measured current, the specific activity (0.68 mA cm^{-2} Pt) and mass activity (0.44 mA μg^{-1} Pt) of core-shell AgPd-Pt/C at 60 °C and 0.90 V are higher than those of commercial Pt/C (0.24 mA cm^{-2} Pt and 0.17 mA μg^{-1} Pt, respectively). The former values are also close to the 2015 targets set by the U.S. Department of Energy (DoE) at 80 °C and 0.9 V (0.72 mA cm^{-2} Pt and 0.44 mA μg^{-1} Pt, respectively). Even on the basis of total mass of Pd and Pt, the core-shell AgPd-Pt/C catalyst shows greater mass activities at room temperature and 60 °C (0.15 mA μg^{-1} Pt + Pd and 0.28 mA μg^{-1} Pt + Pd, respectively) than those of the commercial Pt catalyst (0.11 mA μg^{-1} Pt, and 0.17 mA μg^{-1} Pt, respectively).

It should be noted that the catalytic activities of AgPd alloy and core-shell Ag-Pt nanoparticles towards ORR were found to be much lower than that of the core-shell AgPd-Pt nanoparticles. It is well known that most Pd-based catalysts are lower in intrinsic ORR activities than Pt [58, 59]. Therefore, it was not unexpected that the AgPd alloy nanoparticles show a lower activity than that of core-shell Ag-Pt and AgPd-Pt nanoparticles. For the core-shell Ag-Pt nanoparticles, the tensile effect of Ag core on the Pt shell would result in an upward shift of the d-band center of Pt [60]. As has been demonstrated [61], the surface of a Pt-based catalyst with a high value in d-band center would bind adsorbents too strongly. Compared to that of the core-shell Ag-Pt nanoparticles, the superior catalytic activity of core-shell AgPd-Pt nanoparticles toward ORR is possibly because the latter has a more suitable d-band center to balance the breaking of O–O bonds and the formation of O–H bonds, two usual reactions in a common ORR process [9, 62, 63].

The synthesis of Pt-containing heterogeneous nanoparticles with core-shell structure as efficient electrocatalysts is still a very hot topic in current scientific communities. In 2012, Chen and co-workers prepared core-shell Ru-Pt nanoparticles through a polyol reduction with a sequence controlling processes [32]. The core-shell particles have a Pt shell with a precisely controlled thickness around 1.5 atomic layers. The authors found that the activity of the core–shell Ru-Pt nanoparticles in methanol oxidation reaction (MOR) is significantly dependent on the crystalline structure of the core and the lattice strain at the core–shell interface. Core-shell Ru-Pt nanoparticles delivered 6.1-fold peak MOR current density at -135 mV than Pt particles. The current density is improved by the compressive lattice strain of the surface that is caused by the lattice mismatch between the Pt shell and the Ru core. In addition, the Ru-core nanoparticles maintained a steady current density of 0.11 mA cm^{-2} at 500 mV in a half-cell system for 2 h, which is 100-fold higher than that of Pt particles. These results provide mechanistic information for the development of fuel cell catalysts along with reduced Pt utilization and programmable electrochemical performance.

Also in 2012, the Adzic group described a route to the development of novel PtNiN core–shell catalysts with low Pt content shell and inexpensive NiN core, which have high activity and stability for the ORR [33]. The PtNiN synthesis involves nitriding Ni nanoparticles and simultaneously encapsulating them by 2–4 monolayer-thick Pt shell. The experimental data and the density functional theory calculations indicate nitride has the bifunctional effect that facilitates formation of the core—shell structures and improves the performance of the Pt shell by inducing both geometric and electronic effects‘. More importantly, the synthesis of inexpensive NiN cores opens up possibilities for designing of various transition metal nitride based core–shell nanoparticles for a wide range of applications in energy conversion processes.

To improve the catalytic activity of Pt-based nanomaterials for MOR, the Wang and Yamauchi group prepared core-shell Au-Pt nanocolloids with nanostructured dendritic Pt shells through chemically reducing both H_2PtCl_6 and $HAuCl_4$ species in the presence of a low-concentration surfactant (Pluronic F127) solution [64]. By applying an ultrasonic treatment, the particle size of the core-shell Au-Pt

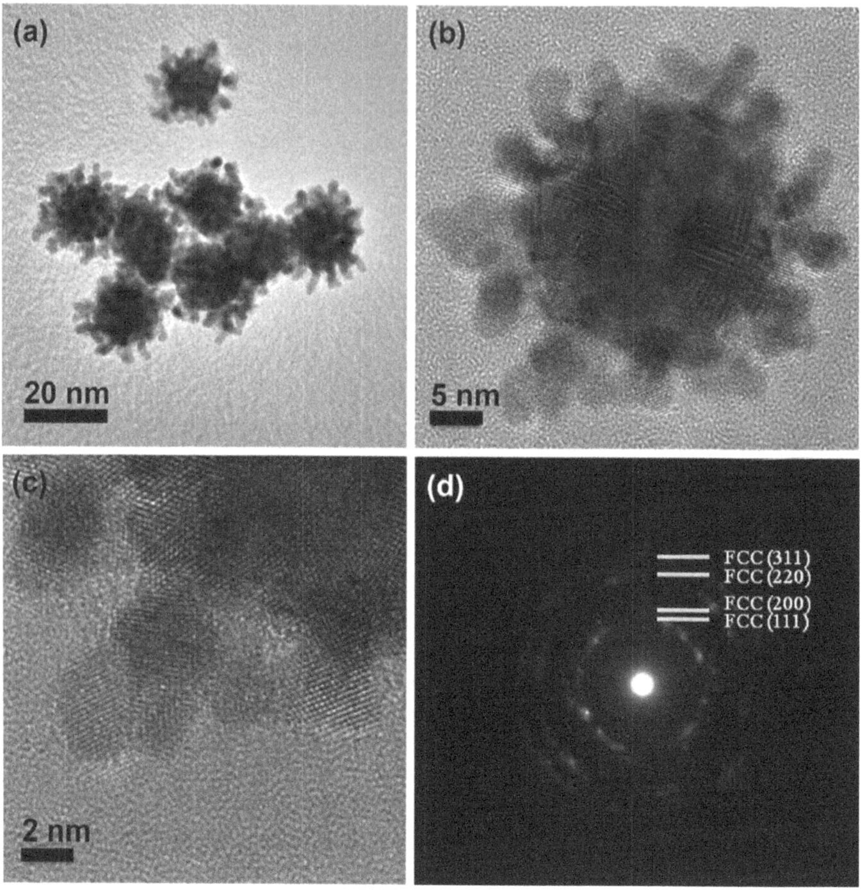

Fig. 3.6 (**a**) Low-magnification and (**b**) high-magnification TEM images of Au@Pt nanocolloids prepared from a typical solution with a Pt/Au molar ratio of 1.0. (**c**) High-magnification TEM image of the Pt shell. (**d**) Selected-area electron diffraction (SAED) patterns taken from the Pt shell region. Reproduced with permission from [64] Copyright American Chemical Society

nanocolloids is dramatically decreased and their size distribution becomes very narrow. The difference in reduction potentials of the two soluble metal salts (Au (III) and Pt(IV) species) plays a key role in the one-step synthesis of the core-shell structure. Because of the different reduction potentials, the reduction of Au ions preferentially occurs over a short time to form the Au seeds. It is followed by overgrowth of Pt nanodendritic nanowires on the Au seeds, as confirmed by the TEM images of the final products (Fig. 3.6). The thickness of the Pt shell on the Au cores can be easily tuned by controlling the Pt/Au molar ratios in the starting precursor solutions. Through the optimization of the Pt shell thicknesses, the core-shell Au-Pt nanocolloids exhibit enhanced activity as an electrocatalyst for MOR.

Subsequently, Kang et al. demonstrated a facile one-pot aqueous approach for the controllable synthesis of trimetallic Au-PdPt core-shell nanoparticles with a well-defined octahedral Au core and a highly crystalline dendritic PdPt alloy shell [35]. The simultaneous reduction of multiple metal precursors with dual reducing agents, namely, ascorbic acid and hydrazine, gave a fine control over the nucleation and growth kinetics of nanoparticles, resulting in the formation of novel core-shell Au-PdPt nanoparticles. The prepared core-shell particles display excellent catalytic performance for methanol electrooxidation, which can be attributed to their optimized binding strength toward adsorbate molecules due to the improved charge transfer between core and shell of the particles.

Most recently, the Sun group reported a core-shell nanoparticle with dual electrocatalytic activities for both methanol oxidation and oxygen reduction reactions [36]. They firstly prepared Au seed particles of 5 nm in diameter in a mixture of tetralin and oleylamine, and then coated these Au seeds with alloy CuPt shell in 1-octadecene at elevated temperature. The thickness of the alloy CuPt shell is ca. 1.5 nm, and the final product was re-dispersed in hexane. The CuPt compositions ($Pt_{65}Cu_{35}$, $Pt_{50}Cu_{50}$, and $Pt_{36}Cu_{64}$) in the alloy shell could be controlled by the molar ratio of Cu and Pt metal precursors. The CuPt alloy effect and interactions between core and shell make these core-shell Au-CuPt nanoparticles a promising catalyst for both ORR and MOR in 0.1 M $HClO_4$ solution. As shown in Fig. 3.7, the specific (mass) reduction and oxidation activities of these core-shell particles reach

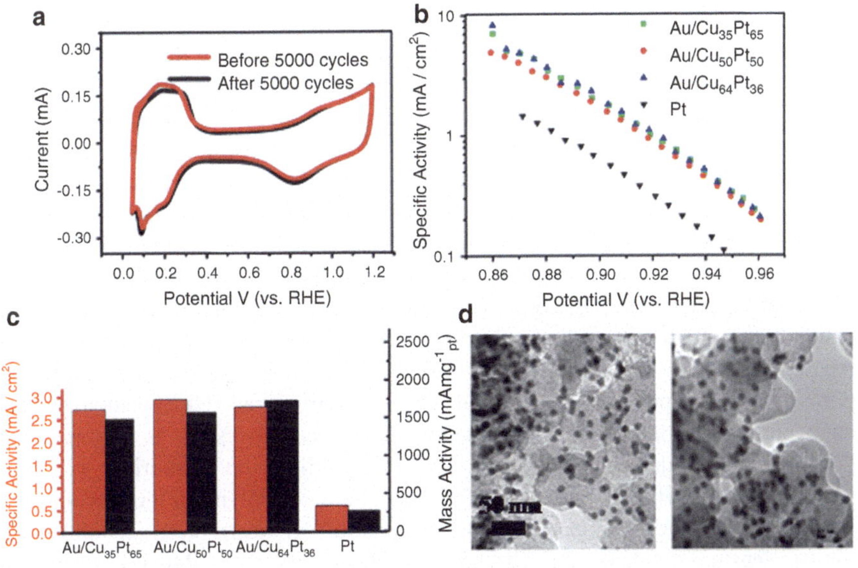

Fig. 3.7 (**a**) CVs of core-shell Au-$Cu_{35}Pt_{65}$/C catalysts before and after 5000 potential cycles between 0.6 and 1.0 V; (**b**) activity–potential plots of Au-Cu_xPt_{100-x}/C and Pt/C catalysts; (**c**) specific and Pt mass activities of Au-Cu_xPt_{100-x}/C and Pt/C catalysts at 0.9 V. (**d**) TEM of Au-$Cu_{35}Pt_{65}$/C catalysts before (*left*) and after (*right*) 5000 potential cycles. Reproduced with permission from [36] Copyright American Chemical Society

2.72 mA cm^{-2} (1500 mA mg^{-1} Pt) at 0.9 V and 0.755 mA cm^{-2} (441 mA mg^{-1} Pt) at 0.8 V (vs RHE), respectively. Their studies show that the existence of an Au core not only minimizes the Pt usage but also improves the stability of the core-shell Au-CuPt catalyst for fuel cell reactions. The results suggest that the core-shell design is indeed effective for optimizing nanoparticle catalysis. The same concept may be extended to other multimetallic nanoparticle systems, making it possible to tune nanoparticle catalysis for many different chemical reactions.

Although great successes have been achieved in preparation and electrochemical application of core-shell structured nanomaterials, to date, as summarized by Yang in a mini-review [28], the reports on the synthesis of well-defined multi-metallic core-shell nanoparticles with sizes below ca. 10 nm are still relatively uncommon. Besides the intrinsic challenges in the synthesis, one important reason perhaps lies in the difficulty to characterize the core-shell structure in multi-metallic nanoparticles in detail. In addition, in view of the profoundly effect of morphology of nanoparticles on the catalytic activity, the performance of core-shell structured nanomaterials for MOR and ORR would be dramatically increased if their overall shape could also be finely controlled at this length scale.

3.3 Pt-Based Heterogeneous Nanomaterials with Hollow Interiors or Cage-Bell Structures

Preparation of Pt-based nanomaterials with hollow interiors is the most common practice to enhance their electrocatalytic property for MOR and ORR [65–72]. For instance, Pt hollow nanospheres are found to be twice as active as solid Pt nanoparticles of roughly the same size for methanol oxidation [65, 66]. The increase in activity could be attributed mainly to the larger surface area of the hollow structure, where the porous shell allows the internal surface of the catalyst to be accessible to the reactants. The continuing research efforts in this area have also given rise to the recent possibility of creating hybrids of core-shell and hollow structures, or a new class of core-shell structures with a distinctive core-void-shell configuration, which are called cage-bell, yolk-shell or rattle-type structures. With the unique properties of a movable core, interior void spaces, and controlled porosity and composition of the shell, cage-bell structured (CBS) nanomaterials have great potential for a diverse range of applications, such as nanoreactors [73–75], drug delivery systems [76–78], lithium-ion batteries [79–83], photocatalysis [84], and photonics [85]. For example, polymeric hollow spheres with a movable Au nanoparticle core have been synthesized, which allow the optical sensing of chemicals diffused into the cavity [86]. Recently, Fan and co-workers developed a photocatalytic approach using densely packed optically active porpyrins to template the synthesis of well-defined hollow Pt nanostructures which were excellent catalyst for the methanol oxidation reaction [87]. The authors also demonstrated reusability of the template by an acid-base treatment. A general strategy for the fabrication of hollow or CBS structures is template-assisted selective etching of core-shell particles. The core particle in this case is overlaid with a single or double shell of a different material. The core or the inner shell is then selectively removed by calcination or with a solvent.

Over the past 20 years, various methods based on sacrificial templates of polymer and inorganic spheres [72, 88–93], liquid droplets [94], vesicles [95–97], and microemulsion droplets [98–100], have been developed for the synthesis of hollow and CBS nanomaterials, and were reviewed recently by the Lou and Jiang groups [101, 102]. These preparative methods are however system specific, more successful for metal oxides than for noble metals. Specific methods based on other principles such as galvanic replacement [65, 68, 103–110], Kirkendall effect [111], Ostwald ripening [112, 113], layer-by-layer assembly [91, 114], copper sulfide leaching [115], and structure transformation [116], have also been used for the synthesis of hollow or cage-bell nanomaterials. These methods are only applicable to a small number of metals (mostly Au and Pt) and there are significant limitations. For example, the core and the shell of cage-bell nanostructures prepared this way are usually made of the same material, and there is no effective control of the shell thickness and shell structure [117, 118]. In summary, a general approach to rationally fabricate hollow and cage-bell structures for a sufficiently wide spectrum of noble metals is still lacking and it poses significant challenges.

In this section, we focus on a facile generic approach reported by the Yang group in 2012, which is based on the inside-out diffusion of Ag in core-shell structures, for the fabrication of hollow or cage-bell noble metal nanomaterials [119]. In this strategy, core-shell Ag-M nanoparticles or core-shell-shell M_A-Ag-M_B nanoparticles are first prepared in an organic solvent, followed by the removal of Ag from the core or the inner shell with bis(p-sulfonatophenyl)phenylphosphane dihydrate (BSPP), which binds strongly with Ag(I)/Ag(0) to promote the inside-out diffusion of Ag in the core-shell structure, which is typically completed in 24~48 h. The hollow and CBS nanoparticles prepared as such have relatively lower densities, which usually translate to a higher surface area than their solid counterparts, hence a higher degree of metal utilization (for example, in catalysis) can be expected. In particular, CBS Pt-Ru nanoparticles were found to exhibit outstanding methanol tolerance for the cathode reaction of the direct methanol fuel cell (DMFC). This may be attributed to the difference in the diffusion of methanol and oxygen through the shell of the CBS nanoparticles. Hence a diffusion-limited design, rather than the intrinsic properties of the catalytic metals, may be used to deliver the desired catalyst selectivity for the mitigation of the methanol crossover problem in DMFCs.

3.3.1 Inside-Out Diffusion of Ag in Ag-Containing Single or Double Shell Core-Shell Metal Nanoparticles

The inside-out diffusion of Ag in Ag-containing single or double shell core-shell metal nanoparticles is the basis for synthesis of noble metal nanomaterials with hollow interiors or cage-bell structures [119]. Typically, core-shell Ag-Pt nanoparticles were used to illustrate this interesting diffusion phenomenon. The TEM images in Fig. 3.8a, b show the initial uniform core-shell Ag-Pt nanoparticles. After storage in toluene for seven months at room temperature, Ag diffuses out from the interior of the core-shell Ag-Pt nanoparticles and TEM shows a product very different from the original nanoparticles (Fig. 3.8a, c): There is increasing

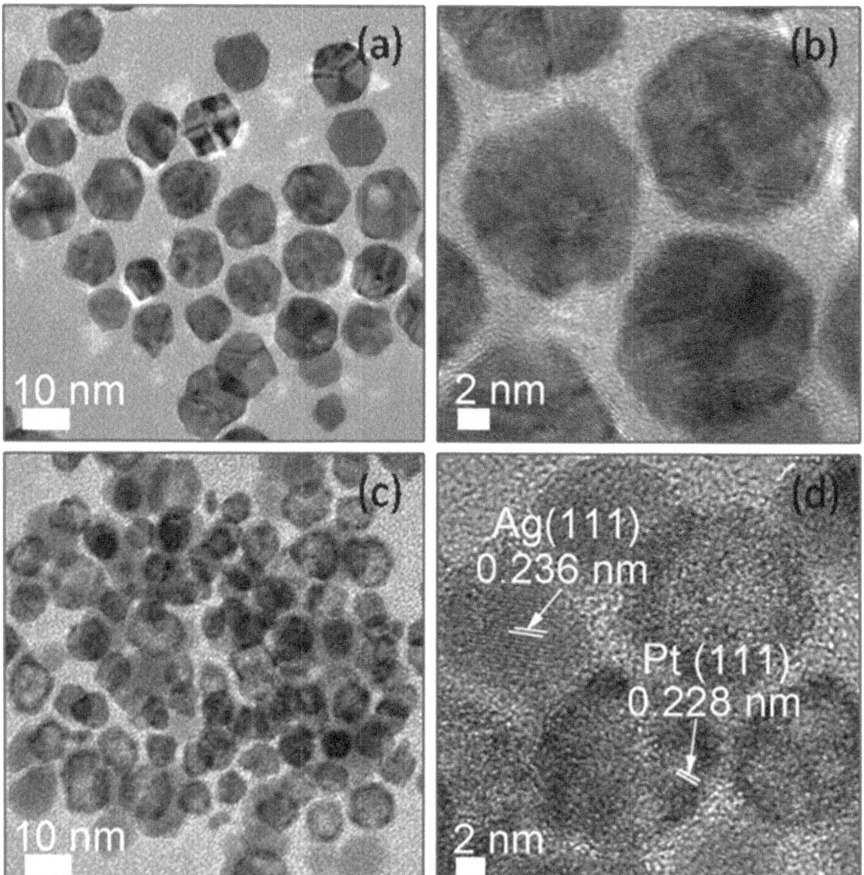

Fig. 3.8 Inside-out diffusion of Ag in core-shell Ag-Pt nanoparticles. (**a**) TEM and (**b**) HRTEM images of original core-shell Ag-Pt nanoparticles; (**c**) TEM and (**d**) HRTEM images of core-shell Ag-Pt nanoparticles after aging in toluene for 7 months at room temperature. Reproduced with permission from [119] Copyright American Chemical Society

presence of isolated Ag nanoparticles and hollow Pt nanoparticles in a mixture of nanoparticles. The HRTEM image of Fig. 3.8d reveals the structures of the isolated Ag and hollow Pt nanoparticles: Two different sets of lattice fringes with separations of 0.236 and 0.228 nm corresponding well with the (1 1 1) planes of face-centered cubic (fcc) Ag and Pt, are found in the solid and hollow nanoparticles, respectively. The inside-out diffusion of Ag in core-shell Ag-Ru and Ag-Rh nanoparticles were also observed and confirmed by TEM characterizations [119].

The inside-out diffusion of Ag is promoted by the structure of the Ag seeds. The Ag seed particles synthesized in oleylamine usually have multiply twinned decahedral structure [119]. The twinned structure has a strong influence on the stability of the Ag nanoparticles and the overlaid Pt shell. The decahedral particles are mosaic structures consisting of five single crystal tetrahedrons oriented radially about a central axis so that all five tetrahedrons share a common edge in the center, and each tetrahedron has two sides in contact with its neighbors. However, this model is not a space filling one. Since the angle between two (1 1 1) planes of a tetrahedron is theoretically 70.53°, five tetrahedrons with adjoining (1 1 1) planes will leave a gap of 7.35°. To make up for this difference, significant lattice distortions and defects must occur [24, 120]. These imperfections in the Ag seed nanoparticles are disruptive to the epitaxial deposition of Pt atoms on the Ag seeds, resulting in roughness and discontinuity in the Pt shell subsequently formed. The twinned Ag nanoparticles are also inherently unstable and could slowly be etched by dissolved O_2 and Cl^- dissociated from the Pt precursor. The crystallographic defects in the twinned particles might be the active sites for the oxidative dissolution of the nanoparticles. The Ag^+ ions released by the O_2/Cl^- etching of twinned Ag nanoparticles could be re-reduced by excess oleylamine in the solution to single crystalline Ag nanoparticles. In principle, the transformation from twinned to single crystalline particles is reversible since O_2/Cl^- etching should also operate on single crystalline Ag nanoparticles. However, as pointed out by Xia and co-workers [120], single crystalline nanoparticles are less soluble and would continue to grow slowly even in the presence of etchant at the expense of the twinned nanoparticles. Thus, the inside-out diffusion of Ag in core-shell Ag-Pt nanoparticles can be rationalized by the scheme shown in Fig. 3.9. Because of the concentration gradient between the interior of the core-shell particles and the surrounding solution, the Ag^+ ions generated from the O_2/Cl^- etching of twinned Ag seeds diffuse out through the discontinuous Pt shell, and are reduced by oleylamine to form isolated single crystalline Ag nanoparticles in the colloidal solution. With the progress of time the core-shell Ag-Pt nanoparticles would eventually disappear due to the etching and outward diffusion of Ag to leave behind a physical mixture of hollow Pt nanoparticles and single crystalline Ag nanoparticles. Oleylamine is an indispensable multifunctional agent in this case: It was simultaneously a particle stabilizing agent and the reducing agent for the etched Ag ions. Without it the nucleation and growth of isolated single crystal Ag nanoparticles from dissolved Ag twined particles would not be possible.

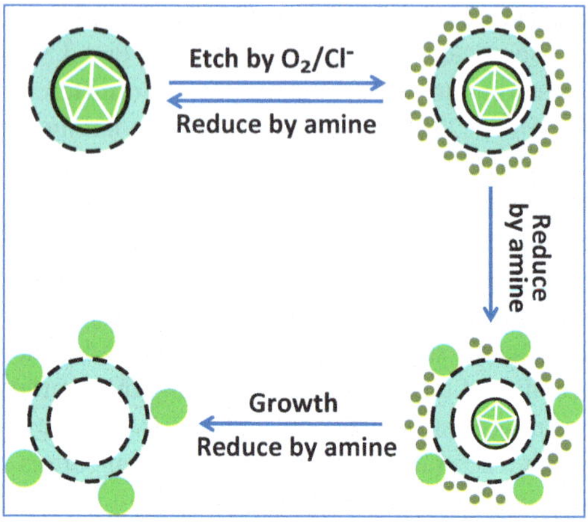

Fig. 3.9 Schematic for the inside-out diffusion of Ag in core-shell Ag-Pt nanoparticles. Reproduced with permission from [119] Copyright American Chemical Society

3.3.2 Monometallic and Alloy Nanoparticles with Hollow Interiors

The inside-out diffusion of Ag in core-shell nanoparticles with Ag residing in the core region could be used to generate hollow metal nanoparticles. In this protocol, the preparation of core-shell Ag-M nanoparticles is an important step preceding the fabrication of hollow noble metal nanoparticles. In addition, the inside-out diffusion process could also be accelerated by introducing a chemical reagent BSPP, which binds strongly with Ag as Ag^+-BSPP coordination complexes to drive the outward diffusion of Ag in the core-shell nanoparticles [66, 121], enabling the completion of the inside-out diffusion of Ag in a much shorter timeframe of 24~28 h (with agitation) after mixing the core-shell Ag-M organosol with an aqueous BSPP solution. The BSPP-Ag^+ coordination compounds are water soluble and their continuous formation left behind an organosol of hollow noble metal nanoparticles, shown by TEM (Fig. 3.10a–d for Ru, Rh, Os, and Pt, respectively) as an increase in the image contrast between the core and shell regions, and the development of visible discontinuity in the metal shell. The removal of Ag was confirmed by EDX analysis of the final products [119].

With a slight modification, the protocol can be used to synthesize hollow alloy nanoparticles. In this case, co-reduction in the presence of preformed Ag seeds involves two or more shell precursor metals in order to form an alloy shell on the Ag core. The second step remains the same—removal of the Ag core by BSPP treatment. TEM images of the hollow PtRu, PtRh, PtOs, PtRuOs, and PtRhPOs nanoparticles are given in Fig. 3.10e, f, g, h and i, respectively. In comparison with their core-shell templates, the particle size and morphology were virtually unchanged by the BSPP treatment, suggesting that the removal of the Ag core from the core-shell nanoparticles did not cause the collapse of the particle geometry [119].

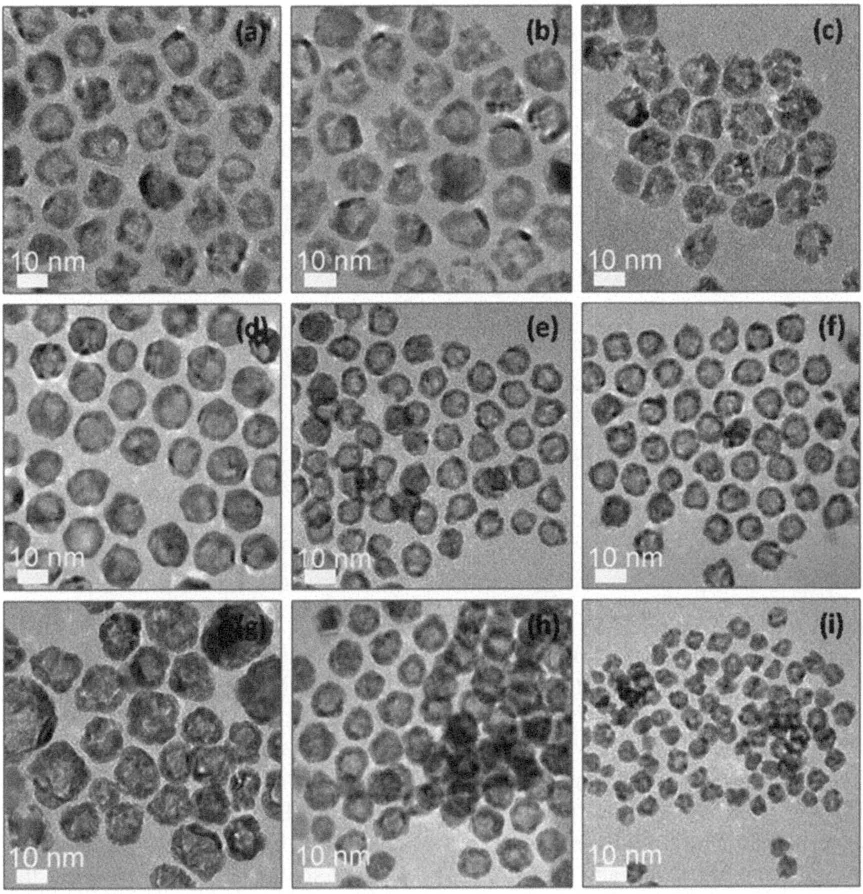

Fig. 3.10 TEM images of hollow structured Ru (**a**), Rh (**b**), Os (**c**), Pt (**d**), PtRu (**d**), PtRh (**f**), PtOs (**g**), PtRuOs (**h**), and PtRhOs nanoparticles (**i**) prepared upon the inside-out diffusion of Ag in core-shell nanostructures. Reproduced with permission from [119] Copyright American Chemical Society

3.3.3 Noble Metal Nanoparticles with Cage-Bell Structures

The protocol can be further modified to synthesize noble metal nanoparticles with a cage-bell structure, which refers to a movable core enclosed by a shell with nano-channels. In principle, the core and shell components may be made of metals or alloys. To achieve the synthesis of cage-bell structured (CBS) nanoparticles, the formation of core-shell-shell M_A-Ag-M_B nanoparticles, in which the inner Ag shell serves as the sacrificial component, is most critical. The Yang group chose Au and Pt as the core component (M_A) in their study, with Ru, Os, Ir, Pt and alloy PtRh as the outermost shell component (M_B). The seed nanoparticles (M_A, Au and Pt) were overlaid with Ag first, followed by the growth of another metal (M_B) shell to form

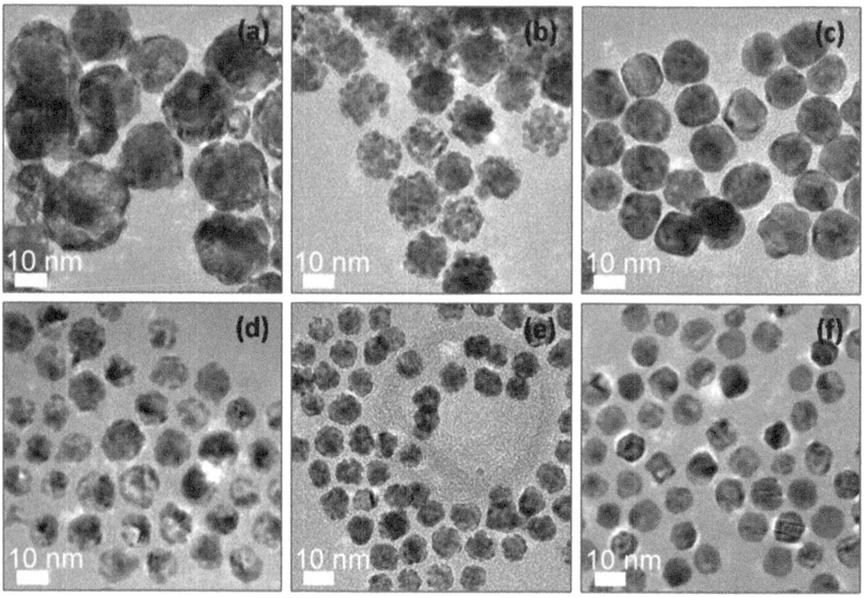

Fig. 3.11 TEM images of CBS Au-Os (**a**), Au-Ir (**b**), Au-Pt (**c**), Pt-Ru (**d**), Pt-Os (**e**), and Pt-Pt nanoparticles (**f**) prepared upon the inside-out diffusion of Ag in core-shell nanostructures. Reproduced with permission from [119] Copyright American Chemical Society

M_A-Ag-M_B nanoparticles with the requisite core-shell-shell structure. After BSPP treatment, the inner Ag layer was removed from the M_A-Ag-M_B nanoparticles, leaving behind Au-Os, Au-Ir, Au-Pt, Pt-Ru, Pt-Os, and Pt-Pt nanoparticles with the cage bell structure, as shown by the TEM images in Fig. 3.11, which show the preservation of the size and morphology of the core-shell-shell nanoparticles in the CBS nanoparticles. The void space between the core and the outer shell regions, formed upon the elimination of the Ag inner shell by BSPP, is discernible by the strong brightness contrast in TEM images.

Most of the materials reported by the Yang group are valuable and important catalytic metals (Pt, Rh, Ru, PtRh, PtRu, etc), thus one would expect the hollow and CBS metal nanoparticles to yield significant applications for catalysis due to their advantage of low density, which translated to a higher surface area and would be beneficial to the catalytic reaction.

An unique application was raised by the CBS nanoparticles. The anode and cathode catalysts of current DMFCs are commonly based on Pt. These catalysts are not selective to methanol oxidation reaction (MOR) or oxygen reduction reaction (ORR) and hence any methanol crossover from the anode to the cathode through the proton exchange membrane (PEM) can be oxidized by the cathode catalyst. This results in the creation of a mixed potential at the cathode which degrades the fuel cell performance [122–124]. Yang and co-workers discovered that the CBS Pt-Ru nanoparticles have none of the failing of common Pt-based catalysts, and are effectively methanol tolerant in oxygen reduction [119]. As a proof, Fig. 3.12a shows the polarization curves of ORR on CBS Pt-Ru nanoparticles in the presence of different methanol concentrations. Even with the presence of methanol in concentrations as high as 1.0 M in the electrolyte, the catalytic reduction of oxygen on the CBS Pt-Ru nanoparticles was hardly affected, demonstrating the effective inhibition of methanol oxidation on the CBS Pt-Ru nanoparticles. For comparison, oxygen reduction on a commercial E-Tek Pt/C catalyst (3~5 nm nanoparticles on carbon powder) with and without methanol was also measured (Fig. 3.12b). The ORR polarization curve in this case was clearly affected even at a low methanol concentration of 0.1 M: a valley was formed at the potential for methanol oxidation. In the CBS Pt-Ru catalyst, the catalytically active metal, i.e. Pt, is located in the core region shielded by a porous Ru shell. Reactants must diffuse through the porous shell of the CBS nanoparticles to access the active core for catalysis to occur. In this case the selectivity for ORR may be caused by the porous shell permitting only the passage of small molecule reactants. The situation is depicted by the scheme in Fig. 3.12c. Methanol and oxygen must diffuse into the CBS nanoparticle interior through the porous Ru shell for MOR and ORR to occur. However, a methanol molecule is larger than an oxygen molecule (the diameters of

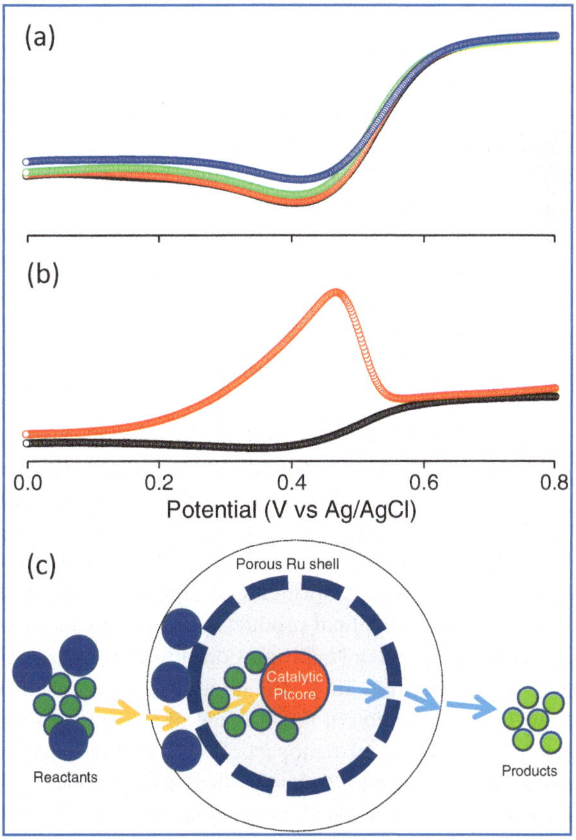

Fig. 3.12 Inhibition of methanol oxidation by CBS Pt-Ru nanoparticles: (**a**) Polarization curves of ORR on CBS Pt-Ru nanoparticles in the presence of (*black circle*) 0.0 M methanol, (*red circle*) 0.1 M methanol, (*green circle*) 0.5 M methanol, and (*blue circle*) 1.0 M methanol; (**b**) polarization curves of ORR on commercial Pt/C catalyst in the presence of (*black circle*) 0.0 M methanol and (*red circle*) 0.1 M methanol; (**c**) Schematic illustration of the differential diffusion and reaction of reactants in CBS Pt-Ru nanoparticles. Reproduced with permission from [119] Copyright American Chemical Society

methanol and oxygen molecules are 0.44 nm and 0.34 nm, respectively). Hence the diffusion of O_2 is faster than the diffusion of methanol in CBS Pt-Ru nanoparticles, rendering the oxidation of methanol on CBS Pt-Ru a non-competitive event. By tailoring the structures (e.g. the size of Pt core and the porosity of Ru shell) of the CBS nanoparticles, one would expect the ORR catalytic activity and methanol-tolerant property of CBS Pt-Ru nanoparticles could be further enhanced.

3.3.4 Carbon-Supported Noble Metal Nanoparticles with Hollow Interiors

The protocol based on the inside-out diffusion of Ag in core-shell nanostructures and BSPP treatment is very general for the fabrication of noble metal nanomaterials with hollow interiors [119]. Unfortunately, the expensive nature of BSPP is a major obstacle to the application of this protocol for large-scale production of hollow structured metal nanomaterials. Therefore, a universal and cost-effective approach to rationally fabricate hollow structures for a sufficiently wide spectrum of noble metals is undoubtedly significant for given technological applications.

Aiming at finding a more cost-effective alternative to BSPP to produce hollow structured metal nanomaterials, the Yang group reported an improved approach to the fabrication of carbon-supported hollow structured noble metal nanoparticles, labeled as hNMNPs/C [125]. Analogous to the protocols they developed previously, this strategy also starts with the synthesis of core-shell noble metal nanoparticles with Ag residing in the core regions, which are then loaded on the carbon supports and refluxed in acetic acid to remove the original organic surfactant. The refluxing in acetic acid also increased the aqueous affinity of the core-shell nanoparticles. The carbon-supported core-shell Ag-noble metal nanoparticles are subsequently agitated in saturated Na_2S or NaCl solution to eliminate Ag from the core region, leading to the formation of noble metal nanoparticles with hollow interiors.

This strategy is universal for the generation of carbon-supported monometallic and alloy noble metal nanoparticles with hollow interiors. As displayed by Fig. 3.13, the strong brightness contrasts in the TEM and HRTEM images of hNMNPs/C including hPt/C (a_1 and a_2), hPtRh/C (b_1 and b_2), hPtRu/C (c_1 and c_2), hPtIr/C (d_1 and d_2), hPtOs/C (e_1 and e_2), hPtRhRu/C (f_1 and f_2), and hPtIrOs/C (g_1 and g_2) show that only carbon-supported nanoparticles with hollow interiors were obtained in the final products after treatment with aqueous Na_2S (Fig. 3.13a_1, b_1, c_1, d_1, e_1, f_1, g_1) or NaCl solution (Fig. 3.13a_2, b_2, c_2, d_2, e_2, f_2, g_2). Analogous to the results observed in BSPP treatment, these electron microscopy images also show the preservation of the size and morphology of the core-shell nanoparticles in the hollow structured nanoparticles. In comparison with the hollow noble metal nanostructures formed through other selective etching process [72, 119, 126], the present synthesis by Na_2S or NaCl shows remarkable simplicity and universality,

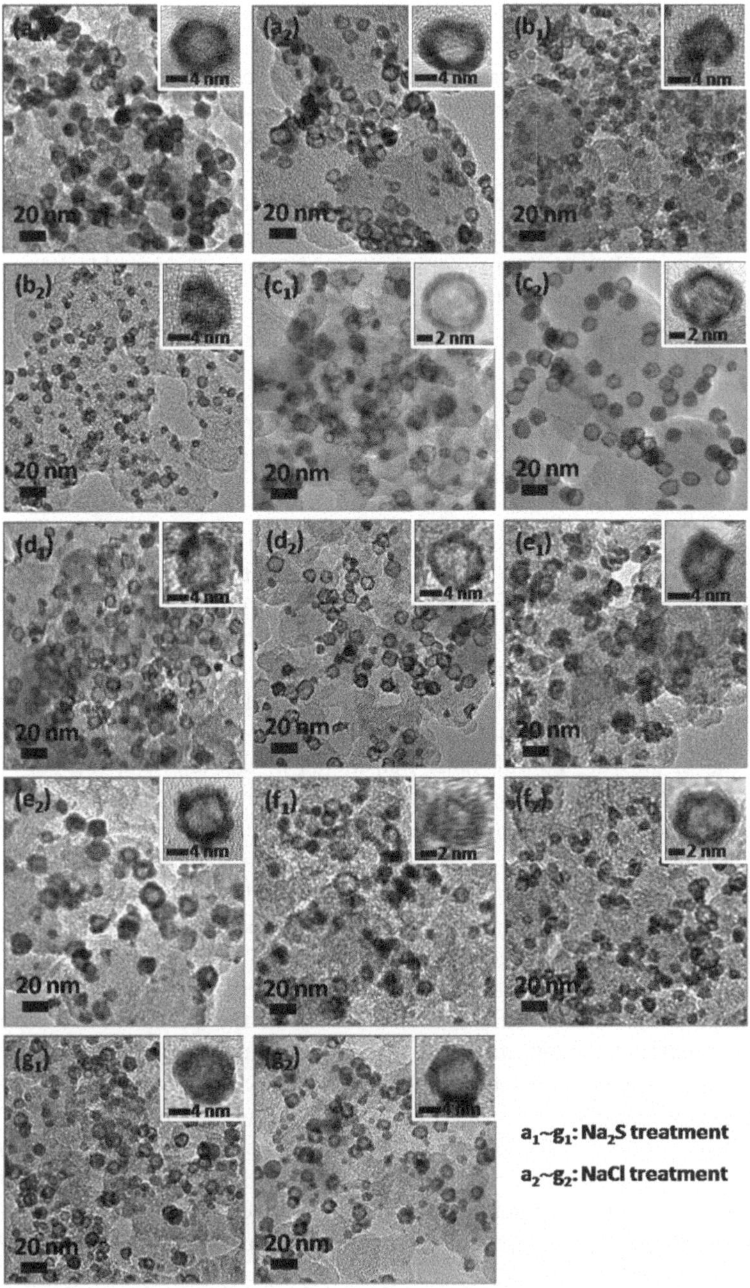

Fig. 3.13 TEM images of hPt/C (**a**$_1$, **a**$_2$), hPtRh/C (**b**$_1$, **b**$_2$), hPtRu/C (**c**$_1$, c$_2$), hPtIr/C (**d**$_1$, **d**$_2$), hPtOs/C (**e**$_1$, **e**$_2$), hPtRhRu/C (**f**$_1$, **f**$_2$), and hPtIrOs/C (**g**$_1$, **g**$_2$) obtained by Na_2S (**a**$_1$, **b**$_1$, **c**$_1$, **d**$_1$, **e**$_1$, **f**$_1$, **g**$_1$) or NaCl treatment (**a**$_2$, **b**$_2$, **c**$_2$, **d**$_2$, **e**$_2$, **f**$_2$, **g**$_2$). Reproduced with permission from [125] Copyright American Chemical Society

which might be used to design and generate other hollow nanostructures for wide application explorations.

In particular, the electrochemical measurements demonstrate that the carbon-supported Pt-containing hollow structured noble metal nanoparticles prepared by NaCl treatment exhibit higher catalytic activity for MOR than that of their core-shell counterparts due to the increase of surface area induced by the hollowing process, whereas those obtained from Na_2S treatment display very poor activity for MOR at room temperature, suggesting that S^{2-} anions have strong poisonousness for the surface of Pt-containing metal shells, which induce serious blockage of the surface areas of the catalysts [125].

3.3.5 *Hollow Pt and Ru Assemblies* via *Electrostatic Interaction*

Although the mastery over the structure and composition of metal nanoparticles is an effective way to improve the electrocatalytic activity on a mass basis, it is generally accepted that Pt alone has rather poor activity for methanol oxidation. At room and moderate temperatures, Pt catalysts could be readily poisoned by carbon monoxide (CO), an intermediate product of methanol oxidation [7, 127, 128]. Without a second oxophilic metal, Pt itself would have to dissociate water to react away CO intermediate at higher potentials.

Instead of alloying Pt with other transitional metals (e.g. Ru, Rh, Ni, or Os) [129–133], the Yang group reported a facile, aqueous route to fabricate the assemblies of hollow Pt (hPt) and ultrafine Ru nanoparticles through electrostatic interaction, and their applications for catalyzing methanol oxidation, the anodic reaction in DMFCs at room temperature [134]. In this approach, negatively charged hollow Pt nanospheres with an average size of 12 nm and positively charged ultrafine Ru nanoparticles with an average size of 0.9 nm were first prepared, respectively. Subsequently, they were mixed together and assemblies of hPt and Ru are formed upon the electrostatic attraction between the particles with opposite charges. Figure 3.14 shows the TEM and HRTEM images of as-prepared hPt-Ru assemblies at different Pt/Ru molar ratios. As indicated, the assembly with ultrafine Ru nanoparticles does not lead to apparent change in particle size of hPt nanospheres. However, the comparison among these microscope images clearly illustrates that the assembly with ultrafine Ru nanoparticles using electrostatic interaction results in the significant change in surface roughness of hPt particles, and the degree of the roughness increased with increasing the molar ratio of Ru in the hPt-Ru assemblies.

Compared with the commercial PtRu alloy nanoparticles, the electrochemical measurements show that the hPt-Ru assemblies have superior catalytic activity toward MOR, of which the assemblies with Pt/Ru molar ratio of 2/1 are the best [134]. XPS analyses confirm that most of the Ru presents as Ru oxides in hPt-Ru

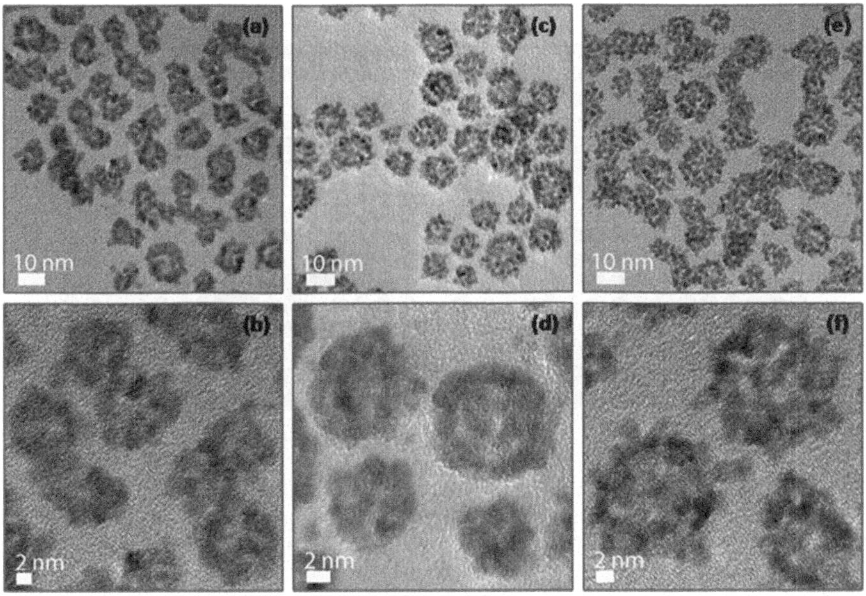

Fig. 3.14 TEM (**a**, **c**, **e**) and HRTEM images (**b**, **d**, **f**) of electrostatic interaction based hPt-Ru assemblies at Pt/Ru molar ratio of 2:1 (**a**, **b**), 1:1 (**c**, **d**), and 1:2 (**e**, **f**), respectively. Reproduced from [134] with permission from the Royal Society of Chemistry

assemblies, which would be favorable for methanol oxidation upon the investigations reported by Rolison and co-workers [135, 136]. They found the homogeneous PtRu alloy catalysts are orders of magnitude less active than mixed-phase electrocatalysts consisting of Pt metal and hyfrous ruthenium oxides. The high activity of a mixed-phase electrocatalysts is derived from the presence of hydrous ruthenium oxides, the electron and proton conducting materials, which are intimately mixed with Pt. Their experimental observations were subsequently confirmed by many other researchers [137–143]. For example, Qiu and co-workers deposited Pt on preformed hydrous Ru oxide and observed enhanced activity for catalyzing MOR [140]. The comparison between alloy PtRu catalysts prepared from co-reduction of the metal precursors and from the mixing of preformed metal colloids was carried out by Lamy and co-workers [144]. They concluded that alloyed catalysts had the lowest activity because of the migration of Ru to the alloy surface under the operating conditions, thereby reducing the Pt surface area available for methanol adsorption.

The less surface dilution effect may also account for the enhanced catalytic activity of hPt-Ru assemblies toward methanol oxidation. It has been generally accepted that methanol oxidation commences by methanol adsorption on multiple Pt sites (3-fold methanol adsorption sites, donated as Pt_3) [122]. The surface conditions of commercial PtRu/C catalysts and hPt-Ru assemblies could be schematically described by Fig. 3.15a, b, respectively. For the alloy PtRu/C catalysts, the atomic mixing and uniform distribution of Pt and Ru on the surface may inhibit

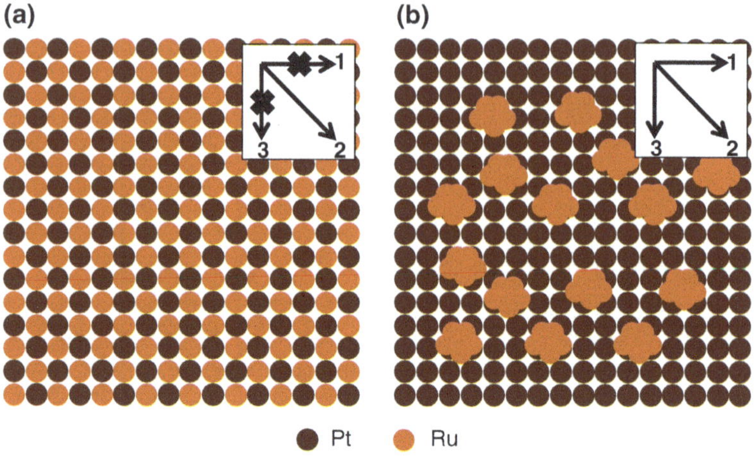

Fig. 3.15 Schematic illustration for the surface conditions of alloy PtRu catalysts (a) and hPt-Ru assemblies (b). Inlets show the possible direction for methanol adsorption on the surface of catalysts. Reproduced from [134] with permission from the Royal Society of Chemistry

the adsorption of methanol at the directions labeled as 1 and 3 (inlet of Fig. 3.15a), and therefore lead to negative effect on the catalytic activity for methanol oxidation. However, for the hPt-Ru assemblies, several closely neighboring Pt atoms share a ultrafine Ru cluster on the surface, not only making use of the oxygen-containing species brought by Ru cluster, but also allowing the adsorption of methanol at all directions (inlet of Fig. 3.15b). Compared with the surface dilution of Pt induced by alloying with Ru, the hPt-Ru assemblies can maintain good catalytic activity toward methanol oxidation at suitable Pt/Ru molar ratios (for example, Pt/Ru of 2/1), although the coherent interfaces between hPt and Ru nanoparticles in the assemblies resulted in some blockage of the surface area of the Pt domains.

3.4 Pt-Based Heterogeneous Nanomaterials with Stellated/Dendritic Morphologies

Extensive research efforts have been devoted towards the development of bimetallic nanomaterials with heterogeneous strtucture, also called hybrid structure or particle-on-particle structure, in which one metal is distributed at single site or at intervals on the surface of another metal particles [6, 15, 17, 145, 146]. These heterogeneously nanostructured materials usually exhibit catalytic properties different from the each one of the constituent materials due to the synergistic effect between the two component metals. For example, the Pt nanoparticles modified by Au clusters were found to display excellent stability for the electrocatalysis of oxygen reduction reaction (ORR) at room temperature [16]. The electronic interaction between Au and Pt can account for the observed stabilization of Pt. When clusters of the softer Au metal are placed on the surface of considerably harder Pt, the surface alloying of Au with Pt, although unlikely, also would modify the Pt electronic structure toward a lower Pt surface energy, or lower-lying Pt *d*-band state, which is helpful to increase the oxidation potential of Pt electrocatalysts against dissolution under potential cycling regimes.

3.4.1 Pt-Containing Heterogeneous Nanomaterials via *the Seed-Mediated Growth*

Analogous to the synthesis of core-shell nanostructures, the general strategy for the fabrication of heterogeneously bimetallic nanoparticles is also the seed-mediated growth method, whereby previously formed seed particles of one metal serve as the sites for nucleation and growth of another metal [17, 147–151]. The advantage of seed-mediated growth is a number of physical/chemical parameters, including the mismatch of lattice structures and constants, the correlation of surface and interface energies, the difference in electronegativity between the two metals, the reducing and stabilizing agent, and even the reaction kinetics, could be selected or tuned for directing the heterogeneous nucleation and growth mode of the second metal on the surface of seed particles in a wet-chemistry preparation [6, 150, 152, 153]. Typical examples include the recent reported bimetallic Pt-on-Pd hetero-nanostructures or nanodendrites with enhanced catalytic activity for ORR [154–157]. The synthesis of such electrocatalysts starts with the preparation of truncated Pd cores through wet chemistry. A dense array of Pt branches are formed subsequently onto the Pd cores from the metal precursors *via* a facet-selective growth. These Pt-on-Pd heterogeneous nanostructures show two and a half times more activity on the basis of equivalent Pt mass for the ORR than the state-of-the-art Pt/C catalyst and five times more activity than the first-generation supportless Pt-black catalyst (Fig. 3.16).

Recently, using the seed-mediated growth method, Kim and co-workers reported the synthesis of AuPt bimetallic hetero-nanostructures consisting of Au nanocrystal cores with well-defined morphologies and dendritic shells of Pt [158]. They first prepared Au seed particles with controlled morphologies (cube, rod, or octahedron) in aqueous phase using ascorbic acid and cetyltrimethylammonium bromide (CTAB) as reducing agent and stabilizer, respectively. Then the aqueous Pt precursors (K_2PtCl_4) was added, and the Pt atoms reduced from K_2PtCl_4 by ascorbic acid deposit on the surface of Au seeds, forming Pt multibranches on Au nanocrystal cores. The final AuPt bimetallic heterogeneous nanostructures would follow the original morphologies of the Au seeds, as schematically illustrated by Fig. 3.17. The authors found that the AuPt hetero-nanostructures exhibit higher electrocatalytic activity and durability for ORR than those of the monometallic Pt catalyst, indicating that the formation of heterostructures can provide higher active

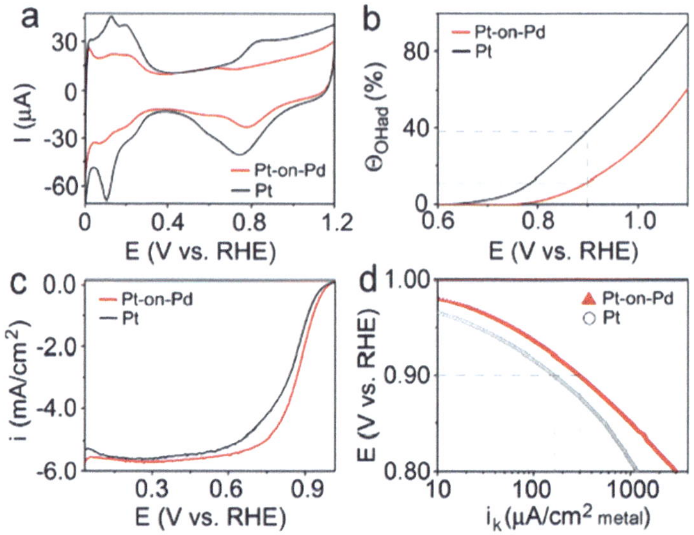

Fig. 3.16 (**a**) CV, (**b**) hydroxyl surface coverage (Θ_{OH}), (**c**) ORR polarization curves, and (**d**) specific kinetic current densities (i_k) for carbonsupported Pt-on-Pd and Pt catalysts. Reproduced with permission from [155] Copyright American Chemical Society

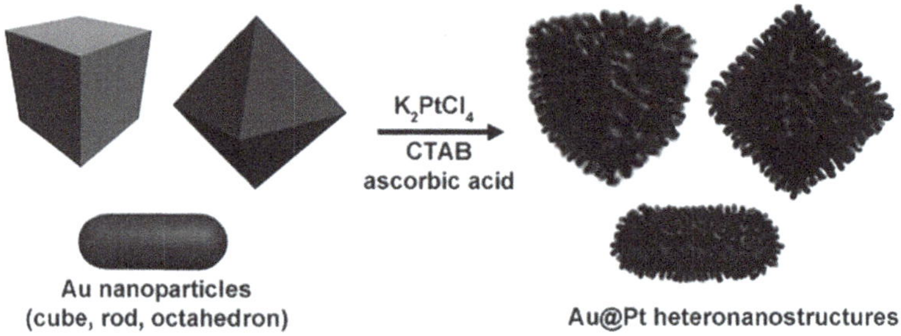

Fig. 3.17 Schematic illustration to show the synthesis of AuPt bimetallic hetero-nanostructures consisting of Au nanocrystal cores with well-defined morphologies and dendritic shells of Pt. Reproduced from [158] with permission from Wiley-VCH

catalytic surfaces. Interestingly, the ORR activities are highly dependent on the shape of the cores. The usage of Au nanooctahedron core would result in the largest improvement of the ORR activity. Their data thus reveal the importance of the core structure for further enhancing the activity of core–shell type nanocatalysts.

To sufficiently increase the catalytic activity but reduce the amount of Pt metals, Wang and Yamauchi reported a facile synthesis of Pt-Pd metallic nanocages, which possess a hollow interior and porous dendritic shell [126]. They prepared dendritic Pt-on-Pd nanoparticles as the starting materials using an one-pot approach they developed before [157, 159]. Then in a typical synthesis of metallic nanocages, the

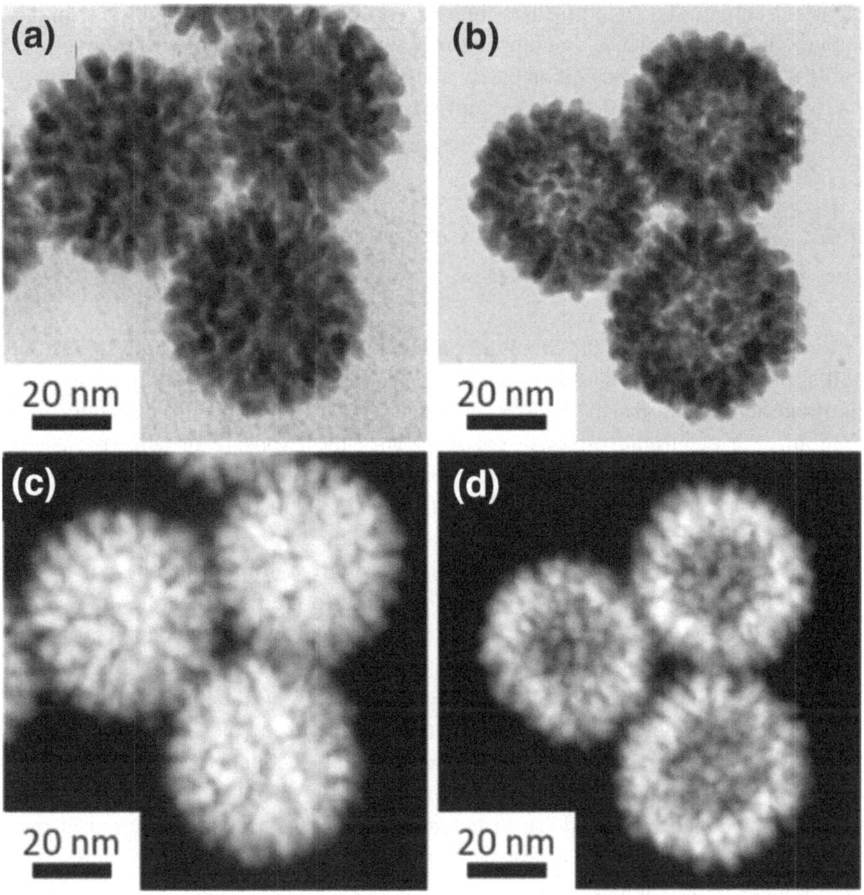

Fig. 3.18 TEM (**a**, **b**) and HAADF-STEM images (**c**, **d**) for dendritic Pt-on-Pd nanoparticles before chemical etching (**a**, **c**) and dendritic nanocages after chemical etching (**b**, **d**). Reprinted with permission from [126] Copyright American Chemical Society

starting Pt-on-Pd dendritic nanoparticles were mixed with an excess amount of concentrated nitric acid, and then the mixture solution was stirred for 5 days at room temperature, allowing selective dissolution of the Pd core. The product was finally collected by consecutive washing/centrifugation cycles with water. As shown by the TEM images in Fig. 3.18, the nanocages are strikingly uniform in shape. The average particle size is ca. 42 nm, which is almost the same size as the starting Pt-on-Pd dendritic nanoparticles before chemical etching. In addition, the resultant dendritic nanocages have well-developed hollow interiors, which quite differed from the starting sample with solid interiors. The exteriors of the dendritic nanocages are well covered by Pt nanoarms with an average diameter of 2 nm. The Pt nanoarms are highly branched in various directions, resulting in the porous wall in each individual entity. Because these metallic nanocages have sufficient catalytic active sites at both their interior and exterior surfaces, they are highly active catalysts for MOR in comparison with other Pt materials. The developed facile synthesis can be scaled up easily and should be highly valuable for routinely producing new metallic nanocages with large surface areas.

3.4.2 *Pt Stellates with an Ag Core or a Hollow Interior* via *an One-Pot Approach*

After a careful review of the literature on recent designs of electrocatalysts for fuel cell applications [6, 147–149, 151, 154–157], Yang and co-workers proposed that the heterogeneous growth of Pt on a metal with different electronegativity may combine the electronic effect in catalysis with the exposure of selective active facets through nanoparticle morphology engineering to increase the diversity in the control of Pt catalytic properties. They have demonstrated this concept with the design and preparation of stellated Pt nanoparticles with a Ag core (SPtNPs-A) [160]. The bimetallic stellated nanoparticles could be formed in one-pot by the competing growth of Pt on the twin sites of in-situ formed decahedral Ag seeds, as shown by Fig. 3.19. Although both Ag and Pt precursors were heated in one-pot, the Ag precursor was preferentially reduced because the reduction of Ag-oleylamine complexes is kinetically more facile. The fast kinetics results in the formation of multiply twinned decahedral Ag nanoparticles, which seed the subsequent deposition of Pt. The twin structure of the Ag seeds, which accounts for many interesting electronic, optical, and catalytic properties [50, 161, 162], are essential for the development of the stellated Ag-Pt heterogeneous nanostructure. The anisotropic growth of Pt on the surface of Ag seeds is due to the selective deposition of Pt at high-energy twin boundary sites.

TEM was used to follow the particle growth process. The results are shown in Fig. 3.20. The TEM (Fig. 3.20a) and high-resolution TEM (HRTEM) images (Fig. 3.20b) show only multiply twinned decahedral Ag nanoparticles with an

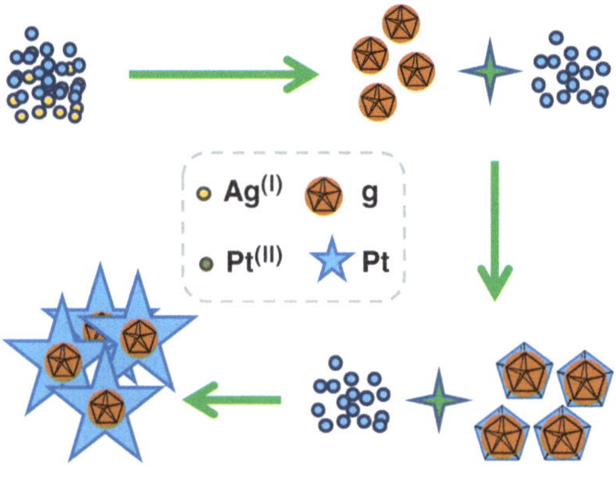

Fig. 3.19 Schematic illustration of the mechanism for the formation of stellated Ag-Pt nanoparticles in the one-pot synthesis. Reproduced from [160] with permission from MacMillan

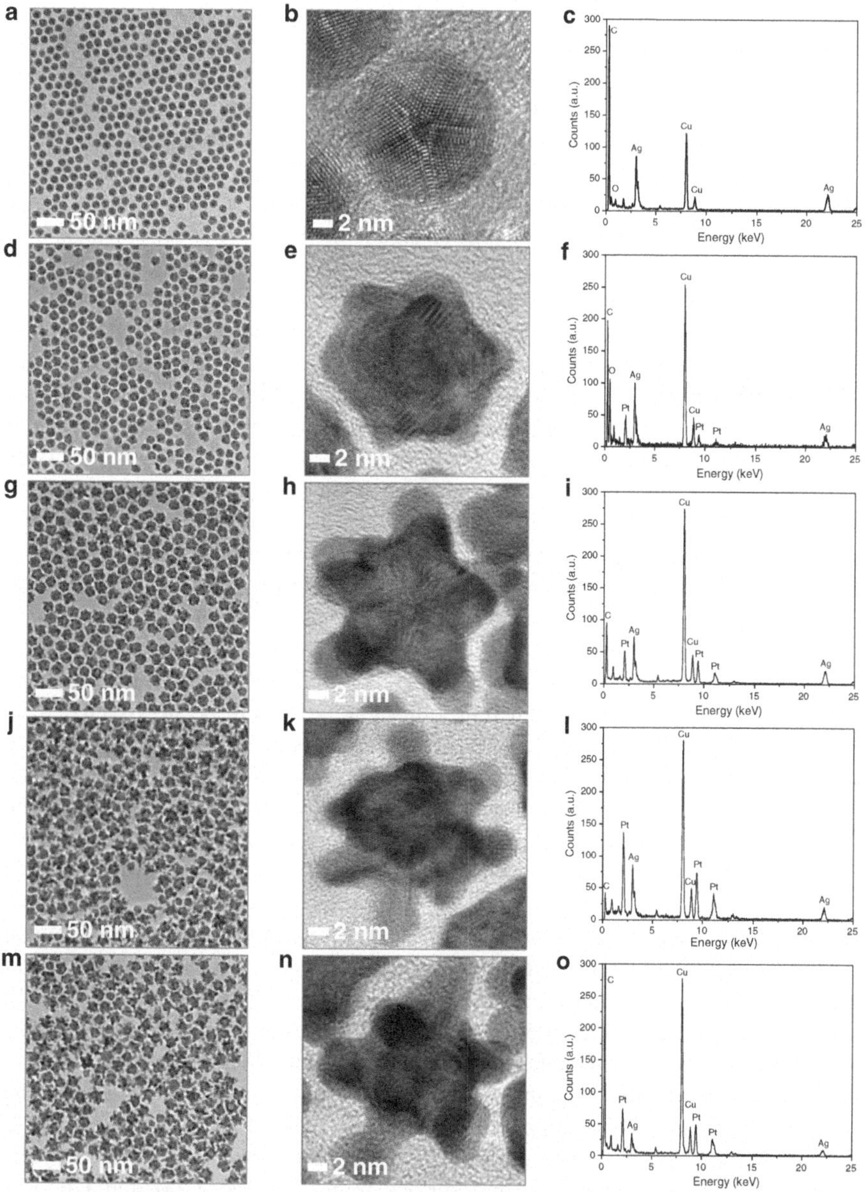

Fig. 3.20 Representative TEM images (**a**, **d**, **g**, **j**, **m**), HRTEM images (**b**, **e**, **h**, **k**, **n**) and corresponding EDX spectra (**c**, **f**, **i**, **l**, **o**) of the stellated Ag-Pt nanoparticles made after the reaction for 30 min (**a**, **b**, **c**), 60 min (**d**, **e**, **f**), 90 min (**g**, **h**, **i**), 120 min (**j**, **k**, **l**), and 180 min (**m**, **n**, **o**), respectively. Reproduced from [160] with permission from MacMillan

average diameter of 12.2 nm after reaction for 30 min, although both Ag and Pt precursors are introduced to oleylamine simultaneously. The absence of the Pt component at this stage is confirmed by EDX analysis (Fig. 3.20c). The Pt signal is only detected by EDX of the nanoparticles recovered from the reaction mixture after 60 min, as shown in Fig. 3.20f. There are also significant changes in the morphology and the size of the particles in the TEM and HRTEM images (Fig. 3.20d, e). With the deposition of Pt on the in-situ formed Ag nanoparticles, the average diameter is increased to 16.6 nm. Core-shell nanoparticles with a number of Pt humps on the shell are the majority product at this time. At 90 min, the Pt humps developed into short branches on the surface of the core-shell Ag-Pt nanoparticles (Fig. 3.20g, h), and the Pt signal in the EDX spectrum increased accordingly (Fig. 3.20i). This stellating growth of Pt continued until the Pt precursor in the reaction mixture is completely exhausted (120 min). The final product is SPtNPs-A with an average size of 25.2 nm (Fig. 3.20j, k) which is morphologically distinct from the spherical nanoparticles formed in the early stages of the reaction. EDX analysis (Fig. 3.20l) of the nanoparticles measured the Ag/Pt atomic ratio to be approximately 1.0/1.9, which is in good agreement with the Ag/Pt atomic ratio used in the preparation (1/2). This is suggestion that all of the Ag and Pt precursors had been completely reduced by oleylamine under the prevailing experimental conditions. The completion of reaction could also be confirmed by extending the reaction time to 180 min, where no further changes in the particle size, morphology, and Pt/Ag atomic ratio were detected by TEM, HRTEM, and EDX (Fig. 3.20o), respectively.

The twinned structure of the Ag seeds provides an expeditious means to modify the internal structure of SPtNPs-A. As discussed in previous section, the multiply-twinned Ag nanoparticles are inherently unstable; known to be slowly etched by dissolved O_2 and Cl^- dissociated from the Pt precursor (K_2PtCl_4), and influence strongly on the continuity and compactness of the overlaid Pt shell [119, 120]. The Ag^+ released from the O_2/Cl^- etching of twinned Ag seeds diffuses out through the discontinuous Pt shell due to the prevailing Ag^+ concentration gradient between the core of the Ag-Pt nanoparticles and the surrounding solution. This diffusion has been developed into a general protocol for the fabrication of noble metal nanoparticles with a hollow or cage-bell structure [119]. A variation of this protocol could be to completely remove the Ag core from SPtNPs-A, resulting in hollow Pt nanoparticles with an intact stellation morphology (SPtNPs-H, or stellated Pt nanoparticles with a hollow interior). After aging the mixture of SPtNPs-A colloidal solution in toluene with aqueous solution of BSPP for 48 h at room temperature, the disappearance of the Ag signal in the EDX analysis of SPtNPs-A evidences the complete elimination of the Ag core from SPtNPs-A (Fig. 3.21d). The void space in the core region formed by the removal of Ag by BSPP could be discerned by the strong contrast between central and surface regions in the TEM and HRTEM images of Fig. 3.21a–c. The particle size and the stellation morphology are practically unchanged by the BSPP treatment. Hence the BSPP treatment is highly selective and did not cause the collapse of the SPtNPs-A structure.

Specifically, the electrochemical measurements demonstrate that the SPtNPs-A are highly active for MOR due to presence of the Ag core which alters the Pt activity through the Pt-Ag electronic coupling effects, while the SPtNPs-H leverages on the abundance of atomic steps, edges, corner atoms, as well as dominant

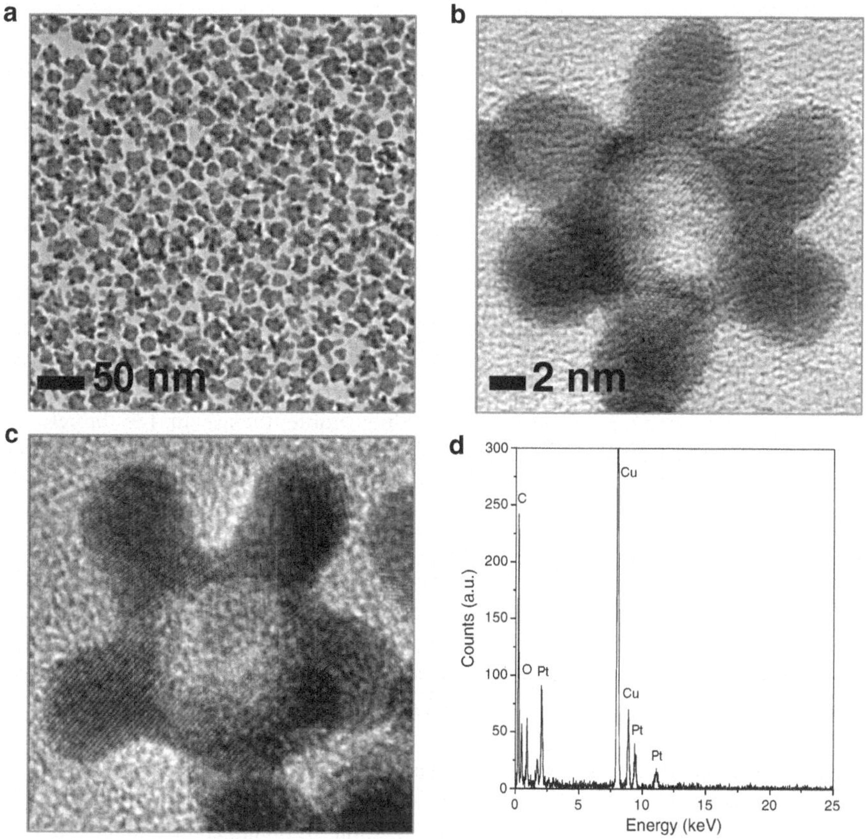

Fig. 3.21 TEM (**a**), HRTEM (**b**, **c**), and corresponding EDX spectrum (**d**) of stellated Pt nanoparticles with a hollow interior prepared from the stellated Ag-Pt bimetallic nanoparticles. Reproduced from [160] with permission from MacMillan

{1 1 1} facets on the stellating geometry to increase the ORR activity of the Pt surface [160]. The stellated morphology offers a convenient platform to construct heterogeneous nanoparticles with very varied catalytic properties quite easily.

3.4.3 Bimetallic Au-Noble Metal Nanodendrites

Considering the great potential of the hybrid nanostructures in many catalytic applications, the production of heterogeneously bimetallic nanoparticles consisting of a wide spectrum of noble metals would be undoubtedly important. Employing Au as a substitute for Ag seeds, the Yang group developed a more generic approach, which is also based on preferential nucleation and growth of noble metals on the active twins of Au seed particles, for the fabrication of bimetallic nanodendrites

[163]. Analogous to the synthesis of bimetallic Ag-Pt stellates, this strategy starts with the preparation of the Au nanoparticles with multiple twins in oleylamine at elevated temperature. These multi-twinned Au particles are served as seeds for the subsequent nucleation and growth of other noble metals, resulting in the formation of bimetallic Au-noble metal nanoparticles with dendritic morphologies.

This strategy is general enough for the generation of dendritic bimetallic nanoparticles of Au and other noble metals. The synthesis follows the common seed-mediated growth protocol besides the temperature, which is 165 °C for Au-Pd and Au-Pt, 320 °C for Au-Ru and Au-Ir, respectively [163]. The TEM (Fig. 3.22a, d, g, j) and HRTEM images (Fig. 3.22b, e, h, k) of bimetallic Au-Pt (Fig. 3.22a, b), Au-Pd (Fig. 3.22d, e), Au-Ru (Fig. 3.22g, h), and Au-Ir systems (Fig. 3.22j, k) show that only heterogeneous nanoparticles with dendritic morphologies were obtained in the final products, which are very different from the starting spherical Au seed particles. The presence of the corresponding noble metals in the bimetallic nanodendrites (Au and Pt for bimetallic Au-Pt, Au and Pd for bimetallic Au-Pd, Au and Ru for bimetallic Au-Ru, Au and Ir for bimetallic Au-Ir) could be confirmed by the results of EDX analyses (Fig. 3.22c, f, i, l) of these four bimetallic nanosystems. In comparison with the dendritic or particle-on-particle nanostructures formed through either Stranski-Krastanov or Volmer-Weber mode [72], the present seed-mediated growth synthesis shows remarkable universality. In addition, the mechanistic understanding for the fabrication of dendritic bimetallic nanoparticles from multiply twinned seeds might be used to design and generate other heterogeneous nanostructures for wide application explorations.

As reported by Yang and co-workers, the heterogeneous Au-Pt bimetallic nanodendrites display superior catalytic activity toward MOR for the presence of unique dendritic structures, as compared with that of commercial Pt/C catalysts [163]. Therefore their study offers a vivid example to demonstrate the tailoring of the catalyst properties by means of a heterogeneous construction, and provides a method with remarkable universality for creating heterogeneously structured nanomaterials for given technological applications.

3.4.4 Oleylamine Synthesis of Dendritic Noble Metal Nanomaterials

Among the different strategies for synthesis of metal nanoparticles with controlled morphologies, the particle synthesis in oleylamine is of particular interest. Oleylamine is an amine of the fatty acid (oleic acid), and is a common reagent in the chemical synthesis of nanoparticles [164]. It can function both as a reducing agent and as a capping ligand to stabilize the surface of the particles. In addition, oleylamine has very high boiling point, and can be a suitable solvent for the reaction mixture. This means amine-based nanoparticle synthesis can usually be conducted in a single-phase reaction system. Compared with that in two-phase system, such as

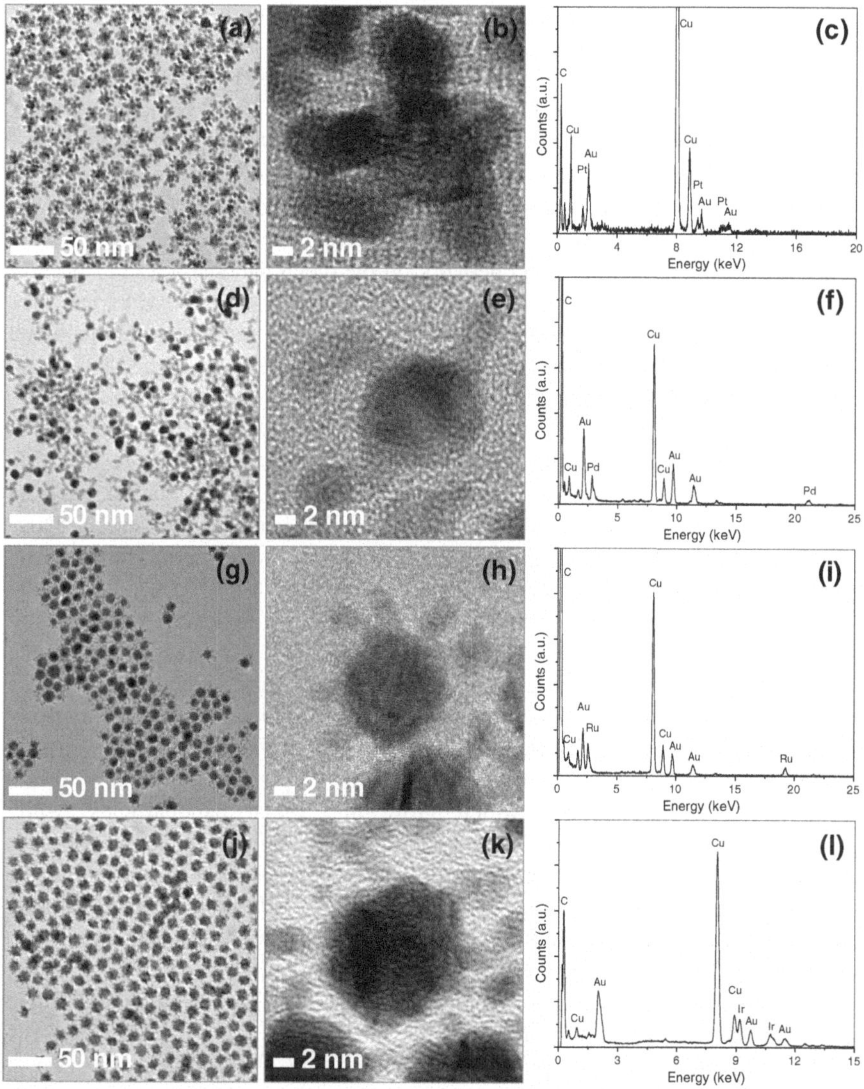

Fig. 3.22 TEM (**a**, **d**, **g**, **j**), HRTEM (**b**, **e**, **h**, **k**), and corresponding EDX spectra (**c**, **f**, **i**, **l**) of the dendritic Au-Pt (**a**, **b**, **c**), Au-Pd (**d**, **e**, **f**), Au-Ru (**g**, **h**, **i**), and Au-Ir bimetallic nanoparticles (**j**, **k**, **l**) as-prepared by seed-mediated growth method. Reproduced from [163] with permission from the Royal Society of Chemistry

the case of Brust-Schiffrin reaction [165], the particle synthesis in single-phase could typically offer the ease of kinetic control. This poses significant advantages to the understanding of nanoparticle growth kinetics.

A number of literatures have been reported for the synthesis of metal nanoparticles in or using oleylamine, particularly the work from the Sun group

[166–178]. For example, in 2009, they reported a simplified approach using oleylamine as both reducing agent and stabilizer for the synthesis of Fe_3O_4 nanoparticles *via* the thermal decomposition of $Fe(acac)_3$ [175]. By varying the volume ratio of oleylamine in benzyl, the Fe_3O_4 nanoparticles from 7 to 10 nm, which show reasonably large magnetization, could be easily produced. In an earlier study, Huo and co-workers investigated the mechanism of gold nanoparticle synthesis using oleylamine as both reducing agent and protecting ligand. They found that amine ligands are oxidized to amides in metal precursor reduction and form a protecting layer of hydrogen bond network on the metal nanoparticles [179].

By reducing the metal precursors in oleylamine at elevated temperature (160 °C), the Yang group recently found that a large variety of noble metal nanodendrites (NMNDs) including monometallic (e.g. Pt, Pd, Ru, Rh, Ir) and alloy (e.g. PtPd, PtRh, PtIr, PtPdRu, PtPdRh, PtRuOs, and PtRhOs) could be synthesized *via* a facile one-pot approach [180]. No extra reducing agent and stabilizers are needed to achieve these syntheses. After reaction, the dendritic products could be purified by precipitation with methanol, centrifugation, washing with methanol, and re-dispersed in toluene. The TEM images of these NMNDs were shown in Fig. 3.23, which indicate that the NMNDs were very uniform in size.

By tracking the growth process of the NMNDs using TEM, Yang and co-workers proposed a competing mechanism to interpret the one-pot synthesis of NMNDs in oleylamine [180]. According to Huo and co-workers [179], amine ligands are oxidized to amides during the metal precursor reduction, forming a protecting layer of hydrogen bond network on the metal nanoparticles. Therefore, the mechanism for the fabrication of NMNDs in oleylamine could be summarized by the scheme in Fig. 3.24. At initial stage, the metal precursors are reduced into metal atoms by oleylamine, which grow into metal nanoparticles, while oleylamides are simultaneously generated from oleylamine as capping agents to stabilize the nanoparticles. Then the particle aggregation would compete with the oleylamide passivation, resulting in the formation of a large number of particle aggregates in the solution (step (a) in Fig. 3.24). Finally, the particle aggregates continuously grow with the expense of small nanoparticles in solution *via* a ripening process, e.g. Ostwald ripening [181], to form larger and more stable nanodendrites (step (b) and (c) in Fig. 3.24).

The competition between particle aggregation and oleylamide passivation at initial stage of the synthesis is essential for the formation of NMNDs, which might be affected by the type of metal precursor and reaction temperature since the ligand in different metal precursors and the temperature would have significant influence on the reduction kinetics of the noble metal ions. For examples, concerning the Pt $(acac)_2$ and K_2PtCl_4 as metal precursors, the acetylacetonate ion in $Pt(acac)_2$ and Cl^- ions in K_2PtCl_4 may have different effect on the reduction of Pt(II) ions, and therefore affect the competition between the aggregation and oleylamide passivation of Pt particles, resulting in Pt with different morphologies. Analogously, since the particle aggregation and oleylamide generation are varied at different temperature. The temperature could influence the morphology of the final metal products

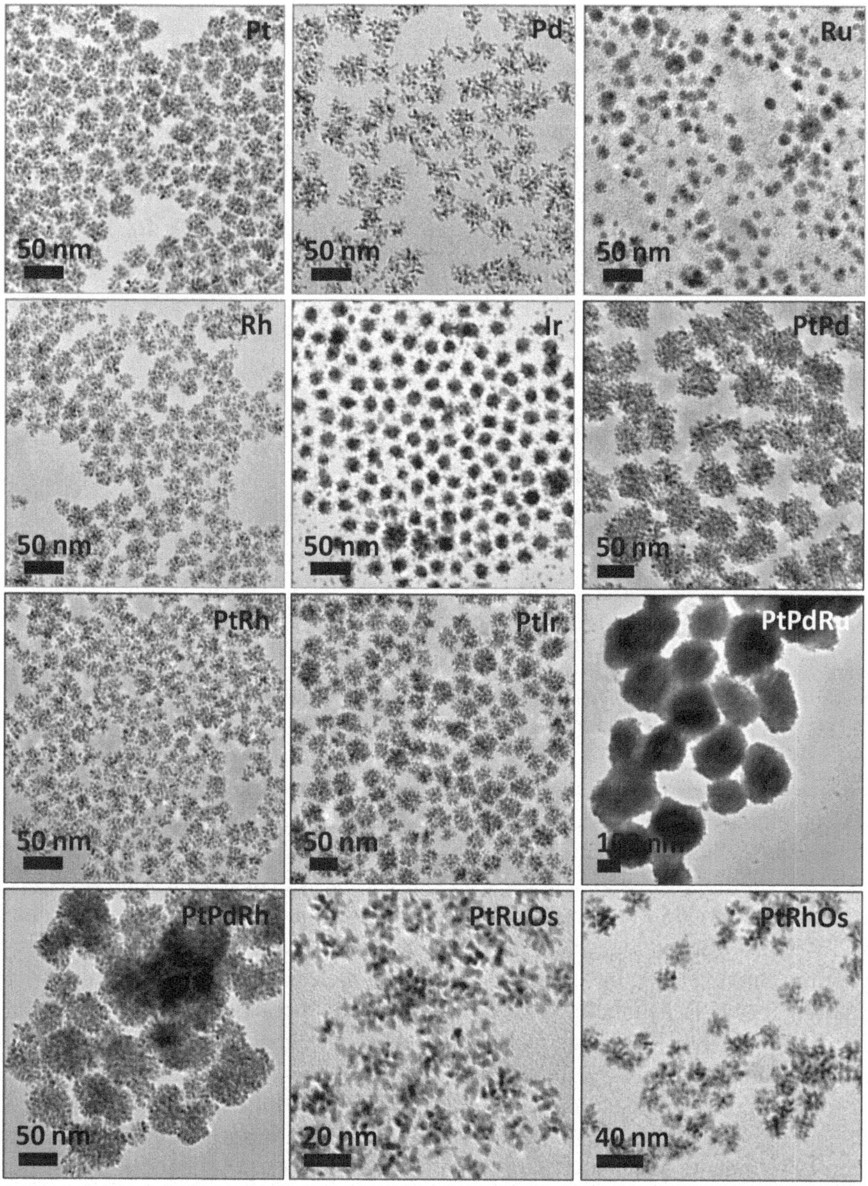

Fig. 3.23 TEM images of NMNDs including Pt, Pd, Ru, Rh, Ir, PtPd, PtRh, PtIr, PtPdRu, PtPdRh, PtRuOs, and PtRhOs synthesized in oleylamine at temperature of 160 °C. Reproduced from [180] with permission from the Royal Society of Chemistry

through affecting the competition between the aggregation and oleylamide passivation of noble metal particles.

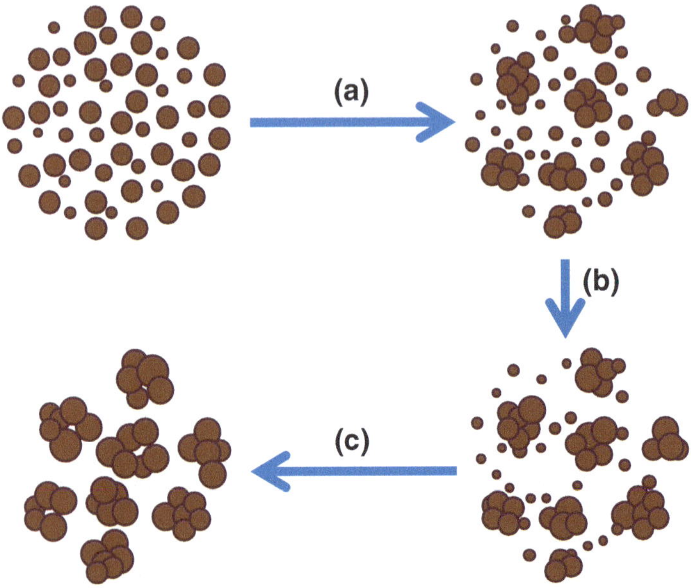

Fig. 3.24 Schematic illustration to demonstrate the mechanism for the formation of NMNDs *via* the reduction of metal precursors in oleylamine at elevated temperature: (**a**) Competition between particle aggregation and oleylamide passivation results in the formation of particle aggregates; (**b**, **c**) particle aggregates grow into nanodendrites *via* Ostwald ripening. Reproduced from [180] with permission from the Royal Society of Chemistry

Yang and co-workers characterized the electrocatalytic activities of a number of Pt-containing NMNDs (Pt, PtPd, PtIr, PtPdRu, and PtRuOs) toward the room temperature MOR and benchmarked with that of commercial PtRu/C catalysts (E-TEK, alloy PtRu nanoparticles with ca. 3.5 nm in diameter on Vulcan XC-72 carbon support) [180]. From the comparison of current densities, the Pt-containing alloy NMNDs show higher catalytic activities than that of Pt nanodendrites. It is worthy to note that the PtRuOs nanodendrites particles display the highest catalytic activity for MOR, definitely supporting that the introduction of Os to the traditional Pt-containing alloy nanomaterials is an effective way to enhance the activity of a MOR catalyst [182, 183], although the effect of the dendritic structure on the catalytic activity of the alloy nanomaterials cannot be ruled out. Os is supposed to modify the adsorptive properties of alloy PtRu towards oxygen-containing species at potentials slightly more negative than for Ru [184]. According to the bi-functional catalysis [185, 186], a timely supply of oxygen-containing species would be helpful to reduce the adsorption of CO-like species, which are the intermediate products during MOR and are the catalyst poisons, on the Pt particles, therefore PtRuOs nanoparticles are superior to pure Pt or other Pt-containing alloy electrocatalysts.

3.5 Pt-Based Heterogeneous Nanomaterials with Dimetric Structures

Bimetallic heterodimers, in which one metal is distributed at single site on the surface of another metal particles, might exhibit unique and superior properties different from the each one of the constituent materials due to the synergistic effect between the two component metals. A number of strategies including localized overgrowth [148, 150], high temperature reduction [187–189], plasmon-mediated synthesis [146], manipulation of reaction kinetics [153], and selective transformation [190], have been developed for the synthesis of bimetallic heterodimers. Recently, Puntes and coworkers reported the fabrication of Au-Pt heterodimers at room temperature using oleylamine as reducing agent (Fig. 3.25), and found that the Pt seeds play an important role for the nucleation and growth of Au region in the heterodimer structures [191]. The Pt seeds markedly accelerate the reaction, serving both as nucleation platforms and as an initial catalytic reducer of the Au ions in solution, as supported by the initial shrinking of the Pt seeds.

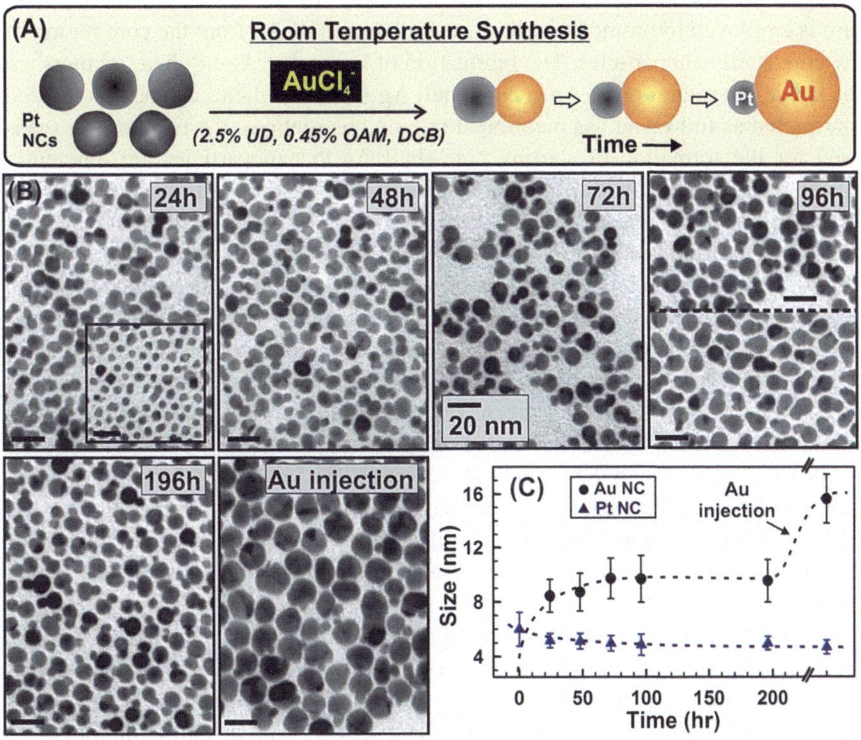

Fig. 3.25 (**a**) Schematic for the heterodimer formation, (**b**) TEM images and (**c**) plot of diameter vs. time for the heterodimer evolution. (*Inset*) TEM of the Pt seeds. All scale bars correspond to 20 nm. Reproduced from [191] with permission from the Royal Society of Chemistry

Unfortunately, up to date, unlike the abundant studies on the semiconductor/metal oxide-noble metal heterodimers, which will be discussed in next section, the literatures reported on heterodimers consisting of two chemically distinct metals are still very limited due to the lack of synthetic inaccessibility. The most commonly used seed-mediated growth method usually results in the formation of bimetallic nanoparticles with core-shell or dendritic aggregates [149, 151, 154–156, 192–195]. Considering the further exploration of the properties and consequent applications of these heterogeneous nanostructures, the development of effective approaches for the production of bimetallic heterodimers would be undoubtedly important and pose significant challenges.

In their recent work, Yang and co-workers presented a facile and high-yield approach to the synthesis of bimetallic heterodimers consisting of Ag and hollow structured Pt nanoparticles, labeled as Ag-hollow Pt (Ag-hPt), which is based on an unique diffusion phenomenon of Ag they observed in core-shell Ag-Pt nanoparticles with Ag residing in the core region [196]. In their strategy, core-shell Ag-Pt nanoparticles are first prepared in oleylaminc and re-dispersed in an organic medium (e.g. toluene) as starting templates. The core-shell Ag-Pt templates are subsequently converted into Ag-hPt heterodimers *via* the inside-out migration of Ag in core-shell Ag-Pt nanoparticles. A heating treatment at elevated temperature is employed to promote the inside-out diffusion of Ag from the core region of the core-shell nanoparticles. The fabrication of bimetallic Ag-hPt heterodimers *via* the inside-out migration of Ag in core-shell Ag-Pt nanoparticles can be rationalized interpreted as following. As mentioned in previous sections, the twinned Ag seeds used for the formation of starting core-shell Ag-Pt nanoparticles are inherently unstable and could slowly be etched by dissolved O_2 and Cl^- dissociated from the Pt precursor to generate Ag^+ ions [119, 120]. These Ag^+ ions are then diffused out through the discontinuous Pt shells, which are produced during the epitaxial deposition of Pt atoms on the Ag seeds with a large number of imperfections on their surfaces, due to the Ag^+ concentration gradient between the interior of core-shell Ag-Pt nanoparticles and the surrounding solution. At suitable temperature (e.g. 80 °C), the Ag^+ ions diffused out from core-shell Ag-Pt nanoparticles are re-reduced by excess reducing agent (e.g. oleylamine) in toluene to form single crystalline Ag nanoparticles decorated on the outer surface of the Pt shell. The Ag nanoparticles grew with the continued outward diffusion of Ag^+ ions until the Ag core is completed depleted, leaving behind a colloidal solution of bimetallic Ag-hPt with heterogeneous nanostructures. Finally, the Ag nanoparticles on the surface of the Pt shell undergo a ripening process to form larger and more stable domains on the surface of Pt shell.

After heating the core-shell Ag-Pt colloidal solution in toluene at 80 °C for 72 h under air, Ag is migrated to the surface of Pt shells from the interior of core-shell Ag-Pt nanoparticles, resulting in the formation of bimetallic heterodimers consisting of Ag solid particles and Pt nanoparticles with a hollow interior. The TEM and STEM images (Fig. 3.26a, b) after heating treatment in toluene show that only heterodimers with strong imaging contrast in different domains are observed in the final products, which are clearly distinct from the starting core-shell

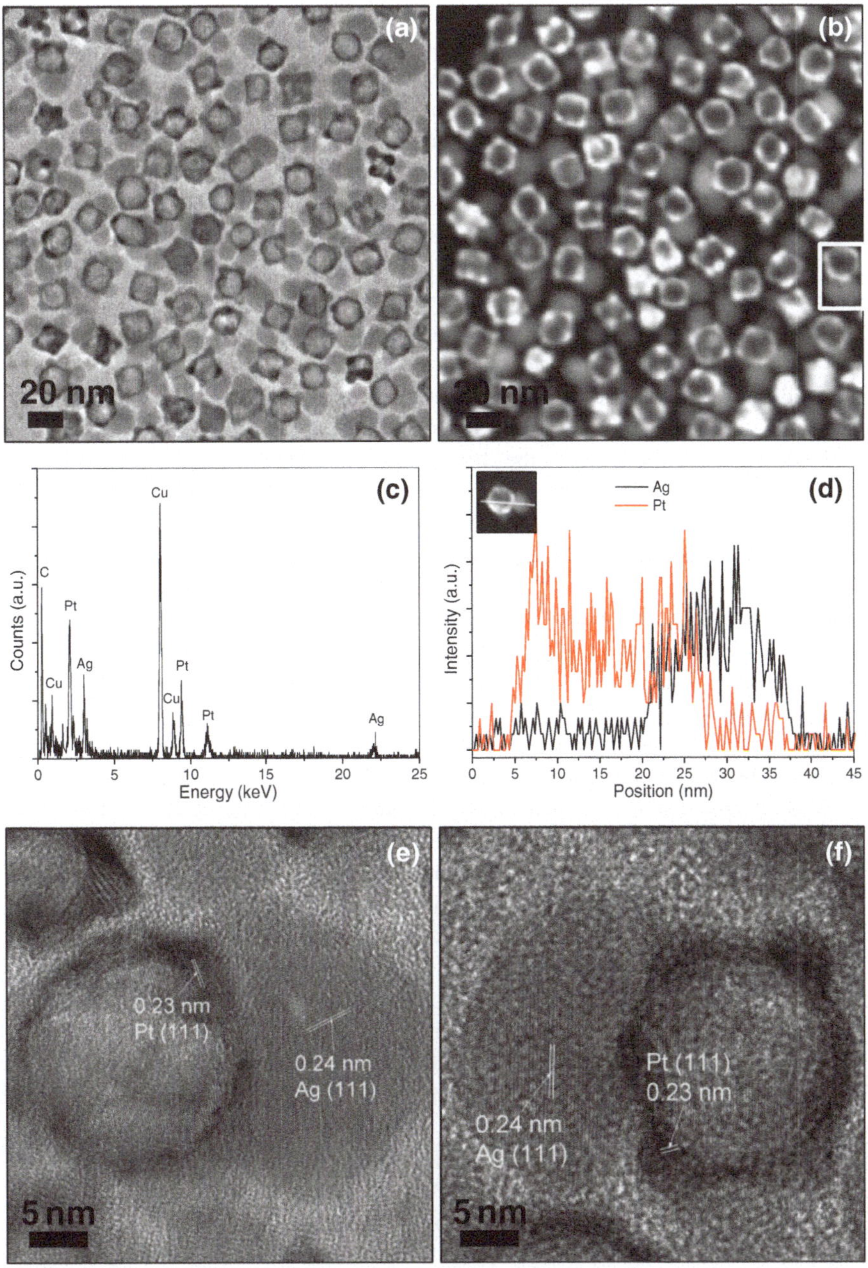

Fig. 3.26 (**a**) TEM image, (**b**) STEM image, (**c**) STEM-EDX spectrum, (**d**) elemental profiles, and (**e**, **f**) HRTEM images of bimetallic Ag-hPt heterodimers synthesized *via* the inside-out migration of Ag in core-shell Ag-Pt nanoparticles at elevated temperature. Reproduced from [196] with permission from the Royal Society of Chemistry

nanoparticles. The void space between the core and the outer Pt shell regions, formed upon the inside-out migration of the Ag cores by heating treatment, is discernible by the strong brightness contrast in TEM and STEM images, while Ag appears as lighter domains sharing the solid-state interfaces with the Pt domains. As indicated by Fig. 3.26c, the EDX analysis of an arbitrary single dimer boxed in the high-angle annular dark-field STEM image (Fig. 3.26b) confirms the presence of Ag and Pt after heating treatment of core-shell Ag-Pt nanoparticles. A more direct evidence for the formation of Ag-hPt heterodimers is provided by the line scanning analysis of an arbitrary single dimer in the high-angle annular dark-field STEM mode. As shown in Fig. 3.26d, the Pt signal is present at one side throughout the heterodimer whereas the Ag signal is detected only at the other side. The interplanar spacings of approximate 0.24 and 0.23 nm indicated in the HRTEM images of the biemtallic Ag-hPt heterodimers (Fig. 3.26e, f) correspond to the (1 1 1) planes of face-centered cubic (fcc) Ag and Pt, respectively.

To demonstrate the effect of the structure of the bimetallic Ag-Pt systems on the catalytic properties, the core-shell Ag-Pt nanoparticles and bimetallic Ag-hPt heterodimers were loaded on Vulcan carbon and examined for their electrocatalytic activities towards the room-temperature ORR. Cyclic voltammograms of core-shell Ag-Pt/C and bimetallic Ag-hPt/C in argon-purged 0.1 M $HClO_4$ at room temperature are used to obtain the electrochemical active surface areas (ECSAs) from the hydrogen adsorption/desorption regions (−0.2 to 0.1 V vs Ag/AgCl). As shown in Fig. 3.27a, the ECSAs normalized by the mass of Pt for core-shell Ag-Pt nanoparticles and bimetallic Ag-hPt heterodimers are 40.6 $m^2\ g_{Pt}^{-1}$ and 48.8 $m^2\ g_{Pt}^{-1}$, respectively. The inside-out migration of Ag from core-shell Ag-Pt nanoparticles may lead to the increase of ECSAs by releasing the inner surface of Pt shell, whereas the growth of Ag domains on the outer surface of Pt shell would result in the decrease of ECSAs due to the contact interfaces between hollow Pt and Ag in the heterodimers, which may induce some blockage of the surface area of the Pt shells. These two effects might have offset each other, such that the ECSA of bimetallic Ag-hPt heterodimers is only slightly higher than that of core-shell Ag-Pt nanoparticles.

Figure 3.27b shows the ORR polarization curves in the potential range of 0.8–0 V for core-shell Ag-Pt nanoparticles and bimetallic Ag-hPt heterodimers in oxygen-saturated 0.1 M $HClO_4$ at room temperature. The half-wave potentials for the core-shell Ag-Pt nanoparticles and bimetallic Ag-hPt heterodimers are 518 mV and 606 mV, respectively. It has been noted that the half-wave potential for bimetallic Ag-hPt heterodimers is much more positive than that of core-shell Ag-Pt nanoparticles, indicating that the Pt shells in the Ag-hPt heterodimers have higher catalytic activity for ORR in comparison with that of core-shell Ag-Pt nanoparticles under the experimental conditions. Furthermore, chronoamperometry of core-shell Ag-Pt nanoparticles and bimetallic Ag-hPt heterodimers at 0.6 V in oxygen-saturated 0.1 M $HClO_4$ solution is used to obtain some indications of the long-term performance of the catalysts in ORR. As shown in Fig. 3.27c, the "steady-state" specific activity of bimetallic Ag-hPt heterodimers is higher than

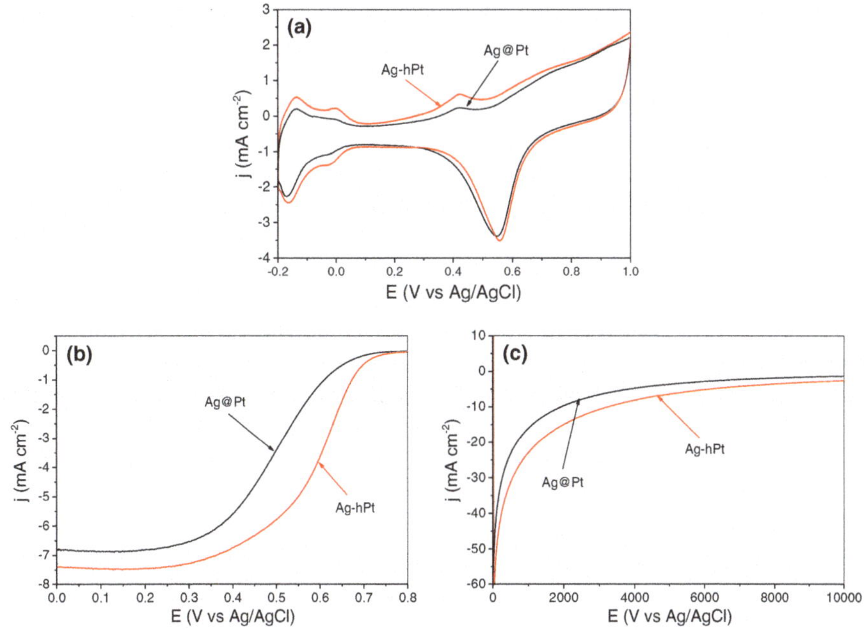

Fig. 3.27 (**a**) Cyclic voltammograms of core-shell Ag-Pt nanoparticles and bimetallic Ag-hPt heterodimers in argon-purged $HClO_4$ (0.1 M) at room temperature, scan rate: 50 mV s^{-1}; (**b**) ORR polarization curves for core-shell Ag-Pt nanoparticles and bimetallic Ag-hPt heterodimers, recorded at room temperature in an O_2-saturated $HClO_4$ solution (0.1 M) at a sweep rate of 20 mV s^{-1} and a rotating speed of 1600 rpm; (**c**) chronoamperograms of core-shell Ag-Pt nanoparticles and bimetallic Ag-hPt heterodimers at 0.6 V in an O_2-saturated $HClO_4$ solution (0.1 M) at room temperature and a rotating speed of 1600 rpm. Reproduced from [196] with permission from the Royal Society of Chemistry

that of core-shell Ag-Pt nanoparticles after 2.5 h, indicating the greater stability of the bimetallic heterodimers.

Considering the electronic coupling between Ag and Pt are present both in core-shell Ag-Pt nanoparticles and bimetallic Ag-hPt heterodimers although it is asymmetric in the latter, the lower catalytic activity of the core-shell Ag-Pt nanoparticles in comparison with that of the bimetallic Ag-hPt heterodimers could be attributed to the tensile effect imposed on the Pt shell by the Ag core in core-shell structures. Since Ag has larger lattice parameter (0.4090 nm) than that of Pt (0.3923 nm), the Ag cores would exert a tensile effect on the Pt atoms deposited on their surfaces [52]. In a common ORR process, i.e. the series 4 electron pathway, the reactions involve both the breaking of an O–O bond and the formation of O–H bonds [9, 62, 63]. The most active Pt-based catalyst should have the d-band center with an intermediate value since the optimal ORR catalyst needs to facilitate both bond-breaking and bond-making steps without hindering one or the other [197, 198]. However, the tensile effect of Ag core on Pt shell, which leads to narrower d-band, results in an up-shift of the d-band center of the Pt shell. As has been discussed in previous section (Sect. 3.2), the surface of a Pt-based catalyst with high value of d-band center tends to bind adsorbents too strongly, thereby enhancing the kinetics of dissociation reactions producing these adsorbents. In this case, the Pt shells may suffer from hindered bond-making step, therefore are less active for ORR. On the other hand, the bimetallic Ag-hPt heterodimers shows good catalytic activity toward ORR possibly because they have a more suitable d-band center to balance the breaking of O-O bonds and the formation of O-H bonds.

The purpose of Yang and co-workers is to provide an unique strategy (inside-out migration of Ag in core-shell Ag-Pt nanoparticles) to fabricate bimetallic Ag-hollow Pt nanoparticles with heterodimeric structures and to demonstrate the physical/chemical properties of a bimetallic system might be tuned by tailoring its structure (by relegating Ag to the core or on the shell region in bimetallic Ag-Pt nanoparticles). Their study offers a vivid example to demonstrate the tuning of the material properties by means of a structural tailoring. In addition, the mechanistic understanding of the structural transformation from core-shell to heterogeneous nanoparticles might be used to design and fabricate other heteronanostructures for a given application.

3.6 Pt-Based Composite Nanomaterials

Besides the more conventional heterogeneous nanomaterials, e.g. core-shell, hollow, and bimetallic heterostructures, there has been increasing interest devoted towards the development of semiconductor-metal nanocomposites that consist of different classes of materials with solid-state interfaces [71, 199–225]. This type of nanostructures combines materials with distinctly different physical and chemical properties to yield a unique hybrid nanosystem with multifunctional capabilities, and tunable or enhanced properties that may not be attainable otherwise. The early studies of semiconductor-metal nanocomposites involved different metals (e.g. Au, Ag, and Pt) deposited on or doped in TiO_2 powders for photocatalytic applications [226–232]. In these structures, the metal domain induces the charge equilibrium in photoexcited TiO_2 nanocrystals, to affect the energetics of the nanocomposites by shifting the Fermi level to more negative potentials. The shift in Fermi level is indicative of improved charge separation in TiO_2-metal systems, and is effective towards enhancing the efficiency of photocatalysis [233, 234].

3.6.1 Nanocomposites Consisting of Semiconductors and Gold

The nanocomposites constructed by depositing gold on the surface of semiconductor nanocrystals are the topics being studied most sufficiently [71, 199–202, 205,

209, 235–242]. Within the last decade have wet-chemistry methods blossomed and become a powerful approach toward the synthesis of semiconductor-Au nanocomposites. The deposition of noble metals on the surface of semiconductor nanocrystals is usually conducted in nonpolar organic solvent *via* seed-mediated growth at room temperature. In detail, the semiconductor nanocrystals with high quality are firstly prepared in organic solvent at elevated temperature, and simultaneously, the metal precursors are also transferred into nonpolar organic solvent using established methods. The semiconductor organosol and the metal precursor solution in organic solvent are then mixed together, followed by addition of weak reducing agent (usually dodecylamine, DDA). The DDA reduces the metal precursors into metal atoms in the presence of semiconductor nanocrystals, which nucleate and grow on the surface of semiconductor nanocrystals, resulting in formation semiconductor-noble metal nanocomposites. Noble metal precursors usually cannot dissolve in nonpolar organic solvents. However, for gold, there are a number of approaches, e.g. Brust-Schiffrin method [165], ethanol-mediated general phase transfer protocol [71], and dodecyldimethylammonium (DDAB) facilitated strategy developed by Banin and coworkers [199], could be used to transfer its precursors, e.g. commonly used $HAuCl_4$, from aqueous phase to nonpolar organic solvents. In addition, the successful deposition of noble metals on the surface of semiconductor nanocrystals has some specific requirements on the reducing agent. For example, too strong reducing agent, e.g. $NaBH_4$, usually leads to the separated nucleation of noble metal atoms in the solution instead of growth on semiconductor nanocrystals. Therefore, although the ethanol-mediated general phase transfer protocol could make a wide variety of metal precursors dissolve in nonpolar organic solvents [71], only the deposition of gold on the surface of semiconductor nanocrystals gain most sufficient study over the past decade.

Banin and co-workers are pioneers in nanocomposites consisting of semiconductor and gold, and they made a breakthrough in 2004 [199]. They demonstrated a solution synthesis for nanocomposites *via* the selective growth of gold tips on the apexes of hexagonal-phase CdSe nanorods at room temperature (Fig. 3.28). The authors prepared CdSe rods and tetrapods of different dimensions by high-temperature pyrolysis of suitable precursors in a coordinating solvent containing a mixture of trioctylphosphineoxide and phosphonic acids. They dissolved $AuCl_3$ in toluene with the addition of dodecyldimethylammonium bromide (DDAB) and dodecylamine, which serves as reducing agent. Then they mixed this solution at room temperature with a toluene solution of the colloidal-grown CdSe nanorods or tetrapods. After the reaction, the composite products were precipitated by addition of methanol, separated by centrifugation, and re-dispersed in toluene for further characterizations. The novel nanostructures display modified optical properties due to the strong coupling between the gold and semiconductor components. The gold tips show increased conductivity, as well as selective chemical affinity for forming self-assembled chains of rods. The architecture of these nanostructures is qualitatively similar to bifunctional molecules such as dithiols, which provide two-sided chemical connectivity for self-assembly and for electrical devices, and contacting

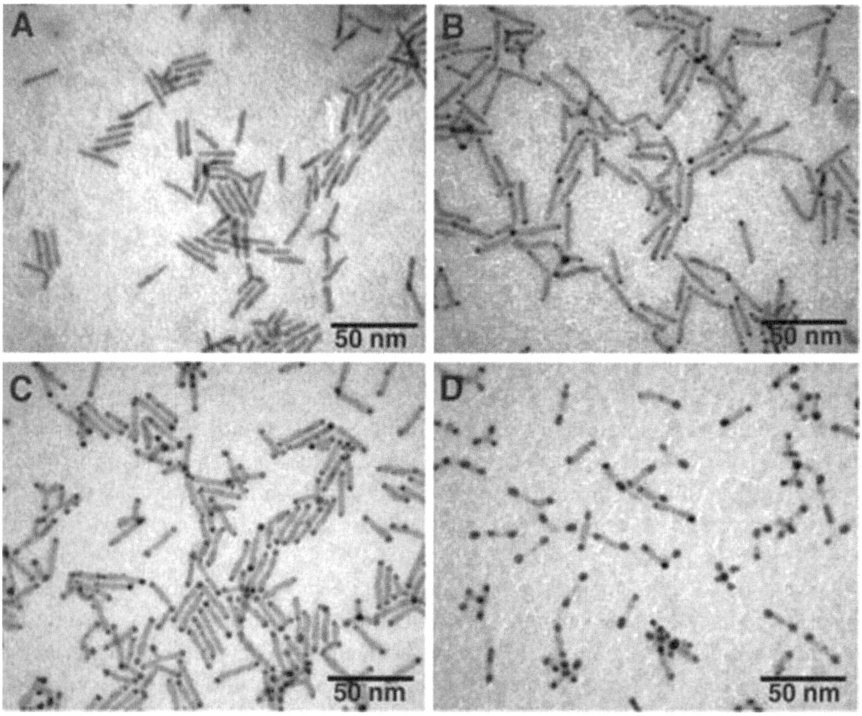

Fig. 3.28 TEM images showing controlled growth of Au onto the tips of CdSe quantum rods. (**a**) Original rod sample, 29 × 4 nm. (**b–d**) Rod samples after Au treatment using gradually increased $AuCl_3$ concentrations; increased Au tip sizes are visible. Reproduced from [199] with permission from the American Association for the Advancement of Science

points for colloidal nanorods and tetrapods. The Banin group later reported the synthesis of asymmetric semiconductor–metal heterostructures whereby gold was grown on one side of CdSe nanocrystalline rods and dots. Theoretical modeling and experimental analysis showed that the one-sided nanocomposites were transformed from the two-sided architectures through a ripening process [235].

Following the breakthrough made by Banin group, various strategies were developed for the synthesis of semiconductor–Au nanocomposites by anisotropic growth of gold on the surface of semiconductor nanocrystals through chemical reduction, physical deposition or photochemistry. The structure displayed in Fig. 3.29 for nanocomposites composed of PbS and gold was reported by Yang and coworkers [238]. They chose PbS prepared in oleylamine as the target semiconductor nanocrystal because it has a typical cubic crystal structure and can easily be produced in controlled shapes and sizes. They dissolved gold precursors into toluene using Brust-Schiffrin method [165], and then mixed them together, followed by addition of dodecylamine served as reducing agent, resulting in formation of PbS-Au nanocomposites. High quality and monodisperse PbS-Au_1, PbS-Au_4 nanostructures and PbS-Au_n nanocubes could be obtained by controlling

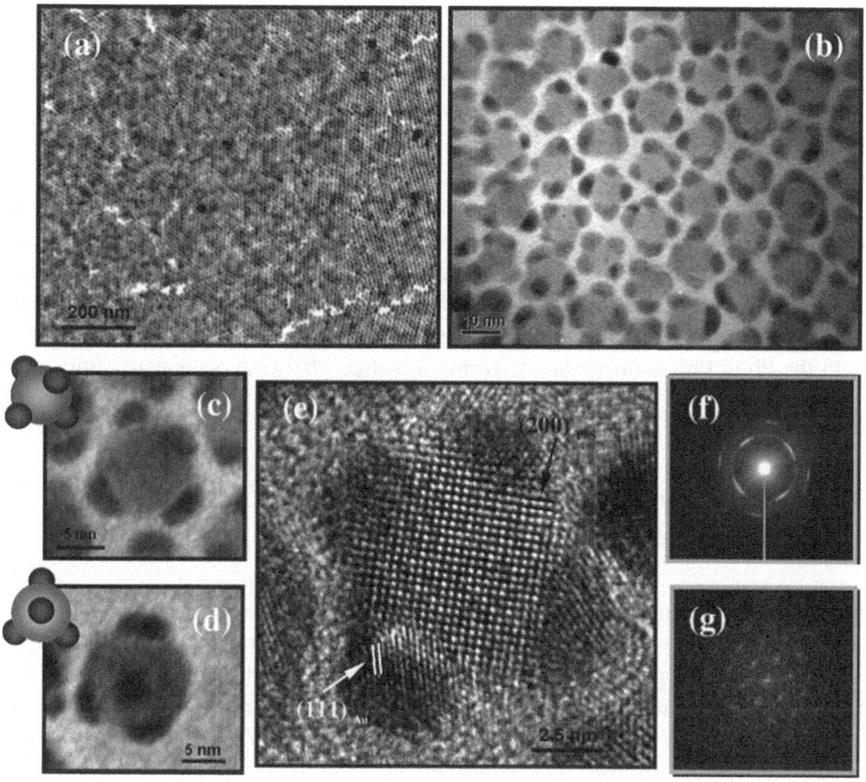

Fig. 3.29 (**a**, **b**) TEM images of the ordered array of PbS-Au_4 heterogeneous nanostructures; (**c**, **d**) TEM images of a typical PbS-Au_4 nanostructure; (**e**) HRTEM image of a typical PbS-Au_4 nanostructure; (**f**) SAED pattern and (**g**) FFT pattern obtained from the ordered array of PbS-Au_4 heterogeneous nanostructures. Reprinted with permission from [238]. Copyright American Chemical Society

the molar ratio of PbS and Au precursors. Their method provides a reasonable method to control the size, number, and distribution of gold nanocrystals on semiconductors, which is important to the design and fabrication of composite nanomaterials for catalysis, optoelectronic devices, and chemical/biological sensors. Owing to their narrow size distribution and intrinsic high symmetry, the as-synthesized PbS-Au_4 nanocomposites could easily self-assemble into hexagonal arrays. Nonlinear optical measurements show that the PbS-Au_4 nanostructure could alter the nonlinear response time of PbS nanocrystals, thereby confirming the applicability of metal-decorated semiconductor nanocrystals in the development of saturable absorber devices.

In a recent study, Yang and co-workers systematically conducted the investigation into the deposition of Au/Ag on the surface of metal sulfide nanocrystals [243]. Upon the characterizations using TEM, HRTEM, STEM, and EDX, they found that both the reaction time and metal/semiconductor precursor ratio have apparent effect on the deposition of Au on the surface of semiconducting metal sulfide nanocrystals. With the increase of reaction time or metal/semiconductor precursor ratio, both the number and the size of the Au domains on each metal sulfide nanocrystal are increased. The deposition of Au on the metal sulfide could be fulfilled in 30 min, suggesting that the surface of semiconducting metal sulfide nanocrystal has catalytic property for the reduction of gold. The deposition mode of Au on metal sulfide nanocrystals might be related to the structural similarity of these two different type of nanoparticles. With the increase of the structural similarity, the number of the sites on the surface of metal sulfide nanocrystals for Au deposition is increased. Specifically, for the PbS nanocrystals, Au nucleates and grows at multiple sites on the surface of PbS nanocrystals, while Ag exchanged with the Pb in PbS nanocrystals to form core-shell PbS@Ag_2S nanostructures.

3.6.2 Nanocomposites Consisting of Semiconductors and Other Noble Metals

Motivated by the unique property and great potentials in wide variety of applications, the syntheses of nanocomposites consisting of semiconductor and noble metals other than gold have also received significant attention in recent years [203, 204, 244]. Actually, early in 80s of last century, researchers had loaded platinum (Pt) or Rhodium (Rh) on the surface of CdS or ZnS semiconductors for producing hydrogen [245–247]. However, the overall size of the semiconductor substrates in those reports have exceeded the scope of nanotechnology.

The Alivisatos group conducted a pioneer study on the photo-deposition of Pt on the surface of CdS and CdSe/CdS nanocrystals [244]. Different from the early aqueous-based deposition of Pt on the surface of semiconductors using photoreduction approach, they performed the deposition reaction in organic phase, which includes CdS nanorods, a organic soluble Pt precursor—(1,5-cyclooctadiene) dimethylplatinum (II) ($(CH_3)_2PtCOD$), and excess of a tertiary amine, which is used as hole-scavenging gent. An excitation wavelength of 458 nm was chosen by the researchers to ensure that CdS is the only light-absorbing component, thus preventing homogeneous nucleation of platinum. After irradiating the reaction mixture for a while under an inert atmosphere, the solution turns from translucent yellow to translucent brown, and the fluorescence of CdS is quenched. The TEM images before and after irradiation show the formation of heterostructures consisting of small nanoparticles positioned along the length of the CdS nanorods (Fig. 3.30). The average diameters of the nanoparticles range from 1.5 to 2.7 nm, depending on the reaction conditions. The results obtained by the Alivisatos group is in contrast to CdSe-Au heterostructures synthesized by thermal methods, where the gold deposition often occurs preferentially on the nanorod ends [199, 235]. The authors also found that the photodeposition of Pt on CdS shows a strong dependence on the nature of the amine used. Highest yields are observed with bulky

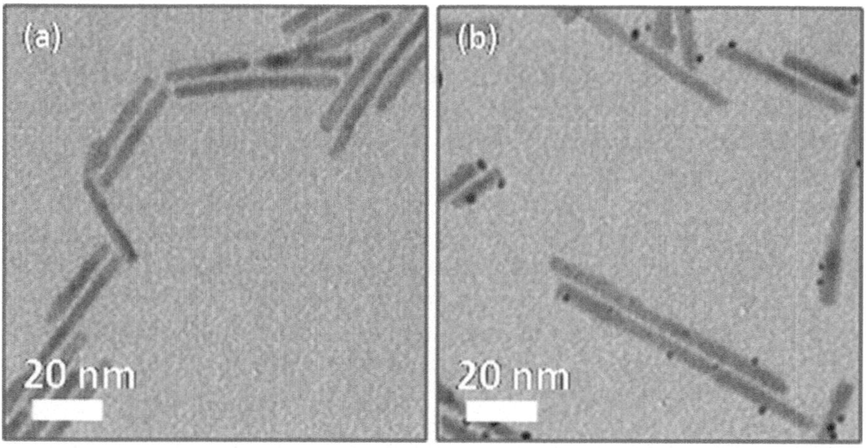

Fig. 3.30 Photodeposition of Pt on CdS nanorods. (**a**) TEM images of CdS nanorods before irradiation and (**b**) of the photodeposition product. After exposure to light, Pt nanoparticles appear along the length of the nanorods. Reproduced from [244] with permission from John Wiley and Sons

tertiary amines. Comparison of octylamine, dioctylamine, and trioctylamine reveals no photodeposition for the primary amine, very little for the secondary, and the most for trioctylamine.

Subsequently, the Banin group reported the synthesis of rod-like CdSe-Pt nanocomposites and their photocatalytic property toward methylene blue [203]. They produced CdSe nanorods (70 nm × 8 nm) using the previously established method [248]. After synthesis the CdSe nanorods were transferred to an aqueous solution by exchanging the alkylphosphine surface ligands with mercapto-undecanoic acid (MUA) [249]. In brief, the CdSe nanorods are dissolved in chloroform along with MUA and the solution is stirred until the MUA dissolved completely. In a different vial KOH is dissolved in triple distilled water (TDW) to give a pH 14 solution. The stock KOH solution is added to the nanorod/MUA solution, followed by vigorous stirring until a white emulsion is formed. The aqueous and organic phases are separated by centrifuging. After centrifugation the water phase is clear and brown colored, indicating the presence of phase-transferred CdSe nanorods, while the organic phase remains nearly colorless. The aqueous nanorod solution contained excess MUA, which is removed by precipitating the rods with methanol and centrifuging. The nanorods were then re-dissolved in TDW for next utilization.

The deposition of Pt on the surface of CdSe nanorods could be achieved through a facile route. The platinum precursor ($PtCl_4$) is dissolved in water and mixed vigorously with the aqueous CdSe nanorod solution at room temperature for 2 days. A dark brown/black precipitate is formed and collected by centrifugation to yield the CdSe–Pt composite particles. No additional reducing agent is needed during the deposition process. A comparison between TEM images of samples taken before and after the reaction with platinum shows that the mean length of the CdSe nanorods is reduced—from 70 to 55 nm.

The above-mentioned method developed by the Banin group for growth of Pt on the surface of CdSe nanorods is very complex and time consuming. More importantly, although the rod-like CdSe-Pt nanocomposites they prepared have superior visible light photocatalytic activity for the reduction of methylene blue, the nanocomposites are easily aggregated in aqueous phase, and this lead to a significant decrease of the surface area available for the photocatalytic reduction of methylene blue. On the other hand, the authors apply phase transfer to avoid the use of expensive organic metal precursors for producing semiconductor-noble metal nanocomposites in aqueous phase, and hence offer a good strategy to deposit noble metals on the surface of semiconductor nanocrystals.

Mokari and coworkers prepared CdS nanorods in organic solvent, and they did not transfer them into aqueous phase. Instead, they directly deposited Pt on the surface of CdS nanorods using platinum acetylacetonate ($Pt(acac)_2$) as metal precursor [204]. The method they used to synthesize CdS nanorods is very simple: A mixture of CdO, trioctylphosphine oxide, and tetradecylphosphonic acid is heated at 300 °C for 15 min. The temperature is then raised to 320 °C and a solution of S dissolved in trioctylphosphine is injected before decreasing the temperature to 300 °C. The reaction mixture is heated for 45 min, cooled, and then precipitated with methanol and separated by centrifugation.

For the deposition of Pt metal, oleic acid, oleylamine, and 1,2-hexadecanediol are heated in diphenyl ether at 80 °C under vacuum for 30 min to remove traces of water. The Pt acetylacetonate is added to a suspension of CdS rods in dichlorobenzene and heated at 65 °C for 10 min to promote dissolution of the Pt precursor. The mixture of surfactants and diphenyl ether is purged with nitrogen and heated to 200 °C before injecting the Pt precursor and semiconductor rods. After several minutes, the reaction is removed from heat and quenched in a water bath. The CdSe-Pt composite product is washed twice by precipitation in ethanol followed by centrifugation, and then separated twice by centrifugation.

Analogous to the deposition of Au on the CdSe nanorods, as reported by Banin and coworkes [199], upon the concentration of Pt precursor, Pt is also selectively grown at one or two tips of the CdS nanorods since the reactivity of the nanorods is higher at the tips than along the body of the rod due to the increased surface energy, as shown by Fig. 3.31. The method developed by Mokari and coworkers could be easily extended to deposit bimetallic PtNi or PtCo on the tips of CdS nanorods, and these heterostructures might be of special interest for a variety of applications including photocatalysis, water splitting, and magnetic applications.

3.6.3 Nanocomposites Consisting of Ag_2S and Noble Metals

The first example to apply the semiconductor-noble metal nanocomposites for catalyzing the reactions (MOR and ORR) in DMFCs is the work reported by

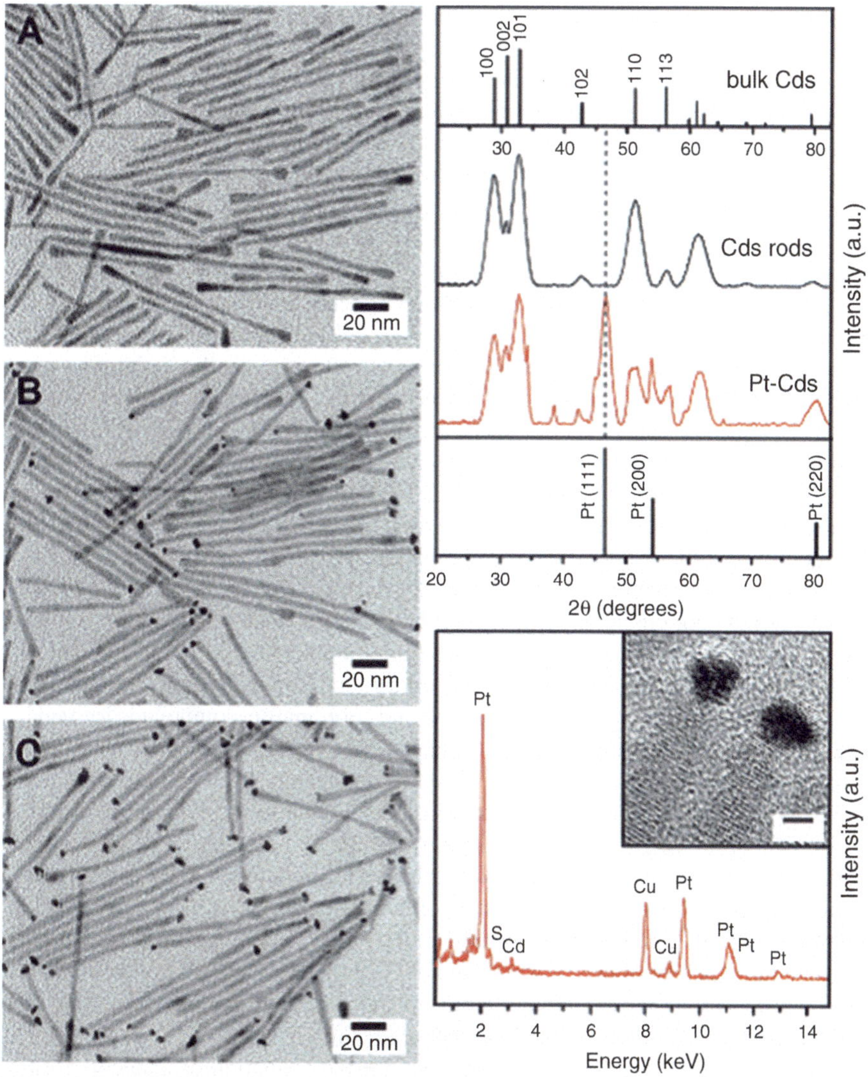

Fig. 3.31 Selective growth of Pt nanoparticles with different sizes on CdS nanorods, (**a**) CdS rods (120 nm × 4 nm), (**b**) CdS with small single Pt tips (4.3 nm), (**c**) CdS with larger double Pt tips (5.7 nm), (**d**) XRD patterns of CdS rods and Pt-CdS hybrid structures with corresponding CdS and Pt bulk patterns (stick patterns shown above and below, respectively), and (**e**) selected area EDS spectrum of a single Pt tip, with inset HRTEM image of two Pt-CdS hybrids. Reproduced with permission from [204] Copyright American Chemical Society

Yang and Ying in 2011. They demonstrated an aqueous route for the synthesis of nanocomposites consisting of Ag_2S and different noble metals [213]. By reducing various noble metal precursors using citrate in aqueous phase in the presence of preformed Ag_2S nanocrystals, uniform semiconductor-noble metal heterogeneous nanostructures are obtained as the dominant product. In addition to binary nanocomposites, ternary and quaternary hybrid systems are also achieved *via* the successive deposition of different noble metals on the surface of Ag_2S nanocrystals. In particular, the Pt-containing nanocomposites are found to exhibit superior catalytic activity for MOR, the key reaction in direct methanol fuel cell (DMFC), due to the electronic coupling effect between the ultrafine Pt crystallites and the semiconductor domains.

A large number of literatures have been devoted to the synthesis of Ag_2S nanocrystals in organic media [250–256]. For example, the Zhao group reported an one-step route to control the organization of semiconductor Ag_2S nanocrystals [253]. In a typical synthetic process, they mixed aqueous $AgNO_3$ solution and toluene solution of dodecylthiol to form a yellowy micellar solution under inHowever, the tensile effect of Ag ively stirring at room temperature. After 30 min, aqueous Na_2S solution was dropped into the microemulsion for the formation of Ag_2S nanocrystals, which were usually self-assemble into hexagonal, quasi-orthogonal, and chain-like arrays on the TEM grid. There are also a few of studies reported on the synthesis of Ag_2S nanocrystals in aqueous phase, either use protein [257, 258], or polymeric membrane [259] to passivate the surface of Ag_2S nanocrystals for preventing them from aggregation in solution. However, their self-assembly and poor quality in term of morphology, size, and size distribution enable them not suitable substrates for the deposition of noble metals.

Alternatively, Yang and Ying developed a room-temperature method based on the reaction between sodium sulfide (Na_2S) and coordinating compounds formed by Ag^+ ions and BSPP to derive aqueous dispersible Ag_2S nanocrystals [213]. Typically, 600 mg of BSPP were added to 300 mL of aqueous $AgNO_3$ solution (1 mM) in a 1000-mL beaker. The mixture was stirred for 1 h to form BSPP-Ag^+ complexes [66, 121], followed by the prompt addition of 10 mL of aqueous Na_2S solution (50 mM), which results in a series of color changes before finally arriving at a brown Ag_2S hydrosol. A TEM image of the as-prepared Ag_2S nanocrystals is shown in Fig. 3.32a. The nanocrystals are spherical, nearly monodispersed, and have an average size of ca. 7.2 nm. The HRTEM image (Fig. 3.32b) illustrates the lattice planes in these nanocrystals, showing an interplanar spacing of ca. 0.26 nm, which corresponds to the $(\bar{1}21)$ planes of monoclinic Ag_2S.

The Ag_2S nanocrystals are used as seeds for the formation of nanocomposites with different metals. With sodium citrate as reducing agent, under the experimental conditions (aqueous phase and elevated temperature), noble metals nucleated preferentially on the existing Ag_2S nanocrystals, rather than homogeneously. The TEM images of binary Ag_2S-Au, Ag_2S-Pt, Ag_2S-Os, and Ag_2S-Pd nanocomposites are displayed in Fig. 3.33a, b, c, and d, respectively. The deposition of noble metals on the Ag_2S nanocrystals could be clearly identified *via* brightness contrast, and confirmed by the EDX analysis of an arbitrary single particle with HAADF-STEM [213].

The final morphology of the nanocomposites depended on whether the surface of substrate particles allows for only a single nucleation site or multiple ones. Unlike

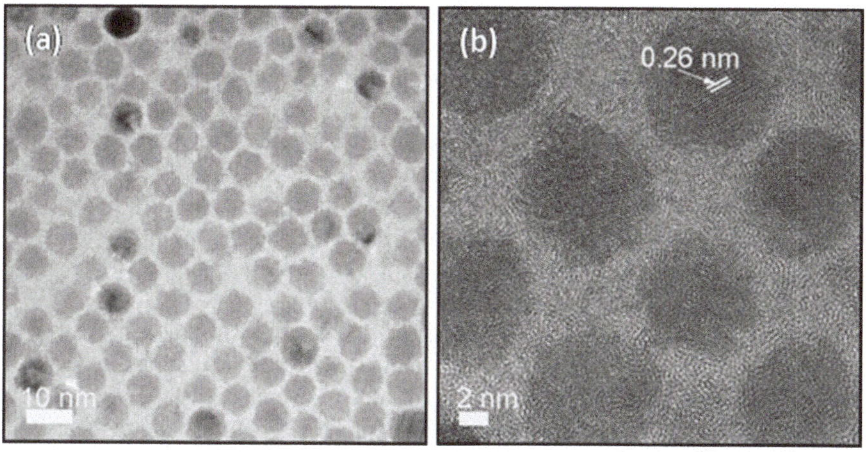

Fig. 3.32 (**a**) TEM and (**b**) HRTEM images of the as-prepared Ag_2S nanocrystals. $\bar{d} = 7.2$ nm, $\sigma = 0.6$ nm, $\bar{\sigma} = 8.4\,\%$. σ and $\bar{\sigma}$ are the standard derivation and relative standard derivation, respectively. Reproduced from [213] with permission from John Wiley and Sons

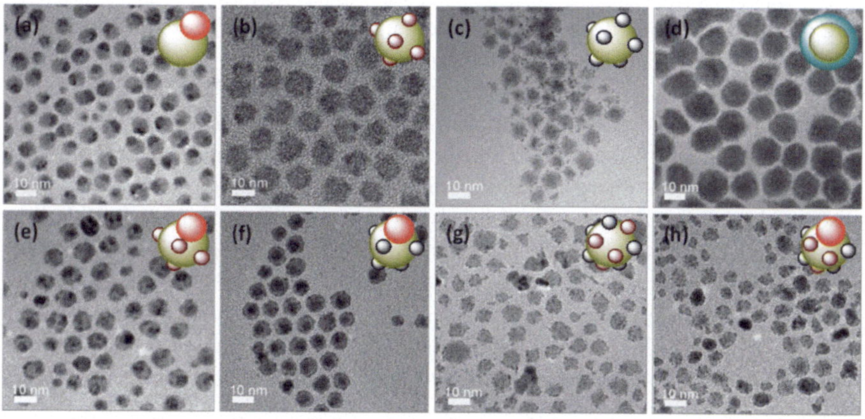

Fig. 3.33 TEM images of binary Ag_2S-Au (**a**), Ag_2S-Pt (**b**), Ag_2S-Os (**c**), Ag_2S-Pd (**d**), multiple Ag_2S-Au-Pt (**e**), Ag_2S-Au-Os (**f**), Ag_2S-Pt-Os (**g**), and Ag_2S-Au-Pt-Os nanocomposites (**h**). Reproduced from [213] with permission from John Wiley and Sons

the face-centered cubic (fcc) or hexagonal materials, monoclinic Ag_2S has very complicated crystal structures. It has many facets with different lattice spacing, which provides favorable sites to match the lattice planes of various noble metals for their epitaxial growth on the substrate seeds. The experimental results show that only a single site on the Ag_2S nanocrystal surface is suitable for the nucleation of gold clusters, but multiple sites existed over the surface of Ag_2S seed for Pt and Os noble metals. An extreme case is observed in the Ag_2S-Pd system (Fig. 3.33d). Numerous sites are provided for the nucleation of Pd nanoparticles, which could grow and eventually coalesce to form a continuous shell on each Ag_2S nanocrystal.

The species-dependent features of noble metal deposition on Ag_2S nanocrystal are further employed to derive multiple semiconductor-metal nanocomposites. The complex semiconductor-noble metal nanocomposites including Ag_2S-Au-Pt, Ag_2S-Au-Os, Ag_2S-Pt-Os, and Ag_2S-Au-Pt-Os, have been prepared by successive reduction of noble metal precursors using citrate in the presence of preformed Ag_2S nanocrystals, as characterized by TEM images in Fig. 3.33e, f, g, h. The growth of more than one type of metal on each Ag_2S nanocrystal demonstrates that the nucleation sites on the surface of Ag_2S nanocrystal are specific to different metals, providing for the possibility of creating complex nanocomposites between Ag_2S and multiple metals *via* a simple and flexible route.

The Pt-containing Ag_2S-noble metal nanocomposites are attractive for catalytic application due to their relative small Pt domain size in the nanocomposite systems. The smaller Pt crystallite size usually translates to a higher surface area, which would be beneficial to the catalytic reaction. The electrochemically active surface area (ECSA) of Pt in Pt-containing Ag_2S-noble metal nanocomposites could be determined using cyclic voltammetry. Although the contact interfaces between Pt and Ag_2S in the nanocomposites result in some blockage of the surface area of the Pt domains, the average ECSA of the Pt in Pt-containing Ag_2S-metal nanocomposites is 84.5 $m^2\ g^{-1}$, which is 16 % higher than that of the commercial Pt/C (72.9 $m^2\ g^{-1}$) due to the smaller Pt domain size of the former (ca. 1 nm) [213].

The other important feature of the nanocomposites is the electronic coupling between the metal and semiconductor domains. The XPS analyses of the Pt-containing nanocomposites demonstrate that electrons are transferred to Pt from other domains of the nanocomposites due to the intra-particle charge transfer (see Fig. 3.34 for the energy level diagram). Analogous charge transfer has been observed in the Au@PbS system, whereby the electrons transfer from PbS shell to the inner Au core resulted in the n-type to p-type change in hydrazine-treated PbS [260]. The electron transfer from Ag_2S to Pt could also be described with the generation of a hole in the Ag_2S domain. In the presence of Au domain (work function = 5.1 eV) [261], the alignment of energy levels in Au and Ag_2S would be favorable for electron transfer from Au to Ag_2S to fill the hole generated by the

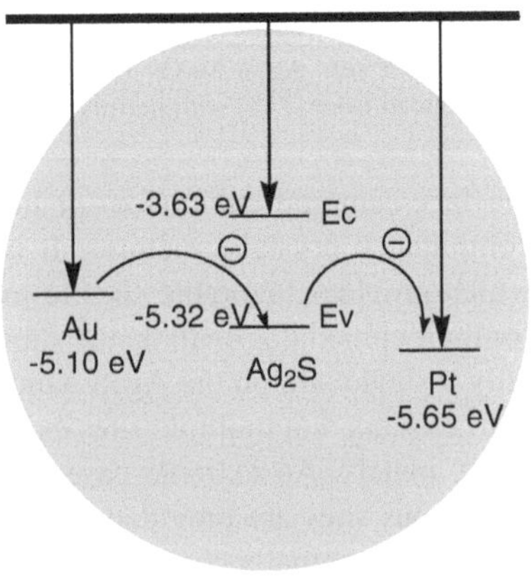

Fig. 3.34 Energy level diagram for Ag_2S-noble metal nanocomposites predicts intraparticle charge transfer among different domains. Reproduced from [213] with the permission from Wiley-VCH

electron transfer to Pt domains, further promoting the electron transfer from Ag_2S to Pt in order for the Fermi levels to match at the interface. The electron transfer from other domains to Pt in Pt-containing nanocomposites increases the electron density around the Pt sites, causing the weakening of CO chemisorption and hence the increase their electrochemical activity for MOR, a key reaction in direct methanol fuel cell [213].

On the other hand, the Pt-containing Ag_2S-noble metal nanocomposites exhibit poor activity for ORR, another key reaction in DMFC [9, 11, 262, 263]. This is not surprising based on a report by Watanabe and co-workers [264], which discussed alloying of Fe, Ni or Co with Pt to enhance the electro-catalytic activity for oxygen reduction. Fe, Ni or Co has more 5d vacancies than Pt and could withdraw electrons from the latter. This withdrawing effect induced an increase in 5d vacancies in Pt, increasing 2π electron donation from O_2 to the Pt surface, and resulting in enhanced O_2 adsorption to favor oxygen reduction. However, for the Pt-containing Ag_2S-noble metal nanocomposites, the electron donation from the semiconductor to the Pt domains would decrease the 5d vacancies in Pt. Thus, the adsorption of O_2 on the nanocomposites would be too weak for the O_2 dissociation reaction [62, 63, 197, 198]. This would account for the poor activity of the Pt-containing nanocomposites towards oxygen reduction.

3.6.4 Cadmium Selenide-Platinum Nanocomposites with a Core-Shell Construction

The role of surface strain in catalysis, particularly as a means of tuning the catalytic activity, has attracted great interest in recent years [265–267]. When a metal is deposited on a substrate with different lattice parameters, a compressive or tensile strain in the surface of the metal layer usually occurs to fulfill the requirement of epitaxial growth, and often affects the overlap of the electron orbitals between the metal atoms and therefore changes the electronic properties of the surface and its reactivity, which will influence the bond strength of an adsorbate [268]. Therefore, the application of compressive strain or tensile stain to a surface can be an effective means to influence the surface reactivity [269]. Strain and the associated shift of the d-band can be brought about by growing the desired metal on other materials with a different lattice constant. The over-layer may thereby be strained or compressed depending on the lattice mismatch between the two materials. The lateral strain has been studied in a number of pseudomorphic metal monolayers formed on electrode substrates [268, 270–273], or core-shell nanoparticles constructed by different metals [274–276]. However, the investigations of lateral strain in semiconductor-metal nanocomposites, a type of nanostructure that combine materials with distinctly different physical and chemical properties, are scarcely reported up to date.

Analogous to the deposition of Pd on Ag_2S nanocrystals [213], the Yang group found that in aqueous media, numerous sites on CdSe nanocrystals are also

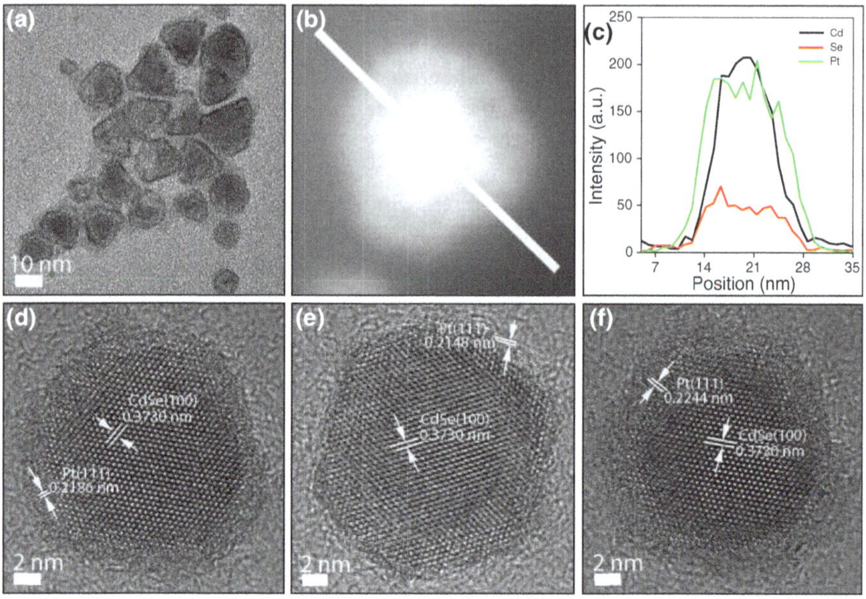

Fig. 3.35 Core-shell CdSe-Pt nanocomposites synthesized in aqueous phase using 10 nm CdSe cores: (**a**) TEM image, (**b**) STEM image, (**c**) element profile, and (**d**) HRTEM image of core-shell CdSe-Pt nanocomposites at CdSe/Pt molar ratio of 1/1; (**e**, **f**) HRTEM images of core-shell CdSe@Pt nanocomposites at CdSe/Pt molar ratio of 2/1 and 1/2, respectively. Reproduced from [214] with permission from the Royal Society of Chemistry

provided for the nucleation of Pt nanoparticles, which then grow and eventually coalesce to form a continuous shell on each CdSe nanocrystal, as displayed by the TEM, HRTEM and STEM in Fig. 3.35 [214]. The thickness of the Pt shell could be controlled by varying the CdSe/Pt molar ratio in the synthesis. Figures 3.35d, e, f illustrate the core-shell CdSe@Pt nanocomposites synthesized at CdSe/Pt molar ratios of 1/1, 2/1, and 1/2, respectively. The thickness of Pt shell could be varied as shown by comparing those HRTEM images.

The XRD analyses of the core-shell CdSe@Pt nanocomposites at different CdSe/Pt molar ratios suggest that the interplanar spacing of Pt is compressed to match the lattice plane of CdSe for its epitaxial growth on the CdSe substrates. The compression of Pt lattice spacing can also be verified by the HRTEM images in Fig. 3.35. The digital analyses of images show that the spacing of Pt(1 1 1) decreases from 0.2244 nm to 0.2148 nm with the decrease of Pt molar ratio in the core-shell nanocomposites. The compressive strain effect of the CdSe core on the deposited Pt shell might be a welcome feature for the core-shell composite nanomaterials in a given catalytic reaction.

In a specific CdSe/Pt molar ratio, e.g. 1/1, the core-shell CdSe@Pt nanocomposites display superior catalytic activity toward oxygen reduction reaction (ORR) [214]. ORR is the key reaction at the cathode of direct methanol fuel cell (DMFC), and the low activity of the cathode catalysts at room temperature is one of the significant challenges need to be resolved before the commercialization of DMFC [62, 155, 277]. In order to overcome this problem, it is necessary to maximize the activity of Pt-based catalysts by engineering their structure, morphology and/or composition.

The observed enhanced catalytic activity for ORR of core-shell CdSe@Pt nanocomposites could be resulted from the reasonably compressive strain effect of the CdSe core on the Pt shell. As discussed in Sect. 3.5, the common ORR process involves both the breaking of an O–O bond and the formation of O–H bonds [9, 62, 63]. The most active Pt-based catalyst should have the d-band center with an intermediate value since the optimal ORR catalyst needs to facilitate both bond-breaking and bond-making steps without hindering one or the other [197, 198]. The surface of a Pt-based catalyst with high value of d-band center tends to bind adsorbents more strongly, thereby enhancing the kinetics of dissociation reactions producing these adsorbents, whereas a surface with lower d-band center value tends to bind adsorbents more weakly and facilitates the formation of bonds amongst them. The compressive stains of CdSe core on Pt shell leading to broader d-band result in a down-shift of the d-band center of the Pt shell. At an appropriate CdSe/Pt molar ratio (1/1) in core-shell nanocomposites, the downward shift of the d-band center sufficiently balances the bond-breaking and bond-making steps of the ORR process, thus offering optimum catalytic activity.

It is worthy to note that the core-shell CdSe@Pt nanocomposites also exhibit excellent catalytic activity toward MOR at CdSe/Pt molar ratio of 1/2 [214]. The enhanced MOR catalytic activity of the core-shell CdSe@Pt nanocomposites could also be attributed to the reasonably compressive strain effect of the CdSe core on the deposited Pt shell, which results in the appropriate shift of the d-band center of Pt shell and improvement of the adsorption of methanol on the surface of Pt catalysts. Interestingly, the CdSe/Pt molar ratio for core-shell CdSe@Pt nanocomposites offering highest activity for MOR is different from that for ORR. This is a welcome feature for these core-shell nanocomposites to be used in DMFC. Comparing with the DMFC using Pt/C catalysts both at anode and cathode, the inhibition of methanol crossover in certain level could be expected when the core-shell nanocomposites with optimal CdSe/Pt molar ratio for ORR are used at cathode. This is important for the commercialization of DMFC since crossover of methanol from the anode to the cathode through the polymer electrolyte membrane (PEM) interferes with oxygen reduction at the cathode, resulting in the creation of a mixed potential and drastically decrease in DMFC performance [122–124].

3.6.5 Nanocomposites of Silver Sulfide and Noble Metals with Controlled Nanostructure

The interest in nanostructured composites is the possibility of integrating materials with vastly different physical and chemical properties into a hybrid nanosystem with multiplexing capability and tunable physical and chemical properties that may not be obtainable otherwise [278, 279]. The interactions between the nanoscale metal and semiconductor components, if synergistic, can lead to significant improvements of application performance. On the other hand, noble metal nanoparticles with controlled internal structures have been often used to modify the noble metal properties in a diverse range of applications, such as catalysis [280–282], nanoreactors [75], and drug delivery systems [76–78]. For example, Hyeon and co-workers reported good catalytic activity for Pd nanoparticles with a hollow interior in Suzuki cross-coupling reactions. The hollow Pd nanoparticle catalyst could be reused seven times without the loss of catalytic activity [280].

Hence, the integration of semiconductor and noble metal nanoparticles into a nanosystem with controlled architecture may lead to further improvements of the application performance of noble metal-semiconductor hybrid materials. Upon the inside-out diffusion of Ag in core-shell nanoparticles [119], the Yang group developed a general approach to the fabrication of silver sulfide and noble metal nanoparticles with an overall hollow or cage-bell structure [283]. In their strategy, core-shell nanoparticles with Ag residing in the core or inner shell region are first prepared as starting templates. Sulfur is then used to promote the inside-out diffusion of Ag from the core or the inner shell region of the core-shell nanoparticles, and to convert the diffused Ag into Ag_2S. The overall result is the conversion of the Ag-containing core-shell nanoparticles into silver sulfide-noble metal nanoparticles with hollow or cage-bell structures.

The inside-out diffusion of Ag in core-shell nanoparticles has been developed into a general protocol for the fabrication of noble metal nanoparticles with a hollow or cage-bell structure [119]. The protocol begins with the synthesis of core-shell Ag-M or core-shell-shell M_A-Ag-M_B nanoparticles in an organic solvent. Ag is then removed from the core or from the inner shell by BSPP, which binds strongly with Ag^+ ions to promote the inside-out diffusion process and to allow the complete removal of Ag in 24~48 h, leaving behind an organosol of hollow or cage-bell structured metal nanomaterials.

Subsequently, Yang and co-workers attempted to find a more cost-effective alternative to produce hollow/cage-bell structured metal nanomaterials, and they employed elemental sulfur (S) to replace expensive BSPP to remove the Ag component from the core-shell particles. However, they experimentally found that the inside-out diffusion of Ag in core-shell nanoparticles could be further developed as a general protocol to fabricate nanocomposites of Ag_2S and metal nanoparticles with a hollow or cage-bell structure [283]. As illustrated by the scheme in Fig. 3.36, after mixing core-shell nanoparticles and sulfur in toluene, the Ag^+ ions released by the O_2/Cl^- etching of the twinned Ag seeds diffuse out through the discontinuous metal shell, and react with sulfur to form Ag_2S nanocrystals decorated on the outer shell of the metal nanoparticles (Fig. 3.36a). The Ag_2S nanocrystals grow with the continued outward diffusion of Ag^+ ions until the Ag core is completely removed, leaving behind a colloidal solution of nanocomposites consisting of Ag_2S and metal nanoparticles with a hollow interior (Fig. 3.36b). Finally, the Ag_2S nanocrystals on the surface of the metal shell

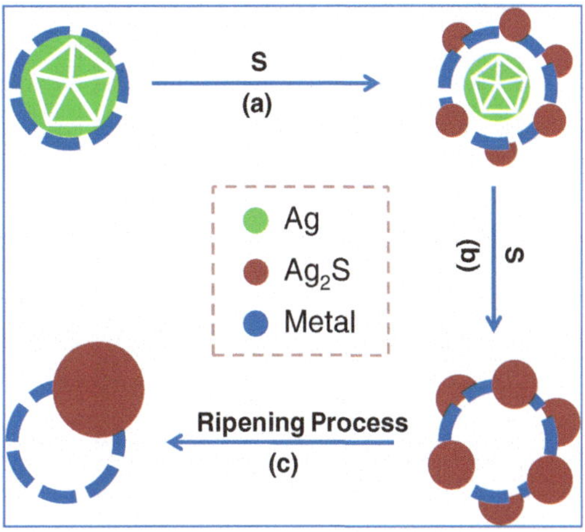

Fig. 3.36 A schematic illustration showing the formation of nanocomposites consisting of Ag_2S and metal nanoparticles with a hollow interior based on the inside-out diffusion of Ag^+ in Ag-containing core-shell nanoparticles. Reproduced from [283] with permission from the Royal Society of Chemistry

underwent a ripening process, e.g. Ostwald ripening or electrochemical ripening [181, 235], to form larger and more stable domains on the surface of the metal shell (Fig. 3.36c).

With a slight modification, the protocol above could be used to prepare nanocomposites of Ag_2S and noble metal nanoparticles with a cage-bell structure. As a typical example, the core-shell-shell Au-Ag-Pt nanoparticles are prepared first, and after treatment with elemental sulfur, the internal Ag shell is removed from the core-shell-shell Au-Ag-Pt nanoparticles by the aforementioned inside-out diffusion of Ag^+ ions. Reaction of Ag^+ ions with sulfur results in the formation of Ag_2S patches on the surface of the now cage-bell Au-Pt nanoparticles, finalizing the fabrication of nanocomposites of Ag_2S and Au-Pt nanoparticles with a cage-bell structure. The morphologies of the starting core-shell-shell Au-Ag-Pt nanoparticles and final nanocomposite were shown by the TEM images in Fig. 3.37a, b. The cage-bell structure of the metallic domains is revealed by the appearance of void space between the core and the outer shell regions. The void space is formed by the removal of the Ag inner shell by elemental sulfur. The Ag_2S patches on the surface of cage-bell metal domains display a stronger contrast than the void space, as can be seen in the TEM image in Fig. 3.37b. The HRTEM image in Fig. 3.37c of a single composite Ag_2S-cage-bell Au-Pt nanoparticle shows three different sets of lattice fringes with separations of 0.26 nm in the light patches, 0.24 nm in the core, and 0.23 nm in the shell regions. These fringes correspond well with the $(\bar{1}21)$ planes of monoclinic Ag_2S, and (1 1 1) planes of face-centered cubic Au and Pt, respectively. The TEM characterizations confirm that the size and morphology of the remaining metallic domains in the nanocomposites are virtually unchanged after treatment with sulfur, suggesting that the diffusion of the Ag from the inner shell of the core-shell-shell nanoparticles does not cause the collapse of the particle geometry.

The strategy for the fabrication of nanocomposites of Ag_2S and hollow structured noble metal nanoparticles could be further tuned to produce ternary nanocomposites consisting of Ag_2S, Au, and Pt nanoparticles with hollow interiors,

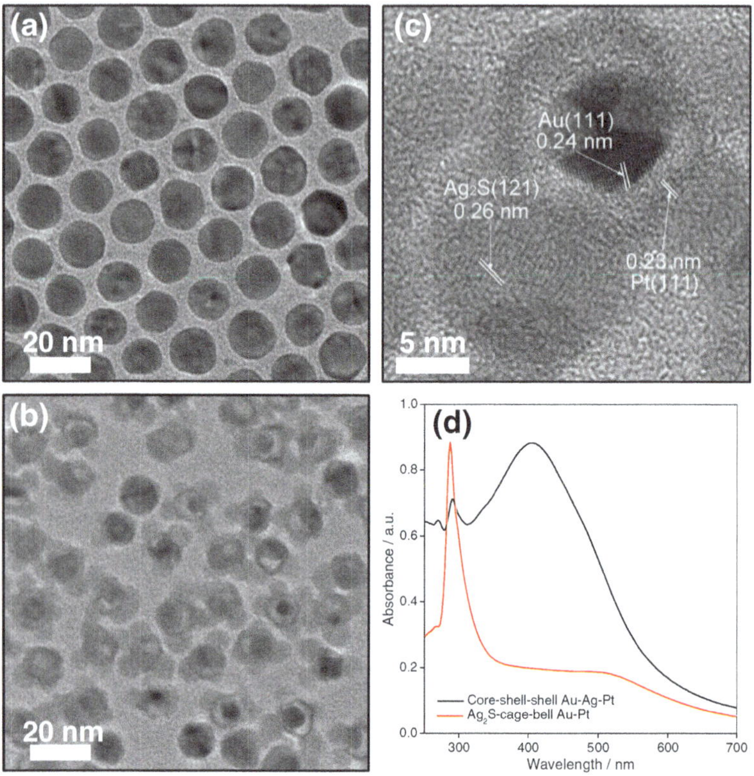

Fig. 3.37 TEM images of core-shell-shell Au-Ag-Pt (**a**), Ag_2S-cage-bell Au-Pt (**b**); HRTEM images of nanocomposites of Ag_2S-cage-bell Au-Pt (**c**); UV-visible spectra of core-shell-shell Au-Ag-Pt nanoparticles before and after treatment with element sulfur (**d**). Reproduced from [283] with permission from the Royal Society of Chemistry

labeled as Ag_2S-Au-hPt [284]. In brief, the gold precursors ($HAuCl_4$) are firstly transferred from aqueous to toluene using the ethanol-mediated transfer protocol [71], and then mixed with Ag_2S-hollow Pt (Ag_2S-hPt) nanocomposite organosol in toluene. Ternary Ag_2S-Au-hPt nanocomposites are found as the dominant product after aging the mixture of Ag_2S-hPt hetero-dimers and $HAuCl_4$ in toluene for 2 h, as indicated by the TEM and HRTEM images in Fig. 3.38. Isolated Au nanoparticles are not observed, indicating that Au nucleates preferentially on the existing Ag_2S-hPt heterodimers under the experimental conditions. In most cases, Au is deposited only at a single site on the Ag_2S domain in each Ag_2S-hPt hetero-dimers. The average diameter of the deposited gold patches is ca. 8.3 nm, which could be discernible by the strong brightness contrast in TEM and HRTEM images.

The direct evidences for the formation of ternary Ag_2S-Au-hPt nanocomposites are provided by the line scanning analysis and elemental mapping of an arbitrary single composite nanoparticle (Fig. 3.38c) in the high-angle annular dark-field STEM mode. As shown in Fig. 3.38d for the line-scanning analysis, the Au and

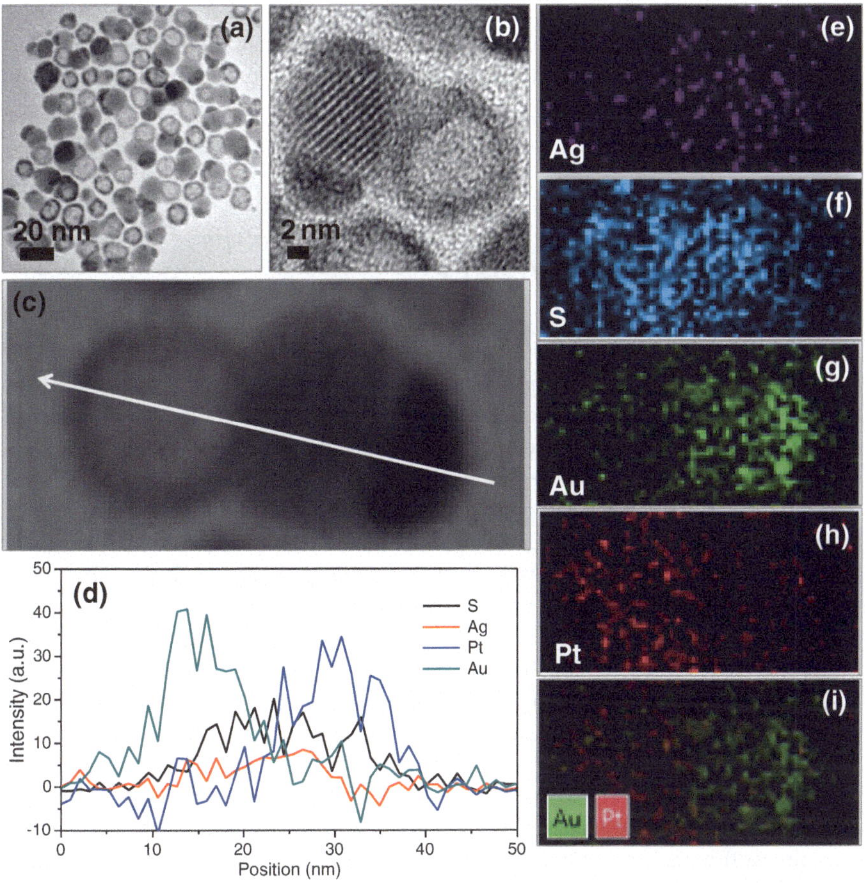

Fig. 3.38 TEM image (**a**), HRTEM image (**b**), STEM image (**c**), line-scan analysis (**d**), and elemental mapping (**e–i**) of a single particle (**c**) of the ternary Ag_2S-Au-hPt nanocomposites as-prepared by depositing Au on the Ag_2S-hPt heterodimers in toluene at room temperature. Reproduced from [284] with the permission from the Nature Publishing Group

Pt signals are present at left- and right-hand side, respectively, whereas the Ag and S signals are only concentrated at the core region. The line-scanning analysis is well in accord with the nanoscale mapping results (Fig. 3.38e–i), which also manifest that the Au and Pt components are respectively distributed on the two sides of the ternary nanocomposites. Further, HRTEM image revealed that the crystal planes of Au are not parallel to those of Ag_2S in each heterodimerNafion membrane nanoparticle (Fig. 3.38b), indicating that the growth of Au on the surface of Ag_2S domain takes place in different orientations.

Analogous to the Ag_2S-Au-Pt ternary nanocomposites discussed previously [213], the ternary Ag_2S-Au-hPt nanocomposites also exhibit enhanced electrocatalytic property toward MOR due to the strong electronic coupling between Pt and other domains in the hybrid particles [284]. The study by Yang and co-workers offers a vivid example to exhibit the enhancement of the material properties by means of a structural tailoring, and the concept might be used toward the design and synthesis of other hetero-nanostructures for catalytic applications other than methanol oxidation.

3.6.6 Pt-Based Selective Catalysts for a Propotype of Membraneless DMFCs

In current DMFCs, the anode and cathode electrocatalysts are commonly based on Pt [3–6, 285, 286]. These catalysts are not selective to MOR at anode or ORR at cathode, and hence any methanol crossover from the anode to the cathode through the proton exchange membrane can be oxidized by the cathode catalyst. This results in the creation of a mixed potential at the cathode, which degrades the fuel cell performance [8, 287, 288]. Although a number of efforts have been devoted toward the modification on the proton exchange membrane to overcome this key obstacle for the commercialization of DMFCs, it is generally thought that the commonly used Nafion membrane has an unacceptably high rate of methanol crossover [289–295]. In this sense, synthesis of electrocatalysts with high selectivity for MOR and ORR represents an alternative for solving this problem in DMFCs.

Upon the deep understanding of the mechanisms of the electrocatalytic reactions, the Yang group reported the design and fabrication of Pt-based nanomaterials with enhanced catalytic activity and high selectivity towards DMFC reactions by sufficiently making use of the structural uniqueness and electronic coupling effects among the different domains of the electrocatalysts so that the DMFCs can be operated well without or with little dependence on the proton exchange membrane [296]. In particular, the ternary Au@Ag_2S-Pt nanocomposites display superior MOR selectivity due to the electronic coupling effect among different domains of the nanocomposites (Fig. 3.39a), while the cage-bell structured Pt-Ru nanoparticles exhibit excellent methanol tolerance for ORR at the cathode because of the differential diffusion of methanol and oxygen in the porous Ru shell of the cage-bell nanoparticles (Fig. 3.39b). The good catalytic selectivity of these Pt-based nanomaterials *via* structural construction enables a DMFC to be built without a proton exchange membrane between the fuel electrode and the oxygen electrode (Fig. 3.39c).

The performance of the membraneless DMFC has been evaluated in term of the open circuit voltage (OCV), the current-voltage (*I–V*), and the current-power (*I–P*) characteristics under ambient conditions [296]. The membraneless DMFC with the selective MOR catalyst at the anode and ORR catalyst at the cathode could maintain an open-circuit voltage of ca. 0.38 V for more than 120 min. It is of great importance that reasonable power with the maximum value of ca.15 μW is obtained without a separate membrane. The OCV of the membraneless DMFC is still rather low, at only ca. 32 % of the theoretical cell voltage of DMFC (ca. 1.18 V)

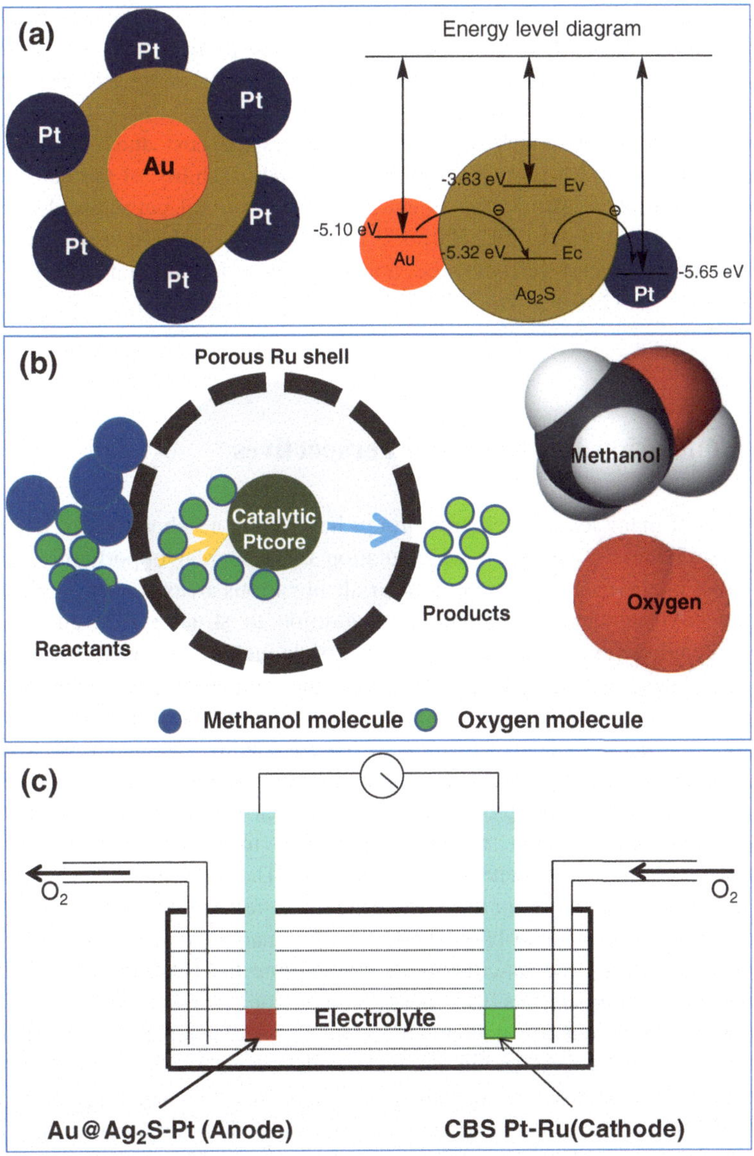

Fig. 3.39 Schematic illustration of the construction of ternary Au@Ag_2S-Pt nanocomposites and the energy level alignment in Ag_2S-Au-Pt nanocomposites (**a**), schematic illustration of the differential diffusion and reaction of methanol and oxygen in CBS Pt-Ru nanoparticles (**b**), and schematic illustration for the membraneless DMFC (**c**). Reproduced from [296] with the permission from the Nature Publishing Group

[297]. The performance of the prototype could be limited by the use of dissolved oxygen as the source of oxygen. Despite the limited performance, the prototype does demonstrate the viability of using selective MOR and ORR catalysts to construct a DMFC without the proton exchange membrane. Since the exemption of a proton exchange membrane not only reduces the cost and volume of the DMFC, it also provides more flexibility and miniaturizability of the cell design. Its implementation is also simpler than membraneless fuel cells that depend on non-mixing laminar flows [298, 299] or 3D anode [300–302]. The performance of the membraneless fuel cell may be improved by using the on-chip fuel cell developed by the Osaka group, which can directly make use of the oxygen from the air [303].

3.7 Conclusion Remarks and Perspectives

The efforts of many leading research groups have led to a rich variety of heterogeneous nanomaterials, and their accumulation creates great opportunities also a tremendous challenge to apply these materials in various areas, e.g. energy conversion and environmental remediation. In addition to structural advantages, the welcome features for a heterogeneous nanoparticle include the electronic coupling effect and lateral strain effect among different domains of the heterogeneous materials. Future research challenges in heterogeneous nanomaterials may include: (1) develop general strategies to prepare heterogeneous nanomaterials with the control on the size and morphology of each domain in the heterogeneous particles; (2) to understand the synergetic effect and interface boundary sites in heterogeneous nanosystems, which are of importance to enhance the properties, e.g. catalysis, of the heterogeneous nanomaterials. The mechanistic understanding of the underlying chemistry for the heterogeneous systems might be valuable for the development of more metal-based heterogeneous nanomaterials with interesting architectures and tailored functionalities; (3) explore the catalytic activity of the heterogeneous nanomaterials for energy conversion. The as-synthesized noble metal-based heterogeneous nanomaterials may possess superior properties for high efficient energy conversion due to its ultrafine size, high stability, ideal morphology, and welcome energy level alignment, some of which are difficult or impossible to achieve by commercial catalysts. The application of these nanomaterials toward DMFC and photocatalysis would be hot topics accordingly. The specific goals include the evaluation of the heterogeneous nanomaterials as catalysts for energy conversion, the devices designed with improved performance, efficiency, durability and reduced costs, and the realization of the commercialization of the catalysts as a long-term objective; (4) Explore other scientific-related issues. Many interesting scientific findings might be derived from the synthesis and characterization of the heterogeneous nanomaterials, which would not only satisfy everlasting human curiosity, but also promise new advances in nanoscience and nanotechnology.

Acknowledgements Financial support from the 100 Talents Program of the Chinese Academy of Sciences, National Natural Science Foundation of China (Grant Nos: 21173226, 21376247, and 21476246), and State Key Laboratory of Multiphase Complex Systems, Institute of Process Engineering, Chinese Academy of Sciences (MPCS-2012-A-11) is gratefully acknowledged.

References

1. Antolini E (2003) Formation of carbon-supported PtM alloys for low temperature fuel cells: a review. Mater Chem Phys 78:563–573
2. Liu H, Song C, Zhang L, Zhang J, Wang H, Wilkinson DP (2006) A review of anode catalysis in the direct methanol fuel cell. J Power Sources 155:95–110
3. Steele BCH, Heinzel A (2001) Materials for fuel-cell technologies. Nature 414:345–352
4. Perry ML, Fuller TF (2002) A historical perspective of fuel cell technology in the 20th century. J Electrochem Soc 149:S59–S67
5. Chen J, Lim B, Lee EP, Xia Y (2009) Shape controlled synthesis of platinum nanocrystals for catalytic and electrocatalytic applications. Nano Today 4:81–95
6. Peng Z, Yang H (2009) Designer platinum nanoparticles: control of shape, composition in alloy, nanostructure and electrocatalytic property. Nano Today 4:143–164
7. Parsons R, VanderNoot T (1988) The oxidation of small organic molecules: a survey of recent fuel cell related research. J Electroanal Chem 257:9–45
8. Antolini E, Lopes T, Gonzalez ER (2008) An overview of platinum-based catalysts as methanol-resistant oxygen reduction materials for direct methanol fuel cells. J Alloys Compd 461:253–262
9. Marković NM, Ross PN (2002) Surface science studies of model fuel cells electrocatalysts. Surf Sci Rep 45:117–229
10. de Brujin FA, Dam VAT, Janssen GJM (2008) Review: durability and degradation issues of PEM fuel cell components. Fuel Cells 8:3–22
11. Chen W, Kim J, Sun S, Chen S (2008) Electrocatalytic reduction of oxygen by FePt alloy nanoparticles. J Phys Chem C 112:3891–3898
12. Wang C, Daimon H, Onodera T, Koda T, Sun S (2008) A general approach to the size- and shape-controlled synthesis of platinum nanoparticles and their catalytic reduction of oxygen. Angew Chem Int Ed 47:3588–3591
13. Kang Y, Murray CB (2010) Synthesis and electrocatalytic properties of cubic Mn-Pt nanocrystals (nanocubes). J Am Chem Soc 132:7568–7569
14. Chen A, Holt-Hindle P (2010) Platinum-based nanostructured materials: synthesis, properties, and applications. Chem Rev 110:3767–3804
15. Cozzoli PD, Pellegrino T, Manna L (2006) Synthesis, properties and perspectives of hybrid nanocrystal structures. Chem Soc Rev 35:1195–1208
16. Zhang J, Sasaki K, Adzic RR (2007) Stabilization of platinum oxygen-reduction electrocatalysts using gold clusters. Science 315:220–222
17. Zhang L-F, Zhong S-L, Xu A-W (2013) Highly branched concave Au/Pd bimetallic nanocrystals with superior electrocatalytic activity and highly efficient SERS enhancement. Angew Chem Int Ed 52:645–649
18. Murray CB, Kagan CR, Bawendi MG (2000) Synthesis and characterization of monodisperse nanocrystals and close-packed nanocrystal assemblies. Annu Rev Mater Sci 30:545–610
19. Daniel M-C, Astruc D (2004) Gold nanoparticles: assembly, supramolecular chemistry, quantum-size-related properties, and applications toward biology, catalysis, and nanotechnology. Chem Rev 104:293–346
20. Bruda C, Chen X, Narayanan R, El-Sayed MA (2005) Chemistry and properties of nanocrystals of different shapes. Chem Rev 105:1025–1102

21. Ferrando R, Jellinek J, Johnston RL (2008) Nanoalloys: from theory to applications of alloy clusters and nanoparticles. Chem Rev 108:845–910
22. Laurent S, Forge D, Port M, Roch A, Robic C, Elst LV, Muller RN (2008) Magnetic iron oxide nanoparticles: synthesis, stabilization, vectorization, physicochemical characterizations, and biological applications. Chem Rev 108:2064–2110
23. Murray RW (2008) Nanoelectrochemistry: metal nanoparticles, nanoelectrodes, and nanopores. Chem Rev 108:2688–2720
24. Xia Y, Xiong Y, Lim B, Skrabalak SE (2009) Shape-controlled synthesis of metal nanocrystals: simple chemistry meets complex physics? Angew Chem Int Ed 48:60–103
25. Novet T, Johnson DC (1991) New synthetic approach to extended solids: selective synthesis of iron suicides via the amorphous state. J Am Chem Soc 113:3398–3403
26. Suryanarayana C (2001) Mechanical alloying and milling. Prog Mater Sci 46:1–184
27. Mazumder V, Chi M, More KL, Sun S (2010) Core/shell Pd/FePt nanoparticles as an active and durable catalyst for the oxygen reduction reaction. J Am Chem Soc 132:7848–7849
28. Yang H (2011) Platinum-based electrocatalysts with core–shell nanostructures. Angew Chem Int Ed 50:2674–2676
29. Wang A, Peng Q, Li Y (2011) Rod-shaped Au–Pd core–shell nanostructures. Chem Mater 23:3217–3222
30. Weber MJ, Mackus AJM, Verheijen MA, van der Marel C, Kessels WMM (2012) Supported core/shell bimetallic nanoparticles synthesis by atomic layer deposition. Chem Mater 24:2973–2977
31. Cardinal MF, Mongin D, Crut A, Maioli P, Rodríguez-González B, Pérez-Juste J, Liz-Marzán LM, Fatti ND, Vallée F (2012) Acoustic vibrations in bimetallic Au@Pd core–shell nanorods. J Phys Chem Lett 3:613–619
32. Chen T-Y, Luo T-JM, Yang Y-W, Wei Y-C, Wang K-W, Lin T-L, Wen T-C, Lee CH (2012) Core dominated surface activity of core–shell nanocatalysts on methanol electrooxidation. J Phys Chem C 116:16969–16978
33. Kuttiyiel KA, Sasaki K, Choi Y, Su D, Liu P, Adzic RR (2012) Nitride stabilized PtNi core–shell nanocatalyst for high oxygen reduction activity. Nano Lett 12:6266–6271
34. Zhang Q, Lee I, Joo JB, Zaera F, Yin Y (2013) Core–shell nanostructured catalysts. Acc Chem Res 46:1816–1824
35. Kang SW, Lee YW, Park Y, Choi B-S, Hong JW, Park K-H, Han SW (2013) One-pot synthesis of trimetallic Au@PdPt core–shell nanoparticles with high catalytic performance. ACS Nano 7:7945–7955
36. Sun X, Li D, Ding Y, Zhu W, Guo S, Wang ZL, Sun S (2014) Core/shell Au/CuPt nanoparticles and their dual electrocatalysis for both reduction and oxidation reactions. J Am Chem Soc 136:5745–5749
37. Tedsree K, Li T, Jones S, Chan CWA, Yu KMK, Bagot PAJ, Marquis EA, Smith GDW, Tsang SCE (2011) Hydrogen production from formic acid decomposition at room temperature using a Ag–Pd core–shell nanocatalyst. Nat Nanotechnol 6:302–307
38. Alayoglu S, Nilekar AU, Mavrikakis M, Eichhorn B (2008) Ru–Pt core–shell nanoparticles for preferential oxidation of carbon monoxide in hydrogen. Nat Mater 7:333–338
39. Serpell CJ, Cookson J, Ozkaya D, Beer PD (2011) Core@shell bimetallic nanoparticle synthesis via anion coordination. Nat Chem 3:478–483
40. Turkevich J, Kim G (1970) Palladium: preparation and catalytic properties of particles of uniform size. Science 169:873–879
41. Paz-Borbon LO, Mortimer-Jones TV, Johnston RL, Posada-Amarillas A, Barcaro G, Fortunelli A (2007) Structures and energetics of 98 atom Pd–Pt nanoalloys: potential stability of the Leary tetrahedron for bimetallic nanoparticles. Phys Chem Chem Phys 9:5202–5208
42. Wang C, van der Vliet D, More KL, Zaluzec NJ, Peng S, Sun S, Daimon H, Wang G, Greeley J, Pearson J, Paulikas AP, Karapetrov G, Strmcnik D, Markovic NM, Stamenkovic VR (2011) Multimetallic Au/$FePt_3$ nanoparticles as highly durable electrocatalyst. Nano Lett 11:919–926

43. Narayanan R, El-Sayed MA (2004) Shape-dependent catalytic activity of platinum nanoparticles in colloidal solution. Nano Lett 4:1343–1348
44. Habas SE, Lee H, Radmilovic V, Somorjai GA, Yang P (2007) Shaping binary metal nanocrystals through epitaxial seeded growth. Nat Mater 6:692–697
45. Wang C, Daimon H, Lee Y, Kim J, Sun S (2007) Synthesis of monodisperse Pt nanocubes and their enhanced catalysis for oxygen reduction. J Am Chem Soc 129:6974–6975
46. Narayanan R, El-Sayed MA (2004) Changing catalytic activity during colloidal platinum nanocatalysis due to shape changes: Electron-transfer reaction. J Am Chem Soc 126:7194–7195
47. Tian N, Zhou Z, Sun S, Ding Y, Wang ZL (2007) Synthesis of tetrahexahedral platinum nanocrystals with high-index facets and high electro-oxidation activity. Science 316:732–735
48. Li C, Sato R, Kanehara M, Zeng HB, Bando Y, Teranishi T (2009) Controllable polyol synthesis of uniform palladium icosahedra: effect of twinned structure on deformation of crystalline lattices. Angew Chem Int Ed 48:6883–6887
49. Wang JX, Inada H, Wu L, Zhu Y, Choi Y, Liu P, Zhou WP, Adzic RR (2009) Oxygen reduction on well-defined core-shell nanocatalysts: particle size, facet, and Pt shell thickness effects. J Am Chem Soc 131:17298–17302
50. Tang Y, Ouyang M (2007) Tailoring properties and functionalities of metal nanoparticles through crystallinity engineering. Nat Mater 6:754–759
51. Zhang Q, Xie J, Yang J, Lee JY (2009) Monodisperse icosahedral Ag, Au, and Pd nanoparticles: size control strategy and superlattice formation. ACS Nano 3:139–148
52. Yang J, Yang J, Ying JY (2012) Morphology and lateral strain control of Pt nanoparticles via core–shell construction using alloy AgPd core toward oxygen reduction reaction. ACS Nano 6:9373–9382
53. Zhang Q, Lee JY, Yang J, Boothroyd C, Zhang JX (2007) Size and composition tunable Ag–Au alloy nanoparticles by replacement reactions. Nanotechnology 18:245605
54. Reyes-Gasga J, Tehuacanero-Nunez S, Montejano-Carrizales JM, Gao XX, Jose-Yacaman M (2007) Analysis of the contrast in icosahedral gold nanoparticles. Top Catal 46:23–30
55. Ling T, Xie L, Zhu J, Yu HM, Ye HQ, Yu R, Cheng Z, Liu L, Yang GW, Cheng ZD, Wang YJ, Ma XL (2009) Icosahedral face-centered cubic Fe nanoparticles: facile synthesis and characterization with aberration-corrected TEM. Nano Lett 9:1572–1576
56. Gasteiger HA, Kocha SS, Sompalli B, Wagner FT (2005) Activity benchmarks and requirements for Pt, Pt-alloy, and non-Pt oxygen reduction catalysts for PEMFCs. Appl Catal B 56:9–35
57. Markovic N, Gasteiger H, Ross PN (1997) Kinetics of oxygen reduction on Pt(hkl) electrodes: implications for the crystallite size effect with supported Pt electrocatalysts. J Electrochem Soc 144:1591–1597
58. Li HQ, Xin Q, Li WZ, Zhou ZH, Jiang LH, Yang SH, Sun GQ (2004) An improved palladium-based DMFCs cathode catalyst. Chem Commun 23:2776–2777
59. Hayre RO, Cha SW, Colella W, Prinz FB (2006) Fuel cell fundamentals. Wiley, New York
60. Bligaard T, Norskov JK (2007) Ligand effects in heterogeneous catalysis and electrochemistry. Electrochim Acta 52:5512–5516
61. Xu Y, Ruba AV, Mavrikakis M (2004) Adsorption and dissociation of O_2 on Pt-Co and Pt-Fe alloys. J Am Chem Soc 126:4717–4725
62. Stamenkovic VR, Mun BS, Arenz M, Mayrhofer KJJ, Lucas CA, Wang GF, Ross PN, Markovic NM (2007) Trends in electrocatalysis on extended and nanoscale Pt-bimetallic alloy surfaces. Nat Mater 6:241–247
63. Stamenkovic VR, Fowler B, Mun BS, Wang GF, Ross PN, Lucas CA, Markovic NM (2007) Improved oxygen reduction activity on $Pt_3Ni(111)$ via increased surface site availability. Science 315:493–497
64. Ataee-Esfahani H, Wang L, Nemoto Y, Yamauchi Y (2010) Synthesis of bimetallic Au@Pt nanoparticles with Au core and nanostructured Pt shell toward highly active electrocatalysts. Chem Mater 22:6310–6318

65. Liang H, Zhang H, Hu J, Guo Y, Wan L, Bai C (2004) Pt hollow nanospheres: facile synthesis and enhanced electrocatalysts. Angew Chem Int Ed 43:1540–1543
66. Yang J, Lee JY, Too H-P, Valiyaveettil S (2006) A bis(*p*-sulfonatophenyl)phenyl-phosphine-based synthesis of hollow Pt nanospheres. J Phys Chem B 110:125–129
67. Chen G, Xia D, Nie Z, Wang Z, Wang L, Zhang L, Zhang J (2007) Facile synthesis of Co–Pt hollow sphere electrocatalyst. Chem Mater 19:1840–1844
68. Chen HM, Liu RS, Lo MY, Chang SC, Tsai LD, Peng YM, Lee JF (2008) Hollow platinum spheres with nano-channels: synthesis and enhanced catalysis for oxygen reduction. J Phys Chem C 112:7522–7526
69. Guo S, Dong S, Wang E (2008) A general method for the rapid synthesis of hollow metallic or bimetallic nanoelectrocatalysts with urchinlike morphology. Chem Eur J 14:4689–4695
70. Xu C, Wang L, Wang R, Wang K, Zhang Y, Tian F, Ding Y (2009) Nanotubular mesoporous bimetallic nanostructures with enhanced electrocatalytic performance. Adv Mater 21:2165–2169
71. Yang J, Sargent EH, Kelley SO, Ying JY (2009) A general phase-transfer protocol for metal ions and its application in nanocrystal synthesis. Nat Mater 8:683–689
72. Peng Z, Wu J, Yang H (2010) Synthesis and oxygen reduction electrocatalytic property of platinum hollow and platinum-on-silver nanoparticles. Chem Mater 22:1098–1106
73. Lee J, Park JC, Song H (2008) A nanoreactor framework of a Au@SiO_2 yolk/shell structure for catalytic reduction of *p*-nitrophenol. Adv Mater 20:1523–1528
74. Gough DV, Wolosiuk A, Braun PV (2009) Mesoporous ZnS nanorattles: programmed size selected access to encapsulated enzymes. Nano Lett 9:1994–1998
75. Liu J, Qiao SZ, Hartono SB, Lu GQ (2010) Monodisperse yolk-shell nanoparticles with a hierarchical porous structure for delivery vehicles and nanoreactors. Angew Chem Int Ed 49:4981–4985
76. Gao J, Liang G, Zhang B, Kuang Y, Zhang X, Xu B (2007) FePt@CoS_2 yolk-shell nanocrystals as a potent agent to kill hela cells. J Am Chem Soc 129:1428–1433
77. Zhao W, Chen H, Li Y, Li L, Lang M, Shi J (2008) Uniform rattle-type hollow magnetic mesoporous spheres as drug delivery carries and their sustained-release property. Adv Funct Mater 18:2780–2788
78. Zhu Y, Ikoma T, Hanagata N, Kaskel S (2010) Rattle-type Fe_3O_4@SiO_2 hollow mesoporous spheres as carrier for drug delivery. Small 6:471–478
79. Zhang W-M, Hu J-S, Guo Y-G, Zheng S-F, Zhong L-S, Song W-G, Wan L-J (2008) Tin-nanoparticles encapsulated in elastic hollow carbon spheres for high-performance anode material in lithium-ion batteries. Adv Mater 20:1160–1165
80. Lou XW, Li CM, Archer LA (2009) Designed synthesis of coaxial SnO_2@carbon hollow nanospheres for highly reversible lithium storage. Adv Mater 21:2536–2539
81. Chen JS, Li CM, Zhou WW, Yan QY, Archer LA, Lou XW (2009) One-pot formation of SnO_2 hollow nanospheres and α-Fe_2O_3@SnO_2 nanorattles with large void space and their lithium storage properties. Nanoscale 1:280–285
82. Chem JS, Luan D, Li CM, Boey FYC, Qiao S, Lou XW (2010) TiO_2 and SnO_2@TiO_2 hollow spheres assembled from anatase TiO_2 nanosheets with enhanced lithium storage properties. Chem Commun 46:8252–8254
83. Chen JS, Wang Z, Dong XC, Chen P, Lou XW (2011) Graphene-wrapped TiO_2 hollow structures with enhanced lithium storage capabilities. Nanoscale 3:2158–2161
84. Li H, Bian Z, Zhu J, Zhang D, Li G, Huo Y, Li H, Lu Y (2007) Mesoporous titania spheres with tunable chamber structure and enhanced photocatalytic activity. J Am Chem Soc 129:8406–8407
85. Benabid F, Couny F, Knight JC, Birks TA, Russell PJ (2005) Compact, stable and efficient all-fibre gas cells using hollow-core phoyonic crystal fibres. Nature 434:488–491
86. Kamata K, Lu Y, Xia Y (2003) Synthesis and characterization of monodispersed core-shell spherical colloids with movable cores. J Am Chem Soc 125:2384–2385

87. Bai F, Sun Z, Wu H, Haddad RE, Xiao X, Fan H (2011) Templated photocatalytic synthesis of well-defined platinum hollow nanostructures with enhanced catalytic performance for methanol oxidation. Nano Lett 11:3759–3762
88. Caruso F, Caruso RA, Möhwald H (1998) Nanoengineering of inorganic and hybrid hollow spheres by colloidal templating. Science 282:1111–1114
89. Caruso F, Shi X, Caruso RA, Susha A (2001) Hollow titania spheres from layered precursor deposition on sacrificial colloidal core particles. Adv Mater 13:740–744
90. Yang Z, Niu Z, Lu Y, Hu Z, Han CC (2003) Templated synthesis of inorganic hollow spheres with a tunable cavity size onto core-shell gel particles. Angew Chem Int Ed 42:1943–1945
91. Lou XW, Yuan C, Archer LA (2007) Shell-by-shell synthesis of tin oxide hollow colloids with nanoarchitectured walls: cavity size tuning and functionalization. Small 3:261–265
92. Wang Z, Luan D, Boey FYC, Lou XW (2011) Fast formation of SnO_2 nanoboxes with enhanced lithium storage capability. J Am Chem Soc 133:4738–4741
93. Lai X, Li J, Korgel BA, Dong Z, Li Z, Su F, Du J, Wang D (2011) General synthesis and gas-sensing properties of multiple-shell metal oxide hollow microspheres. Angew Chem Int Ed 50:2738–2741
94. Schmidt HT, Ostafin AE (2002) Liposome directed growth of calcium phosphate nanoshells. Adv Mater 14:532–535
95. Hubert DHW, Jung M, German AL (2000) Vesicle templating. Adv Mater 12:1291–1294
96. Gao X, Zhang J, Zhang L (2002) Hollow sphere selenium nanoparticles: their in-vitro anti hydroxyl radical effect. Adv Mater 14:290–293
97. Nakashima T, Kimizuka N (2003) Interfacial synthesis of hollow TiO_2 microspheres in ionic liquids. J Am Chem Soc 125:6386–6387
98. Qi L, Li J, Ma J (2002) Biomimetic morphogenesis of calcium carbonate in mixed solutions if surfactants and double-hydrophilic block copolymers. Adv Mater 14:300–303
99. Zhang D, Qi L, Ma J, Cheng H (2002) Synthesis of submicrometer-sized hollow silver spheres in mixed polymer-surfactant solutions. Adv Mater 14:1499–1502
100. Wong MS, Cha JN, Choi K-S, Deming TJ, Stucky GD (2002) Assembly of nanoparticles into hollow spheres using block copolypeptides. Nano Lett 2:583–587
101. Lou XW, Archer LA, Yang Z (2008) Hollow micro-/nanostructures: synthesis and applications. Adv Mater 20:3987–4019
102. Zhao Y, Jiang L (2009) Hollow micro/nanomaterials with multilevel interior structures. Adv Mater 21:3621–3638
103. Sun Y, Mayers BT, Xia Y (2002) Template-engaged replacement reaction: a one-step approach to the large-scale synthesis of metal nanostructures with hollow interiors. Nano Lett 2:481–485
104. Sun Y, Xia Y (2004) Mechanistic study on the replacement reaction between silver nanostructures and chloroauric acid in aqueous medium. J Am Chem Soc 126:3892–3901
105. Chen J, Wiley B, McLellan J, Xiong Y, Li Z-Y, Xia Y (2005) Optical properties of Pd-Ag and Pt-Ag nanoboxes synthesized via galvanic replacement reactions. Nano Lett 5:2058–2062
106. Chen M, Gao L (2006) Synthesis and characterization of Ag nanoshells by a facile sacrificial template route through in situ replacement reaction. Inorg Chem 45:5145–5149
107. Skrabalak SE, Chen J, Sun Y, Lu X, Au L, Cobley CM, Xia Y (2008) Gold nanocages: synthesis, properties, and applications. Acc Chem Res 41:1587–1595
108. Zhang H, Jin M, Wang J, Li W, Camargo PHC, Kim MJ, Yang D, Xie Z, Xia Y (2011) Synthesis of Pd-Pt bimetallic nanocrystals with a concave structure through a bromide-induced galvanic replacement reaction. J Am Chem Soc 133:6078–6089
109. Mohl M, Dobo D, Kukovecz A, Konya Z, Kordas K, Wei J, Vajtai R, Ajayan PM (2011) Formation of CuPd and CuPt bimetallic nanotubes by galvanic replacement reaction. J Phys Chem C 115:9403–9409
110. González E, Arbiol J, Puntes VF (2011) Carving at the nanoscale: sequential galvanic exchange and Kirkendall growth at room temperature. Science 334:1377–1380

111. Yin Y, Rioux RM, Erdonmez CK, Hughes S, Somorjai GA, Alivisatos PA (2004) Formation of hollow nanocrystals through the nanoscale Kirkendall effect. Science 304:711–714
112. Zeng HC (2006) Synthetic architecture of interior space for inorganic nanostructures. J Mater Chem 16:649–662
113. Zeng HC (2007) Ostwald ripening: a synthetic approach for hollow nanomaterials. Curr Nanosci 3:177–181
114. Caruso F (2000) Hollow capsule processing through colloidal templating and self-assembly. Chem Eur J 6:413–419
115. Macdonald JE, Sadan MB, Houben L, Popov I, Banin U (2010) Hybrid nanoscale inorganic cages. Nat Mater 9:810–815
116. Chen C, Kang Y, Huo Z, Zhu Z, Huang W, Xin HL, Snyder JD, Li D, Herron JA, Mavrikakis M, Chi M, More KL, Li Y, Markovic NM, Somorjai GA, Yang P, Stamenkovic VR (2014) Highly crystalline multimetallic nanoframes with three-dimensional electrocatalytic surfaces. Science 343:1339–1343
117. Wu X-J, Xu D (2009) Formation of yolk/SiO_2 shell structure using surfactant mixtures as template. J Am Chem Soc 131:2774–2775
118. Li G, Shi Q, Yuan SJ, Neoh KG, Kang ET, Yang X (2010) Alternating silica/polymer multilayer hybrid microspheres templates for double-shelled polymer and inorganic hollow microstructures. Chem Mater 22:1309–1317
119. Liu H, Qu J, Chen Y, Li J, Ye F, Lee JY, Yang J (2012) Hollow and cage-bell structured nanomaterials of noble metals. J Am Chem Soc 134:11602–11610
120. Wiley B, Herricks T, Sun Y, Xia Y (2004) Polyol synthesis of silver nanoparticles: use of chloride and oxygen to promote the formation of single-crystal, truncated cubes and tetrahedrons. Nano Lett 4:1733–1739
121. Tan Y-N, Yang J, Lee JY, Wang DIC (2007) Mechanistic study on the Bis(*p*-sulfonatophenyl)-phenylphosphine synthesis of monometallic Pt hollow nanoboxes using Ag*-Pt core-shell nanocubes as sacrificial templates. J Phys Chem C 111:14084–14090
122. Gasteiger HA, Markovic N, Ross PN, Cairns EJ (1993) Methanol electrooxidation on well-characterized platinum-ruthenium bulk alloys. J Phys Chem 97:12020–12029
123. Wang JM, Brankovic SR, Zhu Y, Hanson JC, Adzic RR (2003) Kinetic characterization of PtRu fuel cell anode catalysts made by spontaneous Pt deposition on Ru nanoparticles. J Electrochem Soc 150:A1108–A1117
124. Antolini E, Salgado JRC, Gonzalez ER (2005) Carbon supported $Pt_{75}M_{25}$ (M = Co, Ni) alloys as anode and cathode electrocatalysts for direct methanol fuel cells. J Electroanal Chem 580:145–154
125. Liu H, Ye F, Yang J (2014) A universal and cost-effective approach to the synthesis of carbon-supported noble metal nanoparticles with hollow interiors. Ind Eng Chem Res 53:5925–5931
126. Wang L, Yamauchi Y (2013) Metallic nanocages: synthesis of bimetallic Pt–Pd hollow nanoparticles with dendritic shells by selective chemical etching. J Am Chem Soc 135:16762–16765
127. Bagotzky VS, Vassiliev YB, Khazova OA (1977) Generalized scheme of chemisorption, electrooxidation and electroreduction of simple organic-compounds on platinum group metals. J Electroanal Chem 81:229–238
128. Goodenough JB, Hamnett A, Kennedy BJ, Manoharan R, Weeks SA (1990) Porous carbon anodes for the direct methanol fuel-cell. 1. The role of the reduction method for carbon supported platinum electrodes. Electrochim Acta 35:199–207
129. Jusys Z, Schmidt TJ, Dubau L, Lasch K, Jorissen L, Garche J, Behm RJ (2002) Activity of $PtRuMeo_x$ (Me = W, Mo or V) catalysts towards methanol oxidation and their characterization. J Power Sources 105:297–304
130. Park KW, Choi JH, Ahn KS, Sung YE (2004) PtRu alloy and PtRu-WO_3 nanocomposite electrodes for methanol electrooxidation fabricated by a sputtering deposition method. J Phys Chem B 108:5989–5994

131. Park KW, Sung YE, Han S, Yan Y, Hyeon T (2004) Origin of the enhanced catalytic activity of carbon nanocoil-supported PtRu alloy electrocatalysts. J Phys Chem B 108:939–944
132. Park KW, Choi JH, Lee SA, Pak C, Chang H, Sung YE (2004) PtRuRhNi nanoparticle electrocatalyst for methanol electrooxidation in direct methanol fuel cell. J Catal 224:236–242
133. Huang J, Yang H, Huang Q, Tang Y, Lu T, Akins DL (2004) Methanol oxidation on carbon-supported Pt-Os bimetallic nanoparticle electrocatalysts. J Electrochem Soc 151:A1810–A1815
134. Ye F, Yang J, Hu W, Liu H, Liao S, Zeng J, Yang J (2012) Electrostatic interaction based hollow Pt and Ru assemblies toward methanol oxidation. RSC Adv 2:7479–7486
135. Rolison DR, Hagans PL, Swider KE, Long JW (1999) Role of hydrous ruthenium oxide in Pt–Ru direct methanol fuel cell anode electrocatalysts: the importance of mixed electron/proton conductivity. Langmuir 15:774–779
136. Long JW, Stroud RM, Swider KE, Rolison DB (2000) How to make electrocatalysts more active for direct methanol oxidation—avoid PtRu bimetallic alloys! J Phys Chem B 104:9772–9776
137. Villullas HM, Mattos-Costa FI, Bulhões LOS (2004) Electrochemical oxidation of methanol on Pt nanoparticles dispersed on RuO_2. J Phys Chem B 108:12898–12903
138. Chen Z, Qiu X, Lu B, Zhang S, Zhu W, Chen L (2005) Synthesis of hydrous ruthenium oxide supported platinum catalysts for direct methanol fuel cells. Electrochem Commun 7:593–596
139. Scheiba F, Scholz M, Cao L, Schafranek R, Roth C, Cremers C, Qiu X, Stimming U, Fuess H (2006) On the suitability of hydrous ruthenium oxide supports to enhance intrinsic proton conductivity in DMFC anodes. Fuel Cells 6:439–446
140. Cao L, Scheiba F, Roth C, Schweiger F, Cremers C, Stimming U, Fuess H, Chen L, Zhu W, Qiu X (2006) Novel nanocomposite $Pt/RuO_2 \cdot xH_2O$/carbon nanotube catalysts for direct methanol fuel cells. Angew Chem Int Ed 45:5315–5319
141. Profeti LPR, Simões FC, Olivi P, Kokoh KB, Coutanceau C, Léger J-M, Lamy C (2006) Application of $Pt + RuO_2$ catalysts prepared by thermal decomposition of polymeric precursors to DMFC. J Power Sources 158:1195–1201
142. Huang S-Y, Chang C-M, Wang K-W, Yeh C-T (2007) Promotion of platinum-ruthenium catalyst for electrooxidation of methanol by crystalline ruthenium dioxide. ChemPhysChem 8:1774–1777
143. Peng F, Zhou C, Wang H, Yu H, Liang J, Yang J (2009) The role of RuO_2 in the electrocatalytic oxidation of methanol for direct methanol fuel cell. Catal Commun 10:533–537
144. Dubau L, Hahn F, Coutanceau C, Léger J-M, Lamy C (2003) On the structure effects of bimetallic PtRu electrocatalysts towards methanol oxidation. J Electroanal Chem 554–555:407–415
145. Beard KD, Borrelli D, Cramer AM, Blom D, Van Zee JW, Monnier JR (2009) Preparation and structural analysis of carbon-supported Co core/Pt shell electrocatalysts using electroless deposition methods. ACS Nano 3:2841–2853
146. Langille MR, Zhang J, Mirkin CA (2011) Plasmon-mediated synthesis of heterometallic nanorods and icosahedra. Angew Chem Int Ed 50:3543–3547
147. Zhou S, Mcllwrath K, Jackson G, Eichhorn B (2006) Enhanced CO tolerance for hydrogen activation in Au-Pt dendritic heteroaggregate nanostructures. J Am Chem Soc 128:1780–1781
148. Lee H, Habas SE, Somorjai GA, Yang P (2008) Localized Pd overgrowth on cubic Pt nanocrystals for enhanced electrocatalytic oxidation of formic acid. J Am Chem Soc 130:5406–5407
149. Peng Z, Yang H (2009) PtAu bimetallic heteronanostructures made by post-synthesis modification of Pt-on-Au nanoparticles. Nano Res 2:406–415
150. Lim B, Kobayashi H, Yu T, Wang J, Kim MJ, Li Z-Y, Rycenga M, Xia Y (2010) Synthesis of Pd-Au bimetallic nanocrystals via controlled overgrowth. J Am Chem Soc 132:2506–2507

151. Wang F, Li C, Sun L-D, Wu H, Ming T, Wang J, Yu JC, Yan C-H (2011) Heteroepitaxial growth of high-index-faceted palladium nanoshells and their catalytic performance. J Am Chem Soc 133:1106–1111
152. Fan F-R, Liu D-Y, Wu Y-F, Duan S, Xie Z-X, Jiang Z-Y, Tian Z-Q (2008) Epitaxial growth of heterogeneous metal nanocrystals: From gold nano-octahedra to palladium and silver nanocubes. J Am Chem Soc 130:6949–6951
153. Zeng J, Zhu C, Tao J, Jin M, Zhang H, Li Z-Y, Zhu Y, Xia Y (2012) Controlling the nucleation and growth of silver on palladium nanocubes by manipulating the reaction kinetics. Angew Chem Int Ed 51:2354–2358
154. Lim B, Jiang M, Camargo PHC, Cho EC, Tao J, Lu X, Zhu Y, Xia Y (2009) Pd-Pt bimetallic nanodendrites with high activity for oxygen reduction. Science 324:1302–1305
155. Peng Z, Yang H (2009) Synthesis and oxygen reduction electrocatalytic property of Pt-on-Pd bimetallic heteronanostructures. J Am Chem Soc 131:7542–7543
156. Guo S, Dong S, Wang E (2010) Three-dimensional Pt-on-Pd bimetallic nanodendrites supported on graphene nanosheet: Facile synthesis and used as an advanced nanoelectrocatalyst for methanol oxidation. ACS Nano 4:547–555
157. Wang L, Nemoto Y, Yamauchi Y (2011) Direct synthesis of spatially-controlled Pt-on-Pd bimetallic nanodendrites with superior electrocatalytic activity. J Am Chem Soc 133:9674–9677
158. Kim Y, Hong JW, Lee YW, Kim M, Kim D, Yun WS, Han SW (2010) Synthesis of AuPt heteronanostructures with enhanced electrocatalytic activity toward oxygen reduction. Angew Chem Int Ed 49:10197–10201
159. Wang L, Yamauchi Y (2009) Block copolymer mediated synthesis of dendritic platinum nanoparticles. J Am Chem Soc 131:9152–9153
160. Liu H, Ye F, Yao Q, Cao H, Xie J, Lee JY, Yang J (2014) Stellated Ag-Pt bimetallic nanoparticles: an effective platform for catalytic activity tuning. Sci Rep 4:3969
161. Wang ZL (2000) Transmission electron microscopy of shape-controlled nanocrystals and their assemblies. J Phys Chem B 104:1153–1175
162. Brodersen SH, Grønbjerg U, Hvolbæk B, Schiøtz J (2011) Understanding the catalytic activity of gold nanoparticles through multi-scale simulations. J Catal 284:34–41
163. Feng Y, Liu H, Yang J (2014) Bimetallic nanodendrites via selective overgrowth of noble metals on multiply twinned Au seeds. J Mater Chem A 2:6130–6137
164. Mourdikoudis S, Liz-Marzán LM (2013) Oleylamine in nanoparticle synthesis. Chem Mater 25:1465–1476
165. Brust M, Walker M, Bethell D, Schiffrin DJ, Whyman R (1994) Synthesis of thoil-derivatised gold nanoparticle in a two-phase Liquid-Liqiud synthem. J Chem Soc Chem Commun 7:801–802
166. Hiramatsu H, Osterloh F (2004) A simple large-scale synthesis of nearly monodisperse gold and silver nanoparticles with adjustable sizes and with exchangeable surfactants. Chem Mater 16:2509–2511
167. Aslam M, Fu L, Su M, Vijayamohanan K, Dravid VP (2004) Novel one-step synthesis of amine-stabilized aqueous colloidal gold nanoparticles. J Mater Chem 14:1795–1797
168. Yu H, Chen M, Rice PM, Wang SX, White RL, Sun S (2005) Dumbbell-like bifunctional Au–Fe_3O_4 nanoparticles. Nano Lett 5:379–382
169. Wang C, Hou Y, Kim J, Sun S (2007) A general strategy for synthesizing FePt nanowires and nanorods. Angew Chem Int Ed 46:6333–6335
170. Peng S, Sun S (2007) Synthesis and characterization of monodisperse hollow Fe_3O_4 nanoparticles. Angew Chem Int Ed 46:4155–4158
171. Chaubey GS, Barcena C, Poudyal N, Rong C, Gao J, Sun S, Liu JP (2007) Synthesis and stabilization of FeCo nanoparticles. J Am Chem Soc 129:7214–7215
172. Kim J, Rong C, Lee Y, Liu JP, Sun S (2008) From core/shell structured FePt/Fe_3O_4/MgO to ferromagnetic FePt nanoparticles. Chem Mater 20:7242–7245

173. Zeng H, Sun S (2008) Syntheses, properties and potential applications of multicomponent magnetic nanoparticles. Adv Funct Mater 18:391–400
174. Lu X, Tuan H-Y, Korgel BA, Xia Y (2008) Facile synthesis of gold nanoparticles with narrow size distribution by using AuCl or AuBr as the precursor. Chem Eur J 14:1584–1591
175. Xu Z, Shen C, Hou Y, Gao H, Sun S (2009) Oleylamine as both reducing agent and stabilizer in a facile synthesis of magnetite nanoparticles. Chem Mater 21:1778–1780
176. Kim J, Rong C, Liu P, Sun S (2009) Dispersible ferromagnetic FePt nanoparticles. Adv Mater 21:906–909
177. Wang C, Peng S, Lacroix L-M, Sun S (2009) Synthesis of high magnetic moment CoFe nanoparticles via interfacial diffusion in core/shell structured Co/Fe nanoparticles. Nano Res 2:380–385
178. Lacroix L-M, Huls NF, Ho D, Sun X, Cheng K, Sun S (2011) Stable single-crystalline body centered cubic Fe nanoparticles. Nano Lett 11:1641–1645
179. Liu X, Atwater M, Wang J, Dai Q, Zou J, Brennan JP, Huo Q (2007) A study on gold nanoparticle synthesis using oleylamine as both reducing agent and protecting ligand. J Nanosci Nanotechnol 7:3126–3133
180. Feng Y, Ma X, Han L, Peng Z, Yang J (2014) A universal approach to the synthesis of nanodendrites of noble metals. Nanoscale 6:6173–6179
181. Ostwald WF (1897) Studien uber die bildung und umwandlung fester korper. Z Phys Chem 22:289–330
182. Liu RX, Iddir H, Fan QB, Hou GY, Bo AL, Ley KL, Smotkin ES, Sung YE, Kim H, Thomas S (2000) Potential-dependent infrared absorption spectroscopy of adsorbed CO and X-ray photoelectron spectroscopy of arc-melted single-phase Pt, PtRu, PtOs, PtRuOs, and Ru electrodes. J Phys Chem B 104:3518–3531
183. Jiang J, Kucernak A (2009) Electrodeposition of highly alloyed quaternary PtPdRuOs catalyst with highly ordered nanostructure. Electrochem Commun 11:1005–1008
184. Ley KL, Liu RX, Pu C, Fan QB, Leyarovska N, Segre C, Smotkin ES (1997) Methanol oxidation on single-phase Pt-Ru-Os ternary alloys. J Electrochem Soc 144:1543–1548
185. Watanabe M, Motoo S (1975) Electrocatalysis by ad-atoms. 2. Enhancement of oxidation of methanol on platinum by ruthenium ad-atoms. J Electroanal Chem 60:267–273
186. Watanabe M, Motoo S (1975) Electrocatalysis by ad-atoms. 3. Enhancement of oxidation of carbon-monoxide on platinum by ruthenium ad-atoms. J Electroanal Chem 60:275–283
187. Pellegrino T, Fiore A, Carlino E, Giannini C, Cozzoli PD, Ciccarella G, Respaud M, Palmirotta L, Cingolani R, Manna L (2006) Heterodimers based on $CoPt_3$-Au nanocrystals with tunable domain size. J Am Chem Soc 128:6690–6698
188. Choi J-S, Choi HJ, Jung DC, Lee J-H, Cheon J (2008) Nanoparticle assisted magnetic resonance imaging of the early reversible stages of amyloid β self-assembly. Chem Commun 19:2197–2199
189. Wang C, Tian W, Ding Y, Ma Y-Q, Wang ZL, Markovic NM, Stamenkovic VR, Daimon H, Sun S (2010) Rational synthesis of heterostructured nanoparticles with morphology control. J Am Chem Soc 132:6524–6529
190. Bai L, Kuang Y, Luo J, Evans DG, Sun X (2012) Ligand-manipulated selective transformations of Au-Ni bimetallic heteronanostructures in an organic medium. Chem Commun 48:6963–6965
191. Lim SI, Varon M, Ojea-Jiménez I, Arbiol J, Puntes V (2011) Pt nanocrystal evolution in the presence of Au(III)-salts at room temperature: spontaneous formation of AuPt heterodimers. J Mater Chem 21:11518–11523
192. Yang J, Lee JY, Chen LX, Too H-P (2005) A phase-transfer identification of core-shell structures in Ag-Pt nanoparticles. J Phys Chem B 109:5468–5472
193. Luo J, Wang L, Mott D, Njoki PN, Lin Y, He T, Xu Z, Wanjala BN, Lim IS, Zhong C (2008) Core/shell nanoparticles as electrocatalysts for fuel cell reactions. Adv Mater 20:4342–4347
194. Zhou W-P, Yang X, Vukmirovic MB, Koel BE, Jiao J, Peng G, Mavrikakis M, Adzic RR (2009) Improving electrocatalysts for O_2 reduction by fine-tuning the Pt–support interaction:

Pt monolayer on the surfaces of a $Pd_3Fe(111)$ single-crystal alloy. J Am Chem Soc 131:12755–12762
195. Yang J, Chen X, Yang X, Ying JY (2012) Stabilization and compressive strain effect of AuCu core on Pt shell for oxygen reduction reaction. Energy Environ Sci 5:8976–8981
196. Liu H, Yang J (2014) Bimetallic Ag–hollow Pt heterodimers *via* insideout migration of Ag in core–shell Ag–Pt nanoparticles at elevated temperature. J Mater Chem A 2:7075–7081
197. Zhang J, Vukmirovic MB, Sasaki K, Nilekar AU, Mavrikakis M, Adzic RR (2005) Mixed-metal Pt monolayer electrocatalysts for enhanced oxygen reduction kinetics. J Am Chem Soc 127:12480–12481
198. Zhang J, Vukmirovic MB, Xu Y, Mavrikakis M, Adzic RR (2005) Controlling the catalytic activity of platinum-monolayer electrocatalysts for oxygen reduction with different substrates. Angew Chem Int Ed 44:2132–2135
199. Mokari T, Rothenberg E, Popov I, Costi R, Banin U (2004) Selective growth of metal tips onto semiconductor quantum rods and tetrapods. Science 304:1787–1790
200. Shi WL, Zeng H, Sahoo Y, Ohulchanskyy TY, Ding Y, Wang ZL, Swihart M, Prasad PN (2006) A general approach to binary and ternary hybrid nanocrystals. Nano Lett 6:875–881
201. Yang J, Levina L, Sargent EH, Kelley SO (2006) Heterogeneous deposition of noble metals on semiconductor nanoparticles in organic or aqueous solvents. J Mater Chem 16:4025–4028
202. Costi R, Saunders AE, Elmalem E, Salant A, Banin U (2008) Visible light-induced charge retention and photocatalysis with hybrid CdSe-Au nanodumbbells. Nano Lett 8:637–641
203. Elmalem E, Saunders AE, Costi R, Salant A, Banin U (2008) Growth of photocatalytic CdSe-Pt nanorods and nanonets. Adv Mater 20:4312–4317
204. Habas SE, Yang P, Mokari T (2008) Selective growth of metal and binary metal tips on CdS nanorods. J Am Chem Soc 130:3294–3295
205. Yang J, Ying JY (2009) Room-temperature synthesis of nanocrystalline Ag_2S and its nanocomposites with gold. Chem Commun 22:3187–3189
206. Yang J, Ying JY (2010) Diffusion of gold from the inner core to the surface of Ag_2S nanocrystals. J Am Chem Soc 132:2114–2115
207. Zhang J, Tang Y, Lee K, Ouyang M (2010) Tailoring light–matter–spin interactions in colloidal hetero-nanostructures. Nature 466:91–95
208. Zhang J, Tang Y, Lee K, Ouyang M (2010) Nonepitaxial growth of hybrid core-shell nanostructures with large lattice mismatches. Science 327:1634–1638
209. Zhao N, Li L, Huang T, Qi L (2010) Controlled synthesis of PbS-Au nanostar-nanoparticle heterodimers and cap-like Au nanoparticles. Nanoscale 2:2418–2423
210. Carbone L, Cozzoli PD (2010) Colloidal heterostructured nanocrystals: Synthesis and growth mechanisms. Nano Today 5:449–493
211. Li M, Yu X-F, Liang S, Peng X-N, Yang Z-J, Wang Y-L, Wang Q-Q (2011) Synthesis of Au-CdS core-shell hetero-nanorods with efficient exciton-plasmon interactions. Adv Funct Mater 21:1788–1794
212. Qu J, Liu H, Wei Y, Wu X, Yue R, Chen Y, Yang J (2011) Coalescence of Ag_2S and Au nanocrystals at room temperature. J Mater Chem 21:11750–11753
213. Yang J, Ying JY (2011) Nanocomposites of Ag_2S and noble metals. Angew Chem Int Ed 50:4637–4643
214. Yang J, Chen X, Ye F, Wang C, Zheng Y, Yang J (2011) Core-shell CdSe@Pt nanocomposites with superior electrocatalytic activity enhanced by lateral strain effect. J Mater Chem 21:9088–9094
215. Hu W, Liu H, Ye F, Ding Y, Yang J (2012) A facile solution route for the synthesis of PbSe–Au nanocomposites with different morphologies. CrystEngComm 14:7049–7054
216. Haldar KK, Sinha G, Lahtinen J, Patra A (2012) Hybrid colloidal Au-CdSe pentapod heterostructures synthesis and their photocatalytic properties. ACS Appl Mater Interfaces 4:6266–6272
217. Ding X, Zou Y, Jiang J (2012) Au-Cu_2S heterodimer formation *via* oxidation of AuCu alloy nanoparticles and in situ formed copper thiolate. J Mater Chem 22:23169–23174

218. Motl NE, Bondi JF, Schaak RE (2012) Synthesis of colloidal Au-Cu_2S heterodimers via chemically triggered phase segregation of AuCu nanoparticles. Chem Mater 24:1552–1554
219. Vinokurov K, Macdonald JE, Banin U (2012) Structures and mechanisms in the growth of hybrid Ru–Cu_2S nanoparticles: from cages to nanonets. Chem Mater 24:1822–1827
220. Wang D, Li X, Li H, Li L, Hong X, Peng Q, Li Y (2013) Semiconductor–noble metal hybrid nanomaterials with controlled structures. J Mater Chem A 1:1587–1590
221. Ding X, Zou Y, Ye F, Yang J, Jiang J (2013) Pt–CuS heterodimers by sulfidation of CuPt alloy nanoparticles and their selective catalytic activity toward methanol oxidation. J Mater Chem A 1:11880–11886
222. Sitt A, Hadar I, Banin U (2013) Band-gap engineering, optoelectronic properties and applications of colloidal heterostructured semiconductor nanorods. Nano Today 8:494–513
223. Zhao Q, Ji M, Qian H, Dai B, Weng L, Gui J, Zhang J, Ouyang M, Zhu H (2014) Controlling structural symmetry of a hybrid nanostructure and its effect on efficient photocatalytic hydrogen evolution. Adv Mater 26:1387–1392
224. Vinokurov K, Bekenstein Y, Gutkin V, Popov I, Millo O, Banin U (2014) Rhodium growth on Cu_2S nanocrystals yielding hybrid nanoscale inorganic cages and their synergistic properties. CrystEngComm 16:9506–9512
225. Banin U, Ben-Shahar Y, Vinokurov K (2014) Hybrid semiconductor–metal nanoparticles: from architecture to function. Chem Mater 26:97–110
226. Kraeutler B, Bard AJ (1978) Heterogeneous photocatalytic synthesis of methane from acetic acid-new Kolbe reaction pathway. J Am Chem Soc 100:4317–4318
227. Baba R, Nakabayashi S, Fujishima A, Kenichi H (1985) Investigation of the mechanism of hydrogen ecolution during photocatalytic water decomposition on metal-loaded semiconductor powders. J Phys Chem 89:1902–1905
228. Pastoriza-Santos I, Koktysh DS, Mamedov AA, Giersig M, Kotov NA, Liz-Marzan LM (2000) One-pot synthesis of Ag@TiO_2 core-shell nanoparticles and their layer-by-layer assembly. Langmuir 16:2731–2735
229. Chandrasekharan N, Kamat PV, Hu JQ, Jones G (2000) Improving the photoelectrochemical performance of nanostructured TiO_2 films by adsorption of gold nanoparticles. J Phys Chem B 104:10851–10857
230. Subramanian V, Wolf E, Kamat PV (2001) Semiconductor-metal composite nanostructures. To what extent do metal nanoparticles improve the photocatalytic activity of TiO_2 films? J Phys Chem B 105:11439–11446
231. Kamat PV (2002) Photophysical, photochemical and photocatalytic aspects of metal nanoparticles. J Phys Chem B 106:7729–7744
232. Hirakawa T, Kamat PV (2005) Charge separation and catalytic activity of Ag@TiO_2 core-shell composite clusters under UV-irradiation. J Am Chem Soc 127:3928–3934
233. Jakob M, Levanon H, Kamat PV (2003) Charge distribution between UV-irradiated TiO_2 and gold nanoparticles: determination of shift in the Fermi level. Nano Lett 3:353–358
234. Subramanian V, Wolf E, Kamat PV (2004) Catalysis with TiO_2/gold nanocomposites. Effect of metal particle size on the Fermi level equilibration. J Am Chem Soc 126:4943–4950
235. Mokari T, Sztrum CG, Salant A, Rabani E, Banin U (2005) Formation of asymmetric one-side metal-tipped semiconductor nanocrystal dots and rods. Nat Mater 4:855–863
236. Saunders AE, Popov I, Banin U (2006) Synthesis of hybrid CdS-Au colloidal nanostructures. J Phys Chem B 110:25421–25429
237. Mokari T, Aharoni A, Popov I, Banin U (2006) Diffusion of gold into InAsnanocrystals. Angew Chem Int Ed 45:8001–8005
238. Yang J, Elim HI, Zhang Q, Lee JY, Ji W (2006) Rational synthesis, self-assembly, and optical properties of PbS-Au heterogeneous nanostructures *via* preferential deposition. J Am Chem Soc 128:11921–11926
239. Talapin DV, Yu H, Shevchenko EV, Lobo A, Murray CB (2007) Synthesis of colloidal PbSe/PbS core-shell nanowires and PbS/Au nanowire-nanocrystal heterostructures. J Phys Chem C 111:14049–14054

240. Menagen G, Mocatta D, Salant A, Popov I, Dorfs D, Banin U (2008) Selective gold growth on CdSe seeded CdS nanorods. Chem Mater 20:6900–6902
241. Huang SS, Huang JM, Yang JA, Peng JJ, Zhang QB, Peng F, Wang HJ, Yu H (2010) Chemical synthesis, structure characterization, and optical properties of hollow PbSx-solid Au heterodimer nanostructures. Chem Eur J 16:5920–5926
242. Yang J, Lee JY, Ying JY (2011) Phase transfer and its applications in nanotechnology. Chem Soc Rev 40:1672–1679
243. Liu H, Ye F, Ma X, Cao H, Yang J (2013) The growth of metal sulfide–Au/Ag nanocomposites in a nonpolar organic solvent. CrystEngComm 15:7740–7747
244. Dukovic G, Merkle MG, Nelson JH, Hughes SM, Alivisatos AP (2008) Photodeposition of Pt on colloidal CdS and CdSe/CdS semiconductor nanostructures. Adv Mater 20:4306–4311
245. Harbour JR, Wolkow R, Hair ML (1981) Effect of platinization on the photoproperties of CdS pigments in dispersion. Determination by H_2 evolution, O_2 uptake, and electron spin resonance spectroscopy. J Phys Chem 85:4026–4029
246. Reber J-F, Rusek M (1986) Photochemical hydrogen production with platinized suspensions of cadmium sulfide and cadmium zinc sulfide modified by silver sulfide. J Phys Chem 90:824–834
247. Frank AJ, Goren Z, Willner I (1986) Photohydrogenation of acetylene and ethylene by Pt and Rh supported on CdS semiconductor particles. J Chem Soc Chem Commun 15:1029–1030
248. Peng XG, Manna L, Yang WD, Wickham J, Scher E, Kadavanich A, Alivisatos AP (2000) Shape control of CdSe nanocrystals. Nature 404:59–61
249. Salant A, Amitay-Sadovsky E, Banin U (2006) Direct self-assembly of gold-tipped CdSe nanorods. J Am Chem Soc 128:10006–10007
250. Wang X, Zhuang J, Peng Q, Li Y (2005) A general strategy for nanocrystal synthesis. Nature 437:121–124
251. Shi H, Fu X, Zhou X, Hu Z (2006) Preparation of organic fluids with high loading concentration of Ag_2S nanoparticles using the extractant Cyanex301. J Mater Chem 16:2097–2101
252. Motte L, Pileni MP (1998) Influence of length of alkyl chains used to passivate silver sulfide nanoparticles on two- and three-dimensional self-organization. J Phys Chem B 102:4104–4109
253. Gao F, Lu Q, Zhao D (2003) Controllable assembly of ordered semiconductor Ag_2S nanostructures. Nano Lett 3:85–88
254. Liu J, Raveendran P, Shervani Z, Ikushima Y (2004) Synthesis of Ag_2S quantum dots in water-in-CO_2 microemulsions. Chem Commun 2582–2583Please provide the volume number for Ref. [254].This journal has no volume number.
255. Wang X, Liu W, Hao J, Fu X, Xu B (2005) A simple large-scale synthesis of well-defined silver sulfide semiconductor nanoparticles with adjustable sizes. Chem Lett 34:1664–1665
256. Gao F, Lu Q, Komarneni S (2005) Interface reaction for the self-assembly of silver nanocrystals under microwave-assisted solvothermal conditions. Chem Mater 17:856–860
257. Brelle MC, Zhang JZ, Nguyen L, Mehra RK (1999) Synthesis and ultrafast study of cysteine- and glutathione-capped Ag_2S semiconductor colloidal nanoparticles. J Phys Chem A 103:10194–10201
258. Yang L, Xing R, Shen Q, Jiang K, Ye F, Wang J, Ren Q (2006) Fabrication of protein-conjugated silver sulfide nanorods in the bovine serum albumin solution. J Phys Chem B 110:10534–10539
259. Mo X, Krebs MP, Yu SM (2006) Directed synthesis and assembly of nanoparticles using purple membrane. Small 2:526–529
260. Lee JS, Shevchenko EV, Talapin DV (2008) Au-PbS core-shell nanocrystals: plasmonic absorption enhancement and electrical doping *via* intra-particle charge transfer. J Am Chem Soc 130:9673–9675
261. Michaelson HB (1977) The work function of the elements and its periodicity. J Appl Phys 48:4729–4733
262. Yeager E (1984) Electrocatalysts for O_2 reduction. Electrochim Acta 29:1527–1537

263. Aricò AS, Srinivasan S, Antonucci V (2001) DMFCs: from fundamental aspects to technology development. Fuel Cells 1:133–161
264. Toda I, Igarashi H, Uchida H, Watanabe M (1999) Enhancement of the electroreduction of oxygen on Pt alloys with Fe Ni and Co. J Electrochem Soc 146:3750–3756
265. Xu C, Goodman DW (1996) Adsorption and reaction of formic acid on a pseudomorphic palladium monolayer on Mo(110). J Phys Chem 100:245–252
266. Baldauf M, Kolb DM (1996) Formic acid oxidation on ultrathin Pd films on Au(*hkl*) and Pt (*hkl*) electrodes. J Phys Chem 100:11375–11381
267. Kibler LA, Kolb DM (2003) Physical electrochemistry: recent developments. Z Phys Chem 217:1265–1280
268. Kibler LA, El-Aziz AM, Hoyer R, Kolb DM (2005) Tuning reaction rates by lateral strain in a palladium monolayer. Angew Chem Int Ed 44:2080–2084
269. Kumar S, Zou S (2007) Electrooxidation of carbon monoxide and methanol on platinum-overlayer-coated gold nanoparticles: effects of film thickness. Langmuir 23:7365–7371
270. Naohara H, Ye S, Uosaki K (2000) Electrocatalytic reactivity for oxygen reduction at epitaxially grown Pd thin layers of various thickness on Au(111) and Au(100). Electrochim Acta 45:3305–3309
271. Naohara H, Ye S, Uosaki K (2001) Thickness dependent electrochemical reactivity of epitaxially palladium thin layers on Au(111) and Au(110) surfaces. J Electroanal Chem 500:435–445
272. El-Aziz AM, Kibler LA (2002) Influence of steps on the electrochemical oxidation of CO adlayers on Pd(111) and on Pd films electrodeposited onto Au(111). J Electroanal Chem 534:107–114
273. Kibler LA, El-Aziz AM, Kolb DM (2003) Electrochemical behaviour of pseudomorphicoverlayers: Pd on Au(111). J Mol Catal A Chem 199:57–63
274. Yang JH, Cheng CH, Zhou W, Lee JY, Liu Z (2010) Methanol-tolerant heterogeneous PdCo@PdPt/C electrocatalyst for the oxygen reduction reaction. Fuel Cells 10:907–913
275. Yang JH, Lee JY, Zhang Q, Zhou W, Liu Z (2008) Carbon-supported pseudo-core-shell Pd-Pt nanoparticles for ORR with and without methanol. J Electrochem Soc 155:B776–B781
276. Yang JH, Zhou W, Cheng CH, Lee JY, Liu Z (2010) Pt-decorated PdFe nanoparticles as methanol-tolerant oxygen reduction electrocatalyst. ACS Appl Mater Interfaces 2:119–126
277. Adzic RR, Zhang J, Sasaki K, Vukmirovic MB, Shao M, Wang JX, Nilekar AU, Mavrikakis M, Valerio JA, Uribe F (2007) Platinum monolayer fuel cell electrocatalysis. Top Catal 46:249–262
278. Shemesh Y, Macdonald JE, Menagen G, Banin U (2011) Synthesis and photocatalytic properties of a family of CdS-Pd_X hybrid nanoparticles. Angew Chem Int Ed 123:1217–1221
279. Shaviv E, Schubert O, Alves-Santos M, Goldoni G, Felice RD, Vallée F, Fatti ND, Banin U, Sönnichsen C (2011) Absorption properties of metal-semiconductor hybrid nanoparticles. ACS Nano 5:4712–4719
280. Kim SW, Kim M, Lee WY, Hyeon T (2002) Fabrication of hollow palladium spheres and their successful application to the recyclable heterogeneous catalyst for Suzuki coupling reactions. J Am Chem Soc 124:7642–7643
281. Li Y, Zhou P, Dai Z, HuZ SP, Bao J (2006) A facile synthesis of PdCo bimetallic hollow nanospheres and their application to Sonogashira reaction in aqueous media. New J Chem 30:832–837
282. Cheng F, Ma H, Li Y, Chen J (2007) $Ni_{1-x}Pt_x$ ($x = 0$-0.12) hollow spheres as catalysts for hydrogen generation from ammonia borane. Inorg Chem 46:788–794
283. Liu H, Ye F, Cao H, Ji G, Lee JY, Yang J (2013) A core–shell templated approach to the nanocomposites of silver sulfide and noble metal nanoparticles with hollow/cage-bell structures. Nanoscale 5:6901–6907
284. Feng Y, Liu H, Wang P, Ye F, Tan Q, Yang J (2014) Enhancing the electrocatalytic property of hollow structured platinum nanoparticles for methanol oxidation through a hybrid construction. Sci Rep 4:6204

285. Wang X, Waje M, Yan Y (2004) Methanol resistant cathodic catalyst for direct methanol fuel cells. J Electrochem Soc 151:A2183–A2188
286. Xia BY, Wu HB, Yan Y, Low XW, Wang X (2013) Ultrathin and ultralong single-crystal platinum nanowire assemblies with highly stable electrocatalytic activity. J Am Chem Soc 135:9480–9485
287. Liu F, Lu G, Wang C-Y (2006) Low crossover of methanol and water through thin membranes in direct methanol fuel cells. J Electrochem Soc 153:A543–A555
288. Du CY, Zhao TS, Yang WW (2007) Effect of methanol crossover on the cathode behavior of a DMFC: a half-cell investigation. Electrochim Acta 52:5266–5271
289. Jia N, Lefebvre MC, Halfyard J, Qi Z, Pickup PG (2000) Modification of Nafion proton exchange membranes to reduce methanol crossover in PEM fuel cells. Electrochem Solid-State Lett 3:529–531
290. Gurau B, Smotkin ES (2002) Methanol crossover in direct methanol fuel cells: a link between power and energy density. J Power Sources 112:339–352
291. Kim Y-M, Park K-W, Choi J-H, Park I-S, Sung Y-E (2003) A Pd-impregnated nanocomposite Nafion membrane for use in high-concentration methanol fuel in DMFC. Electrochem Commun 5:571–574
292. Sahu AK, Pitchumani S, Sridhar P, Shukla AK (2009) Nafion and modified-Nafion membranes for polymer electrolyte fuel cells: An overview. Bull Mater Sci 32:285–294
293. Zhang H, Huang H, Shen PK (2012) Methanol-blocking Nafion composite membranes fabricated by layer-by-layer self-assembly for direct methanol fuel cells. Int J Hydrogen Energy 37:6875–6879
294. Zhang Y, Cai W, Si F, Ge J, Liang L, Liu C, Xing W (2012) A modified Nafion membrane with extremely low methanol permeability *via* surface coating of sulfonated organic silica. Chem Commun 48:2870–2872
295. Beauger C, Lainé G, Burr A, Taguet A, Otazaghine B, Rigacci A (2013) Nafion®-sepiolite composite membranes for improved proton exchange membrane fuel cell performance. J Membr Sci 130:167–179
296. Feng Y, Yang J, Liu H, Ye F, Yang J (2014) Selective electrocatalysts toward a prototype of the membraneless direct methanol fuel cell. Sci Rep 4:3813
297. Cameron DS, Hards GA, Harrison B, Potter RJ (1987) Direct methanol fuel cells. Platinum Metals Rev 31:173–181
298. Ferrigno R, Stroock AD, Clark TD, Mayer M, Whitesides GM (2002) Membraneless vanadium redox fuel cell using laminar flow. J Am Chem Soc 124:12930–12931
299. Jayashree RS, Egas D, Spendelow JS, Natarajan D, Markoski LJ, Kenis PJA (2006) Air-breathing laminar flow-based direct methanol fuel cell with alkaline electrolyte. Electrochem. Solid State Lett 9:A252–A256
300. Lam A, Wilkinson DP, Zhang J (2008) Novel approach to membraneless direct methanol fuel cells using advanced 3D anodes. Electrochim Acta 53:6890–6898
301. Lam A, Wilkinson DP, Zhang J (2009) Control of variable power conditions for a membraneless direct methanol fuel cell. J Power Sources 194:991–996
302. Lam A, Dara MS, Wilkinson DP, Fatih K (2012) Aerobic and anaerobic operation of an active membraneless direct methanol fuel cell. Electrochem Commun 17:22–25
303. Tominaka S, Ohta S, Obata H, Momma T, Osaka T (2008) On-chip fuel cell: micro direct methanol fuel cell of an air-breathing, membraneless, and monolithic design. J Am Chem Soc 130:10456–10457

Chapter 4
Synthesis and Electrocatalysis of Pt-Pd Bimetallic Nanocrystals for Fuel Cells

Ruizhong Zhang and Wei Chen

4.1 Introduction

Over the past few decades, research on novel environment-friendly energy conversion devices and their potential applications has attracted extensive interest due to the depletion of fossil fuel reserves and thus the growing demand for efficient but low-cost renewable energy [1, 2]. Fuel cells, such as proton-exchange membrane fuel cells (PEMFCs), direct methanol fuel cells (DMFCs), and direct formic acid fuel cells (DFAFCs), have been considered as the most promising power sources because of high power density, high energy-conversion efficiency, and zero or low emission of pollutants [3–6]. For fuel cells, platinum (Pt) has received unremitting interest as an electrocatalyst because of the highest catalytic activity among the studied catalysts for electro-oxidation of small organic fuels on the anode and oxygen reduction on the cathode [7–10]. However, several issues arise from using pure Pt as fuel cell catalysts. First, Pt surfaces are easily self-poisoned by the strong adsorption of CO intermediates originated from small organic fuel oxidation, leading to a severe decrease in the catalytic performance [11, 12]. Second, using Pt as cathodic catalysts in DMFCs, methanol molecules crossover from anode to cathode may lower the ORR performance because of the mixed potentials formed from the simultaneous methanol oxidation and oxygen reduction [13–15]. Third, the limited reserve in nature and the resulting sky-rocketing price of Pt has become

R. Zhang
State Key Laboratory of Electroanalytical Chemistry, Changchun Institute of Applied Chemistry, Chinese Academy of Sciences, Changchun, Jilin 130022, China

University of Chinese Academy of Sciences, Beijing 100039, China

W. Chen (✉)
State Key Laboratory of Electroanalytical Chemistry, Changchun Institute of Applied Chemistry, Chinese Academy of Sciences, Changchun, Jilin 130022, China
e-mail: weichen@ciac.ac.cn

K.I. Ozoemena, S. Chen (eds.), *Nanomaterials for Fuel Cell Catalysis*, Nanostructure Science and Technology, DOI 10.1007/978-3-319-29930-3_4

one of the major barriers for the wide commercialization of fuel cells [16]. To improve the catalytic activities and lower the costs of catalysts in fuel cells, much effort has been devoted to combining Pt with another metal that is less constrained in terms of reserve and economic dependence, to form bimetallic nanocatalysts. As compared to monometallic Pt, bimetallic Pt-based nanocrystals are expected to display not only a combination of the properties associated with two distinct metals, but also new and unexplored properties because of a possible bi-functional mechanism or the so-called ligand effect [17–19].

Among the metals around Pt in the periodic table, palladium (Pd) is probably the best candidate to generate bimetallic electrocatalysts with Pt due to the following obvious advantages. First, Pd and Pt share the same face-centered cubic (fcc) structure and almost identical lattice constants (with a mismatch of only 0.77%), both of which are beneficial to the formation of Pt-Pd bimetallic nanocrystals with single crystallinity. Second, the combination of Pt with Pd has a crucial impact on the electronic structure of Pt, resulting in superior electrocatalytic activities for a specific reaction owing to the formation of Pt-Pd bonds [20, 21]. According to the studies by Nørskov and co-workers [22–25], the strain and electronic coupling presented in a catalyst play a key role in determining the catalytic properties because both of the two parameters can result in shifts of the d-band center (calculated with respect to the Fermi level), which is directly related to the adsorption energies of reactants on a catalyst as well as their activation barriers. Of them, a compressive strain tends to down-shift the energy of the d-band center, causing adsorbates to bind less strongly to the catalyst, while a tensile strain has the opposite effect. On the other hand, the electronic coupling can result in shifts for the d-band center due to a change in the density of states near the Fermi level. As for Pt-Pd bimetallic catalysts, a small compressive strain resulting from their weak lattice contract can cause a downshift of the d-band center, thus lowering the binding strength for the adsorbed intermediates. As a consequence, the changes in the d-band properties of Pt caused by its combination with Pd, and the small mismatch in the lattice constants between Pt and Pd, makes Pt-Pd bimetallic nanocrystals attractive catalysts for various reactions. Taking ORR as an example, the rate-limiting step at high potential is the desorption of intermediates (O, OH, etc.) produced during the reaction. The weakened binding strength of the intermediates on a Pt-Pd bimetallic catalyst can activate the Pt sites required for the adsorption of O_2 and then the splitting of the O-O bond, therefore accelerating the kinetics of oxygen reduction [26]. In addition, a strong interaction between Pt and Pd due to the formation of Pt-Pd bonds can also change the electronic structure of Pt, causing the enhanced amount of O_2 adsorbed on the surface of Pt and thus improving ORR catalytic activity.

Moreover, it has been identified that the dissolution of Pt is a major reason for the degradation of catalysts in proton-exchange membrane (PEM) fuel cells due to the presence of dissolved molecular oxygen, highly aggressive condition in terms of acidic pH, and the highly positive potential for device operation [27]. Therefore, compared to other metals such as Ag, Cu, Co or Ni, the use of Pd may also help minimize the corrosion and loss of catalysts in an acidic environment such as the medium of a PEM fuel cell [28]. Especially, the introduction of Pd can prevent the electrocatalysts from degradation to a certain extent by up-shifting the dissolution potential of Pt and thus assure long-term stability [29]. From these perspectives, it has been actively explored Pt-Pd bimetallic catalysts for a variety of reactions with enhanced performance.

In addition to elemental compositions, the size, shape and structure of Pt-Pd bimetallic system can also be finely manipulated to further enhance their catalytic performance. Since electrocatalytic reactions are very sensitive to the exposed crystal facets and the proportion of atoms located at the facets, edges or corners of catalysts, recent years tremendous efforts have been devoted to the syntheses of Pt-Pd bimetallic nanocrystals with well-defined shapes in high yields and purity by tuning various experimental parameters [30]. By using Pd nanocrystals as the seeds for overgrowth or sacrificial templates for galvanic replacement, Pt-Pd alloys with different structures/shapes including alloys, core-shell, dendrities, alternating multi-shells, and atomic monolayer, can be easily synthesized by different strategies. These preparation methods include co-chemical reduction and its combination with galvanic replacement to generate alloy nanocrystals, galvanic replacement between Pd nanocrystals and a Pt salt precursor for generating dendritic nanostructures, seed-mediated overgrowth for generating core-shell, multi-shell, and dendritic nanostructures, and a combination of electrochemical deposition and galvanic replacement for generating Pt monolayer on Pd nanocrystals, etc.

Besides the catalytic activity and costs, the stability and lifetime of an electrocatalyst are also critical issues for its practical applications in fuel cells. When Pt-Pd bimetallic nanocrystals are used as fuel cell catalysts, the catalyst support also plays an important role in determining the catalytic properties. For a catalyst support, it should have a high surface area for catalyst dispersion, a strong affinity to immobilize the catalyst particles, a high electrical conductivity to accelerate electron transfer in redox reactions, and a high electrochemical stability in acidic or alkaline electrolytes to ensure a stable structure. Up to now, various carbon materials including traditional carbon materials (e.g., Vulcan XC-72R carbon black) and nanocarbon materials (e.g., carbon nanotubes, graphene and ordered mesoporous carbon) have been used for the dispersion of catalyst particles. However, earlier studies have shown that except for the dissolution and aggregation of metal nanoparticles, the severe corrosion and oxidation of carbon support materials in the harsh operating environment could also lead to quick degradation of the electrocatalytic performance [31, 32]. Therefore, in recent years great efforts have also been devoted to addressing the challenges of catalyst supports via developing non-carbon support materials such as metal oxides, electronically conductive polymer, nitrides and carbides [33].

Herein, we will first summarize recent progress in the development of experimental techniques for the preparation of unsupported/supported Pt-Pd bimetallic nanocrystals with unique structures, and then focus on their applications in fuel cells as anode and cathode catalysts.

4.2 Synthesis of Pt-Pd Bimetallic Nanocrystals

It is well-known that the catalytic activities of nanoparticle electrocatalysts are strongly dependent on their composition, shapes, size, exposed surface planes and the interactions between nanocrystals and catalyst supports. In the past few decades, various methods have been applied in the synthesis of Pt-Pd bimetallic nanocrystals with structures in the form of alloy, core-shell, dendrities, multi-shells, and supported monolayer. These well-defined Pt-Pd nanostructures can be generally prepared through co-chemical reduction, galvanic replacement, seed-mediated growth and electrochemical deposition synthetic routes. Typically, co-chemical reduction synthesis of Pt-Pd bimetallic nanocrystals refers to the reduction of Pt and Pd precursors in the presence of capping agents and/or stabilizers. Pt-Pd nanoalloys can also be obtained through *in situ* oxidation and dissociation of Pd nanocrystals by galvanic replacement with a Pt salt precursor. For a seed-mediated growth approach, the well-defined shapes of Pt or Pd as seeds for epitaxial growth of a Pd or Pt-shell. It is also feasible to control the shell by using electrochemical deposition methods. In this section, we summarize various leading synthetic techniques and the formation mechanisms for the preparation of Pt-Pd nanocrystals.

4.2.1 Co-chemical Reduction Method

In the past decades, co-chemical reduction has proved to be a straightforward strategy for the facile synthesis of bimetallic nanocrystals. With this technique, it is feasible to achieve simultaneous reduction of both Pt and Pd salt precursors in the presence of a capping agent and/or a stabilizer due to their similar electrochemical potentials of 0.74 V (versus a reversible hydrogen electrode) for $PtCl_6^{2-}$/Pt and 0.62 V for $PdCl_4^{2-}$/Pd. The obtained Pt-Pd bimetallic nanocrystals are interesting for various electrocatalytic reactions due to the co-existence of Pt and Pd atoms on the nanocrystal surfaces [34]. To date, various reducing agents have been used to co-reduce Pd and Pt salt precursors for the synthesis of Pt-Pd alloy nanocrystals, including sodium borohydride, alcohol, formic acid, formaldehyde, hydrazine, ascorbic acid, etc. By using sodium borohydride ($NaBH_4$) as a reductant, Crooks and coworkers [35] prepared Pt-Pd bimetallic nanoparticles via the co-reduction of K_2PtCl_4 and K_2PdCl_4 in a poly(amidoamine) dendrimer (G6-OH) aqueous solution. Two steps are involved in the process: (i) loading Pt and Pd ions into the dentrimers by fully complexing with the interior amines of dendrimers, and (ii) co-reducing the complexed Pd and Pt ions by $NaBH_4$. Interestingly, the synthesized bimetallic Pt-Pd nanoparticles by this process have almost the same diameter (~1.8 nm) containing an average of 180 atoms but with seven different Pt/Pd ratios. Here, the dendrimer template plays a vital role in controlling the size of Pt-Pd nanoparticles (smaller than 3 nm) and the composition variation (molar ratios adjustable from 1:5 to 5:1). TEM and single-particle energy-dispersive spectroscopy (EDS) indicated that the calculated particle diameter and Pt/Pd ratios in the Pt and Pd precursors are very consistent with the measured ones. Therefore this dendrimer-templating is a unique method for preparing nanoparticles having particular Pt/Pd ratios, uniform size and composition. In another work, one-dimensional ultrathin Pt-Pd alloy nanowires were synthesized in a cetyltrimethylammonium (CTAB)-assisted water-chloroform micelles system [36]. In this synthesis, a mixed aqueous solution of Pt and Pd salts was mixed with a chloroform solution of CTAB, followed by the addition of a $NaBH_4$ aqueous solution. In the mixed solution, the reduction of precursor ions and metal growth occurred in the swollen wormlike micelle networks of chloroform droplets with the CTAB molecules. Because of the strict limitation of the wormlike micelle networks, the obtained Pt-Pd nanowires showed an average size of 2.5 nm with a narrow diameter distribution, and both Pt and Pd can be co-reduced to form alloy with an atomic ratio of about 1:1, as confirmed by EDX measurements.

When Pt-Pd bimetallic materials are used as fuel cell catalysts, they are usually dispersed on a catalyst support. Among the used catalyst supports, Vulcan XC-72 carbon black is the most popular carbon support for immobilizing and stabilizing Pt-Pd nanoparticles [37–44], while carbon nanotubes [45–48], conducting polymer composite matrix [49], and tungsten carbide [50] have also been explored as potential supporting materials. For instance, Zhang et al. [51] synthesized a series of Pd_xPt_{1-x} nanoparticles dispersed on carbon black by reducing the mixture of Pd (II) and Pt(II) precursors by $NaBH_4$ in the presence of Vulcan XC-72 carbon. In the synthesis, EDTA was also used as a chelator for Pd and Pt ions to ensure the co-reduction of Pd(II) and Pt(II) species. The highly dispersed Pd_xPt_{1-x} nanoparticles on carbon black exhibited composition-dependent catalytic activity for formic acid electro-oxidation and the $Pd_{0.9}Pt_{0.1}$ nanoparticles with a mean size of 3.2 nm showed the best performance among the series.

In addition to strong reducing agents, Pt-Pd bimetallic nanostructures have also been prepared by a polyol process, in which mild reductants such as ethylene glycol [52–68], methanol [69], glycerol [70, 71], and 1,4-butanediol [53], have been used. It is well-known that Pt nanocrystals with different exposed surfaces have different electronic structures and surface atomic arrangements, and the appropriate crystal phase and/or composition can greatly enhance reaction kinetics due to the minimized surface energy and total excess free energy. Therefore, much work has been done to prepare shape-controlled metallic nanostructures with desired exposed crystal facets. For instance, Lee and co-workers [72] reported a glycerol reduction method to synthesize Pt-Pd nanoparticles exhibiting dominantly exposed (1 1 1) facets in octahedral shape with complete alloy formation between Pt and Pd metallic phases. During the fast reduction process, the thermodynamically minimized crystalline surface energy and thus the formation of crystal facets with a low surface energy in the Pt-Pd structure can improve the electrocatalytic performance of the obtained carbon supported Pt-Pd nanoparticles composite toward methanol oxidation. By using a hollow-core mesoporous shell (HCMS) carbon as support, Berker et al. [65] reported a rapid method to synthesize Pt-Pd/HCMS composite by the co-reduction of H_2PtCl_6 and $PdCl_2$ using ethylene glycol as a reducing agent under microwave irradiation. It was proposed that the HCMS carbon facilitated the diffusion of hydrogen and oxygen as well as the water transport within fuel cells, leading to significantly improved fuel cell performance. Formic acid has also been used to synthesize Pt-Pd bimetallic nanocrystals [73–75]. Guo et al. [76] demonstrated a simple procedure to synthesize Pt/Pd hybrid supported on polyaniline (PANI) nanofibers with high conductivity and surface-to-volume ratios. In this method, the PANI nanofibers were first synthesized by a wet-chemical approach and then the as-obtained 1D PANI nanofibers were added into a mixture containing H_2PtCl_6 and H_2PdCl_4. HCOOH was finally injected into the above mixture to reduce the precursors at room temperature. The SEM and TEM images shown in Fig. 4.1a, b indicate that a large number of PANI nanofibers with a diameter of 60–100 nm have been obtained. Compared with the smooth surface of pristine PANI, a rougher surface of the as-prepared PANI/Pt and PANI/PtPd indicates that Pt (Fig. 4.1c, d) and PtPd nanoparticles (Fig. 4.1e, f) have been successfully deposited on the surface of PANI fibers. Moreover, the corresponding magnified images further revealed that the small PtPd nanoparticles formed a network and thus leading to many nanoporous structures on the surface of hybrid nanofibers, which favor a high electrocatalytic activity.

Most recently, by using graphene as support, Wang and co-workers [77] used the formic acid method to prepare graphene-supported 1D PtPd nanorods (G-PtPd NRs) by co-reducing H_2PtCl_6 and $Pd(NO_3)_2$ in the presence of HCOOH. In comparison with the carbon-supported PtPd NRs and graphene-supported Pt NRs, the G-PtPd NRs showed a larger diameter of about 4.4 nm and a longer length of about 35 nm measured from the corresponding TEM images. Extended X-ray absorption fine structure (EXAFS) studies confirmed the formation of G-PtPd alloy with a Pt-rich inner core and a Pd-rich outer shell. XRD patterns indicated that the growth of (1 1 1) and (2 2 0) planes of G-PtPd NRs was promoted for the alloying of Pt and Pd on graphene support. In addition, the results of X-ray absorption near edge spectroscopy (XANES) also showed that using graphene as a support and alloying with Pd synergistically modified the d-band of Pt, and the total number of unoccupied d-states for G-PtPd was reduced to as low as 0.295. All these results suggested that the G-PtPd had lower unfilled d-states and more d-band electrons were transferred from Pd to Pt, resulting in enhanced ORR performance.

Meanwhile, a simple microemulsion method was also developed to construct Pt-Pd nanoparticles [78, 79]. With this technique, the preparation of metal nanoparticles is realized by mixing two different micro-emulsions carrying the specific reactants (metallic salts and reducing agent) dissolved in aqueous phase. Microemulsion method has been accepted to be a unique method for the production of metal particles with a very narrow size distribution. For example, Escudero and coworkers [80] prepared alloyed Pt-Pd nanoparticles by reducing H_2PtCl_6 and $PdCl_2$ with hydrazine in a water-in-oil micro-emulsion of water/Berol 050/*iso*-octane. The obtained Pt-Pd nanoparticles were smaller than 5 nm and exhibited potential application in fuel cells.

In recent years, microwave-assisted technique has also been applied to the synthesis of metal nanoparticles [81]. Compared with the traditional chemical

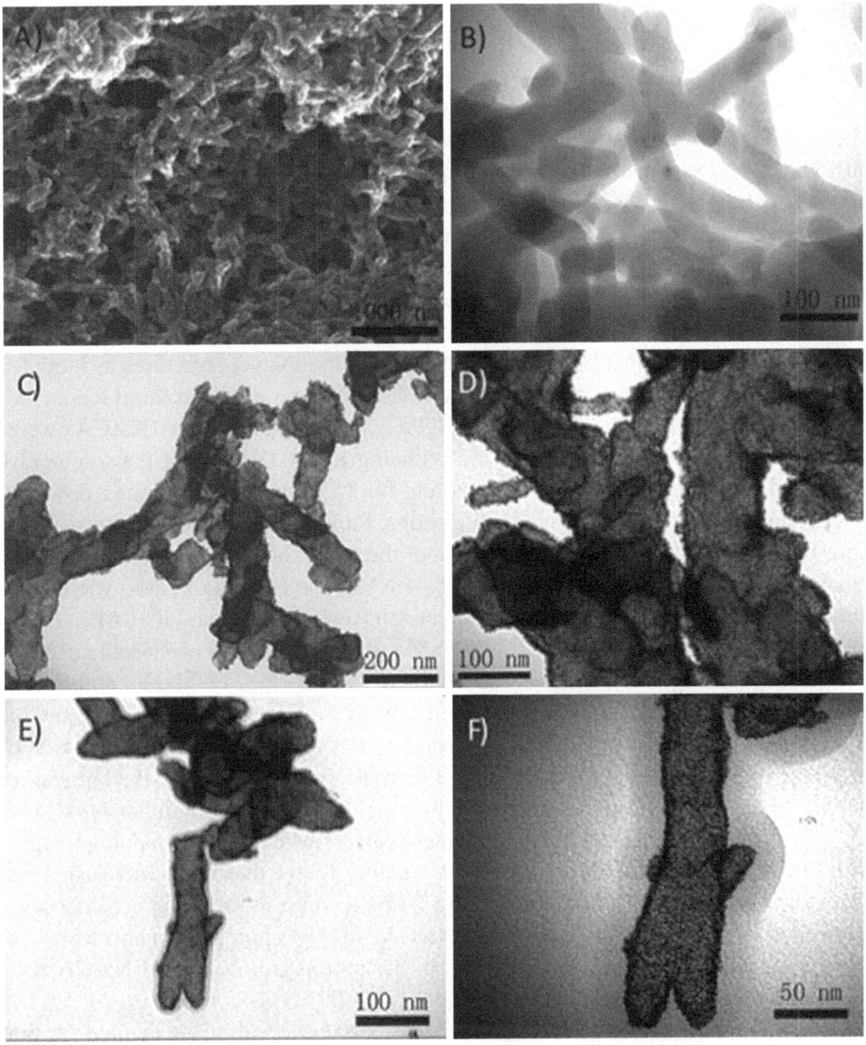

Fig. 4.1 (**a**) SEM and (**b**) TEM images of PANI nanofibers. (**c**, **d**) TEM images of PANI nanofiber-supported supra-high density Pt NPs. (**e**, **f**) TEM images of PANI nanofiber-supported Pt/Pd NPs at different magnifications. Reprinted with permission from [76]. Copyright 2009 Wiley-VCH

methods, the microwave heating provides homogeneous reaction conditions and a fast kinetic process. Zhang and coworkers [82] reported a simple method to synthesize three-dimensional Pd@Pt core-shell nanostructures with a high yield via the co-reduction of K_2PtCl_4 and $PdCl_2$ in the presence of CTAB by using ascorbic acid as a reducing agent under microwave irradiation. It was shown that the long alkyls of CTAB served as a shape-control agent to tailor the structure and improve the dispersion of nanoparticles in the synthesis. TEM and SEM characterizations showed that the morphology of the synthesized Pd@Pt nanostructures can be easily controlled by changing the molar ratio between Pt and Pd precursors. And the Pd@Pt morphology was changed from cubic to spherical shapes by decreasing the Pd/Pt molar ratio (3:1 to 1:3). The authors also studied the shape effect on the eletrocatalytic activity for MOR and ORR, and among the different Pd@Pt nanostructures, the sample at a Pd/Pt molar ratio of 1:3 exhibited the best catalytic activity.

Interfacial self-assembly was found to be a simple and effective strategy for the synthesis of noble metal nanomaterials, and the liquid-liquid, liquid-air and oil-water-air interfaces have showed promise as flexible 2D platforms for nanoparticles assembly [83–87]. Wu et al. [88] prepared macroscopic free-standing Pd/Pt nanomembranes (Pd/Pt-FNMs) with well-defined monolayer structures as large as several square centimeters by one-step self-assembly at a water-air interface. In the synthesis, aqueous solutions of $PdCl_2$, K_2PtCl_6 and sodium citrate (Na_3CA) were mixed in a beaker. Subsequently, the reducing agent $NH_2OH{\cdot}HCl$ was quickly added into the above solution. After boiling for 15 min and then cooling down to room temperature, a monolayer of macroscopic Pd/Pt-FNMs floated at the water-air interface and covered the solution surface of the whole beaker. The composition of the Pd/Pt-FNMs ($Pd_{53}Pt_{47}$-FNMs, $Pd_{33}Pt_{67}$-FNMs, and $Pd_{15}Pt_{85}$-FNMs) was easily tuned by adding a different amount of precursors and confirmed by ICP-AES. From the TEM images, the obtained Pd/Pt-FNMs exhibited similar well-defined morphologies, and Fig. 4.2a–c shows the details of the $Pd_{33}Pt_{67}$-FNMs. It can be seen that the Pd/Pt nanoparticles have a uniform size with an average diameter of 9.25 nm. Although no obvious grain boundaries between Pd and Pt are observed from the high-resolution TEM (HRTEM) measurements (Fig. 4.2d), the bimetallic nature of the FNMs can be clearly seen from the high-angle annular dark-field (HDDAF-STEM) image and the nanometer-scale TEM elemental mapping images (Fig. 4.2e). In the mechanistic study, the authors found that when increasing the dosage of Na_3CA, Pd/Pt nanomembranes failed to form at the water-air interface. However, with decreasing dosage of Na_3CA, highly aggregated nanostructures were produced. Therefore, the presence of an appropriate dosage of Na_3CA was believed to be the key to achieve high-quality Pd/Pt-FNMs.

In another study, Sun and coworkers [89] reported an oil-phase method for the synthesis of polyhedral Pd-Pt alloy nanocrystals with a controlled size (3.5–6.5 nm) and composition ($Pd_{88}Pt_{12}$ to $Pd_{34}Pt_{66}$). This synthesis involves co-reduction of Pd $(acac)_2$ and $Pt(acac)_2$ with morpholine borane (MB) in oleylamine (OAm) at different temperatures. The TEM and HAADF-STEM images confirmed the formation of single-crystal nanoparticles in high yield with a uniform size. Both Pd and Pt are uniformly distributed throughout each particle from EDX mapping measurements. The linear dependence with a nearly unity slope implies that the two precursors were co-reduced at the same rate in generating the bimetallic nanocrystals. The formation of a single complex between the Pt and Pd precursors with the capping agent facilitates the simultaneous reduction in an elevated oil-phase. In a typical synthesis, the composition of the Pt-Pd alloy nanocrystals could be adjusted by controlling the amounts of $Pd(acac)_2$ and $Pt(acac)_2$ added into

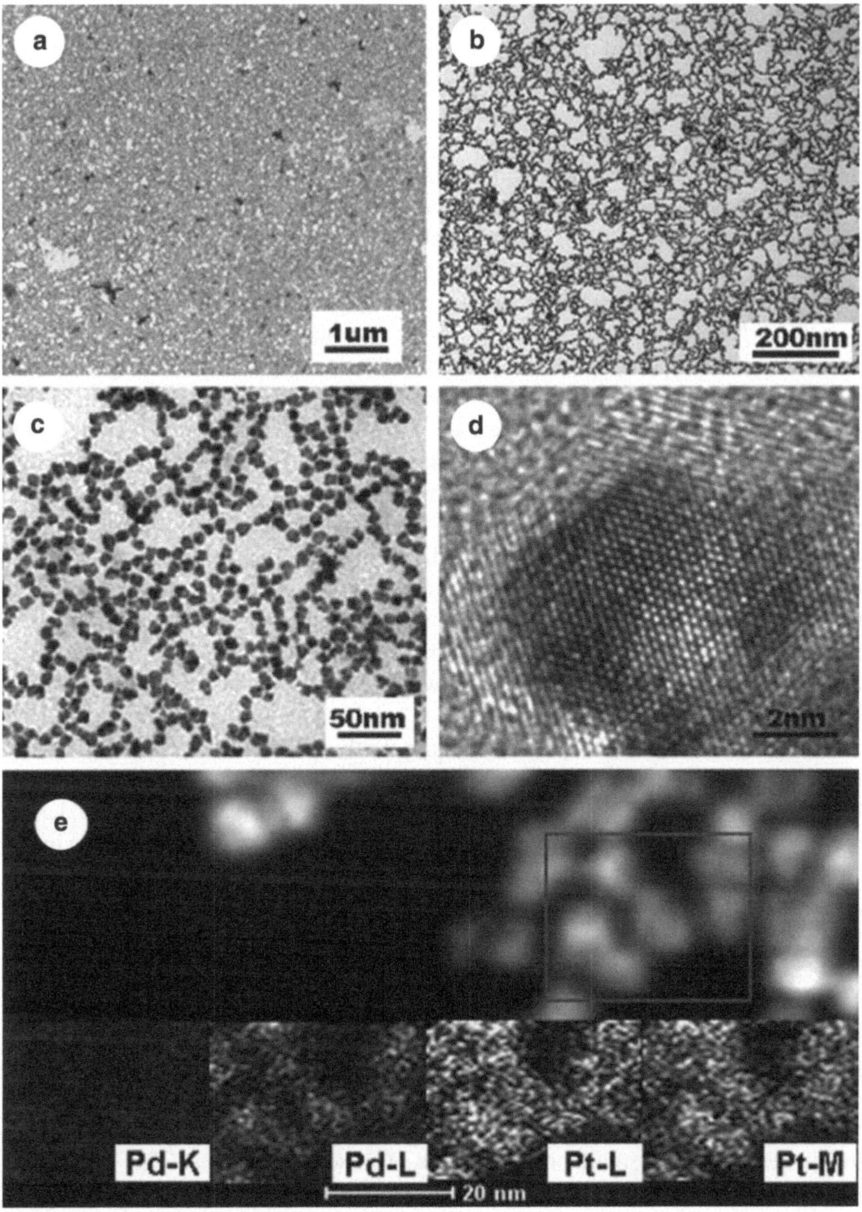

Fig. 4.2 (**a**–**c**) TEM images, (**d**) HRTEM image, and (**e**) HAADF-STEM image and the corresponding nanometer-scale TEM elemental mapping (*inset*) of the $Pd_{33}Pt_{67}$-FNMs. Reprinted with permission from [88]. Copyright 2012 Wiley-VCH

the reaction solution. More importantly, at a constant composition, the size of the Pt-Pd nanocrystals could be easily tuned by varying the temperature at which MB was injected. When MB was injected at 40, 60 and 80 °C, 6.5, 5 and 4.5 nm $Pd_{67}Pt_{33}$ nanoparticles can be formed, respectively. The as-synthesized Pt/Pd nanoparticles exhibited highly composition-dependent catalytic activities and high stability for methanol oxidation reaction in acid media.

Although great success has been achieved in the synthesis of Pt-Pd alloy nanoparticles via the afore-mentioned techniques, the products from these syntheses were largely restricted to small nanoparticles (typically less than 5 nm in size) with poorly defined crystallinity and morphology. By heating an aqueous solution containing Na_2PdCl_4, K_2PtCl_4 and poly (vinyl pyrrolidone) (PVP) at 80 °C for 18 h, Xia and coworkers [90] demonstrated a co-reduction approach for the synthesis of Pt-Pd alloy nanocrystals with well-defined shapes and twinned structures. In this process, the commercially available PVP was employed as a weak reducing agent to manipulate the reduction kinetics owing to its hydroxy (OH)-end groups. It was found that the slow reduction rate associated with the weak reducing power of PVP is the prerequisite for the formation of Pt-Pd alloy nanocrystals with twinned structures. From the TEM characterizations in Fig. 4.3a, b, the formed Pt-Pd nanocrystals are mainly star-shaped decahedra with an average size of 40 nm and triangular nanoplates with lateral dimensions of 30–50 nm, as well as a small fraction of other shapes such as octahedra. The compositional line profiles of Pd and Pt on a star-shaped decahedron shown in Fig. 4.3c indicate the formation of Pt-Pd alloy. The HRTEM image (Fig. 4.3d) confirms the presence of fivefold twins from the center of a star-shaped decahedron. In this work, the PVP-mediated slow reduction rate could help retain the particles at small sizes for a long period of time before nucleation. During the period, the small particles easily coalesced into larger particles to reduce surface-to-volume ratio, leading to the formation of twinned structures. When the reaction was conducted using a relatively high-rate reducing agent of ethylene glycol, Pt-Pd nanocrystals with a truncated, octahedral shape were produced, and this fast reduction process favored the formation of Pt-Pd nanocrystals with a single-crystal structure.

Hydrothermal method is another useful and frequently used technique for the preparation of Pt-Pd nanocrystals with the advantages of simplicity, free of templates, and easy shape-control of metal nanocrystals [91–93]. For instance, Yan and coworkers [94] demonstrated a shape-selective synthesis of Pt-Pd nanotetrahedrons (NTs) and nanocubes (NCs) with less than 10 nm in size via a one-step hydrothermal process by using small ions as efficient facet-selective agents (Fig. 4.4a). With a combination of $Na_2C_2O_4$ and formaldehyde as the (1 1 1) facet-selective agent and reducing agent, single-crystalline Pt-Pd NTs enclosed by four (1 1 1) facets with a shape selectivity of ~70% and an average size of 4.9 nm were produced (Fig. 4.4b). The selective adsorption of $C_2O_4^{2-}$ species on the (1 1 1) facets was found to be a critical factor in directing the formation of Pt-Pd NTs. In comparison, Pt-Pd NCs (~8.5 nm) with a shape selectivity of about 88% were produced in the presence of both large amount of Br^- and tiny amount of I^- anions owing to their selective capping for the (100) facet (Fig. 4.4c). Furthermore, reduction rate dependent on

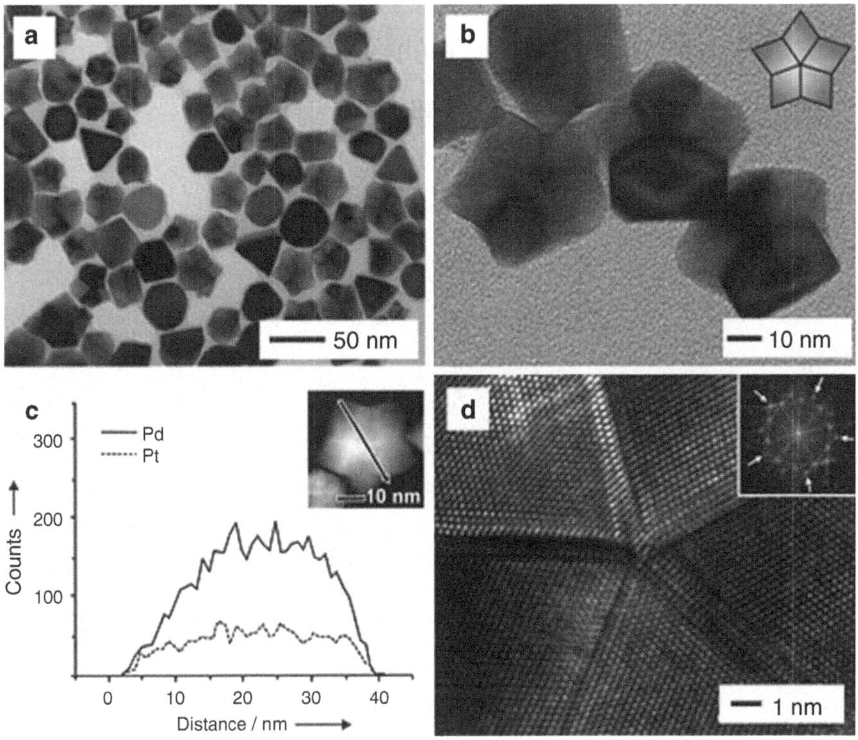

Fig. 4.3 (**a**, **b**) TEM images, (**c**) EDX line-scan, and (**d**) HRTEM image of Pt-Pd alloy nanocrystals synthesized by co-reduction of Na_2PdCl_4 and K_2PtCl_4 with PVP in an aqueous solution. The inset in (**b**), (**c**) and (**d**) correspond to schematic illustration, the HAADF-STEM image, and the Fourier transform pattern of the star-shaped decahedron, respectively. Reprinted with permission from [90]. Copyright 2009 Wiley-VCH

the type of reductant also played a vital role in determining the shape of the Pt-Pd products. A fast enough reducing rate by using a certain amount of formaldehyde could improve the shape- and size-uniformity of the obtained Pt-Pd NTs, while a slow reduction rate (with PVP instead of formaldehyde as the reducing agent) was found to be beneficial to the formation of regularly shaped Pt-Pd NCs. The electrocatalytic studies showed that the obtained alloy nanocrystals exhibit enhanced and facet-dependent catalytic activity and stability for methanol electrooxidation in the order of NCs > NTs > commercial Pt/C.

Besides the intrinsic catalytic properties of catalysts, the support used also plays important roles in determining their catalytic performance. Due to the high electronic conductivity and large surface area, graphene has been recently used as a support material for Pt-Pd nanocrystals dispersion. Our group developed a facile hydrothermal method for the one-pot fabrication of reduced graphene oxide (rGO)-supported Pt-Pd alloy nanocubes (PtPd/rGO) [95]. In a typical procedure, $Pd(acac)_2$ and $Pt(acac)_2$ were mixed with PVP, and NaI in DMF solution of graphene oxide.

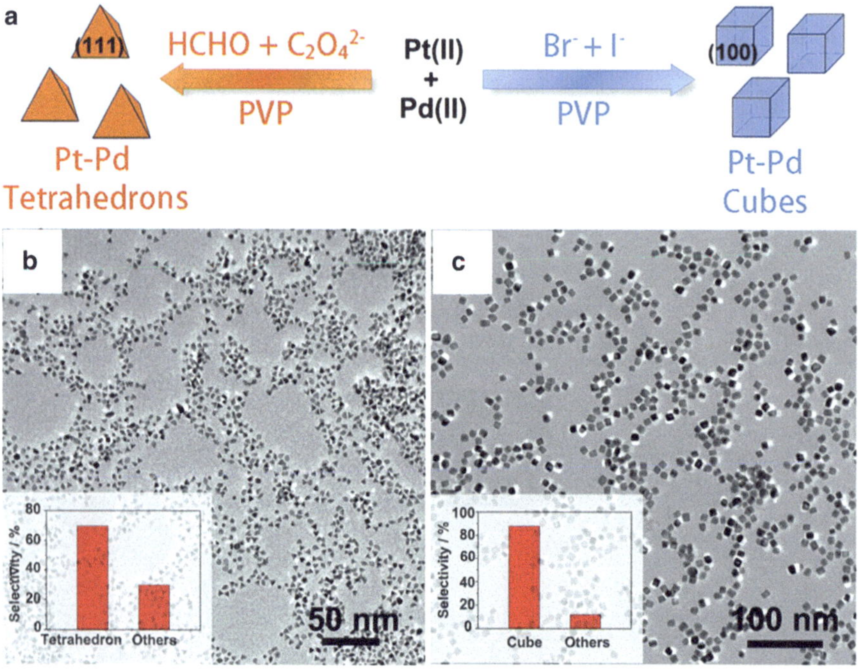

Fig. 4.4 (**a**) Schematic illustration of shape-controlled synthesis of Pt-Pd alloy nanocrystals with tetrahedral and cubic shapes. (**b**, **c**) TEM images of Pt-Pd tetrahedrons and cubes, respectively. The insets in (**b**) and (**c**) show the percentages of the alloy tetrahedrons and cubes, respectively. Reprinted with permission from [94]. Copyright 2011 American Chemical Society

After ultrasonic treatment, the mixed solution was then transferred to a Teflon-lined stainless steel autoclave and heated at 150 °C for 8 h. In this process, two key steps were included: (1) the reduction of graphene oxide (GO) and the nucleation of nanocrystals attached onto the surface of rGO, and (2) the gradual nuclei growth into cubic shapes under the protection of PVP. From HRTEM images and elemental mapping shown in Fig. 4.5a–g, single-crystalline Pt-Pd nanocubes with shape selectivity of 82% and an average size of 8.5 nm were uniformly distributed on the rGO surface. More recently, using GO as both a support material and a structure- and/or morphology-directing agent, rGO-supported PtPd concave nanocubes have also been successfully synthesized through a simple hydrothermal process [96]. In sharp contrast, only cubic PtPd alloy nanocrystals were obtained in the absence of GO. The as-prepared PtPd concave nanocubes exhibited enhanced electrocatalytic activity and durability toward methanol oxidation owing to the exposed high-index facets of {730} and the strong interaction between the catalysts and graphene support.

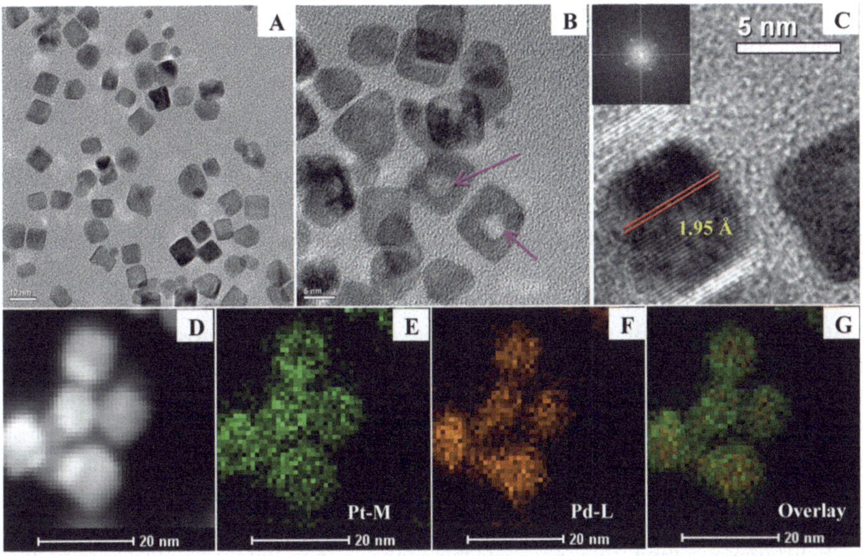

Fig. 4.5 (**a**–**c**) High-resolution TEM micrographs of the PtPd alloy nanocubes supported on rGO at different magnifications. The inset in (**c**) shows the FFT pattern of an individual PtPd nanocrystal. (**d**) The high-angle annular dark-field (HAADF)-STEM image of PtPd/rGO and the corresponding elemental mapping of (**e**) Pt, (**f**) Pd, and (**g**) the overlay. Reprinted with permission from [95]. Copyright 2013 American Chemical Society

4.2.2 Galvanic Replacement and Its Combination with Chemical Reduction

Galvanic replacement is a simple and popular route for controllably constructing various types of bimetallic nanocrystals. Compared with the co-chemical reduction method, galvanic replacement is based on an etching process without using hazardous reducing agents. Therefore, this technique is considered to be a green method for nanocrystals preparation. A galvanic replacement reaction is driven by the different electrochemical potentials between a sacrificial metal template and another metal ion in a solution phase. Typically, this process involves oxidation and dissolution of the template accompanied by reduction of another metal ions and deposition of the resultant atoms on the surface of template. Moreover, the size and morphology of the final product can be easily manipulated by using sacrificial templates with different size and shape and/or by controlling the extent of replacement reaction.

In recent years, galvanic replacement has been applied to the synthesis of supported- and unsupported-PtPd bimetallic nanocrystals [97–100]. By using Ag nanowires as sacrificial templates, Chen et al. [101] synthesized Pt-Pd nanotubes (PtPd NTs) with an average diameter of 45 nm, wall thickness of 7 nm and length of 10 μm. In this synthesis, Ag nanowires were first prepared via a polyol method, and the PtPd NTs were subsequently obtained by a galvanic replacement reaction between Ag nanowires and the mixture of $Pt(CH_3COO)_2$ and $Pd(CH_3COO)_2$ in an aqueous solution. The as-prepared supportless-PtPd NTs exhibited enhanced catalytic activity and much improved durability for ORR compared to Pt NTs and the commercial Pt/C, which can be ascribed to their unique dimensions (i.e. micrometer-sized length) and anisotropic morphologies.

In theory, galvanic replacement reaction can occur between any pair of metals with appropriate difference in their redox potentials. In spite of the large difference in redox potential between $PtCl_6^{2-}$/Pt and $PdCl_4^{2-}$/Pd, it was found that no obvious galvanic replacement reaction occurs between Pd nanocrystals and $PtCl_6^{2-}$ ions unless under special reaction conditions. Xia and co-workers [102, 103] reported that the presence of Br^- ions could promote the initiation of galvanic replacement between $PtCl_6^{2-}$ ions and Pd nanocrystals. Interestingly, the Br^--induced galvanic replacement exhibited a preferential selectivity towards the (100) facets of Pd nanocrystals, resulting in the formation of Pt-Pd bimetallic nanocrystals with a concave structure due to the simultaneous dissolution of Pd atoms from the (100) facets and deposition of Pt atoms on the (111) facets. Figure 4.6 shows the typical TEM images of Pt-Pd nanocrystals synthesized via galvanic replacement reaction with different reaction times using Pd nanocubes as templates. As illustrated by the schematic drawings in the insets, the Pt-Pd nanocrystals evolved from Pd truncated cubes to Pt-Pd concave cubes and Pt-Pd octapods. The Pt-Pd nanocrystals can also be manipulated in terms of both morphology and size by using Pd templates with different sizes and shapes [102]. More interestingly, when a reducing agent of citric acid (CA) was introduced into the above system, Pt-Pd bimetallic nanocages rather than concave nanocrystals were generated [103]. In the process, it was suggested that two important stages were included: (1) Br^--induced selective galvanic replacement reaction between (100) facets of Pd nanocubes and $PtCl_4^{2-}$, resulting in the formation of Pt-Pd concave nanocubes; (2) co-reduction of Pd^{2+} ions evolved from the galvanic replacement, together with $PtCl_4^{2-}$ ions remained in the solution by CA into atoms and the subsequent deposition of these atoms on the side faces of the concave nanocubes. Thus, Pt-Pd alloy nanocages with cubic morphology could be easily obtained *via* a combination of Pd dissolution (related to galvanic replacement) and the following Pt-Pd overgrowth (due to co-reduction).

In a recent study [104], our group also investigated the galvanic replacement between Pd nanowires and $PtCl_6^{2-}$ in aqueous solution. PtPd alloy nanorods with a porous structure were successfully synthesized through this bromide-induced galvanic replacement route. The obtained PtPd nanorods showed much larger electrochemically accessible surface area compared with the Pd nanowires and commercial Pt/C, making them promising for application in fuel cells as cathode catalysts.

Despite great success achieved in the synthesis of Pt-Pd nanostructures *via* the afore-mentioned strategies, a key procedure for the synthesis of Pd templates is required. Zheng and co-workers [105] reported a one-pot fabrication of hollow Pt-Pd nanocubes by using a mixed precursors of $Pd(acac)_2$ and $Pt(acac)_2$, with a PVP and NaI solution in DMF. In this synthesis, because of the strong coordination of I^- ions to Pd^{2+} ions, the addition of I^- ions into the mixture of $Pd(acac)_2$ and

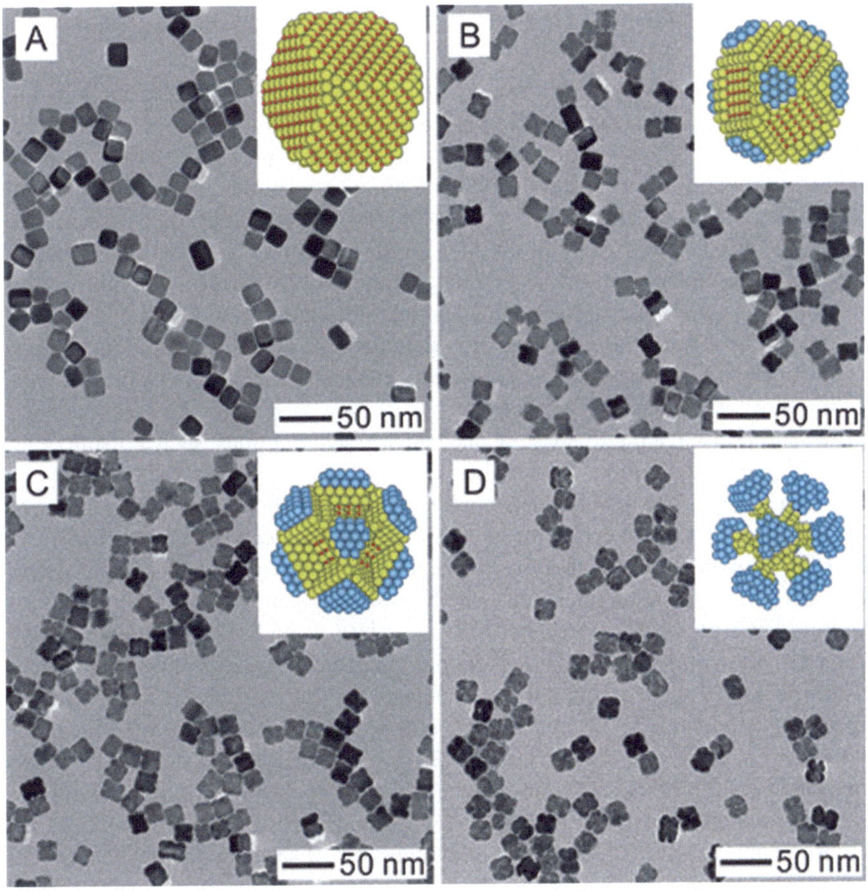

Fig. 4.6 TEM images of Pt-Pd nanocrystals in the form of nanocubes, concave nanocubes, and octapods that were formed through bromide-induced galvanic replacement at various reaction times: (**a**) 0.5, (**b**) 4, (**c**) 9, and (**d**) 20 h. The *yellow*, *blue* and *red* balls represent Pd atoms, Pt atoms, and Br^- ions, respectively. Reprinted with permission from [102]. Copyright 2011 American Chemical Society

$Pt(acac)_2$ can generate the new dominating precursors of PdI_4^- and $Pt(acac)_2$. In the DMF solution, PdI_4^- is more favorably reduced to form Pd nanocubes, and the galvanic replacement between temporal Pd nanocubes and Pt^{2+} species occurred subsequently to produce hollow Pt/Pd nanocubes. In addition to the use of iodide ions as the shape controller, the authors also demonstrated that acetylacetonate precursors can alter the reduction kinetics of metal cations and thus control the one-pot synthesis of Pt-Pd hollow nanocubes. Compared to solid Pt-Pd nanocubes, the hollow Pt-Pd nanocubes with increased accessible surface area exhibited improved catalytic activity towards formic acid electrooxidation.

4.2.3 Seed-Mediated Growth

Morphological control of nanocrystals has become increasingly important, as many of their physical and chemical properties are highly shape-dependent. As a convenient and versatile synthesis method, seed-mediated growth is probably the most powerful route for the synthesis of bimetallic heterostructures with controlled morphology. In a seed-mediated growth process, a pre-synthesized seed of one metal is significant to serve as initial sites for the nucleation and growth of a second metal. During nucleation, the second metal can follow two different pathways by homogeneous and heterogeneous nucleation. Because the activation energy for nucleation on a pre-synthesized seed is prominently lower than that in a homogeneous nucleation process, the heterogeneous nucleation is always thermodynamically more favorable than homogeneous nucleation as long as the seed-mediated growth proceeds at a relatively slow rate under mild conditions, such as using weak reducing agent and at low temperature.

For seed-mediated growth, there are three major growth modes corresponding to the products with three distinct structures: (i) Frank-vander Merwe mode (or layer-by-layer growth) for core-shell nanocrystals, (ii) Volmer-Weber mode (or island growth) for hybrid structures, and (iii) Stranski-Krastanow mode (or island-on-wetting-layer growth) for branched structures [106]. According to the previously demonstrated results [107], the nucleation and growth mode of the second metal over the seed metal are mainly manipulated by various physical parameters, including lattice match, difference in bond dissociation energy and electronegativity between these two metals. As for Pt and Pd, due to the negligible lattice mismatch (only 0.77%) and the lack of galvanic replacement unless under the modified conditions (the presence of Br^-, I^- ions, etc.), the growth mode of Pd atoms on preformed Pt metal seeds (or Pt atoms on Pd seeds) is mainly determined by the bond dissociation energy.

Controllable synthesis of Pt-Pd bimetallic nanocrystals has been attracting increasing attention due to their novel catalytic properties which are distinctly different from those of their monometallic counterparts. Particularly, recent advances reveal that Pt-Pd bimetallic nanoelectrocatalysts with a core-shell structure have been recognized as a promising alternative to commercial catalysts for effectively improving the catalytic activity and durability for fuel cell application. To this end, a lot of studies have been performed to synthesize high-efficiency bimetallic Pt-Pd nanoelectrocatalysts with well-defined core-shell structures [108–115]. By manipulating the reaction kinetics, the growth and nucleation of Pt-Pd core-shell nanocrystals can be directed to a layer-by-layer epitaxial mode, leading to the formation of Pt-Pd core-shell nanocrystals. Yang and coworkers [116] developed a facile method for the synthesis of Pt-Pd core-shell cubes, cuboctahedra and octahedra through epitaxial growth of Pd on Pt cubic seeds. This process proceeded through reducing K_2PdCl_4 by ascorbic acid with Pt nanocubes (9.5 nm in edge length) as seeds and tetradecyltrimethylammonium bromide (TTAB) as a surfactant agent in an aqueous solution. Interestingly, in

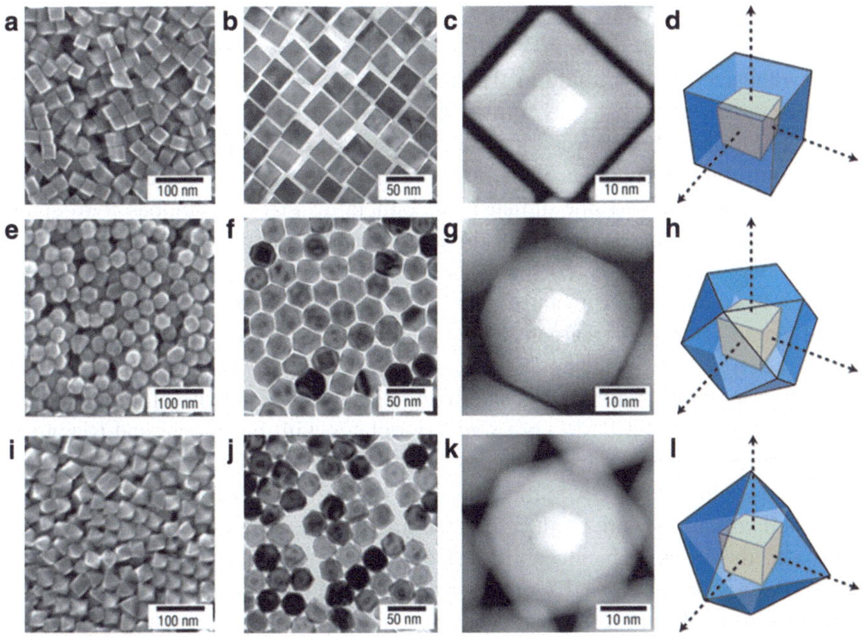

Fig. 4.7 SEM (the first column), TEM (the second column), HAADF-STEM (the third column) images of Pt-Pd core–shell nanocubes (**a**, **b**, **c**), cuboctahedra (**e**, **f**, **g**) and octahedral (**i**, **j**, **k**), and the modeled orientation (the fourth column) of the core and shell of Pt-Pd nanocubes (**d**), Pt-Pd cuboctahedra (**h**) and octahedral (**l**), respectively. Reprinted with permission from [116]. Copyright 2007 Nature Publishing Group

this synthesis, the authors found that by using different controlled facets of Pt cubic nanocrystals as nucleation centers for the overgrowth of Pd metal, shape-controlled Pt-Pd heterostructures can be obtained. The epitaxial growth of cubic Pd shells on cubic Pt seeds along the (100) and (111) directions resulted in the formation of Pt-Pd core-shell cubes. However, cuboctahedrally and octahedrally shaped Pd shells were formed upon addition of increasing amount of NO_2 which can alter the growth rates along the (100) and (111) directions to produce Pt-Pd core-shell cuboctahedron and octahedron. Both scanning electron microscopy (SEM) (Fig. 4.7a, e, i) and the corresponding transmission electron microscopy (TEM) (Fig. 4.7b, f, j) images clearly show the overall cubic, cuboctahedral and octahedral morphologies and monodispersity of the Pt-Pd products. Moreover, from the HAADF-STEM images, the Pt cores and the shaped Pd shells are discernible, demonstrating the layer-by-layer epitaxial growth on Pt nanocubes. As illustrated by the schematic drawings in Fig. 4.7d, h, l, the products evolved from Pt nanocubes to Pt-Pd cubes, cuboctahedra and octahedra along the (100) and (111) directions with different growth rates. This method and concept could also be used to synthesize other metal nanostructures such as FePt, CoPt, with desirable morphology.

In another study, Wang and coworkers [117] prepared Pt-Pd petal-like nanotubes via a wet-chemical strategy, in which the Pt nanotubes with petal-like surface was first synthesized using ultrathin Te nanowires as sacrificial templates and an effective epitaxial growth was further employed to deposit thin Pd nanoshells on the novel Pt nanotubes. It was found that the thickness of the Pd nanoshells can be easily controlled through the synthetic parameters (the amount of added Pd precursor, etc.). The obtained one-dimensional bimetallic Pt-Pd nanotubes with small diameter and nanometer-sized wall thickness demonstrated promising application in fuel cells as effective electrocatalysts.

Since Pt is extremely rare and expensive, it has been shown that deposition of Pt on Pd single-crystal surface can reduce the cost of materials while enhance their catalytic activity. By reducing H_2PtCl_6 with citric acid (CA) in the presence of Pd nanoplates as seeds and PVP as a stabilizing agent in an aqueous solution, Xia and coworkers synthesized Pd-Pt core-shell nanoplates with hexagonal and triangular shapes through layer-by-layer epitaxial growth of Pt on Pd nanoplates [118]. When AA was used instead of CA in this process, Pt-Pd nanodendrities rather than Pt-Pd core-shell structures were produced, indicating that the slow reduction rate associated with the weak reducing ability of CA played a vital role in achieving the epitaxial growth of Pt shells on Pd nanoplates. Furthermore, Pd-Pt core-shell structures with different shapes, such as regular octahedra, truncated octahedra and cubes, could also be obtained from the epitaxial growth of Pt on well-defined Pd nanocrystals [119]. The epitaxial growth of Pt shells on regular and truncated octahedra of Pd at slow reducing resulted in the formation of Pd-Pt core-shell octahedra. However, an incomplete octahedral Pt shell was formed when the Pd cube was used as a seed.

Most recently, by using rGO as a support, Bai et al. [120] developed a unique synthetic approach to prepare core-shell-like Pt-Pd-rGO stack structures. Two important steps (Fig. 4.8a) were suggested to be involved in the synthesis: (i) In situ growth of Pd nanocubes on rGO sheets via the co-reduction of K_2PdCl_4 and GO nanosheets by using ascorbic acid as a reducing agent, and (ii) Pt shells were coated onto the Pd nanocrystals by reducing H_2PtCl_6 in DMF. As shown in Fig. 4.8b, c, cubic nanocrystals with an average shell thickness of about 2 nm are dispersed on rGO sheets. HRTEM images in Fig. 4.8d show that the Pt shell is a single crystal enclosed by (100) facets, forming a perfect interface with the Pd nanocrystal. STEM and EDX mapping studies further confirm that Pt and Pd are enriched in the shells and cores, respectively. Importantly, the thickness of the Pt shell in the Pd-Pt-rGO stack structure can be controlled by simply changing the ratio of rGO-Pd to Pt precursors. Moreover, Pt can be selectively deposited on Pd nanocubes rather than on rGO sheets. In the synthesis of Pd-Pt-rGO structure, the potential difference of rGO and Pd causes the electrons accumulation on Pd surfaces, and then Pt can be preferentially reduced on the Pd surfaces with a relatively high electron density. Meanwhile, the more negative potential of rGO (0.38 V vs. SHE) than Pd (0.62 V vs. SHE) could provide a steady electron supply to prevent the oxidation of Pd in the redox reactions. All these results clearly demonstrate that rGO played a key role

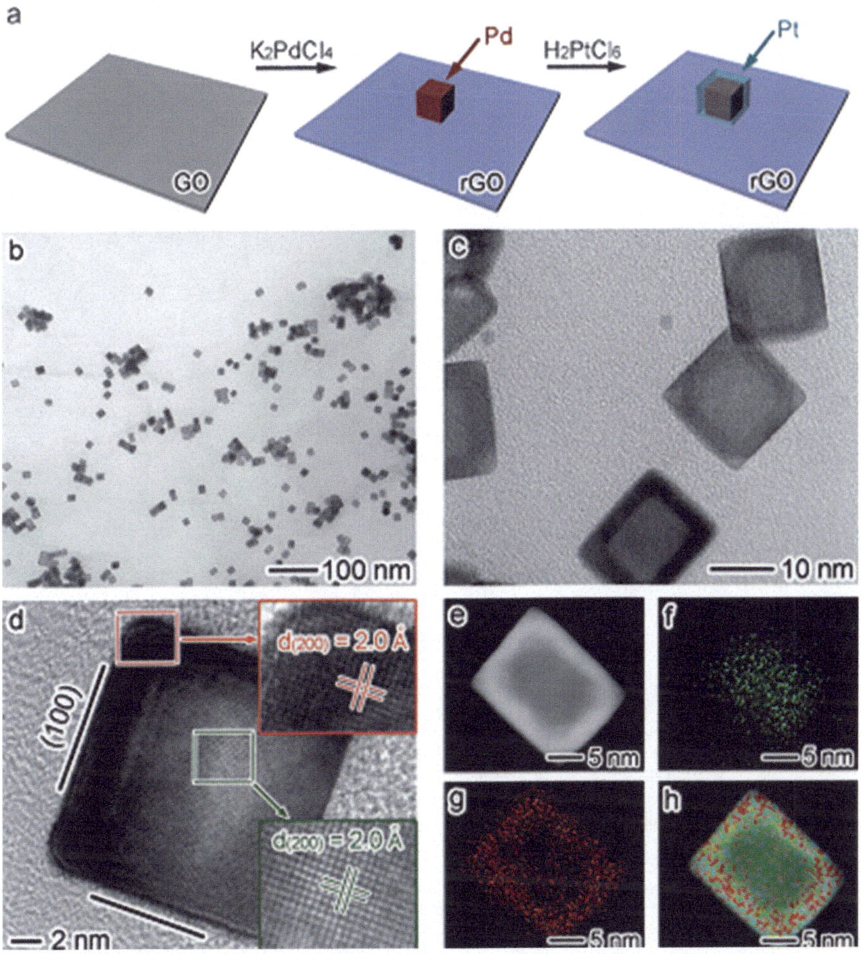

Fig. 4.8 (**a**) Schematic illustration of the synthesis of Pt-Pd-rGO structures. (**b**, **c**) TEM images of the Pt-Pd-rGO structure at different magnifications. (**d**) HRTEM image of a Pt-Pd nanocrystal supported on rGO. (**e**) STEM image and (**f–h**) EDS mapping profiles of a single Pt-Pd nanocrystal on rGO: (**f**) Pd (*green*), (**g**) Pt (*red*), and (**h**) Pt-Pd-STEM overlay. Reprinted with permission from [120]. Copyright 2014 Wiley-VCH

in manipulating the reaction kinetics to generate layer-by-layer epitaxial growth of Pt on Pd nanocrystals.

The layer-by-layer growth strategy was also extended to prepare Pt-Pd multi-shelled nanocrystals with alternating shells of Pt and Pd. Xia and coworkers [121] demonstrated a facile method for the heteroepitaxial growth of Pt-Pd nanocrystals with multi-shelled structures by sequentially adding Pt and Pd salt precursors into an aqueous solution containing Pt or Pd seeds with CA as both capping and reducing agents. Figure 4.9a shows a schematic of this synthesis starting from a

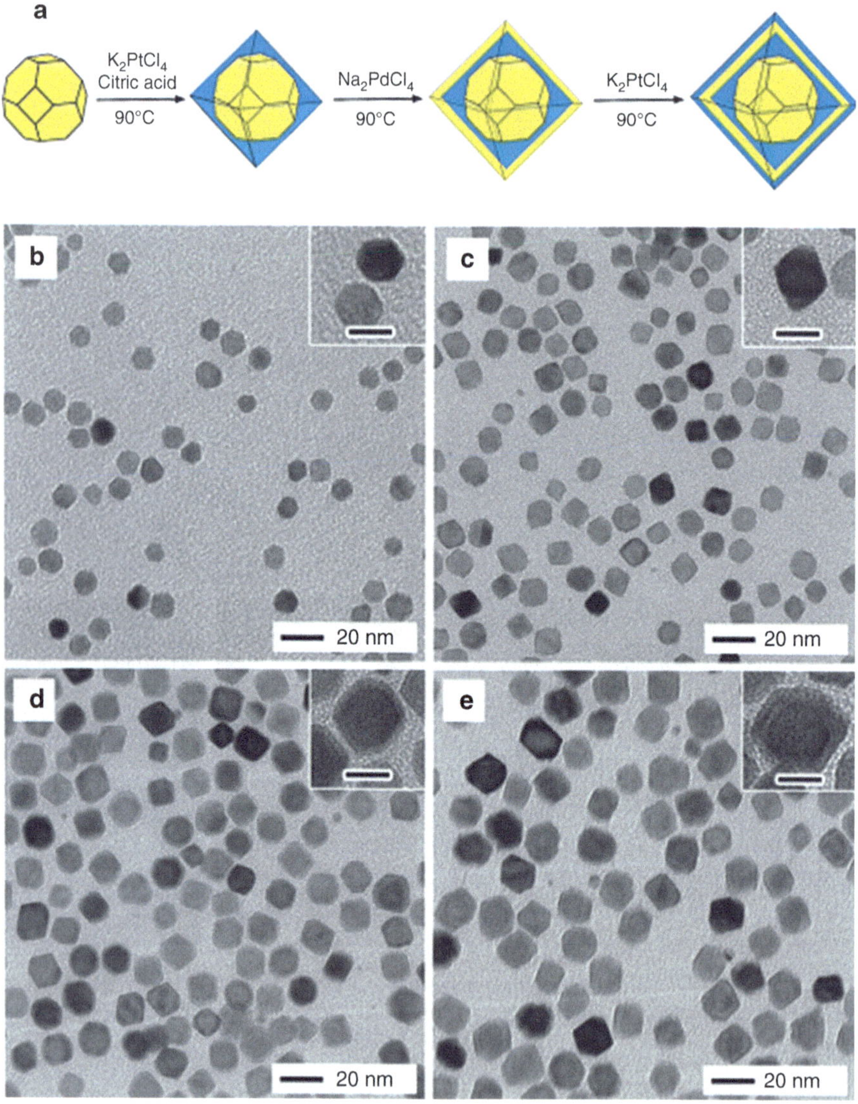

Fig. 4.9 (**a**) Schematic illustration of layer-by-layer epitaxial growth of Pt-Pd multi-shelled nanocrystals on a Pd cuboctahedral seed. (**b**) TEM image of Pd cuboctahedra of 9 nm in size that serve as seeds for the overgrowth steps. (**c–e**) TEM images of Pd@Pt (**c**), Pd@Pt@Pd (**d**), and Pd@Pt@Pd@Pt (**e**) nanocrystals prepared by reducing K_2PtCl_4 with citric acid (CA) as a reducing agent in the presence of cuboctahedral seeds of Pd. The insets show TEM images of individual nanocrystals at a higher magnification. Reprinted with permission from [121]. Copy right 2011 American Chemical Society

cuboctahedral Pd seed. The TEM images in Fig. 4.9b–e clearly shows a set of products obtained at different stages of this heteroepitaxial growth process, with the shape evolving from cuboctahedra to octahedra. The morphological transition from cuboctahedra to octahedra derived from the preferential adsorption of CA on (1 1 1) facets of Pt, resulting in a faster growth rate along the Pt(1 0 0) direction than the (1 1 1) facets. The thickness of Pd and Pt shells could be independently manipulated by varying the amounts of Pd and Pt salt precursors added in the reaction solution. The core-shell nanocrystals with Pt and Pd shells can be repeated more times to generate larger and more complex Pd-Pt multi-shell nanocrystals. In addition to the use of Pd cuboctahedra as seeds for the alternating deposition of Pt and Pd shells, the authors found that Pd octahedra/plates and Pt cubes can also be employed as seeds to prepare Pt-Pd multi-shelled nanocrystals with other different shapes. For instance, starting from seeds of cubic Pt nanocrystals, Pt-Pd multi-shell nanocrystals composing of alternating Pd and Pt shells were also obtained, except for a morphological transition from cubes to octahedra owing to extensive overgrowth. All of these studies clearly demonstrated that the use of CA as both reducing and capping agents is the key to the successful synthesis of the multi-shelled nanocrystals. As a relatively weak reducing agent, CA can produce Pt and Pd atoms at the right place to ensure layer-by-layer epitaxial growth of both metals. The synthetic technique presented here can be used to prepare multi-shelled nanostructures with other compositions and morphologies for various applications.

Except for the afore-mentioned layer-by-layer growth mode, the island-on-wetting-layer mode is another preferred process for the growth of Pd on Pt seeds due to the following order in bond dissociation energy: $E_{\text{Pt-Pt}}$ (307 kJ/mol) > $E_{\text{Pt-Pd}}$ (191 kJ/mol) > $E_{\text{Pd-Pd}}$ (136 kJ/mol), resulting in the formation of Pt-Pd branched structures. Among various Pt-Pd heteronanostructures, Pt-on-Pd nanodendrites with highly branched shapes have received great interest because of their unique properties originating from the electronic coupling between the metals and a wide variety of promising applications in catalysis. Xia and coworkers [122] developed a facile, seed-mediated approach to the synthesis of Pt-Pd nanodendrites consisting of a dense array of Pt branches on a Pd nanocrystal core. In the synthesis, Pd truncated octahedra with an average size of 9 nm were used as seeds to direct the dendritic growth of Pt upon the reduction of K_2PtCl_4 by L-ascorbic acid (AA) in an aqueous solution containing PVP. It was proposed that the high initial supersaturation of Pt atoms associated with fast reduction by AA was probably responsible for the branched growth of Pt. In the reaction system, once Pt has nucleated on the surface of a Pd nanocrystal upon fast reduction by AA, the Pt nuclei can serve as catalytic sites for further reduction of the Pt precursor and create favorable sites for atomic addition. Growth preferentially occurs on the Pt nuclei, and deposition proceeds along the developing Pt branches. And the spatially separated Pt branches structures could be generated due to the multiple nucleation sites provided by truncated octahedral Pd seeds. The TEM images of the product showed that numerous Pt branches have grown from each Pd core, resulting in the formation of Pt-Pd nanodendrites. HRTEM characterizations clearly revealed the continuous lattice fringes from the Pd core to the Pt branches, further indicating the Pt branches were epitaxially nucleated and grown on the Pd seeds. The authors also found that the generated product has a high surface area and particularly active facets, providing a promising application in fuel cells. Furthermore, by using differently shaped Pd seeds and mediated growth mechanism, other Pt-Pd dendritic nanostructures have also been generated [123, 124]. For example, Wang and coworkers [125] synthesized ultralong Pt-Pd bimetallic nanowires with a 100% yield by employing Pd nanowires as seeds to direct the dendritic growth of Pt upon the reduction of K_2PtCl_4 by AA in aqueous solution. Interestingly, the as-prepared Pt-Pd nanowires have the cores of Pd nanowires and shells of dendritic Pt, and the small single-crystal Pt nanobranches interweave with each other, resulting in nanopores on the surface of Pd nanowires. Due to the unique nanostructure, the synthesized Pt-Pd nanowires exhibited a high surface area and enhanced electrocatalytic activity towards methanol oxidation reaction.

In another study, Yang et al. [126] demonstrated an oil-phase synthetic approach for the synthesis of Pt-Pd branched nanostructures by reducing $Pt(acac)_2$ in a mixture of diphenyl ether and oleylamine with 5 nm Pd nanoparticles as seeds under an argon atmosphere at 180 °C. As shown in Fig. 4.10a, the branched arms of Pt with an average diameter of 3 nm are distributed evenly on the surface of Pd

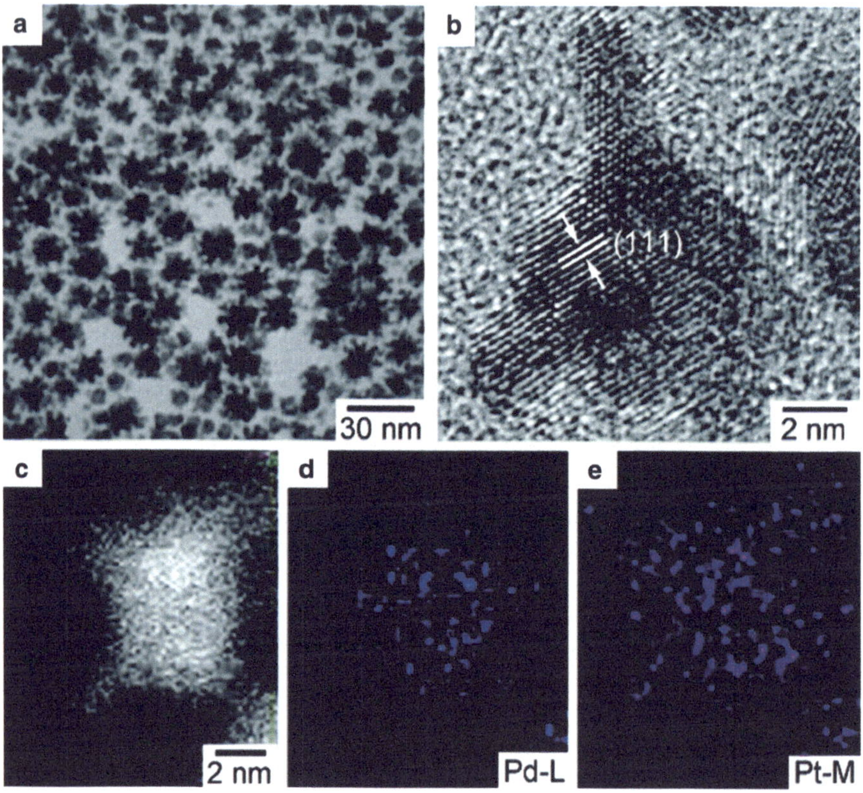

Fig. 4.10 Representative (**a**) TEM, (**b**) HRTEM (**c**) HAADF-STEM image and (**d**, **e**) EDX mapping of Pd-Pt bimetallic nanodendrites synthesized by reducing $Pt(acac)_2$ in a mixture of diphenyl ether and oleyamine in the presence of preformed Pd nanoparticles. Reprinted with permission from [126]. Copyright 2009 American Chemical Society

nanoparticles, generating Pt-Pd nanodendrite structures. From the HRTEM image (Fig. 4.10b), the synthesized nanoparticles show well-defined crystalline fringes and the Pt branches grew along the (1 1 1) crystal planes on the Pd seed. From the representative STEM and the corresponding EDX maps (Fig. 4.10c–e), Pt is dispersed throughout the entire particle, including the branches, whereas Pd could only be detected in the core region, indicating the formation of Pt-Pd dendritic structures.

Graphene nanosheets have been studied extensively owing to its unique electronic, thermal, mechanical properties arising from its strictly 2D structure. Particularly, its unique structure and the resulting properties endow it to be a promising 2D supporting material to load metal nanoparticles for application in fuel cells. By using graphene sheets as support, Wang and coworkers [127] constructed high-quality 3D Pt-on-Pd bimetallic nanodendrites supported on PVP-functionalized graphene nanosheets. In the synthetic process, the PVP-functionalized graphene was first obtained under the reduction of hydrazine. Pd/graphene seeds were then synthesized using HCOOH as reducing agent at room temperature. Finally, Pt-on-Pd nanodendrites supported on graphene sheets were produced by using graphene sheets-supported Pd nanoparticles as seeds to direct the dendritic growth of Pt upon the reduction of K_2PtCl_4 by ascorbic acid in an aqueous solution. The TEM results indicated that the Pt-on-Pd bimetallic nanodendrites with an average size of 15 nm were dispersed on graphene sheets, in which Pt branches with an average diameter of about 3–5 nm were distributed evenly on the surface of a Pd nanoparticle. Importantly, the number of Pt branches could be easily controlled through simply manipulating the reaction parameters. For example, lower concentrations of graphene can lead to relatively lower amounts of Pd nanoparticles adsorbed on the surface of graphene nanosheets, thus resulting in Pt-on-Pd bimetallic nanodentrites with more Pt branches supported on the graphene sheets. More importantly, because the small single-crystal Pt nanobranches with porous structure and good dispersion were directly grown onto the surface of graphene nanosheets, the obtained hybrids exhibited an enlarged electrochemical surface area as high as 81.6 m^2/g. All these unique structural features together with the synergetic effects of the Pt-Pd and the enhanced electron transfer stemming from graphene support are highly favorable for the application of the graphene nanosheets-supported 3D Pt-on-Pd bimetallic nanodendrites in fuel cells, with a much higher catalytic activity than conventional E-TEK Pt/C electrocatalysts for methanol electro-oxidation.

In spite of the significant achievement in the synthesis of Pt-on-Pd nanodendritic structures *via* the aforementioned seed-mediated growth strategy, the involved processes of the syntheses were strongly dependent on the use of faceted Pd seeds to direct the subsequent growth of Pt branches. Without the well-defined Pd seeds, both the particle size and shape of the Pt-on-Pd nanodendrites are usually uncontrollable. Recently, without using any pre-synthesized Pd seeds, organic solvent and high temperature, Wang et al. [128] proposed a simple approach for one-step direct synthesis of Pt-on-Pd nanodendrites in aqueous solution at room temperature. In the synthesis, a block copolymer (Pluronic P123) was employed to mediate the reduction of K_2PtCl_4 and Na_2PdCl_4 by using AA as a reducing agent for 30 min at room temperature. It was found that the as-prepared product consists of well-dispersed nanodendrites with Pd interior and dendritic Pt exterior, in which the Pt nanoarms have widths of 3 nm branching in various directions. The preferential reduction of the Pd precursor by AA caused by the different reduction kinetics of Pd and Pt complex with AA was found to a key factor in directing the formation of Pt-on-Pd nanodendrites. The formed Pd nanoparticles can serve as in situ seeds for the subsequent deposition of Pt. Moreover, the use of Pluronic P123 and the selected concentration (0.87 mM) also played critical roles in the synthesis of Pt-on-Pd nanodendrites. In comparison, aggregated Pt-Pd nanoclusters were obtained with an increase/decrease of the Pluronic P123 concentration or replacing Pluronic P123 with Pluronic F68. A surfactant of PVP was found to be unfavorable for the formation of Pt-on-Pd nanodendrites. Most importantly, by a simple control of the mole ratios of the Pt and Pd precursors in the reaction solution, Pt-on-Pd nanodendrites with a designed Pt and Pd content could be obtained. In comparison with the two-step seed-mediated methods, this rational block copolymer-mediated synthesis could trigger the facile creation of novel bimetallic heterostructures with designed compositions and desired properties.

In addition to seed-mediated growth involving direct reduction of a second metal onto pre-formed seeds, Beer and co-workers [129] recently reported a novel anion coordination route to control the formation of bimetallic core-shell nanoparticles for any two noble-metals including Pt and Pd. This method uses ligand-based supramolecular forces to ensure surface segregation of the shell metal onto the pre-formed core before its reduction. And four different types of bimetallic core-shell nanoparticles (Au-Pd, Pd-Au, Pt-Pd and Pd-Pt) with an average size of less than 5 nm have been synthesized by using this new protocol. The success of this synthesis was based on the ability to anchor metal ions to the pre-formed seeds through amides-chlorometallates anion coordination by hydrogen bonding before reduction occurred. This work not only provides novel core-shell nanoparticles with small size (<5 nm), but also offers an impetus for the exploitation of supramolecular interactions in the design and synthesis of structured nanoparticles with controlled composition.

4.2.4 Electrochemical Deposition and Its Combination with Galvanic Replacement

In recent years, to improve the mass activity of shaped nanocrystals, much work has focused on the crystallographic orientation of metal atoms at the surface of nanocrystals [30, 130]. Owing to the simplicity and no need of templates, electrochemical deposition is a useful approach to prepare Pt-Pd bimetallic nanocrystals with decorated surface by Pt or Pd adatoms and consequently with enhanced electrochemical properties [131–144]. For example, Feliu and coworkers [145] have successfully electrodeposited Pd adatom on cubic Pt nanoparticles as an anode electrocatalyst toward formic acid oxidation in fuel cells. The authors proposed that the amounts of Pd on the Pt surfaces can be easily monitored in situ by observing the voltammetric changes during the deposition process. Compared with the Pd-modified quasispherical Pt nanoparticle, the Pd adatoms-decorated Pt(100) nanoparticles exhibited enhanced catalytic activity for formic acid electrooxidation. In the Pd adatoms-decorated Pt(100) nanoparticles, the high fraction of Pt(100) sites can decrease surface poising, lower onset potential and thus greatly improve the kinetics of formic acid oxidation over Pt surface.

In addition, the nanostructure of Pd nanocrystals covered by a Pt monolayer shell has also attracted much attention because in such structure all Pt atoms located on the surface of Pd core can be sufficiently utilized and thus improved Pt mass activity can be achieved. DFT calculations demonstrated that the strong interaction between the Pt monolayer and the base metal also plays a vital role in determining the structure and properties of the Pt monolayer. In this case, the electrocatalytic properties of the Pt monolayer can be easily manipulated by changing the base metal. For example, the electrocatalytic activity of a Pt monolayer on different base metals for ORR exhibited a volcano-type dependence [146]. Pd substrate was thought not only to apply a compressive strain upon the Pt monolayer but also to impart a so-called "ligand-effect", leading to lowered energy of the weighted center of the Pt d-band. Therefore, the Pd(1 1 1) decorated with Pt monolayer is located at the top of the volcano curve with the highest ORR activity. The extremely low Pt loading together with a perfect catalytic activity make the Pt-monolayer decorated Pd nanocrystals very attractive for practical applications in fuel cells.

Recently, much work has been done to prepare Pt monolayer-decorated Pd nanocrystals with improved electrocatalytic properties [147–150]. For example, Adzic and coworkers [151] reported a simple approach for the synthesis of a Pt monolayer on a Pd base metal by first electrochemically depositing a monolayer of Cu atoms on Pd cores through Cu underpotential deposition (UPD), followed by the controlled displacement of these adatoms with Pt *via* galvanic exchange. Figure 4.11a schematically illustrates the key steps for the growth of a Pt monolayer on a Pd core. The Z-contrast image from HAADF-STEM (Fig. 4.11b) demonstrates the formation of a bright shell on the relatively darker nanoparticle core. The EDX line-scan analysis (Fig. 4.11c) shows that the Pt trace has two peaks in two sides, while the Pd trace has one peak in the center, further confirming the formation of a Pt monolayer on the Pd core. In addition, the thickness of the Pt shell could be gradually increased with repeated UPD and galvanic replacement processes.

Although outstanding ORR activities have been achieved with the aforementioned 0D core-shell catalysts, 1D nanostructures are characterized by their uniquely anisotropic nature, which imparts advantageous structural and electronic factors in the catalytic reduction of oxygen. In particular, Koenigsmann et al. [152] designed a Pd nanowire core-Pt monolayer shell structure with enhanced electrocatalytic activity and durability by successfully combining the uniquely advantageous core-shell motif with the electrocatalytic advantages of ultrathin 1D nanostuctures. In their synthesis, the ultrathin Pd nanowires (~2 nm) were first

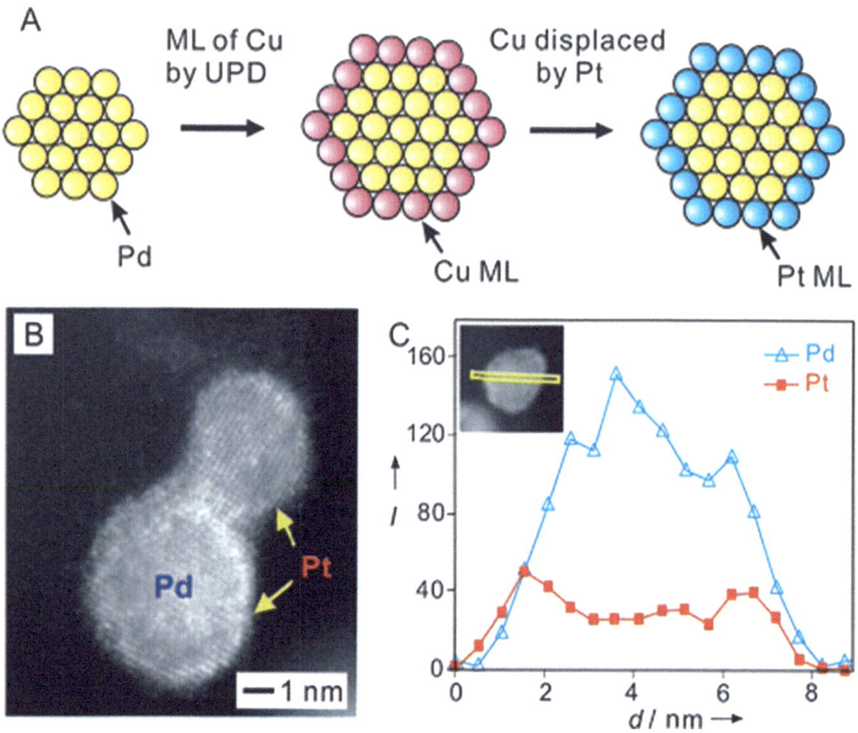

Fig. 4.11 (**a**) A schematic of the two major steps involved in the synthesis of a Pt monolayer on a Pd core through a combination of electrochemical deposition and galvanic replacement. (**b**) HAADF-STEM image and (**c**) the EDX line-scanning profile showing the formation of a Pt monolayer on a Pd nanoparticle. Reprinted with permission from [151]. Copyright 2010 Wiley-VCH

synthesized, and the deposition of Pt onto the surface of the treated Pd nanowires was achieved by Cu UPD followed by galvanic displacement of the Cu adatoms with Pt^{2+}. Importantly, the use of ultrathin nanowires with a diameter below 5 nm in this work could maximize the surface area-to-volume ratio, achieving higher mass activity compared with conventional commercial Pt nanoparticles, and core-shell nanoparticles. Moreover, in the obtained Pd nanowire core-Pt monolayer shell structure, a contraction of the core nanowire surface would be highly advantageous for ORR electrocatalysis, because this would enhance the strain-induced contraction of the Pt monolayer and therefore improve the inherent ORR activity. In addition, by varying the shape of the Pd cores (Pd cubes, octahedra or rods), the morphology-tailored core-shell nanoparticles could also be obtained by the same way [150, 153, 154].

In addition to the methods listed above, the synthesis of Pt-Pd (or supported Pt-Pd) bimetallic nanocrystals can also be achieved by other techniques, such as thermal treatment [155–160], plasma sputtering [161, 162], electroless deposition [163, 164], and supercritical CO_2 deposition [165, 166].

4.3 Applications of Unsupported/Supported Pt-Pd Bimetallic Nanocrystals as Electrocatalysts in Fuel Cells

4.3.1 *Alcohol Oxidation Reaction*

Among the various fuel cell technologies, direct methanol fuel cells (DMFCs) have attracted special interests because of their ability to utilize methanol as a liquid fuel, which can be easily and safely stored as well as transport comparing with hydrogen-based counterparts [167, 168]. Moreover, methanol can be directly prepared either from natural gas or renewable biomass and thus ideally meet the future power needs with a high energy efficiency of 600 Wh/kg [169]. More importantly, DMFCs have the unique advantage of operation at near ambient temperatures between 40 and 80 °C, which is much lower than the normal operating temperatures for solid oxide fuel cells (800–1000 °C). Therefore, DMFCs technology represents a potentially effective fuel cell candidate for future applications in the automotive, portable power generating, and electronics industries [170]. However, it is significant to note that large portions of the cost of DMFCs can be attributed to the high loading of expensive electrocatalysts that are at the heart of the device itself. Meanwhile, the slow anode reaction and methanol crossover reaction at the cathode are also the vital limitation to the widespread application of DMFCs. Hence, lowering the costs and improving efficiency of the catalysts have become two critical technological issues towards the development of practical and inexpensive DMFCs.

The process of methanol oxidation reaction (MOR) at the anode in DMFCs includes the methanol adsorption and the subsequent dissociation into adsorbed intermediates [171]. According to the dual-pathway mechanism, CO is a poisoning intermediate species, which can largely reduce the catalytic activity of catalysts, especially Pt-based catalysts. To eliminate the CO poisoning to catalysts, oxygen-containing surface species (e.g, OH) formed on adjacent catalyst sites are usually needed to remove CO adsorbed on the catalyst surface (CO_{ad}). Thus, to catalyze the methanol oxidation efficiently, catalysts with multiple active sites are required for the adsorption of methanol and formation of OH species. Therefore, a significant amount of work has been dedicated to the synthesis of Pt-based nanostructures combining with another metal so as to improve CO-tolerance. Among these, Pt-Pd bimetallic nanocrystals represent an active and durable catalysts for methanol oxidation.

Effectively controlling the size and shape of Pt-Pd bimetallic nanocrystals can provide a great opportunity to improve their catalytic properties and increase their mass or specific activity [57, 100, 110]. Huang and coworkers [115] employed a seed-mediated growth method to synthesize Pt-Pd core-shell nanocrystals with different Pd shell thicknesses. The catalytic activities of the different sizes (i.e. Pd shell thickness) of Pt-Pd nanocrystals at Pd/Pt ratios from 1/6 to 2/3 for MOR in alkaline media were compared in CV measurements. It was found that the Pt-Pd core-shell nanocrystals with a Pd/Pt ratio of 1/3 (near monolayer Pd shell) yield the highest current density and the most negative potential for the oxidation peak in the forward sweep due to the highest tolerance of the sample to CO poisoning. Such results demonstrate the importance of Pd shell thickness of the core-shell nanocrystals in the manipulation of the catalytic performance for fuel cell applications.

To enhance the mass activity of Pt, the core-shell type of nanocrystal catalysts with a Pt shell have also been developed to remarkably reduce Pt loading. Yamauchi and coworkers [128] synthesized Pt-on-Pd nanodendrites and studied their catalytic properties for methanol oxidation reaction in acid condition. The electrochemical measurements showed that the forward peak current density of MOR on Pt-on-Pd (0.49 A/mg_{Pt}) is much higher than that of Pt nanodendrites (0.21 A/mg_{Pt}) and Pt black (0.11 A/mg_{Pt}), and at any oxidation current density, the corresponding oxidation potentials on the Pt-on-Pd nanodendrites are obviously lower than those on the Pt dendrites and Pt black, indicating that Pt-on-Pd nanodendrites has the highest activity for methanol oxidation. It was found that the formation of the inserted pseudo-Pd-Pt alloy heterointerface plays a critical role in reducing the electronic binding energy in Pt and facilitating the C-H cleavage reaction in methanol decomposition. Furthermore, the various atomic steps exposed on the Pt branch surface can act as highly active sites for the methanol oxidation reaction. Therefore, superior catalytic activity was exhibited through this open dendritic structure with the designed Pt and Pd ratios. In addition, nanostructures with high aspect ratios such as nanowires, nanotubes and nanorods can provide improved mass transport and higher electrochemical active surface area than those of low aspect-ratio nanoparticles [172]. For example, Guo et al. [125]. investigated the activity of the ultralong Pt-Pd bimetallic nanowires for methanol oxidation reaction. By comparing with other catalysts, the Pt-Pd nanowires exhibited much larger electrochemical surface area (ECSA) (90.7 m^2/g) than those of E-TEK Pt/C catalyst (53.5 m^2/g) and mesoporous Pt with giant mesocages (74 m^2/g). Compared to the commercial E-TEK catalyst, a significant enhancement of the peak current and a negative shift of the onset potential of methanol oxidation were achieved on the Pt-Pd nanowires. Moreover, it should be noted that the Pt-Pd nanowires also exhibited much higher mass activity and stability than some state-of-art Pt-based nanomaterials.

It has been well-documented that the catalytic activities of bimetallic nanocrystals are strongly dependent on their compositions [139, 173]. Sun and co-workers [89] studied polyhedral Pt-Pd bimetallic nanoparticles with the compositions ranging from $Pd_{88}Pt_{12}$ to $Pd_{34}Pt_{66}$ as anode catalysts for methanol oxidation reaction. As shown in Fig. 4.12a, as compared to pure Pd nanoparticles, both peak potentials and current densities of MOR change with the Pt content in the Pt-Pd nanoparticles. The plots of peak potentials and peak current densities versus Pt atomic % in the nanoparticles (Fig. 4.12b) showed a composition-dependent MOR activity and the catalyst having 40–60% Pt exhibited the maximum activity plateau. Cai and coworkers [82] synthesized three-dimensional Pt-Pd core-shell nanostructures by using a one-step microwave heating method. It was found that the Pt-Pd nanostructure with a Pd/Pt molar ratio of 1:3 exhibited the highest

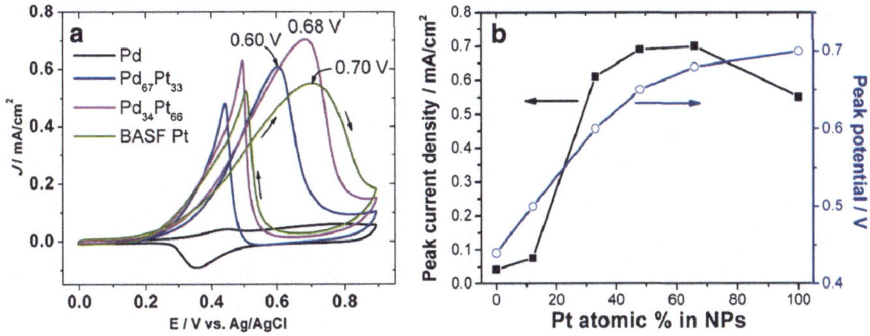

Fig. 4.12 (**a**) J-V curves of MOR on Pd, BASF Pt and Pd-Pt catalysts with different compositions in 0.1 M $HClO_4$ and 0.1 M methanol. (**b**) Methanol oxidation peak current density and peak potential vs the Pt atomic % in the Pt-Pd nanoparticles. Reprinted with permission from [89]. Copyright 2011 American Chemical Society

electrocatalytic activity toward methanol oxidation as compared to Pt, Pd and other Pt-Pd nanostructures. Moreover, the Pt-Pd nanostructure with a Pd/Pt molar ratio of 1:3 also exhibited high stability due to its enhanced tolerance to intermediate species during the methanol oxidation. Jin and coworkers [88] investigated the catalytic activity of the assembled free-standing Pt-Pd nanomembranes (PdPt-FNMs) for methanol oxidation in alkaline condition. By using the hydrogen adsorption-desorption method, the PdPt-FNMs have much higher ECSA values ($Pd_{53}Pt_{47}$–FNMs (48.01 m^2/g), $Pd_{33}Pt_{67}$–FNMs (52.04 m^2/g), $Pd_{15}Pt_{85}$–FNMs (50.55 m^2/g)) than commercial Pt black (22.94 m^2/g), indicating the enhanced active sites on PdPt–FNMs for the electrooxidation of methanol. By comparing the CVs of methanol oxidation, the $Pd_{33}Pt_{67}$–FNMs exhibited the best catalytic performance with the most negative onset potential and the largest current density among the PdPt-FNMs with different compositions and commercial Pt black. Moreover, the PtPd-FNMs showed higher stability than commercial Pt black after CV cycling tests, and the $Pd_{33}Pt_{67}$–FNMs exhibited the highest stability with the least loss of the electrocatalytic activity among the studied PdPt-FNMs. The enhanced MOR activity of all the above-mentioned Pt-Pd bimetallic nanocrystals can be explained by the bifunctional methanol oxidation mechanism. In the Pt-Pd nanocrystals, Pd is mainly responsible for the water dehydrogenation to form Pd-OH, while Pt catalyzes the methanol dehydrogenation to form Pt-CO. The reaction between Pd-OH and Pt-CO produces CO_2 to regain the active metal surfaces. However, without Pd, the water dehydrogenation on Pt occurs at a higher potential, making the oxidation process sluggish. The activity also decreases with the excessive Pd due to the lack of Pt for methanol dehydrogenation.

In the synthesis, metal nanoparticles are usually formed with certain facets to minimize surface energy and the total free energy. On different crystal surfaces of Pt-Pd materials, the electronic structures and atomic arrangements are quite different, and the appropriate crystal phase is able to greatly enhance reaction kinetics

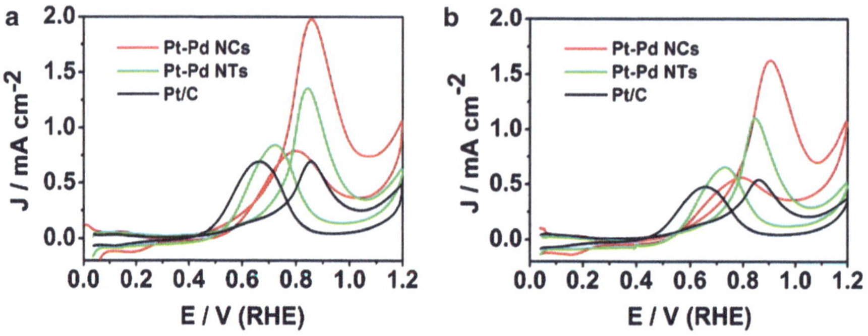

Fig. 4.13 (**a**) Stable CV curves obtained from the Pt-Pd NCs, NTs, and Pt/C in 0.1 M $HClO_4$ and 1 M CH_3OH at a sweep rate of 50 mV/s. (**b**) CV curves obtained from different electrocatalysts after 4000 additional cycles. Reprinted with permission from [94]. Copyright 2011 American Chemical Society

[113]. Yin et al. [94] have successfully prepared the mono-disperse single-crystalline sub-10 nm Pt-Pd nanotetrahedra (NTs) and nanocubes (NCs). The single-crystalline Pt-Pd NTs are enclosed by four (1 1 1) facets, while the Pt-Pd NCs are enclosed by (1 0 0) facets. As shown in Fig. 4.13a, both the Pt-Pd NCs and NTs exhibited high catalytic activities towards methanol electrooxidation in acid electrolyte. As compared the peak voltage (E_f=0.85 V) and the peak current density (J_f=0.51 mA/cm^2) for Pt/C, more negative E_f and much larger J_f in a forward scan were obtained with 0.85 V and 1.49 mA/cm^2 for Pt-Pd NCs, and with 0.84 V and 1.12 mA/cm^2 for Pt-Pd NTs, respectively. Moreover, the ratio of the current density values in two sequential forward (positive) and backward (negative) sweeps (J_f/J_b) is considered to be an important indicator of the catalyst tolerance to poisoning species. The different J_f/J_b values for Pt-Pd NCs (2.5) and NTs (1.4) confirmed that different reaction pathways might be adopted on the (1 0 0) or (1 1 1) surfaces. Meanwhile, due to the more durable nature of the (1 1 1) facet of Pt-based nanocrystals, the Pt-Pd NTs remained higher activities than both Pt-Pd NCs and Pt/C after repeating the CV sweeps for over 4000 cycles (Fig. 4.13b). The different electrocatalytic performances of the Pt-Pd NCs and NTs demonstrate the facet-sensitive nature of methanol oxidation on Pt-Pd nanocrystals.

In another study, Lee et al. [72] synthesized octahedral Pt-Pd nanoparticles with exposed (1 1 1) facets as anode catalysts for methanol oxidation. The octahedral Pt-Pd alloy with dominantly exposed (1 1 1) facets exhibited an enhanced catalytic performance with lower onset and peak potentials, and higher current density as compared to polycrystalline Pt/C for MOR. Moreover, a higher ratio of the forward to the reverse anodic peak current density than that of commercial Pt/C indicates less accumulation of residues on the Pt-Pd octahedron during the forward anodic scan. In addition, the remained size and morphology of the well-defined Pt-Pd alloy octahedron compared to the massive agglomeration of Pt/C after the stability test further confirmed the superior electrocatalytic activity and stability of the Pt-Pd bimetallic nanocrystals for MOR. All these studies indicate that shape-controlled synthesis of facet-sensitive multi-metal nanocrystals can open up new opportunities for Pt-Pd nanocrystals with potential applications.

To further improve the catalytic activity of the Pt-Pd bimetallic nanocrystals and lower the usage of noble metals, loading of catalysts on the surface of suitable supporting materials is highly desirable [43, 54, 81, 96, 174, 175]. Wang and coworkers [127] investigated the catalytic activity of three-dimensional Pt-on-Pd bimetallic nanodendrites supported on graphene nanosheets (TP-BNGN). By comparing with the carbon nanofiber- or CNT-supported Pt nanoparticles, CNT/Pt composite and commercial E-TEK Pt/C catalysts, the TP-BNGN exhibited much improved catalytic activity for MOR with much higher mass current density, more negative onset oxidation potential and more efficient removal of the poisoning species on the catalyst surface. Recently, we studied the catalytic activities of rGO-supported Pt-Pd nanocubes (PtPd/rGO) for methanol oxidation in acid medium [95]. As shown in Fig. 4.14, compared with the unsupported PtPd alloy nanocubes and commercial Pt/C catalysts, the PtPd/rGO showed enhanced

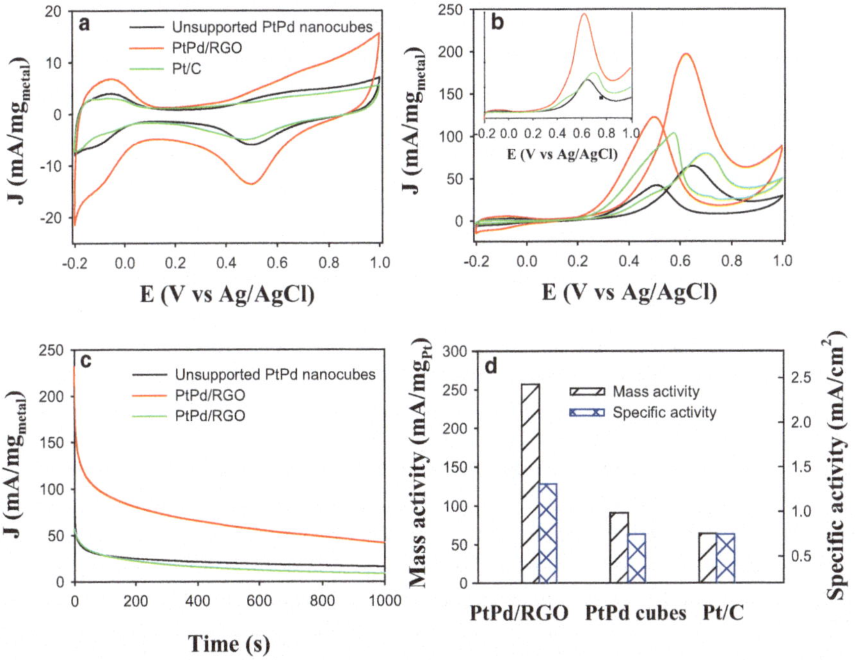

Fig. 4.14 CVs of the unsupported and rGO-supported (PtPd/rGO) PtPd alloy nanocubes, and the commercial Pt/C catalysts in (**a**) 0.1 M $HClO_4$ solution, and (**b**) 0.1 M $HClO_4$+1.0 M CH_3OH solution. (**c**) Chronoamperometric curves of methanol oxidation at 0.62 V in 0.1 M $HClO_4$+1.0 M CH_3OH solution after the CO stripping treatment. Potential scan rate 50 mV/s. All currents were normalized to the total mass of noble metals (Pt and Pd). (**d**) Comparison of mass and specific activities of the three catalysts for methanol oxidation. Reprinted with permission from [95]. Copyright 2013 American Chemical Society

electrocatalytic performance with increased ECSA, lower negative onset potential and higher current density for methanol oxidation. Furthermore, the PtPd/rGO exhibited high durability during the methanol oxidation. All these studies indicate that graphene sheets can be used as excellent catalyst support for enhancing the catalytic performance and improving the stability of Pt-Pd bimetallic catalysts.

With a high surface area, good electrical conductivity, light weight, high chemical stability and inherent size and hollow geometry, carbon nanotubes (CNTs) are one of the most attractive nanomaterials used as an efficient catalyst support in fuel cells. Wai and coworkers [166] studied the catalytic activity of Pt-Pd nanoparticles supported on multi-walled carbon nanotubes (MWCNT) for methanol electro-oxidation. From the CVs of methanol oxidation, compared to MWCNT-supported monometallic nanoparticles, the Pt-Pd/MWCNTs exhibited much lower onset potential and higher ratio of the forward scan peak current over the backward scan peak current, indicating the Pt-Pd/MWCNTs catalysts are more efficient at lowering the adsorbed carbon monoxide and thus show an enhanced catalytic activity for MOR.

In addition, conducting polymer nanostructures have recently received special attention as supporting nanomaterials owing to their highly π-conjugated polymeric chains, unique electrical properties, and controlled morphologies. Recently, Wang and coworkers [76] reported a 1D nanoelectrocatalyst based on polyaniline (PANI) nanofiber-supported Pt-Pd hybrid nanoparticles for methanol oxidation reaction. From the CV measurements, the current density of MOR on the PANI nanofiber-supported Pt-Pd nanoparticles catalyst was 2.99 times higher than that of E-TEK Pt/C catalyst for MOR. And the hybrid catalyst exhibited a higher ratio (1.98) of the forward oxidation current peak to the reverse current peak compared to commercial E-TEK Pt/C (1.08), revealing the more effective removal of the poisoning species on the surface of PANI nanofiber-supported Pt-Pd nanoparticles. The enhanced catalytic activity of PANI nanofiber-supported Pt-Pd can be ascribed to the high conductivity of PANI support in acidic conditions and the supra-high density distribution of small Pt-Pd nanoparticles on the surface of PANI nanofibers. In another work, highly dispersed Pt-Pd nanoparticles supported on polypyrrole (Ppy) film were fabricated by electrochemical deposition method [137]. Because of the uniform dispersion of Pt-Pd nanoparticles in the Ppy film, the synergistic effect of the Pt and Pd and the effective reduction of CO species, the Ppy-supported Pt-Pd nanoparticles showed remarkably improved electrocatalytic performance for methanol oxidation reaction. Galal et al. [140] synthesized Pt-Pd nanoparticles supported on poly(3-methylthiophene) (PMT) films via electro-deposition method. The voltammetric measurements indicated a higher activity of PMT-supported Pt-Pd nanoparticles than other types of deposited substrates such as Pt and glassy carbon, for methanol oxidation. Moreover, it was also found that the thickness of PTM in the Pt-Pd/PTM hybrid has an obvious effect on the catalytic performance, and the best activity was obtained at the relatively thinner films.

Direct ethanol fuel cells (DEFCs) have also attracted enormous attention due to the higher theoretical energy density but less toxicity of ethanol as compared to methanol. Moreover, ethanol can be synthesized directly from biomass, whereas methanol is most commonly generated from natural gas, an abundant but non-renewable source [176]. For ethanol oxidation on Pt-based electrocatalysts, several reaction mechanisms have been proposed with complex processes [177, 178]. According to the mechanisms, it is widely accepted that the electro-oxidation of ethanol can only occur with a multifunctional electrocatalyst. Platinum itself is known to be rapidly poisoned on its surface by strongly adsorbed species coming from the dissociative adsorption of the ethanol molecule. So other metals modified with Pt can activate water at low potentials to produce oxygenated species (e.g. adsorbed OH), and these OH species are necessary to oxidize the species from the dissociation of ethanol to carbon dioxide. Moreover, unlike methanol oxidation, the oxidation of ethanol is more difficult with the necessity to break the C-C bond to obtain its complete oxidation. Except for the formation of adsorbed CO, acetaldehyde, acetic acid, or CO_2 can be formed through different reaction pathways during ethanol electro-oxidation reaction. In the past years, Pt-Pd bimetallic nanocrystals have been recognized as a promising alternative to commercial catalysts for effectively improving the activity and durability for ethanol electro-oxidation [64, 142, 155]. For example, Guo et al. [117] synthesized Pt-Pd bimetallic nanotubes with petal-like surface and further studied their catalytic activity for ethanol oxidation in alkaline media. Figure 4.15a compares the CVs of Pt-Pd bimetallic nanotubes with petal-like surfaces (PBNPSs, curve a), Pt nanotubes with petal-like surfaces (PNPSs, curve b) and E-TEK Pd/C (curve c) electrodes in a N_2-saturated 0.5 M H_2SO_4 solution. It can be seen that the reduction peak potential of $Pd(OH)_2$ shifts positively on the PBPNSs (0.48 V vs. Ag/AgCl) as compared with E-TEK Pd/C (0.38 V vs. Ag/AgCl), indicating the reduced oxophilicity and weakened chemisorpotion with oxygen-containing species on PBNPSs catalyst. Furthermore, from the CVs of ethanol oxidation shown in Fig. 4.15b (specific activity), c (mass activity), the PBNPSs (curve a) exhibit much higher specific and mass activities, which are about 3.7 and 3.1 times larger than PNPSs (curve b), demonstrating the importance of thin Pd shells for enhancing ethanol electro-oxidation. Meanwhile, from Fig. 4.15b, c, the PBNPSs also show higher specific (3.05 times) and mass (3.7 times) activities than the E-TEK Pd/C catalyst. In addition, after 1800 s continuous operating (Fig. 4.15d), the specific activities of PBNPSs (curve a) are much higher than those of the PNPSs (curve b) and commercial Pd/C (curve c), suggesting that the PBNPSs catalyst has great stability in ethanol oxidation reaction.

Recently, Lee and co-workers [144] prepared bimetallic Pt-Pd alloy nanoparticles on a Nafion-graphene film by a simple electrochemical deposition method. Due to the synergistic effects of the PtPd nanoparticles and the enhanced electron transfer originating from graphene, the hybrid exhibited efficient catalytic activity and stability toward ethanol electro-oxidation in alkaline media. Datta et al. [49] successfully incorporated Pt-Pd bimetallic nanocrystallites into polyvinyl carbazole (PNVC) cross-linked with V_2O_5 (PNVC-V_2O_5) and investigated its electrocatalytic activity for ethanol oxidation in alkaline condition. As compared to the Pt/C, PtPd/C and Pt/PNVC-V_2O_5, the PtPd/PNVC-V_2O_5 catalyst exhibited enhanced electrocatalytic performance with increased electrochemical active

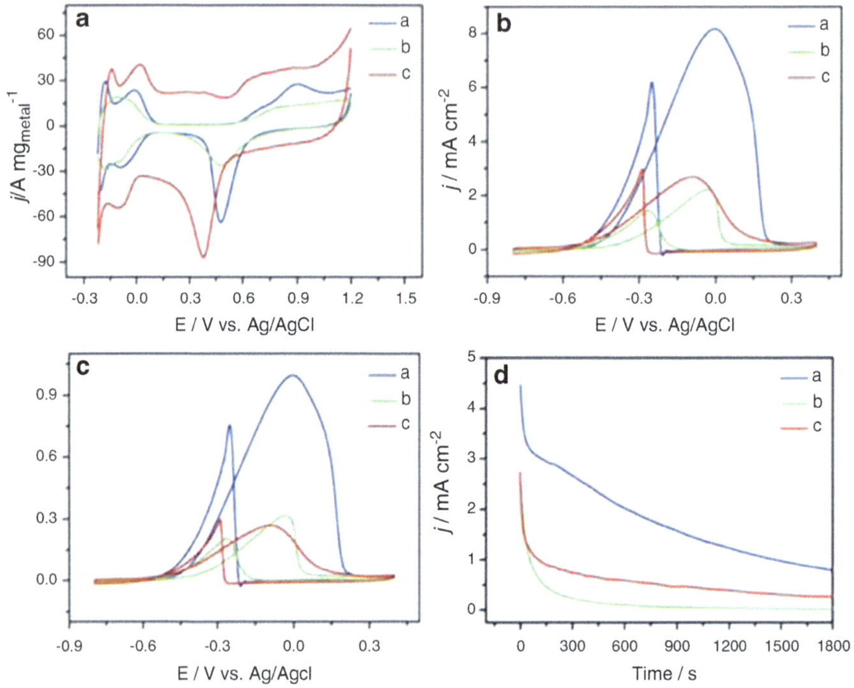

Fig. 4.15 (**a**) CVs of the PBNPSs (curve a), PNPSs (curve b) and E-TEK Pd/C (curve c) electrodes in a 0.5 M H_2SO_4 solution; CVs ((**b**): specific activity; (**c**): mass activity) of the PBNPSs (curve a), PNPSs (curve b) and E-TEK Pd/C (curve c) electrodes in a 0.5 M H_2SO_4+ 1 M ethanol solution at the scan rate of 50 mV/s; (**d**) Current density-time curves (specific activity) of the PBNPSs (curve a), PNPSs (curve b) and E-TEK Pd/C (curve c) electrodes in a 0.5 M NaOH +1 M ethanol solution at −0.2 V. Reprinted with permission from [117]. Copyright 2010 Royal Society of Chemistry

surface area, lower onset potential, and larger current density of ethanol oxidation. This study indicates that PNVC-V_2O_5 polymer-metal composite can serve as a promising catalyst supporting material for fuel cells.

4.3.2 Formic Acid Oxidation Reaction (FAOR)

Formic acid is a liquid at room temperature, and dilute formic acid is recognized as a safe fuel used in low-temperature fuel cells. As compared to traditional hydrogen fuel, aqueous solutions of formic acid exhibit the advantages of ease of transportation, storage and handling. Moreover, direct formic acid fuel cells (DFAFCs) have a smaller crossover of formic acid than that of methanol fuels in DMFCs, which simultaneously improves fuel utilization and enables the use of thinner membranes. In addition, DFAFCs has a theoretical open circuit voltage (OCV) of about 1.48 or 1.45 V, which is higher than hydrogen-oxygen (1.23 V) and methanol-oxygen (1.18 V) fuel cells [179–181]. For formic acid oxidation, a so-called dual pathway mechanism has been commonly accepted: (i) direct pathway (dehydrogenation pathway), which involves the removal of two hydrogen atoms from formic acid molecule to directly produce CO_2 ($HCOOH \rightarrow CO_2+2H^++2e^-$), and (ii) indirect pathway (dehydration pathway), which involves the dehydration of formic acid to yield a poisonous intermediate CO species (CO_{ads}) and then further oxidation to CO_2 ($HCOOH \rightarrow CO_{ads}+H_2O \rightarrow CO_2+2H^+ +2e^-$). On the basis of such a mechanism, to enhance fuel cell efficiency and avoid poisoning, the materials on which the direct dehydrogenation pathway predominantly occurs or have high CO tolerance are ideal anode catalysts for formic acid oxidation. Pt-Pd bimetallic catalysts are considered as a substitute of platinum in fuel cells due to their attractive performance and better CO tolerance during formic acid oxidation than that of Pt catalysts [69, 135, 182].

By controlling the crystallographic orientation of the surface atoms, Pt-Pd bimetallic nanocrystals with enhanced electrochemical properties can be designed [124, 173]. For instance, by using highly faceted cubic Pt as seeds to direct the epitaxial overgrowth of Pd, Yang and coworkers [116] prepared shape-controlled Pt-Pd heterostructures and studied their catalytic activities for formic acid electro-oxidation. From the CVs of formic acid oxidation shown in Fig. 4.16, (100)-facets terminated cubic Pt-Pd core-shell particles (Fig. 4.16a) exhibit a much higher peak current that is five times of that from the (1 1 1)-facets terminated Pt-Pd octahedra (Fig. 4.16c), whereas the octahedral Pt-Pd shows a lower peak potential (0.15 V) than the cubic Pt-Pd (0.36 V). Moreover, the Pt-Pd cubes show a sharper peak at 0.47 V in the negative scan than the octahedra, and the Pt-Pd cuboctahedra show intermediate catalytic properties between the cubes and octahedra (Fig. 4.16b). The surface-dependent properties can be ascribed to the different arrangement of atoms at the surface of different shaped Pt-Pd nanocrystals, and thus the corresponding surface energy affects adsorption, surface diffusion, intermediate formation, chemical rearrangements and finally desorption of the products in the formic acid oxidation.

In another work, Pt-Pd binary nanoparticles were fabricated by decorating Pd on the well-defined Pt(100) single crystal surface via a seed-mediated growth process [123]. Because the presence of adsorbed Pd on Pt(100) can minimize CO formation at low potential, and thus decrease the self-poisoning and lower the oxidation potential, the prepared Pt-Pd nanoparticles showed remarkably enhanced electrocatalytic performance for formic acid oxidation. Feliu and coworkers [145] prepared Pd adatom-modified (100) preferentially oriented cubic Pt nanoparticles by an electrochemical deposition approach using a reductant- and surfactant-free process. In electrochemical measurements, the formic acid oxidation current was strongly inhibited in the positive scan on the bare cubic Pt nanoparticles owing to the almost complete poisoning of (100) surface sites by CO. After the incorporation of Pd, the formic acid oxidation was enhanced. However, due to the lower fraction of Pt(100) sites in the quasispherical Pt nanoparticles, the current density obtained from the cubic Pt nanoparticles is higher than that from quasispherical Pt

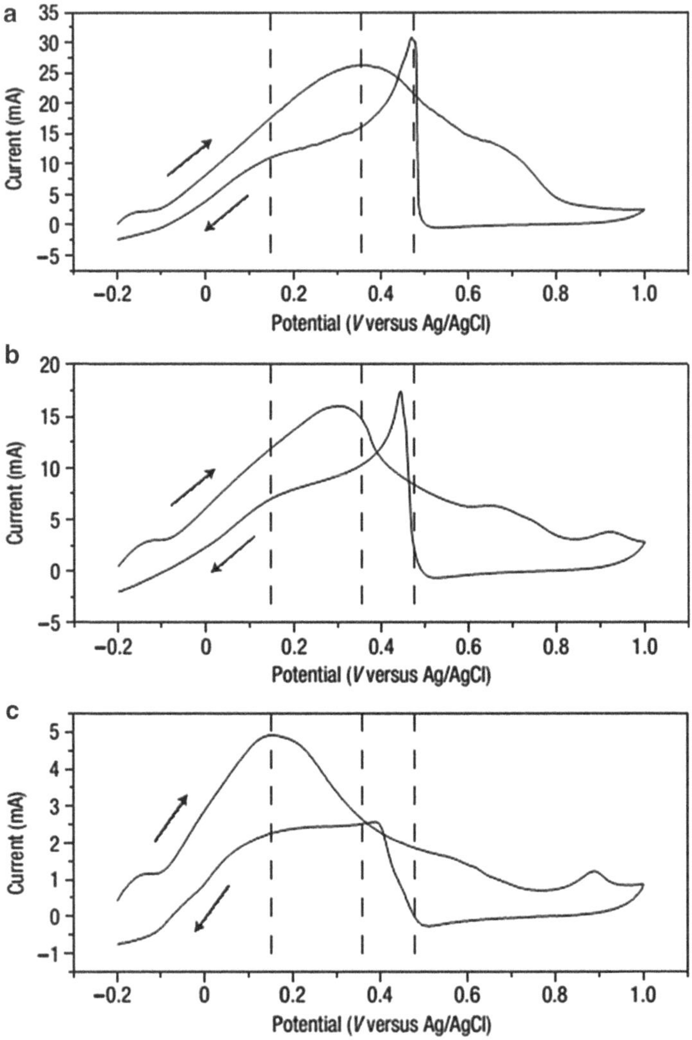

Fig. 4.16 Electrocatalytic activities of the Pt-Pd binary metal nanocrystals for formic acid oxidation. (**a**–**c**) Cyclic voltammograms for cubes (**a**), cuboctahedra (**b**) and octahedra (**c**) in 0.1 M H_2SO_4 and 0.2 M formic acid at a sweep rate of 50 mV/s. Reprinted with permission from [116]. Copyright 2007 Nature Publishing Group

nanoparticles. Moreover, the Pd adatom-decorated cubic Pt nanoparticles exhibited high stability during the formic acid oxidation. Such results indicate that modification of shape-controlled Pt by surface adatoms is an efficient strategy to enhance their electrocatalytic activities.

It was also found that the catalytic activities of Pt-Pd bimetallic nanocrystals towards formic acid electro-oxidation depend on their compositions [67, 183]. Zheng and coworkers [111] reported that Pt can be alloyed into the Pd

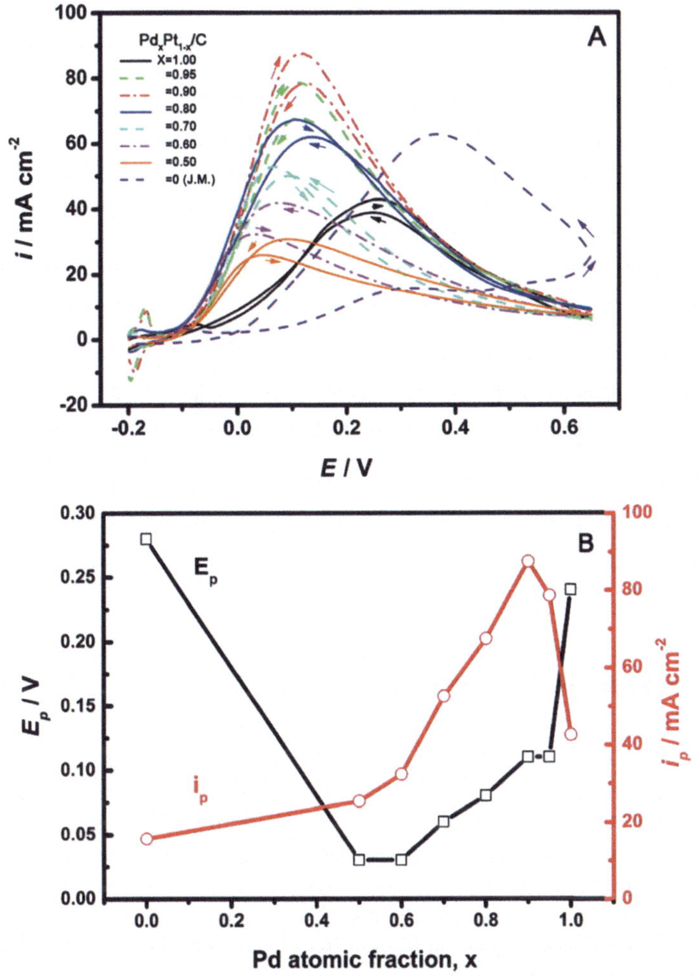

Fig. 4.17 (**a**) Cyclic voltammograms recorded at 50 mV/s for Pd_xPt_{1-x}/C coated on GC electrodes in 0.5 M H_2SO_4 solution containing 0.5 M formic acid. (**b**) Variation of E_p and i_p with Pd atomic fraction (x). Reprinted with permission from [51]. Copyright 2010 American Chemical Society

nanospheres and the Pt-Pd nanoparticles with a molar ratio of Pd/Pt = 3:1 exhibited the highest activity for electrocatalytic oxidation of formic acid. Cai and coworkers [51] synthesized a series of Pd_xPt_{1-x} (x = 0.5–1) nanoparticles dispersed on carbon black support and studied their catalytic activities for formic acid oxidation. Figure 4.17a compares the CVs of formic acid oxidation on different compositions of Pd_xPt_{1-x}/C catalysts in 0.5 M H_2SO_4 solution containing 0.5 M formic acid (FA). Obviously, the peak current density (i_p) of FA oxidation is x-dependent, and the $Pd_{0.9}Pt_{0.1}/C$ shows the highest i_p which is nearly twice as high as that of Pd/C and six times of that of commercial Pt/C. The larger gap of negatively shifted peak potentials (Fig. 4.17b) between Pd/C and $Pd_{0.9}Pt_{0.1}$/C but smaller difference between Pd_xPt_{1-x}/C catalysts demonstrate the synergistic effect of Pd and Pt sites in the FA electro-oxidation. Moreover, after 1000 s stability test, the oxidation current density on $Pd_{0.9}Pt_{0.1}$/C is the highest among the Pd_xPt_{1-x}/C catalysts, which is 2.4- and 3.6-fold larger than those of commercial Pd/C and Pt/C, respectively.

Carbon nanotubes have also exhibited their potential applications as catalyst supports in fuel cells [71]. Chen and coworkers [99] studied the catalytic properties of hollow Pt-Pd nanospheres dispersed on carbon nanotubes. The electrochemical measurements showed that the prepared hybrids have enhanced activity with a high peak current and steady-state current toward formic acid oxidation as compared to commercial Pt/C and E-TEK PtRu/C. In another study, Pt_xPd_y alloy nanoparticles supported on carbon nanotubes were also studied as anode catalysts for FAOR [59]. It was found that the catalytic activity and stability of the CNT-supported catalysts increase with more Pd content in the catalysts. In addition, titanium [92] and conductive polymer (polythiophene) [138] were also employed as support materials to disperse Pt-Pd bimetallic nanocrystal with enhanced electrocatalytic activities for formic acid oxidation.

4.3.3 Oxygen Reduction Reaction (ORR)

Oxygen reduction reaction (ORR) on cathode plays a crucial role in manipulating the performance of a fuel cell. Whereas much progress has been achieved on cathodic catalysts in the past few decades, the inherently slow reaction kinetics of ORR and the large overpotential still prevent the fuel cells from being scaled-up for commercial applications. The ORR has complicated reaction pathways and involves a multi-electron transfer process in which O_2 is converted into H_2O or OH^-, depending on the solution used in the electrochemical studies. In an acidic solution, O_2 can be reduced in an efficient $4e^-$ process and converted into H_2O: $O_2+ 4e^- \rightarrow 2H_2O$. Meanwhile, O_2 may also undergo a less efficiently partial $2e^-$ reduction to form hydrogen peroxide, H_2O_2, followed by another $2e^-$ reduction to convert H_2O_2 into H_2O: $O_2+2H^+ + 2e^- \rightarrow H_2O_2$; $H_2O_2+2H^+ + 2e^- \rightarrow 2H_2O$. In alkaline solution, O_2 can be reduced by a $4e^-$ process to form hydroxide, OH^-: $O_2+2H_2O+ 4e^- \rightarrow 4OH^-$, or by two $2e^-$ processes to form HO_2^- and then OH^-: $O_2+H_2O+2e^- \rightarrow HO_2^- +OH^-$; $HO_2^- +H_2O+2e^- \rightarrow 3OH^-$. In addition, under common ORR conditions, oxygen may be converted into different intermediates, such as oxygenated (O), hydroxyl (OH) and superhydroxyl (OOH) species, which are very difficult to be detected experimentally [184].

The kinetic parameters of ORR can be determined using the rotating disk electrode (RDE) technique. The kinetic current density and the electron transfer numbers can be derived from the following equations: [14]

$$\frac{1}{j} = \frac{1}{j_K} + \frac{1}{j_L} = \frac{1}{j_K} + \frac{1}{B^{1/2}} \tag{4.1}$$

$$B = 0.2nF(D_0)^{2/3}\,{}^{1/6}C_0 \tag{4.2}$$

$$j_K = nFkC_0 \tag{4.3}$$

where j is the measured current density, j_K and j_L are the kinetic and diffusion limitation current densities, ω is the electrode rotating rate, n represents the number of electrons transferred per oxygen molecule, F is the Faraday constant ($F=$ 96,485 C/mol), D_0 is the diffusion coefficient of O_2, ν is the kinetic viscosity of the electrolyte, and C_0 is the bulk concentration of O_2 dissolved in the electrolyte. The constant 0.2 is used when the rotation speed is expressed in rpm. According to Eqs. 4.1–4.3, the number of electron transfer (n) and kinetic current density (j_K) can be obtained from the slope and intercept of the Koutecky-Levich plots, respectively. In addition, rotating ring-disk electrode (RRDE) can also be used to evaluate the ORR performance of a catalyst. Electron transfer number (n) and the H_2O_2 percentage ($H_2O_2\%$) generated at the disk electrode can be calculated from the ring and disk current as the following equations (i_D and i_R represent the disk and ring current, and N is the RRDE collection efficiency) [185]:

$$n = \frac{4i_D}{i_D + i_R/N} \tag{4.4}$$

$$H_2O_2\% = \frac{200i_R/N}{i_D + i_R/N} \tag{4.5}$$

At present, because of the outstanding catalytic properties, Pt has been the most widely used electrocatalyst to speed up oxygen reduction at cathode. However, despite extensive research efforts, the sluggish oxygen reduction kinetics at the cathode, high cost, low stability and limited supply of platinum still largely restrict the wide spread commercialization of fuel cells. Moreover, the state-of-art commercial catalysts are usually prepared by dispersing Pt nanoparticles (2–5 nm) on carbon supports, whose electrochemical surface areas and catalytic efficiency degrade over time owing to the corrosion of the carbon supports and the dissolution, aggregation of Pt nanoparticles during the catalysis processes. This has posed tremendous challenges in the development of Pt-based catalysts for practical fuel-cell applications. The combination of Pd with Pt can downshift the d-band center of Pt catalyst, leading to a lower degree of adsorption of oxygenated spectator species (such as OH^-) and an increase of the number of active sites accessible to oxygen, Pt-Pd bimetallic nanocrystals have been accepted as one type of the most active and effective cathodic catalysts for oxygen reduction reaction.

The enhancement of the Pt mass or specific activity for ORR can be achieved by controlling the composition of the alloyed Pt-Pd nanomaterials [38–40, 56, 60, 73]. Cooks and coworkers [35] prepared Pt-Pd bimetallic nanoparticles (~1.8 nm) containing an average of 180 atoms with seven different Pt:Pd ratios

(G6-OHPt$_n$Pd$_{180-n}$DENs, n = 180, 150, 120, 90, 60, 30 and 0) by a dendrimer templating method. From the results of CV and rotating disk voltammetry measurements shown in Fig. 4.18, as compared to the equivalent monometallic Pt DENs catalyst, the maximum relative mass activity enhancement of 240 % was obtained from the bimetallic PtPd DENs containing 17–33 % Pd. And the G6-OH (Pt$_{150}$Pd$_{30}$) DENs exhibited the most active ORR performance with the most positive onset potential, the highest kinetic current density and an efficient 4e$^-$ oxygen reduction process. Huang and coworkers studied the ORR activity of the Pt-Pd bimetallic nanocrystals with different compositions [115]. It was found that the bimetallic Pt-Pd nanocrystals have higher ORR activity than that of commercially available Pt black and Pd black in acid solution, and the bimetallic nanocrystals with Pd/Pt ratio of 1/3 gave the highest ORR activity. Lee et al. [70] reported that the Pt-Pd nanocatalysts deposited on Vulcan XC-72R carbon support with a molar ratio of 1:3 exhibited much enhanced catalytic activity toward oxygen reduction reaction in comparison with commercial Pt/C. By using carbon nanotubes as supports for electrocatalyst, Ghosh et al. [55] investigated the catalytic properties of multiwalled

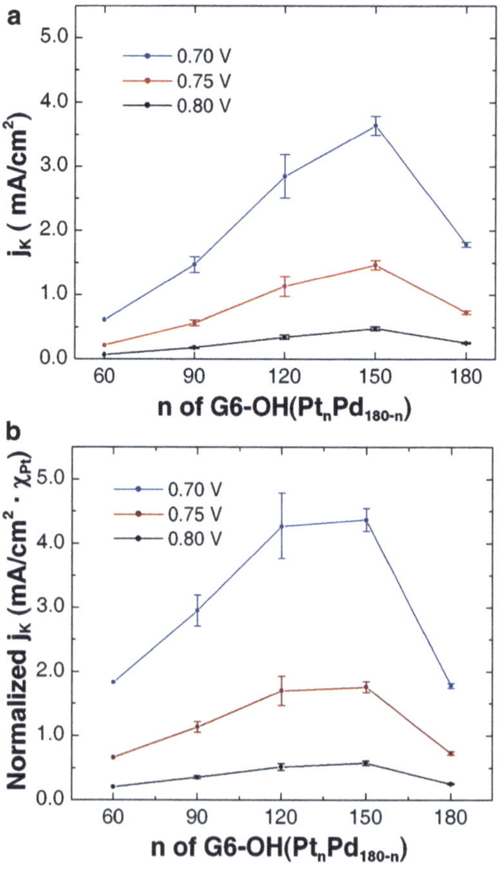

Fig. 4.18 Plots of (**a**) kinetic current density and (**b**) kinetic current density normalized to the fraction amount of Pt contained in each nanoparticle composition at different potentials. Error bars represent the span of data for two independently executed experiments (only one experiment was carried out for G6-OH (Pt$_{60}$Pd$_{120}$). Reprinted with permission from [35]. Copyright 2007 American Chemical Society

carbon nanotubes (MCNTs)-supported Pt-Pd nanoparticles with different compositions. The electrochemical measurements showed that the $Pt_{46}Pd_{54}$ nanoparticles supported on MCNTs have an enhanced electrocatalytic activity toward ORR with specific and mass activities of 378 $\mu A/cm^2{}_{Pt\text{-}Pd}$ and 64 $\mu A/\mu g_{Pt\text{-}Pd}$ at 0.8 V, as compared to 150 $\mu A/cm^2{}_{Pt\text{-}Pd}$ and 26.5 $\mu A/\mu g_{Pt\text{-}Pd}$ for $Pt_{64}Pd_{36}$, 33 $\mu A/cm^2{}_{Pt\text{-}Pd}$ and 4.5 $\mu A/\mu g_{Pt\text{-}Pd}$ for $Pt_{28}Pd_{72}$, and 142 $\mu A/cm^2{}_{Pt\text{-}Pd}$ and 45.7 $\mu A/\mu g_{Pt\text{-}Pd}$ for commercial Pt black. Moreover, the $Pt_{46}Pd_{54}$/MCNTs exhibited high stability and tolerance to methanol during the oxygen reduction. The synergistic effect between Pt and Pd and the strong MCNT-catalyst interaction was believed to enhance the overall catalytic performance of the catalysts.

In addition to composition, the catalytic activities of Pt-Pd bimetallic nanocrystals toward ORR are also dependent on their shapes. Pt-Pd core-shell nanocrystals are a class of electrocatalysts for ORR with an enhanced performance. For example, Iwasawa and coworkers [114] investigated the catalytic activities of three types of bimetallic Pt-Pd nanoparticles, Pt/C, Pd (core)-Pt (shell), and Pt (core)-Pd (shell) for ORR. By comparing the CVs, the Pt/C, PdPt alloy/C, and Pd (core)-Pt (shell)/C catalysts with Pt-enriched surfaces showed much higher ORR specific activities than the Pd/C and Pt (core)-Pd (shell)/C catalysts with Pd-enriched surfaces. And the Pt surface-enriched bimetallic catalysts with a core-shell structure showed higher Pt-based mass activity than the Pt monometal catalyst. Moreover, after 1000 cycles of CV test, the loss of ECSA is less than 20% for Pd (core)-Pt (shell)/C and PdPt alloy/C with Pt-enriched surfaces, while drop as large as 53.8% and 41.6% for Pd/C and Pt (core)-Pd (shell)/C with Pd-enriched surfaces, suggesting that Pt surface-enriched catalysts have great stability in ORR.

Pt monolayer deposited on Pd provides another promising core-shell structure to enhance the ORR activity through the strong interaction between these two metals, as well as the reduction of Pt loading down to a minimum level [110, 146, 148, 150, 186, 187]. Adzic and coworkers [153] reported a synthesis of well-defined Pd core-Pt monolayer shell nanocatalysts for ORR. It was found that the optimized Pt monolayer on commercial Pd nanoparticles exhibited outstanding mass and area-specific ORR activities of 0.96 A/mg_{Pt} and 0.5 mA/cm^2, respectively. The significant enhancement in activity could be attributed to the uniquely advantageous core-shell motif. In the structure, the Pd core imparts a compressive strain and a so-called "ligand-effect" upon the Pt monolayer, which can effectively lower the energy of the Pt d-band and thus directly weaken its interaction with OH_{ads} group, thereby yielding improved ORR activity. Sasaki et al. [151] investigated the catalytic performance of a Pd core-protected Pt monolayer shell catalyst for ORR. Compared to commercial Pt/Ketjen black and Pt/C, the Pd nanoparticles covered by a Pt monolayer (Pt ML/Pd/C) exhibited much higher electrocatalytic durability. The mass activity of the Pt ML/Pd/C decreased by 37% after 100,000 cycles of stability tests, while the mass activity of Pt/C lost almost 70% after 60,000 potential cycles and the Pt/Ketjen black carbon experienced more than 40% loss after only 10,000 cycles. Surprisingly, due to the effect of potential cycling on the particle size distribution, the Pt mass activity of PtML/Pd/C

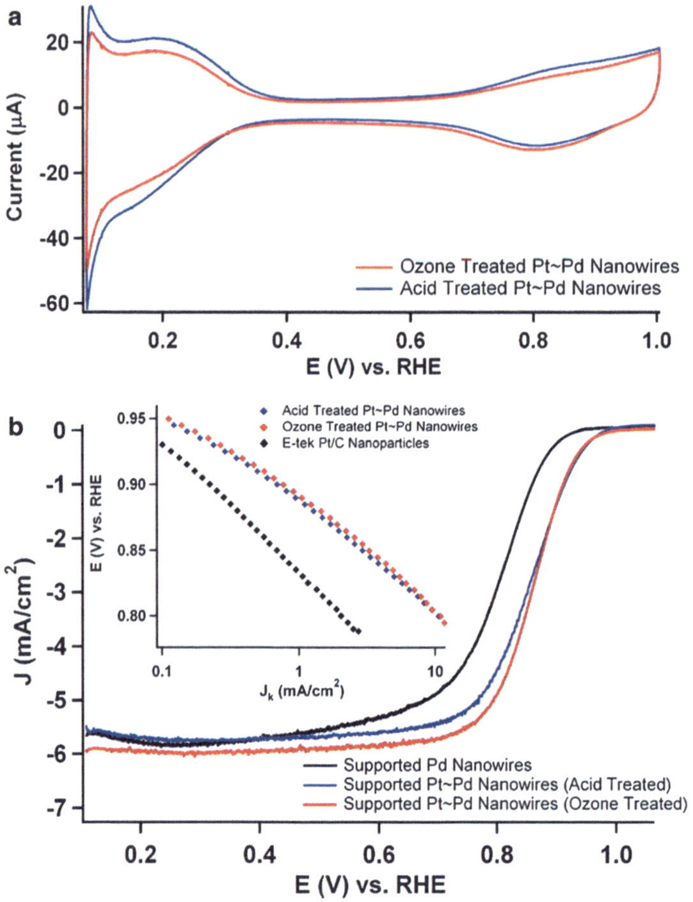

Fig. 4.19 Cyclic voltammograms (**a**) obtained for the ozone- and acid-treated Pt-PdNW/C core-shell nanowires, after Pt monolayer deposition, loaded separately onto a GCE in a 0.1 M $HClO_4$ solution at 20 mV/s. (**b**) The polarization curves for the treated Pt-Pd/C core-shell nanowire composites were both obtained on a glassy carbon rotating disk electrode. Curves (anodic sweep direction) were collected using a rotation rate of 1600 rpm in a 0.1 M $HClO_4$ solution at 20 °C. Kinetic current vs potential plots (inset) of treated Pt-Pd/C composites and commercial carbon supported Pt nanoparticles are also presented. Reprinted with permission from [152]. Copyright 2011 American Chemical society

increased from the initial 3-fold to 5-fold enhancement over that of Pt/C after 60,000 cycles.

In another work, Adzic and coworkers [152] reported the deposition of a monolayer quantity of Pt onto the surface of the treated Pd nanowires/C (Pd NW/C) composites by Cu UPD followed by galvanic replacement of the Cu adatoms with Pt^{2+}, and their electrocatalytic activity for ORR in acid condition. As shown in Fig. 4.19a, b, the ultrathin Pt monolayer shell-Pd nanowire core catalyst (Pt-Pd NW/C) displayed dramatically enhanced ORR activity compared to the Pd

NW/C and commercial Pt nanoparticles. Moreover, the ozone-treated Pt-Pd NW/C exhibited much higher area-specific (0.77 mA/cm^2) and mass-specific (1.83 A/mg$_{Pt}$) activities than those of acid-treated Pt-Pd nanowires (0.70 mA/cm^2 and 1.47 A/mg$_{Pt}$). Such results indicate that the ozone treatment for Pd NW before Cu UPD can lead to the exceptional enhancement in the catalytic activity of Pt-Pd nanowires. Most importantly, after the accelerated durability half-cell test, the ozone-treated Pt-Pd nanowires still maintained excellent electrochemical durability and the area-specific activity increased by 1.5-fold.

Pt-Pd nanodendrites are another novel structure imparting its enhanced ORR activity by maximizing the expression of certain highly active facets and improving surface area. Xia and coworkers [122] synthesized Pt-Pd bimetallic nanodendrites consisting of a dense array of Pt branches on a Pd core, and investigated their catalytic activity for ORR in acid conditions. The calculated ECSA of the Pt-Pd nanodendrites (57.1 m^2/g$_{Pt}$) is much larger than that of commercial Pt black catalyst (19.1 m^2/g$_{Pt}$), but is lower than that of Pt/C catalyst (74 m^2/g$_{Pt}$). It was found that the mass activity from the Pt-Pd nanodendrites (0.241 mA/μg$_{Pt}$) is 2.5- and 5-fold larger, respectively, than those of state-of-art Pt/C and the first-generation supportless Pt-black catalyst at room temperature. Moreover, at 60 °C, the Pt mass activity of the Pt-Pd nanodendrites (0.433 mA/μg$_{Pt}$) was still greater than that of the Pt/C catalyst (0.204 mA/μg$_{Pt}$) and the Pt black (0.078 mA/μg$_{Pt}$). In addition to mass activity, Pt-Pd nanodendrites exhibited a specific activity of 3.1–3.4-fold that of the Pt/C catalyst and 1.7–2.0-fold that of the Pt black depending on the temperature, indicating a superior capability of the Pt-Pd nanodendrites in accelerating ORR. The high ORR activity of the Pt-Pd nanodendrites can be attributed to the high and accessible surface area of the dendritic morphology and the preferential exposure of (1 1 1) facets along with some (1 1 0) and high-index (3 1 1) facets on the Pt branches that are particularly active towards ORR. In another work, Yang and coworkers [126] reported the synthesis of Pt-on-Pd heteronanostructure supported on carbon black. The as-synthesized nanostructure showed much larger electrochemically active surface area as compared to E-TEK Pt/C (20 wt%Pt) catalyst (Fig. 4.20a). From Fig. 4.20b, it can be seen that the incorporation of Pd also greatly altered the ability to absorb hydroxyl species (OH$_{ads}$, E>0.6 V) compared to E-TEK Pt/C (20 wt%Pt) catalyst, and the Pt-on-Pd catalyst exhibited much faster hydroxyl desorption ability with positively shifted onset and peak potentials in the backward sweep. As shown in Fig. 4.20c, d, the Pt-on-Pd catalyst shows a more positive onset potential and higher activity than the commercially pure Pt. Moreover, after 30,000 cyclic voltammetry cycling, the Pt-on-Pd catalyst lost only 12% of the initial ECSA and showed a small degradation of 9 mV in the half-wave potential, while a loss of 39% of the initial ECSA and a large decrease of 35 mV in the half-wave potential for Pt/C catalyst, suggesting the enhanced stability of Pt-on-Pd/C catalyst for ORR.

In addition, the ORR catalytic activities of Pt-Pd bimetallic nanostructures with other specific size and morphologies such as hollow urchin-like [97], bowl-like [100], and rods [77], were also investigated. Yan et al. [101] found that Pt-Pd nanotubes exhibited much enhanced ORR mass/specific activity, and improved

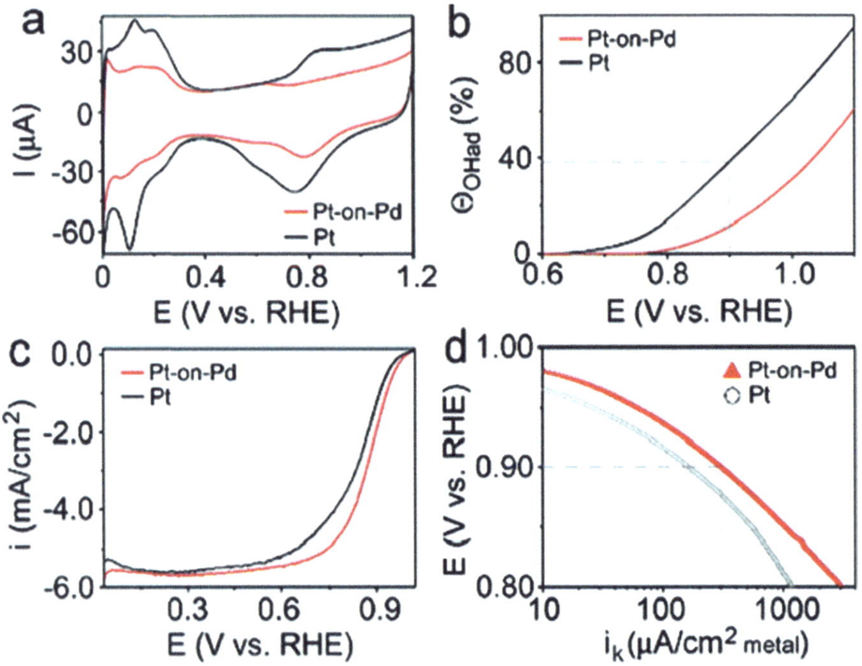

Fig. 4.20 (**a**) CVs, (**b**) hydroxyl surface coverage (Θ_{OH}), (**c**) ORR polarization curves, and (**d**) specific kinetic current densities (i_k) for carbon-supported Pt-on-Pd and Pt catalysts. Reprinted with permission from [126]. Copyright 2009 American Chemical Society

stability in acid electrolyte as compared to commercial Pt/C and Pt black catalysts. As reported by Xia and coworkers in another work [102], Pt-Pd concave cubes with different weight percentages of Pt can be prepared by using bromide-induced galvanic replacement method. It was found that the ORR activities of the Pt-Pd concave nanocubes on the basis of Pt mass monotonically increase with decreasing Pt weight percentage. And the Pt-Pd concave nanocubes with a Pt weight percentage of 3.4 showed the highest mass activity for ORR, which is 4 times higher than that of the Pt/C catalyst in terms of equivalent Pt mass. In a recent work from our group [104], Pt-Pd porous nanorods (PtPd PNRs) were successfully synthesized through a bromide-induced galvanic replacement reaction between Pd nanowires (Pd NWs) and K_2PtCl_6. The unique porous Pt-Pd nanorods showed greatly enhanced ORR activity with the highest limiting current densities and the most positive onset potential compared to those obtained from the original Pd NWs and commercial Pt/C catalyst. In addition to the improved catalytic activity, the PtPd PNRs also showed excellent durability in ORR (Fig. 4.21a–d) with only 5.88 % loss of the initial ECSA after the accelerated durability tests (1000 potential cycles), whereas 21.6 % and 40.4 % of their original ECSA were lost for the Pd NWs and commercial Pt/C catalyst, respectively. The enhanced catalytic activity of PtPd

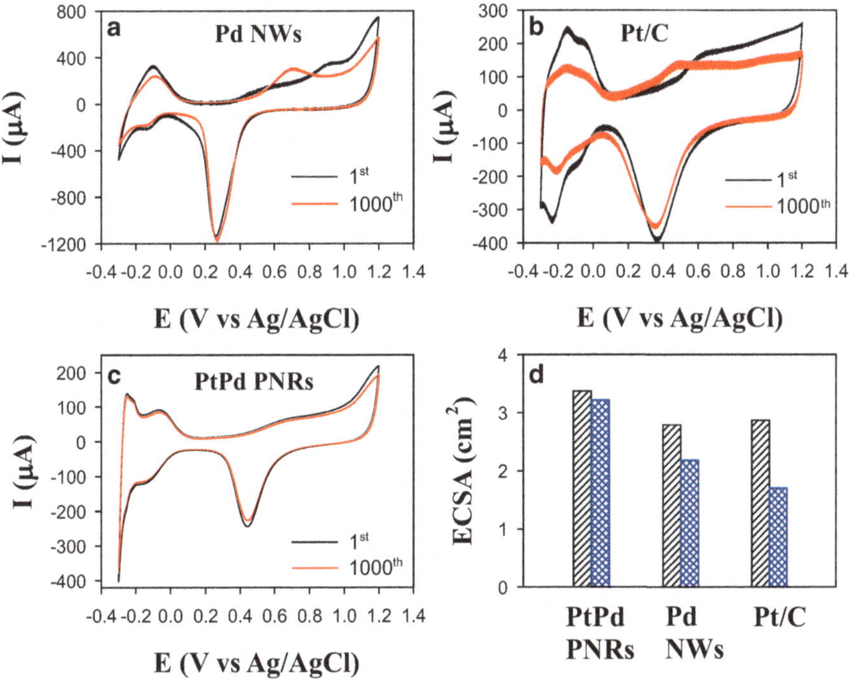

Fig. 4.21 Cyclic voltammograms recorded on Pd NWs (**a**), Pt/C (**b**) and PtPd PNRs (**c**) before (*black curves*) and after 1000 cycles (*red curves*) of the accelerated durability test, in 0.1 M $HClO_4$ solution with a potential scan rate of 0.1 V/s. (**d**) Comparison of the electrochemical surface area (ECSA) before and after 1000 cycles of the accelerated durability tests on the PtPd PNRs, Pd NWs, and commercial Pt/C catalyst, respectively. Reprinted with permission from [104]. Copyright 2013 Elsevier

PNRs toward ORR with efficient $4e^-$ pathway can be ascribed to its high specific surface area, unique one-dimensional morphology and the alloy structure.

4.4 Conclusion

Over the past several years, Pt-Pd bimetallic nanocrystals have received great attention owing to their attractive properties and extensive applications in fuel cells. In this chapter, the recent development related to the synthesis of Pt-Pd bimetallic nanocrystals with different structures, including Pt-Pd alloys, core-shell, multi-shells, dendrites and supported monolayer has been extensively reviewed. Furthermore, a number of strategies based on co-chemical reduction, galvanic displacement, seed-mediated growth and the combination of these methods used to control the formation of Pt-Pd bimetallic nanocrystals with a specific structure have been discussed. Finally, the application of Pt-Pd (or supported Pt-Pd) bimetallic nanocrystals in fuel cells as catalysts for fuels oxidation at anode and oxygen reduction reaction at cathode was summarized.

Despite substantial progress that has been made in designing Pt-Pd bimetallic electrocatalysts and understanding their electrocatalytic mechanism, the development of new methodologies, the rational control on size, shape, structure and composition of bimetallic nanocrystals, in-depth understanding of the structure-catalytic property relationship, and the exploration of more extensive applications of bimetallic nanocrystals in fuel cells still need to be continuously studied.

Acknowledgements This work was supported by the National Natural Science Foundation of China with the Grant Numbers of 21275136, 21043013 and the Natural Science Foundation of Jilin Province, China (No. 201215090).

References

1. Potocnik J (2007) Renewable energy sources and the realities of setting an energy agenda. Science 315:810–811
2. Armand M, Tarascon JM (2008) Building better batteries. Nature 451:652–657
3. Steele BCH, Heinzel A (2001) Materials for fuel-cell technologies. Nature 414:345–352
4. Russell AE, Rose A (2004) X-ray absorption spectroscopy of low temperature fuel cell catalysts. Chem Rev 104:4613–4635
5. Lamy C, Lima A, LeRhun V, Delime F, Coutanceau C, Leger JM (2002) Recent advances in the development of direct alcohol fuel cells (DAFC). J Power Sources 105:283–296
6. Vigier F, Rousseau S, Coutanceau C, Leger JM, Lamy C (2006) Electrocatalysis for the direct alcohol fuel cell. Top Catal 40:111–121
7. Tian N, Zhou ZY, Sun SG, Ding Y, Wang ZL (2007) Synthesis of tetrahexahedral platinum nanocrystals with high-index facets and high electro-oxidation activity. Science 316:732–735
8. Chen W, Kim J, Sun SH, Chen SW (2006) Electro-oxidation of formic acid catalyzed by FePt nanoparticles. Phys Chem Chem Phys 8:2779–2786
9. Chen W, Kim JM, Sun SH, Chen SW (2007) Composition effects of FePt alloy nanoparticles on the electro-oxidation of formic acid. Langmuir 23:11303–11310
10. Yang HZ, Zhang J, Sun K, Zou SZ, Fang JY (2010) Enhancing by weakening: electrooxidation of methanol on Pt_3Co and Pt nanocubes. Angew Chem Int Ed 49:6848–6851
11. Chen YX, Miki A, Ye S, Sakai H, Osawa M (2003) Formate, an active intermediate for direct oxidation of methanol on Pt electrode. J Am Chem Soc 125:3680–3681
12. Tripkovic AV, Popovic KD, Grgur BN, Blizanac B, Ross PN, Markovic NM (2002) Methanol electrooxidation on supported Pt and PtRu catalysts in acid and alkaline solutions. Electrochim Acta 47:3707–3714
13. Qiao Y, Li CM (2011) Nanostructured catalysts in fuel cells. J Mater Chem 21:4027–4036
14. Zhang RZ, Chen W (2013) Non-precious Ir-V bimetallic nanoclusters assembled on reduced graphene nanosheets as catalysts for the oxygen reduction reaction. J Mater Chem A 1:11457–11464
15. Liu MM, Zhang RZ, Chen W (2014) Graphene-supported nanoelectrocatalysts for fuel cells: synthesis, properties, and applications. Chem Rev 114:5117–5160
16. Zhang H, Jin MS, Xia YN (2012) Enhancing the catalytic and electrocatalytic properties of Pt-based catalysts by forming bimetallic nanocrystals with Pd. Chem Soc Rev 41:8035–8049
17. Gasteiger HA, Markovic N, Ross PN, Cairns EJ (1994) Temperature-dependent methanol electrooxidation on well-characterized Pt-Ru alloys. J Electrochem Soc 141:1795–1803

18. Dinh HN, Ren XM, Garzon FH, Zelenay P, Gottesfeld S (2000) Electrocatalysis in direct methanol fuel cells: in-situ probing of PtRu anode catalyst surfaces. J Electroanal Chem 491:222–233
19. Wang DS, Li YD (2011) Bimetallic nanocrystals: liquid-phase synthesis and catalytic applications. Adv Mater 23:1044–1060
20. Calvo SR, Balbuena PB (2007) Theoretical analysis of reactivity on Pt(111) and Pt-Pd(111) alloys. Surf Sci 601:4786–4792
21. Calvo SR, Balbuena PB (2007) Density functional theory analysis of reactivity of Pt_xPd_y alloy clusters. Surf Sci 601:165–171
22. Norskov JK, Bligaard T, Rossmeisl J, Christensen CH (2009) Towards the computational design of solid catalysts. Nat Chem 1:37–46
23. Norskov JK, Abild-Pedersen F (2009) Catalysis bond control in surface reactions. Nature 461:1223–1225
24. Kitchin JR, Norskov JK, Barteau MA, Chen JG (2004) Role of strain and ligand effects in the modification of the electronic and chemical properties of bimetallic surfaces. Phys Rev Lett 93:156801–156804
25. Hammer B, Norskov JK (2000) Theoretical surface science and catalysis-calculations and concepts. Adv Catal 45:71–129
26. Norskov JK, Rossmeisl J, Logadottir A, Lindqvist L, Kitchin JR, Bligaard T, Jonsson H (2004) Origin of the overpotential for oxygen reduction at a fuel-cell cathode. J Phys Chem B 108:17886–17892
27. de Bruijn FA, Dam VAT, Janssen GJM (2008) Durability and degradation issues of PEM fuel cell components. Fuel Cells 8:3–22
28. Stamenkovic VR, Fowler B, Mun BS, Wang GF, Ross PN, Lucas CA, Markovic NM (2007) Improved oxygen reduction activity on $Pt_3Ni(111)$ via increased surface site availability. Science 315:493–497
29. Ramirez-Caballero GE, Hirunsit P, Balbuena PB (2010) Shell-anchor-core structures for enhanced stability and catalytic oxygen reduction activity. J Chem Phys 133:134705–134713
30. Xia YN, Xiong YJ, Lim B, Skrabalak SE (2009) Shape-controlled synthesis of metal nanocrystals: simple chemistry meets complex physics? Angew Chem Int Ed 48:60–103
31. Shao YY, Yin GP, Gao YZ (2007) Understanding and approaches for the durability issues of Pt-based catalysts for PEM fuel cell. J Power Sources 171:558–566
32. Ferreira PJ, La O' GJ, Shao-Horn Y, Morgan D, Makharia R, Kocha S, Gasteiger HA (2005) Instability of Pt/C electrocatalysts in proton exchange membrane fuel cells—a mechanistic investigation. J Electrochem Soc 152:A2256–A2271
33. Wang YJ, Wilkinson DP, Zhang JJ (2011) Noncarbon support materials for polymer electrolyte membrane fuel cell electrocatalysts. Chem Rev 111:7625–7651
34. Greeley J, Stephens IEL, Bondarenko AS, Johansson TP, Hansen HA, Jaramillo TF, Rossmeisl J, Chorkendorff I, Norskov JK (2009) Alloys of platinum and early transition metals as oxygen reduction electrocatalysts. Nat Chem 1:552–556
35. Ye HC, Crooks RM (2007) Effect of elemental composition of PtPd bimetallic nanoparticles containing an average of 180 atoms on the kinetics of the electrochemical oxygen reduction reaction. J Am Chem Soc 129:3627–3633
36. Yang SC, Hong F, Wang LQ, Guo SW, Song XP, Ding BJ, Yang ZM (2010) Ultrathin Pt-based alloy nanowire networks: synthesized by CTAB assistant two-phase water-chloroform micelles. J Phys Chem C 114:203–207
37. Basu D, Basu S (2012) Performance studies of Pd-Pt and Pt-Pd-Au catalyst for electro-oxidation of glucose in direct glucose fuel cell. Int J Hydrogen Energy 37:4678–4684
38. Thanasilp S, Hunsom M (2011) Preparation of a high-performance Pt-Pd/C-electrocatalyst-coated membrane for ORR in PEM fuel cells via a combined process of impregnation and seeding: effect of electrocatalyst loading on carbon support. Electrochim Acta 56:1164–1171

39. Thanasilp S, Hunsom M (2011) Effect of Pt: Pd atomic ratio in Pt-Pd/C electrocatalyst-coated membrane on the electrocatalytic activity of ORR in PEM fuel cells. Renew Energy 36:1795–1801
40. Tang YF, Zhang HM, Zhong HX, Xu T, Jin H (2011) Carbon-supported Pd-Pt cathode electrocatalysts for proton exchange membrane fuel cells. J Power Sources 196:3523–3529
41. Maghsodi A, Hoseini MRM, Mobarakeh MD, Kheirmand M, Samiee L, Shoghi F, Kameli M (2011) Exploration of bimetallic Pt-Pd/C nanoparticles as an electrocatalyst for oxygen reduction reaction. Appl Surf Sci 257:6353–6357
42. Li XW, Zhu Y, Zou ZQ, Zhao MY, Li ZL, Zhou Q, Akins DL, Yang H (2009) Simple complexing-reduction synthesis of Pd-Pt/C alloy electrocatalysts for the oxygen reduction reaction. J Electrochem Soc 156:B1107–B1111
43. Kim IT, Lee HK, Shim J (2008) Synthesis and characterization of Pt-Pd catalysts for methanol oxidation and oxygen reduction. J Nanosci Nanotechnol 8:5302–5305
44. Joo JB, Kim YJ, Kim W, Kim ND, Kim P, Kim Y, Yi J (2008) Methanol-tolerant PdPt/C alloy catalyst for oxygen electro-reduction reaction. Korean J Chem Eng 25:770–774
45. Golikand AN, Asgari M, Lohrasbi E (2011) Study of oxygen reduction reaction kinetics on multi-walled carbon nano-tubes supported Pt-Pd catalysts under various conditions. Int J Hydrogen Energy 36:13317–13324
46. Golikand AN, Lohrasbi E, Asgari M (2010) Enhancing the durability of multi-walled carbon nanotube supported by Pt and Pt-Pd nanoparticles in gas diffusion electrodes. Int J Hydrogen Energy 35:9233–9240
47. Morales-Acosta D, Arriaga LG, Alvarez-Contreras L, Luna SF, Varela FJR (2009) Evaluation of $Pt_{40}Pd_{60}$/MWCNT electrocatalyst as ethylene glycol-tolerant oxygen reduction cathodes. Electrochem Commun 11:1414–1417
48. Golikand AN, Lohrasbi E, Maragheh MG, Asgari M (2009) Carbon nano-tube supported Pt-Pd as methanol-resistant oxygen reduction electrocatalyts for enhancing catalytic activity in DMFCs. J Appl Electrochem 39:2421–2431
49. Datta J, Dutta A, Biswas M (2012) Enhancement of functional properties of PtPd nano catalyst in metal-polymer composite matrix: application in direct ethanol fuel cell. Electrochem Commun 20:56–59
50. Wu M, Shen PK, Wei ZD, Song SQ, Nie M (2007) High activity PtPd-WC/C electrocatalyst for hydrogen evolution reaction. J Power Sources 166:310–316
51. Zhang HX, Wang C, Wang JY, Zhai JJ, Cai WB (2010) Carbon-Supported Pd-Pt nanoalloy with low Pt content and superior catalysis for formic acid electro-oxidation. J Phys Chem C 114:6446–6451
52. Godinez-Garcia A, Gervasio DF (2014) Pd-Pt nanostructures on carbon nanofibers as an oxygen reduction electrocatalyst. RSC Adv 4:42009–42013
53. Tao F, Grass ME, Zhang YW, Butcher DR, Renzas JR, Liu Z, Chung JY, Mun BS, Salmeron M, Somorjai GA (2008) Reaction-driven restructuring of Rh-Pd and Pt-Pd core-shell nanoparticles. Science 322:932–934
54. Arikan T, Kannan AM, Kadirgan F (2013) Binary Pt-Pd and ternary Pt-Pd-Ru nanoelectrocatalysts for direct methanol fuel cells. Int J Hydrogen Energy 38:2900–2907
55. Ghosh S, Sahu RK, Raj CR (2012) Pt-Pd alloy nanoparticle-decorated carbon nanotubes: a durable and methanol tolerant oxygen reduction electrocatalyst. Nanotechnology 23:385602–285610
56. Zhao JA, Manthiram A (2011) Preleached Pd-Pt-Ni and binary Pd-Pt electrocatalysts for oxygen reduction reaction in proton exchange membrane fuel cells. Appl Catal B Environ 101:660–668
57. Long NV, Hien TD, Asaka T, Ohtaki M, Nogami M (2011) Synthesis and characterization of Pt-Pd alloy and core-shell bimetallic nanoparticles for direct methanol fuel cells (DMFCs): enhanced electrocatalytic properties of well-shaped core-shell morphologies and nanostructures. Int J Hydrogen Energy 36:8478–8491

58. Zhou ZM, Shao ZG, Qin XP, Chen XG, Wei ZD, Yi BL (2010) Durability study of Pt-Pd/C as PEMFC cathode catalyst. Int J Hydrogen Energy 35:1719–1726
59. Winjobi O, Zhang ZY, Liang CH, Li WZ (2010) Carbon nanotube supported platinum-palladium nanoparticles for formic acid oxidation. Electrochim Acta 55:4217–4221
60. He W, Liu JY, Qiao YJ, Zou ZQ, Zhang XG, Akins DL, Yang H (2010) Simple preparation of Pd-Pt nanoalloy catalysts for methanol-tolerant oxygen reduction. J Power Sources 195:1046–1050
61. He W, Chen M, Zou ZQ, Li ZL, Zhang XG, Jin SA, You DJ, Pak C, Yang H (2010) Oxygen reduction on Pd_3Pt_1 bimetallic nanoparticles highly loaded on different carbon supports. Appl Catal B Environ 97:347–353
62. Wang WM, Huang QH, Liu JY, Zou ZQ, Zhao MY, Vogel W, Yang H (2009) Surface and structure characteristics of carbon-supported Pd_3Pt_1 bimetallic nanoparticles for methanol-tolerant oxygen reduction reaction. J Catal 266:156–163
63. Kadirgan F, Kannan AM, Atilan T, Beyhan S, Ozenler SS, Suzer S, Yorur A (2009) Carbon supported nano-sized Pt-Pd and Pt-Co electrocatalysts for proton exchange membrane fuel cells. Int J Hydrogen Energy 34:9450–9460
64. Kadirgan F, Beyhan S, Atilan T (2009) Preparation and characterization of nano-sized Pt-Pd/C catalysts and comparison of their electro-activity toward methanol and ethanol oxidation. Int J Hydrogen Energy 34:4312–4320
65. Ficicilar B, Bayrakceken A, Eroglu I (2009) Effect of Pd loading in Pd-Pt bimetallic catalysts doped into hollow core mesoporous shell carbon on performance of proton exchange membrane fuel cells. J Power Sources 193:17–23
66. Li HQ, Sun GQ, Li N, Sun SG, Su DS, Xin Q (2007) Design and preparation of highly active Pt-Pd/C catalyst for the oxygen reduction reaction. J Phys Chem C 111:5605–5617
67. Blair S, Lycke D, Iordache CA (2006) Palladium-platinum alloy anode catalysts for direct formic acid fuel cells. ECS Trans 3:1325–1332
68. Zhou WJ, Zhou ZH, Song SQ, Li WZ, Sun GQ, Tsiakaras P, Xin Q (2003) Pt based anode catalysts for direct ethanol fuel cells. Appl Catal B Environ 46:273–285
69. Li X, Hsing I (2006) Electrooxidation of formic acid on carbon supported Pt_xPd_{1-x} (x = 0-1) nanocatalysts. Electrochim Acta 51:3477–3483
70. Lee YW, Ko AR, Kim DY, Han SB, Park KW (2012) Octahedral Pt-Pd alloy catalysts with enhanced oxygen reduction activity and stability in proton exchange membrane fuel cells. RSC Adv 2:1119–1125
71. Selvaraj V, Grace AN, Alagar M (2009) Electrocatalytic oxidation of formic acid and formaldehyde on nanoparticle decorated single walled carbon nanotubes. J Colloid Interface Sci 333:254–262
72. Lee YW, Ko AR, Han SB, Kim HS, Park KW (2011) Synthesis of octahedral Pt-Pd alloy nanoparticles for improved catalytic activity and stability in methanol electrooxidation. Phys Chem Chem Phys 13:5569–5572
73. Nishanth KG, Sridhar P, Pitchumani S, Shukla AK (2011) A DMFC with methanol-tolerant-carbon-supported-Pt-Pd-alloy cathode. J Electrochem Soc 158:B871–B876
74. Antolini E, Zignani SC, Santos SF, Gonzalez ER (2011) Palladium-based electrodes: a way to reduce platinum content in polymer electrolyte membrane fuel cells. Electrochim Acta 56:2299–2305
75. Lopes T, Antolini E, Gonzalez ER (2008) Carbon supported Pt-Pd alloy as an ethanol tolerant oxygen reduction electrocatalyst for direct ethanol fuel cells. Int J Hydrogen Energy 33:5563–5570
76. Guo SJ, Dong SJ, Wang EK (2009) Polyaniline/Pt hybrid nanofibers: high-efficiency nanoelectrocatalysts for electrochemical devices. Small 5:1869–1876
77. Chen HS, Liang YT, Chen TY, Tseng YC, Liu CW, Chung SR, Hsieh CT, Lee CE, Wang KW (2014) Graphene-supported Pt and PtPd nanorods with enhanced electrocatalytic performance for the oxygen reduction reaction. Chem Commun 50:11165–11168

78. Solla-Gullon J, Rodes A, Montiel V, Aldaz A, Clavilier J (2003) Electrochemical characterisation of platinum-palladium nanoparticles prepared in a water-in-oil microemulsion. J Electroanal Chem 554:273–284
79. Chen CH, Hwang BJ, Wang GR, Sarma LS, Tang MT, Liu DG, Lee JF (2005) Nucleation and growth mechanism of Pd/Pt bimetallic clusters in sodium bis(2-ethylhexyl)sulfosuceinate (AOT) reverse micelles as studied by in situ X-ray absorption spectroscopy. J Phys Chem B 109:21566–21575
80. Escudero MJ, Hontanon E, Schwartz S, Boutonnet M, Daza L (2002) Development and performance characterisation of new electrocatalysts for PEMFC. J Power Sources 106:206–214
81. Zhang H, Xu XQ, Gu P, Li CY, Wu P, Cai CX (2011) Microwave-assisted synthesis of graphene-supported Pd_1Pt_3 nanostructures and their electrocatalytic activity for methanol oxidation. Electrochim Acta 56:7064–7070
82. Zhang H, Yin YJ, Hu YJ, Li CY, Wu P, Wei SH, Cai CX (2010) Pd@Pt core-shell nanostructures with controllable composition synthesized by a microwave method and their enhanced electrocatalytic activity toward oxygen reduction and methanol oxidation. J Phys Chem C 114:11861–11867
83. Xia H, Wang D (2008) Fabrication of macroscopic freestanding films of metallic nanoparticle monolayers by interfacial self-assembly. Adv Mater 20:4253–4256
84. Liang HW, Cao XA, Zhou F, Cui CH, Zhang WJ, Yu SH (2011) A free-standing Pt-nanowire membrane as a highly stable electrocatalyst for the oxygen reduction reaction. Adv Mater 23:1467–1471
85. Jin YD, Dong SJ (2002) Diffusion-limited, aggregation-based, mesoscopic assembly of roughened core shell bimetallic nanoparticles into fractal networks at the air-water interface. Angew Chem Int Ed 41:1040–1044
86. Shi HY, Hu B, Yu XC, Zhao RL, Ren XF, Liu SL, Liu JW, Feng M, Xu AW, Yu SH (2010) Ordering of disordered nanowires: spontaneous formation of highly aligned, ultralong Ag nanowire films at oil-water-air interface. Adv Funct Mater 20:958–964
87. Kang YJ, Ye XC, Chen J, Cai Y, Diaz RE, Adzic RR, Stach EA, Murray CB (2013) Design of Pt-Pd binary superlattices exploiting shape effects and synergistic effects for oxygen reduction reactions. J Am Chem Soc 135:42–45
88. Wu HX, Li HJ, Zhai YJ, Xu XL, Jin YD (2012) Facile synthesis of free-standing Pd-based nanomembranes with enhanced catalytic performance for methanol/ethanol oxidation. Adv Mater 24:1594–1597
89. Liu Y, Chi MF, Mazumder V, More KL, Soled S, Henao JD, Sun SH (2011) Composition-controlled synthesis of bimetallic PdPt nanoparticles and their electro-oxidation of methanol. Chem Mater 23:4199–4203
90. Lim B, Wang JG, Camargo PHC, Cobley CM, Kim MJ, Xia YN (2009) Twin-induced growth of palladium-platinum alloy nanocrystals. Angew Chem Int Ed 48:6304–6308
91. Papageorgopoulos DC, Keijzer M, Veldhuis JBJ, de Bruijn FA (2002) CO tolerance of Pd-rich platinum palladium carbon-supported electrocatalysts-Proton exchange membrane fuel cell applications. J Electrochem Soc 149:A1400–A1404
92. Yi QF, Huang W, Liu XP, Xu GR, Zhou ZH, Chen AC (2008) Electroactivity of titanium-supported nanoporous Pd-Pt catalysts towards formic acid oxidation. J Electroanal Chem 619:197–205
93. Huang D, Yuan Q, Wang HH, Zhou ZY (2014) Facile synthesis of PdPt nanoalloys with sub-2.0 nm islands as robust electrocatalysts for methanol oxidation. Chem Commun 50:13551–13554
94. Yin AX, Min XQ, Zhang YW, Yan CH (2011) Shape-selective synthesis and facet-dependent enhanced electrocatalytic activity and durability of monodisperse sub-10 nm Pt-Pd tetrahedrons and cubes. J Am Chem Soc 133:3816–3819

95. Lu YZ, Jiang YY, Wu HB, Chen W (2013) Nano-PtPd cubes on graphene exhibit enhanced activity and durability in methanol electrooxidation after CO stripping-cleaning. J Phys Chem C 117:2926–2938
96. Lu YZ, Jiang YY, Chen W (2014) Graphene nanosheet-tailored PtPd concave nanocubes with enhanced electrocatalytic activity and durability for methanol oxidation. Nanoscale 6:3309–3315
97. Guo SJ, Dong SJ, Wang EK (2008) A general method for the rapid synthesis of hollow metallic or bimetallic nanoelectrocatalysts with urchinlike morphology. Chem Eur J 14:4689–4695
98. Yang JH, Lee JY, Zhang QB, Zhou WJ, Liu ZL (2008) Carbon-supported pseudo-core-shell Pd-Pt nanoparticles for ORR with and without methanol. J Electrochem Soc 155:B776–B781
99. Liu B, Li HY, Die L, Zhang XH, Fan Z, Chen JH (2009) Carbon nanotubes supported PtPd hollow nanospheres for formic acid electrooxidation. J Power Sources 186:62–66
100. Zhou XM, Fan LZ (2010) Pt/Pd alloy nanoparticles composed of bimetallic nanobowls Synthesis, characterization and electrocatalytic activities. Electrochim Acta 55:8111–8115
101. Chen ZW, Waje M, Li WZ, Yan YS (2007) Supportless Pt and PtPd nanotubes as electrocatalysts for oxygen-reduction reactions. Angew Chem Int Ed 46:4060–4063
102. Zhang H, Jin MS, Wang JG, Li WY, Camargo PHC, Kim MJ, Yang DR, Xie ZX, Xia YA (2011) Synthesis of Pd-Pt bimetallic nanocrystals with a concave structure through a bromide-induced galvanic replacement reaction. J Am Chem Soc 133:6078–6089
103. Zhang H, Jin MS, Liu HY, Wang JG, Kim MJ, Yang DR, Xie ZX, Liu JY, Xia YN (2011) Facile synthesis of Pd-Pt alloy nanocages and their enhanced performance for preferential oxidation of CO in excess hydrogen. ACS Nano 5:8212–8222
104. Lu YZ, Jiang YY, Chen W (2013) PtPd porous nanorods with enhanced electrocatalytic activity and durability for oxygen reduction reaction. Nano Energy 2:836–844
105. Huang XQ, Zhang HH, Guo CY, Zhou ZY, Zheng NF (2009) Simplifying the creation of hollow metallic nanostructures: One-pot synthesis of hollow palladium/platinum single-crystalline nanocubes. Angew Chem Int Ed 48:4808–4812
106. Bauer E, Vandermerwe JH (1986) Structure and growth of crystalline superlattices-from monolayer to superlattice. Phys Rev B 33:3657–3671
107. Fan FR, Liu DY, Wu YF, Duan S, Xie ZX, Jiang ZY, Tian ZQ (2008) Epitaxial growth of heterogeneous metal nanocrystals: From gold nano-octahedra to palladium and silver nanocubes. J Am Chem Soc 130:6949–6951
108. Beard KD, Van Zee JW, Monnier JR (2009) Preparation of carbon-supported Pt-Pd electrocatalysts with improved physical properties using electroless deposition methods. Appl Catal B Environ 88:185–193
109. Long NV, Asaka T, Matsubara T, Nogami M (2011) Shape-controlled synthesis of Pt-Pd core shell nanoparticles exhibiting polyhedral morphologies by modified polyol method. Acta Mater 59:2901–2907
110. Long NV, Ohtaki M, Hien TD, Randy J, Nogami M (2011) A comparative study of Pt and Pt-Pd core-shell nanocatalysts. Electrochim Acta 56:9133–9143
111. Zhang L, Zhang JW, Jiang ZY, Xie SF, Jin MS, Han XG, Kuang Q, Xie ZX, Zheng LS (2011) Facile syntheses and electrocatalytic properties of porous Pd and its alloy nanospheres. J Mater Chem 21:9620–9625
112. Jung DH, Bae SJ, Kim SJ, Nahm KS, Kim P (2011) Effect of the Pt precursor on the morphology and catalytic performance of Pt-impregnated on Pd/C for the oxygen reduction reaction in polymer electrolyte fuel cells. Int J Hydrogen Energy 36:9115–9122
113. Nguyen VL, Ohtaki M, Matsubara T, Cao MT, Nogami M (2012) New experimental evidences of Pt-Pd bimetallic nanoparticles with core-shell configuration and highly fine-ordered nanostructures by high-resolution electron transmission microscopy. J Phys Chem C 116:12265–12274
114. Liu L, Samjeske G, Nagamatsu S, Sekizawa O, Nagasawa K, Takao S, Imaizumi Y, Yamamoto T, Uruga T, Iwasawa Y (2012) Enhanced oxygen reduction reaction activity

and characterization of Pt-Pd/C bimetallic fuel cell catalysts with Pt-enriched surfaces in acid media. J Phys Chem C 116:23453–23464
115. Li YJ, Wang ZW, Chiu CY, Ruan LY, Yang WB, Yang Y, Palmer RE, Huang Y (2012) Synthesis of bimetallic Pt-Pd core-shell nanocrystals and their high electrocatalytic activity modulated by Pd shell thickness. Nanoscale 4:845–851
116. Habas SE, Lee H, Radmilovic V, Somorjai GA, Yang P (2007) Shaping binary metal nanocrystals through epitaxial seeded growth. Nat Mater 6:692–697
117. Guo SJ, Dong SJ, Wang EK (2010) Pt/Pd bimetallic nanotubes with petal-like surfaces for enhanced catalytic activity and stability towards ethanol electrooxidation. Energy Environ Sci 3:1307–1310
118. Lim BW, Lu XM, Jiang MJ, Camargo PHC, Cho EC, Lee EP, Xia YN (2008) Facile synthesis of highly faceted multioctahedral Pt nanocrystals through controlled overgrowth. Nano Lett 8:4043–4047
119. Jiang MJ, Lim B, Tao J, Camargo PHC, Ma C, Zhu YM, Xia YN (2010) Epitaxial overgrowth of platinum on palladium nanocrystals. Nanoscale 2:2406–2411
120. Bai S, Wang C, Deng M, Gong M, Bai Y, Jiang J, Xiong YJ (2014) Surface polarization matters: enhancing the hydrogen-evolution reaction by shrinking Pt Shells in Pt–Pd–Graphene stack structures. Angew Chem Int Ed 126:12316–12320
121. Zhang H, Jin MS, Wang JG, Kim MJ, Yang DR, Xia YN (2011) Nanocrystals composed of alternating shells of Pd and Pt can be obtained by sequentially adding different precursors. J Am Chem Soc 133:10422–10425
122. Lim B, Jiang MJ, Camargo PHC, Cho EC, Tao J, Lu XM, Zhu YM, Xia YN (2009) Pd-Pt bimetallic nanodendrites with high activity for oxygen reduction. Science 324:1302–1305
123. Lee HJ, Habas SE, Somorjai GA, Yang PD (2008) Localized Pd overgrowth on cubic Pt nanocrystals for enhanced electrocatalytic oxidation of formic acid. J Am Chem Soc 130:5406–5407
124. Lim B, Jiang MJ, Yu T, Camargo PHC, Xia YN (2010) Nucleation and growth mechanisms for Pd-Pt bimetallic nanodendrites and their electrocatalytic properties. Nano Res 3:69–80
125. Guo SJ, Dong SJ, Wang EK (2010) Ultralong Pt-on-Pd bimetallic nanowires with nanoporous surface: nanodendritic structure for enhanced electrocatalytic activity. Chem Commun 46:1869–1871
126. Peng ZM, Yang H (2009) Synthesis and oxygen reduction electrocatalytic property of Pt-on-Pd bimetallic heteronanostructures. J Am Chem Soc 131:7542–7543
127. Guo SJ, Dong SJ, Wang EK (2010) Three-dimensional Pt-on-Pd bimetallic nanodendrites supported on graphene nanosheet: facile synthesis and used as an advanced nanoelectrocatalyst for methanol oxidation. ACS Nano 4:547–555
128. Wang L, Nemoto Y, Yamauchi Y (2011) Direct synthesis of spatially-controlled Pt-on-Pd bimetallic nanodendrites with superior electrocatalytic activity. J Am Chem Soc 133:9674–9677
129. Serpell CJ, Cookson J, Ozkaya D, Beer PD (2011) Core@shell bimetallic nanoparticle synthesis via anion coordination. Nat Chem 3:478–483
130. Lee H, Habas SE, Kweskin S, Butcher D, Somorjai GA, Yang PD (2006) Morphological control of catalytically active platinum nanocrystals. Angew Chem Int Ed 45:7824–7828
131. Climent V, Markovic NM, Ross PN (2000) Kinetics of oxygen reduction on an epitaxial film of palladium on Pt(111). J Phys Chem B 104:3116–3120
132. Arenz M, Schmidt TJ, Wandelt K, Ross PN, Markovic NM (2003) The oxygen reduction reaction on thin palladium films supported on a Pt(111) electrode. J Phys Chem B 107:9813–9819
133. Babu PK, Kim HS, Chung JH, Oldfield E, Wieckowski A (2004) Bonding and motional aspects of CO adsorbed on the surface of Pt nanoparticles decorated with Pd. J Phys Chem B 108:20228–20232

134. Zhao MC, Rice C, Masel RI, Waszczuk P, Wieckowski A (2004) Kinetic study of electro-oxidation of formic acid on spontaneously-deposited Pt/Pd nanoparticles-CO tolerant fuel cell chemistry. J Electrochem Soc 151:A131–A136
135. Jayashree RS, Spendelow JS, Yeom J, Rastogi C, Shannon MA, Kenis PJA (2005) Characterization and application of electrodeposited Pt, Pt/Pd, and Pd catalyst structures for direct formic acid micro fuel cells. Electrochim Acta 50:4674–4682
136. Grace AN, Pandian K (2006) Pt, Pt-Pd and Pt-Pd/Ru nanoparticles entrapped polyaniline electrodes—a potent electrocatalyst towards the oxidation of glycerol. Electrochem Commun 8:1340–1348
137. Selvaraj V, Alagar M, Hamerton I (2006) Electrocatalytic properties of monometallic and bimetallic nanoparticles-incorporated polypyrrole films for electro-oxidation of methanol. J Power Sources 160:940–948
138. Selvaraj V, Alagar M, Hamerton I (2007) Nanocatalysts impregnated polythiophene electrodes for the electrooxidation of formic acid. Appl Catal B Environ 73:172–179
139. Xu Y, Lin X (2007) Facile fabrication and electrocatalytic activity of $Pt_{0.9}$ $Pd_{0.1}$alloy film catalysts. J Power Sources 170:13–19
140. Galal A, Atta NF, Darwish SA, Ali SM (2008) Electrodeposited metals at conducting polymer electrodes. II: Study of the oxidation of methanol at poly(3-methylthiophene) modified with Pt-PdCo-catalyst. Top Catal 47:73–83
141. Zhou ZL, Kang TF, Zhang Y, Cheng SY (2009) Electrochemical sensor for formaldehyde based on Pt-Pd nanoparticles and a Nafion-modified glassy carbon electrode. Microchim Acta 164:133–138
142. Mahapatra SS, Dutta A, Datta J (2010) Temperature effect on the electrode kinetics of ethanol oxidation on Pd modified Pt electrodes and the estimation of intermediates formed in alkali medium. Electrochim Acta 55:9097–9104
143. Mahapatra SS, Dutta A, Datta J (2011) Temperature dependence on methanol oxidation and product formation on Pt and Pd modified Pt electrodes in alkaline medium. Int J Hydrogen Energy 36:14873–14883
144. Yang X, Yang QD, Xu J, Lee CS (2012) Bimetallic PtPd nanoparticles on Nafion-graphene film as catalyst for ethanol electro-oxidation. J Mater Chem 22:8057–8062
145. Vidal-Iglesias FJ, Solla-Gullon J, Herrero E, Aldaz A, Feliu JM (2010) Pd adatom decorated (100) preferentially oriented Pt nanoparticles for formic acid electrooxidation. Angew Chem Int Ed 49:6998–7001
146. Zhang JL, Vukmirovic MB, Xu Y, Mavrikakis M, Adzic RR (2005) Controlling the catalytic activity of platinum-monolayer electrocatalysts for oxygen reduction with different substrates. Angew Chem Int Ed 44:2132–2135
147. Zhang J, Mo Y, Vukmirovic M, Klie R, Sasaki K, Adzic R (2004) Platinum monolayer electrocatalysts for O_2 reduction: Pt monolayer on Pd (111) and on carbon-supported Pd nanoparticles. J Phys Chem B 108:10955–10964
148. Adzic RR, Zhang J, Shao M, Sasaki K, Vukmirovic M, Uribe FA (2006) Platinum and mixed platinum-metal monolayer fuel cell electrocatalysts: design, activity and long-term performance stability. ECS Trans 3:31–36
149. Vukmirovic MB, Zhang J, Sasaki K, Nilekar AU, Uribe F, Mavrikakis M, Adzic RR (2007) Platinum monolayer electrocatalysts for oxygen reduction. Electrochim Acta 52:2257–2263
150. Bliznakov ST, Vukmirovic MB, Yang L, Sutter EA, Adzic RR (2012) Pt Monolayer on electrodeposited Pd nanostructures: advanced cathode catalysts for PEM fuel cells. J Electrochem Soc 159:F501–F506
151. Sasaki K, Naohara H, Cai Y, Choi YM, Liu P, Vukmirovic MB, Wang JX, Adzic RR (2010) Core-protected platinum monolayer shell high-stability electrocatalysts for fuel-cell cathodes. Angew Chem Int Ed 49:8602–8607
152. Koenigsmann C, Santulli AC, Gong KP, Vukmirovic MB, Zhou WP, Sutter E, Wong SS, Adzic RR (2011) Enhanced electrocatalytic performance of processed, ultrathin, supported

Pd-Pt core-shell nanowire catalysts for the oxygen reduction reaction. J Am Chem Soc 133:9783–9795
153. Wang JX, Inada H, Wu LJ, Zhu YM, Choi YM, Liu P, Zhou WP, Adzic RR (2009) Oxygen reduction on well-defined core-shell nanocatalysts: particle size, facet, and Pt shell thickness effects. J Am Chem Soc 131:17298–17302
154. Shao MH, He GN, Peles A, Odell JH, Zeng J, Su D, Tao J, Yu T, Zhu YM, Xia YN (2013) Manipulating the oxygen reduction activity of platinum shells with shape-controlled palladium nanocrystal cores. Chem Commun 49:9030–9032
155. Seweryn J, Lewera A (2012) Electrooxidation of ethanol on carbon-supported Pt-Pd nanoparticles. J Power Sources 205:264–271
156. Ignaszak A, Song C, Zhu W, Wang YJ, Zhang J, Bauer A, Baker R, Neburchilov V, Ye S, Campbell S (2012) Carbon–$Nb_{0.07}Ti_{0.93}O_2$ composite supported Pt–Pd electrocatalysts for PEM fuel cell oxygen reduction reaction. Electrochim Acta 75:220–228
157. Lopes PP, Ticianelli EA, Varela H (2011) Potential oscillations in a proton exchange membrane fuel cell with a Pd-Pt/C anode. J Power Sources 196:84–89
158. Kim KH, Yu JK, Lee HS, Choi JH, Noh SY, Yoon SK, Lee CS, Hwang TS, Rhee YW (2007) Preparation of Pt-Pd catalysts for direct formic acid fuel cell and their characteristics. Korean J Chem Eng 24:518–521
159. Parinyaswan A, Pongstabodee S, Luengnaruemitchai A (2006) Catalytic performances of Pt-Pd/CeO_2 catalysts for selective CO oxidation. Int J Hydrogen Energy 31:1942–1949
160. Koutsopoulos S, Johannessen T, Eriksen KM, Fehrmann R (2006) Titania-supported Pt and Pt-Pd nanoparticle catalysts for the oxidation of sulfur dioxide. J Catal 238:206–213
161. Mougenot M, Caillard A, Brault P, Baranton S, Coutanceau C (2011) High performance plasma sputtered PdPt fuel cell electrodes with ultra low loading. Int J Hydrogen Energy 36:8429–8434
162. Schmidt TJ, Markovic NM, Stamenkovic V, Ross PN, Attard GA, Watson DJ (2002) Surface characterization and electrochemical behavior of well-defined Pt-Pd{111} single-crystal surfaces: a comparative study using Pt{111} and palladium-modified Pt{111} electrodes. Langmuir 18:6969–6975
163. Ohashi M, Beard KD, Ma SG, Blom DA, St-Pierre J, Van Zee JW, Monnier JR (2010) Electrochemical and structural characterization of carbon-supported Pt-Pd bimetallic electrocatalysts prepared by electroless deposition. Electrochim Acta 55:7376–7384
164. Wang C, Peng B, Xie HN, Zhang HX, Shi FF, Cai WB (2009) Facile fabrication of Pt, Pd and Pt-Pd alloy films on Si with tunable infrared internal reflection absorption and synergetic electrocatalysis. J Phys Chem C 113:13841–13846
165. Cangul B, Zhang LC, Aindow M, Erkey C (2009) Preparation of carbon black supported Pd, Pt and Pd-Pt nanoparticles using supercritical CO_2 deposition. J Supercrit Fluids 50:82–90
166. Yen CH, Shimizu K, Lin YY, Bailey F, Cheng IF, Wai CM (2007) Chemical fluid deposition of Pt-based bimetallic nanoparticles on multiwalled carbon nanotubes for direct methanol fuel cell application. Energy Fuel 21:2268–2271
167. Arico AS, Srinivasan S, Antonucci V (2001) DMFCs: from fundamental aspects to technology development. Fuel Cells 1:133–161
168. Choi WC, Jeon MK, Kim YJ, Woo SI, Hong WH (2004) Development of enhanced materials for direct-methanol fuel cell by combinatorial method and nanoscience. Catal Today 93–95:517–522
169. Koenigsmann C, Wong SS (2011) One-dimensional noble metal electrocatalysts: a promising structural paradigm for direct methanol fuel cells. Energy Environ Sci 4:1161–1176
170. Shukla AK, Ravikumar MK, Gandhi KS (1998) Direct methanol fuel cells for vehicular applications. J Solid State Electrochem 2:117–122
171. Leger JM (2001) Mechanistic aspects of methanol oxidation on platinum-based electrocatalysts. J Appl Electrochem 31:767–771
172. Zhang XY, Lu W, Da JY, Wang HT, Zhao DY, Webley PA (2009) Porous platinum nanowire arrays for direct ethanol fuel cell applications. Chem Commun 2:195–197

173. Arenz M, Stamenkovic V, Ross PN, Markovic NM (2004) Surface (electro-)chemistry on Pt (111) modified by a pseudomorphic Pd monolayer. Surf Sci 573:57–66
174. Qian W, Hao R, Zhou J, Eastman M, Manhat BA, Sun Q, Goforth AM, Jiao J (2013) Exfoliated graphene-supported Pt and Pt-based alloys as electrocatalysts for direct methanol fuel cells. Carbon 52:595–604
175. Feng JX, Zhang QL, Wang AJ, Wei J, Chen JR, Feng JJ (2014) Caffeine-assisted facile synthesis of platinum@palladium core-shell nanoparticles supported on reduced graphene oxide with enhanced electrocatalytic activity for methanol oxidation. Electrochim Acta 142:343–350
176. Bianchini C, Shen PK (2009) Palladium-based electrocatalysts for alcohol oxidation in half cells and in direct alcohol fuel cells. Chem Rev 109:4183–4206
177. Lamy C, Rousseau S, Belgsir EM, Coutanceau C, Leger JM (2004) Recent progress in the direct ethanol fuel cell: development of new platinum-tin electrocatalysts. Electrochim Acta 49:3901–3908
178. Simoes FC, dos Anjos DM, Vigier F, Leger JM, Hahn F, Coutanceau C, Gonzalez ER, Tremiliosi G, de Andrade AR, Olivi P, Kokoh KB (2007) Electroactivity of tin modified platinum electrodes for ethanol electrooxidation. J Power Sources 167:1–10
179. Yu XW, Pickup PG (2008) Recent advances in direct formic acid fuel cells (DFAFC). J Power Sources 182:124–132
180. Rees NV, Compton RG (2011) Sustainable energy: a review of formic acid electrochemical fuel cells. J Solid State Electrochem 15:2095–2100
181. Rees NV, Compton RG (2012) Sustainable energy: a review of formic acid electrochemical fuel cells. J Solid State Electrochem 16:419–419
182. Grigoriev SA, Lyutikova EK, Martemianov S, Fateev VN (2007) On the possibility of replacement of Pt by Pd in a hydrogen electrode of PEM fuel cells. Int J Hydrogen Energy 32:4438–4442
183. Baranova EA, Miles N, Mercier PH, Le Page Y, Patarachao B (2010) Formic acid electro-oxidation on carbon supported $Pd_xPt_{1-x}(0 \geq x \geq 1)$ nanoparticles synthesized via modified polyol method. Electrochim Acta 55:8182–8188
184. Guo SJ, Zhang S, Sun SH (2013) Tuning nanoparticle catalysis for the oxygen reduction reaction. Angew Chem Int Ed 52:8526–8544
185. Jiang YY, Lu YZ, Lv XY, Han DX, Zhang QX, Niu L, Chen W (2013) Enhanced catalytic performance of Pt-free iron phthalocyanine by graphene support for efficient oxygen reduction reaction. ACS Catal 3:1263–1271
186. Bliznakov ST, Vukmirovic MB, Yang L, Sutter EA, Adzic RR (2011) Pt monolayer on electrodeposited Pd nanostructures-advanced cathode catalysts for PEM fuel cells. ECS Trans 41:1055–1066
187. Sasaki K, Wang JX, Naohara H, Marinkovic N, More K, Inada H, Adzic RR (2010) Recent advances in platinum monolayer electrocatalysts for oxygen reduction reaction: Scale-up synthesis, structure and activity of Pt shells on Pd cores. Electrochim Acta 55:2645–2652

Chapter 5
Integrated Studies of Au@Pt and Ru@Pt Core-Shell Nanoparticles by *In Situ* Electrochemical NMR, ATR-SEIRAS, and SERS

Dejun Chen, Dianne O. Atienza, and YuYe J. Tong

5.1 Introduction

Nanoscale noble metal materials, mostly in the form of nanoparticles (NPs), have been subjected to intensive research in the context of their applications to fuel cell electrocatalysis [1] traditionally and also to lithium-air batteries [2, 3] recently. Among others, despite decades-long continuous efforts in discovering viable replacements for the expensive and earthly-scant platinum (Pt), Pt-based electrocatalysts [4, 5] are still the materials of choice in terms of having high activity for CO oxidation reaction (COR) [6, 7], methanol oxidation reaction (MOR) [8], ethanol oxidation reaction (EtOR) [9, 10], formic acid oxidation reaction (FAOR) [11, 12], hydrogen evolution/oxidation reaction (HER/HOR) [13, 14], oxygen reduction/evolution reaction (ORR/OER) [15–17], and CO_2 reduction reaction (CO_2R) [18, 19], just to name a few widely studied reactions. Benefited from the fertile knowledge gained from the decades-long intensive investigations of model electrocatalysts, such as single crystal surfaces, fundamental research on discovering, understanding and utilizing novel electronic, geometric, and/or bifunctional properties of Pt-based NPs [20–22] for the aforementioned reactions continues to flourish and push the boundary of research in an environment as close to that of practical applications as possible. Constant advance of the sophistication of many *in situ* spectroelectrochemical techniques has been a critical part of the latter. In this Chapter, we will discuss integrated studies of Au@Pt and Ru@Pt NPs by *in situ* electrochemical (EC) nuclear magnetic resonance (NMR), attenuated-total-reflection surface-enhanced IR reflection adsorption spectroscopy (ATR-SEIRAS), and surface-enhanced Raman scattering spectroscopy (SERS)

D. Chen • D.O. Atienza • Y.Y.J. Tong (✉)
Department of Chemistry, Georgetown University, 37th and O Streets, NW, Washington, DC 20057, USA
e-mail: yyt@georgetown.edu

K.I. Ozoemena, S. Chen (eds.), *Nanomaterials for Fuel Cell Catalysis*, Nanostructure Science and Technology, DOI 10.1007/978-3-319-29930-3_5

spectroscopy, with a focus on how an organic integration of these techniques can enhance the investigative power brought to bear by their complementary nature. That is, an organically integrated whole is better than the sum of its parts.

The strength of NMR is in its unique ability to probe the electronic and structural properties of Pt-based NPs [23]. Particularly, ^{195}Pt NMR can distinguish, at least semi-quantitatively, the Pt atoms on the surface of from those inside the NPs and the Pt atoms in metallic from non-metallic (e.g. oxidized) state [24, 25]. Moreover, using adsorbed ^{13}CO as a molecular probe, ^{13}C NMR can offer complementary electronic and diffusional information of the surface [26–28]. However, NMR continues to suffer from its intrinsic low mass detection sensitivity so it is generally not a method of choice for detecting reaction intermediates during the reaction. On the other hand, SEIRAS and SERS are more suited for the latter [29, 30].

Since its inception [31], EC-SEIRAS has been applied extensively to study electrocatalysis mainly on conductive metal thin-films (~50 to 100 nm thick) used as working electrodes [32]. Single metallic films, such as Au [32–34], Pt [35–43], Pd [44, 45], Cu [46], Ag [47], and bimetallic PtRu [43, 48], PtPd [44] alloy film as well as additional ultrathin-film (bi)metallic layers deposited on Au substrate film [49, 50], have been the subjects of studies of COR [34, 41], MOR [40, 42, 43, 48], EtOR [36], FAOR [39, 44, 45, 50], HER/HOR [35, 38], ORR [51, 52] and CO_2R [37]. Numerous surface species that are important to the reactions under investigation, such as adsorbed CO (CO_{ads}) in linearly/bridge/multiple bonding configurations [37, 43, 45, 50], formate [39, 40, 45, 53], adsorbed anion [33, 45] and interfacial water [32, 54, 55], have been identified and studied [30]. On the other hand, while SERS may not be as versatile and sensitive as SEIRAS, EC-SERS does offer some complementary advantages, such as observing species that only have symmetric vibrational modes, for instance O_2^- [56], accessing to lower wavenumber range (100~1000 cm^{-1}) in which the vibrations of metal-adsorbate, such as Pt-CO [57] or Pt-O [58], locate and not worrying about strong absorption by water. With the enhancement offered by roughened bulk surfaces [29, 59] or deposited NPs [60–62], EC-SERS has become increasingly used to study catalytic reactions, such as COR [57, 60, 63, 64], MOR [57, 65], FAOR [62], ORR [56] and H_2O_2 electro-reduction reaction [66, 67], on the respective electrocatalyst surface.

Despite their complementary nature, EC-NMR, -SEIRAS or -SERS has been largely used alone for various reasons. But an integrated approach of combining them together to investigate the same system is expected to provide a fuller mechanistic picture of the system under investigation. With this in mind, and as already briefly mentioned above, we will focus here on two samples, Ru@Pt and Au@Pt NPs, which have been studied in various degree of detail by *in situ* EC-NMR, -SEIRAS, and -SERS. This chapter will be structured as follows: we will first present briefly the physical aspects of the samples (Sect. 5.2), then move to NMR (Sect. 5.3), SEIRAS (Sect. 5.4), and SERS (Sect. 5.5) sequentially, which will be followed by an integrated discussion (Sect. 5.6) of the results presented in the previous sections. We will conclude the chapter with some summative comments.

5.2 Ru@Pt and Au@Pt NPs Samples

Ru@Pt NPs. The synthesis was adopted from the wet homogeneous reaction-like procedure developed by Du et al. [68]. Briefly, the ethylene glycol (EG) was used as both reaction medium and reducing agent of Johnson-Matthey (J-M) Ru Black (~3 nm) and $H_2PtCl_6{\cdot}6H_2O$. The first step is the complete reduction of Ru NPs at 140 °C for 2 h with vigorous stirring. This step is critical for the success of controlling the surface Pt deposition where it has been determined experimentally that Ru readily oxidizes to RuO_x in the atmosphere [69]. The cleaned Ru NPs were then collected and rinsed with ethanol and re-dispersed ultrasonically into a fresh EG solution. The calculated volume of $H_2PtCl_6{\cdot}6H_2O$ solution according to the experimentally determined linear relationship [68] gave the desired ≈40 % Pt packing density (PD). This mixture was ultrasonicated and heated at 120 °C for 4 h to reduce Pt(IV) to Pt(0) on the surface of cleaned Ru NPs. This synthetic method in a lower temperature homogeneous-reaction like media is advantageous in preventing sintering effects. A representative TEM image is shown in Fig. 5.1a from which the average particle size was determined to be 3.2 nm.

Au@Pt NPs. Au colloid NPs (3.5 nm) were first prepared from $HAuCl_4{\cdot}3H_2O$ in aqueous solution using sodium citrate and sodium borohydride as reductant and then supported spontaneously on Vulcan XC-72R that gave a metal loading of 30 wt %. The Pt shell was deposited by reducing Pt^{4+} onto the surface of dispersed carbon-supported Au NPs by using ascorbic acid with target Pt coverage of 100 % (atomic ratio of Pt/Au = 1). The detail of the synthesis can be found in the original paper [70]. A representative TEM image is shown in Fig. 5.1b that gives an average particle size of 4.6 nm.

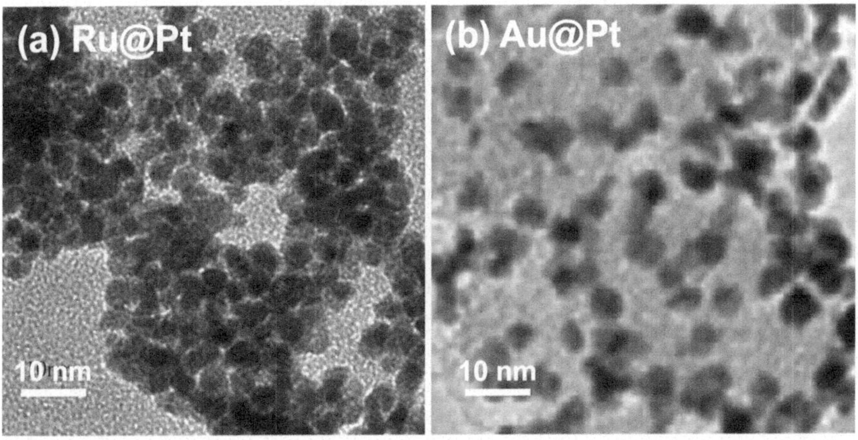

Fig. 5.1 TEM images of Ru@Pt (**a**) and Au@Pt (**b**). The average particle size is 3.2 nm for the former and 4.6 nm for the latter. (**b**) is adapted with permission from [70] (© 2007 Elsevier)

5.3 *In Situ* EC-NMR Study of Ru@Pt and Au@Pt NPs [71]

For EC-NMR measurements, the NMR samples (~50–80 mg) were treated at room temperature by an EC cleaning process in a three-electrode setup with the NPs as working electrode by holding the potential at 50 mV (vs. Ag/AgCl). For CO adsorption from dissociation of ^{13}C-labeled methanol (^{13}C-MeOH) onto the pre-clean NPs surface was done with the same cell but in 0.1 M $HClO_4 + 0.5$ M MeOH solution by holding the potential at 250 mV. The EC setup was under a continuous stream of ultrapure Argon gas during the entire period of the above treatments. After the EC cleaning or CO adsorption, the NMR sample cell filled with the supporting electrolyte (0.1 M $HClO_4$ or 0.1 M $HClO_4 + 0.5$ M ^{13}C-MeOH) was sealed immediately with a one-to-one fitted ground-glass stopper, which was then inserted into the NMR probe and lowered down to the cryostat that was pre-cooled at 80 K. After the NMR measurements, the cell was immediately reconnected back to the EC cleaning setup. The open-circuit potential was re-checked to make sure that its stability was within a 20 mV difference before and after the NMR measurements. The ^{13}C NMR spectra of the adsorbed ^{13}CO and the corresponding inversion-recovery spin-lattice T_1 measurements were obtained at various temperature ranges of 80–120 K at a "home-assembled" spectrometer equipped with an active-shielded 9.395 T widebore superconducting magnet, an Oxford Spectrostat-CF cryostat (Oxford Instrument, U.K.), an AMT (Lancaster, PA) 1 kW power amplifier, a Tecmag (Houston, TX) Apollo data acquisition system, and a home-built single channel solenoid probe with a coil of 5 mm inner diameter and 28 mm length. Both the ^{13}C and ^{195}Pt spectra were acquired using the conventional "$\pi/2$-τ-π-τ-echo" Hahn spin-echo sequence with the $\pi/2$ pulse length and τ of 3 μs and 25 μs respectively but for the ^{195}Pt spectra the frequency was manually changed between ca. 82–87 MHz with a 0.047 G/kHz increment between each resonance position at 80 K because of the extremely broadness of the spectra. The slow beats of the ^{195}Pt signal at a given frequency (see Fig. 5.2e, f) are obtained by varying the τ in the spin-echo sequence and the corresponding T_1 measurements were done with the conventional saturation-recovery method.

Figure 5.2a, b shows the point-by-point ^{195}Pt NMR spectra of the as-received (red) and EC-cleaned (blue) samples of Au@Pt (with Pt packing density (PD) = 3.8, A) and of Ru@Pt (with Pt PD = 0.4, D) core-shell NPs in comparison with that of the pure Pt/C NPs [73] (black dashed lines and open circles). The vertical dashed line indicates the surface peak position (1.1000 G/kHz,) of the EC-cleaned pure 2.5 nm Pt/C NPs [73]. Although the frozen electrolyte at 80 K prevented active potentiostatic control, the tightly vacuum-sealed sample cell ensured that the sample surface potential would not change during the NMR measurements. This was verified by measuring open circuit potentials before and after each NMR measurement that always indicated that a constant potential was indeed achieved.

The as-received Au@Pt NPs had a shelf life of more than 48 months [70], therefore, the surface was heavily oxidized, as indicated by the peak of Pt oxide at 1.089 G/kHz (Fig. 5.2a). Notice that distinguishing a Pt surface covered by

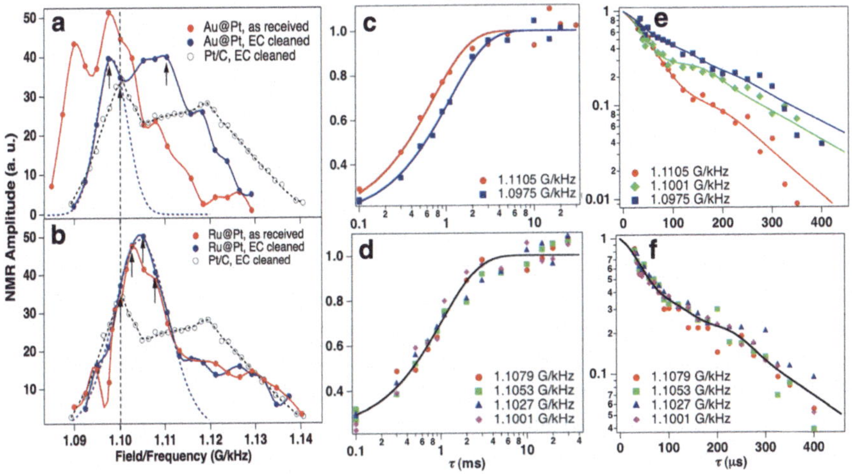

Fig. 5.2 Point-by-point ^{195}Pt NMR spectra (**a**, **b**), spin–lattice relaxation curves (**c**) and (**d**), spin-spin relaxation (*J*-coupling modulation or slow-beat) curves (**e**, **f**) for the Au@Pt (**a**, **c**, **e**) and Ru@Pt (**b**, **d**, **f**) NPs, respectively. The calculated Pt PD (packing density) was 3.8 for the Au@Pt and 0.4 for the Ru@Pt NPs. According to the straight line in Fig. 3 of [72], the expected Pt surface fraction would be 1.3, i.e., a fully Pt covered Au core. The % of the total deposited Pt on the surface would be ~1.3/3.8 = 0.34, in an excellent agreement with the value of 33 % determined by ^{195}Pt NMR in a. Adapted from [71] with permission (© 2012 American Chemical Society)

adsorbed oxygen from that by Pt oxide is straightforward. After the EC cleaning, the Pt oxide peak disappeared and a clean-surface peak appeared at 1.0975 G/kHz. The latter amounts to a 2278-ppm *positive* shift with respect to the clean-surface peak position (1.1000 G/kHz) of the pure Pt NPs, manifesting a strong electronic effect of Au on Pt. The significant portion of the ^{195}Pt atoms in the Au@Pt NPs resonating at higher fields (above 1.1000 G/kHz) indicates that many deposited Pt atoms went inside the Au NPs. Indeed, using a Gaussian de-convolution of the surface peak (the dashed blue curve in Fig. 5.2a), we found that only ≈33 % of all Pt deposited atoms were on the surface of the Au@Pt NPs, which is in agreement with expected atomic exchange between deposited Pt atoms and the underlying Au atoms [72]. Notice that such a ~33 % value is actually predicted by the straight line in Fig. 3 of [72] (see caption of Fig. 5.2).

The as-received Ru@Pt NPs were freshly synthesized with a shelf life of only a few hours [68]. Under the protection of ethylene glycol, little surface oxidation was expected in such a short time. Indeed, no significant surface oxidation was observed in the corresponding ^{195}Pt NMR spectrum in Fig. 5.2b (red). The clean-surface peak (blue line) now appeared at 1.1053 G/kHz, corresponding to a 4795-ppm *negative* shift with respect to the clean-surface peak position (1.1000 G/kHz) of the pure Pt NPs, revealing also a strong electronic effect of Ru on Pt as well, which is *opposite* to that of Au. A Gaussian de-convolution of the surface peak gave a surface atom fraction of ≈74 %, indicating a dominant Ru core Pt shell structure.

The nuclear spin-lattice (T_1) and spin-spin (T_2) relaxation measurements are presented in Fig. 5.2c, e for the EC-cleaned Au@Pt and in 2d and 2f for the EC-cleaned Ru@Pt, respectively, at the spectral positions listed in the respective figures. Detailed discussion of the relaxation results is out of scope here due to the limited space, but it suffices to demonstrate the most important difference here. Both the T_1 (Fig. 5.2d) and T_2 (Fig. 5.2f) at different spectral positions on the Ru@Pt fall into the *same* respective relaxation curves, which is a manifestation of the same chemical environment, i.e., the majority of the Pt atoms were on the surface. Particularly, a J-coupling modulation analysis [25] of the data in Fig. 5.2f by fitting to the J-coupling induced slow beats equation [74] $S(\tau)/S_0 = \exp(-2\tau/T_2)(P_0 + \exp(-(\tau/T_{2J})^2)[P_1\cos(J\tau) + P_2\cos^2(J\tau)])$ with J for J-coupling constant, $1/T_{2J}$ for spread in J, and $P_0 + P_1 + P_2 = 1$ (solid line) gives a Pt atomic fraction of 0.37. Since for a full pseudo-morphic Pt monolayer on a Ru(0 0 0 1) surface, the Pt atomic fraction among all the possible next nearest neighbors is 6/9 = 0.67, a local Pt atomic fraction of 0.37 is thus consistent with a Pt PD of 0.4 of mono-atomic Pt islands. In contrast, both the T_1 (Fig. 5.2c) and T_2 (Fig. 5.2e) at different spectral positions on the Au@Pt were different, which indicates a wider distribution of Pt atoms on surface and inside the NPs as well.

From the T_1 value, the known spectral position and temperature at which the T_1 is measured, one can use the two-band model to calculate the s- and d-like E_f–LDOSs, denoted here as $D_s(E_f)$ and $D_d(E_f)$ respectively [75]. These values for the surface Pt atoms in Au@Pt and Ru@Pt samples are collected in Table 5.1. *As can be clearly seen, Au and Ru cores exert different (opposing) electronic influence on Pt shell: the former lowers more the d-like while the latter lowers more the s-like E_f-LDOSs.*

Also collected (in the parentheses) in Table 5.1 are the average surface values calculated by using the OpenMX DFT package [76] on Pt13@Pt42 (Pt55), Ru13@Pt42, and Au13@Pt42 cubo-octahedral clusters. Considering the very simplified cluster models, it is remarkable that the experiments and theoretical calculations have achieved a good agreement in numerical values and in trend for both $D_s(E_f)$ and $D_d(E_f)$.

How changes in the surface Pt s- and d-like E_f-LDOS induced by the Au and Ru cores influence metal-adsorbate bonding was probed and corroborated by ^{13}C NMR of ^{13}CO adsorbed onto the Au@Pt and Ru@Pt NPs via dissociative adsorption of $^{13}CH_3OH$ (MeOH). The ability to probe the surface electronic properties by

Table 5.1 E_f-LDOS values deduced from ^{195}Pt and ^{13}C NMR data (1 Ry = 13.6 eV and mol = molecule)

Samples	$D_s(E_f)/Ry^{-1}$ atom^{-1}	$D_d(E_f)/Ry^{-1}$ atom^{-1}	$D_{5\sigma}(E_f)/Ry^{-1}$ mol^{-1}	$D_{2\pi*}(E_f)/Ry^{-1}$ mol^{-1}
Ru@Pt	3.7 (2.9)	13.3 (12.9)	0.7	4.9
Au@Pt	4.3 (3.0)	10.1 (12.7)	0.6	5.8
Pt	5.1 (3.4)	13.5 (15.2)	0.7	6.7

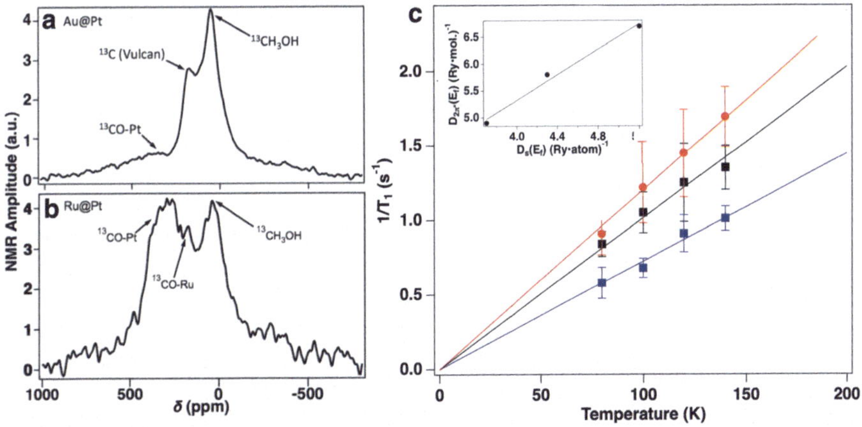

Fig. 5.3 ^{13}C NMR spectra (9.39 T, 80 K) of chemisorbed CO on (**a**) Au@Pt and (**b**) Ru@Pt NPs. Electro-deposition of CO was performed by holding the potential at 0.35 V vs. Ag/AgCl (3 M) while immersed in 0.5 M D_2ClO_4 with $^{13}CH_3OH$ [13]. CH_3OH at 49 ppm was also used as an internal shift reference. (**c**) Temperature dependent T_1 relaxation measurements obtained at 333 ppm for CO-Pt/Ru@Pt and CO-Pt/Au@Pt and at 219 ppm for CO-Ru/Ru@Pt. Adapted from [71] with permission (© 2012 American Chemical Society). Inset in (**c**): Linear correlation between the $D_{2\pi*}(E_f)$ and $D_s(E_f)$

using surface-adsorbed molecules, such as adsorbed ^{13}CO, is among the unique investigative strengths of *in situ* EC-NMR. Figure 5.3 shows the ^{13}C NMR spectra of ^{13}CO on the Au@Pt (a) and Ru@Pt (b) NPs, with the assigned peaks indicated by the arrows. For the Au@Pt sample, despite the dominant peaks from the remaining MeOH in the electrolyte and the carbon support, the broad and weak—yet still discernable—^{13}C NMR peak at ≈333 ppm was assigned to the adsorbed CO, an assignment corroborated by the temperature dependent T_1 measurements (*vide infra*). For the Ru@Pt sample, the peaks at ≈300 ppm and ≈219 ppm can be assigned to CO on Pt and on Ru sites, respectively, based on the literature values [27, 77]. *The difference in the signal strength of the adsorbed CO on the Au@Pt and Ru@Pt NPs may reflect the difference in dissociative adsorption of MeOH on the respective surfaces.*

The results of temperature dependent T_1 measurements are presented in Fig. 5.3c. For the CO on Pt, the T_1's were measured at 333 ppm for both Au@Pt ($T_1T = 83$ s K) and Ru@Pt ($T_1T = 99$ s K) samples to facilitate comparison, although the peak position for the latter was at ≈ 300 ppm. The pass-through-origin straight lines are the hallmark of the Korringa relaxation behavior [78], which indicates that the adsorbed CO molecules on three different sites all acquired metallic characteristics through surface bonding. For CO on Ru sites, the shift (219 ppm) and Korringa constant $T_1T = 138$ s K (where T is the absolute temperature at which T_1 is measured) are very close to those observed on pure Ru [77] and on Ru deposited on Pt NPs [27], which lends strong support to our peak assignment that is also consistent with the expected exposure of Ru core for a Pt PD of 40 %. On the other hand, the Korringa constants measured at the same spectral position (333 ppm) were different for CO on the surface Pt atoms of the Au@Pt and on those of the Ru@Pt NPs. The former showed a faster relaxation (i.e., smaller Korringa constant T_1T), and thus a stronger metal-adsorbate bonding. In fact for CO on Pt, we can also deduce the values of the respective 5σ- and 2π*-like E_f–LDOS, i.e., $D_5\sigma(E_f)$ and $D_2\pi_*(E_f)$, at ^{13}C by using a different two-band model developed for adsorbed CO on Pt [26].

As can be seen from the data in Table 5.1, the $D_{5\sigma}(E_f)$ was essentially constant, while the $D_{2\pi^*}(E_f)$ decreased proportionally (inset in Fig. 5.3c) with the $D_s(E_f)$ for CO on Pt from pure Pt to Au@Pt to Ru@Pt. This suggests that the electronic alterations on metal-CO bonding, i.e., changes in metal to CO 2π* back-donation [79], and therefore in bonding strength, were mainly caused by the variations in the $D_s(E_f)$ of the surface Pt atoms. On the other hand, it has been observed that the ability to dissociatively adsorb MeOH that leads to adsorbed CO follows the order Pt ≈ Ru@Pt > Au@Pt, which shares the same order in the respective $D_d(E_f)$ as shown in Table 5.1. These results suggest that the *d*-like electrons are probably responsible for agostic interaction that activates the three methyl protons in MeOH during its dissociative adsorption.

5.4 *In Situ* EC ATR-SEIRAS of Ru@Pt and Au@Pt NPs [80]

Figure 5.4 shows the schematic of the EC IR cell that was used for *in situ* EC ATR-SEIRAS measurements of the Ru@Pt and Au@Pt samples. A gold film was first electrolessly deposited onto the reflecting plane of a Si attenuation total reflection (ATR) prism of a triangular shape, which was polished with successively

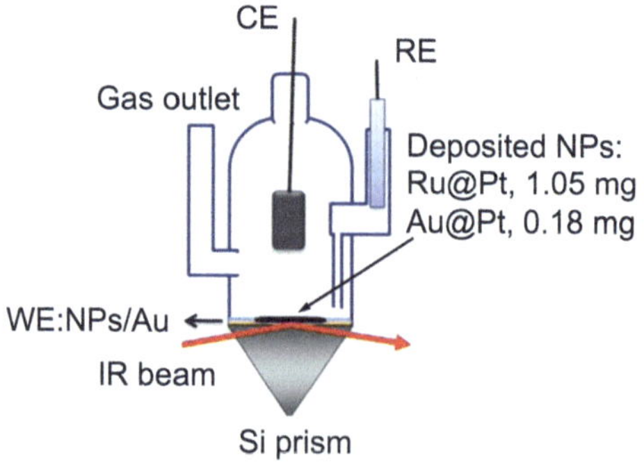

Fig. 5.4 Schematic of the *in situ* EC ATR-SEIRAS cell used for ATR-SEIRAS measurements. WE, RE and CE stand for working, reference and counter electrode respectively. A ~50-nm Au film was first deposited onto the pre-polished Si prism onto which the Ru@Pt or Au@Pt NPs were then drop-casted and dried with nitrogen flow

finer grade alumina slurries down to 0.3 μm and cleaned by sonication in the Milli-Q water. The detailed deposition procedure can be found elsewhere [81].

For the *in situ* EC-SEIRAS measurements, Ru@Pt (1.05 mg) and Au@Pt/C (0.18 mg) were dispersed onto the as-prepared gold film. The catalytic adlayer of the core-shell NPs was then dried in a gentle nitrogen flow at room temperature. No Nafion® solution was used in order to avoid any optical interference. The working electrode surface (Au film plus NPs) was cleaned by repetitive potential cycling until stable and reproducible cyclic voltammograms (CV) were achieved in 0.1 M $HClO_4$ at 50 mV s^{-1} between −0.3 V and 0.7 V for the Ru@Pt, and −0.25 V and 1.0 V for the Au@Pt/C NPs, respectively. Notice that the Ru@Pt had an expected Pt PD or coverage of 40 % while that for the Au@Pt was 100 %.

Figure 5.5a, b present the SEIRAS spectra of two sequential COR potential scans after gaseous CO adsorption on the Ru@Pt (a-1 and a-2) and the Au@Pt (b-1 and b-2) NPs. The reference spectra were taken at 0.7 V for the former and 1.0 V for the latter at which all adsorbed CO had been fully oxidized according to the corresponding CO stripping voltammograms. While the same IR band of linearly-bound CO-Pt was observed on the Au@Pt NPs for the two sequential COR runs (Fig. 5.5b, d), there were changes for the CO spectra on the Ru@Pt NPs. In (a-1), the spectra show a series of a *single*, potential-dependent CO IR band that varied between 1998 and 2047 cm^{-1}, which can be assigned to a linearly-bonded CO_L [43, 82]. The peak position of this single band at −0.15 V (spectrum (a) in the inset of Fig. 5.5c) is 2005 cm^{-1}, a value in agreement with those of CO adsorbed on the Ru sites of Ru-decorated Pt(1 1 1) surfaces (from 2001 to 2008 cm^{-1}) [82–84] but between those of CO_L on pure Ru (~1995 cm^{-1}) [43, 82] and on PtRu alloy (~2044 cm^{-1}) [43, 85] under the similar conditions. For CO absorbed on the Ru-modified Pt(1 1 1) surfaces, two distinct CO stretching bands corresponding to Pt-CO_L and Ru-CO_L respectively were observed; but on the Ru@Pt of the first COR in this study and on PtRu alloys [43, 85, 86] in general, only one band was observed.

On the other hand, the SEIRAS spectra obtained between −0.3 to 0.3 V of the CO during the second COR on the Ru@Pt can be reasonably de-convoluted into three peaks whose values at −0.15 V are 1948 cm^{-1}, 2000 cm^{-1} and 2031 cm^{-1} respectively (spectrum b in the inset of Fig. 5.5c). Based on the available literature values and assignments of the C-O stretching frequencies of adsorbed CO on Ru [43], Ru-decorated Pt [82, 83, 87], and PtRu alloy [43, 88, 89] surfaces, we assigned the three peaks to CO_L bonded to Ru-coordinated-to-Ru (or Ru sites distal to the Pt islands–Ru-like sites, 1948 cm^{-1}), Ru-coordinated-to-Pt (or Ru sites proximal to the Pt islands–PtRu sites, 2000 cm^{-1}), and Pt-islands-on-Ru-core (or Pt adlayer sites that are most close to Pt sites in PtRu alloys–Pt-like sites, 2031 cm^{-1}) sites on the surface of the Ru@Pt NPs respectively. The middle-frequency sites are most likely around the peripheries of the Pt islands formed on the surface of Ru core [90].

Because Ru is highly oxophilic, the exposure of the Ru@Pt NPs to air after the synthesis can easily lead to an enrichment of Ru species to the surface [91]. Moreover, neither repetitive CVs between −0.3 V and 0.7 nor holding electrode potential at −0.3 V could readily modify such a Ru enrichment on the surface of the Ru@Pt NPs. Consequently, only one C-O stretching band (Fig. 5a-1) similar to that of

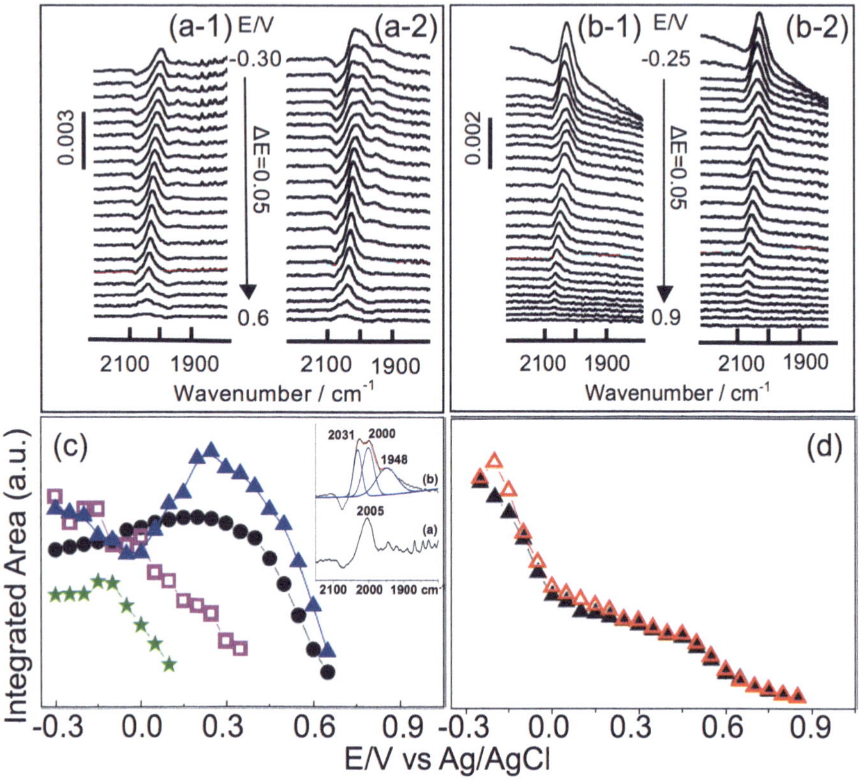

Fig. 5.5 The potential dependent EC-SEIRAS spectra of the pre-adsorbed CO during the first (**a-1**/**b-1**) and the second (**a-2**/**b-2**) COR on the Ru@Pt/Au@Pt/C NPs respectively in 0.1 M $HClO_4$. The spectra taken at 0.7 V for the Ru@Pt and 1.0 V for the Au@Pt/C were used as the corresponding reference spectra. The integrated areas of the spectra in (**a**) and (**b**) are plotted in (**c**) and (**d**) respectively. The inset in (**c**) compares fine spectral features of (**a-1**) and (**a-2**) taken at −0.15 V. The spectra in (**a-2**) can be de-convoluted into three bands: 2031-cm^{-1} (*green stars*), 2000-cm^{-1} (*blue triangles*) and 1948-cm^{-1} (*pink squares*) bands, which should be compared with those in (**a-1**) where only one band (2005 cm^{-1}, *black circles*) was observed. In contrast, only one same IR band that varied between 2025 and 2070 cm^{-1} was observed in (**b-1**, *black triangles* in **d**) and (**b-2**, *red triangles* in **d**). Adapted from [80] with permission (© 2011 American Chemical Society)

CO_L-Ru of Ru-modified Pt(1 1 1) surface [82] was observed, suggesting a dominant CO-Ru species on the surface.

On the other hand, the first COR acted like a CO annealing process and was able to reduce (at least partially) the surface Ru oxide (RuO_x or RuO_xH_y) to metallic Ru and to bring the segregated Pt sites back to the surface of the Ru@Pt NPs so that CO could adsorb onto different surface sites that led to the appearance of the three different C-O stretching peaks as observed in the inset of Fig. 5.5c. The potential dependent integrated IR band intensity of each de-convoluted band is presented in Fig. 5.5c. What is intriguing is that the integrated intensity of each band behaves differently. The 2031-cm^{-1} band (the Pt-islands-on-Ru-core sites, green stars) shows the most negative onset potential of the COR, i.e., ~−0.1 V, at which the CO IR intensity starts decreasing. The 2000-cm^{-1} band (the Ru sites proximal to the Pt islands, blue solid triangles) shows little intensity changes from −0.3 to 0.0 V, a noticeable intensity increase from 0.0 to 0.25 V, and then a rapid intensity decease beyond 0.25 V. The latter can be considered the COR onset potential of these sites. Lastly, the CO IR intensity of the 1948-cm^{-1} band (the Ru sites distal to the Pt islands) decreases slowly until ~0.15 V beyond which a shaper decrease starts. Thus, 0.15 V can also be considered as its COR onset potential.

The above observations led us to conclude that, in terms of the COR, the Pt-islands-on-Ru-core sites were the most active; the proximal Ru sites were the least active, and the distal Ru sites were in between. Although these differences in the reaction rate of the COR may reflect differences in reaction barrier and/or CO diffusion rate at these sites, their relatively lower onset potentials for the COR should be contrasted to that of the first COR whose value is about 0.35 V (Fig. 5.5c, black circles). This indicates that the more segregated surface structures of the Ru@Pt NPs generated by the CO annealing, i.e., Pt islands on metallic Ru core, is much more active than the (largely ruthenated) surface structure before the CO annealing of which surface Ru oxides dominated, an observed was further confirmed by our recent study of the chemistry related to the activation of commercial PtRu alloy electrocatalysts [92].

Figure 5.5b-1 and b-2 depict the potential dependent EC-SEIRAS spectra for the two sequential CORs on the Au@Pt/C NPs. As briefly mentioned above and in contrast to the cases of the Ru@Pt NPs presented in Fig. 5.5a-1 and a-2, no obvious differences were observed in terms of the SEIRAS of the pre-adsorbed CO between the two sequential CORs, as can been seen in Fig. 5.5b, d: only one CO IR stretching band varying from 2025 to 2070 cm^{-1} was observed over the entire potential range, which can be assigned to a linearly bonded CO_L-Pt. This infers rather constant surface structure and properties of the Au@Pt/C, an indication of stability.

What is puzzling but also intriguing is the rather precipitous decrease in the IR intensity below 0.0 V where adsorbed CO is supposed to be stable. Coincidently, the potential dependence of the vibrational frequency in this region also behaved oddly (not shown here but can be consulted in the original paper [80]). Since the linearly increasing current during CO stripping positive-going potential scan was largely negative in this potential regime [80], we speculated that an ongoing protonation-like process converted adsorbed CO into a different species whose exact identity is still unknown. However, the species-conversion hypothesis is consistent with the observation of the non-monotonic frequency variation observed in the same potential regime mentioned above [93] and with the observation of co-adsorbed hydrogen on Pt electrodes by visible-infrared SFG [94] and by SEIRAS [95–98]. Moreover, if this large decrease in IR intensity were caused by a much earlier COR, it would be hard to reconcile with that no enhanced MOR was observed on the Au@Pt NPs as will be discussed below.

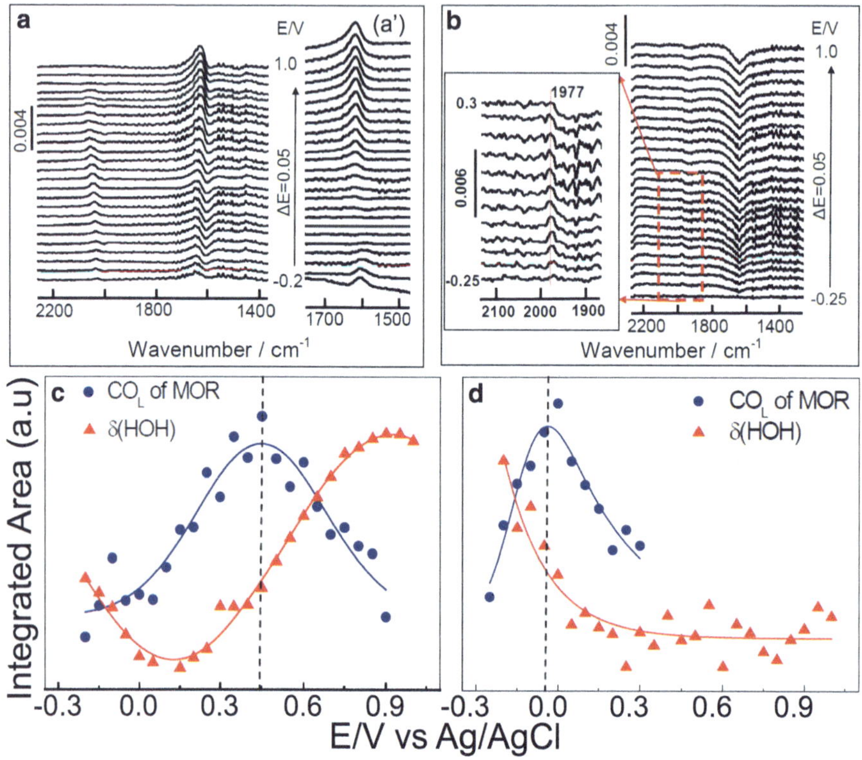

Fig. 5.6 The *in situ* SEIRAS spectra acquired in 0.5 M MeOH + 0.1 M $HClO_4$ during positive potential scan of MOR on the Au@Pt (**a** and **a**′) and Ru@Pt (**b**) NPs, with the spectra taken at −0.25 and −0.3 V as the respective references. The spectra in (**a**′) are the water bending bands for the Au@Pt/C NPs obtained using the spectrum recorded at 0.1 V as the reference while the inset in (**b**) amplifies the very weak CO spectra. (**c**) and (**d**) show the corresponding potential dependent integrated IR intensity of the adsorbed CO_L (*blue*) and the surface water bending δ(HOH) (*red*). The *solid lines* are for eye-guiding purpose. Adapted from [80] with permission (© 2011 American Chemical Society)

The MOR on the Ru@Pt and Au@Pt/C NPs after the two sequential CORs discussed above was followed by the *in situ* stair-step potential dependent EC-SEIRAS in 0.5 M MeOH + 0.1 M $HClO_4$. On the Au@Pt/C NP surface a weak Pt-CO_L band around 2040 cm^{-1} was observed as shown in Fig. 5.6a whose corresponding integrated intensity is shown in Fig. 5.6c (solid blue circles). Notice that while the COR of the pre-adsorbed gaseous CO (Fig. 5.5b) started with a fully CO-covered surface, the MOR started with a CO free surface. Therefore, the increase in IR intensity of the CO_L during the MOR before 0.45 V is a manifestation of the accumulation of the surface-bound poisonous CO generated from the dissociative adsorption of MeOH. Notice that the CO intensity started decreasing at 0.45 V, which is consistent with the main-current peak potential of the CO stripping [80].

During the potential dependent MOR, a somewhat bipolar water bending vibration δ(HOH) band was observed at ~1620 cm^{-1} with the negative going signal appearing at the lower frequency side. If the spectra were re-referenced by the spectrum obtained at 0.1 V, the original bipolar spectral shape largely disappeared (Fig. 5.6a′). This is because 0.1 V is around the potential of zero charge, PZC [99], at which IR bending vibration of the absorbed water is expected to be weakest [32]. The now dominantly mono-polar spectra in Fig. 5.6a′ reveals a more clear trend: The peak intensity decreased first gradually and the bending frequency moved from 1608 cm^{-1} at −0.2 V to 1600 cm^{-1} just below 0.1 V; it then started increasing continuously with a constant bending frequency at 1622 cm^{-1} above 0.1 V (red triangles in Fig. 5.6c). Such a behavior of the δ(HOH) is very similar to that observed on Pt film [95], which was interpreted as due to changes in the hydrogen-bonding-associated average orientation of water molecules with oxygen-up or down on the negative or positive charged Pt surface separated by the PZC.

Below the PZC, the Au@Pt/C NP surface was negatively charged so it attracted the hydrogen end of the water molecules. As the electrode potential became less negative, the attraction to hydrogen became weaker, so less hydrogen-down water molecules would bond to the surface and led to weaker IR intensity of the water-bending band. As the electrode potential moved further positively beyond the PZC, the Au@Pt/C NP surface became more and more positively charged, leading to attracting more and more water molecules via the oxygen-end of the water molecules. Consequently, the IR intensity of the water-bending band became stronger. Although the electro-oxidation of MeOH and adsorbed CO must have consumed surface water, the continuous increase in water adsorption above 0.1 V indicates that the MOR on the Au@Pt/C NPs was overall slow. The oxidation of the surface above 0.75 V led to the observed level-off and then decrease in the IR intensity.

The band of the adsorbed CO generated during the MOR on the Au@Pt/C NPs was much weaker than the pre-adsorbed gaseous CO for the COR. This is consistent with the EC results that showed the degree of the suppression of hydrogen adsorption was much lower for the former than that for the latter. Yet, an even weaker linearly bound CO band at about 1977 cm^{-1} (at −0.05 V) was observed for the MOR on the Ru@Pt NPs, whose integrated intensity, although about three times smaller than that of the CO generated during the MOR on Au@Pt NPs, followed the same increase-then-decrease pattern with the peak at −0.05 V (Fig. 5.6d). It is highly likely that this CO band corresponds to the poisonous CO generated during the MOR on the Pt-islands-on-Ru-core sites, for the Ru sites are essentially inactive for MOR. Moreover, this onset is almost identical with that of CO_L on Pt islands of Ru@Pt for second COR (green stars in Fig. 5.5c). The very low CO coverage was probably responsible for the significant red shift in stretching frequency from the full-coverage value of ~2030 cm^{-1} to the low-coverage value of 1977 cm^{-1}.

Notice that the MOR generated CO-band peak potential of −0.05 V for the Ru@Pt NPs was 0.5 V more negative than that (0.45 V) for the Au@Pt/C NPs. This observation implies a much less CO poisoning and faster MOR on the Ru@Pt than on the Au@Pt/C NPs. Indeed, the amplitude of the corresponding surface water bending vibration δ(HOH) band at 1641 cm^{-1} (Fig. 5.6b) did not show an increase after 0.1 V but instead reached a constant level (Fig. 5.6d), signaling that the MOR on the Ru@Pt NPs was much faster than on the Au@Pt/C NPs such that it led to a higher rate of surface water consumption, therefore a steady state of surface water adsorption.

As activated surface water is a necessary surface reactant for both COR and MOR [100, 101], its *in situ* IR spectroscopic investigation has proven informative and revealing [7, 41, 102]. The IR vibrational frequency, band shape, and intensity of the surface water at the EC interface are very sensitive to several parameters that include the type of metal surface, the strength and direction of interfacial electric field, and the co-adsorbed species. In addition to the bending mode δ(HOH) discussed above, three distinguishable ν(O-H) stretching bands of interfacial water have been identified and reported: [32, 95, 103, 104] the non-hydrogen bonded water monomer at ~3660 cm^{-1}, the strongly hydrogen-bonded ice-like water at ~3040 cm^{-1}, and the disordered weakly hydrogen-bonded water at ~3400 cm^{-1}.

For the MOR, two water bands were observed (Fig. 5.7a): the weakly hydrogen-bonded at 3380 cm^{-1} and strongly hydrogen-bonded at 3075 cm^{-1}. Their potential dependent, integrated IR intensity obtained by peak de-convolution are presented in Fig. 5.7b. The initial intensity decrease in both bands can be again rationalized by the reduced attraction to the hydrogen of the water molecule as the surface became

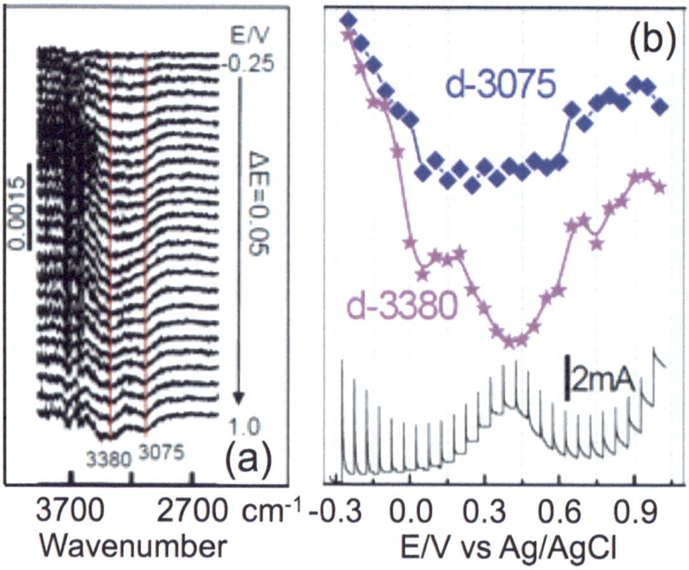

Fig. 5.7 (**a**) The potential dependent EC-SEIRAS spectra of the ν(O-H) bands observed on the Ru@Pt NPs during MOR and (b) the corresponding integrated IR intensity of the 3380-cm^{-1} (*pink stars*) and 3075-cm^{-1} (*blue diamonds*) bands as indicated by the *vertical red lines* in (**a**). Also presented in (**b**) is the current transient recorded during the stair-step IR measurements (*black curve with spikes*). Adapted from [80] with permission (© 2011 American Chemical Society)

less negatively charged. The second large decrease in the integrated intensity for the weakly hydrogen-bonded water at 3380 cm^{-1} that peaked at ~0.45 V as compared to the flat amplitude of the strongly hydrogen-bonded water band at 3075 cm^{-1} over the same potential region strongly suggests that the former type of water was at least more actively involved in the MOR and therefore was consumed more. This activity correlation is further supported by the simultaneously recorded current transients over the potential-step changes (black curve in Fig. 5.7b). The clear overlap of the MOR current peak with the IR intensity dip of the weakly hydrogen-bonded water is the strongest experimental evidence showing that the weakly hydrogen-bonded water was the active water species for the MOR on the Ru@Pt NPs.

The *in situ* EC-SEIRAS spectra of the ν(O–H) band of the interfacial water on the Au@Pt/C NPs during the COR in 0.1 M $HClO_4$ and MOR in 0.1 M $HClO_4$ + 0.5 M MeOH are presented in Fig. 5.8a, b. Very similar behaviors were observed for both cases, consisting of a broad band at 3245 cm^{-1} with a shoulder at ~3400 cm^{-1} over the entire potential range studied. Additionally, a small yet visible and also narrower peak at ~3580 cm^{-1} was observed for potentials more positive than ~0.3 V for the COR and than ~−0.1 V for the MOR, respectively. The potential dependent integrated band intensities obtained via de-convolution were plotted in Fig. 5.8c, d (the blue circles for the COR and red triangles for the MOR) for the 3435-cm^{-1} (weakly hydrogen-bonded) and 3245-cm^{-1} (strongly hydrogen-bonded) water bands, respectively.

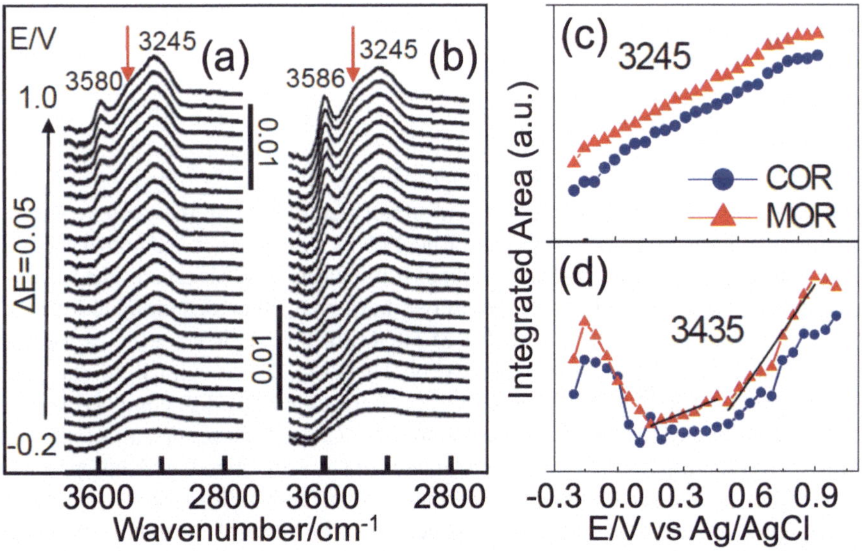

Fig. 5.8 Potential dependent EC-SEIRAS spectra of ν(O-H) stretch vibration for the Au@Pt/C NPs during the COR (**a**) and MOR (**b**). The integrated IR intensity of the 3245-cm^{-1} (**c**) and 3580-cm^{-1} (**d**) bands as a function of potential for the COR (*blue circle*) and MOR (*red triangle*) obtained by spectral de-convolution. Adapted from [80] with permission (© 2011 American Chemical Society)

Among the three different types of water observed, only the weakly hydrogen-bonded water in both the COR and MOR (Fig. 5.8d) show a variation pattern in the potential-dependent, integrated IR band intensity that has three different potential regimes: -0.25 to 0.1 V, 0.1–0.45 V, and 0.45–1.0 V, which closely match those of IR band of the adsorbed CO (Fig. 5.5d). Interestingly, two distinct increasing rates of the IR intensity were observed in the latter two potential regimes: 0.28, 0.12 a.u./V below and 1.01, 0.79 a.u./V above ~0.45 V (see the black straight lines in Fig. 5.8d) for MOR and COR, respectively, of which the electrode potential of 0.45 V coincides with the peak potential of the COR and the poisonous CO generated during the MOR (Fig. 5.6c). Thus, the much bigger increasing rate of the IR intensity above 0.45 V is most likely due to the oxidation of adsorbed CO_L in COR and MOR that rendered more surface sites available for water adsorption. Such a pattern match strongly suggests that the weakly hydrogen-bonded water species most likely occupied the surface sites freed by the CO oxidation and might also be the source of oxygen-containing species for the COR and MOR as suggested by the observations made on the Ru@Pt NPs (Fig. 5.7b), although more direct evidence supporting this assignment is still needed. Moreover, since no ν(OH) band corresponding to adsorbed Pt-OH_{ads} (~3700 cm^{-1}) [103] was observed during both the COR and MOR and on both samples, the dissociative adsorption of water on Pt (Pt-$OH_2 \rightarrow$ Pt-$OH_{ads} + H^+ + e^-$) was probably the rate-determining step for both the COR and MOR [7].

5.5 *In Situ* SERS of Pt and Ru@Pt

Figure 5.9 shows the Schematic of our *in situ* EC-SERS cell. The cell body is made of Teflon with a quartz window through which the laser is shined onto the surface of working electrode, which can be a rough Au film electrode, a roughened Pt disk electrode, or NPs deposited on the substrate electrode. For the current study, commercial Pt black or the synthesized Ru@Pt NPs were drop-casted onto a normal Pt disk electrode from which Raman signal is too weak to be observed. The surface enhancement came from the NPs themselves but it was time-consuming to locate a "hot" spot. For the measurements, 0.1 M $HClO_4$ or 0.1 M $HClO_4$ + 0.5 M MeOH were used for COR and MOR respectively. The EC-SERS spectra were obtained using a confocal Raman microscope system (Renishaw RM1000) equipped with a deep depletion CCD peltier cooled down to -70 °C. The microscope attachment is based on an Olympus BH2-UMA system and uses a $\times 50$ objective. A holographic notch filter was used to filter the excitation line, and 1200 g mm^{-1} selective holographic grating was employed depending on the spectral resolution required. The excitation wavelength was 785 nm from a Renishaw diode laser with a maximum power of 100 mW. For each spectrum, exposure time was 120 s; each potential step was held for 253 s during the *in situ* EC SERS measurements.

Figure 5.10 compares the *in situ* EC SERS spectra of the adsorbed CO at -0.25 and 0.0 V acquired during COR and MOR on Pt black and Ru@Pt respectively. It shows

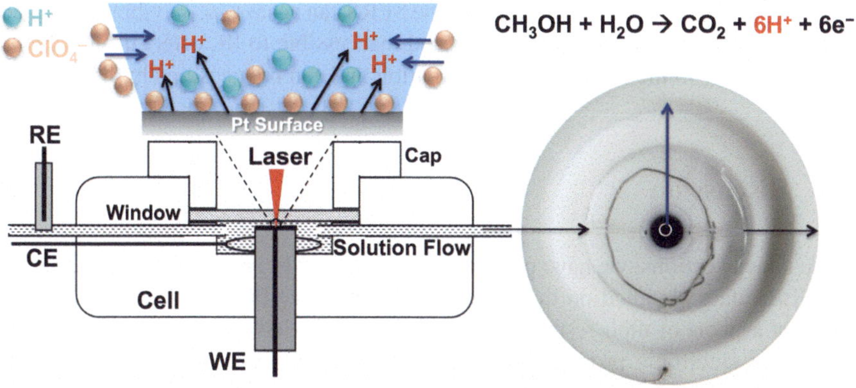

Fig. 5.9 The schematic of the *in situ* EC SERS cell with a photo of the working electrode encased in the cell body as seen from the top. Adapted from [93] with permission (© 2014 Royal Society of Chemistry)

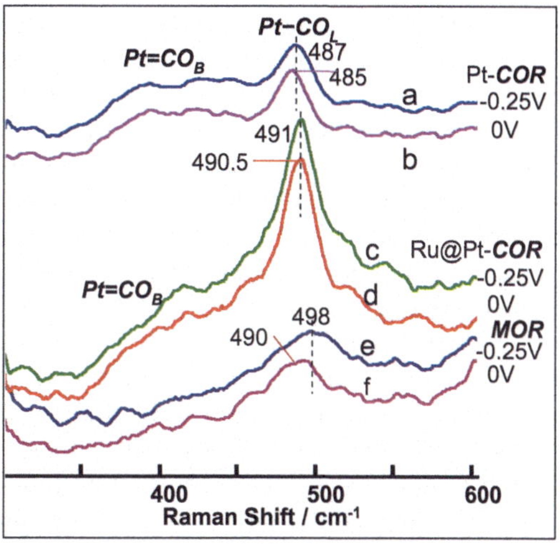

Fig. 5.10 *In-situ* EC SERS spectra of the bonding of Pt–CO_{ads} during the COR (**a–d**) and MOR (**e**, **f**) at −0.25 and 0.0 V on the Pt black and the Ru@Pt NPs

that the gaseous adsorption generated Pt–CO bands on Pt and Ru@Pt at 0.0 V appeared at 485 cm^{-1} and 490.5 cm^{-1}, respectively, with a difference of ~+5 cm^{-1}. On the other hand, the ν(C–O) on Pt and Ru@Pt at the same potential presents bands at 2052 cm^{-1} and 2006 cm^{-1}, respectively, with a difference of ~−46 cm^{-1}. In other words, in contrast to the large red shift (~−46 cm^{-1}) for C-O stretching on pure Pt vs on Ru@Pt, the corresponding difference in the Pt–CO vibrational frequency shows a small but still significant blue shift (~+5 cm^{-1}). Such opposing shifts in $\nu_{Pt\text{–}CO}$ vs $\nu_{C\text{–}O}$ indicates that the addition of Ru makes the Pt sites more electron-accepting or having higher Lewis acidity [105]. For MeOH generated CO on the Ru@Pt, the band intensity was much weaker, indicating improved CO tolerance. But it also showed the largest negative Stark tuning slope: −32 cm^{-1}/V (the other two were much smaller though still negative). That the SERS can access $\nu_{Pt\text{–}CO}$ vibration, which is directly related to metal–CO bonding and inaccessible to IR, offers insight that is complementary to the intra-molecular vibration that can be measured by IR.

5.6 Discussion

In the above sections, we have presented *in situ* EC ^{195}Pt and ^{13}CO NMR, SEIRAS and SERS data obtained on the same Ru@Pt and Au@Pt NPs, shedding light on different aspects of their respective electrocatalytic behaviors. For instance, NMR probes the electronic properties of the metal NPs, IR exams the intra-molecular vibrations and by which identifies reaction intermediates, and Raman accesses to the metal–adsorbate vibrations. Altogether, they can provide a fuller picture and deeper understanding of the system under investigation.

NMR's ability to provide quantitative information on electronic partition at the Fermi level (Table 5.1) can be of good use to connect with some recent theoretical development. For instance, recent quantum calculations on oxide-supported Pt model systems [106] have (re-) unearthed a potentially important and also more chemically intuitive surface bonding descriptor–the electronic partition, i.e., 6sp vs 5d electrons, at Pt that has been less (if not at all) investigated in electrocatalysis. The calculations predict that the Pt–H bond would be stronger at a Pt site with higher 6sp partition, while the Pt–O bond would be stronger at a Pt site with higher 5d partition, with Pt–CH_3 and Pt–CH_2 in between. The relatively recent application of the *C*rystal *O*rbital *H*amiltonian *P*opulation (COHP) formalism [107] within the extended Hückel *m*olecular *o*rbital (MO) theory to chemisorption of CO, hydrogen, methyl, and ethyl to metal surfaces by Hoffman and co-workers [108] also highlight the revealing power of such a chemistry-based *e*lectronic-*o*rbital-*s*pecific (EOS) formalism. For instance, the COHP analysis revealed that the metal *sp* orbitals actually contribute significantly (much more than previously believed) to the overall CO-Ni and CO-Pt chemisorption [109–111]. Also, based on the COHP analysis, strong agostic interactions between the C–H bonds of methyl and ethyl groups and the Pt sites of high symmetries were proposed for C–H bond activation [108]. This may find useful application in further delineating the mechanism of the formic acid oxidation reaction (FAOR) and MOR on metal (Pt) surfaces, where C–H bond-breaking is a necessary reaction step. This line of reasoning may also be useful in designing suitable catalysts for partial oxidation of methane.

Indeed, the aforementioned EOS description has long been embodied in the organometallic analogy of surface bonding advocated by Somorjai [112–114] and in the frontier-orbital formalism by Hoffmann [115, 116], which has been very successful in rationalizing many reaction mechanisms at solid/gas interfaces of heterogeneous catalysis. It is somewhat surprising that it has not found wide-spread use in electrocatalysis lately, which might have to do with the fact that the powerful valence-electron-orbital-probing ultraviolet photoelectron spectroscopy is not applicable to a solid/liquid interface, so a direct experimental connection with the chemically intuitive EOS description is thus lost. Nonetheless, we strongly believe that an EOS description as briefly mentioned above can complement the highly successful *d*-band center theory by offering more chemical specificity in terms of surface-bonding-involved electronic orbitals by which the chemistry taking place at an EC interface may be better nuanced. Moreover, FAOR, MOR, and oxygen reduction reaction (ORR) at an electrode surface, all involve elementary reaction steps that will necessarily have Pt–H, Pt–O, Pt-CH_3, and/or Pt–CH_2 bond formation and C–H, O–H, and/or O=O bond breakings, either sequentially or simultaneously. They can thus serve as a natural and fertile testing ground for the EOS description. Therefore, the EOS description can help parse more insightfully the chemistry of M@Pt core-shell NPs for fuel cell electrocatalysis as a function of core element M at an electronic and molecular level.

Figure 5.11 illustrates some EOS information obtained by the ^{195}Pt and ^{13}CO NMR: the pure Pt has the highest s-like electrons at the Fermi level, Ru@Pt lowest, with Au@Pt in between. Now if higher availability of s-like electrons leads to stronger Pt-H bonding, the ability to abstract hydrogen would be strongest on Pt, weakest on Ru@Pt, with Au@Pt in between. For MOR, stronger hydrogen abstracting ability would lead to more generation of CO [117], which would rationalize why only very low amount of CO was observed on the Ru@Pt (Fig. 5.6b). At the same time, the availability of *d*-like electrons is not reduced at all as compared to pure Pt so it would retain its ability to bind oxygen-containing species, which would facilitate the formation of methoxy therefore enhance the direct reaction pathway [117]. In other words, the higher MOR activity on Ru@Pt

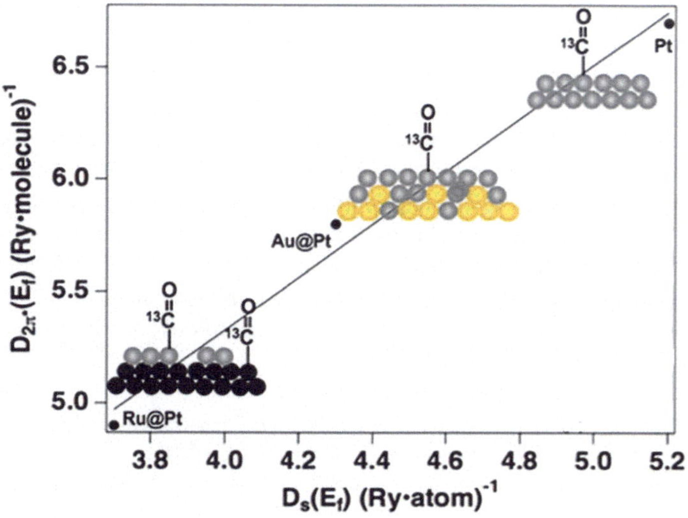

Fig. 5.11 Correlation between the $D_{2\pi^*}(E_f)$ and $D_s(E_f)$ deduced by using the two-band model from the ^{13}C NMR Data of CO_{ads} on the Pt, Ru@Pt and Au@Pt NPs as well as ^{195}Pt NMR of them. Adapted from [71] with permission (© 2012 American Chemical Society)

might arise from its ability to enhance the direct reaction pathway for MOR. By the same token, the substantially reduced d-like electrons at the Fermi level for the Au@Pt would account for its inferior ability to do MOR because of its worsened ability to bond oxygen-containing species.

As to the blue shift of the Pt–CO vibration from the pure Pt to Ru@Pt observed by *in situ* EC-SERS, one could rationalize it by the reduction in *s*-like electrons at the Fermi level as this correlates to a lower $D_{2\pi^*}(E_f)$ (Fig. 5.11), an indication of stronger Pt–CO bonding [105]. Consequently, ν_{Pt-CO} became higher on the Ru@Pt NPs.

5.7 Summary and Future Outlooks

We have shown in this chapter that integrating *in situ* EC NMR, SEIRAS and SERS studies of the Ru@Pt and Au@Pt NPs has enabled us to achieve a better mechanistic understanding of the two systems. *In situ* ^{195}Pt and ^{13}C NMR were able to provide some quantitative EOS information based on which some SEIRAS and SERS observations can be rationalized. The more relevant information one can garner, the better one can understand the catalyst's performance through which better performing electrocatalysts can be designed and developed. For instance, for Ru@Pt NPs, the SEIRAS data have suggested that segregated Pt and Ru ensembles seem to help enhance the MOR reactivity. Also, lowering the s-like but increasing the d-like electrons' availability may guide MOR through the direct reaction pathway without generating poisonous CO. Such specific information can be fed to the next round of designing and developing better MOR electrocatalysts.

In terms of developing further *in situ* techniques, there are several promising ones on the horizon that will become mature in the next decade. The first is the synchrotron light source and free-electron laser [118] based *in situ* X-ray spectroscopic methods [119]. As the synchrotron light source becomes more intense, the sensitivity of the technique will improve as well. It is also expected that the spatial resolution of X-ray based microscope and imaging will also improve substantially over the next decade. The second is the simple and straightforward high-resolution *in situ* EC NMR [120] that can explore all the analytical power of routine NMR. The third is nitrogen vacancy based scanning NMR microscope [121–123]. The key features of these new developments are chemical specificity, sensitivity, and spatial resolution.

Acknowledgments The authors gratefully acknowledge the financial supports provided by DOE (DE-FG02-07ER15895), NSF (CHE-1413429) and ARO (66191-CH).

References

1. Markovic NM, Ross PN (2000) New electrocatalysts for fuel cells from model surfaces to commercial catalysts. CATTECH 4:110–126
2. Steele BCH, Heinzel A (2001) Materials for fuel-cell technologies. Nature 414 (6861):345–352
3. Girishkumar G, McCloskey B, Luntz AC, Swanson S, Wilcke W (2010) Lithium–air battery: promise and challenges. J Phys Chem Lett 1(14):2193–2203. doi:10.1021/jz1005384
4. Shao Y, Park S, Xiao J, Zhang J-G, Wang Y, Liu J (2012) Electrocatalysts for nonaqueous lithium–air batteries: status, challenges, and perspective. ACS Catal 2(5):844–857. doi:10.1021/cs300036v
5. Zhang S, Yuan X-Z, Hin JNC, Wang H, Friedrich KA, Schulze M (2009) A review of platinum-based catalyst layer degradation in proton exchange membrane fuel cells. J Power Sources 194(2):588–600. http://dx.doi.org/10.1016/j.jpowsour.2009.06.073
6. Sealy C (2008) The problem with platinum. Mater Today 11(12):65–68
7. Arenz M, Mayrhofer KJJ, Stamenkovic V, Blizanac BB, Tomoyuki T, Ross PN, Markovic NM (2005) The effect of the particle size on the kinetics of CO electrooxidation on high surface area Pt catalysts. J Am Chem Soc 127(18):6819–6829. doi:10.1021/ja043602h
8. Lebedeva NP, Koper MTM, Feliu JM, van Santen RA (2002) Mechanism and kinetics of the electrochemical CO adlayer oxidation on Pt(111). J Electroanal Chem 524–525:242–251. http://dx.doi.org/10.1016/S0022-0728(02)00669-1
9. Iwasita T (2002) Electrocatalysis of methanol oxidation. Electrochim Acta 47:3663–3674
10. Wang H, Jusys Z, Behm RJ (2006) Ethanol electro-oxidation on carbon-supported Pt, PtRu and Pt_3Sn catalysts: a quantitative DEMS study. J Power Sources 154(2):351–359. http://dx.doi.org/10.1016/j.jpowsour.2005.10.034
11. Xia XH, Liess HD, Iwasita T (1997) Early stages in the oxidation of ethanol at low index single crystal platinum electrodes. J Electroanal Chem 437(1–2):233–240. http://dx.doi.org/10.1016/S0022-0728(97)00404-X
12. Kang Y, Qi L, Li M, Diaz RE, Su D, Adzic RR, Stach E, Li J, Murray CB (2012) Highly active Pt3Pb and core–shell Pt3Pb–Pt electrocatalysts for formic acid oxidation. ACS Nano 6 (3):2818–2825. doi:10.1021/nn3003373
13. Chen DJ, Zhou ZY, Wang Q, Xiang DM, Tian N, Sun SG (2010) A non-intermetallic PtPb/C catalyst of hollow structure with high activity and stability for electrooxidation of formic acid. Chem Commun 46(24):4252–4254. doi:10.1039/c002964e
14. Schmidt TJ, Ross Jr PN, Markovic NM (2002) Temperature dependent surface electrochemistry on Pt single crystals in alkaline electrolytes: Part 2. The hydrogen evolution/oxidation reaction. J Electroanal Chem 524–525:252–260. http://dx.doi.org/10.1016/S0022-0728(02)00683-6
15. Esposito DV, Hunt ST, Stottlemyer AL, Dobson KD, McCandless BE, Birkmire RW, Chen JG (2010) Low-cost hydrogen-evolution catalysts based on monolayer platinum on tungsten monocarbide substrates. Angew Chem Int Ed 49(51):9859–9862. doi:10.1002/anie.201004718
16. Lim B, Jiang M, Camargo PHC, Cho EC, Tao J, Lu X, Zhu Y, Xia Y (2009) Pd-Pt bimetallic nanodendrites with high activity for oxygen reduction. Science 324(5932):1302–1305. doi:10.1126/science.1170377
17. Zhang J, Sasaki K, Sutter E, Adzic RR (2007) Stabilization of platinum oxygen-reduction electrocatalysts using gold clusters. Science 315(5809):220–222
18. Lu YC, Xu ZC, Gasteiger HA, Chen S, Hamad-Schifferli K, Shao-Horn Y (2010) Platinum-gold nanoparticles: a highly active bifunctional electrocatalyst for rechargeable lithium-air batteries. J Am Chem Soc 132(35):12170–12171. doi:10.1021/ja1036572
19. Kerbach I, Climent V, Feliu JM (2011) Reduction of CO_2 on bismuth modified Pt(110) single-crystal surfaces. Effect of bismuth and poisoning intermediates on the rate of hydrogen

evolution. Electrochim Acta 56(12):4451–4456. http://dx.doi.org/10.1016/j.electacta.2011.02.027
20. Qu J, Zhang X, Wang Y, Xie C (2005) Electrochemical reduction of CO_2 on RuO_2/TiO_2 nanotubes composite modified Pt electrode. Electrochim Acta 50(16–17):3576–3580. http://dx.doi.org/10.1016/j.electacta.2004.11.061
21. Climent V, Garcia-Araez N, Feliu JM (2009) Clues for the molecular-level understanding of electrocatalysis on single-crystal platinum surfaces modified by p-block adatoms. Fuel cell catalysis a surface science approach. Wiley, Hoboken
22. Liu P, Nørskov JK (2001) Ligand and ensemble effects in adsorption on alloy surfaces. Phys Chem Chem Phys 3(17):3814–3818. doi:10.1039/b103525h
23. Tong YY, Wieckowski A, Oldfield E (2002) NMR of electrocatalysts. J Phys Chem B 106 (10):2434–2446. doi:10.1021/jp0129939
24. Du B, Danberry AL, Park I-S, Sung Y-E, Tong Y (2008) Spatially resolved ^{195}Pt NMR of carbon-supported PtRu electrocatalysts: local electronic properties, elemental composition, and catalytic activity. J Chem Phys 128(5):052311. doi:10.1063/1.2830952
25. Tan F, Du B, Danberry AL, Park I-S, Sung Y-E, Tong Y (2008) A comparative in situ 195Pt electrochemical-NMR investigation of PtRu nanoparticles supported on diverse carbon nanomaterials. Faraday Discuss 140:139–153. doi:10.1039/b803073a
26. Tong RC, Wieckowski A, Oldfield E (2000) A detailed NMR-based model for CO on Pt catalysts in an electrochemical environment: shifts, relaxation, back-bonding, and the fermi-level local density of states. J Am Chem Soc 122(6):1123–1129. doi:10.1021/ja9922274
27. Tong KHS, Babu PK, Waszczuk P, Wieckowski A, Oldfield E (2002) An NMR investigation of CO tolerance in a Pt/Ru fuel cell catalyst. J Am Chem Soc 124(3):468–473. doi:10.1021/ja011729q
28. Kobayashi T, Babu PK, Gancs L, Chung JH, Oldfield E, Wieckowski A (2005) An NMR determination of CO diffusion on platinum electrocatalysts. J Am Chem Soc 127 (41):14164–14165. doi:10.1021/ja0550475
29. Tian ZQ, Ren B (2004) Adsorption and reaction at electrochemical interfaces as probed by surface-enhanced Raman spectroscopy. Annu Rev Phys Chem 55:197–229. doi:10.1146/annurev.physchem.54.011002.103833
30. Osawa M (2006) Diffraction and spectroscopic methods in electrochemistry: in-situ surface-enhanced infrared spectroscopy of the electrode/solution interface, vol 9, Advances in electrochemical science and engineering. Wiley-VCH, New York
31. Osawa M, Ataka K, Yoshii K, Yotsuyanagi T (1993) Surface-enhanced infrared ATR spectroscopy for in situ studies of electrode/electrolyte interfaces. J Electron Spectrosc Relat Phenom 64:371–379. uuid: CE0FC3F0-7DA6-4A00-8375-39A57E55959A
32. Ataka K, Yotsuyanagi T, Osawa M (1996) Potential-dependent reorientation of water molecules at an electrode/electrolyte interface studied by surface-enhanced infrared absorption spectroscopy. J Phys Chem 100(25):10664–10672. doi:10.1021/jp953636z
33. Garcia-Araez N, Rodriguez P, Bakker HJ, Koper MTM (2012) Effect of the surface structure of gold electrodes on the coadsorption of water and anions. J Phys Chem C 116 (7):4786–4792. doi:10.1021/jp211782v
34. Sun SG, Cai WB, Wan LJ, Osawa M (1999) Infrared absorption enhancement for CO adsorbed on Au films in perchloric acid solutions and effects of surface structure studied by cyclic voltammetry, scanning tunneling microscopy, and surface-enhanced IR spectroscopy. J Phys Chem B 103(13):2460–2466
35. Yoshida M, Yamakata A, Takanabe K, Kubota J, Osawa M, Domen K (2009) ATR-SEIRAS investigation of the fermi level of Pt cocatalyst on a GaN photocatalyst for hydrogen evolution under irradiation. J Am Chem Soc 131(37):13218–13219. doi:10.1021/ja904991p
36. Shao MH, Adzic RR (2005) Electrooxidation of ethanol on a Pt electrode in acid solutions: in situ ATR-SEIRAS study. Electrochim Acta 50(12):2415–2422. doi:10.1016/j.electacta.2004.10.063

37. Smolinka T, Heinen M, Chen YX, Jusys Z, Lehnert W, Behm RJ (2005) CO_2 reduction on Pt electrocatalysts and its impact on H-2 oxidation in CO_2 containing fuel cell feed gas—a combined in situ infrared spectroscopy, mass spectrometry and fuel cell performance study. Electrochim Acta 50(25–26):5189–5199. doi:10.1016/j.electacta.2005.02.082
38. Kunimatsu K, Uchida H, Osawa M, Watanabe M (2006) In situ infrared spectroscopic and electrochemical study of hydrogen electro-oxidation on Pt electrode in sulfuric acid. J Electroanal Chem 587(2):299–307. doi:10.1016/j.jelechem.2005.11.026
39. Osawa M, K-i K, Samjeské G, Uchida T, Ikeshoji T, Cuesta A, Gutiérrez C (2011) The role of bridge-bonded adsorbed formate in the electrocatalytic oxidation of formic acid on platinum. Angew Chem Int Ed 50(5):1159–1163. doi:10.1002/anie.201004782
40. Chen YX, Miki A, Ye S, Sakai H, Osawa M (2003) Formate, an active intermediate for direct oxidation of methanol on Pt electrode. J Am Chem Soc 125(13):3680–3681. doi:10.1021/ja029044t
41. Samjeske G, Komatsu K, Osawa M (2009) Dynamics of CO oxidation on a polycrystalline platinum electrode: a time-resolved infrared study. J Phys Chem C 113(23):10222–10228. doi:10.1021/jp900582c
42. Kunimatsu K, Hanawa H, Uchida H, Watanabe M (2009) Role of adsorbed species in methanol oxidation on Pt studied by ATR-FTIRAS combined with linear potential sweep voltammetry. J Electroanal Chem 632(1–2):109–119
43. Yajima T, Uchida H, Watanabe M (2004) In-situ ATR-FTIR spectroscopic study of electro-oxidation of methanol and adsorbed CO at PtRu alloy. J Phys Chem B 108(8):2654–2659. doi:10.1021/jp037215q
44. Wang C, Peng B, Xie H-N, Zhang H-X, Shi F-F, Cai W-B (2009) Facile fabrication of Pt, Pd and Pt–Pd alloy films on Si with tunable infrared internal reflection absorption and synergetic electrocatalysis. J Phys Chem C 113(31):13841–13846. doi:10.1021/jp9034562
45. Miyake H, Okada T, Samjeske G, Osawa M (2008) Formic acid electrooxidation on Pd in acidic solutions studied by surface-enhanced infrared absorption spectroscopy. Phys Chem Chem Phys 10(25):3662–3669. doi:10.1039/b805955a
46. Wang HF, Yan YG, Hu SJ, Cai WB, Xu QH, Osawa M (2007) Seeded growth fabrication of Cu-on-Si electrodes for in situ ATR-SEIRAS applications. Electrochim Acta 52 (19):5950–5957. doi:10.1016/j.electacta.2007.03.042
47. Delgado JM, Orts JM, Rodes A (2007) A comparison between chemical and sputtering methods for preparing thin-film silver electrodes for in situ ATR-SEIRAS studies. Electrochim Acta 52(14):4605–4613. doi:10.1016/j.electacta.2006.12.045
48. Yajima T, Wakabayashi N, Uchida H, Watanabe M (2003) Adsorbed water for the electro-oxidation of methanol at Pt-Ru alloy. Chem Commun 7:828–829. doi:10.1039/b212197b
49. Yan LQ-X, Huo S-J, Ma M, Cai W-B, Osawa M (2005) Ubiquitous strategy for probing ATR surface-enhanced infrared absorption at platinum group metal-electrolyte interfaces. J Phys Chem B 109(16):7900–7906. doi:10.1021/jp044085s
50. Wang J-Y, Zhang H-X, Jiang K, Cai W-B (2011) From HCOOH to CO at Pd electrodes: a surface-enhanced infrared spectroscopy study. J Am Chem Soc 133:14876–14879. doi:10.1021/ja205747j
51. Vassilev P, Koper MTM (2007) Electrochemical reduction of oxygen on gold surfaces: a density functional theory study of intermediates and reaction paths. J Phys Chem C 111 (6):2607–2613. doi:10.1021/jp064515+
52. Kunimatsu K, Yoda T, Tryk DA, Uchida H, Watanabe M (2010) In situ ATR-FTIR study of oxygen reduction at the Pt/Nafion interface. Phys Chem Chem Phys 12(3):621–629. doi:10.1039/B917306D
53. Cuesta A, Cabello G, Hartl FW, Escudero-Escribano M, Vaz-Domínguez C, Kibler LA, Osawa M, Gutiérrez C (2013) Electrooxidation of formic acid on gold: an ATR-SEIRAS study of the role of adsorbed formate. Catal Today 202:79–86. http://dx.doi.org/10.1016/j.cattod.2012.04.022

54. Shiroishi H, Ayato Y, Kunimatsu K, Okada T (2005) Study of adsorbed water on Pt during methanol oxidation by ATR-SEIRAS (surface-enhanced infrared absorption spectroscopy). J Electroanal Chem 581(1):132–138. doi:10.1016/j.jelechem.2005.04.027
55. Futamata M, Luo LQ (2007) Adsorbed water and CO on Pt electrode modified with Ru. J Power Sources 164(2):532–537. doi:10.1016/j.jpowsour.2006.10.079
56. Li X, Gewirth AA (2005) Oxygen electroreduction through a superoxide intermediate on Bi-modified Au surfaces. J Am Chem Soc 127(14):5252–5260. doi:10.1021/ja043170a
57. Yang H, Yang Y, Zou S (2007) In situ surface-enhanced raman spectroscopic studies of CO adsorption and methanol oxidation on Ru-modified Pt surfaces. J Phys Chem C 111 (51):19058–19065. doi:10.1021/jp075929l
58. Xu B, Park I-S, Li Y, Chen D-J, Tong YJ (2011) An in situ SERS investigation of the chemical states of sulfur species adsorbed onto Pt from different sulfur sources. J Electroanal Chem 662:52–56. doi:10.1016/j.jelechem.2011.02.031
59. Tian ZQ, Ren B, Wu DY (2002) Surface-enhanced Raman scattering: from noble to transition metals and from rough surfaces to ordered nanostructures. J Phys Chem B 106 (37):9463–9483. doi:10.1021/jp0257449
60. Gómez R, Pérez JM, Solla-Gullón J, Montiel V, Aldaz A (2004) In situ surface enhanced raman spectroscopy on electrodes with platinum and palladium nanoparticle ensembles. J Phys Chem B 108(28):9943–9949. doi:10.1021/jp038030m
61. Gómez R, Solla-Gullón J, Pérez JM, Aldaz A (2005) Surface-enhanced raman spectroscopy study of ethylene adsorbed on a Pt electrode decorated with Pt nanoparticles. ChemPhysChem 6(10):2017–2021. doi:10.1002/cphc.200500168
62. Solla-Gullón J, Gómez R, Aldaz A, Pérez JM (2008) A combination of SERS and electrochemistry in Pt nanoparticle electrocatalysis: promotion of formic acid oxidation by ethylidyne. Eletrochem Commun 10(2):319–322. http://dx.doi.org/10.1016/j.elecom.2007.12.010
63. Pu Zhang JC, Chen Y-X, Tang Z-Q, Dong C, Yang JL, Wu D-Y, Ren B, Tian Z-Q (2010) Potential-dependent chemisorption of carbon monoxide at a gold core-platinum shell nanoparticle electrode: a combined study by electrochemical in situ surface-enhanced raman spectroscopy and density functional theory. J Phys Chem C 114:403–411
64. Park I-S, Chen D-J, Atienza DO, Tong YYJ (2013) Enhanced CO monolayer electro-oxidation reaction on sulfide-adsorbed Pt nanoparticles: a combined electrochemical and in situ ATR-SEIRAS spectroscopic study. Catal Today 202:175–182. http://dx.doi.org/10.1016/j.cattod.2012.05.045
65. Park I-S, Atienza DO, Hofstead-Duffy AM, Chen D, Tong YJ (2011) Mechanistic insights on sulfide-adsorption enhanced activity of methanol electro-oxidation on Pt nanoparticles. ACS Catal 2(1):168–174. doi:10.1021/cs200546f
66. Li X, Gewirth AA (2003) Peroxide electroreduction on bi-modified au surfaces: vibrational spectroscopy and density functional calculations. J Am Chem Soc 125(23):7086–7099. doi:10.1021/ja034125q
67. Li X, Heryadi D, Gewirth AA (2005) Electroreduction activity of hydrogen peroxide on Pt and Au electrodes. Langmuir 21(20):9251–9259. doi:10.1021/la0508745
68. Du B, Rabb SA, Zangmeister C, Tong Y (2009) A volcano curve: optimizing methanol electro-oxidation on Pt-decorated Ru nanoparticles. Phys Chem Chem Phys 11 (37):8231–8239. uuid:F4A2892D-47E4-478B-825C-8EB0584C7F0F
69. Le Rhun V, Garnier E, Pronier S, Alonso-Vante N (2000) Electrocatalysis on nanoscale ruthenium-based material manufactured by carbonyl decomposition. Electrochem Commun 2(7):475–479
70. Park IS, Lee KS, Jung DS, Park HY, Sung Y-E (2007) Electrocatalytic activity of carbon-supported Pt-Au nanoparticles for methanol electro-oxidation. Electrochim Acta 52:5599–5605. uuid:84E52FA9-41B4-4917-93F1-E2222D793214

71. Atienza DO, Allison TC, Tong YJ (2012) Spatially resolved electronic alterations as seen by in situ 195Pt and 13CO NMR in Ru@Pt and Au@Pt core–shell nanoparticles. J Phys Chem C 116(50):26480–26486. doi:10.1021/jp310313k
72. Du B, Zaluzhna O, Tong YJ (2011) Electrocatalytic properties of Au@Pt nanoparticles: effects of Pt shell packing density and Au core size. Phys Chem Chem Phys 13 (24):11568–11574
73. Tong RC, Godbout N, Wieckowski A, Oldfield E (1999) Correlation between the knight shift of chemisorbed CO and the fermi level local density of states at clean platinum catalyst surfaces. J Am Chem Soc 121(13):2996–3003. doi:10.1021/ja9830492
74. Stokes HT, Rhodes HE, Wang PK, Slichter CP, Sinfelt JH (1982) NMR of platinum catalysts. III. Microscopic variation of the Knight shifts. Phys Rev B 26(7):3575–3581. uuid:572642F8-60BF-49D7-83A6-B7AE0A58B890
75. Bucher J, van der Klink J (1988) Electronic properties of small supported Pt particles: NMR study of 195Pt hyperfine parameters. Phys Rev B Condens Matter 38(16):11038–11047. doi:10.1103/PhysRevB.38.11038
76. Adekunle AS, Ozoemena KI (2008) Electron transfer behaviour of single-walled carbon nanotubes electro-decorated with nickel and nickel oxide layers. Electrochim Acta 53 (19):5774–5782
77. Wang PK, Ansermet JP, Rudaz SL, Wang Z, Shore S, Slichter CP, Sinfelt JH (1986) NMR studies of simple molecules on metal surfaces. Science 234(4772):35–41. doi:10.1126/science.234.4772.35
78. Korringa J (1950) Nuclear magnetic relaxation and resonance line shift in metals. Physica XVI(7–8):601–610
79. Blyholder G (1964) Molecular orbital view of chemisorbed carbon monoxide. J Phys Chem 68(10):2772–2777. doi:10.1021/j100792a006
80. Chen D-J, Hofstead-Duffy AM, Park I-S, Atienza DO, Susut C, Sun S-G, Tong YJ (2011) Identification of the most active sites and surface water species: a comparative study of CO and methanol oxidation reactions on core–shell M@Pt (M = Ru, Au) nanoparticles by in situ IR spectroscopy. J Phys Chem C 115(17):8735–8743. doi:10.1021/jp200557m
81. Chen D-J, Xu B, Sun S-G, Tong YJ (2012) Electroless deposition of ultrathin Au film for surface enhanced in situ spectroelectrochemistry and reaction-driven surface reconstruction for oxygen reduction reaction. Catal Today 182:46–53. doi:10.1016/j.cattod.2011.08.052
82. Lin WF, Zei MS, Eiswirth M, Ertl G, Iwasita T, Vielstich W (1999) Electrocatalytic Activity of Ru-Modified Pt(111) Electrodes toward CO Oxidation. J Phys Chem B 103 (33):6968–6977. doi:10.1021/jp9910901
83. Lu GQ, White JO, Wieckowski A (2004) Vibrational analysis of chemisorbed CO on the Pt (111)/Ru bimetallic electrode. Surf Sci 564(1–3):131–140
84. Spendelow JS, Babu PK, Wieckowski A (2005) Electrocatalytic oxidation of carbon monoxide and methanol on platinum surfaces decorated with ruthenium. Curr Opin Solid State Mater Sci 9(1–2):37–48
85. Watanabe M, Sato T, Kunimatsu K, Uchida H (2008) Temperature dependence of co-adsorption of carbon monoxide and water on highly dispersed Pt/C and PtRu/C electrodes studied by in-situ ATR-FTIRAS. Electrochim Acta 53(23):6928–6937. doi:10.1016/j.electacta.2008.02.023
86. Ianniello R, Schmidt VM, Stimming U, Stumper J, Wallau A (1994) CO adsorption and oxidation on Pt and Pt Ru alloys: dependence on substrate composition. Electrochim Acta 39 (11–12):1863–1869
87. Friedrich KA, Geyzers KP, Dickinson AJ, Stimming U (2002) Fundamental aspects in electrocatalysis: from the reactivity of single-crystals to fuel cell electrocatalysts. J Electroanal Chem 524–525:261–272
88. Zheng MS, Sun SG, Chen SP (2001) Abnormal infrared effects and electrocatalytic properties of nanometer scale thin film of PtRu alloys for CO adsorption and oxidation. J Appl Electrochem 31(7):749–757

89. Lin WF, Iwasita T, Vielstich W (1999) Catalysis of CO electrooxidation at Pt, Ru, and PtRu alloy. An in situ FTIR study. J Phys Chem B 103(16):3250–3257
90. Brankovic SR, Wang JX, Adzic RR (2001) Pt submonolayers on Ru nanoparticles—a novel low Pt loading, high CO tolerance fuel cell electrocatalyst. Electrochem Solid State Lett 4 (12):A217–a220
91. Vogel W, Le Rhun V, Garnier E, Alonso-Vante N (2001) Ru clusters synthesized chemically from dissolved carbonyl: in situ study of a novel electrocatalyst in the gas phase and in electrochemical environment. J Phys Chem B 105(22):5238–5243. doi:10.1021/jp0100654
92. Chen DJ, Tong YYJ (2015) In situ Raman spectroscopic measurement of near-surface proton concentration changes during electrochemical reactions. Chem Commun 51(26):5683–5686. doi:10.1039/C5CC00427F
93. Zhu Y, Uchida H, Watanabe M (1999) Oxidation of carbon monoxide at a platinum film electrode studied by Fourier transform infrared spectroscopy with attenuated total reflection technique. Langmuir 15(25):8757–8764. doi:10.1021/la990835r
94. Peremans A, Tadjeddine A (1994) Vibrational spectroscopy of electrochemically deposited hydrogen on platinum. Phys Rev Lett 73(22):3010
95. Osawa M, Tsushima M, Mogami H, Samjeske G, Yamakata A (2008) Structure of water at the electrified platinum-water interface: a study by surface-enhanced infrared absorption spectroscopy. J Phys Chem C 112(11):4248–4256. doi:10.1021/jp710386g
96. Kunimatsu K, Senzaki T, Samjesk G, Tsushima M, Osawa M (2007) Hydrogen adsorption and hydrogen evolution reaction on a polycrystalline Pt electrode studied by surface-enhanced infrared absorption spectroscopy. Electrochim Acta 52(18):5715–5724
97. Kunimatsu K, Uchida H, Osawa M, Watanabe M (2006) In situ infrared spectroscopic and electrochemical study of hydrogen electro-oxidation on Pt electrode in sulfuric acid (vol 587, p 299, 2006). J Electroanal Chem 596(2):169–169. doi:10.1016/j.jelechem.2006.07.015
98. Futamata M, Luo L, Nishihara C (2005) ATR-SEIR study of anions and water adsorbed on platinum electrode. Surf Sci 590(2–3):196–211
99. Climent V, Gomez R, Feliu JM (1999) Effect of increasing amount of steps on the potential of zero total charge of Pt(111) electrodes. Electrochim Acta 45(4–5):629–637
100. Bergelin M, Herrero E, Feliu JM, Wasberg M (1999) Oxidation of CO adlayers on Pt(111) at low potentials: an impinging jet study in H_2SO_4 electrolyte with mathematical modeling of the current transients. J Electroanal Chem 467(1–2):74–84
101. Vidal-Iglesias FJ, Solla-Gullon J, Campina JM, Herrero E, Aldaz A, Feliu JM (2009) CO monolayer oxidation on stepped Pt(S) [(n-1)(100)*(110)] surfaces. Electrochim Acta 54 (19):4459–4466
102. Roth C, Benker N, Buhrmester T, Mazurek M, Loster M, Fuess H, Koningsberger DC, Ramaker DE (2005) Determination of O[H] and CO coverage and adsorption sites on PtRu electrodes in an operating PEM fuel cell. J Am Chem Soc 127(42):14607–14615. doi:10.1021/ja050139f
103. Coker DF, Miller RE, Watts RO (1985) The infrared predissociation spectra of water clusters. J Chem Phys 82(8):3554–3562
104. Richmond GL (2002) Molecular bonding and interactions at aqueous surfaces as probed by vibrational sum frequency spectroscopy. Chem Rev 102(8):2693–2724. doi:10.1021/cr0006876
105. Wasileski SA, Koper MTM, Weaver MJ (2001) Field-dependent chemisorption of carbon monoxide on platinum-group (111) surfaces: relationships between binding energetics, geometries, and vibrational properties as assessed by density functional theory. J Phys Chem B 105(17):3518–3530. doi:10.1021/jp003263o
106. Oudenhuijzen MK, van Bokhoven JA, Ramaker DE, Koningsberger DC (2004) Theoretical study on Pt particle adsorbate bonding: influence of support ionicity and implications for catalysis. J Phys Chem B 108(52):20247–20254. uuid:FEE96E5D-8FA4-4172-95A9-A6C6AC66EE05

107. Dronskowski R, Bloechl PE (1993) Crystal orbital Hamilton populations (COHP): energy-resolved visualization of chemical bonding in solids based on density-functional calculations. J Phys Chem 97(33):8617–8624. doi:10.1021/j100135a014
108. Papoian G, Norskov JK, Hoffmann R (2000) A comparative theoretical study of the hydrogen, methyl, and ethyl chemisorption on the Pt(111) surface. J Am Chem Soc 122 (17):4129–4144. doi:10.1021/ja993483j
109. Glassey WV, Hoffmann R (2001) A comparative study of the p (2×2)-CO/M (111), M = Pt, Cu, Al chemisorption systems. J Phys Chem B 105(16):3245–3260. doi:10.1021/jp003922x
110. Glassey WV, Hoffmann R (2001) A molecular orbital study of surface–adsorbate interactions during the oxidation of CO on the Pt(111) surface. Surf Sci 475(1–3):47–60. doi:10.1016/S0039-6028(00)01062-1
111. Glassey WV, Papoian GA, Hoffmann R (1999) Total energy partitioning within a one-electron formalism: a Hamilton population study of surface–CO interaction in the c (2 × 2)-CO/ Ni(100) chemisorption system. J Chem Phys 111(3):893–910. doi:10.1063/1.479200
112. Somorjai GA, Aliaga C (2010) Molecular studies of model surfaces of metals from single crystals to nanoparticles under catalytic reaction conditions. Evolution from prenatal and postmortem studies of catalysts. Langmuir 26(21):16190–16203. doi:10.1021/la101884s
113. Somorjai GA, Contreras AM, Montano M, Rioux RM (2006) Clusters, surfaces, and catalysis. Proc Natl Acad Sci U S A 103:10577–10583
114. Somorjai GA (1994) Introduction to surface chemistry and catalysis. Wiley, New York
115. Hoffmann R (1993) A chemical and theoretical approach to bonding at surfaces. J Phys Condens Matter 5:A1–A16
116. Hoffmann R (1971) Interaction of orbitals through space and through bonds. Acc Chem Res 4 (1):1–9
117. Housmans THM, Wonders AH, Koper MTM (2006) Structure sensitivity of methanol electrooxidation pathways on platinum: an on-line electrochemical mass spectrometry study. J Phys Chem B 110(20):10021–10031. doi:10.1021/jp055949s
118. Huang Z, Kim K-J (2007) Review of X-ray free-electron laser theory. Phys Rev ST Accel Beams 10(3):034801–034826. doi:10.1103/PhysRevSTAB.10.034801
119. Shearing P, Wu Y, Harris SJ, Brandon N (2011) In situ X-ray spectroscopy and imaging of battery materials. Interface 20:43–47. uuid:27465B6F-DD87-45E1-8587-699AFDB658FF
120. Huang L, Sorte EG, Sun SG, Tong YYJ (2015) A straightforward implementation of in situ solution electrochemical 13C NMR spectroscopy for studying reactions on commercial electrocatalysts: ethanol oxidation. Chem Commun 51(38):1–3. doi:10.1039/C5CC00862J
121. DeVience SJ, Pham LM, Lovchinsky I, Sushkov AO, Bar-Gill N, Belthangady C, Casola F, Corbett M, Zhang H, Lukin M, Park H, Yacoby A, Walsworth RL (2015) Nanoscale NMR spectroscopy and imaging of multiple nuclear species. Nat Nanotechnol 10(2):129–134. doi:10.1038/nnano.2014.313
122. Häberle T, Schmid-Lorch D, Reinhard F, Wrachtrup J (2015) Nanoscale nuclear magnetic imaging with chemical contrast. Nat Nanotechnol 10(2):125–128. doi:10.1038/nnano.2014.299
123. Rugar D, Mamin HJ, Sherwood MH, Kim M, Rettner CT, Ohno K, Awschalom DD (2015) Proton magnetic resonance imaging using a nitrogen-vacancy spin sensor. Nat Nanotechnol 10(2):120–124. doi:10.1038/nnano.2014.288

Chapter 6
Recent Development of Platinum-Based Nanocatalysts for Oxygen Reduction Electrocatalysis

David Raciti, Zhen Liu, Miaofang Chi, and Chao Wang

6.1 Introduction

Fed by hydrogen and oxygen gases, a fuel cell generates electricity from the hydrogen oxidation reaction at the anode and the ORR at the cathode. As it has water as the only product, fuel cell is considered to be environmentally friendly. With hydrogen produced from water splitting with solar electricity and oxygen from the air, it is also a renewable solution for chemical-electrical energy conversion. This has led to a broad range of applications of fuel cells, including stationary power systems, mobile electronics and vehicles [1]. However, the need for platinum (Pt) as catalysts for the electrochemical reactions, especially the ORR at the cathode, has limited large-scale applications of fuel cells [2]. The sluggish kinetics of ORR gives high overpotentials at the cathode, typically several hundred millivolts on Pt catalysts, and thereby large amounts of Pt are needed for practical applications. In order to lower the system cost, substantial improvement of the catalyst's performance is demanded for reduced use of Pt. It is expected that a five-fold enhancement of the ORR activity versus state-of-the-art Pt/C catalysts is required to enable the commercial implementation of fuel cells in transportation [3].

Various approaches have been developed to improve the performance of Pt electrocatalysts for the ORR, including optimization of Pt particle size and shape, alloying Pt with 3*d* transition metals and development of composite nanostructures.

D. Raciti • Z. Liu • C. Wang (✉)
Department of Chemical and Biomolecular Engineering, Johns Hopkins University, Baltimore, MD 21212, USA
e-mail: chaowang@jhu.edu

M. Chi
Center for Nanophase Materials Sciences, Oak Ridge National Laboratory, Oak Ridge, TN, USA

K.I. Ozoemena, S. Chen (eds.), *Nanomaterials for Fuel Cell Catalysis*, Nanostructure Science and Technology, DOI 10.1007/978-3-319-29930-3_6

Throughout these approaches the catalyst surface is maintained to be Pt, as Pt is likely the only element that is both active and stable under PEMFC conditions. Therefore, the design of advanced electrocatalysts beyond Pt usually goes through the manipulation of subsurface nanoscale architectures to induce modifications to the surface geometry and/or electronic structures, and thus alter the surface adsorption and catalytic properties of Pt.

Here we provide a focused review of the recent advancement of Pt-based electrocatalysts for the ORR. We will discuss the various efforts on the development of Pt and Pt-based alloy and composite nanostructures for electrocatalytic applications, especially how to tailor surface structures towards enhanced catalytic activity and stability for the ORR. Due to the presence of various protocols for ORR measurements, we will focus our discussion on relative activities, typically expressed in terms of improvement factors versus Pt bench marker. Though our emphasis will be placed on high-surface-area nanocatalysts, we will also try to link the structure-property relationships of nanocatalysts to those established on extended-surface model catalysts. Possible approaches towards further improvement of the Pt-based nanocatalysts will also be discussed.

6.2 Single-Component Pt Catalysts

Albeit the focus of research efforts is placed on alloy and composite nanostructures, most of the catalysts employed in present fuel cells are still single-component Pt particles supported on high-surface-area carbon. The Pt/C catalysts have been extensively studied for more than half a century, with the ORR catalytic activity found to be correlated to the particle size and shape [4]. The presence of extensive studies in history, however, has not impeded the continual effort on this simple system, where new insights and nanostructures with substantial catalytic improvement have been developed.

6.2.1 Particle Size Effect

One purpose of using nanoparticulated materials as catalysts is to harvest the large surface/volume ratio and number of surface sites. For many reactions, it is also found that the catalytic activity of smaller particles, in terms of turnover frequency (TOF), is higher than larger particles or extended surfaces due to the reduced coordination of surface atoms [5–9]. These two trends ask for decreasing particle size in catalytic improvement. The scenery of ORR on Pt, however, is different. It has long been realized that the catalytic activity of Pt for the ORR is dependent on particle size, with smaller particles giving lower specific activity (current normalized by the real surface area of the catalyst) than larger ones. It is believed that the size-dependent ORR activity originates from the varying ratios of ordered surface facets, i.e., (1 1 1) and (1 0 0) for cubo-octahedral particles, and edge and corner sites on the surface as the particle size changes [4].

However, debates still exist in the literature. For example, Watanabe et al. argued that the specific activity of ORR on Pt does not depend on the particle size but rather on the interparticle distance [11]. In their studies of Pt clusters for the ORR Yamamoto et al. found the Pt clusters of sizes smaller than 2 nm have higher specific activities than commercial Pt/C catalysts (2.5 nm) [12]. On the other side, a recent study by Shao et al. [10] confirmed the size-dependent ORR catalytic activity with Pt/C catalysts prepared by layer-by-layer growth on the same seeds with underpotential deposition methods, which can be considered to be with reasonable control over the interparticle distance. They observed an increase of specific activity with particle size and the maximum mass activity (current normalized by the mass of Pt) at an intermediate particle size (Fig. 6.1a), consistent with the previous trends established by Kinoshita [4] except that mass activity achieves the maximum at ~2.2 nm (versus ~3 nm in Kinoshita's report). Moreover, they calculated the oxygen binding energies for the particles of various sizes and was able to clearly assign the increase in ORR specific activity with particle size to the increase of Pt–O binding energy (Fig. 6.1b, due to the diminishing occupation of edge sites on larger particles) [10], as the rate of ORR on Pt is limited by the desorption of OH.

It is noticed by this review that size-*independent* ORR activities have also been reported recently. Arenz et al. did not observe a clear difference in ORR activity among commercial Pt/C catalysts of 1–5 nm, though substantially different ORR activities were found between unsupported Pt black (~30 nm) and the 1–5 nm Pt/C catalysts [13]. Shao-Horn et al. revealed size-independent ORR activities on Pt catalysts below 5 nm, which was ascribed to the similarity of surface compositions and surface electronic structures of Pt particles below 5 nm as well as comparable OH_{ad} coverage at the potential at which ORR was measured, though the catalysts with different particle sizes in this study were produced by thermal treatment which had caused increase of size dispersion [14].

From above one can see that the particle size effect, one of the most studied topics of relevance to the ORR, still remains elusive for Pt electrocatalysts. While the different experimental protocols with varying potential windows, catalyst loading, electrolytes, etc. may all contribute to the controversial observations, the materials that have been studied, the various Pt/C catalysts, could be of particular importance. Commercial catalysts prepared by impregnation methods usually possess relatively wide distributions of particle size and shape, which may not be considered as appropriate starting materials for fundamental studies. Development of better defined catalysts, with monodisperse (but variable) and uniform (of the same shape) particles by controlled synthesis is expected to resolve these issues and help uncover the intrinsic particle size effect of Pt electrocatalysts for the ORR.

While the studies on particle size effect have been focused on the ORR activity, it is necessary to point out that the size of Pt particles also plays an important role in the stability of ORR electrocatalysts. In their studies of Pt/C catalysts of different sizes Shao-Horn et al. also observed increasing catalyst stability upon potential cycling as the particle size increases (Fig. 6.2) [14]. It is suggested that the

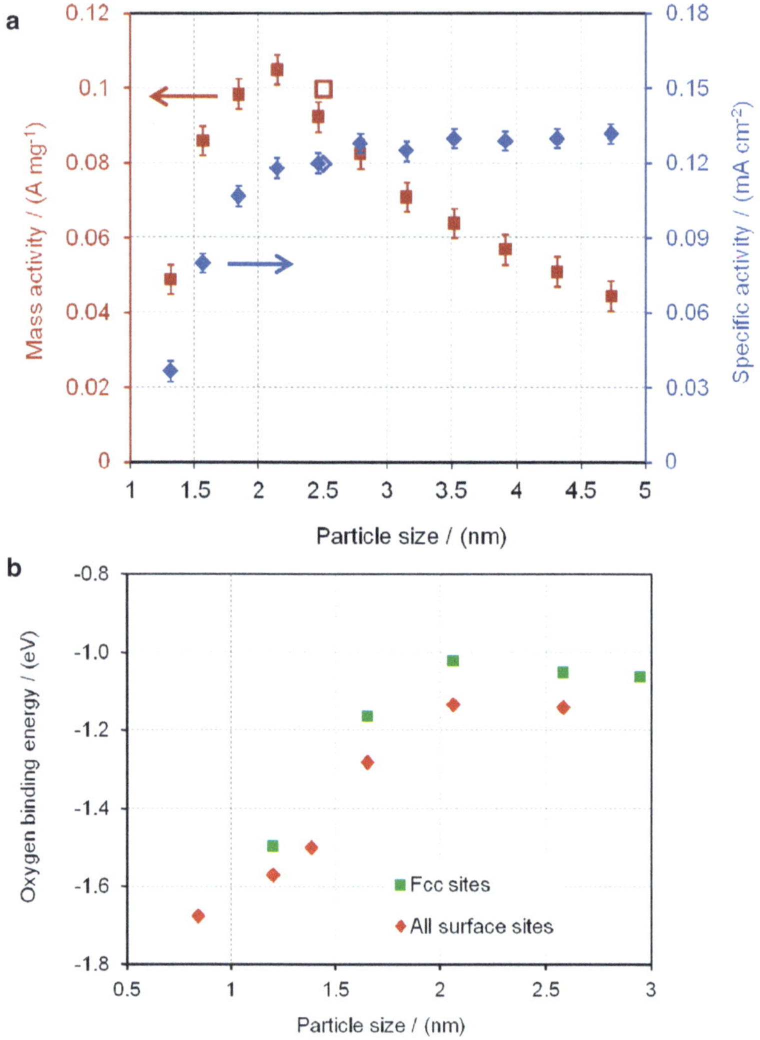

Fig. 6.1 (**a**) Size dependence of specific activity (*blue diamond*) and mass activity (*red square*) of Pt/C for oxygen reduction reaction at 0.93 V; (**b**) averaged oxygen binding energy calculated as a function of particle size for cubo-octahedrons (Reprinted with permission from [10]. Copyright 2011 ACS)

degradation of Pt/C catalysts in PEMFCs can undergo different routes, including Pt dissolution and migration, Ostwald ripening and particle detachment from carbon support [15]. Larger Pt particles are more stable as their surface atoms possess lower chemical potential for dissolution or movement [16] which could be due to the higher average coordination and presence of fewer defects. Therefore, a balance between the both size-dependent activity and stability is desired for the design of

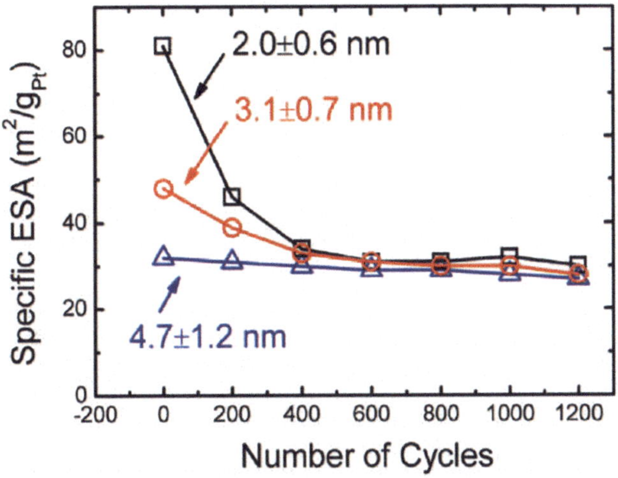

Fig. 6.2 Specific electrochemical surface area (ESA) of Pt NPs as a function of the number of potential cycles for Pt-46 % (2.0 ± 0.6 nm, *black squares*), Pt-46 %-900 °C-1 m (3.1 ± 0.7 nm, *red circles*), and Pt-46 %-900 °C-2 h (4.7 ± 1.2 nm, *blue triangles*) (Reprinted with permission from [14]. Copyright 2012 ECS)

ORR electrocatalysts and the optimal particle size could be different from that established by simply considering the maximal mass activity.

6.2.2 Particle Shape Effect

The ORR is a structure-sensitive reaction on Pt catalysts. Studies of single-crystal Pt extended surfaces have revealed that the rate of ORR follows the trend (1 1 0) > (1 1 1) > (1 0 0) [17, 18]. It is thus natural to pursue shape control over Pt catalysts for enhanced electrocatalytic activity. However, except for a few reports [19–21], the studies of shape-dependent ORR catalysis are very limited for Pt nanocatalysts (in sharp contrast to the case of Pt-based alloy catalysts to be discussed in the following). The challenges are mainly present in the difficulty of selectively growing Pt nanocrystals enclosed by (1 1 0) or (1 1 1) facets, whereas synthesis of Pt nanocubes with (1 0 0) surface has been found to be relatively easy and reported by many groups [22–25]. In this sense, substantial effort may still be needed in elucidating the particle shape effect in the ORR and correlate the electrocatalytic performance of Pt nanoctalysts to their bulk counterparts.

Besides controlling the surface structure by directly growing Pt nanocrystals with shape control, many other studies have focused on the development of Pt electrocatalysts of alternated morphologies beyond simple NPs for the ORR (Table 6.1). These include the syntheses of single-crystalline Pt nanowires on carbon spheres (Fig. 6.3a) [26], porous Pt nanotubes by galvanic displacement of

Table 6.1 Comparison of ORR activities of Pt catalysts of different morphologies

Ref	Pt nanostructure	Specific surface area ($cm^2\ mg^{-1}$)	Specific surface activity at 0.9 V vs. RHE (A/g)	Measuring conditions
a	Nanowires	436	120	In a single cell PEMFC at 80 °C
b	Porous nanotubes	704	88	Argon saturated 0.1 M $HClO_4$ electrolyte
c	Hollow nanoparticles	600	1150	0.1 M $HClO_4$ at 80 °C, 0.9 V in 10 mV/s potential sweeps
d	Mesoporous double gyroid networks	560	170	N2-purged 0.1M $HClO_4$ (aq), 23 °C, sweep rate 50 mV/s

Pd nanowires (Fig. 6.3b) [27], hollow Pt NPs by galvanic displacement of Ni NPs (Fig. 6.3c) [28], and mesoporous double gyroid Pt networks (Fig. 6.3d) [29]. High mass activity has generally been achieved in these studies by manipulation of the nanoscale architectures to increase the exposure of surface sites, whereas the catalyst stability is typically superior than simply reducing the particle size due to the formation of more continuous surface. While these nanostructures possess unique properties and can potentially lead to unconventional advanced electrocatalysts, their scaling up and incorporation into membrane electrode assemblies must also be considered which may reduce their potential for practical applications.

6.3 Pt-Based Alloy Catalysts

Bimetallic alloys of Pt and 3*d* transition metals such as Cr, Mn, Fe, Ni, Co and Cu have been extensively studied as electrocatalysts for the ORR [18, 30–35]. In these alloys, the 3*d* transition metals are not directly involved in the ORR, but rather enhance the catalytic activity of surface Pt atoms through electronic and geometric modifications. Such modifications can lead to weaker Pt–O binding and reduced adsorption of oxygenated spectator species, e.g., OH^-, on the surface, provide more active sites accessible for molecular oxygen and thus improve the catalytic activity [18, 31–33].

Although having promising catalytic activity, challenges still exist for the employment of Pt-alloy catalysts for the ORR. The main concern is that the 3*d* metals have much lower oxidation potential than Pt and they may be leached out under the running conditions in PEMFCs [36]. Therefore, the architecture of surface structures becomes critical for the design of Pt-alloy catalysts, which do not only determine the catalytic activity, but also long-term durability of the catalysts.

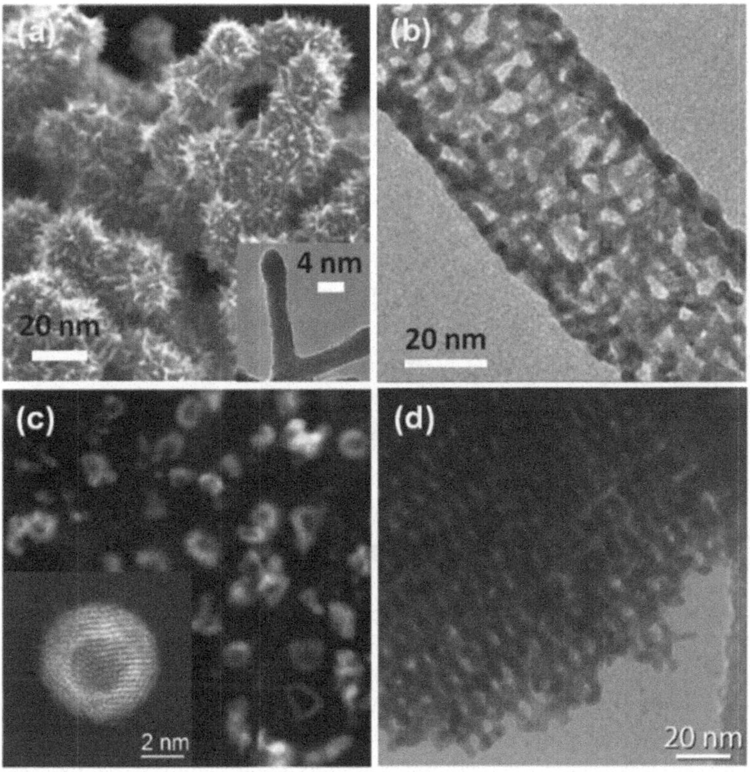

Fig. 6.3 (**a**) SEM micrographs of the as-synthesized Pt nanowire/C nanostructure, with 60 wt% Pt on carbon (inset: HRTEM micrograph of a branched Pt nanowire grown on carbon support); (Reprinted with permission from [26]. Copyright 2008 Wiley) (**b**) HRTEM image of Pt nanotubes (Reprinted with permission from [27]. Copyright 2010 Wiley); (**c**) STEM images of Pt hollow nanoparticles (inset: High-resolution STEM images of Pt hollow particles) (Reprinted with permission from [28]. Copyright 2011 ACS); (**d**) TEM image of the Pt mesoporous double gyroid networks (Reprinted with permission from [29]. Copyright 2012 ACS)

Conventionally Pt-alloy catalysts are prepared by impregnation of 3*d* transition metals into Pt/C catalysts [37, 38]. This type of synthesis usually involves high-temperature thermal treatment to improve alloy homogeneity. The induced particle sintering and agglomeration cause loss in surface area and control over particle size and size distribution, making it challenging for the systematical study of the Pt-alloy catalysts. Recently progress has been made on the synthesis of Pt alloy nanoparticles (NPs) based on organic solution synthesis [39–48]. Monodisperse and homogeneous colloidal NPs of the Pt-bimetallic alloys have been made by coupling the reduction of Pt salts with the decomposition of organometallic precursors of 3*d* transition metals [42–45] or by subsequent addition of the metal salt precursors [44, 46–48], to address the different reduction and growth rates between Pt and 3*d* metals. Pretreatments for surfactant removal and surface cleaning are usually applied [45, 49–52] either directly on the as-synthesized NPs or after their

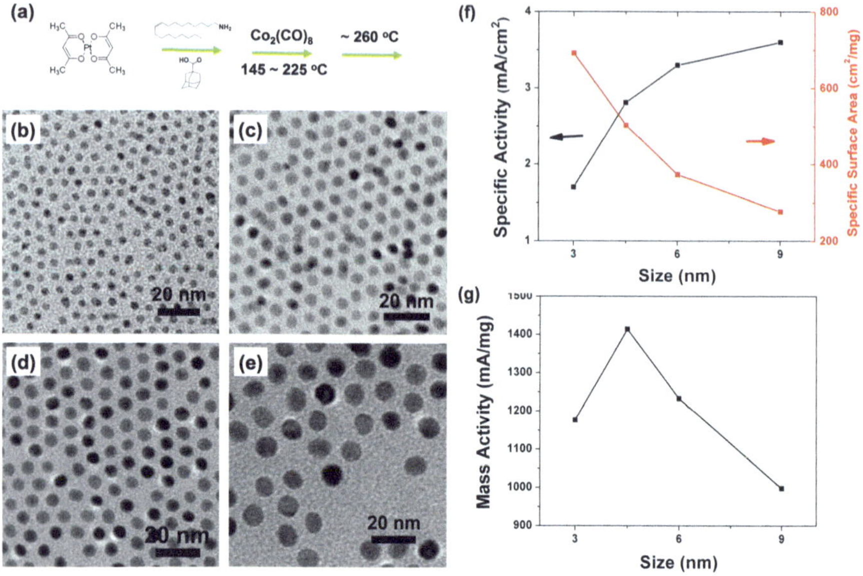

Fig. 6.4 Size dependent specific activity and specific surface area for the Pt_3Co/C catalysts of various particle sizes (Reprinted with permission from [45]. Copyright 2011 ACS)

deposition onto high-surface-area carbon, before the application as catalysts. By separation of the nanoparticle growth and particle deposition onto support, control over particle size, shape and composition has been achieved by the colloidal approach, which has enabled systematic studies of the Pt-alloy catalysts for the ORR.

6.3.1 Particle Size Effect

Based on monodisperse and homogeneous Pt_3Co NPs obtained from organic solution synthesis, the effect of particle size on electrocatalytic activity has been investigated for Pt-bimetallic catalysts for the ORR (Fig. 6.4) [45]. It was found that while the specific surface area (surface area normalized by the mass of Pt) decreases as the particle size increases, the specific activity for the ORR increases due to less oxophilic properties of the surface for large particles [3, 4, 53]. NPs of an intermediate size, ~4.5 nm, were found to give the maximum in mass activity, with an improvement factor of about 3 versus Pt/C of similar particle size. This volcano-type dependence is similar to the size-dependence activity of Pt/C (Fig. 6.1), where the difference in the particle size maximizing mass activity implies the dissimilar surface structures between these two systems (see the following section).

6.3.2 Particle Shape Effect

Previous studies of Pt_3Ni(h k l) single crystals have shown that the ORR activity is an order of magnitude higher on Pt_3Ni(1 1 1) than on Pt(1 1 1) and other low-index Pt surfaces (Fig. 6.5) [18]. This has inspired significant efforts on the synthesis of Pt alloy nanocatalysts of octahedral shape aiming at achieving the same level of catalytic enhancement [54–56]. Pt_3Ni NPs of octahedral shape with (1 1 1) faceted surfaces were found to be ~4 and ~6 times more active than Pt_3Ni nanocubes and Pt nanocubes with (1 0 0) surfaces of comparable sizes, respectively (Fig. 6.6) [54]. However, in another study Yang et al. only observed an improvement factor of ~1.5 for the particle shape evolved from cube to octahedron [55] and ~2.2 to icosahedron [57] for Pt_3Ni. Such a divergence indicates that any trivial difference in synthetic and catalyst preparation methods and/or particle sizes could lead to substantially different catalytic performance, pointing to the lack of fundamental understanding for the surface structure-property relationship of alloy nanocatalysts.

Although the single-crystal work [18] is usually cited for the dependence of ORR activity on surface structure, it is necessary to point out that the nanoscale alloy catalysts could possess different surface structures from the well-defined extended surfaces and several factors might be present to impede the achievement of similar level of catalytic improvement. First and most importantly, the ORR activity of Pt_3Ni(1 1 1) shown in Fig. 6.5 was recorded on single crystals with a *Pt-skin* surface. This structure is formed by surface segregation via

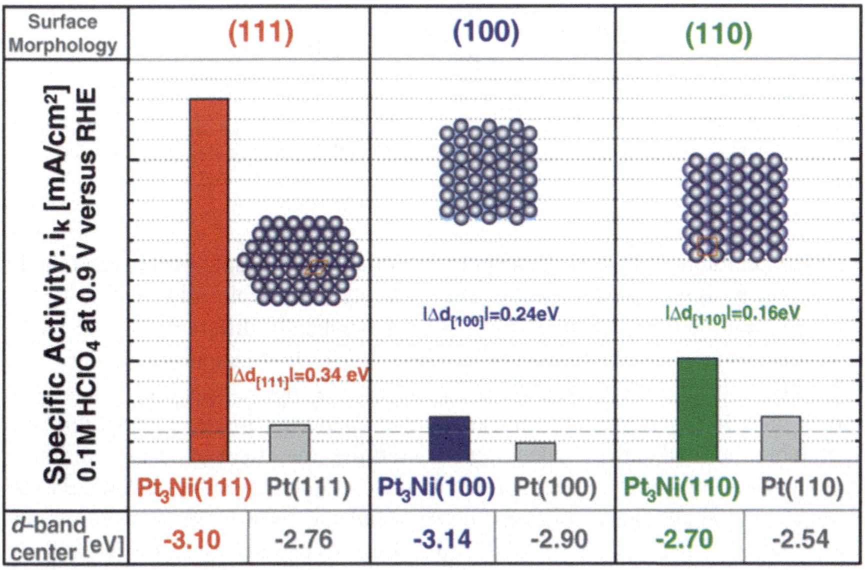

Fig. 6.5 Catalytic activity trends of single-crystal Pt and Pt_3Ni (h k l) surfaces for the ORR in correlation to the *d*-band center positions (Reprinted with permission from [18]. Copyright 207 AAAS)

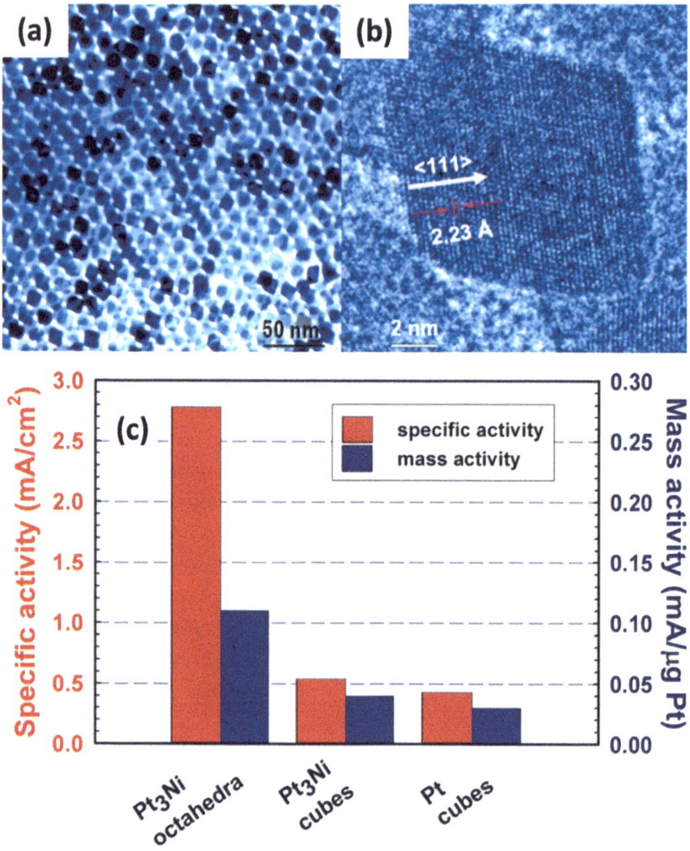

Fig. 6.6 (**a**) TEM and (**b**) HRTEM images of Pt_3Ni nanoctahedra; (**c**) Shape dependent electrocatalytic activities of Pt_3Ni nanocatalysts for the ORR (Reprinted with permission from [60]. Copyright 2010 ACS)

high-temperature (~1000 K) annealing in ultrahigh vacuum, which is unlikely present in as-synthesized Pt alloy nanocatalysts. The same alloy catalysts but with different surface structure could behave drastically differently in the ORR, as will be discussed in the following (see Sect. 6.3.4). In principle, annealing could bring about similar structuring processes in alloy nanocatalysts, but it has been shown that high-temperature annealing also induces particle aggregation and agglomeration, making it challenging to determine the effect of surface structure in alloy nanocatalysts [58]. Secondly, alloy nanoparticles, even those claimed to be in octahedral shape, have truncated corners and the surface comprises a certain portion of other facets such as (1 0 0) and (1 1 0), which have much lower catalytic activity than (1 1 1) surface of Pt alloys (see Fig. 6.5). An additional but not the last factor is the presence of more defects on alloy nanocatalysts than on single-crystal surfaces. Surfaces of nanoparticles could be enriched with steps, vacancies and

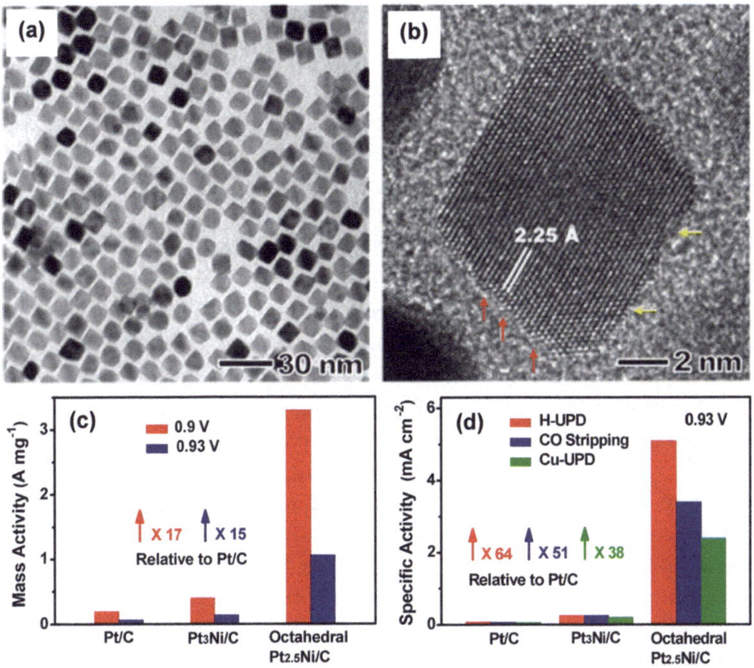

Fig. 6.7 (**a**) TEM and (**b**) HRTEM images of $Pt_{2.5}Ni$ octahedral nanoparticles; (**c, d**) Comparison of the ORR activities of $Pt_{2.5}Ni/C$ nanocatalysts with commercial cubo-octahedral Pt/C and Pt_3Ni/C catalysts (Reprinted with permission from [56]. Copyright 2013 ACS)

adatoms owing to the fast kinetics of nanocrystal nucleation and growth in typical wet-chemical methods (as labelled by arrows in Fig. 6.7b). Similar to the corner and edge sites discussed in the particle size effect, such defect sites have smaller coordination number, higher oxophilicity and lower ORR activity than their counterparts on ordered extended surfaces (which is more likely the case on single crystals).

Besides the intrinsic structure difference, the method of catalyst preparation, especially the way of cleaning and removing organic surfactants, could also affect the catalytic performance of nanomaterials derived from solution synthesis [59]. A recent study of octahedral Pt-Ni alloy nanocatalysts shows more than one order of magnitude of enhancement in mass activity compared to commercial Pt (Fig. 6.7) [56], with the more substantial enhancement than previous studies [54, 55] ascribed to the adoption of acetic acid washing for surface cleaning. Although surface defects are still visible in their HRTEM images, this approach is believed to produce clean octahedral particles with better preserved (1 1 1) facets than the other methods available in the literature [56].

6.3.3 Composition Effect and Dealloying

Despite their enhanced catalytic activity, concerns are present about the stability of Pt-alloy catalysts for implementation into PEMFCs. Owing to their low oxidation potentials, the 3*d* transition metals exposed on the surface of Pt-alloy catalysts dissolve almost instantaneously in acidic electrochemical environments, as demonstrated by low energy ion scattering (LEIS) studies on extended Pt-alloy surfaces (Fig. 6.8) [61]. It has also been found in Pt-alloy nanocatalysts that a Pt-rich shell is formed as a result of the depletion of non-precious metals in the near-surface regions (also called dealloying) [48, 62–67]. Systematic studies based on Pt_xNi_{1-x} ($0 < x < 1$) alloy NPs from organic solution synthesis show that the extent of depletion, or the thickness of the formed Pt-rich shell, is dependent on the initial alloy composition of the catalysts (Fig. 6.9) [48]. For the same particle size of ~5 nm, a maximum ratio of ~27 % was preserved for $Pt_{0.5}Ni_{0.5}$ among various compositions, for which the Pt-rich shell thickness found to be 0.7~1 nm, corresponding to about two atomic layers. Moreover, the dependence of ORR catalytic activity on the composition for the Pt_xNi_{1-x} catalysts was found to follow the same trend for the ratio of Ni left after surface depletion, with $Pt_{0.5}Ni_{0.5}$ gave the highest activity among the investigated series [48]. Other studies on the Pt-alloy catalysts of different compositions also revealed the dependence of ORR activity on the initial alloy composition, albeit the formation of different nanostructures after dealloying depending on the particle size and initial element distribution [67–70].

Strasser et al. studied the core/shell structures formed by dealloying of PtCu alloy catalysts with greater details [71–74]. It was argued that the formed Pt-rich shell is under compressive strain due to the smaller atomic size of Cu in comparison to Pt. The composition-dependent ORR catalytic activity was thus ascribed to the varying surface strain depending on the shell thickness, which is further determined by the initial alloy compositions (Fig. 6.10) [71].

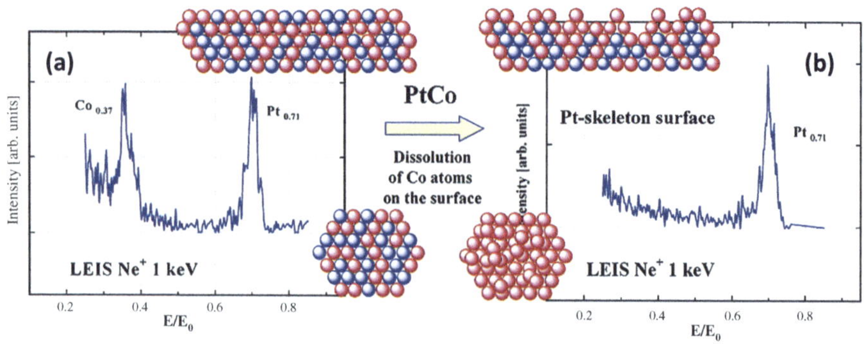

Fig. 6.8 Schematic and LEIS studies for the formation of Pt skeleton after exposure of Pt-alloy surfaces to the electrochemical environment (Reprinted with permission from [61]. Copyright 2006 ACS)

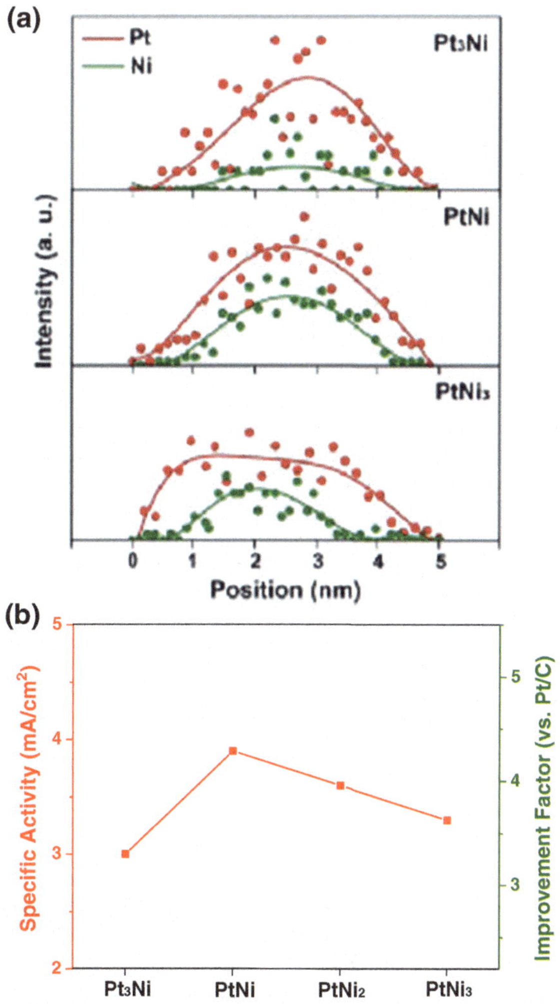

Fig. 6.9 (**a**) Composition line profiles for the Pt_xNi_{1-x}/C nanocatalysts after electrochemical depletion of surface Ni; (**b**) summary of the specific activities for the ORR at 0.9 V and improvement factors versus Pt/C of he Pt_xNi_{1-x}/C catalysts (Reprinted with permission from [48]. Copyright 2011 Wiley)

After surface depletion, the preserved 3*d* metal in the particle core may still not be stable under the running conditions of PEMFCs. Two mechanisms can lead to further degradation of Pt-bimetallic catalysts on the course of potential cycling, even after pretreatment by acid. Redox reactions induced by potential cycling can cause the dissolution of surface Pt atoms, exposing the non-precious atoms in the subsurface [68, 75]. The embedded 3*d* metal atoms in the subsurface can also diffuse to the surface and dissolve under the oxidative and acidic conditions in the ORR, driven by their higher oxophilicity than Pt. The latter effect was demonstrated by potential cycling of acid treated Pt_3Co catalysts in alkaline electrolytes, where the diffusion of 3*d* metal atoms onto surface was visualized by the appearance and growth of Co^{2+}/Co^{3+} redox peaks (Fig. 6.11) [36]. The observed fast

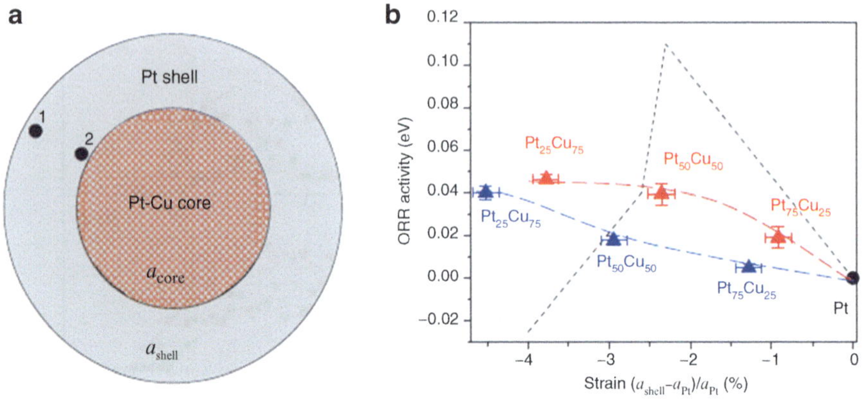

Fig. 6.10 (**a**) A simple structural two-phase core/shell model for the dealloyed PtCu NPs; (**b**) Experimentally observed (*solid triangles*) and predicted (*black dashed line*) relationships between ORR activity and lattice strain for the dealloyed PtCu catalysts Reprinted by permission from Macmillan Publishers Ltd. [Nature Chemistry] [71] Copyright 2010

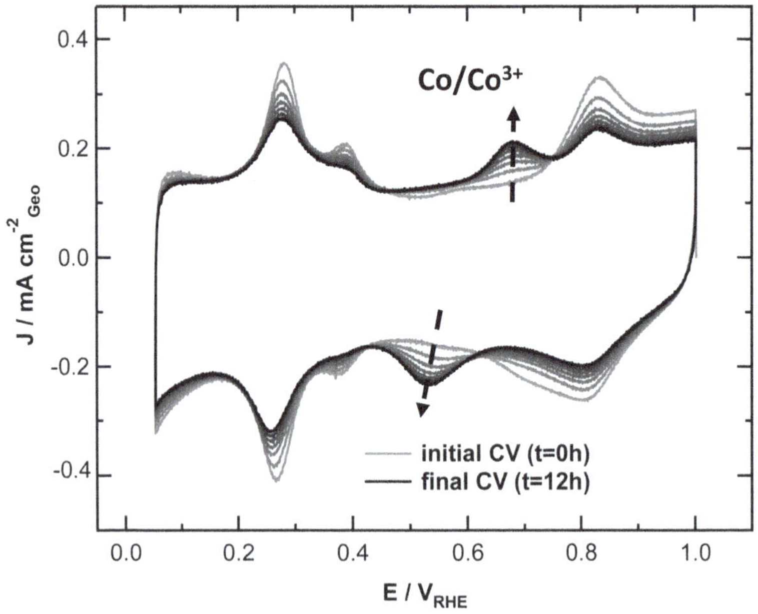

Fig. 6.11 CV series (50 mV/s scan rate at room temperature) recorded for an acid leached Pt_3Co catalyst in Ar saturated 0.1 M KOH electrolyte. The gradual segregation of Co atoms to the surface due to potential cycling was indicated by appearance of Co^{2+}/Co^{3+} peaks (labelled by *arrows*) (Reprinted with permission from [36]. Copyright 2009 ACS)

degradation of Pt-alloy catalysts under potential cycling is likely associated with the disordered surface structure of dealloyed catalysts. The Pt-skeleton shell formed after 3*d* metal leaching is usually enriched by defects, such as undercoordinated atoms and vacancies (Fig. 6.8) [68, 76]. The surface atoms with coordination numbers smaller than their counterparts on ordered surfaces are less stable and can be easier to dissolve [77]. The vacancies present in the Pt-skeleton shell could provide the pathway for diffusion of the 3*d* metal atoms from the core to the surface. Therefore, the Pt-skeleton surfaces formed by either acid pretreatment or *in situ* electrochemical leaching may not be able to protect the nonprecious metal atoms that remain in the particles.

6.3.4 Alloy Catalysts with Pt-Skin Surface

As discussed above (in particle size and shape effects), ORR electrocatalysis is structure-sensitive and the absence of a Pt-skin surface on octahedral nanocatalysts has impeded the achievement of the same level of catalytic enhancement as demonstrated on single-crystal extended surfaces. Meanwhile, from the prior section it is noticed that the continual loss of 3*d* metals causes the degradation of Pt-bimetallic catalysts towards Pt-like catalytic performance. Because of these undesirable factors, improvement of only 2–3 times in mass activity was observed in Pt ally nanocatalysts [63, 64, 68, 78–80], which is substantially lower than that expected based on fundamental studies of extended surfaces (Fig. 6.5). Moreover, the dissolved 3*d* metal ions from alloy catalysts were found to be detrimental to the cell performance of MEA, e.g., causing accumulation of protons at the cathode [1]. Therefore, it is of utmost importance to protect the non-precious component for the employment of Pt-alloy catalysts in PEMFCs.

While using large particles can generally preserve more non-precious metal contents than smaller ones, considering that a thicker Pt shell can sustain longer and also slow down the diffusion of 3*d* metals [67], it results in the sacrifice of specific surface area and reduces the gain in mass activity. A more desirable solution would be modification of the surface structure towards smooth and ordered Pt surfaces which can better protect the subsurface 3*d* metals [18]. Studies of extended surfaces of Pt alloys in ultrahigh vacuum (UHV) show that thermal annealing (~1000 K) induces the segregation of Pt to the topmost atomic layer, which is counterbalanced by the depletion of Pt in the next two or three subsurface layers. The formed pure Pt surface, termed *Pt skin* (Fig. 6.8) [61], exhibits distinctive electronic and adsorption properties, and much higher catalytic activity than pure Pt and Pt-skeleton surfaces. Moreover, the formed Pt-skin surface is found to be able to protect subsurface 3*d* metals from leaching out and preserve the beneficial surface and catalytic properties under electrochemical conditions [18, 32, 61].

Inspired by the findings on extended surfaces, efforts have been dedicated to modify the surface structures of Pt-alloy nanocatalysts towards Pt-skin surfaces [66, 76, 81–85]. For example, in order to induce surface segregation, thermal annealing was applied to commercial Pt_3Co/C nanocatalysts. The observation of a bright outermost atomic layer followed by a dark second layer was ascribed to the formation of Pt-skin surface induced by surface segregation (Fig. 6.12) [66]. The

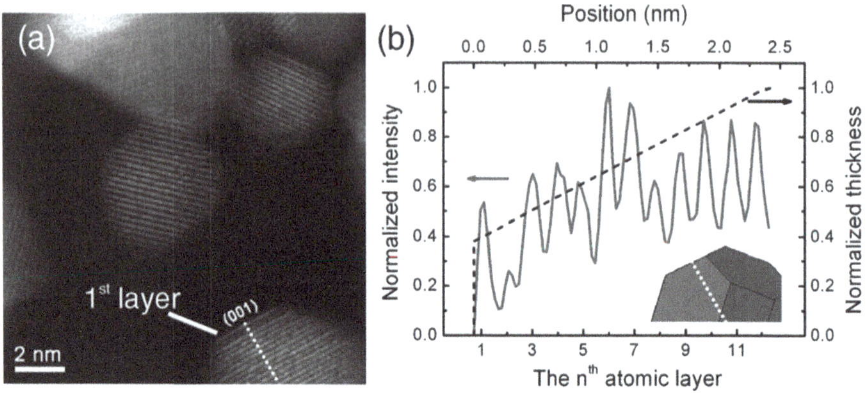

Fig. 6.12 Electron microscopic studies of thermally annealed Pt_3Co/C catalysts. (**a**) Cs-corrected HAADF-STEM image. (**b**) Normalized intensity (*solid curve*) of the particle in (**a**) along the *dotted line*, versus normalized thickness variations for an ideal truncated octahedron particle (Reprinted with permission from [66]. Copyright 2008 ACS)

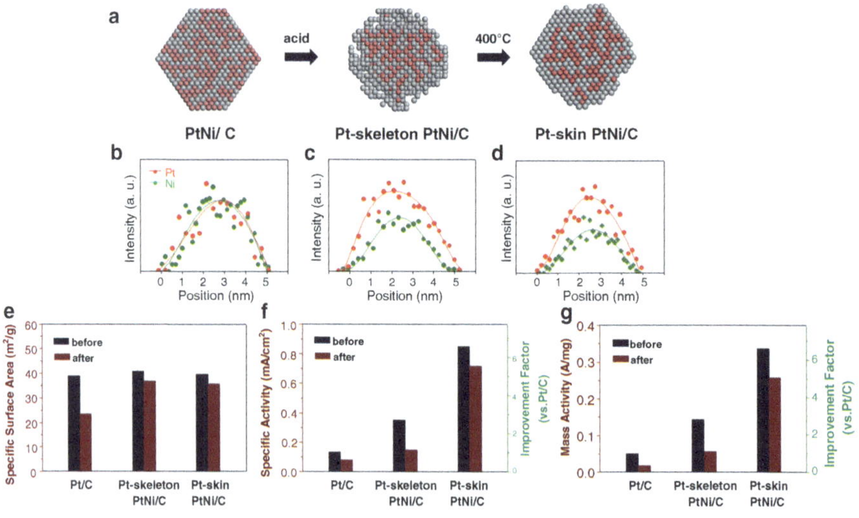

Fig. 6.13 (**a**) Schematic and (**b–d**) composition line profile analysis for the nanostructure evolution in the synthesis of PtNi/C catalyst with multilayered Pt-skin surface; Summary of electrochemical durability studies obtained by RDE before and after 4000 potential cycles between 0.6 and 1.1 V for the Pt/C and PtNi/C catalysts in 0.1M$HClO_4$ at 0.95 V and 60 °C: (**e**) specific surface area, (**f**) specific activity, and (**g**) mass activity (Reprinted with permission from [76]. Copyright 2011 ACS)

annealed catalyst also exhibited ~4 times enhancement in ORR catalytic activity, versus ~2 times for the acid treated catalyst.

A more promising approach towards the Pt-skin surface other than simple thermal annealing is a two-step process by successive acid and thermal treatments of Pt-alloy catalysts (Fig. 6.13a) [76]. Starting with homogeneous $Pt_{0.5}Ni_{0.5}$ NPs derived from organic solution synthesis, acid leaching was first applied to form a Pt-rich skeleton shell. The acid treated catalyst was then annealed at ~400 °C. This mild-temperature treatment may not be sufficient to induce complete surface segregation to form monolayer Pt-skin, but was found to be able to smooth the surface, reduce under coordinated defect sites, and form an ordered Pt skin without inducing particle size change. Composition line profile analysis shows that the formed Pt-skin structure has a thickness of ~0.5 nm, corresponding to about two atomic layers (Fig. 6.13b–d) [76]. This multilayered Pt-skin surface was further found to be capable of preserving the beneficial properties of subsurface Ni, which exhibited an enhanced ORR catalytic activity ~6 times greater than Pt/C particles with similar size (~5 nm). Moreover, the multilayered Pt-skin surface also shows improved durability than the Pt-skeleton surface (Fig. 6.13e–g), preserving more than 80 % of the specific activity after extensive potential cycling versus less than 50 % for the Pt-skeleton surface. The improved durability was further confirmed by the elemental analysis of the catalysts after electrochemical studies showing that more Ni was preserved by the Pt-skin than the Pt-skeleton surface.

6.4 Composite Nanostructures

Besides alloys of Pt and 3*d* transition metals, composite nanostructures, such as core/shell [86–92], heterodimer [93, 94], nanotube [95, 96], and nanoporous [70, 97], have also been explored for electrocatalytic applications. Some of these nanostructures are still based on bimetallic systems of Pt and 3*d* transition metals [77, 87, 88, 90], but with more sophisticated nanostructures. By going beyond simple particulated catalysts, these composite nanostructures provide additional dimensions for tuning the surface catalytic properties and have brought about additional catalytic improvement.

6.4.1 Pt-Based Core/Shell Nanostructures

Core/shell nanostructures are of particular interest for the development of ORR electrocatalysts considering the potential of reducing the cost of Pt electrocatalysts by replacing catalytically inaccessible inner Pt atoms by cheaper, less noble metals. For example, Pd/Pt core/shell nanocatalysts have been synthesized by galvanic displacement of under potentially deposited (UPD) Cu with Pt on Pd and other metal or alloy seeds, with the ORR catalytic activity dependent on the shell thickness (Fig. 6.14) [86, 98–102]. Similar to the case in core/shell nanostructures formed by dealloying [71], the Pt surface on the Pd/Pt core/shell particles was also believed to be under compressive strain due to the lattice mismatch between the core and shell materials, leading to the improved ORR catalytic activity [86].

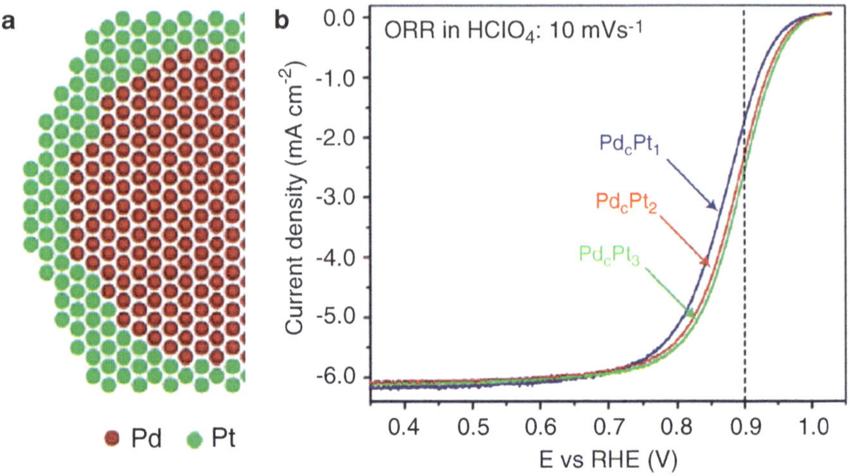

Fig. 6.14 (**a**) Projection of the structure model of Pd/Pt core/shell particles formed by galvanic displacement of UPD Cu; (**b**) ORR polarization curves measured for Pd/Pt core/shell catalysts with one- to three-monolayer of Pt shell (Reprinted with permission from [86]. Copyright 2009 ACS)

On the other side, with an electrochemically more stable noble gold metal as the core, multimetallic Au/$FePt_3$ core/shell NPs were found to exhibit high durability under potential cycling in the ORR-relevant potential regions (Fig. 6.15) [91]. While significant changes to the particle size and shape were observed for Pt/C and Pt-alloy ($FePt_3$/C) catalysts after extensive potential cycling, the changes to Au/$FePt_3$ were almost negligible. The stabilization of particle morphology and thus electrocatalytic durability by the Au core was further correlated to a counter balance between the segregation of Au to the surface and stronger binding of Pt–O vs. Au–O. It is worth pointing out that stabilization of Pt electrocatalysts by Au was also observed previously in Au clusters-modified Pt/C catalysts, where the enhanced durability was ascribed to the raised oxidation potential of Pt [93].

6.4.2 *Pt-Based Nanostructured Thin-Films (NSTFs)*

Pt-based NSTFs represents another group of electrocatalysts with unique composite nanostructures. Developed by 3M researchers, NSTF catalysts were synthesized by sputtering Pt-based metallic films over arrays of polymer nanowires [103, 104]. The concept of NSTF catalysts is drawn from the fact that the ORR activities on high-surface-area nanocatalysts are much lower than on their extended-surface counterparts, with typical differences of up to one order of magnitude [13]. While extended surfaces, even planar thin films, cannot be directly adopted as electrocatalysts for PEMFCs, NSTFs can combine the high surface area of nanocatalysts and high

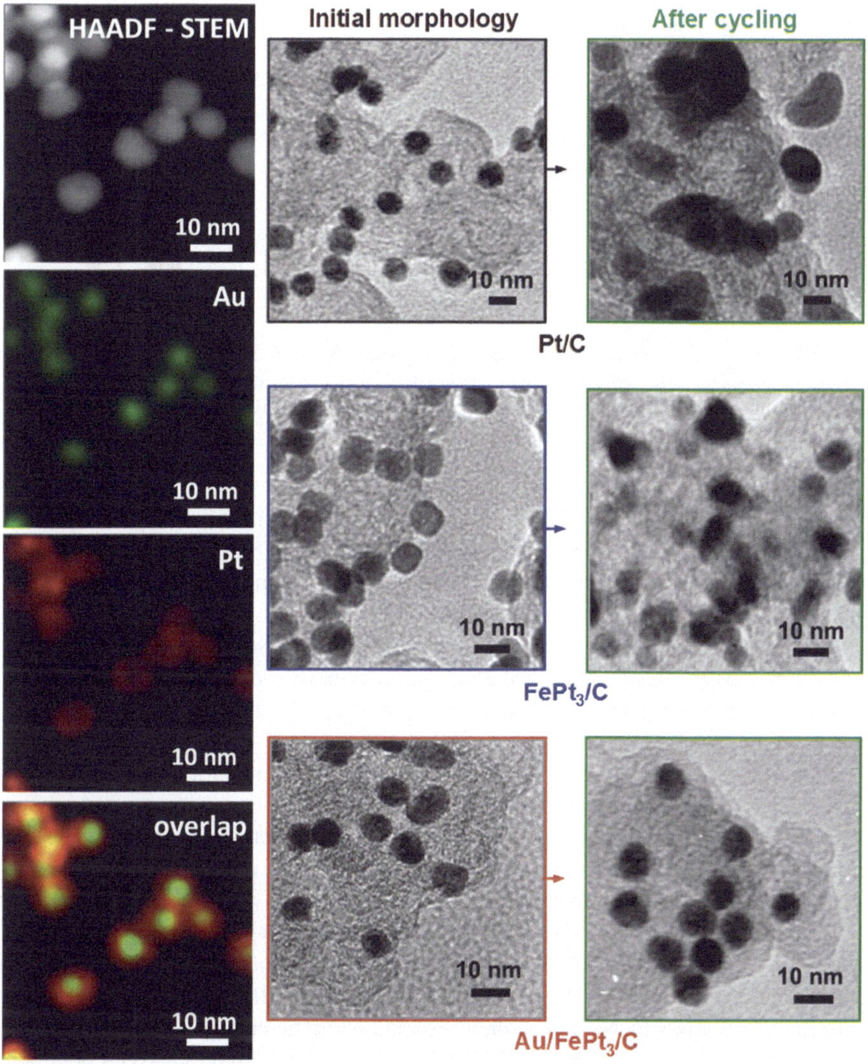

Fig. 6.15 (**a**) HAADF-STEM characterization and elemental mapping of Au (*green*), Pt (*red*) and their overlap; (**b**) TEM characterization of the catalysts before and after the potential cycling (Reprinted with permission from [91]. Copyright 2011 ACS)

catalytic activity of extended surfaces, achieving advanced catalytic performance that could be hard to achieve by regular nanoparticulated catalysts.

The as-synthesized NSTFs by sputtering, though already exhibit much high ORR activities than nanoparticulated Pt/C, still have rough surfaces which are full of defects (Fig. 6.16a). It was recently found that further annealing of these NSTFs can smoothen the surface, induce surface segregation and further improve

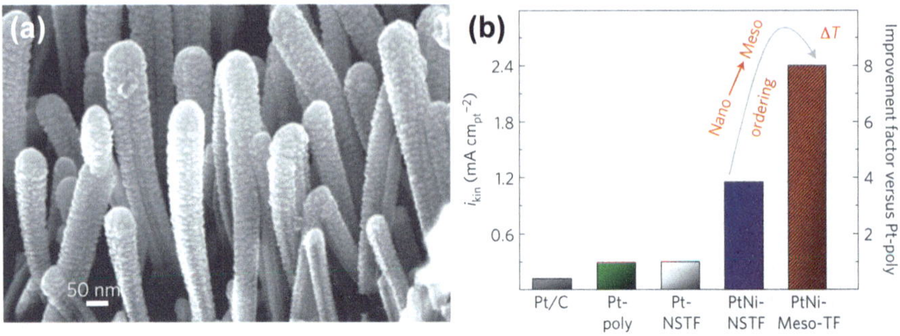

Fig. 6.16 (**a**) High-resolution SEM image of Pt-based NSTFs; (**b**) Specific activities of ORR measured by RDE in 0.1 M $HClO_4$ with 1600 rpm, 20 mV s^{-1} at 0.95 V with corresponding improvement factors versus polycrystalline Pt. Reprinted by permission from Macmillan Publishers Ltd: *Nature Materials* [96] Copyright 2012

the catalytic activity by another factor of two (Fig. 6.16b) [96]. The ORR specific activity obtained on the formed mesostructured NSTFs reaches 2.4 mA/cm^2 at 0.95 V vs. RHE, representing improvement factors of 8 and 20 versus polycrystalline Pt and commercial Pt/C catalysts [18].

Porous materials with nanoscale pore structures have wide applications in heterogeneous catalysis. Among the various nanoporous catalysts zeolite is probably the most important system which has been employed in petrochemical industry for catalytic cracking, isomerization and reforming [105–107]. Nanoporous gold (NPG) has also been studied as catalysts for selective oxidation [108–110] and hydrogenation [110–112] reactions. The advantages of nanoporous materials as catalysts do not only include the distinctive surface structures exposed inside the pores, but also the unique nanoscale confinement effects which can be tailored to improve catalytic activity and selectivity.

Erlebacher et al. developed the synthesis of nanoporous Pt_3Ni foams (Fig. 6.17a) [97] as well as NPs (Fig. 6.17b) [70] by controlled electrochemical dealloying. By impregnation of a hydrophobic and protic ionic liquid with high oxygen solubility into the nanoporous Pt_3Ni foams, they reported a kinetic current density as high as 18.2 mA/cm^2 at 0.9 V vs. RHE [97], which approaches the highest activity obtained on single-crystal extended surfaces [18]. More recently, Stamenkovic et al. applied this approach to 3D Pt_3Ni nanoframes derived from Ni-rich Pt-Ni alloy polyhedral nanoparticles and reported improvement factors of >20 in specific activity and >30 in mass activity versus commercial Pt/C (Fig. 6.18) [113]. Such improvement factors are already in line with the previous results obtained on extended surfaces (Fig. 6.5). Although the catalytic enhancement also arises from the boosted oxygen mass transport by using the ionic liquid, it is believed that the edges in the nanoframes possess highly crystalline surfaces that are good mimic of ordered facets on extended surfaces, likely due to the unique 1D morphology in the hierarchical nanoscale architecture.

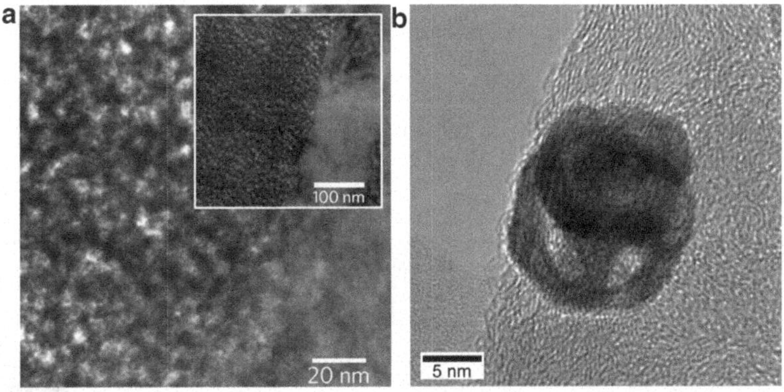

Fig. 6.17 (**a**) Cross-sectional TEM image of a nanoporous-Pt_3Ni foam. The inset shows a lower magnification view of a sharp interface between the porous dealloyed region (*left*) and undealloyed metal (*right*). Reprinted by permission from Macmillan Publishers Ltd: Nature Materials [97] copyright 2010; (**b**) HRTEM image of a 15 nm nanoporous Pt_3Ni NP (Reprinted with permission from [70]. Copyright 2012 ACS)

6.5 Summary and Future Directions

Over the past two decades significant progress has been made in the development of Pt-based nanomaterials for electrocatalytic applications in fuel cells. We briefly reviewed the recent development of Pt-based ORR electrocatalysts in terms of single-component Pt, bimetallic alloys and composite nanostructures. While these directions may only be a few of the many towards advanced electrocatalysts, we believe they represent the major efforts in the literature. Further improvements along these directions could potentially achieve the performance required for the commercial implementation of PEMFCs in stationary and transportation applications [1, 3].

Despite the progress that has been made, challenges still exist in Pt-based ORR electrocatalysts. These include: (*i*) while the improvement in catalytic activity has reached or been above the PEMFC's target [3], the catalyst's stability has to be further improved for practical applications, namely much prolonged tests under PEMFC conditions; (*ii*) the developed various catalysts need to be incorporated into the membrane electrode assemblies (MEAs) for fuel cells with structural and property integrity while maintaining their catalytic performance that are measured by rotating disk electrodes (RDEs); (*iii*) scale-up production of the catalysts with novel nanostructures at gram, or even kilogram scales. To address these challenges it would require fine balances between catalytic activity and stability, between sophisticated nanostructures and simple processes easy for scaling up, as well as between improvement in catalytic performance and cost effectiveness. In this sense, interdisciplinary efforts combining insights from physics, chemistry, materials science and chemical engineering will be particularly important.

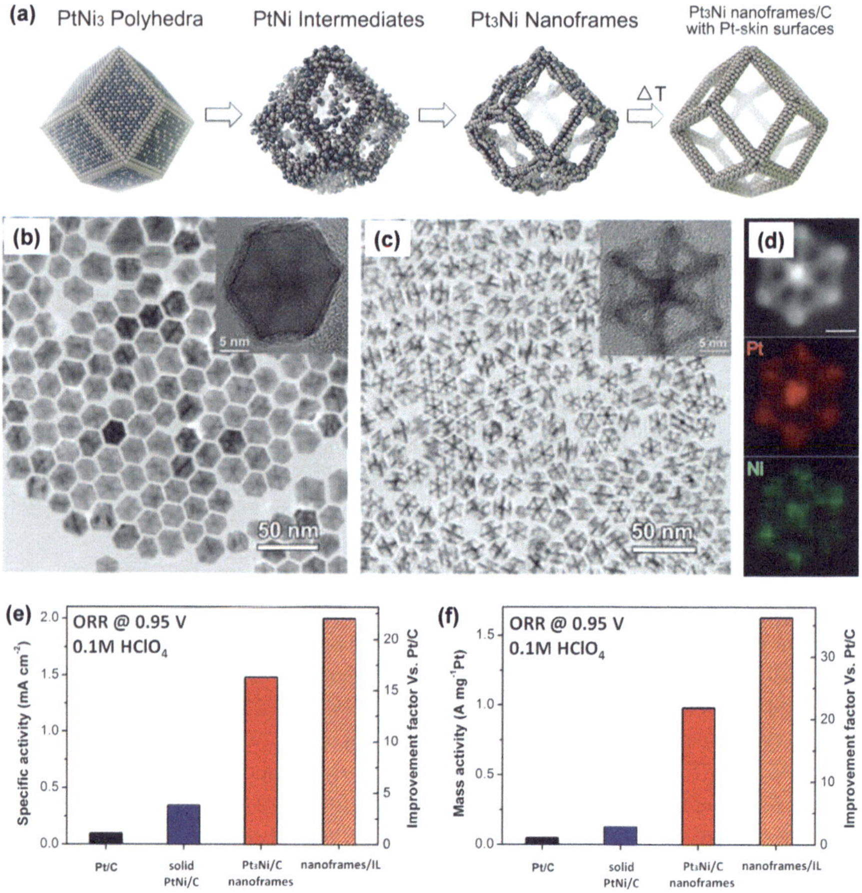

Fig. 6.18 (**a**) Synthetic scheme for the Pt_3Ni nanoframe/C catalysts; (**b**, **c**) TEM analyses of the seeding PtNi3 polyhedral NPs (**b**) and Pt3Ni nanoframes produced by aging in air (**c**); (**d**) element mapping for the generated Pt_3Ni nanoframes; (**e**, **f**) summary of the ORR catalytic performance for the Pt_3Ni nanoframe and nanoframe + ionic liquid composite catalysts (Reprinted with permission from [113]. Copyright 2014 AAAS)

Acknowledgements The authors at JHU thank the start-up support from the Whiting School of Engineering, Johns Hopkins University and funding support from NSF/DMR.

References

1. Vielstich W, Lamm A, Gasteiger HA (2003) Handbook of fuel cells: fundamentals, technology, and applications. Wiley, Hoboken
2. Richter B, Goldston D, Crabtree G, Glicksman L, Goldstein D, Greene D, Kammen D, Levine M, Lubell M, Savitz M, Sperling D, Schlachter F, Scofield J, Dawson J (2008) How

America can look within to achieve energy security and reduce global warming. Rev Mod Phys 80(4):S1–S107. doi:10.1103/RevModPhys.80.S1
3. Gasteiger HA, Kocha SS, Sompalli B, Wagner FT (2005) Activity benchmarks and requirements for Pt, Pt-alloy, and non-Pt oxygen reduction catalysts for PEMFCs. Appl Catal B Environ 56(1–2):9–35
4. Kinoshita K (1990) Particle-size effects for oxygen reduction on highly dispersed platinum in acid electrolytes. J Electrochem Soc 137(3):845–848
5. Valden M, Lai X, Goodman DW (1998) Onset of catalytic activity of gold clusters on titania with the appearance of nonmetallic properties. Science 281(5383):1647–1650
6. Arenz M, Mayrhofer KJJ, Stamenkovic V, Blizanac BB, Tomoyuki T, Ross PN, Markovic NM (2005) The effect of the particle size on the kinetics of CO electrooxidation on high surface area Pt catalysts. J Am Chem Soc 127(18):6819–6829
7. Bezemer GL, Bitter JH, Kuipers HPCE, Oosterbeek H, Holewijn JE, Xu XD, Kapteijn F, van Dillen AJ, de Jong KP (2006) Cobalt particle size effects in the Fischer-Tropsch reaction studied with carbon nanofiber supported catalysts. J Am Chem Soc 128(12):3956–3964. doi:10.1021/Ja058282w
8. Hvolbaek B, Janssens TVW, Clausen BS, Falsig H, Christensen CH, Norskov JK (2007) Catalytic activity of Au nanoparticles. Nano Today 2(4):14–18. doi:10.1016/S1748-0132(07)70113-5
9. Lei Y, Mehmood F, Lee S, Greeley J, Lee B, Seifert S, Winans RE, Elam JW, Meyer RJ, Redfern PC, Teschner D, Schlogl R, Pellin MJ, Curtiss LA, Vajda S (2010) Increased silver activity for direct propylene epoxidation via subnanometer size effects. Science 328 (5975):224–228. doi:10.1126/science.1185200
10. Shao MH, Peles A, Shoemaker K (2011) Electrocatalysis on platinum nanoparticles: particle size effect on oxygen reduction reaction activity. Nano Lett 11(9):3714–3719. doi:10.1021/Nl2017459
11. Yano H, Inukai J, Uchida H, Watanabe M, Babu PK, Kobayashi T, Chung JH, Oldfield E, Wieckowski A (2006) Particle-size effect of nanoscale platinum catalysts in oxygen reduction reaction: an electrochemical and Pt-195 EC-NMR study. PCCP 8(42):4932–4939. doi:10.1039/B610573d
12. Yamamoto K, Imaoka T, Chun WJ, Enoki O, Katoh H, Takenaga M, Sonoi A (2009) Size-specific catalytic activity of platinum clusters enhances oxygen reduction reactions. Nat Chem 1(5):397–402. doi:10.1038/Nchem.288
13. Nesselberger M, Ashton S, Meier JC, Katsounaros I, Mayrhofer KJJ, Arenz M (2011) The particle size effect on the oxygen reduction reaction activity of pt catalysts: influence of electrolyte and relation to single crystal models. J Am Chem Soc 133(43):17428–17433. doi:10.1021/Ja207016u
14. Sheng WC, Chen S, Vescovo E, Shao-Horn Y (2012) Size influence on the oxygen reduction reaction activity and instability of supported Pt nanoparticles. J Electrochem Soc 159(2): B96–B103. doi:10.1149/2.009202jes
15. Shao-Horn Y, Sheng WC, Chen S, Ferreira PJ, Holby EF, Morgan D (2007) Instability of supported platinum nanoparticles in low-temperature fuel cells. Top Catal 46(3–4):285–305. doi:10.1007/s11244-007-9000-0
16. Tang L, Han B, Persson K, Friesen C, He T, Sieradzki K, Ceder G (2010) Electrochemical stability of nanometer-scale Pt particles in acidic environments. J Am Chem Soc 132 (2):596–600
17. Markovic NM, Ross PN (2002) Surface science studies of model fuel cell electrocatalysts. Surf Sci Rep 45(4–6):121–229
18. Stamenkovic VR, Fowler B, Mun BS, Wang GF, Ross PN, Lucas CA, Markovic NM (2007) Improved oxygen reduction activity on Pt3Ni(111) via increased surface site availability. Science 315(5811):493–497

19. Wang C, Daimon H, Onodera T, Koda T, Sun SH (2008) A general approach to the size- and shape-controlled synthesis of platinum nanoparticles and their catalytic reduction of oxygen. Angew Chem Int Ed 47(19):3588–3591. doi:10.1002/anie.200800073
20. Chen JY, Lim B, Lee EP, Xia YN (2009) Shape-controlled synthesis of platinum nanocrystals for catalytic and electrocatalytic applications. Nano Today 4(1):81–95. doi:10.1016/j.nantod.2008.09.002
21. Sanchez-Sanchez CM, Solla-Gullon J, Vidal-Iglesias FJ, Aldaz A, Montiel V, Herrero E (2010) Imaging structure sensitive catalysis on different shape-controlled platinum nanoparticles. J Am Chem Soc 132(16):5622–5624. doi:10.1021/Ja100922h
22. Ahmadi TS, Wang ZL, Green TC, Henglein A, ElSayed MA (1996) Shape-controlled synthesis of colloidal platinum nanoparticles. Science 272(5270):1924–1926
23. Song H, Kim F, Connor S, Somorjai GA, Yang PD (2005) Pt nanocrystals: shape control and Langmuir-Blodgett monolayer formation. J Phys Chem B 109(1):188–193. doi:10.1021/Jp0464775
24. Ren JT, Tilley RD (2007) Preparation, self-assembly, and mechanistic study of highly monodispersed nanocubes. J Am Chem Soc 129(11):3287–3291. doi:10.1021/Ja067636w
25. Wang C, Daimon H, Lee Y, Kim J, Sun S (2007) Synthesis of monodisperse Pt nanocubes and their enhanced catalysis for oxygen reduction. J Am Chem Soc 129(22):6974–6975
26. Sun SH, Jaouen F, Dodelet JP (2008) Controlled growth of Pt nanowires on carbon nanospheres and their enhanced performance as electrocatalysts in PEM fuel cells. Adv Mater 20(20):3900–3904. doi:10.1002/adma.200800491
27. Alia SM, Zhang G, Kisailus D, Li DS, Gu S, Jensen K, Yan YS (2010) Porous platinum nanotubes for oxygen reduction and methanol oxidation reactions. Adv Funct Mater 20 (21):3742–3746. doi:10.1002/adfm.201001035
28. Wang JX, Ma C, Choi YM, Su D, Zhu YM, Liu P, Si R, Vukmirovic MB, Zhang Y, Adzic RR (2011) Kirkendall effect and lattice contraction in nanocatalysts: a new strategy to enhance sustainable activity. J Am Chem Soc 133(34):13551–13557. doi:10.1021/Ja204518x
29. Kibsgaard J, Gorlin Y, Chen ZB, Jaramillo TF (2012) Meso-Structured Platinum Thin Films: Active and Stable Electrocatalysts for the Oxygen Reduction Reaction. J Am Chem Soc 134 (18):7758-7765. doi:Doi 10.1021/Ja2120162
30. Toda T, Igarashi H, Uchida H, Watanabe M (1999) Enhancement of the electroreduction of oxygen on Pt alloys with Fe, Ni, and Co. J Electrochem Soc 146(10):3750–3756
31. Stamenkovic V, Mun BS, Mayrhofer KJJ, Ross PN, Markovic NM, Rossmeisl J, Greeley J, Norskov JK (2006) Changing the activity of electrocatalysts for oxygen reduction by tuning the surface electronic structure. Angew Chem Int Ed 45(18):2897–2901
32. Stamenkovic VR, Mun BS, Arenz M, Mayrhofer KJJ, Lucas CA, Wang GF, Ross PN, Markovic NM (2007) Trends in electrocatalysis on extended and nanoscale Pt-bimetallic alloy surfaces. Nat Mater 6(3):241–247
33. Greeley J, Stephens IEL, Bondarenko AS, Johansson TP, Hansen HA, Jaramillo TF, Rossmeisl J, Chorkendorff I, Norskov JK (2009) Alloys of platinum and early transition metals as oxygen reduction electrocatalysts. Nat Chem 1(7):552–556. doi:10.1038/Nchem.367
34. Peng ZM, Yang H (2009) Designer platinum nanoparticles: control of shape, composition in alloy, nanostructure and electrocatalytic property. Nano Today 4(2):143–164. doi:10.1016/j.nantod.2008.10.010
35. Wang C, Markovic NM, Stamenkovic VR (2012) Advanced platinum alloy electrocatalysts for the oxygen reduction reaction. ACS Catal 2(5):891–898
36. Mayrhofer KJJ, Hartl K, Juhart V, Arenz M (2009) Degradation of carbon-supported Pt bimetallic nanoparticles by surface segregation. J Am Chem Soc 131(45):16348–16349
37. Ponec V, Bond GC (1995) Catalysis by metals and alloys, vol 95, Studies in surface science and catalysis. Elsevier, Amsterdam
38. Ertl G, Knözinger H, Weitkamp J (1997) Handbook of heterogeneous catalysis. Weinheim, VCH

39. Murray CB, Kagan CR, Bawendi MG (2000) Synthesis and characterization of monodisperse nanocrystals and close-packed nanocrystal assemblies. Annu Rev Mater Sci 30:545–610
40. Cushing BL, Kolesnichenko VL, O'Connor CJ (2004) Recent advances in the liquid-phase syntheses of inorganic nanoparticles. Chem Rev 104(9):3893–3946
41. Ferrando R, Jellinek J, Johnston RL (2008) Nanoalloys: from theory to applications of alloy clusters and nanoparticles. Chem Rev 108(3):845–910
42. Sun SH, Murray CB, Weller D, Folks L, Moser A (2000) Monodisperse FePt nanoparticles and ferromagnetic FePt nanocrystal superlattices. Science 287(5460):1989–1992
43. Shevchenko EV, Talapin DV, Rogach AL, Kornowski A, Haase M, Weller H (2002) Colloidal synthesis and self-assembly of COPt3 nanocrystals. J Am Chem Soc 124 (38):11480–11485
44. Wang C, Chi MF, Li DG, van der Vliet D, Wang GF, Lin QY, Mitchell JF, More KL, Markovic NM, Stamenkovic VR (2011) Synthesis of homogeneous Pt-bimetallic nanoparticles as highly efficient electrocatalysts. ACS Catal 1(10):1355–1359. doi:10.1021/Cs200328z
45. Wang C, van der Vilet D, Chang KC, You HD, Strmcnik D, Schlueter JA, Markovic NM, Stamenkovic VR (2009) Monodisperse Pt(3)Co nanoparticles as a catalyst for the oxygen reduction reaction: size-dependent activity. J Phys Chem C 113(45):19365–19368
46. Ahrenstorf K, Albrecht O, Heller H, Kornowski A, Gorlitz D, Weller H (2007) Colloidal synthesis of Ni(x)Pt(1-x) nanoparticles with tuneable composition and size. Small 3 (2):271–274
47. Ahrenstorf K, Heller H, Kornowski A, Broekaert JAC, Weller H (2008) Nucleation and growth mechanism of Ni(x)Pt(1-x) nanoparticles. Adv Funct Mater 18(23):3850–3856
48. Wang C, Chi MF, Wang GF, van der Vliet D, Li DG, More K, Wang HH, Schlueter JA, Markovic NM, Stamenkovic VR (2011) Correlation between surface chemistry and electrocatalytic properties of monodisperse Pt(x)Ni(1-x) nanoparticles. Adv Funct Mater 21 (1):147–152
49. Liu ZF, Shamsuzzoha M, Ada ET, Reichert WM, Nikles DE (2007) Synthesis and activation of Pt nanoparticles with controlled size for fuel cell electrocatalysts. J Power Sources 164 (2):472–480
50. Liu ZF, Ada ET, Shamsuzzoha M, Thompson GB, Nikles DE (2006) Synthesis and activation of PtRu alloyed nanoparticles with controlled size and composition. Chem Mater 18 (20):4946–4951
51. Lee YH, Lee G, Shim JH, Hwang S, Kwak J, Lee K, Song H, Park JT (2006) Monodisperse PtRu nanoalloy on carbon as a high-performance DMFC catalyst. Chem Mater 18 (18):4209–4211
52. Chen W, Kim JM, Sun SH, Chen SW (2008) Electrocatalytic reduction of oxygen by FePt alloy nanoparticles. J Phys Chem C 112(10):3891–3898. doi:10.1021/Jp7110204
53. Mayrhofer KJJ, Blizanac BB, Arenz M, Stamenkovic VR, Ross PN, Markovic NM (2005) The impact of geometric and surface electronic properties of Pt-catalysts on the particle size effect in electocatalysis. J Phys Chem B 109(30):14433–14440
54. Zhang J, Yang HZ, Fang JY, Zou SZ (2010) Synthesis and oxygen reduction activity of shape-controlled Pt(3)Ni nanopolyhedra. Nano Lett 10(2):638–644
55. Wu JB, Gross A, Yang H (2011) Shape and composition-controlled platinum alloy nanocrystals using carbon monoxide as reducing agent. Nano Lett 11(2):798–802
56. Choi SI, Xie SF, Shao MH, Odell JH, Lu N, Peng HC, Protsailo L, Guerrero S, Park JH, Xia XH, Wang JG, Kim MJ, Xia YN (2013) Synthesis and characterization of 9 nm Pt-Ni octahedra with a record high activity of 3.3 A/mg(Pt) for the oxygen reduction reaction. Nano Lett 13(7):3420–3425
57. Wu JB, Qi L, You HJ, Gross A, Li J, Yang H (2012) Icosahedral platinum alloy nanocrystals with enhanced electrocatalytic activities. J Am Chem Soc 134(29):11880–11883. doi:10.1021/Ja303950v

58. Wang C, Wang GF, van der Vliet D, Chang KC, Markovic NM, Stamenkovic VR (2010) Monodisperse Pt3Co nanoparticles as electrocatalyst: the effects of particle size and pretreatment on electrocatalytic reduction of oxygen. PCCP 12(26):6933–6939
59. Li DG, Wang C, Tripkovic D, Sun SH, Markovic NM, Stamenkovic VR (2012) Surfactant removal for colloidal nanoparticles from solution synthesis: the effect on catalytic performance. ACS Catal 2(7):1358–1362
60. Zhang J, Yang HZ, Fang JY, Zou SZ (2010) Synthesis and oxygen reduction activity of shape-controlled Pt3Ni nanopolyhedra. Nano Lett 10(2):638–644. doi:10.1021/Nl903717z
61. Stamenkovic VR, Mun BS, Mayrhofer KJJ, Ross PN, Markovic NM (2006) Effect of surface composition on electronic structure, stability, and electrocatalytic properties of Pt-transition metal alloys: Pt-skin versus Pt-skeleton surfaces. J Am Chem Soc 128(27):8813–8819
62. Watanabe M, Tsurumi K, Mizukami T, Nakamura T, Stonehart P (1994) Activity and stability of ordered and disordered Co-Pt alloys for phosphoric-acid fuel-cells. J Electrochem Soc 141(10):2659–2668
63. Ball S, Hudson S, Theobald B, Thompsett D (2006) Enhanced stability of PtCo catalysts for PEMFC. ECS Trans 1(8):141–152
64. Ball S, Hudson S, Theobald B, Thompsett D (2007) PtCo, a durable catalyst for automotive proton electrolyte membrane fuel cells? ECS Trans 11(1):1267–1278
65. Mani P, Srivastava R, Strasser P (2008) Dealloyed Pt-Cu core-shell nanoparticle electrocatalysts for use in PEM fuel cell cathodes. J Phys Chem C 112(7):2770–2778
66. Chen S, Ferreira PJ, Sheng WC, Yabuuchi N, Allard LF, Shao-Horn Y (2008) Enhanced activity for oxygen reduction reaction on "Pt(3)CO" nanoparticles: direct evidence of percolated and sandwich-segregation structures. J Am Chem Soc 130(42):13818–13819
67. Chen S, Sheng WC, Yabuuchi N, Ferreira PJ, Allard LF, Shao-Horn Y (2009) Origin of oxygen reduction reaction activity on "Pt(3)Co" nanoparticles: atomically resolved chemical compositions and structures. J Phys Chem C 113(3):1109–1125
68. Chen S, Gasteiger HA, Hayakawa K, Tada T, Shao-Horn Y (2010) Platinum-alloy cathode catalyst degradation in proton exchange membrane fuel cells: nanometer-scale compositional and morphological changes. J Electrochem Soc 157(1):A82–A97
69. Gan L, Heggen M, Rudi S, Strasser P (2012) Core-shell compositional fine structures of dealloyed PtxNi1-x nanoparticles and their impact on oxygen reduction catalysis. Nano Lett 12(10):5423–5430. doi:10.1021/Nl302995z
70. Snyder J, McCue I, Livi K, Erlebacher J (2012) Structure/processing/properties relationships in nanoporous nanoparticles as applied to catalysis of the cathodic oxygen reduction reaction. J Am Chem Soc 134(20):8633–8645. doi:10.1021/Ja3019498
71. Strasser P, Koh S, Anniyev T, Greeley J, More K, Yu CF, Liu ZC, Kaya S, Nordlund D, Ogasawara H, Toney MF, Nilsson A (2010) Lattice-strain control of the activity in dealloyed core-shell fuel cell catalysts. Nat Chem 2(6):454–460. doi:10.1038/Nchem.623
72. Koh S, Strasser P (2007) Electrocatalysis on bimetallic surfaces: modifying catalytic reactivity for oxygen reduction by voltammetric surface dealloying. J Am Chem Soc 129 (42):12624–12625
73. Strasser P, Koha S, Greeley J (2008) Voltammetric surface dealloying of Pt bimetallic nanoparticles: an experimental and DFT computational analysis. PCCP 10(25):3670–3683. doi:10.1039/B803717e
74. Cui CH, Gan L, Heggen M, Rudi S, Strasser P (2013) Compositional segregation in shaped Pt alloy nanoparticles and their structural behaviour during electrocatalysis. Nat Mater 12 (8):765–771
75. Ferreira PJ, La O' GJ, Shao-Horn Y, Morgan D, Makharia R, Kocha S, Gasteiger HA (2005) Instability of Pt/C electrocatalysts in proton exchange membrane fuel cells—a mechanistic investigation. J Electrochem Soc 152(11):A2256–A2271
76. Wang C, Chi MF, Li DG, Strmcnik D, van der Vliett D, Wang GF, Komanicky V, Chang KC, Paulikas AP, Tripkovic D, Pearson J, More KL, Markovic NM, Stamenkovic VR (2011)

Design and synthesis of bimetallic electrocatalyst with multilayered Pt-skin surfaces. J Am Chem Soc 133(36):14396–14403
77. Wang C, van der Vliet D, More KL, Zaluzec NJ, Peng S, Sun SH, Daimon H, Wang GF, Greeley J, Pearson J, Paulikas AP, Karapetrov G, Strmcnik D, Markovic NM, Stamenkovic VR (2011) Multimetallic Au/FePt(3) nanoparticles as highly durable electrocatalyst. Nano Lett 11(3):919–926
78. Wagner FT, Gasteiger HA, Makharia R, Neyerlin KC, Thompson EL, Yan SG (2006) ECS Trans 3(1):19
79. Yu P, Pemberton M, Plasse P (2005) PtCo/C cathode catalyst for improved durability in PEMFCs. J Power Sources 144(1):11–20
80. Neyerlin KC, Srivastava R, Yu CF, Strasser P (2009) Electrochemical activity and stability of dealloyed Pt-Cu and Pt-Cu-Co electrocatalysts for the oxygen reduction reaction (ORR). J Power Sources 186(2):261–267. doi:10.1016/j.jpowsour.2008.10.062
81. Hwang BJ, Kumar SMS, Chen CH, Monalisa CMY, Liu DG, Lee JF (2007) An investigation of structure-catalytic activity relationship for Pt-Co/C bimetallic nanoparticles toward the oxygen reduction reaction. J Phys Chem C 111(42):15267–15276
82. Yano H, Kataoka M, Yamashita H, Uchida H, Watanabe M (2007) Oxygen reduction activity of carbon-supported Pt-M (M = V, Ni, Cr, Co, and Fe) alloys prepared by nanocapsule method. Langmuir 23(11):6438–6445
83. Malheiro AR, Perez J, Santiago EI, Villullas HM (2010) The extent on the nanoscale of Pt-skin effects on oxygen reduction and its influence on fuel cell power. J Phys Chem C 114 (47):20267–20271
84. Ingram DB, Christopher P, Bauer JL, Linic S (2011) Predictive model for the design of plasmonic metal/semiconductor composite photocatalysts. ACS Catal 1(10):1441–1447
85. Mu RT, Fu QA, Xu H, Zhang HI, Huang YY, Jiang Z, Zhang SO, Tan DL, Bao XH (2011) Synergetic effect of surface and subsurface Ni species at Pt-Ni bimetallic catalysts for CO oxidation. J Am Chem Soc 133(6):1978–1986
86. Wang JX, Inada H, Wu LJ, Zhu YM, Choi YM, Liu P, Zhou WP, Adzic RR (2009) Oxygen reduction on well-defined core-shell nanocatalysts: particle size, facet, and Pt shell thickness effects. J Am Chem Soc 131(47):17298–17302. doi:10.1021/Ja9067645
87. Mazumder V, Chi MF, More KL, Sun SH (2010) Core/shell Pd/FePt nanoparticles as an active and durable catalyst for the oxygen reduction reaction. J Am Chem Soc 132 (23):7848–7849. doi:10.1021/Ja1024436
88. Chen YM, Liang ZX, Yang F, Liu YW, Chen SL (2011) Ni-Pt core-shell nanoparticles as oxygen reduction electrocatalysts: effect of Pt shell coverage. J Phys Chem C 115 (49):24073–24079. doi:10.1021/Jp207828n
89. Kuttiyiel KA, Sasaki K, Choi YM, Su D, Liu P, Adzic RR (2012) Nitride stabilized PtNi core-shell nanocatalyst for high oxygen reduction activity. Nano Lett 12(12):6266–6271. doi:10.1021/Nl303362s
90. Beard KD, Borrelli D, Cramer AM, Blom D, Van Zee JW, Monnier JR (2009) Preparation and structural analysis of carbon-supported Co core/Pt shell electrocatalysts using electroless deposition methods. ACS Nano 3(9):2841–2853. doi:10.1021/Nn900214g
91. Wang C, van der Vliet D, More KL, Zaluzec NJ, Peng S, Sun SH, Daimon H, Wang GF, Greeley J, Pearson J, Paulikas AP, Karapetrov G, Strmcnik D, Markovic NM, Stamenkovic VR (2011) Multimetallic Au/$FePt_3$ nanoparticles as highly durable electrocatalyst. Nano Lett 11(3):919–926. doi:10.1021/Nl102369k
92. Yang H (2011) Platinum-based electrocatalysts with core-shell nanostructures. Angew Chem Int Ed 50(12):2674–2676. doi:10.1002/anie.201005868
93. Zhang J, Sasaki K, Sutter E, Adzic RR (2007) Stabilization of platinum oxygen-reduction electrocatalysts using gold clusters. Science 315(5809):220–222
94. Wang C, Daimon H, Sun SH (2009) Dumbbell-like Pt-Fe_3O_4 nanoparticles and their enhanced catalysis for oxygen reduction reaction. Nano Lett 9(4):1493–1496. doi:10.1021/Nl8034724

95. Alia SM, Jensen KO, Pivovar BS, Yan YS (2012) platinum-coated palladium nanotubes as oxygen reduction reaction electrocatalysts. ACS Catal 2(5):858–863. doi:10.1021/Cs200682c
96. van der Vliet DF, Wang C, Tripkovic D, Strmcnik D, Zhang XF, Debe MK, Atanasoski RT, Markovic NM, Stamenkovic VR (2012) Mesostructured thin films as electrocatalysts with tunable composition and surface morphology. Nat Mater 11(12):1051–1058
97. Snyder J, Fujita T, Chen MW, Erlebacher J (2010) Oxygen reduction in nanoporous metal-ionic liquid composite electrocatalysts. Nat Mater 9(11):904–907. doi:10.1038/Nmat2878
98. Zhang J, Mo Y, Vukmirovic MB, Klie R, Sasaki K, Adzic RR (2004) Platinum monolayer electrocatalysts for O-2 reduction: Pt monolayer on Pd(111) and on carbon-supported Pd nanoparticles. J Phys Chem B 108(30):10955–10964
99. Sasaki K, Naohara H, Cai Y, Choi YM, Liu P, Vukmirovic MB, Wang JX, Adzic RR (2010) Core-protected platinum monolayer shell high-stability electrocatalysts for fuel-cell cathodes. Angew Chem Int Ed 49(46):8602–8607. doi:10.1002/anie.201004287
100. Adzic RR, Zhang J, Sasaki K, Vukmirovic MB, Shao M, Wang JX, Nilekar AU, Mavrikakis M, Valerio JA, Uribe F (2007) Platinum monolayer fuel cell electrocatalysts. Top Catal 46(3–4):249–262. doi:10.1007/s11244-007-9003-x
101. Sasaki K, Mo Y, Wang JX, Balasubramanian M, Uribe F, McBreen J, Adzic RR (2003) Pt submonolayers on metal nanoparticles—novel electrocatalysts for H-2 oxidation and O-2 reduction. Electrochim Acta 48(25–26):3841–3849. doi:10.1016/S0013-4686(03)00518-8
102. Shao M, Sasaki K, Marinkovic NS, Zhang L, Adzic RR (2007) Synthesis and characterization of platinum monolayer oxygen-reduction electrocatalysts with Co-Pd core-shell nanoparticle supports. Electrochem Commun 9(12):2848–2853
103. Debe MK, Schmoeckel AK, Vernstrorn GD, Atanasoski R (2006) High voltage stability of nanostructured thin film catalysts for PEM fuel cells. J Power Sources 161(2):1002–1011. doi:10.1016/j.jpowsour.2006.05.033
104. Debe MK (2012) Electrocatalyst approaches and challenges for automotive fuel cells. Nature 486(7401):43–51. doi:10.1038/Nature11115
105. Marcilly C (2003) Present status and future trends in catalysis for refining and petrochemicals. J Catal 216(1–2):47–62. doi:10.1016/S0021-9517(02)00129-X
106. Vermeiren W, Gilson JP (2009) Impact of zeolites on the petroleum and petrochemical industry. Top Catal 52(9):1131–1161. doi:10.1007/s11244-009-9271-8
107. Armor JN (2011) A history of industrial catalysis. Catal Today 163(1):3–9. doi:10.1016/j.cattod.2009.11.019
108. Fajin JLC, Cordeiro MNDS, Gomes JRB (2011) On the theoretical understanding of the unexpected O-2 activation by nanoporous gold. Chem Commun 47(29):8403–8405. doi:10.1039/C1cc12166a
109. Wittstock A, Zielasek V, Biener J, Friend CM, Baumer M (2010) Nanoporous gold catalysts for selective gas-phase oxidative coupling of methanol at low temperature. Science 327(5963):319–322. doi:10.1126/science.1183591
110. Zielasek V, Jurgens B, Schulz C, Biener J, Biener MM, Hamza AV, Baumer M (2006) Gold catalysts: nanoporous gold foams. Angew Chem Int Ed 45(48):8241–8244. doi:10.1002/anie.200602484
111. Zeis R, Lei T, Sieradzki K, Snyder J, Erlebacher J (2008) Catalytic reduction of oxygen and hydrogen peroxide by nanoporous gold. J Catal 253(1):132–138. doi:10.1016/j.jcat.2007.10.017
112. Yan M, Jin T, Ishikawa Y, Minato T, Fujita T, Chen LY, Bao M, Asao N, Chen MW, Yamamoto Y (2012) nanoporous gold catalyst for highly selective semihydrogenation of alkynes: remarkable effect of amine additives. J Am Chem Soc 134(42):17536–17542. doi:10.1021/Ja3087592
113. Chen C, Kang YJ, Huo ZY, Zhu ZW, Huang WY, Xin HLL, Snyder JD, Li DG, Herron JA, Mavrikakis M, Chi MF, More KL, Li YD, Markovic NM, Somorjai GA, Yang PD, Stamenkovic VR (2014) Highly crystalline multimetallic nanoframes with three-dimensional electrocatalytic surfaces. Science 343(6177):1339–1343

Chapter 7
Enhanced Electrocatalytic Activity of Nanoparticle Catalysts in Oxygen Reduction by Interfacial Engineering

Christopher P. Deming, Peiguang Hu, Ke Liu, and Shaowei Chen

7.1 Introduction

As fuel cells are emerging as an optimal candidate for a renewable, environmentally friendly energy source, the challenge of large scale commercialization remains at the forefront of energy research [1–3]. It is well-known that the sluggish kinetics of electrochemical oxygen reduction is a significant limitation to the efficiency of fuel cells and therefore a catalyst must be utilized [4–6]. Platinum has been recognized as the best single metal catalyst for oxygen reduction due to the stability under the harsh operating conditions of the cell as well as a high intrinsic activity, but must be used in excess of 50 g for a 100 kW vehicle given the state of the art PEMFC (polymer electrolyte membrane fuel cell) [3]. This requirement significantly limits commercial production given the yearly global output of Pt and the yearly demand for motor vehicles. Experts predict that the benchmark activity of 160 A/g and 200 $\mu A/cm^2$ at +0.90 V must be enhanced by 4–10 folds for this technology to emerge in the market as a cost efficient alternative to internal combustion engine vehicles. To improve the intrinsic nature of a metal towards oxygen reduction, it is beneficial to examine the individual steps that comprise the complete reduction.

Many ORR pathways have been proposed [7], and in general, each pathway involves adsorption of oxygen on the catalyst surface followed by electron transfer/proton addition, ending with the release of water. Depending on the metal surface the rate limiting step will either be the initial electron transfer or the final release of product.

C.P. Deming • P. Hu • K. Liu • S. Chen (✉)
Department of Chemistry and Biochemistry, University of California, 1156 High Street, Santa Cruz, CA 95064, USA
e-mail: Shaowei@ucsc.edu

K.I. Ozoemena, S. Chen (eds.), *Nanomaterials for Fuel Cell Catalysis*, Nanostructure Science and Technology, DOI 10.1007/978-3-319-29930-3_7

It has been shown experimentally and computationally that the strength of interactions between different oxygenated intermediates (OIs) and a catalyst surface cannot be independently optimized. This idea is rationalized by the fact that all intermediates bind to the catalyst surface through the oxygen atom and that stronger or weaker surface interactions for one intermediate will yield the same result for all other surface bound intermediates. Therefore, a surface that binds too strongly to oxygen will undergo facile initial adsorption (initial step) of oxygen but will not easily release the final product (final step) resulting in poisoning of active sites and poor activity. Metals that bind to oxygenated species too weakly will allow for the rapid release of the final product but will also result in an unfavorable initial step of oxygen adsorption and consequently will exhibit poor activity. Therefore, the best catalyst will exhibit moderate adsorptive interactions with OIs to maximize the rate of the limiting step. The benchmark activity of 160 A/g and 200 μA/cm^2 at +0.90 V measured for carbon supported platinum nanoparticles is therefore attributed to the moderate binding to OIs.

Although platinum displays the highest reported activity as a single metal catalyst, theoretical predictions indicate platinum still binds to OIs too strongly. This model, in fact, shows a volcano-shaped trend when activity is plotted vs metal/adsorbate binding energy with the peak of the volcano exhibiting an activity over 20 times that of platinum. This trend was predicted by Sabatier over 100 years ago [8] and is graphically represented in Fig. 7.1. Here the *x* axis represents the difference in binding energy of a sample as compared to that of commercial

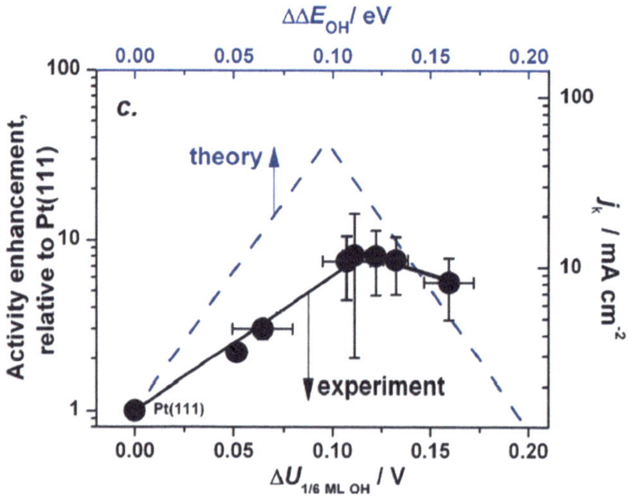

Fig. 7.1 Experimental and theoretical data relating electrocatalytic activity and oxygen binding energy. Reprinted with permission from (Stephens, I. E.; Bondarenko, A. S.; Perez-Alonso, F. J.; Calle-Vallejo, F.; Bech, L.; Johansson, T. P.; Jepsen, A. K.; Frydendal, R.; Knudsen, B. P.; Rossmeisl, J.; Chorkendorff, I.: Tuning the activity of Pt(1 1 1) for oxygen electroreduction by subsurface alloying. *J Am Chem Soc* **2011**, *133*, 5485–91). Copyright (2011) American Chemical Society

platinum while the *y* axis represents the enhancement in activity compared to that of commercial Pt. The blue dashed line represents theoretical calculations while the black dots represent experimental points that will be discussed in detail later.

Furthermore, to quantitatively describe the adsorptive behaviors of metals, a model has been developed relating the d band center of a metal to adsorptive properties. This model states that an upshift of the d band center relative to the Fermi level will create empty anti-bonding states that result in stronger bonding to adsorbates. The converse is also true that a downshift of the *d* band center will result in weaker interactions with adsorbates. This model is particularly useful in predicting how a catalyst surface will respond to structural modifications and when coupled with the dynamics of OIs on the surface, giving a promising outline for future catalysts to meet and/or exceed the theoretical maximum and afford the efficiency required for large scale commercialization.

In order to improve the activity of cathode catalysts or reduce the usage amount of Pt, a number of strategies have been proposed and explored, including manipulation of size [9, 10], morphology [11–13], elemental composition [14, 15], and surface atomic arrangement [16, 17]. Additionally, deliberate engineering of the nanoparticle interface has proven an effective method to alter the adsorptive properties of the particle and thus the electrocatalytic activity [14, 18–22].

Among the pathways of interfacial manipulation, alloying with a second metal serves to alter the electronic structure of surface atoms by imparting lateral force on the particle lattice [23] as well as driving electron rearrangements between neighboring atoms with different d band filling and electronegativity [24, 25]. Both situations will ultimately alter the d band structure of the surface metal and thus the degree of filling for the metal-adsorbate antibonding state and strength of adsorptive interaction. The *d* band model is well established and has allowed for many predictions and rationalizations of enhanced alloy nanoparticles. Consequently, a great amount of efforts have been devoted to alloy cathode catalysts, which can be classified into Pt-based alloy catalysts [24–27] and non-Pt alloy catalysts [15].

Additionally, the functionalization of metal nanoparticles with organic ligands has afforded activity enhancement through direct manipulation of the electronic structure of the metal [20, 21, 28–31]. These interactions, along with other opto/electronic properties, are largely a result of the nature of the interfacial connection between the metal and capping ligand. For example, metal-sulfur bonds have been shown to be highly polarized bonds with strongly localized charge while metal-carbon bonds exhibit more covalent characteristics [21]. Additionally, there are many examples of π overlap between the ligand p orbitals and the metal d orbitals which lead to an extensively conjugated particle-ligand system [32–35]. The different interfacial bonding exhibited by each class of ligands will impart a different effect on the electronic structure of the metal and correspondingly affect the dynamics of OI adsorption. As models for nanoparticle alloys have linked the d band center shift to a particle's characteristics such as composition and type of alloy, the development of a strong model relating how each type of ligands will modify a metal d band is essential for understanding the origin of activity enhancement from organic ligand functionalization and allow for the proper choice of capping ligands for a starting material.

Further studies have demonstrated that surface functionalization of metal oxide nanoparticles will similarly enhance the electrocatalytic activity, rendering them to be viable catalysts for oxygen reduction [22, 36]. Note that transition metal oxides are generally cheap and abundant and therefore represent a possible route for commercially available fuel cell technology, but at present the activity is generally considered subpar compared to that of platinum group metals [37, 38]. The controlled enhancement of such materials is therefore essential for commercial realization of this technology, but it must follow a slightly different approach than with zero-valence metal particles. Metal oxide electrocatalysts are generally tested in alkaline solution due to high corrosion in acidic media, and therefore a different mechanism is needed to describe the reduction process. This mechanism involves the generation of surface OH groups, replacement of these groups with O_2, stepwise reduction through peroxide and oxide intermediates, and the regeneration of surface OH [38, 39]. For this reduction pathway, the kinetics are limited by the strength of interaction between the material surface and oxygen. Specifically, an interaction too weak will limit the initial displacement of OH by O_2 and an interaction too strong would prevent facile regeneration of OH [38, 40]. Again, the optimal surface would possess properties in between the two extremes. To reach this volcano peak, many surface manipulation techniques have been performed including direct manipulation of oxide vacancies [22], surface functionalization with chlorine [36], and alteration of metal components [40].

In this chapter, we will highlight some recent progress in ORR electrocatalysis, including Pt-based nanoparticles, non-Pt metal nanoparticles, organically capped metal nanoparticles, as well as metal oxide-based nanocomposites, with a focus on the impacts of interfacial engineering on the nanoparticle electrocatalytic activity.

7.2 Interfacial Engineering of Metal Nanoparticle Catalysts: Metal Compositions

Metal alloying has been proved to be an effective way to realize surface engineering in the improvement of ORR catalytic activity. It has been widely known that the catalytic activity of Pt can be effectively improved by alloying with 3d metals [18, 26, 27, 41, 42]. For instance, Wu et al. [42] prepared truncated-octahedral Pt_3Ni (*t,o*-Pt_3Ni) crystals with dominant exposure of (1 1 1) facets, which showed remarkably enhanced electrocatalytic activity towards ORR, and the activity increased with increasing (1 1 1) surface fraction. As shown in the inset to Fig. 7.2a, the electrochemical surface areas (ECSAs) obtained from the calculated hydrogen adsorption/desorption charge in the cyclic voltammgrams (CV) were 33.8, 53.7, 62.4, and 65 m^2/g_{Pt} for the 70, 90, 100 % ((1 1 1) surface fraction) *t,o*-Pt_3Ni, and commercial Pt (TKK, 50 wt% Pt, 3 nm diameter on Vulcan carbon)

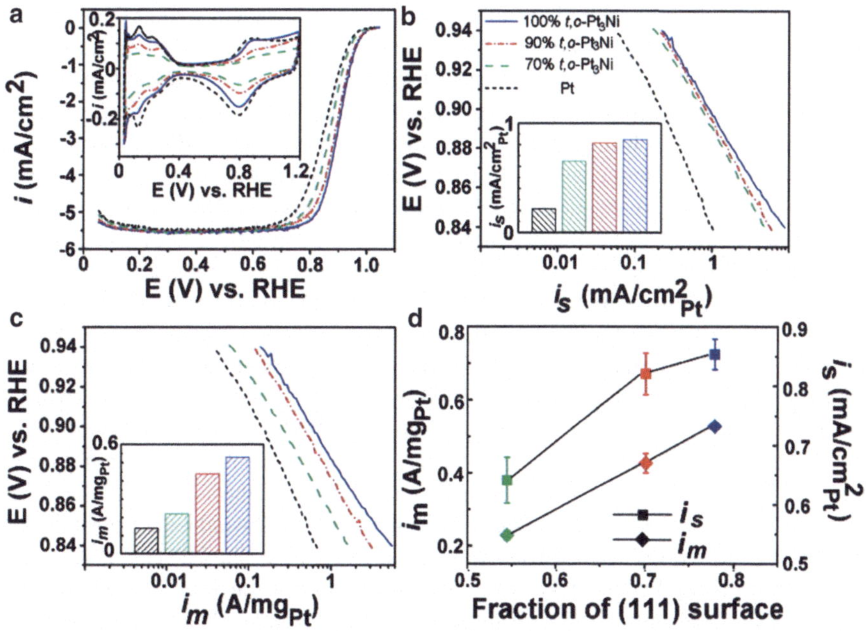

Fig. 7.2 (**a**) Polarization curves (at 1600 rpm) and CV curves (inset). (**b**) area-specific (mA/cm^2_{Pt}), and (**c**) mass-specific (A/mg_{Pt}) ORR activities for the *t,o*-Pt_3Ni and commercial Pt catalysts. (**d**) Correlations of the area-specific and mass-specific activities with the fraction of the (1 1 1) surfaces of these *t,o*-Pt_3Ni catalysts. Reprinted (Adapted) with permission from (Wu, J. B.; Zhang, J. L.; Peng, Z. M.; Yang, S. C.; Wagner, F. T.; Yang, H., Truncated Octahedral Pt_3Ni Oxygen Reduction Reaction Electrocatalysts. *J Am Chem Soc* **2010**, *132* (14), 4984) Copyright (2010) American Chemistry Society

catalysts, respectively. The electrocatalytic activities of the *t,o*-Pt_3Ni and commercial Pt catalysts were then evaluated by RDE (rotating disk electrode) voltammetry, and all three *t,o*-Pt_3Ni catalysts exhibited more positive onset potentials than commercial Pt/C. Furthermore, as shown in Fig. 7.2b, c, within the potential range of +0.84 to +0.94 V (vs. RHE), the area-specific and mass-specific current densities were 850 $\mu A/cm^2$ and 0.53 A/mg_{Pt} respectively at +0.90 V for the 100 % *t, o*-Pt_3Ni catalysts, both 4-times higher than those (215 $\mu A/cm^2$ and 0.14 A/mg_{Pt}) of the commercial Pt/C catalysts. In addition, one can see that the higher (1 1 1) surface fraction of *t,o*-Pt_3Ni, the higher area-specific and mass-specific current density, as manifested in Fig. 7.2d. In another study, Stamenkovic et al. [27] demonstrated that the Pt_3Ni(1 1 1) surface was 10-fold more active on ORR than the corresponding Pt(1 1 1) surface and 90-fold more active than commercial Pt/C catalysts for PEMFC, and further studies were also conducted to explore the mechanism for the significantly improved activity. It was shown by low energy ion scattering (LEIS) that the composition of the outermost atomic layer was pure Pt, a so-called Pt-skin structure. Such first-layer Pt enrichment was suggested to be counterbalanced by its depletion in the next two to three layers so that the concentration profile oscillates around the bulk value. They also showed in background-corrected ultraviolet photoemission spectroscopy (UPS) measurements that such a near-surface compositional change also resulted in distinctive electronic properties for PtNi alloys. It is found that the *d*-band density of states (DOS) is structure-sensitive, i.e., the *d*-band center shifts as the surface structure changes (−2.70 eV on $Pt_3Ni(1\,1\,0)$, −3.10 eV on $Pt_3Ni(1\,1\,1)$, and −3.14 eV on $Pt_3Ni(1\,0\,0)$), which are also quite different from that of Pt single crystals. Furthermore, the average energy of the surface atoms influences the chemisorption energies in surface electrochemistry when the adsorbate binds onto the surface (the ligand effect). Experimentally, on the Pt-skin surface of a Pt_3Ni single crystal the formation potentials were found to be negatively shifted by ca. 0.15 V for H_{upd} (which refers to the underpotentially deposited hydrogen) and positively shifted by ca. 0.1 V for OH_{ad} (which refers to the adsorbed hydroxyl layer), relative to those on the Pt(1 1 1), and correspondingly, the fractional coverages of H_{upd} and OH_{ad} on $Pt_3Ni(1\,1\,1)$ were dramatically reduced by 50 % relative to those on Pt(1 1 1). Both of the observations above were consistent with the large downshift (0.34 eV) of the *d*-band center position on the Pt-skin structure. ORR measurements showed that the ORR kinetics was accelerated on the $Pt_3Ni(1\,1\,1)$ skin, compared with that on Pt(1 1 1), where one can find a positive shift of 100 mV in the half-wave potential. The authors ascribed such improved catalytic activity to the key parameter of reduced coverage of OH_{ad}. Additionally, the lower coverage of H_{upd} substantially attenuated the production of peroxide on the Pt-skin surface.

Further studies were also carried out by Stamenkovic et al. on other PtM (M = Co, Fe, Ti, and V) surfaces in order to investigate the fundamental relationship in catalytic trends between the experimentally determined surface electronic structure (the *d*-band center) and the activity for oxygen reduction reaction [26]. As shown in Fig. 7.3, a volcano-shape relationship was found between the electrocatalytic activity and the *d*-band center position on the Pt-skin of PtM surfaces with the maximum activity from Pt_3Co. Such a trend strongly indicates that two opposite aspects, the strong binding energy of O_2 and reaction intermediates like O_2^-, O_2^{2-}, H_2O_2, and the relatively low coverage by oxygenated species and specifically adsorbed anions should be taken into consideration and counterbalanced to obtain better catalysts than Pt. As a result, for metals or alloys with too high binding energy to oxygen/oxides/anions, like Pt, the *d*-band center is too close to the Fermi level and the availability of OH_{ad}/anion-free Pt sites is the key parameter that limits the ORR kinetic rate. In contrast, in the cases like Pt_3V and Pt_3Ti of which the *d*-band center is too far away from the Fermi level, although the catalyst surface is less occupied by oxygen species and anions, the binding between oxygen/intermediates and metal/alloy is too weak to enable a turnover rate of ORR. Consequently, the volcano curve of the electrocatalytic activity can be interpreted by the Sabatier principle: the reaction is limited by the rate of removing surface oxides and anions on the surface of catalysts binding oxygen too strongly,

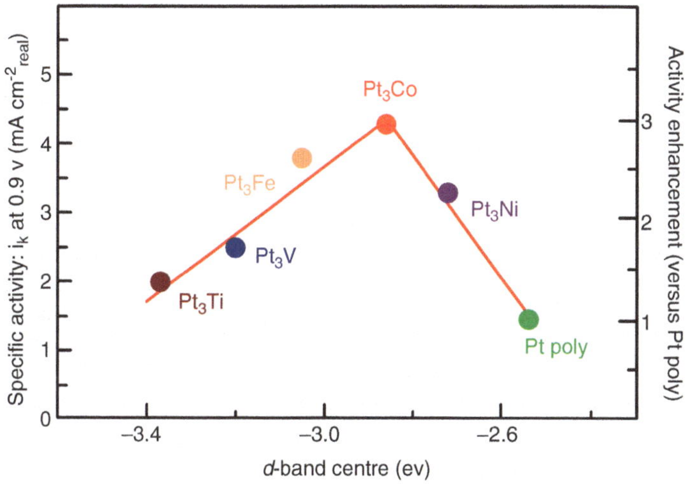

Fig. 7.3 Relationships between the catalytic properties and electronic structure of Pt-skin on Pt_3M alloys. Reprinted by permission from Macmillan Publishers Ltd: Nat Mater (Stamenkovic VR, Mun BS, Arenz M, Mayrhofer KJJ, Lucas CA, Wang GF, Ross PN, Markovic NM, Trends in electrocatalysis on extended and nanoscale Pt-bimetallic alloy surfaces. Nat Mater 2007, 6, 241–247), copyright (2007)

but on the surface of catalysts binding oxygen too weakly it is limited by the rate of electron and proton transfer to the adsorbed oxygen.

Besides ligand effects discussed above, geometrical strain is another important effect in surface engineering by metal alloying [43]. As a matter of fact, ligands and geometric effects generally simultaneously occur and impact the catalytic activity at the same time [44, 45]. In order to understand how geometric strain effect influence the catalytic activity in an electrocatalytic system, a core-shell catalyst structure was designed and prepared by Strasser et al. In the structure, the shell consists of a few atomic monometallic layers, which is supported on the core substrate with different lattice parameters. Such a structure is supposed to be able to isolate geometric strain effects since ligand effect on surface catalytic activity is only limited within two or three atomic layers [23]. To prepare the special catalyst mentioned above, precursor PtCu alloy nanoparticles were synthesized at first, and the core-shell structure was obtained via preferential dissolution/removal of the electrochemically more reactive component (Cu) from the bimetallic alloy, which is often referred to as "dealloying". It was then shown that dealloyed PtCu nanoparticles exhibited uniquely high catalytic activity for ORR.

The ORR catalytic activity of a serial of PtCu dealloyed nanoparticles was both calculated by density functional theory and measured experimentally to quantify the relationship between lattice strain, oxygen binding energy, and electrocatalytic activity, as shown in Fig. 7.4. It is obvious that lattice strain increases as Cu content increases, and both computational analysis and X-ray spectroscopy data showed a linear relationship between lattice strain and oxygen binding energy, which is also

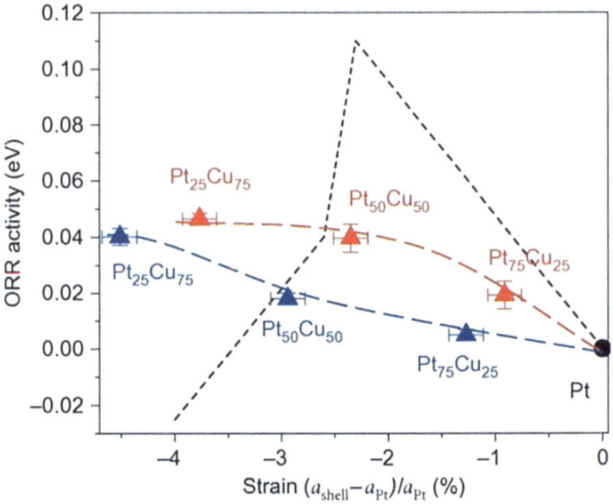

Fig. 7.4 Experimental and predicted relationships between electrocatalytic ORR activity and lattice strain. Reprinted by permission from Macmillan Publishers Ltd: Nat Chem (Strasser P, Koh S, Anniyev T, Greeley J, More K, Yu CF, Liu ZC, Kaya S, Nordlund D, Ogasawara H, Toney MF, Nilsson A, Lattice-strain control of the activity in dealloyed core–shell fuel cell catalysts. Nat Chem 2010, 2, 454–460), copyright (2010)

consistent with computational calculations in other reports [43]. Based on a microkinetic model developed for ORR [46], the strain-binding energy relationship was combined, and a volcano plot was derived for the relationship between the predicted ORR rate and lattice strain. The volcano plot implies that lattice strain first accelerates the ORR rate by weakening the bond between surface Pt atoms and intermediate oxygenated adsorbates, lowering the energy barriers for electron and proton transfer, and subsequently reducing the coverage of adsorbed oxygen species. However, the ORR rate is predicted to decrease after a critical lattice strain, which can be understood by the fact that the binding energy becomes too weak between surface Pt atoms and oxygen, leading to an increased energy barrier for oxygen dissociation and the following formation of oxygenated intermediates. Experimentally, the ORR catalytic activity of PtCu dealloys, in consistence with calculation analysis, was also found to be enhanced at first with the increased lattice strain. However, unlike the prediction in computational studies, the ORR catalytic activity did not decrease on the left side of the volcano plot, which was rationalized by the compressive strain relaxation in the Pt shell. That is, Pt atoms adjacent to the PtCu core adopt the lattice parameters close to that of the core, and in contrast, the Pt atoms in shell generally relax towards the lattice constants of pure Pt. Therefore, the real surface strain is less than the calculated value plotted in Fig. 7.4, which is an average strain in the Pt shell. Thus, it is expected of a data point shift to the right if the real surface strain is used in Fig. 7.4. Moreover, it is believed that due to the surface strain relaxation, it is probably hard to obtain sufficient surface strain by synthesizing dealloyed nanoparticles to achieve the true maximum of the volcano curve.

Further studies were conducted by Greeley et al. [18] and Xiao et al. [47] to quantitatively understand the surface structure-activity relationship for fuel cell cathode catalysts. It turns out that although weakening the surface binding energy to oxygen by alloying Pt with other metals is responsible for the improvement of the

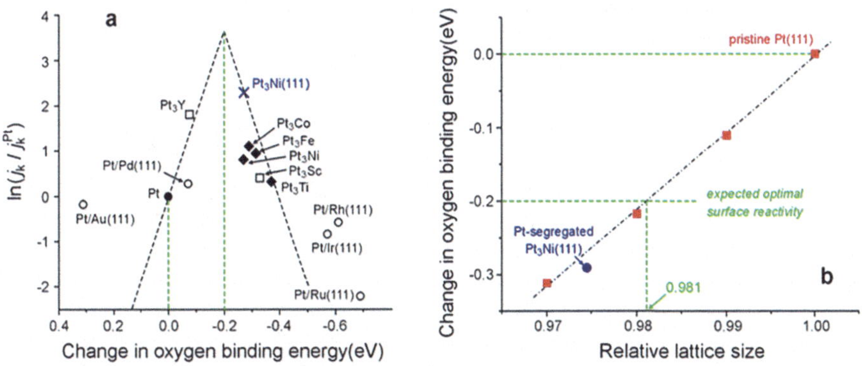

Fig. 7.5 (**a**) The volcano plot between catalytic activity and oxygen binding energy relative to that of Pt(1 1 1); (**b**) the linear relationship between the relative lattice size and oxygen binding energy relative to that of Pt(1 1 1). Reprinted by permission from Macmillan Publishers Ltd: Nat Chem, (Greeley J, Stephens IEL, Bondarenko AS, Johansson TP, Hansen HA, Jaramillo TF, Rossmeisl J, Chorkendorff I, Norskov JK (2009) Alloys of platinum and early transition metals as oxygen reduction electrocatalysts. Nat Chem 1 (7):552–556), Copyright (2009). Reproduced from (Xiao L, Huang B, Zhuang L, Lu JT (2011) Optimization strategy for fuel-cell catalysts based on electronic effects. Rsc Adv 1, 1358–1363.) with permission of The Royal Society of Chemistry

electrocatalytic activity of Pt, the surface catalytic activity of PtNi(1 1 1) is not yet optimal. As shown in Fig. 7.5a, both experimental and theoretical data strongly suggest that a surface with better ORR catalytic activity than Pt should have a binding energy to oxygen 0–0.4 eV weaker than that of Pt, with an optimal value at the binding energy 0.2 eV lower than Pt [18]. According to Fig. 7.5b, it is found by Xiao et al. [47] that the binding energy to oxygen of Pt decreases linearly as the Pt lattice compresses. Taken together with the surface electronic structure-surface reactivity relationship, it is clear of the guideline for catalyst optimization, which is to control the ligand effect and geometric strain effect to achieve an alloy structure with the oxygen binding energy around 0.2 eV lower than that of Pt, and alloy here is not limited to Pt or any other precious metal/alloys, but also includes non-precious metal/alloys.

In addition to achieving better activity than Pt by surface engineering through metal alloying, another aspect that should be taken into consideration is the stability of the surface alloy structures, which means not only the stability in thermodynamics but also the stability under PEMFC working conditions. In order to search for stable alloy catalysts, DFT calculations have also been conducted by Greeley et al. with a number of alloys [18]. As shown in Fig. 7.6 (red points), it is clear that Pt_3Y and Pt_3Sc turn out to be the most stable among the alloys with ΔE_O in the optimal range (0–0.4 eV). The high stability can be rationalized in terms of approximately half filled metal-metal *d* bands, meaning filled bonding states and empty anti-bonding states. The ORR activity of the catalysts was measured and shown in Fig. 7.7. It is clear in Fig. 7.7a that both Pt_3Y and Pt_3Sc exhibit

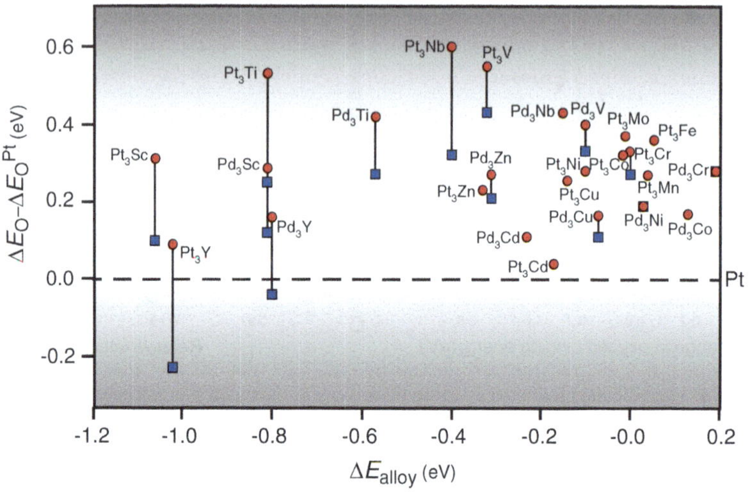

Fig. 7.6 Output of computational screening procedure, showing the oxygen binding energy, relative to that of Pt, on a Pt or Pd skin surface, as a function of alloying energy. ΔE_{alloys} is the calculated energy of formation of the indicated elements from the appropriate bulk elemental precursors, specified on a per atom basis. For some systems we include results for the cases where there are 50 % (*circles*) and 25 % (*squares*) of the alloying element in the second layer. Reprinted by permission from Macmillan Publishers Ltd: Nat Chem (Greeley J, Stephens IEL, Bondarenko AS, Johansson TP, Hansen HA, Jaramillo TF, Rossmeisl J, Chorkendorff I, Norskov JK. Alloys of platinum and early transition metals as oxygen reduction electrocatalysts. Nat Chem 2009, 1, 552–556), Copyright (2009)

significantly enhanced ORR catalytic activity over Pt, which is consistent with the prediction in calculations. Moreover, the alloy catalysts are also very stable under continued cycling conditions, as shown in Fig. 7.7b.

To further reduce the costs of cathode catalysts in fuel cells, a large number of efforts have also been devoted to non-Pt based alloy catalysts. Greeley et al. [48] have conducted a density functional theory (DFT)-based combinatorial search for improved ORR alloy catalysts. In their calculations, oxygen binding energy was used as an ORR descriptor for the prediction of ORR activity trends, since for Pt-based alloys the calculation using oxygen binding energy as a descriptor is in excellent agreement with experimental measurements. According to their calculations, 85 of 736 transitional metal surface alloys are predicted to be at least as active as pure Pt for ORR. However, among the 29 most active and stable alloy candidates for ORR, only 8 (containing primarily Pt, Pd, Ag, or group VA elements in the surface layer) will survive above +0.5 V in acidic environment based on the calculation. What is worse, even this level of stability may not be sufficient to yield robust catalysts with long-term operation stability in fuel cells.

Among non-Pt based catalysts, Pd is considerably less expensive than Pt and also less active than Pt for ORR catalysis, but more active and stable than any other metals [15]. It has been reported that interfacial engineering by alloying can significantly enhance the catalytic activity of Pd [14, 15, 49]. Therefore, Pd-based

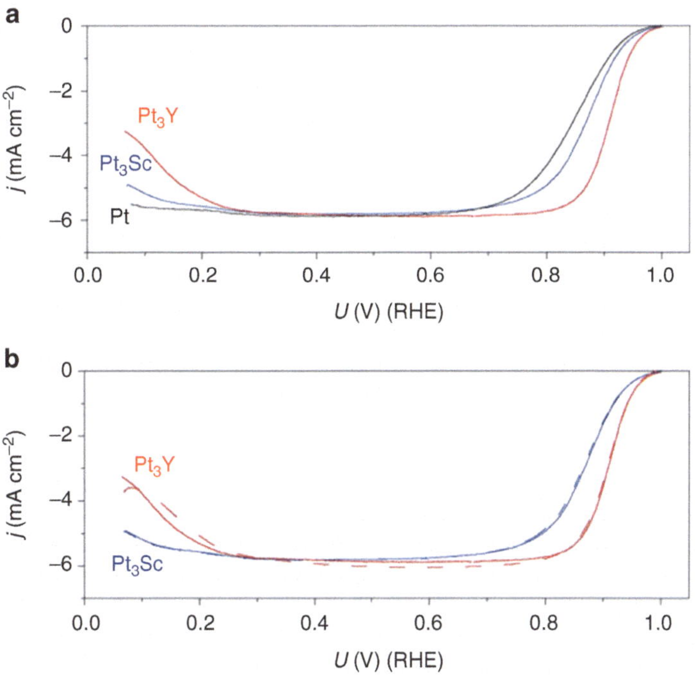

Fig. 7.7 Anodic sweeps of cyclic voltammograms of Pt, Pt_3Sc and Pt_3Y in O_2-saturated electrolyte. (**a**) The first stable sweep. (**b**) A comparison of the first steady cycle (*solid lines*) and after 90 min of cycling (*dashed lines*), for Pt_3Y and Pt_3Sc only, under the same conditions as (**a**). Measurements were taken at 20 mV/s and 1600 rpm in 0.1 M $HClO_4$ at 23 ± 2 °C. Reprinted by permission from Macmillan Publishers Ltd: Nat Chem (Greeley J, Stephens IEL, Bondarenko AS, Johansson TP, Hansen HA, Jaramillo TF, Rossmeisl J, Chorkendorff I, Norskov JK, Alloys of platinum and early transition metals as oxygen reduction electrocatalysts. Nat Chem 2009, 1, 552–556), copyright (2009)

alloy catalysts have been considered as one of the best candidates for non-Pt ORR catalysts. Suo et al. [50] carried out first-principles studies of the improved electrocatalytic activity of Pd-based alloys toward ORR. It is found that there is a volcano relationship (Fig. 7.8) between the degree of alloying (in the case of Pd-Co alloy) and the catalytic activity, which is believed to arise from the contrary influences from lattice strain effect and ligand effect by the interfacial engineering. At a low degree of alloying with Co, the lattice strain effect is predominant because of the negligible content of Co on the surface, which leads to weakened oxygen binding energy and increased activity, while at a high degree of alloying, the ligand effect becomes overwhelming as a result of the high surface content of Co, leading to increased oxygen binding energy and reduced catalytic activity.

In addition, given that the lattice strain effect on the surface was considered as the main reason for the enhanced activity, another significant calculation was also conducted to find the relationship between the oxygen binding energy (denoted as AE in Fig. 7.9) and the lattice strain on the surface described as the ratio of lattice

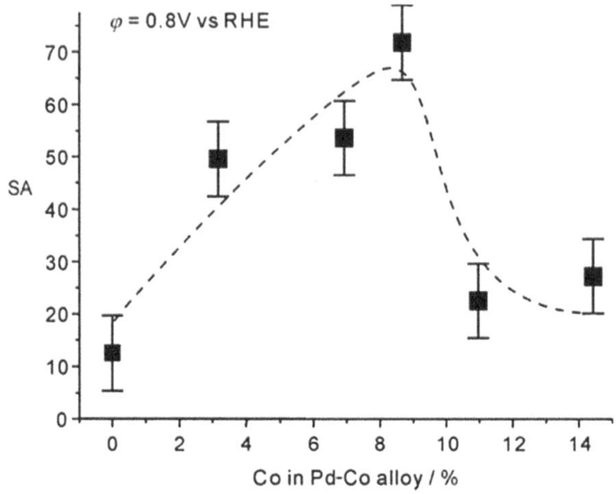

Fig. 7.8 Relationship between surface specific activity (SA) of Pd–Co alloys and degree of alloying. The kinetic current obtained from RDE experiments (25 °C, 5 mV/s, O_2-saturated 0.5 M H_2SO_4, 1600 rpm) was normalized by the electrochemical surface area of Pd–Co alloys to give SA. Reprinted with permission from (Suo YG, Zhuang L, Lu JT: First-principles considerations in the design of Pd-alloy catalysts for oxygen reduction. Angew Chem Int Edit 2007, 46, 2862–2864. Copyright © 2007 WILEY-VCH Verlag GmbH & Co. KGaA, Weinheim

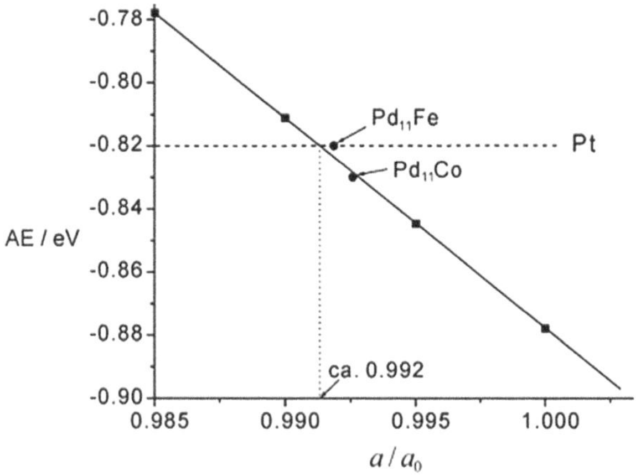

Fig. 7.9 Relationship between the AE of O_{ads} and the strain ratio in the lattice constant of Pd, which may serve as a guideline for the design of Pd-alloy catalysts for ORR. Reprinted with permission from (Suo YG, Zhuang L, Lu JT: First-principles considerations in the design of Pd-alloy catalysts for oxygen reduction. Angew Chem Int Edit 2007, 46, 2862–2864. Copyright © 2007 WILEY-VCH Verlag GmbH & Co. KGaA, Weinheim

constant, as shown in Fig. 7.9. According to Fig. 7.9, the relationship is linear and the reduced lattice ratio on the surface is supposed to benefit the catalytic activity. It is also predicted that Fe would be a better alloy element than Co for interfacial engineering of Pd for a better catalytic performance, and the activity of $Pd_{11}Fe$ alloy may be as comparable to that of Pt since the oxygen binding energy on it is very close to that on Pt, based on the relationship. Such a prediction, as a matter of fact, was confirmed by Shao et al. [15]. They prepared a serial of Pd-Fe alloys with different atomic ratios and found a volcano relationship between the activity and the atomic ratio. It was reported that different alloying degree of Fe resulted in different Pd-Pd bond length, generating different lattice strain, further altering the *d*-band center position of the alloy, which determined the surface activity of the alloy. Moreover, the Pd_3Fe was found to be very stable, methanol tolerated, and exhibit even higher activity than Pt.

Ag and Ag-based catalysts have also been gaining increasing interests as a viable candidate for non-Pt catalysts applied in fuel cells with alkaline electrolytes [51–53]. Ag is relatively inexpensive as compared with Pd, Au, and more importantly, although still inferior to Pt, Ag gives relatively high catalytic activity with a 4-electron ORR pathway [52, 53]. Nevertheless, the application of Ag as a PEMFC catalyst is still limited by both poor activity and stability. Fortunately, in recent years, remarkable progress has been made in alkaline polymer electrolytes (APEs), facilitating the development of alkaline polymer electrolyte fuel cell (APEFC), which is expected to combine the advantages of both PEMFC and AFC (alkaline fuel cell) [54]. The development of APEFC thus enables the use of non-Pt metal/alloy catalysts, which are not compatible with the acidic electrolyte in PEMFC mainly due to the stability problem.

For instance, by simultaneous reduction of Ag and Pd precursors, Slanac et al. [55] prepared Ag-rich AgPd alloy nanoparticles (~5 nm). The ORR mass activity per total metal was found to be 60 % higher for the $AgPd_2$ alloy than that of pure Pd, and the mass activities of Ag_9Pd (340 mA/mg_{metal}) and Ag_4Pd (598 mA/mg_{metal}) were 2.7 and 3.2 times higher, respectively, than those calculated by a linear combination of mass activities of Ag (60 mA/mg_{Ag}) and Pd (799 mA/mg_{Pd}) nanoparticles. The enhancement of the ORR activity was ascribed to interfacial engineering, leading to the combined effects of weak binding energy of Ag and strong binding energy of Pd towards oxygen/oxygen intermediates (ensemble effect). That is, by creating a AgPd alloy surface, Pd atoms facilitate the adsorption of oxygen, while Ag atoms desorb the oxygenated products (like OH^-). Additionally, the electronic structure shifted by the ligand effect, which was confirmed by XPS measurements, is also believed to contribute to the activity enhancement.

7.3 Metal Nanoparticle Catalysts: Ligand-Mediated ORR Activity

In addition to metal compositions, another potent method to improve the performance of a metal surface for ORR is through deliberate surface functionalization with organic ligands. Self-assembled monolayers of organic ligands on metal nanoparticles have been well studied [56, 57] and have been specifically explored for improved electroreduction of oxygen on metal surfaces in the form of thiols [21, 58–60], *para*-substituted phenyls [20, 21, 28], cyanide ions [61], phenyl acetylene derivatives [29], n-alkynes [21, 29], phenyl phosphine derivatives [62], and graphene quantum dots (GQD) [30, 31]. The catalytic performance largely depends on three factors: the chemical nature of the metal, the structures and properties of the capping ligands, and the nature of the interfacial bonding interactions between the metals and the ligands. Among these, the structure of the capping ligands, such as chain length, degree of branching, and functional groups may alter the nature of the catalytic process through steric hindrance while the nature of interfacial bonds may affect the catalytic reactions by manipulation of the electron density in the metal. The primary focus of this section will be to summarize the effects of interfacial bonding on ORR, specifically, how interactions between the ligand and metal may be exploited for the enhancement of the kinetics of oxygen reduction by providing more favorable interactions with OIs.

Thiol ligands have been shown to improve the stability of platinum nanoparticle catalysts in acidic solution. For instance, for platinum nanoparticles functionalized by 4-mercapto aniline then further derivatized with 2-thiophencarbonyl chloride, it was found that no activation process was needed for ligand desorption and the as-prepared nanoparticles exhibited excellent stability after prolonged electrochemical cycling as indicated by cyclic voltammetry and XPS measurements [58].

For platinum nanoparticles functionalized with *para*-substituted phenyl fragments, the ORR activity was markedly enhanced, which was correlated with the electron withdrawing capacity of the phenyl substituents with the most active sample exhibiting 3 times the specific activity of commercial Pt/C catalysts [20]. In this study, the *para*-substituents for the phenyl moiety included $-CH_3$, –F, –Cl, $-OCF_3$, and $-CF_3$ within a large range of Hammet coefficient (σ). The ORR activity was evaluated with rotating ring disk electrode (RRDE) voltammetry in oxygen-saturated 0.1 M $HClO_4$ and it was found that specific activities were very well correlated with the Hammet coefficient, as shown in Fig. 7.10. The trend in activity was ascribed to the weakening interaction between platinum and OIs as the electron density of Pt was diminished by the electron-withdrawing substituting groups.

Similar effects have also been observed with other metal-ligand interfacial bonding interactions. In a more recent study [21], a variety of silver nanoparticles were prepared by surface functionalization with hexanethiol, 1-octyne, and 4-trifluoromethylphenyl fragments (denoted as AgSC6, AgHC8, and AgPhCF3, respectively), and the impacts of the metal-ligand interfacial bonds were examined, with silver nanoparticles supported on carbon (Ag/C) as the benchmark material. The ORR activity was found to increase in the order AgSC6 < Ag/C < AgHC8 < AgPhCF3. The most active sample was identified as AgPhCF3 that exhibited a specific current density 13 times that of bare Ag and an onset potential 200 mV more positive. This was accounted for by the manipulation of the electronic states of silver by the capping ligand, as confirmed by XPS measurements of the binding energy of the Ag 3d electrons.

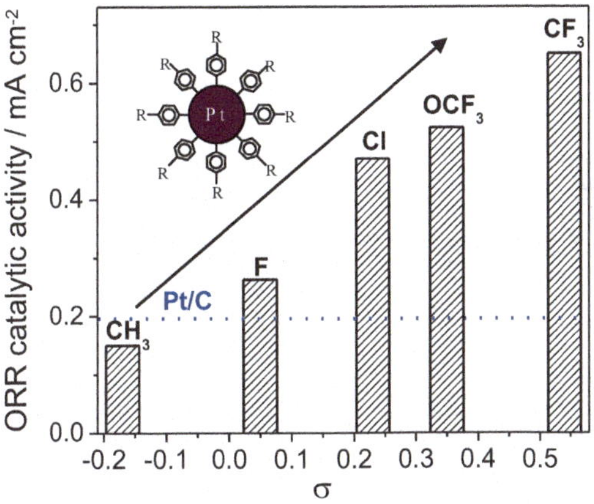

Fig. 7.10 Activity of Pt nanoparticles functionalized with different para substituted phenyl ligands as a function of Hammet coefficient (electron withdrawing nature) Reprinted with permission from (Zhou, Z.-Y.; Kang, X.; Song, Y.; Chen, S.: Ligand-Mediated Electrocatalytic Activity of Pt Nanoparticles for Oxygen Reduction Reactions. *J Phys Chem C* **2012**, *116*, 10592–10598). Copyright (2012) American Chemical Society

Triphenylphosphine triphosphonate (TPPTP) capped platinum particles have also been found to show superior activity as compared to pure Pt particles. The successful synthesis of the nanoparticles was confirmed by ^{1}H and ^{31}P NMR spectroscopy, XPS, TEM, and UV-Vis measurements [63]. Cyclic voltammetry measurements in 0.1 M $HClO_4$ revealed that the surface oxide reduction peak shifted to a more positive potential for the TPPTP functionalized particles than similarly prepared Pt particles without TPPTP functionalization. This suggests weakened interactions between the metal surface and OIs, which is further confirmed by the larger specific activity of the TPPTP-capped particles than that of uncapped particles. This study highlights the impacts of metal-ligand interfacial bonds (in this case Pt-P) that will allow for unique electronic interactions and ultimately result in enhanced electrocatalytic oxygen reduction.

In a recent study, bimetallic AgAu Janus nanoparticles were prepared by galvanic replacement reactions of hexanethiolate-protected silver (AgC6) nanoparticles with Au(I) mercapto-propanediol complex on one hemisphere of the particle, leading to phase segregation of both metals and ligands (Fig. 7.11) [64]. Metallic phase segregation was confirmed by XPS and energy-dispersive X-ray spectroscopy (EDX, Fig. 7.12) measurements while ligand segregation was confirmed by contact angle measurements. Electrochemical measurements showed that the bimetallic AgAu Janus particles exhibited apparent ORR activity that was seven times that of AgC6 particles and over 4 times that of bulk exchanged AgAu particles. This enhancement is attributed to the segregation of polar and non-polar ligands on the particle surface that enhanced the charge transfer from Ag to Au, thus improving the adsorption of oxygen. Since gold is a metal with a low lying *d* band center, it interacts weakly with OIs and is limited by the initial adsorption and

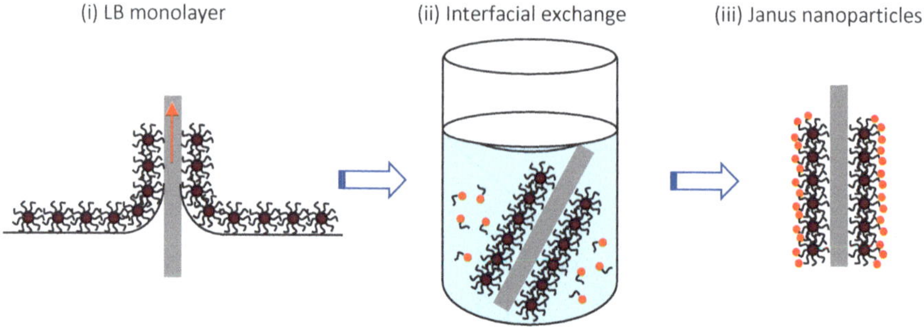

Fig. 7.11 Schematic of bimetallic AgAu Janus nanoparticle synthesis. Reprinted with permission from (Song, Y.; Liu, K.; Chen, S.: AgAu bimetallic Janus nanoparticles and their electrocatalytic activity for oxygen reduction in alkaline media. *Langmuir* **2012**, *28*, 17143–52). Copyright (2012) American Chemical Society

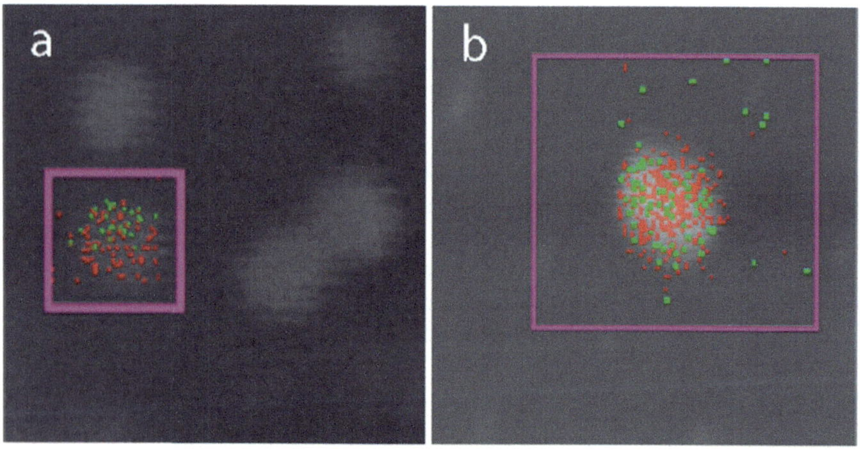

Fig. 7.12 EDX measurements of (**a**) Janus exchanged particles and (**b**) bulk exchange particles. Red represents silver domains while green represents gold domains. Reprinted with permission from (Song, Y.; Liu, K.; Chen, S.: AgAu bimetallic Janus nanoparticles and their electrocatalytic activity for oxygen reduction in alkaline media. *Langmuir* **2012**, *28*, 17143–52). Copyright (2012) American Chemical Society

electron transfer step. Therefore, the charge transfer from silver to gold will raise the *d* band center of gold and result in stronger binding to OIs and is consistent with the previously stated d band model [25, 65].

Another area gaining much attention is the functionalization of metal nanoparticles with high surface area supports such as carbon nanotubes [66] and graphene quantum dots (GQDs) [30, 67]. Research has shown that the function of such supports can extend beyond merely a means to disperse the catalysts and connect them to the electrode, but may also serve to enhance the activity through electronic interactions that result in more favorable binding with OIs

[67–70]. Furthermore, recent DFT calculations have shown that the defects in the graphitic structure of GQDs may diminish electron density of the deposited platinum nanoparticles leading to enhanced activity through weakened adsorption of OIs [7]. As carbon black or activated carbon is commonly used for supports, a Pt/GQD composite is a promising route for improving catalytic activity, for GQDs will offer the same availability, stability, and high surface to volume ratio as the common carbon supports in addition to activity enhancement.

In fact, in a recent study focused on the defect-mediated activity of GQD supported platinum nanoparticles, it was found that the relative amount of GQD defects can be easily manipulated through both temperature and time of heating. Platinum salt was dissolved in water with GQDs and hydrothermally reduced in a Teflon-lined autoclave at controlled temperatures (140, 160, 180, or 200 °C) for a selected period of time (3, 6, or 12 h) to study the relationship between hydrothermal treatment and defect concentration.

These defects were largely oxygen containing groups that disrupted the graphitic nature of GQDs and provided anchoring sites for metal deposition [66] as well as diminishing electron density. The relative amounts of these groups were examined by FTIR and XPS measurements, as depicted in Figs. 7.13 and 7.14. It was found that the C–O–C asymmetric epoxy stretch (1135 cm^{-1}), the C=O stretch (1685 cm^{-1}), and the OH deformation in C–OH (1350 cm^{-1}) that were present in the initial FTIR spectra of the untreated GQDs were markedly diminished and/or

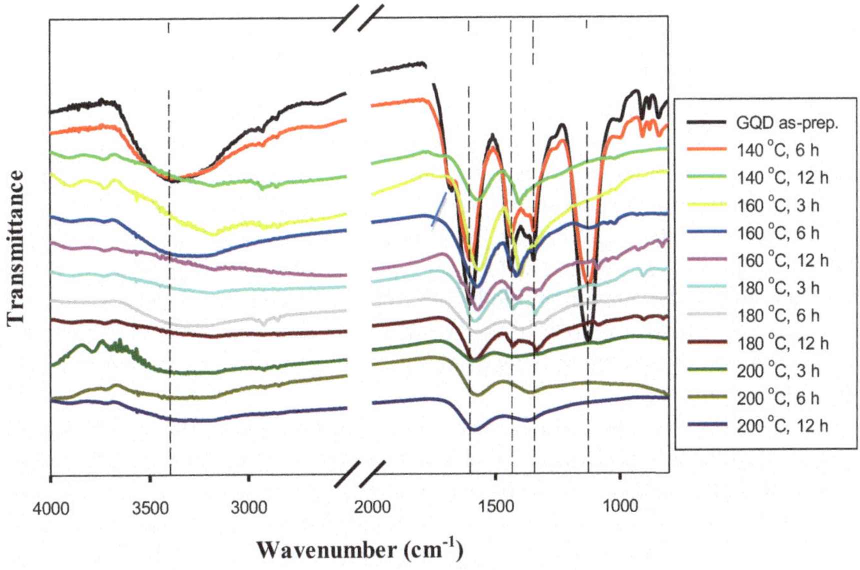

Fig. 7.13 FTIR of platinum functionalized GQDs for different annealing temperatures and times. Reprinted with permission from (Song, Y.; Chen, S.: Graphene quantum-dot-supported platinum nanoparticles: defect-mediated electrocatalytic activity in oxygen reduction. *ACS Appl Mater Interfaces* **2014**, *6*, 14050–60). Copyright (2014) American Chemical Society

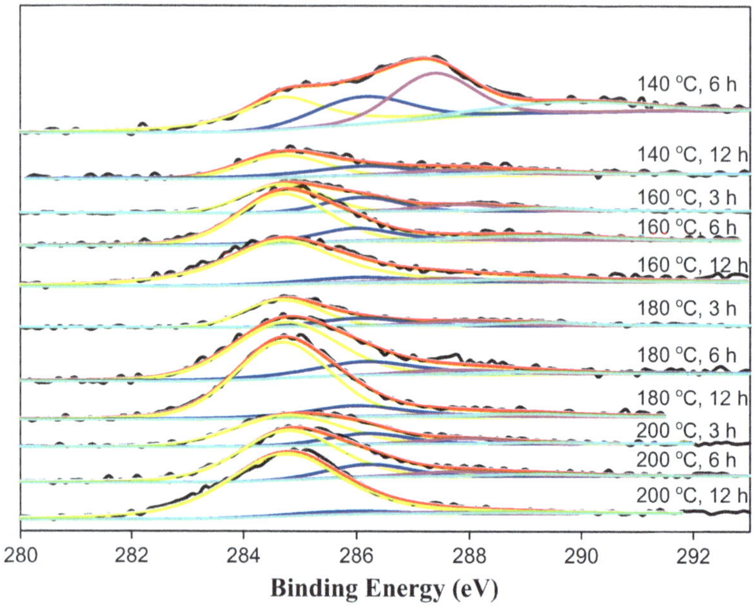

Fig. 7.14 XPS of C 1s electrons of platinum functionalized GQDs for different annealing temperatures and times. Reprinted with permission from (Song, Y.; Chen, S.: Graphene quantum-dot-supported platinum nanoparticles: defect-mediated electrocatalytic activity in oxygen reduction. *ACS Appl Mater Interfaces* **2014**, *6*, 14050–60). Copyright (2014) American Chemical Society

disappeared with increasing temperature and heating time. In addition, deconvolution of the C1s signal from the XPS spectra shows a diminishment of the C=O and COOH components as temperature and time increase with an almost complete removal of these groups for samples when treated at 200 °C for 12 h.

Furthermore, Raman spectroscopic measurements yielded additional insights into the nature of the GQD support by examining the intensity ratio of the defect (D) and graphitic (G) bands. The G band represents the E2g phonon while the D band represents the breathing mode of κ-point phonons with A1g symmetry [71, 72]. The intensity ratio of these two bands is well correlated with the XPS and FTIR results where more defects are removed with a longer heating time and higher temperature.

The samples were then tested for oxygen reduction and it was found that all samples facilitated four-electron reduction of oxygen and the onset potentials were more positive than +0.80 V vs RHE. The best ORR performance was demonstrated by the samples treated at 160 °C for 12 h and at 180 °C for 6 h which had mass activities of 468.1 A/g and 435.7 A/g, and specific activities of 26.2 A/m^2 and 24.6 A/m^2, respectively, as depicted in Fig. 7.15.

Furthermore, the activities exhibit a volcano shaped variation with defect concentration as a result of direct manipulation of the electronic structure of the Pt particles by the GQD defects. This trend is in agreement with the dynamics of

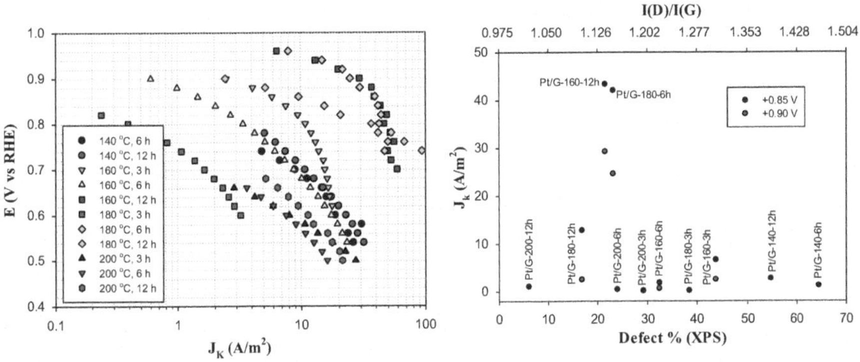

Fig. 7.15 (*Left*) Tafel plot of Pt nanoparticle functionalized GQDs prepared at different annealing temperatures for a different period of times. (*Right*) Variation of the ORR kinetic current density at +0.85 (*solid circles*) and +0.90 V (*empty circles*) with GQD structural defects manifested as the defect concentrations from XPS measurements and I(D)/I(G) ratio in Raman spectroscopic measurements. Reprinted with permission from (Song, Y.; Chen, S.: Graphene quantum-dot-supported platinum nanoparticles: defect-mediated electrocatalytic activity in oxygen reduction. *ACS Appl Mater Interfaces* **2014**, *6*, 14050–60). Copyright (2014) American Chemical Society

oxygen reduction in that a large number of defects will withdraw too much electron density and limit the first step of reduction while samples with a small amount or no defects will not diminish significant electron density resulting in the tight binding between OIs and Pt and an unfavorable release of the final product. Importantly, the peak performance in this study met the DOE target for 2017.

The results of these studies tie in very well with the *d* band model proposed earlier if we allow the downshift of the *d* band center when electrons are withdrawn and an upshift when charge is provided. A physical manifestation of this notion is apparent in Fig. 7.10 where it can be seen that electron withdrawing groups enhance the activity of Pt when compared to a bare surface while the activity diminishes when the group is electron-donating. Although much can be concluded from these results, more research is still needed to bring the level of quantitative understanding concerning the impact on *d* band structure from organic capping ligands to that of the impacts of alloying on *d* band structure.

7.4 Metal Oxide-Based Catalysts

As the battle for reducing the mass and enhancing the performance of platinum electrocatalysts continues, metal oxides offer a cheap, abundant alternative to precious metal catalysis. Although the mechanism of oxygen reduction on a metal oxide surface is slightly different than that on a traditional metal electrocatalyst (Pt, Pd), optimization of binding interactions between the surface and O_2 is the key to significant enhancement. Consequently, there has been much focus on improving activity and many different models proposed for enhancing ORR activity, including direct manipulation of oxide vacancies [22], deliberate surface functionalization with selected chemical ligands [36], and stoichiometry and properties of the metal(s) [40].

For instance, recently it was shown that the removal of surface oxygen sites on MnO_2 was not only facile and controllable, but led to a vast improvement of the ORR activity [22]. Experimentally, β-MnO_2 nanorods were synthesized via a hydrothermal route and were further processed at elevated temperatures for different periods of times in argon or air atmospheres. The samples were denoted as Air-250-2 h, Air-400-2 h, Ar-350-2 h, Ar-450-6 h to represent the treatment conditions. Thermogravimetric analysis (TGA) showed that the loss of lattice oxygen varied with the hydrothermal conditions, with the most vacancies identified in the Ar-450-6 h sample and the least oxygen removal for the samples treated in air. In electrocatalytic testing, the oxides with moderate surface oxygen removal performed the best as predicted with current models of oxygen reduction in alkaline media.

Theoretical studies were additionally performed to assess the nature of the enhancement through oxygen removal as well as to compare samples with different vacancy concentrations. The reduction pathway for pristine MnO_2 and MnO_2 with one and two vacancies was modeled for the strength of adsorption of reaction intermediates. It was found that O_2 adsorbed onto the samples with oxygen vacancies experienced bond elongation which activated the bond and allowed for facile breaking of the bond in subsequent steps. Furthermore the barrier for the rate determining step was 0.47 eV lower for the sample with one vacancy compared that with two. The stick and ball surface models and the reaction pathway can be seen in Fig. 7.16. The intermediate level of defects again shows the best adsorptive properties and best ORR activity, in strong agreement with experimental data.

In another study, functionalization of indium tin oxide (ITO) nanoparticles with chlorine ions has also been found to lead to marked enhancement of the ORR activity. ITO nanoparticles were synthesized through a solvothermal method followed by photo irradiation in *o*-dichlorobenzene. The resulting particles showed an average diameter of 7.51 nm from TEM measurements and about 3.9 at.% chlorine from XPS analysis. The Cl-functionalized particles exhibited enhanced ORR activity, as compared to the bare particles at all potentials. Although the mechanism for such enhancement is not fully understood, it is thought that the improved activity results from the replacement of surface oxygen with chlorine leading to a lower ratio of O/In and thus quasi-oxygen deficient sites as depicted in Fig. 7.17 [36]. These sites facilitate the formation of oxide species and provide for improved reduction kinetics [73].

Additionally, perovskite oxide (ABO_3)-based electrocatalysts have displayed surface property modulation resulting from alteration of the stoichiometry and properties of the metals. A change in metal ratio or substitution for another metal has been found to drastically alter the surface properties and thus activity [40]. Similar to the *d* band theory that successfully predicts the activity of metal nanoparticles, it has been shown than the degree of e_g orbital filling of the active (B) site can accurately describe the catalytic behavior of perovskite oxides. When

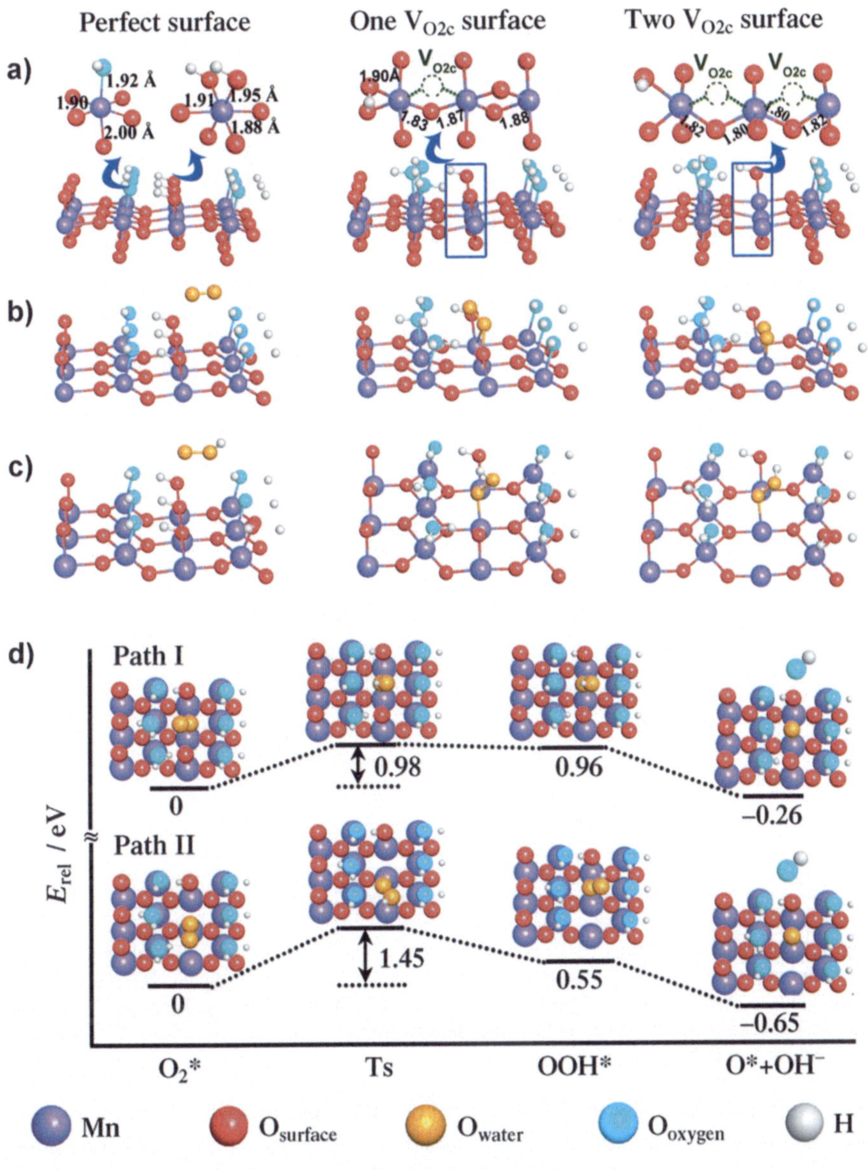

Fig. 7.16 (**a**) Ball and stick models of perfect MnO_2 surfaces and MnO_2 with 1 and 2 defects. (**b**) Adsorption of O_2. (**c**) Adsorption of OOH. (**d**) Energy diagram for reduction process. Reprinted with permission from (Cheng, F.; Su, Y.; Liang, J.; Tao, Z.; Chen, J.: MnO_2-Based Nanostructures as Catalysts for Electrochemical Oxygen Reduction in Alkaline Media. Chem Mater 2010, 22, 898–905) Copyright (2010) American Chemical Society

the e_g orbital filling is 1, there is sufficient destabilization of the B-OH bond and an energetically favorable energy transfer that promotes the facile removal of OH and addition of O_2. Filling greater than one will mean more electrons in the antibonding orbitals and thus weaker interactions and poorer initial swap between OH and O_2.

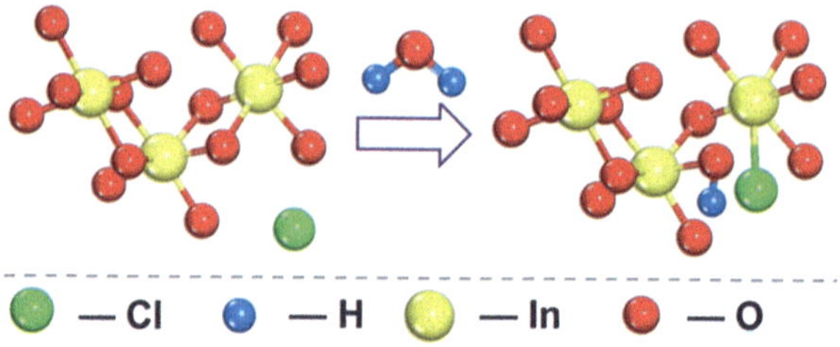

Fig. 7.17 Ball and stick model for surface functionalization of indium tin oxide with chlorine. Reprinted with permission from (Wang, N.; Niu, W.; Li, L.; Liu, J.; Tang, Z.; Zhou, W.; Chen, S.: Oxygen electroreduction promoted by quasi oxygen vacancies in metal oxide nanoparticles prepared by photoinduced chlorine doping. *Chem Commun* **2015**, *51*, 10620–10623.) Copyright (2015) Royal Society of Chemistry

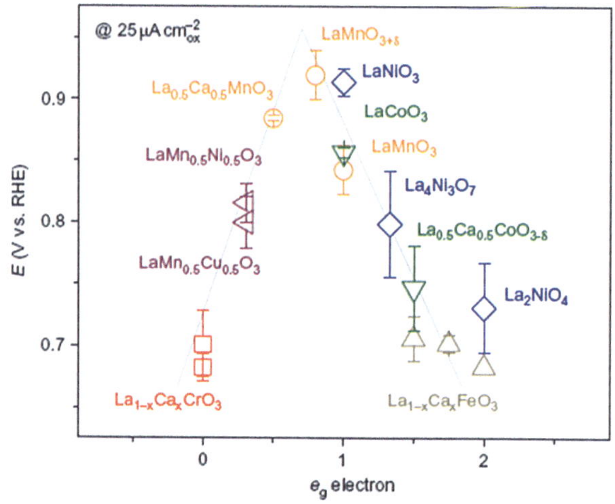

Fig. 7.18 Potentials at 25 μA cm^{-2} as a function of e_g orbital in perovskite-based oxides. Data symbols vary with type of B ions (Cr, red; Mn, orange; Fe, grey; Co, green; Ni, blue; mixed compounds, purple), where $x = 0$ and 0.5 for Cr, and 0, 0.25 and 0.5 for Fe. Error bars represent standard deviations. Reproduced with permission from Nat Chem 2011, *3*, 546–550. Copyright 2011 Macmillan Publishers Ltd

When e_g filling is smaller than 1, there will be little contribution to the formation of antibonding orbitals and the bonding interaction will be too strong and hence limit hydroxide regeneration. In fact, the plot of activity vs e_g filling shows an apparent volcano-shaped trend (Fig. 7.18), with the peak activity 4 times that of the poorest performance in the series.

Interestingly, this model assumes that surface atoms exhibit localized e_g orbitals facing outwards towards the adsorbate rather than a delocalized *d* band model. The localization of e_g orbitals on B sites is seen as a consequence of the different bonding symmetry of surface atoms, as compared to the bulk. The site activity is therefore dependent on the surrounding metals and concentration of each as previously mentioned. This study is yet another example of surface engineering of an electrocatalyst for optimal ORR activity.

7.5 Conclusions and Perspectives

The kinetic limitations of electrocatalytic oxygen reduction has been recognized as the bottleneck for efficient energy extraction in fuel cell technologies and therefore many techniques have been developed to optimize catalytic activity. In this chapter we have reviewed recent examples of interfacial engineering of nanoparticle catalysts, including metal alloys in the forms of homogenous and segregated composition, organic functionalization, as well as oxygen vacancy control and surface orbital manipulation of metal oxides. As the strength of interactions between the catalyst surface and oxygenated intermediates is largely responsible for the kinetic limitations, the common theme for all of these techniques is alteration of the electronic structure of the active metal that will result in more favorable interactions with oxygenated intermediates. Since the binding interaction between a metal surface and each individual reaction intermediate can not be altered independently, an optimal binding interaction exists. This chapter has outlined the experimental techniques utilized to reach this peak in addition to the theoretical models to quantify and predict catalytic behaviors. Specifically, the measure of a metal's *d* band center has been directly linked to the metal adsorbate binding energy. Models relating the effects of alloying on the *d* band center are well developed and are in strong agreement with experimental evidence. Additionally much progress has been made towards understanding the origin of enhancement from organic ligand functionalization and how that may be exploited for further catalytic gain. Finally, the rate limiting steps for ORR on oxide surfaces were recognized as the OH/O_2 displacement and the regeneration of surface OH. To predict and explain the behavior of various perovskite oxides, the filling of the e_g antibonding orbitals serve more than a sufficient indicator. Even as the field of electrocatalysis is rich in systematic activity enhancements and strong theoretical models, much work is still needed to bring this technology to commercial settings.

Acknowledgment The authors thank the National Science Foundation for partial support of the work.

References

1. Wagner FT, Lakshmanan B, Mathias MF (2010) Elecrochemistry and the future of the automobile. J Phys Chem Lett 1:2204–2219. doi:10.1021/jz100553m|J
2. Cano-Castillo U (2013) Hydrogen and fuel cells: potential elements in the energy transition scenario. Rev Mex Fis 59(2):85–92
3. Stephens IEL, Bondarenko AS, Grønbjerg U, Rossmeisl J, Chorkendorff I (2012) Understanding the electrocatalysis of oxygen reduction on platinum and its alloys. Energy Environ Sci 5 (5):6744. doi:10.1039/c2ee03590a
4. Norskov JK, Rossmeisl J, Logadottir A, Lindqvist L, Kitchin JR, Bligaard T, Jonsson H (2004) Origin of the overpotential for oxygen reduction at a fuel-cell cathode. J Phys Chem B 108:17886–17892

5. Song C, Zhang J (2008) PEM fuel cell electrocatalysis and catalyst layers: fundamentals and applications. Electrocatalytic oxygen reduction reaction. Springer, New York
6. Rabis A, Rodriguez P, Schmidt TJ (2012) Electrocatalysis for polymer electrolyte fuel cells: recent achievements and future challenges. ACS Catal 2(5):864–890. doi:10.1021/cs3000864
7. Lim D-H, Wilcox J (2012) Mechanisms of the oxygen reduction reaction on defective graphene-supported Pt nanoparticles from first-principles. J Phys Chem C 116 (5):3653–3660. doi:10.1021/jp210796e
8. Sabatier P (1911) Announcement. Hydrogenation and dehydrogenation for catalysis. Ber Dtsch Chem Ges 44:1984–2001. doi:10.1002/cber.19110440303
9. Kinoshita K (1990) Particle-size effects for oxygen reduction on highly dispersed platinum in acid electrolytes. J Electrochem Soc 137(3):845–848
10. Yano H, Inukai J, Uchida H, Watanabe M, Babu PK, Kobayashi T, Chung JH, Oldfield E, Wieckowski A (2006) Particle-size effect of nanoscale platinum catalysts in oxygen reduction reaction: an electrochemical and Pt-195 EC-NMR study. Phys Chem Chem Phys 8 (42):4932–4939. doi:10.1039/B610573d
11. Lim B, Jiang MJ, Camargo PHC, Cho EC, Tao J, Lu XM, Zhu YM, Xia YN (2009) Pd-Pt bimetallic nanodendrites with high activity for oxygen reduction. Science 324 (5932):1302–1305. doi:10.1126/science.1170377
12. Chen ZW, Waje M, Li WZ, Yan YS (2007) Supportless Pt and PtPd nanotubes as electrocatalysts for oxygen-reduction reactions. Angew Chem Int Ed 46(22):4060–4063. doi:10.1002/Anie.200700894
13. Xiao L, Zhuang L, Liu Y, Lu JT, Abruna HD (2009) Activating Pd by morphology tailoring for oxygen reduction. J Am Chem Soc 131(2):602–608
14. Savadogo O, Lee K, Oishi K, Mitsushima S, Kamiya N, Ota KI (2004) New palladium alloys catalyst for the oxygen reduction reaction in an acid medium. Electrochem Commun 6 (2):105–109. doi:10.1016/J.Elecom.2003.10.020
15. Shao MH, Sasaki K, Adzic RR (2006) Pd-Fe nanoparticles as electrocatalysts for oxygen reduction. J Am Chem Soc 128(11):3526–3527. doi:10.1021/Ja060167d
16. Zhang JL, Vukmirovic MB, Sasaki K, Nilekar AU, Mavrikakis M, Adzic RR (2005) Mixed-metal Pt monolayer electrocatalysts for enhanced oxygen reduction kinetics. J Am Chem Soc 127(36):12480–12481
17. Zhang JL, Vukmirovic MB, Xu Y, Mavrikakis M, Adzic RR (2005) Controlling the catalytic activity of platinum-monolayer electrocatalysts for oxygen reduction with different substrates. Angew Chem Int Ed 44(14):2132–2135
18. Greeley J, Stephens IEL, Bondarenko AS, Johansson TP, Hansen HA, Jaramillo TF, Rossmeisl J, Chorkendorff I, Norskov JK (2009) Alloys of platinum and early transition metals as oxygen reduction electrocatalysts. Nat Chem 1(7):552–556. doi:10.1038/Nchem.367
19. Bing YH, Liu HS, Zhang L, Ghosh D, Zhang JJ (2010) Nanostructured Pt-alloy electrocatalysts for PEM fuel cell oxygen reduction reaction. Chem Soc Rev 39 (6):2184–2202. doi:10.1039/B912552c
20. Zhou Z-Y, Kang X, Song Y, Chen S (2012) Ligand-mediated electrocatalytic activity of pt nanoparticles for oxygen reduction reactions. J Phys Chem C 116(19):10592–10598. doi:10.1021/jp300199x
21. He G, Song Y, Phebus B, Liu K, Deming CP, Hu P, Chen S (2013) Electrocatalytic activity of organically functionalized silver nanoparticles in oxygen reduction. Sci Adv Mater 5 (11):1727–1736. doi:10.1166/sam.2013.1624
22. Cheng FY, Zhang TR, Zhang Y, Du J, Han XP, Chen J (2013) Enhancing electrocatalytic oxygen reduction on MnO_2 with vacancies. Angew Chem Int Ed 52(9):2474–2477. doi:10.1002/anie.201208582
23. Strasser P, Koh S, Anniyev T, Greeley J, More K, Yu CF, Liu ZC, Kaya S, Nordlund D, Ogasawara H, Toney MF, Nilsson A (2010) Lattice-strain control of the activity in dealloyed core-shell fuel cell catalysts. Nat Chem 2(6):454–460. doi:10.1038/Nchem.623

24. Kitchin JR, Norskov JK, Barteau MA, Chen JG (2004) Modification of the surface electronic and chemical properties of Pt(111) by subsurface 3d transition metals. J Chem Phys 120 (21):10240–10246. doi:10.1063/1.1737365
25. Stephens IE, Bondarenko AS, Perez-Alonso FJ, Calle-Vallejo F, Bech L, Johansson TP, Jepsen AK, Frydendal R, Knudsen BP, Rossmeisl J, Chorkendorff I (2011) Tuning the activity of Pt (111) for oxygen electroreduction by subsurface alloying. J Am Chem Soc 133 (14):5485–5491. doi:10.1021/ja111690g
26. Stamenkovic VR, Mun BS, Arenz M, Mayrhofer KJJ, Lucas CA, Wang GF, Ross PN, Markovic NM (2007) Trends in electrocatalysis on extended and nanoscale Pt-bimetallic alloy surfaces. Nat Mater 6(3):241–247
27. Stamenkovic VR, Fowler B, Mun BS, Wang GF, Ross PN, Lucas CA, Markovic NM (2007) Improved oxygen reduction activity on Pt3Ni(111) via increased surface site availability. Science 315(5811):493–497
28. Zhou ZY, Kang XW, Song Y, Chen SW (2012) Enhancement of the electrocatalytic activity of Pt nanoparticles in oxygen reduction by chlorophenyl functionalization. Chem Commun 48 (28):3391–3393. doi:10.1039/C2cc17945h
29. Liu K, Kang XW, Zhou ZY, Song Y, Lee LJ, Tian D, Chen SW (2013) Platinum nanoparticles functionalized with acetylene derivatives: Electronic conductivity and electrocatalytic activity in oxygen reduction. J Electroanal Chem 688:143–150
30. He GQ, Song Y, Liu K, Walter A, Chen S, Chen SW (2013) Oxygen reduction catalyzed by platinum nanoparticles supported on graphene quantum dots. ACS Catal 3(5):831–838. doi:10.1021/Cs400114s
31. Song Y, Chen SW (2014) Graphene quantum-dot-supported platinum nanoparticles: defect-mediated electrocatalytic activity in oxygen reduction. ACS Appl Mater Interfaces 6(16):14050–14060. doi:10.1021/Am503388z
32. Chen W, Chen SW, Ding FZ, Wang HB, Brown LE, Konopelski JP (2008) Nanoparticle-mediated intervalence transfer. J Am Chem Soc 130(36):12156–12162. doi:10.1021/Ja803887b
33. Chen W, Zuckerman NB, Kang XW, Ghosh D, Konopelski JP, Chen SW (2010) Alkyne-protected ruthenium nanoparticles. J Phys Chem C 114(42):18146–18152
34. Kang XW, Chen W, Zuckerman NB, Konopelski JP, Chen SW (2011) Intraparticle charge delocalization of carbene-functionalized ruthenium nanoparticles manipulated by selective ion binding. Langmuir 27(20):12636–12641
35. Zhou ZY, Ren J, Kang X, Song Y, Sun SG, Chen S (2012) Butylphenyl-functionalized Pt nanoparticles as CO-resistant electrocatalysts for formic acid oxidation. Phys Chem Chem Phys 14(4):1412–1417. doi:10.1039/c1cp23183a
36. Wang N, Niu W, Li L, Liu J, Tang Z, Zhou W, Chen S (2015) Oxygen electroreduction promoted by quasi oxygen vacancies in metal oxide nanoparticles prepared by photoinduced chlorine doping. Chem Commun 51:10620–10623. doi:10.1039/C5CC02808F
37. Cheng F, Su Y, Liang J, Tao Z, Chen J (2010) MnO_2-based nanostructures as catalysts for electrochemical oxygen reduction in alkaline media. Chem Mater 22(3):898–905. doi:10.1021/cm901698s
38. Cheng FY, Chen J (2012) Metal-air batteries: from oxygen reduction electrochemistry to cathode catalysts. Chem Soc Rev 41(6):2172–2192. doi:10.1039/c1cs15228a
39. Goodenough JB, Cushing BL (2003) Handbook of fuel cells-fundamentals, technology and applications, vol 2. Wiley, New York
40. Suntivich J, Gasteiger HA, Yabuuchi N, Nakanishi H, Goodenough JB, Shao-Horn Y (2011) Design principles for oxygen-reduction activity on perovskite oxide catalysts for fuel cells and metal-air batteries. Nat Chem 3(7):546–550
41. Xiong L, Kannan AM, Manthiram A (2002) Pt-M (M = Fe, Co, Ni and Cu) electrocatalysts synthesized by an aqueous route for proton exchange membrane fuel cells. Electrochem Commun 4(11):898-903. Pii: S1388-2481(02)00485x

42. Wu JB, Zhang JL, Peng ZM, Yang SC, Wagner FT, Yang H (2010) Truncated octahedral Pt3Ni oxygen reduction reaction electrocatalysts. J Am Chem Soc 132(14):4984–4985. doi:10.1021/ja100571h
43. Mavrikakis M, Hammer B, Norskov JK (1998) Effect of strain on the reactivity of metal surfaces. Phys Rev Lett 81(13):2819–2822
44. Rodriguez JA, Goodman DW (1992) The nature of the metal bond in bimetallic surfaces. Science 257(5072):897–903. doi:10.1126/Science.257.5072.897
45. Chen MS, Kumar D, Yi CW, Goodman DW (2005) The promotional effect of gold in catalysis by palladium-gold. Science 310(5746):291–293. doi:10.1126/Science.1115800
46. Stamenkovic V, Mun BS, Mayrhofer KJJ, Ross PN, Markovic NM, Rossmeisl J, Greeley J, Norskov JK (2006) Changing the activity of electrocatalysts for oxygen reduction by tuning the surface electronic structure. Angew Chem Int Ed 45(18):2897–2901
47. Xiao L, Huang B, Zhuang L, Lu JT (2011) Optimization strategy for fuel-cell catalysts based on electronic effects. Rsc Adv 1(7):1358–1363. doi:10.1039/C1ra00378j
48. Greeley J, Norskov JK (2009) Combinatorial density functional theory-based screening of surface alloys for the oxygen reduction reaction. J Phys Chem C 113(12):4932–4939. doi:10.1021/Jp808945y
49. Fernandez JL, Raghuveer V, Manthiram A, Bard AJ (2005) Pd-Ti and Pd-Co-Au electrocatalysts as a replacement for platinum for oxygen reduction in proton exchange membrane fuel cells. J Am Chem Soc 127(38):13100–13101. doi:10.1021/Ja0534710
50. Suo YG, Zhuang L, Lu JT (2007) First-principles considerations in the design of Pd-alloy catalysts for oxygen reduction. Angew Chem Int Ed 46(16):2862–2864. doi:10.1002/Anie.200604332
51. Sleightholme AES, Varcoe JR, Kucernak AR (2008) Oxygen reduction at the silver/hydroxide-exchange membrane interface. Electrochem Commun 10(1):151–155. doi:10.1016/J.Elecom.2007.11.008
52. Guo JS, Hsu A, Chu D, Chen RR (2010) Improving oxygen reduction reaction activities on carbon-supported Ag nanoparticles in alkaline solutions. J Phys Chem C 114(10):4324–4330. doi:10.1021/Jp910790u
53. Chatenet M, Genies-Bultel L, Aurousseau M, Durand R, Andolfatto F (2002) Oxygen reduction on silver catalysts in solutions containing various concentrations of sodium hydroxide—comparison with platinum. J Appl Electrochem 32(10):1131–1140. doi:10.1023/A:1021231503922
54. Varcoe JR, Slade RCT (2005) Prospects for alkaline anion-exchange membranes in low temperature fuel cells. Fuel Cells 5(2):187–200. doi:10.1002/Fuce.200400045
55. Slanac DA, Hardin WG, Johnston KP, Stevenson KJ (2012) Atomic ensemble and electronic effects in Ag-rich AgPd nanoalloy catalysts for oxygen reduction in alkaline media. J Am Chem Soc 134(23):9812–9819. doi:10.1021/Ja303580b
56. Chen S, Templeton AC, Murray RW (2000) Monolayer-protected cluster growth dynamics. Langmuir 16:3543–3548
57. Templeton AC, Wuelfing MP, Murray RW (2000) Monolayer protected cluster molecules. Acc Chem Res 33(1):27–36
58. Cavaliere S, Fdr R, Etcheberry A, Herlem M, Perez H (2004) Direct electrocatalytic activity of capped platinum nanoparticles toward oxygen reduction. Electrochem Solid-State Lett 7(10): A358. doi:10.1149/1.1792259
59. Baret B, Aubert PH, L'Hermite MM, Pinault M, Reynaud C, Etcheberry A, Perez H (2009) Nanocomposite electrodes based on pre-synthesized organically capped platinum nanoparticles and carbon nanotubes. Part I: Tuneable low platinum loadings, specific H upd feature and evidence for oxygen reduction. Electrochim Acta 54(23):5421–5430. doi:10.1016/j.electacta.2009.04.033
60. Genorio B, Strmcnik D, Subbaraman R, Tripkovic D, Karapetrov G, Stamenkovic V, Pejovnik S, Markovic N (2010) Selective catalysts for the hydrogen oxidation and oxygen

reduction reactions by patterning of platinum with calix[4]arene molecules. Nat Mater 9:998–1003. doi:10.1038/nmat2883

61. Strmcnik D, Escudero-Escribano M, Kodama K, Stamenkovic V, Cuesta A, Markovic N (2010) Enhanced electrocatalysis of the oxygen reduction reaction based on patterning of platinum surfaces with cyanide. Nat Chem 2:880–885. doi:10.1038/nchem.771
62. Pietron JJ, Garsany Y, Baturina O, Swider-Lyons KE, Stroud RM, Ramaker DE, Schull TL (2008) Electrochemical observation of ligand effects on oxygen reduction at ligand-stabilized Pt nanoparticle electrocatalysts. Electrochem Solid-State Lett 11(8):B161. doi:10.1149/1.2937448
63. Kostelansky CN, Pietron JJ, Chen MS, Dressick WJ, Swider-Lyons KE, Ramaker DE, Stroud RM, Klug CA, Zelakiewicz BS, Schull TL (2006) Triarylphosphine-stabilized platinum nanoparticles in three-dimensional nanostructured films as active electrocatalysts. J Phys Chem B 110(43):21487–21496. doi:10.1021/Jp062663u
64. Song Y, Liu K, Chen SW (2012) AgAu bimetallic janus nanoparticles and their electrocatalytic activity for oxygen reduction in alkaline media. Langmuir 28 (49):17143–17152. doi:10.1021/La303513x
65. Lima FHB, Zhang J, Shao MH, Sasaki K, Vukmirovic MB, Ticianelli EA, Adzic RR (2007) Catalytic activity-d-band center correlation for the O_2 reduction reaction on platinum in alkaline solutions. J Phys Chem C 111:404–410
66. Hull RV, Li L, Xing YC, Chusuei CC (2006) Pt nanoparticle binding on functionalized multiwalled carbon nanotubes. Chem Mater 18(7):1780–1788. doi:10.1021/Cm0518978
67. Palaniselvam T, Irshad A, Unni B, Kurungot S (2012) Activity modulated low platinum content oxygen reduction electrocatalysts prepared by inducing nano-order dislocations on carbon nanofiber through N-2-doping. J Phys Chem C 116(28):14754–14763. doi:10.1021/Jp300881p
68. Timperman L, Feng YJ, Vogel W, Alonso-Vante N (2010) Substrate effect on oxygen reduction electrocatalysis. Electrochim Acta 55(26):7558–7563. doi:10.1016/j.electacta.2009.09.076
69. Vogel W, Timperman L, Alonso-Vante N (2010) Probing metal substrate interaction of Pt nanoparticles: structural XRD analysis and oxygen reduction reaction. Appl Catal Gen 377 (1–2):167–173. doi:10.1016/j.apcata.2010.01.034
70. Liu X, Yao KX, Meng CG, Han Y (2012) Graphene substrate-mediated catalytic performance enhancement of Ru nanoparticles: a first-principles study. Dalton Trans 41(4):1289–1296. doi:10.1039/C1dt11186h
71. Ferrari AC, Basko DM (2013) Raman spectroscopy as a versatile tool for studying the properties of graphene. Nat Nanotechnol 8(4):235–246
72. Dresselhaus MS, Terrones M (2013) Carbon-based nanomaterials from a historical perspective. Proc IEEE 101(7):1522–1535
73. Kim J, Yin X, Tsao KC, Fang S, Yang H (2014) Ca(2)Mn(2)O(5) as oxygen-deficient perovskite electrocatalyst for oxygen evolution reaction. J Am Chem Soc 136 (42):14646–14649. doi:10.1021/ja506254g

Chapter 8
Primary Oxide Latent Storage and Spillover for Reversible Electrocatalysis in Oxygen and Hydrogen Electrode Reactions

Milan M. Jaksic, Angeliki Siokou, Georgios D. Papakonstantinou, and Jelena M. Jaksic

8.1 Introduction

Ever since Sir William Grove invented fuel cells (FCs) [1], electrocatalysts of the oxygen electrode (ROE) reactions became the main imperative target, challenge and dream in the whole electrochemical science of aqueous and later PEM media, while its substantiation is the subject matter of the present paper. The main focus has always primarily been posed on noble (Pt, Au, Ru) metals, while the reversible primary (Pt-OH, Au-OH, Ru-OH), and polarizable surface (Pt=O, Au=O, Ru=O) oxides, along with H-adatoms (Pt-H, Au-H, Ru-H) represent interactive species defining the overall electrode behavior and properties. In such a respect potentiodynamic spectra, (Fig. 8.1), usually reveal within a narrow potential range the highly reversible peaks of primary oxide adsorptive growth and its backwards desorptive removal [2–4],

$$Pt + 2H_2O \leftrightarrow Pt\text{-}OH + H_3O^+ + e^- \tag{8.1}$$

as a typical double layer (DL) charging and discharging pseudo-capacitance [5–11], while the former soon later irreversibly disproportionates into the strongly polarizable, more adsorptive and remarkably more stable surface oxide (Pt=O) monolayer [2–4],

M.M. Jaksic
Institute of Chemical Engineering Sciences, CEHT/FORTH, Patras, Greece

Faculty of Agriculture, University of Belgrade, Belgrade, Serbia
e-mail: milan@iceht.forth.gr

A. Siokou • G.D. Papakonstantinou • J.M. Jaksic (✉)
Institute of Chemical Engineering Sciences, CEHT/FORTH, Patras, Greece
e-mail: jelena@iceht.forth.gr

K.I. Ozoemena, S. Chen (eds.), *Nanomaterials for Fuel Cell Catalysis*, Nanostructure Science and Technology, DOI 10.1007/978-3-319-29930-3_8

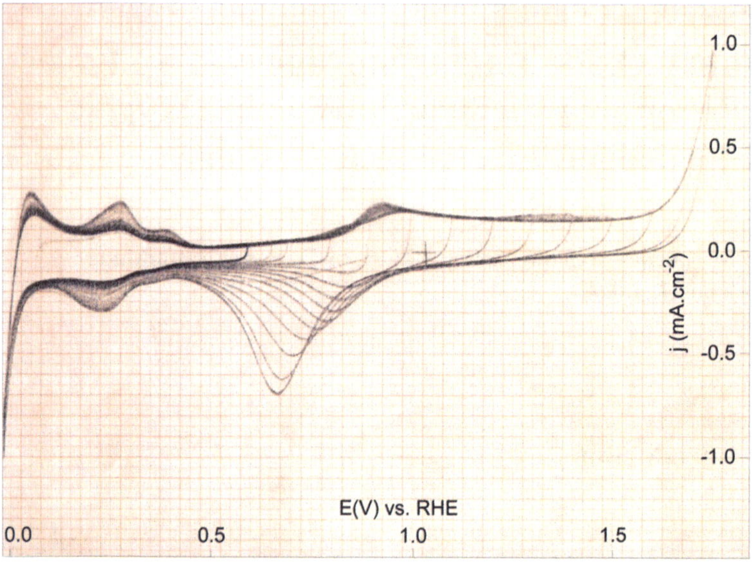

Fig. 8.1 Cyclic voltammograms of polycrystalline Pt scanned in 0.1 M NaOH at sweep rate 100 mV s^{-1}(Reprinted with permission from [7]. Copyright (2014) American Chemical Society)

$$\text{Pt-OH} + H_2O \rightarrow \text{Pt} = O + H_3O^+ + e^- \tag{8.2}$$

With the corresponding pronounced irreversible desorption peak all along the equivalent reverse potential scan, back and forth. The former, (Eq. 8.1), arises as the highly reversible reaction, proceeds independently and extremely fast, while the latter, (Eq. 8.2), as a typical irreversible disproportionation reaction step, is rather slow and highly polarizable. Such state of the experimental evidence reveals and definitely defines electrocatalytic fate and properties of all plain Pt(Au), and non-interactive (Pt/C, Au/C) supported noble and in particular other hypo ($n \leq 5$) and hyper ($n > 5$) d-metals along transition series of elements. In such circumstances, there is no plain individual d-metal which might feature electrocatalytic properties all along the potential axis between hydrogen and oxygen evolving limits, but partially reversible and partially polarizable.

In fact, in the absence of the (Pt-OH $\rightarrow$ 0), therefrom imposes such a critical and typical, very pronounced the ***reaction polarization range*** (Pt=O $\rightarrow$ 1, Eq. 8.2 and Fig. 8.1), all along, after the primary oxide adsorption/desorption peak, until the oxygen potential evolving limits, and in both scan directions, back and forth [5–11]. Such highly reversible (Eq. 8.1), relative to the subsequent strongly irreversible, and substantially polarizable (Eq. 8.2) transient and coexistence of two interrelated reaction steps, has been decisive as concerns the overall electrocatalytic properties and behavior of all d-metals all along transition series, on the first place for the oxygen electrode reactions (ORR and OER).

The present analysis is best reflected and proved by comparison of Figs. 8.1, 8.2 and 8.3, once when missing Pt-OH spillover (Pt-OH $\rightarrow$ 0, Pt=O $\rightarrow$ 1), then when

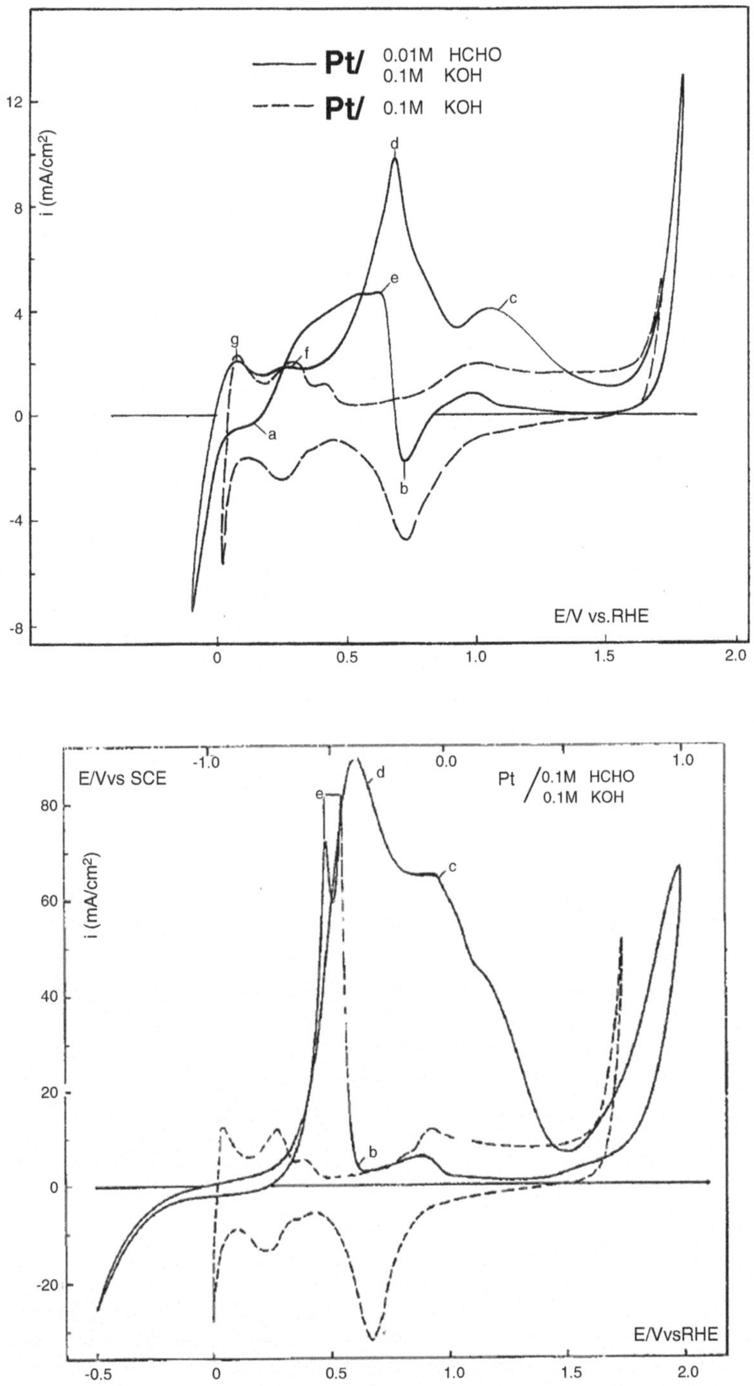

Fig. 8.2 (**a**, **b**) Cyclic voltammograms scanned on a polycrystalline Pt wire electrode in alkaline (0.1 M KOH, dashed lines) solution and in admixture of formaldehyde (0.01 M (**a**) and 0.1 M HCHO (**b**), *full lines*) at 200 mV s^{-1} sweep rate between hydrogen and oxygen potential evolving

there proceeds simple effusion of self-generated Pt-OH, or finally, the enriched latent storage spillover enables enormous primary oxide adsorptive deposition and reverse desorption, (Pt-OH $\rightarrow$ 1, Pt=O $\rightarrow$ 0), exactly as big capacitors do, and when continuously and reversibly present all over the potentiodynamic cycle (see further downwards detailed stepwise discussion).

8.2 Electrocatalysis of Oxygen Electrode Reactions: Initial Ideas

Since the equimolar ratio of the primary and surface oxide concentrations defines the optimal interfering self-catalytic spillover reaction step in cathodic oxygen reduction reaction mechanism (ORR) [12],

$$\text{Pt-OH} + \text{Pt} = \text{O} + 3\text{H}_3\text{O}^+ + 3\text{e}^- \rightarrow 5\text{H}_2\text{O} + 2\text{Pt} \tag{8.3}$$

and in particular along the reversible (low slope, 30 mV/dec, or even lower) parts of Tafel line plots, the irreversible disproportionation (Eq. 8.2) imposes an extremely high reaction polarization barrier (see later downwards) that amounts for even more than 600 mV s and then, in absence of the external Pt-OH supply, makes plain Pt (and Au) and non-interactive supported both (Pt/C or Au/C) platinum and gold irreversible for the oxygen electrode reactions within the broader potential range, back and forth. The thermodynamic definition of irreversibility then would state that plain Pt(Au) by no means can feature the reversibility and/or (electro)catalytic activity all along the potential axis between hydrogen and oxygen evolving limits, or more specifically, within the whole potentiodynamic circle. The strongly adsorptive and thence highly polarizable Pt=O (Au=O), deprived from any local and/or external Pt-OH (Au-OH) source and supply, then defines one of the most pronounced issues of the reaction polarization in the entire electrochemical science: No Pt-OH, means that there is no reversible reaction (Eq. 8.3). Quite another story arises when nanostructured Pt(Au) electrocatalyst is selective interactive grafting bonded on various, in particular mixed valence hypo-d-(f)-oxide supports.

Whereas hydrogen molecules undergo spontaneous adsorptive dissociation on plain Pt (Pt/C) yielding H-adatoms (Pt-H) to establish thermodynamic equilibrium of the RHE (Reversible Hydrogen Electrode, (Pt(H_2)/Pt-H/H_3O^+)), within the

Fig. 8.2 (continued) limits. Labels: (*a*) reversible hydrogen adsorption peak; (*b*) irreversible Pt surface oxide (Pt=O) desorption peak; (*c*, *d*) successive peaks of anodic aldehyde oxidation; (*e*) sudden sharp current jump and reverse peak of repeated HCHO oxidation in the course of successive cathodic scan; (*g*,*f*) reversible H-adatoms oxidation and desorption peaks, respectively (Reprinted with permission from [7]. Copyright (2014) American Chemical Society)

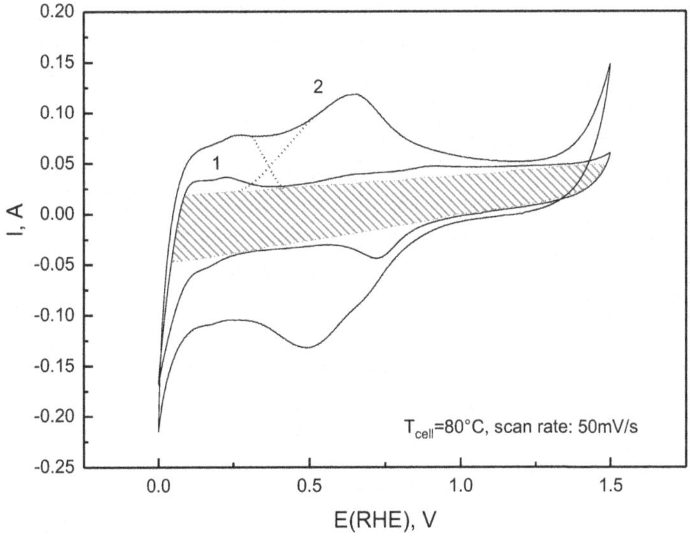

Fig. 8.3 Cyclic voltammograms of mixed valence hypo-d-oxides supported nanostructured Pt electrode (Pt/TiO_2,WO_3/C), scanned in He stream, ones at negligible moisture content (*curve 1*) and at 80 °C water vapor saturation (*curve 2*) (Reprinted with permission from [7]. Copyright (2014) American Chemical Society)

above identified critical potential range Pt (Au) is deprived (Eq. 8.2), from the Pt-OH (Au-OH) to provide reversible properties for the ROE ((Pt(O_2)/Pt-OH, Pt=O)/OH^-), (Eq. 8.3), and both together assemble the reversible hydrogen fueled L&MT PEMFCs. In other words, the Pt-OH (or Au-OH) plays the same thermodynamic constitutional role for establishing the equilibrium for the ROE, as Pt-H does for the RHE, and thence, unavoidably imposes continuous need and requires a permanent source of the former (Pt-OH) to overcome the reaction polarization by the latter (Pt=O), and enables the reversible electrocatalytic properties of the ROE [7]. In such a respect, the first main step in the present concept towards the ROE implies that catalytic hyper-d-metals (Pt, Au, Ru) establish with hypo-d-(f)-oxides ($n \leq 5$, or their mixed valence compounds, like Nb_2O_5, TiO_2, CeO_2; Ta_2O_5, TiO_2, CeO_2; or WO_3, Nb_2O_5, TiO_2, GdO_2) the interactive Brewer type (d-d or d-f) bonding effect [13, 14], otherwise well known in heterogeneous catalysis as Tauster [15–18] SMSI (Strong Metal-Support Interaction), one of the strongest in the entire chemistry [5]). Consequently, this way interactive inter-bonded composite electrocatalysts (example, Pt/Nb_2O_5, TiO_2), while strong bonding, impose remarkably stretched d-orbitals, and thereby exhibit much weaker adsorptive inter-bonding strengths of intermediates (Pt-H, Pt-OH) in the RDSs (Rate Determining Steps), thence correspondingly facilitated cleavage of the latter, and thereby, increased the catalytic activity both for hydrogen and oxygen electrode reactions. Meanwhile, some other accompanying effects, in particular the ones associated with the Pt-OH latent storage and spillover, play even more significant and decisive role for the latter [5–7, 19, 20]. In such a respect, the dramatically pronounced Pt-OH spillover effect has primarily been already noticed as the remarkable increased CO-tolerance [5]. Then, hypo-d-oxides and their mixed valence compounds, as based on typical d-d-(f)-metallic bonds, exhibit extra high stability in both acidic and alkaline media, and many of them pronounced (above 300 S/cm) electron conductivity. Furthermore, majority of hypo-d-(f)-oxides and in particular of higher altervalent numbers, feature prevailingly high percentage of dissociative water molecules adsorption [21, 22], (Eq. (8.1-a)), and thence, the mostly enhanced surface membrane type of hydroxyl ions migration mass transfer [23–25], (Eqs. (8.1-b) and (8.1-c)),

$$Nb_2O_5 + 5H_{s2}O \leftrightarrow 2Nb(OH)_5 \quad (8.1\text{-a})$$

$$Nb(OH)_5 + Pt \rightarrow Nb(OH)^{4+} + Pt\text{-}OH + e^- \quad (8.1\text{-b})$$

$$Nb(OH)^{4+} + 2H_2O \rightarrow Nb(OH)^{5+} + H_3O^+ \quad (8.1\text{-c})$$

$$\sum Pt + 2H_2O \leftrightarrow Pt\text{-}OH + H_3O^+ + e^- \quad (8.1)$$

finally, under anodic polarization, ending up with the prevailing electron transfer to the interactive supported metallic catalyst, so that the Pt-OH behaves as a pronounced dipole species [26], and thus, exhibits the strong spillover surface repulsion, transfer and distribution. At the same time, the highly pronounced reversible potentiodynamic peaks testify for the extremely fast overall spillover (*surface*) reaction (Eq. 8.1), in both directions [2–4], primarily used for DL charging and discharging, and then being ready and available for fast heterogeneous electrocatalytic reactions. In the same context, it would be significant inferring that whenever anodic oxidation reactions of the Pt-OH (Au-OH), exceed at least for an order of magnitude the disproportionation rate (Eq. 8.2), such as the ones with HCHO [5–11], and other aldehydes, simple alcohols and their acids, then this way these succeed highly to suppress and remarkably postpone the Pt=O growth all along the potential axis till the close proximity of anodic oxygen evolving limits, and at the same time consequently show as experimentally confirmed enormous broad extension growth of the Pt-OH (Au-OH) adsorption peak (Fig. 8.2a, b). In fact, since Pt-OH(Au-OH) (electro)catalytically belong to heterogeneous surface reactions and are insoluble in aqueous media, to approach the effect similar to the one just inferred with HCHO, and then partially suppress or completely eliminate the surface oxide adsorptive growth (Eq. 8.2), along with the corresponding reaction polarization, and create the reversible electrocatalysts for the ROE (ORR, OER), (Eq. 8.3), there have been employed interactive hypo-d-(f)-oxide supports yielding the advanced continuous, recoverable and renewable latent storage of primary oxides [5–11, 19, 20]. These dramatically differ from plain Pt (Au) and impose quite another quality and behavior in electrocatalysis for the oxygen electrode reactions (ORR, OER).

In such an overall constellation, individual hypo-d-(f)-oxides, and even more so their mixed valence composite compounds, when in expanded (sol-gel synthesis with thorough liquid CO_2 supercritical drying) hydrated surface state (Fig. 8.3), behave as unlimited latent storage and spillover sources of the primary oxide [5–11, 19, 20], continuously being renewed and recovered simply by water vapor supply and instantaneously proceeding with spontaneous dissociative adsorption of water molecules [21, 22].

8.2.1 Adsorptive Dissociation of Water Molecules and Membrane Type Surface Migration

The first decisive step towards rather fast spillover widespreading phenomena of the Pt-OH is the consequence of a strong first principle thermodynamic confirmed evidence (Density Functional Calculations, DFC) by Vittadini et al. [21], that water molecules undergo prevailing spontaneous dissociative adsorption on anatase (1 0 1), and even rutile titania, and more so on the higher altervalent oxides [22] of tungsten, molybdenum, tantalum, niobium and/or cerium, etc. (Fig. 8.4), as the general oxophilicity properties of hypo-d-metal oxides. In addition, the first-principles molecular-dynamic simulations showed the existence of a mechanism for thermodynamically favored spontaneous dissociation of water molecules even

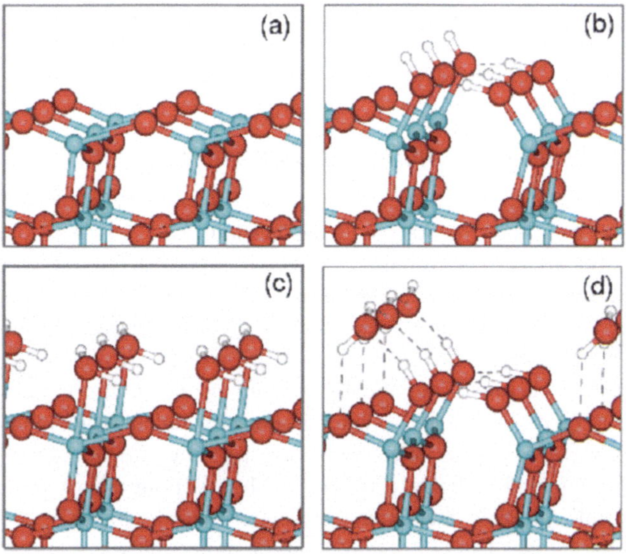

Fig. 8.4 The perspective views of DFT-optimized atomic structures for: (**a**) the clean anatase (ADM) ad-molecule model of unreconstructed (0 0 1) surface; (**b**) the dissociated state of water (0.5 monolayer) on (0 0 1); (**c**) the relaxed geometries of molecular state of adsorbed water (1.0 monolayer of hydroxylated anatase) on (0 0 1); and (**d**) the mixed state of water on (001) with a half-dissociated coverage of adsorbed monolayer water molecules (courtesy of A. Vittadini, cf. [21])

at low coverage of oxygen vacancies of the anatase (1 0 1) surface [22], and consequently at the Magneli phases, as substantially suboxide structure significant both as highly bulk electronic conductive (up to 1000 S/cm), membrane type surface transferring hydroxide species and interactive catalyst support. In fact, this is the status of reversible open circuit dissociative adsorption of water molecules at the equilibrium state, something like capillary phenomena in adsorption after some critical coverage extents. Meanwhile, in the presence of the nano-sized metallic part of the catalyst, and continuous enough moisture supply, directional electric field (or, electrode polarization), further disturbs such an established equilibrium and dynamically imposes further continuous forced dissociation of water molecules, and as the consequence, their membrane transport properties (Livage et al. [23–25], (Eqs. (8.1-a), (8.1-b), (8.1-c) summation ($\sum$) yields Eq. (8.1)), definitely resulting with the Pt-OH dipole spillover features.

Such an oxide mixed valence network, in particular of polyvalent (high altervalent numbers) hypo-d-elements, when in hydrous state, distinctly behaves as an ion exchange membrane even for their own surface hydroxide migration. In fact, gels (aero and xerogels) are biphasic systems in which solvent molecules are trapped inside an oxide network, and such a material can be considered as a water-oxide membrane composite [23–25]. At the same time, the highly pronounced reversible potentiodynamic peaks testify for the extremely fast and independent overall spillover reaction (Eq. 8.1), in both directions [2–4], primarily used for DL charging and discharging pseudo-capacitance [5–8], and being always available for heterogeneous electrode reactions (see further downwards).

8.2.2 Electrocatalytic Spillover Phenomena

The first spillover phenomenon in heterogeneous catalysis was observed and defined by Boudart [27–29] for the interactive supported bronze type (Pt/WO_3) catalyst, initially at high temperature (above 400 °C) for pure solid oxide system. Meanwhile, after the dissociative adsorption of water molecules on hypo-d-oxide supports of Pt [21, 22], the fast interactive effusion of H-adatoms over its hydrated ($W(OH)_6$) surface becomes dramatically sped up even at ambient conditions in the ultimate presence of condensed at least monolayer aqueous precipitate, and *vice versa*, establishing capillary phenomena of speeding up effusion of the main catalytic reaction intermediates (Pt-OH). Such a striking sharp wetness impact upon the overall spillover phenomena associated with hydrated hypo-d-oxides in aqueous media, implies Ertel [30] auto-catalytic molecular water effect, too, which states that the catalytic reaction of hydrogen oxidation upon Pt surface, even at deep low temperatures (140 K), proceeds with remarkable amounts of the Pt-OH, as the decisive and accumulated intermediate, including the self-catalytic step with adsorbed water molecules (equivalent to Eq. 8.1),

$$2Pt + O_{ads} + H_2O_{ads} \rightarrow 2Pt\text{-}OH \tag{8.4}$$

In fact, Ertel has pointed out the substantial overall significance of water molecules in heterogeneous catalysis for oxidation processes that in general proceed over the Pt-OH generation and spillover, and substantially implies the fundamental peak relations (Eq. 8.1) on the same wet way in (electro)catalysis.

This is significant both for the evidence of the extremely fast spillover widespreading and thereby resulting with imposed the reversible hydrated substrate reduction. The latter finally leads to the corresponding form of electrocatalytically active bronze ($Pt/H_{0.35}WO_3$) for cathodic processes, in which non-stoichiometric incorporated hydrogen obeys the same free reactive properties like adsorptive (Pt-H), and is the main source for the electrode or heterogeneous catalytic reactions. In other words, the point is that spontaneous dissociative adsorption of water molecules [21, 22] imposes much smaller activation energy for transformation of the resulting hydrated $W(OH)_6$, into corresponding bronze state, and occurs even at ambient temperature, then behaving remarkably different than the initial solid oxide WO_3, and thereby, dramatically facilitates the overall spillover effect under pronounced wet status (the activation energies thence being in the ratio of 2.2, one with another). The alterpolar interchanges between the bronze type electrocatalyst and its hydrated state are correspondingly approved occurring instantaneously and reversibly fast, exactly because of the substantially facilitated Boudart [27–29] spillover effect and behave as a thermodynamic equilibrium ($Pt/H_{0.35}WO_3 \leftrightarrow Pt/W(OH)_6$ or, $Pt/H_xNbO_5 \leftrightarrow Pt/Nb(OH)_5$). In other words, such state of the art enables to perform (electro)catalytic processes from very high temperature down at ambient conditions, simply by the wetness effect [5–11].

Hypo-d-electronic transition metal oxides usually feature several altervalent states giving rise even to interactive mixed valence compounds, such as, for example, TiO_2/WO_3, TiO_2/Nb_2O_5, TiO_2/Ta_2O_5, or $TiO_2/Nb_2O_5/CeO_2$, and then correspondingly increase the overall latent storage and spillover effect of both H-adatoms and primary oxides (M-OH). The whole spillover and SMSI effect behave typical synergistic electrocatalytic properties and never any individual hypo-d-oxide enables that much as mixed altervalent composites. Such coinciding and interconnected events and phenomena have finally been perfectly and broadly tuned in mutual interfering phase, almost like a lucky concatenation, and enabled us to approach and substantiate the reversible electrocatalysts for the oxygen electrode reactions.

The problem so far was in unattainable nanostructured Pt-bronze, the catalytic activity of which exponentially increases with decreased Pt nano-size approaching maximum at monoatomic dispersion [31]. This requirement has now been fulfilled by the grafting implementation of Pt-acetylacetonate within colloidal particles of peroxopolytungstic acid, niobia (Nb_2O_5), tantalia (Ta_2O_5) and ceria (CeO_2), (Fig. 8.3) [7, 8, 19, 20]. Such homogeneous and even distributions of nanostructured Pt particles, and such SMSI bonding effectiveness, so far were missing the experimental evidence in PEMFCs development.

8.2.3 Anodic HCHO Oxidation and Primary Oxide Spillover

Since the heterogeneous reaction of formaldehyde oxidation with Pt-OH, and in particular Au-OH, proceeds as a fast reversible anodic process mass transfer limited, and since HCHO is soluble in all ratios in aqueous media, the primary oxide generation rate and its yielding spillover have primarily been investigated by potentiodynamic spectra within the broader concentration range and between hydrogen and oxygen potential evolving limits (Fig. 8.2a, b). Such for an order faster anodic reaction is able to postpone within unusually broad potential range the recombination of the primary (Pt-OH), into the more polarizable, more adsorptive and more stable surface oxide (Pt=O), (Eq. 8.2). Formaldehyde oxidation starts at its reversible potential (0.32 V *vs.* RHE), merges with the second UPD desorption peak of H-adatoms, and extends as an exaggerated broad twins peak all along the anodic scan, nearly until the beginning of OER (Fig. 8.2a). In the same sense anodic Pt-OH striping CO oxidation on composite hypo-d-oxides supported Pt or Pt, Ru catalysts takes place even within the usual interval of UPD H-adatoms desorption (see further downwards). In other words, Pt-OH arises available for reaction not only within its nominal reversible adsorption/desorption peak limits in regular mineral acid or alkaline aqueous solutions, but depending on the reactant (HCHO, HCOOH, CO, etc.), its concentration, affinity and the actual reaction rate, along a broad and ***extendable*** potential range. Meanwhile, as the link of DL charging/discharging, it actually extends and appears available all along the whole potential axis of cyclic voltammograms. Such an unusually broader charge capacity area (Fig. 8.2a, b), usually features all the properties of typical under- and over-potential oxidation (UPO, OPO) peaks, in particular when compared with cathodic UPD properties of H-adatoms on various metals. Such specific cyclic voltammograms clearly reveal the interference between Pt-OH and Pt=O, and testify for the reaction polarization of the latter along the broader potential range.

Stepwise extension of positive potential limits toward the oxygen evolution reaction (OER), clearly shows (Fig. 8.2a, b) the absence and/or distinctly reduced and postponed Pt=O growth almost until oxygen starts evolving. Then, since during the reverse potential scan toward the HER, as the result of the former, there arises and finishes the Pt=O desorption much earlier and of dramatically reduced charge capacity, than nominally in the simple acidic or alkaline solutions, the appearance of a characteristic sharp anodic current jump testifies for the hysteretic (cathodic sweep, anodic peak) aldehyde oxidation within the former nominal DL charging/discharging range, and this way reflects and features the specific and highly reactive properties of the Pt-OH. In other words, as a corollary, there is no anodic aldehyde, alcohol, their simple acids, and even CO oxidation, nor the ORR, upon any M=O (Pt=O, Au=O) covered metal surface prior to the potential of molecular oxygen evolution, when the latter becomes broken, and thence, occurs only upon prevailing Pt-OH spillover deposits.

Namely, since aldehydes are often soluble in aqueous media almost in all ratios, their voltammograms at high contents feature imprinted extremely high both charge capacities and limiting currents at their peaks (Fig. 8.2b), and thereby testify for almost unlimited reversible reaction rates (Eq. 8.1), as long as diffusional mass-transfer supply provides enough reacting species. Even more so, for interactive supported Pt and Au upon higher altervalent hypo-d-oxides and their mixed valence compounds, since these behave as highly enriched latent storage capacities of primary oxide spillover sources [32–36].

This is the cause and reason why within the reversible part of Tafel plots electrocatalytic metal (Pt,Au) surface is always covered by the interacting Pt-OH/Pt=O species, and naturally tends to impose the reversible oxygen electrode properties, and these are the experimental evidence that the optimal catalytic rate implies their optimal ratio.

In such a context, meanwhile, one of the most outstanding observations has then been that cyclic voltammograms of both formaldehyde (Fig. 8.3, [37]) and formic (muriatic) acid (Fig. 8.3, [38]) anodic oxidation distinctly differ upon plain (Pt/C) and the same, but more or less enriched supports in specific amounts of interactive hypo-d-oxide (Pt/Ta_2O_5/C [37] and/or hypo-f-oxide supports, Pt/CeO_2/C [38]). In other words, the all parameters and conditions being the same, the reaction rate becomes dramatically different upon various supported electrocatalysts in their hypo-d-oxide amounts and there resulting latent storage capacity, as the result of distinctly different the additional Pt-OH spillover feeding and spreading effects (Fig. 8.5) [32], and as the only distinctly imposed difference. Such a conclusive observation belongs to the main experimental arguments to prove the theory of the M-OH interfering self-catalytic spillover contributions in electrocatalysis of aqueous media, (Eq. 8.3) [5–11], finally providing the ROE behavior and properties, and traced the way and entire approach for its substantiation. In such a respect, cyclic voltammograms deeper enlightened and more clearly revealed the spillover reaction impact properties of the Pt-OH.

8.3 Potentiodynamic Scans of Primary Oxide Latent Storage and Electrocatalytic Spillover Spreading

Some intermolecular compatible hypo-d-oxide mixed valence architecture (Pt/WO_3,TiO_2/C; Pt/Nb_2O_5,TiO_2/C), (Fig. 8.3), as the interactive catalytic submonolayer (and much more so as the multilayer, Fig. 8.5, [32]) supports of high altervalent number or capacity, have been investigated by potentiodynamic scans to reveal both the primary oxide latent storage and resulting Pt-OH spillover yielding properties, along with same for H-adatoms. In this respect, cyclic voltammograms scanned at low moisture content of He stream (just enough to enable basic electrode processes to occur and proceed), insufficient for WO_3 (or TiO_2) hydration, repeatedly reveal similar potentiodynamic spectra characteristic for indifferent carbon supported (Pt/C), or plain Pt itself (Fig. 8.3), but with high double layer charging capacity, because of the accompanying parallel

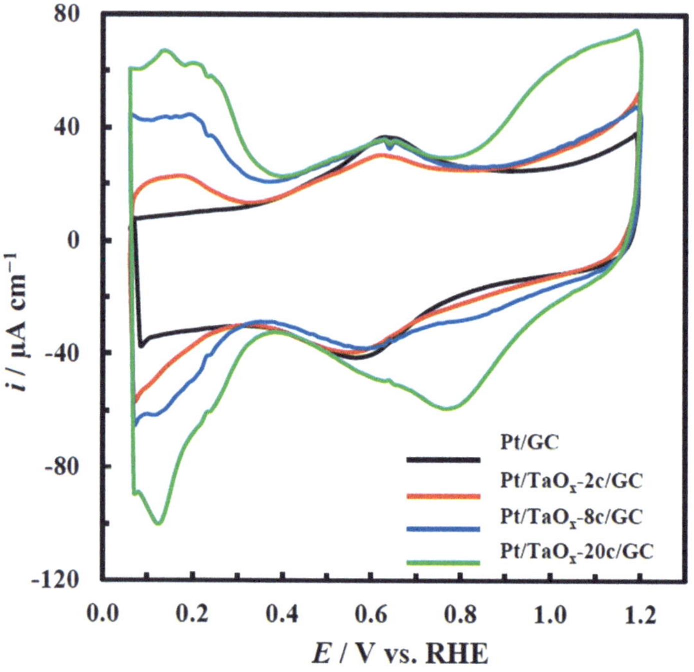

Fig. 8.5 Cyclic voltammograms scanned at the Pt/GC and Pt/TaOx/GC catalytic electrodes with 2c, 8c, and 20c (cycles) in Ar-saturated 0.5 M H_2SO_4 solution at scan rate of 50 mV s^{-1}, revealing the effect of proportional increasing of interactive Pt supporting Ta_2O_5 deposit on the Pt-OH spillover effect and growth for the ORR (courtesy of T. Ohsaka [32])

charging of Vulcan carbon particles beside the metal (with correspondingly large charge value, $Q_{DL} = 1.07$ C).

In contrast to such fairly common occurrences, a continuous supply of saturated water vapor in the He stream at higher temperature (80 °C), imposing condensation (Boudart spillover precondition [27–29]), and leading to the appearance of wet titania-tungstenia mixed valence oxide composite, along with spontaneous dissociative adsorption of water molecules all over its exposed surface [21, 22], as the interactive catalytic support, has been accompanied by the unusual phenomenon of a dramatic blowing expansion of two reversible pairs of peaks of both the primary oxide ($Q_{Pt\text{-}OH(a)} = Q_{Pt\text{-}OH(c)} = 1.453$ C), and H-adatoms (247 versus 47 mC cm^{-2}, or in the ratio of about 5.3:1) chemisorptive deposition and desorption (Fig. 8.3), like a DC capacitance of extremely developed electrode surface charging and discharging. Since these are highly reversible and evidently behave remarkably pronounced the latent Pt-OH storage (cf. [5–7, 21]), they keep the same their extents even after multiple and repeating number of cycles at any other time. The latter have both been of unusually high spillover charge and discharge capacity values and for Pt-OH (UPD and OPD) shifted towards both much more negative and far positive potential limits, in common with Fig. 8.2a, b and discussion thereon. In fact, two distinctly different cyclic voltammogram shapes and charge capacities (Fig. 8.3), appear only as the result of the difference in water vapor supply, all other parameters being unaltered the same, and as the effect of the equivalent dipole (Pt-OH) charging and discharging of the double layer, since nothing else takes place in between. Every cessation in the steam supply instantaneously imposes the sudden reversible shrinkage of both such rather exaggerated pairs of peaks down to the same initial potentiodynamic shape similar to the nanostructured plain and non-interactive supported Pt/C voltammogram spectra themselves. *Vice versa*, the renewed saturate water vapor feeding immediately leads to their former Pt-OH peaks and the same former charge capacities; namely, the effect later already noticed and scanned for formaldehyde [37] and formic acid [38] oxidation. Such an appearance without exception behaves as a typical reversible transient phenomenon by its endless altering repetition [5–7], and never appears upon the plain Pt/C electrocatalyst, both wet and/or dry, nor with small and insufficient amounts of catalytic hypo-d-oxide supports (Fig. 8.5, [32]).

The complementary interactive Ta_2O_5-based electrocatalytic support strongly reinforces just displayed potentiodynamic features of Pt/WO_3,TiO_2/C by coinciding and congenial spectral behavior of their cyclic voltammograms: The distinct growth of Pt-OH and H-adatoms adsorption and desorption peaks, reflecting their different accumulated latent charge storage capacities, as a function of the amount (charge density) of interactive composite hypo-d-oxide deposits per unit electrode surface (Fig. 8.5) [32]. In such a respect, for example, as the effect of much smaller d-ionic radius, Y_2O_3 much more does so, than Nb_2O_5, or WO_3. In fact, such Pt-OH latent storage growth (including the corresponding spillover effect) does not extend endlessly and usually passes over remarkably pronounced maximum in the ORR catalytic rate and activity (see Fig. 8.11, [14]).

What is now the substantial difference between voltammograms in Fig. 8.3, wet state, and Fig. 8.2a, b (or Fig. 8.5)? Even when mostly suppressed in the surface oxide (Pt=O) adsorptive growth (Fig. 8.2a, b), the reversal backward cathodic scans on plain Pt proceed still highly polarized for about 600 mV with negligible or zero current, exactly corresponding to Fig. 8.1. In other words, there is still no stored Pt-OH on plain Pt to start either hysteretic HCHO oxidation peak or the self-catalytic interfering reaction (Eq. 8.3) of the ORR. However, on the hypo-d-oxide interactive supported Pt catalyst, the *a priori* latently accumulated initial storage of the primary oxide from the very beginning is already ready and available, and thence continuously provides and spillover enhances the latter to proceed as the uninterrupted and extra fast reversible electrode reaction, as long as there is continuous water vapor supply. However, for the plain Pt (or Pt/C), the sudden hysteretic sharp anodic current jump in the course of reversal cathodic sweep (Fig. 8.2a, b) rearises at and coincides with the classical position of the reversible peak for the Pt-OH growth, reflects the local interfering self-catalytic effects at such potential range, provided by the repeated Pt-OH spillover growth (Scheme 8.1). This is the striking point and the core substance of the present study, while Fig. 8.3, supported by Fig. 8.5, are the best illustrative issues of the substantiated reversible electrocatalyst for the oxygen electrode reactions (ORR, OER).

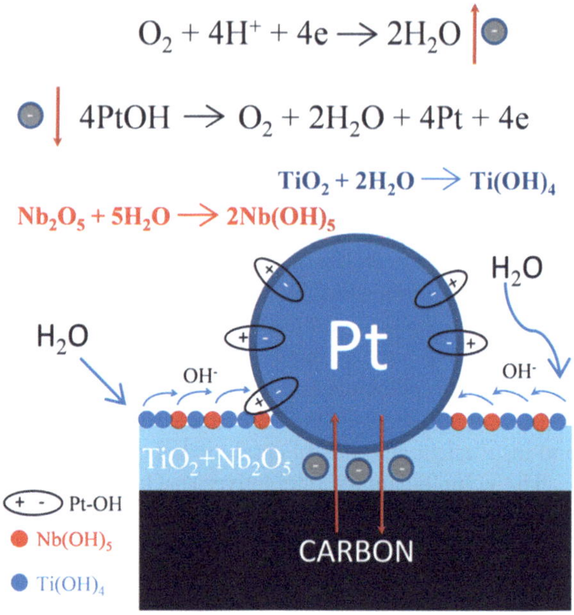

Scheme 8.1 Visual presentation of novel spillover latent storage Pt-OH type interactive supported electrocatalyst for oxygen electrode reactions, exampled for $Pt/Nb_2O_5,TiO_2/Nb(OH)_5,Ti(OH)_4$ issue (Reprinted with permission from [7]. Copyright (2014) American Chemical Society)

8.4 Interchangeable Interfering Reversible and Polarizable Oxygen Electrode Reactions

In general, nanostructured Pt (Au) electrocatalyst selectively interactive grafted upon hypo-d-(f)-oxide supports, when the latter exist in the hydrated external surface and/or internal bulk state (Scheme 8.1), thence consequently exhibits enriched Pt-OH (Au-OH) latent storage, as the feedback oxophilicity effect and property. The corresponding primary oxide then continuously features renewable spillover, simply by water vapor supply (surface phenomena, no concentration polarization), and further continuously yielding spontaneous dissociative adsorption of aqueous species, so that the reversible anodic oxygen evolution (Eq. 8.5) now takes place straight from pronouncedly high and also the highly reversible Pt-OH (Au-OH) adsorptive peak capacities, (Figs. 8.3 and 8.5) [5–7],

$$4\text{Pt-OH} \leftrightarrow O_2 + 2H_2O + 4Pt + 4e^- \tag{8.5}$$

and, thence, at the thermodynamic equilibrium (ROE) potential value. This is the substance, this is the point and evidence, this is the achievement of the substantial reversibility in the oxygen electrode reactions. Quite on the contrary, in classical issues (Fig. 8.1), the latter occurs from the strong irreversibly deposited and highly polarizable monolayer of the Pt=O, (Eq. 8.2), and thereby, while being deprived from primary oxide presence or external supply, at rather high anodic overpotentials,

$$2Pt = O \rightarrow O_2 + 2Pt + 4e^- \quad (8.6)$$

The same, *vice versa*, consequently alternately occurs both in these reversible and/or irreversible states, as concerns the ORR, all along the reverse cathodic scan and polarization (Figs. 8.1, 8.3 and 8.5) [5–7]. The transient between the reversible and irreversible status features some sharp thresholds both as a function of hypo-d-(f)-oxide unit content and/or wetness percentage [5–7], and substantially reflects the Pt-OH (Au-OH) properties (present or absent) for the oxygen electrode reactions. In such a context, plain and non-interactive supported Pt (Pt/C), in the absence of primary oxides, and thereby established or *a priori* existing the broad reaction polarization range, by no means can feature the overall reversible oxygen electrode properties along the whole potential axis, in between of the hydrogen and oxygen evolving limits, or within the entire potentiodynamic circle. Thus, there exists either the reversible behavior in the presence or continuous external supply of primary oxides, or highly irreversible properties in the absence of the latter, (Pt-OH $\rightarrow$ 0, Au-OH $\rightarrow$ 0), and this is the substance for Pt and Au electrocatalytic properties!

In such a context, the simple stoichiometric combination of Eqs. 8.1 and 8.5, reveals that the overall reversible anodic oxygen evolution initiates from water molecules (Eq. 8.1), and obeys the auto- or self-catalytic Ertel framework [30], with the general oxidation mechanism based on and catalyzed by M-OH species, or the primary oxides as the main interfering electrocatalytic species, and substantially takes place at the reversible oxygen potential value; the same in the reverse cathodic ORR scans. However, when oxygen evolving initiates from Pt (Au) monolayer covered by the strongly adsorptive and polarizable surface oxide (Pt=O, Au=O), the anodic, and reverse cathodic reaction have both to overcome remarkable overpotentials (Eq. 8.6, adjoined with Eq. 8.2). This is the substance and difference as concerns the irreversible plain (Pt, Au) and non-interactive supported Pt/C(Au/C), and/or the enriched latent storage and continuous external spillover feeding of the primary oxide for interactive hypo-d-(f)-oxides supported Pt electrode. The fascinating and incredible surprising facts, meanwhile, are that such simple distinct phenomena have been obscure and unknown in electrocatalysis for many decades now, mainly because of longer missing of cyclic voltammetry.

8.5 Electrocatalytic Consequences

Since Pt(Au) electrocatalyst becomes interactive d-d-(f)-bonded with and grafting fixed upon hypo-d-(f)-oxide type catalytic supports in new composite nano-structure, there is no more its metallic nano-particles (cluster) surface diffusion and agglomeration, nor any Nafion membrane cross-over of hydrogen. Even more so, from the same reasons, the life-time of such Pt(Au) (electro)catalysts is at least twice longer and can be guaranteed. Alterpolar spillover interchanges in the revertible, PEMFC *versus* WE, proceed reversibly and instantaneous smoothly, and enable their superior unique operation. There is a great number of parametric and structural variability for the voltage/current optimization and stability, such as hypo-d-oxide radius of small, medium and large d-(f)-metals (Y, Ti, Nb, Ta, W, Ce, Gd, Ho, La, etc.), to keep low Tafel slope within the operating range, 0.0–1.0 A cm^{-2}. Since some of hypo-d-(f)-oxide supports of composite electrocatalysts (Nb_2O_5, Ta_2O_5, WO_3, TiO_2) feature advanced electron conductivity, nano-particulate carbon carriers and current collectors can be completely avoided that is substantially significant for WE and mostly for anodic oxygen evolution. In such a respect, the present paper defines the main substantial and advantageous frameworks in electrocatalysis of reversible electrocatalysts for oxygen electrode reactions (mostly ORR and AOE), primarily for L&MT PEMFCs, WE, and their revertible combination, and as the overall main conclusion, hopefully means one of the most important contribution all along after Sir William Grove. All stated herein for Pt, means even more so for Au when oxygen electrode reactions are in consideration (ORR, OER) [6, 31, 39–42].

Mixed altervalent hypo-d-(f)-oxide supports, depending on their preceding thermal treatment, feature advanced electron conductivity from 1000 S cm^{-1}, for Magneli phases, down to about 300 S cm^{-1} for average composites, and this way enable to replace and even completely remove nanostructured carbon as a non-interactive catalyst support and current collector. However, except for the OER, when carbon becomes partially the subject of anodic oxidation, nano-particulate carbon still quite satisfies requirements for its nominal purposes.

In fact, the main benefit comes from the dissociative adsorption of water molecules upon mixed valence hypo-d-(f)-oxide supports, enabling us to increase the latent storage capacity, and therefrom the yielding Pt-OH (Au-OH) spillover intensity, and/or the higher current density at the more reversible electrode potentials. In the same respect, the supercritical drying within the sol-gel procedure for hypo-d-(f)-oxide supports development enables to extend remarkably the available surface for dissociative water molecules adsorption and consequently enlarged the Pt-OH latent storage and spillover intensity.

The actual attainable target for LT PEMFCs: Tafel slopes of 60 mV/dec, or $\eta = 180$ mV at 1000 mA cm^{-2}, or ideally 30 mV/dec, or 90 mV at 1.0 A cm^{-2}.

8.6 Hypo-Hyper-d-d-Interelectronic Bonding and Interfering Spillover Nature of Electrocatalysis for Hydrogen and Oxygen Electrode Reactions

The majority of physical and chemical properties of elements behave typically symmetric volcano plots along transition series, observed and well known as Friedel d-d-electronic correlations [14, 43–48]. These properties substantially reflect the characteristic periodicity features of d-metals, and when plotted one into another, result with simple straight line dependences of deeper fundamental scientific significance [14, 46–48]. In such a context, one of the most significant linear interdependences arises as the so called *one by sixth law* between surface free and cohesive energies [49], and in a straight forward manner connects them both to the catalytic activity properties, since the former directly correlates with the adsorptive bonding strength (or, the free enthalpy of adsorption as the driving force) of reacting intermediates in the RDS. In the same respect, asymmetric volcano curves of the electrocatalytic activity ($log\ j_o$) for the cathodic hydrogen evolution (HER), (Fig. 8.1, [50]) [43, 51], and/or the free enthalpy of H-adatoms adsorption, and work function (Fig. 8.2, [14]), [46–48], all being with optimal peak values of d^8-electronic configuration, when plotted with each other, also reveal a straight line dependence of most significant and substantial meaning in electrocatalysis. While Trasatti [52] came to such a linear dependence from some theoretical analysis, our approach follows the overall general fundamental properties of volcano plots along the transition series also associated with and based on the d-band state or the d-energy level of individual d-elements [14, 43, 46, 48]. Even more so, since the fundamental Butler-Volmer equation in electrode kinetics obeys the typical Maxwell-Boltzmann law of exponential relation between current density (j), as the reaction rate intensity feature, and polarization or the overvoltage (η), as the electrode reaction driving force, logarithm of the exchange current (j_o) obeys a typical linear function dependence of the free activation enthalpy of adsorption of intermediates at the RDS ($log\ j_o \approx k_o \cdot \Delta H_o$) [46–48, 53]. In other words, in more general form, $log\ j \approx k \cdot \Delta \mathrm{H}$, and when $\eta \Rightarrow 0$, then $j \Rightarrow j_o$, and thence, these interrelate with other volcano plots on quite general basis and simple linearization manner. Actually, the optimal d-electronic configuration peak for asymmetric volcano plots introduces a simple kinetic constant correction term, which shifts the initial (or, real) volcano curves to their nominal symmetric position [14, 43, 46, 48]. In the same respect, Brewer's [13, 14] intermetallic bonding theory predicts and confirmed that every hypo-hyper-d-d-interelectronic combination (or, the local phase diagram within the Periodic Table of transition elements) implies even much more pronounced intermetallic bonding effectiveness, consequently remarkably further extended their common volcano plots and correspondingly extra high intermetallic phase stability at peak values of such cohesive energy curves [14, 43, 44, 46, 48, 53]. In other words, every hypo-hyper-d-d-interelectronic phase diagram (Fig. 8.6), behaves for various physical property plots as a part of the Periodic Table, with local volcano curves composed from intermetallic phases along the axis in between that (energetically or by the average d-electronic configuration) correspond to the set of elements located within the periods of two initial constituents (ingredients) in the local phase diagram consideration. Thence, as the overall general meaning of the hypo-hyper-d-d-interelectronic correlations [44], such local intermetallic volcano curves, when plotted one into another, yield again the same type of straight line interdependence and various characteristic maximum values correspondingly coincide with each other [43, 46–48]. In such a context of coherent mutual consequences, the intermetallic Brewer peak bonding effectiveness values correspond to the maximal stability, the highest interatomic and orbital symmetry, and

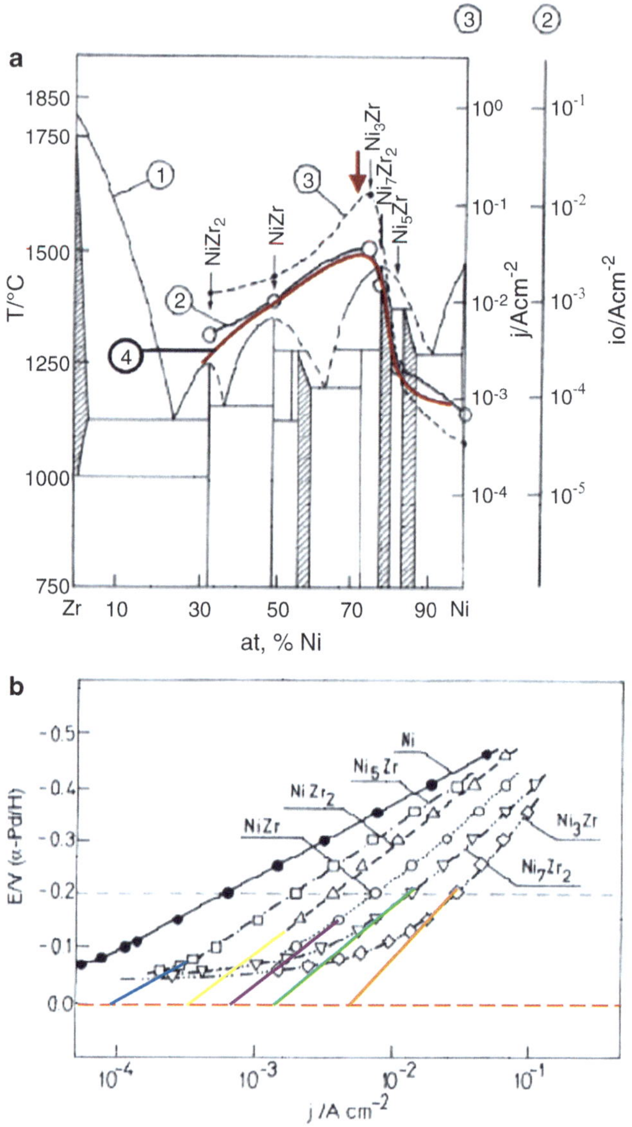

Fig. 8.6 (**a**) Electrocatalytic activities of various intermetallic phases (polished below 1.8 in the roughness factor) along Zr–Ni phase diagram (*curve 1*) for the HER, taken as the exchanged current (j_o, close circles, *curve 3*) and relative current density changes (j, open circuits, *curve 2*) at constant overvoltage (−0.2 V), plotted together with the cohesive energy of intermetallic phases (*curve 4*). (**b**) Tafel plots for the HER with indicated extrapolations to assess the relevant exchange currents, j_o; for (**a**); (**c**) linear interdependence between cohesive energy (E_c) and electrocatalytic activity (*log* j_o) for the HER

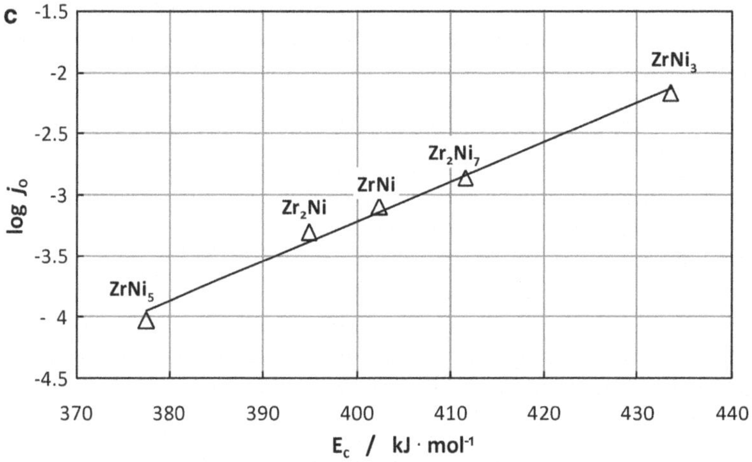

Fig. 8.6 (continued)

consequently, both the maximal work function values and synergistic electrocatalytic activities for the hydrogen electrode reactions. Thus, cohesive and catalytic features are brought in the simple common linear interdependence with the coinciding maximal values: the stronger the bonding, the higher the electrocatalytic activity [14, 43, 46–48]. In such a sense, some specific overall maximum belongs to the $HfPd_3$ symmetric intermetallic phase [13, 14, 43]. Such a theoretical approach yields the basic guide line in electrocatalysis for hydrogen electrode reactions amongst individual and hypo-hyper-d-d-interelectronic and intermetallic composite electrocatalysts, both for macroscopic issues and nano-structured clusters.

Where then lays the link between the hypo-hyper-d-d-interelectronic bonding and SMSI synergistic electrocatalytic effects in the broader interfering spillover sense? The whole electrocatalytic theory [14, 43, 46] relies on the Brewer [13, 14] intermetallic bonding model and Friedel [44–46] volcanic tape hypo-hyper-d-d-electronic correlations. They both inferring that the stronger the hypo-hyper-d-d-intermetallic cohesive bonding, the more stretched, more strengthened and more exposed arise d-orbitals within the ideal symmetric intermetallic phases, like $ZrNi_3$, $HfPd_3$, $MoPt_3$, $LaNi_5$, $La_xCe_{(1-x)}Ni_5$, etc., and thereby, the weaker and easier the cleavage of their adsorptive intermediates (Pt-H, Pt-OH) in the RDS, and consequently, the higher the reaction rate and the overall catalytic activity [5, 9–11, 43, 46, 48]. The same Brewer [13, 14] type d-d-intermetallic bonding model has been much earlier anticipated and means the preceding basis for Tauster [15–18] promotional SMSI effect, with the far-reaching consequences in both heterogeneous catalysis and electrocatalysis. The latter systematically predetermined interactive grafting [9, 10, 19, 20], and the homogeneous, evenly uniform distribution of individual and prevailing hyper-d-metallic catalysts upon hypo-d-oxide supports [5–11]. The same Brewer type [13, 14] of hypo-hyper-d-d-interactive bonding between nanostructured metal particles of such composite catalysts ($MoPt_3$,$TiNi_3$), and their hypo-d-oxide supports, additionally reinforces the entire electrocatalytic activity effect based on the overall interactive d-d-interelectronic bonding strength, similarly as cohesive energy itself defines and advances the electrocatalytic activity for nanostructured mono-phase systems [5–7, 19, 20, 43]. In fact, the interactive hypo-hyper-d-d-interelectronic bonding strengths belong to the strongest bonding effectiveness in the whole chemical science [5–11, 19, 20], sometimes even proceeding so vigorously with exceptionally intense explosion ($HfPd_3$), and thereby resulting with extra high intermetallic phase stability [13, 14, 46, 48], and consequently, high electrocatalytic activity for the HER. In other words, the interelectronic-d-d-(f)-bonding and/or cohesive strength linearly correlates with the (electro)catalytic activity, [46, 48]. The pronounced cathodic and anodic yields of interactive spillover contributions within and based on the SMSI, have additionally been significant for the present theory and its embodiment in electrocatalysis of both hydrogen and oxygen electrode reactions, particularly for Low and Medium Temperature (L&MT) PEMFCs and WE [5–11, 19, 20]. While in aqueous media plain nanostructured Pt (Pt/C) features the catalytic surface properties of the active Pt-H and polarizable Pt=O, missing any effusion of other interacting species, a new generation of composite interactive supported (SMSI) electrocatalysts in condensed wet state primarily characterizes exceptional and extremely fast reversible spillover interplay of either H-adatoms, or more important the primary oxides (Pt-OH, Au-OH), as the unique significant and advanced synergistic interactive electrocatalytic ingredients [5–11, 19, 20].

8.7 Volcano Plots: Their Sense, Properties, Significance and Uses

Volcano plots of various physical and chemical properties along transition series reveal the periodicity features of d-metals [14, 44–48], based on the d-d-electronic correlations [44], with similar symmetric shape, and *vice versa*. Periodicity features group transition series of elements into congenial, parallel and symmetric volcano plots. In the essence, such symmetric volcano plots draw their basic roots from the hypo-hyper-d-d-electronic variations in the elemental configuration and the resulting individual atomic energy level states. The ascending part of such volcano curves comes from hypo-d-elements while filling their semi-d-orbitals, whereas the descending half reflects hyper-d-electron pairing and yielding the anti-bonding d-band along with still remaining unpaired semi-d-orbitals, to balance d-states periodicity on both volcano sides, ascending and descending, along transition series. Substantially this is the basic reproducible d-electronic symmetric volcano plot we always imply and start from. All other physical and chemical volcano type properties simply come as the reflection, reproduction and consequence of such initial and substantially involved basic hypo-hyper-d-d-electronic distribution of individual d-electronic level state of d-metals. Thence, as the consequence of symmetry and similarity, every common plot of one into other periodicity state, together yields various linear interdependences of deeper physical meaning and significance, like cohesive and surface free energy [46–48], or even more so, the cohesive bonding strength into electrocatalytic activity, and the latter straight into the surface and/or adsorptive free energy, etc., further unlimited in the general and universal properties of the matter! As a consequence, the d-band of transition elements has been confirmed to play a crucial decisive role in the overall behavior of transition elements, and thereby, is considered for the bonding, adsorptive, catalytic and electrocatalytic band of transition series [46, 48]. Thus, consequently, any search for advances and synergism amongst the latter, should be based on the d-d-interelectronic correlations and modifications [14, 43–48]. Such a state of theoretical knowledge and experimental evidence, established on the symmetry basis of elemental properties, consequently leads to the general statement and universal conclusion that every hypo-hyper-d-d-interelectronic phase diagram behaves as the part of the Periodic Table between the two initial periods of the interacting ingredients, with their intermetallic phases taken for the axis [43, 46–48], (Fig. 8.6) [14]. The latter means and implies with no exception, so that the partial volcano plots in the small [14, 15, 44–48], bear a broad and deep catalytic, and in the substance general physical sense and meaning. Their intermetallic phases in between two initial ingredients are of the same average d-electronic configuration, (Fig. 8.6), that way replacing the 'missing elements' in between, in their energy state, level and behavior [46, 48], and consequently, have been used to assess the synergistically maximally active electrocatalysts for the HER. Meanwhile, since transition elements themselves establish and reproduce simple individual fundamental symmetric ascending/descending volcano plots, depending on basic individual d-electronic number of evolved d-electrons along transition series and their individual configuration of d-band, hypo-hyper-d-d-interelectronic combinations involve antibonding pairs and consequently their volcano curves are usually asymmetric (Fig. 8.6) These theoretical and experimental facts can then be employed to predict the synergistically active electrocatalysts from the peak values of corresponding local volcano plots along each hypo-hyper-d-d-interelectronic phase diagram [46–48], and then further catalytically advance by the interactive hypo-d-oxide interfering spillover contributions (Pt-H for hydrogen, and Pt-OH for oxygen electrode reactions) [5–11], displayed more systematically and more thoroughly in the present and preceding study [7]. Even more so, maximal synergistically active electrocatalysts can be estimated indirectly with a rater high certainty and probability, straight from alternative secondary volcano plots for various correlating physical properties, on the first place now from cohesive energy, Fig. 8.6 [14, 46, 48]. On the same way Fermi energy plots fundamental elemental simple symmetric volcano curve dependence along transition series, thence on a straightforward linear manner correlates with the electrocatalytic activity for the HER, and since the former is in the linear interdependence with the total cohesive energy, the overall functional correlation results straight one from another. Thus, one of the guiding aims should have to be to assess experimentally or theoretically most promising synergistic electroctalysts, and consequently correlate the actual hypo-hyper-d-d-interelectronic SMSI effect or hypo-hyper-d-d-interbonding effectiveness for new type interactive supported composite ($Pt/Nb_2O_5,TiO_2/C$, Pt/Ta_2O_5, TiO_2/C, or more complex mixed valence compounds for the HER, $HfPd_3/Nb_2O_5$, $TiO_2,CeO_2/C$) electrocatalysts. The formal difference is that interactive bonding is interatomic or intermetallic (Brewer-Friedel correlations), while in the more complex and broader Tauster extension SMSI sense [15–18], they are interionic-intermetallic interactions, or all together being hypo-hyper-d-d-interelectronic bonding correlations in electrocatalysis. Thus, the interfering Brewer-Tauster (SMSI) hypo-hyper-d-d-(f)-interbonding effect, along with the resulting primary oxide (Pt-OH,Au-OH) latent storage and spillover properties, bring electrocatalysis for hydrogen and oxygen electrode reactions to the common point for mutual optimization in (electro)catalytic activity.

All attempts to construct volcano plots in heterogeneous catalysis along any energetic (adsorptive) axis so far failed. Since all (electro)catalytic processes are governed by, and as the subject matter of the Maxwell-Boltzmann general kinetic and distribution law, imply and exceptionally impose the exponential dependence and relations, then consequently, the logarithm of reaction intensity ($log\ j_o = k_o \cdot \Delta H_o$) results as a linear, but by no means volcanic function [14, 43, 46–48]. Volcano plots display without exception the periodicity features along transition series of elements as a function of d-electron configuration, or, d-electronic density of states at the Fermi level, which itself reveals the fundamental symmetric volcanic parabola of hypo-hyper-d-d-electronic structure of individual transition elements, and reproduces endlessly itself in all physical and chemical property plots along the Periodic Table [43–48].

8.8 Electrocatalysis and d-Electronic Density of Sates Versus Fermi Level

The electrocatalytic reaction mechanism by contemporary *ab initio* DFT calculations, both for the HER and ORR, implies that a d-band centered near the Fermi level (E_F) can lower the activation energy as the bonding orbital passes E_F, the critical step for reduction processes taking place when the antibonding orbital passes the Fermi level of the metal from above and picks up electrons to become filled [54]. In electro-oxidation reactions (HOR, OER), the bonding orbital passes the Fermi level from below and gets emptied. Consequently, a good catalyst for these reactions should have a high density of d-states near the Fermi level [40, 55, 56]. Thus, the present concept consists of the proper hypo-hyper-d-d-interelectronic combinations of transition elements both being of rather high densities of d-states at the Fermi level (Fig. 8.7a), in which hypo-d-oxide components (W, Nb, Ta, Ti), besides the substantially high SMSI bonding effect, as typical oxophilic d-metals, involve their pronounced surface type membrane migration properties for the

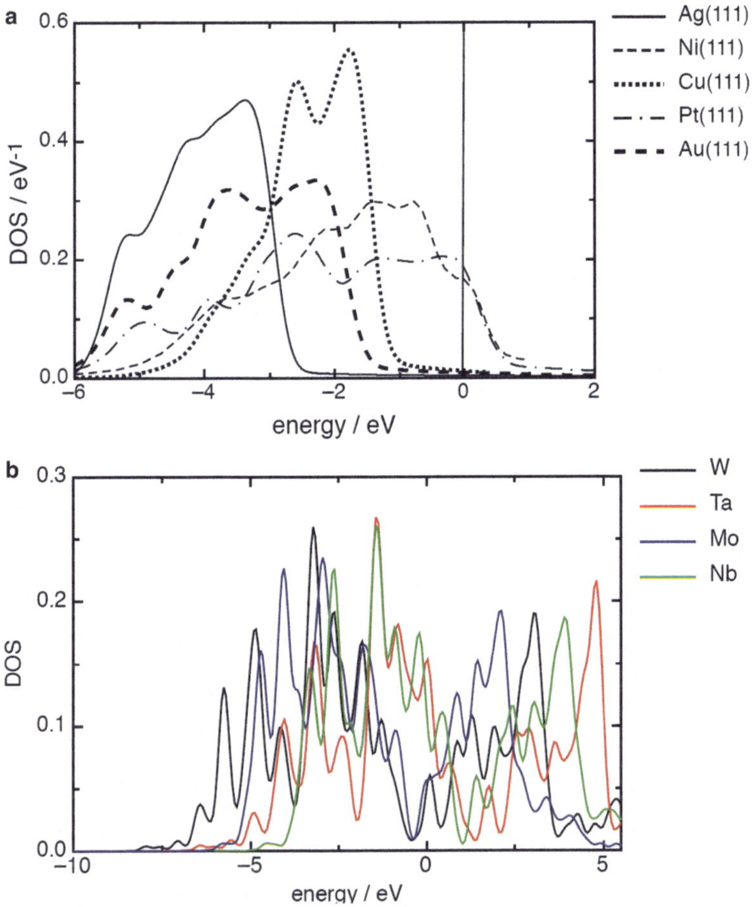

Fig. 8.7 (**a**) Surface densities of d-band states of some selected hyper-d-electronic transition metals. The integral over the densities has been normalized to unity; the vertical line indicates the Fermi level. Indications and labels: *dashed-dotted line*: Pt(1 1 1); *thick line*: Au(1 1 1); *thin line*: Ag (1 1 1); *dotted line*: Cu(111); *dashed line*: Ni(111). (**b**) Surface densities of d-band states (bulk values differing from exact surface densities for less than 3 %) of some selected hypo-d-electronic transition metals (W, Ta, Mo, Nb, color labeled) (Calculated by Professor Wolfgang Schmickler, University of Ulm, Germany)

primary oxide latent storage, transferring and effusion. In such a state, our aim has been to keep the composite transition element ingredients with their initial high densities of d-states, or even to increase the latter by their SMSI interbonding effect, and at the same time to use the benefits of the primary oxide spillover for the overall reaction, in particular for the ORR. In this respect, Fig. 8.3b reveals why Nb and Ta, by the position in the Periodic Table of elements, are electrocatalytically predestined for even higher activity than W, Mo and Ti.

8.9 Significance and Experimental Evidence for the SMSI d-d-(f)-Interbonding Effect

The evaluation of Tauster Strong Metal-Support Interaction properties (SMSI) [15–18], as the d-d-interelectronic bonding effect in thereby advanced catalysis, otherwise necessary to impose membrane ionic transport, too, and finally yielding with spillover of dipole type RDS intermediates in (electro)catalysis, implies some experimental evidence of deeper fundamental theoretical meaning and significance. In such a context, to study, get some insight in, and assess experimentally the SMSI bonding effect itself, and further imposed its resulting spillover phenomena, very fine, nano-sized Au films were deposited by controlled electron beam evaporation of ultrahigh purity gold metal under high vacuum onto stationary, nano-crystalline anatase titania (1 0 1) coated microscopic slides [5, 6]. The XP spectra of the Au 4*f* electrons reveal the remarkable binding energy shift, (Fig. 8.8), that provides the insight for the d-d-interelectronic SMSI within the Au/TiO_2 interphase and this has been one of the first reliable precise experimental evidences of that kind in heterogeneous catalysis. The smaller the nano-particle size or the thickness of the Au nano-layer, the larger the binding energy shift in the XP spectra with titania, and the more pronounced the d-d-SMSI effect, with tendency to its maximal d-d-binding strength at monoatomic dispersion (as extrapolated straight assessed value), and for 1:1 deposition of Au on available Ti atoms, as theoretically predicted in the present and preceding papers [8–11, 19, 20, 31]. Due to the nano-dimensions of the Au layer, the signal originated mainly from Au/TiO_2 interface,

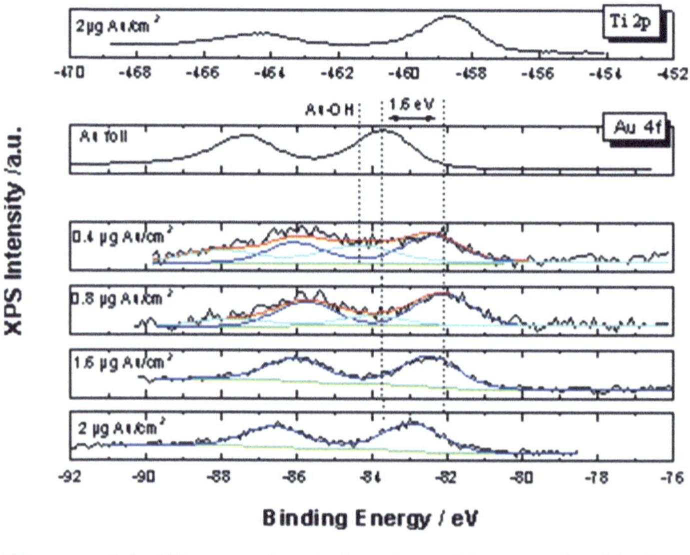

Fig. 8.8 XP spectra of Au 4*f* for vapor deposited nanolayered Au upon a fine thin films of anatase titania (1 0 1) with deconvolution for lower amounts of deposits to reveal the existence of primary oxides (Au-OH an AuOOH) [6]

reflects the bonding status within the latter. In other words, the thinner the nanolayer, the closer to the interphase itself (Au/TiO_2) penetrate spectral beams and thereby, better and more completely reflect the bonding status within the latter, this way enabling to get some insight and feeling as concerns the hypo-hyper-d-d-interelectronic SMSI bonding effect, which is apparently very strong.

The deconvoluted Au 4*f* peaks with lower Au loadings reveal that Au nanoparticles in interactive bonding contact with titania appear partially oxidized [57, 58]. The peak located at 82.15 ± 0.1 eV is attributed to metallic Au, while the peak at 84.05 ± 0.1 eV corresponds to the gold primary (Au-OH or AuOOH) oxides. The latter, in accordance with the present theory, appear as the *a priori* naturally provided primary oxide latent storage spillover species, associated with and promoted by the wet anatase titania interaction (Eq. 8.1), and are in advance, already available and ready for the cathodic ORR and anodic CO, aldehyde group and other oxidation processes.

Starting with their plain combinations (Au/C *versus* Pt/C) of nanostructured electrocatalysts, and in particular with interactive SMSI supported bronze type ($Au/Nb_2O_5,TiO_2/C$ *versus* $Pt/Nb_2O_5,TiO_2/C$) issues, Au is definitely much better and advanced nano-sized catalyst for the ORR than Pt, primarily because of remarkably smaller adsorptive strength of corresponding spillover Au-OH *versus* Pt-OH reacting species [41].

Haruta et al. [41, 59, 60] have, for example, in such a sense shown that the same reactants (propylene in admixture with equimolar amounts of hydrogen and oxygen) yield different products upon different nano-sized Au catalysts supported on anatase titania (Au/TiO_2): (i) propane by hydrogenation at nano-particles <2 nm Au, and (ii) Propylene oxide by epoxidation or oxygen addition for >2 nm Au. Hydrogenation implies H-adatoms adsorption on Au that should not spontaneously occur on any surface of pure massive bulky gold. Haruta [41, 59–64] ascribes such chemisorptive properties to "forced" or strained Au-d-orbitals within smaller (below the critical threshold) nanostructured metal particles, in particular when d-d-SMSI deposited on the interactive anatase titania (cf. [6]). In other words, smaller nanostructured Au particles (<2 nm), interactively d-d-bonded with anatase titania, or in particular when interactive supported upon mixed valence hypo-d-d-compounds ($Au/Nb_2O_5,TiO_2$, $Au/Ta_2O_5,TiO_2$, or $Au/WO_3,TiO_2$), thereby being even more reinforced in their d-d-interbonding effectiveness, and thence exposed with the more strained d-orbitals, are qualitatively different than massive Au. Such highly dispersed nano-particles consequently behave H-adatoms adsorption, and thus feature provided the reversible behavior of hydrogen electrode in the Nernst sense [6], and finally, is thereby able to carry out the hydrogenation processes, the property so far unknown for gold at all. Such defined lower size (<2 nm) Au nanostructure is something else both in heterogeneous catalysis and particularly electrocatalysis, than massive Au ("the most noble and most inactive noble metal" [40–43], primarily because of featuring spontaneous H-adatoms adsorption [6, 43]. In the same context, it has been further shown that self-reconstructed Au electrode surface, after multiple potentiodynamic cycles between hydrogen and oxygen evolving limits, features the pronounced H-adatoms adsorption and even absorption, and consequently, the reversible electrode properties ([6] and references therein). The reconstruction effect has been much more pronounced and faster in heavy water media, because of the stronger interatomic interrelations and deeper penetrating interphase effect of twice larger deuterium than protium ions and atoms [6]. Finally, the advanced electrocatalytic properties of sol-gel prepared nanostructured Au for the HER, and extra superior for the ROE, and even Ag (Ag, $AgNi_3$, $AgNiCo_2$) for alkaline media, after homogeneous interactive SMSI grafting deposition upon carefully super-critical dried hypo-d- and hypo-d-f-mixed valence compounds in liquid CO_2, ($Au/Nb_2O_5,TiO_2/C$, or $Au/Ta_2O_5,TiO_2,CeO_2/C$), follow in the forthcoming studies. The main point, meanwhile, has here substantially been to show that XPS measurements help approaching and the estimation of the hypo-hyper-d-d-(f)-metal (Au) *versus* corresponding oxide SMSI inter-bonding effectiveness, at least in the order and as the limiting values.

In the same context, one of the striking issues and contributions of the present paper is that every hypo-hyper-d-d-interelectronic bonding effectiveness exceeds the entering value of individual transition metal cohesive energy strength, and in particular all hypo-hypo-d-d- and/or hyper-hyper-d-d-interelectronic bonds! This is the point and substance of the advanced interactive bonding of hypo-d-d-(f)-oxide supported hyper-d-electronic (Pt, Au, Ru) electrocatalysts.

8.10 Stepwise Survey of the Theory Development and Experimental Evidence of Its Confirmation

Although Pt best satisfies electrocatalytic requirements of PEMFCs for hydrogen electrode reactions (HER), but many more common transition elements (Ni,Co), too, both common energy chart diagrams (Fig. 8.9), and interconnected cyclic voltammograms (Fig. 8.1) heuristically show that the reversible behavior for cathodic oxygen reduction (ORR), like a broad plateau in the energy conversion, starts practically within the potential range of the reversible peak of the Pt-OH desorption and consequently, overall loses within the preceding critical potential range of the reaction polarization amount for more than half of thermodynamic available values; or about and even more than 600 mVs. The main energy barrier obstacle arises as the imposed ***reaction polarization*** between the reversible adsorption/desorption peaks of the primary oxide growth and removal (Eq. 8.1), and anodic oxygen evolution, in both potential scan directions, as the result of irreversible disproportionation (Eq. 8.2), and the overall transition of Pt-OH (Au-OH) into highly polarizable and strongly adsorptive surface oxide (Pt=O, Au=O). Thus, ever since Sir William Grove [1] invented fuel cells (FCs), the electrocatalytic search for the reversible oxygen electrode (ROE), and most particularly the ORR, became the main challenge and imperative target in electrochemistry of aqueous and PEM media. The substance is to replace the potential range of the strong reaction polarization of the monolayered surface oxide deposit (Figs. 8.1 and 8.9)

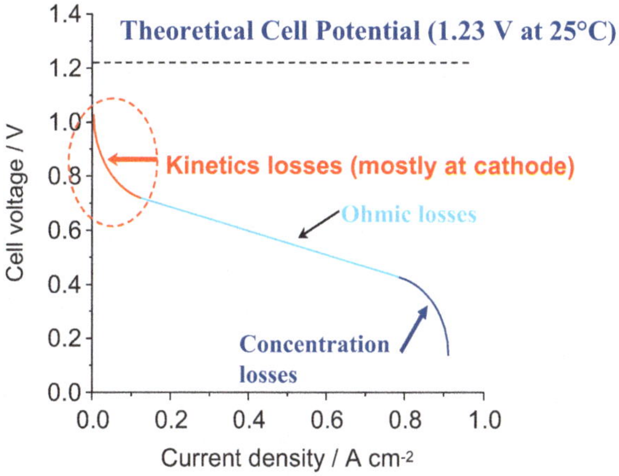

Fig. 8.9 Classical and characteristic energy chart diagram for nanostructured Pt/C electrocatalyst in LT PEMFCs (Reprinted with permission from [7]. Copyright (2014) American Chemical Society)

by the reversible (electro)catalysis of their ideally equimolar concentration ratio (Pt-OH:Pt=O ≈ 1:1, Eq. 8.3), and make Pt(Au) catalytically active all over the potential axis between hydrogen and oxygen evolving limits, back and forth, otherwise Pt(Au) is for a broad potential range substantially and highly polarizable electrode even when with extremely developed surfaces.

8.10.1 General Theory and Its Substantiation

Under such sharply defined polarization circumstances of *a priori* existing and broadly imposed the strong ***reaction polarization*** range of the primary oxide (Pt-OH) disproportionation into the irreversible, rather stable, strongly adsorptive bonded and highly polarizable (Eq. 8.2) surface oxide (Pt=O), electrocatalytic activation certainly implies the external reversible latent storage and continuous spillover supply of the missing sources of the former.

In such a respect, polarization properties of some typical and characteristic novel interactive hypo-d-oxides supported nanostructured Pt electrocatalysts for both the ORR and OER [5–8], enable to select some diagnostic kinetic criterions on the way towards the ROE, revealing from their Tafel plots: (i) nano-dispersed Pt/C clusters (10 wt% Pt) non-interactive adhering upon sol-gel developed indifferent nano-particulate E-tek, Inc., Vulcan-XC-72 carbon (240 sq m/g) carrier and current collecting species, considered for the classical issue of such an electrocatalytic activity comparison; (ii) The interactive supported nanostructured Pt particles upon both the extra stable and electron conductive (300–1000 S/cm), ceramic Magneli phases (Ebonex®), $Ti_nO_{(2n-1)}$, in average Ti_4O_7, usually defined as a shared rutile structure, accommodating the oxygen suboxide deficiency in the structure by the formation of crystal shared planes along the *n*th layerlike plane of octahedron, so that Ti_4O_7, has one TiO for every three TiO_2 layers; (iii) Advanced interactive

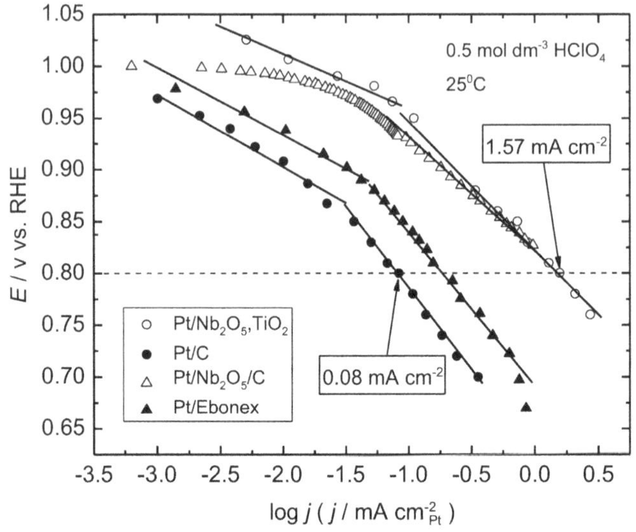

Fig. 8.10 Tafel plots for the cathodic ORR scanned on RDE in 0.5 M $HClO_4$ solution at 25 °C for E-tek, Inc., Pt/C (Vulcan XP-72, *closed circles*); Pt/Ebonex (Magneli phases, *closed triangles*); Pt (10 wt%)/Nb_2O_5 (20 wt%)/C (70 wt%) (*open triangles*); and Pt/Nb_2O_5 (5 mol%),TiO_2 (95 mol%) (*open circles*) (N. V. Krstajic measured and plotted)

selective grafted and homogeneously distributed nanostructured Pt clusters down to the prevailing (2.2 ± 0.2 nm) nano-size, upon the optimized structure of mixed valence hypo-d-oxide compounds (Pt/Nb_2O_5,TiO_2/C), and even further extended (iv) composites with hypo-f-oxides (Pt/Nb_2O_5,CeO_2,TiO_2/C, including doped GdO_2, HoO_2, and/or LaO_2 itself), and their relative combinations of extra high stability and remarkable electron conductivity, too. Such comparable diagnostic Tafel polarization interdependences distinctly displays Fig. 8.10, with differences of more than an order of magnitude in the electrocatalytic activity for the ORR that arise **only** as the result of different hypo-d-f-oxide type of interactive catalyst supports, relative ratios of their amounts *versus* metallic part of the catalyst (Pt), and their corresponding SMSI [6].

8.10.2 Leading Idea and Its Confirmations for the ROE Electrocatalysts

Some subtle potentiodynamic survey of fundamental significance has then been focused on the actual mixed interfering Pt-OH/Pt=O coverages at selected points within the characteristic potentials close to the open circuit value and all along the reversible potential range of low Tafel line slopes (mostly 30 mV/dec, Fig. 8.11). The main conclusions then have been that Pt electrodes, both plain and interactive

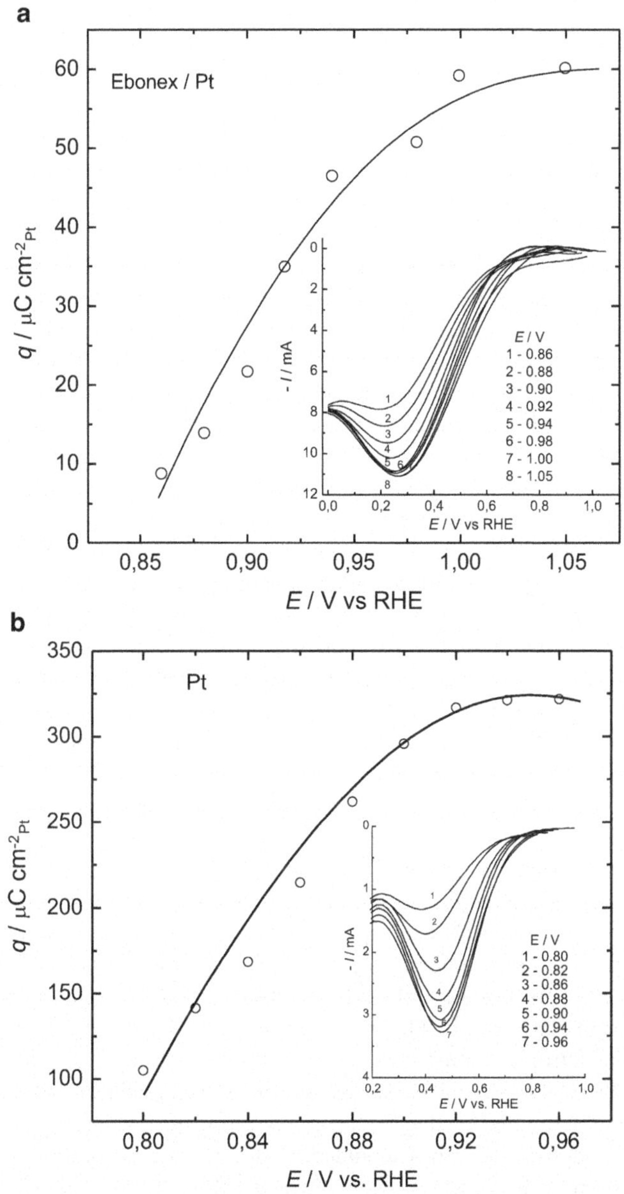

Fig. 8.11 (**a**) Charge density that required oxygen species for their reduction, presented as a function of potential for Pt/(Magneli phases) (5 mg) electrode, same as in Fig. 8.10. The insert shows potentiodynamic I vs. E relations scanned from different initial potentials (hold) with sweep rate of 5 V s^{-1}.) (N. V. Krstajic conceived and plotted). (**b**) The same as in (**a**), but for polycrystalline Pt metal

supported upon hypo-d-oxides, have been covered by mixed (Pt-OH, Pt=O) oxides (besides many other experimental methods, *in situ* and *ex-situ* XPS confirmed [5, 6, 8, 43]), while at higher irreversible polarization (120 mV/dec), they were completely deprived from such adsorptive layers. In other words, while within the reversible range, the ORR is associated with and **proceeds upon mixed oxide covered Pt surface**, as the interfering self-catalytic spillover electrode process, (Eq. 8.3), otherwise known as the fastest reaction step [12], the oxide-free Pt surface imposes much higher polarization for the direct electron transfer reaction to occur [5–11]. Thus, the conclusive observation has now been that ***Pt oxides play the same self-catalytic role to establish the ROE properties***, as the spontaneous hydrogen (Pt-H) adsorption does and means for the RHE. In fact, the leading idea consists now from the extension of the reversible Tafel plot for the ORR all along the potential axis down from or up to the thermodynamic value (1.29 V *vs.* RHE) that implies enriched external latent Pt-OH storage and continuous spillover supply all along in both scan directions, and particularly within the critical potential range (Figs. 8.1 and 8.9). Namely, so far the problem and obstacle were the initial highly reaction polarizable potential range of strongly adsorbed monolayered Pt=O and missing the Pt-OH, which cannot be supplied from aqueous solution, but only as the adsorptive surface species. These rather specific potentiodynamic measurements even more clearly show that the ORR upon Pt/Ebonex Magnely phase starts and finishes at remarkably more positive potential values (1.05 down to 0.86 V, *versus* 0.95 down to 0.8 V, all *versus* the RHE, Fig. 8.11), relative to polycrystalline Pt metal and/or nanostructured Pt/C [5, 6]; whereas the other two congenial issues (iii and iv) initiate with 1.29 V as the completed reversible property of the ROE (Figs. 8.3 and 8.5), and have been the ones of main substantial conclusive remarks on the way towards the ROE (electro)catalysts. These afford and enable the reliable link between the reversible and polarized oxygen electrode properties.

In such a respect, the guiding concept implies homogeneous nanostructured distribution and selective grafting while interactive hypo-hyper-d-d-interelectronic bonding of Pt(Au) nano-clusters upon various individual and preferably mixed valence hypo-d-oxide supports (Fig. 8.8), taken for the reversible external interconnected latent storage and spillover Pt-OH (Au-OH) sources (Scheme 8.1), primarily Nb_2O_5,TiO_2 (or Ta_2O_5,TiO_2), because of their much advanced both the stability and electronic conductivity. In such a constellation, nano-particles of solid oxides and Pt establish the SMSIs, the ones of strongest in the whole chemistry (Fig. 8.8), together with the electron conductive transfer, while external surface of hypo-d-oxide deposit undergoes spontaneous dissociative adsorption of water molecules and thereby becomes, along with continuous further water vapor supply (Scheme 8.1), the renewable and dynamically almost unlimited latent storage and spillover source of the Pt-OH.

8.11 Nanostructured Characterization of Hypo-d-Oxide Supported Electrocatalysts

The correlation of XRD spectra and Ultrahigh-Resolution Transmission Electron Microscopy (UHRTEM) and in particular Scanning Transmission Electron Microscopy in High-Angle Annular Dark Field (STEM-HAADF) mode (also known as Z-contrast) images belongs to the most reliable methods to estimate the nanostructured particle size distribution of Pt electrocatalysts upon interactive hypo-d-d-(f)-oxide catalytic supports, and represents substantially the most significant nano-size analysis manner for the time being now. These along with the SMSI of individual (Pt, Au), or prevailing hyper-d-metal ($MoPt_3$) catalyst interelectronic grafting bonded upon hypo-d-d-(f)-oxide (example, Nb_2O_5, TiO_2, CeO_2) supports are now the most reliable complementary sophisticated methods of prediction and estimation of advanced synergistic electrocatalytic activity effects in particular for contemporary L&MT PEMFCs.

The XRD spectrum for the Pt/(Nb_2O_5,TiO_2)/C catalyst reveals an overlapping region between 35 and 50° (Fig. 8.12). In this particular region one expects the reflection lines for the Pt(1 1 1) and (2 0 0), TiO_2 anatase (0 0 4) and (2 0 0), as well as the graphite (1 0 0) crystallographic planes, indicated with the labeling straight lines drawn therein. Evidently, all the above mentioned reflections are essentially overlapping. In such a respect, it is not possible to estimate exactly the Pt and TiO_2 particles size, while peaks attributable to Nb specimens are at the noise level and

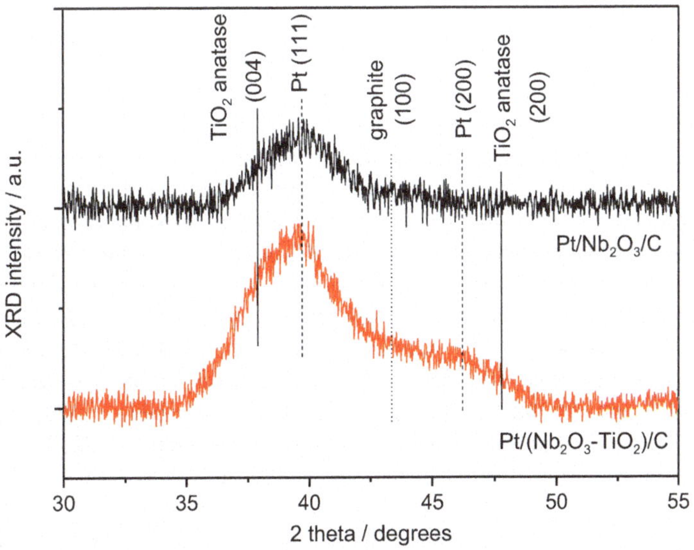

Fig. 8.12 XRD spectra of the Pt supported on Nb_2O_5/C and (Nb_2O_5,TiO_2)/C. The vertical lines drawn show the positions of the respective peak reflections (Reprinted with permission from [7]. Copyright (2014) American Chemical Society)

absent, due to their very small amount. Nb oxide reflection peaks are also absent even from the XRD spectrum of the enriched $Pt/Nb_2O_5/C$ catalyst, though the amount of niobia is now significantly higher as compared with the $Pt/(Nb_2O_5, TiO_2)/C$ sample. The only visible reflection peak in the $Pt/Nb_2O_5/C$ sample is ascribed to the Pt (1 1 1) crystallographic plane. By fitting this particular peak with a Gaussian function, the average Pt particle size estimated by applying the Scherrer equation has been found to be approximately 2.2 nm, this way indicating the highly and uniformly dispersed nature of such a catalyst. From hydrogen gas B.E.T. chemisorption measurements [65], the average Pt particle size was estimated to be 3.7 nm (77 m^2/g of Pt assuming spherical particles, while the electrochemically active surface area, UPD H-adatoms estimated by the corresponding potentiodynamic UPD peaks, is even assessed to be lower, 57.5 m^2/g Pt). The significantly larger Pt particles overestimated by the B.E.T. H_2 chemisorption indicates the interaction between the Pt particles and the oxide support (SMSI) that hinders the hydrogen gas adsorptive properties of Pt crystallites, leading to particle size overestimation. On the other hand, the particle size determination by XRD is usually overestimated (the overlapping effect), and since the size is close to the instrument limits, such a state of experimental evidence convincingly suggests that Pt particles are evenly dispersed and strongly bonded on the hypo-d-oxide catalytic support as shown by many XRD measurements with similar interactive hypo-d-oxide supported electrocatalysts [5–11]. The absence of niobia reflection peaks indicates that Nb-oxides are either highly sub-up to mono-layer dispersed on the carbon particles surface (as essentially the surface species), or less probably in amorphous state. In fact, the present electrocatalyst has been *a priori* planned and calculated by the mass and surface ratio between hypo-d-oxide and Vulcan carbon particles to be deposited as submonolayer (Nb_2O_5 having 170 m^2/g, while carbon particles being in higher amount and of much larger surface area, 260 m^2/g). Such a unique XRD experimental insight convincingly testifies to the existence of high Pt (1 1 1) bronze dispersion upon highly developed hypo-d-oxide support. Such a state relative to the standard Pt/C electrocatalyst, otherwise suffering from Pt surface diffusion and agglomeration, characterizes remarkably increased electrocatalytic activity, a much longer lasting catalyst because of the SMSI bonding effect, while the interactive structure enables Pt metal recovery, which is another high quality achievement of the novel nanostructured bronze type electrocatalysts. Meanwhile, the most significant is the control of homogeneous dispersion of grafted nanostructured Pt size magnitude: The ratio of available exposed hypo-d-oxide surface relative to the Pt 10 wt% defines the latter, and in our issues it is usually fixed between 2.0 and 2.4 nm in average, in particular for NbO_x, WO_x and TiO_2 individual and mixed valence composite hypo-d-oxide supports (SMSI); for TaO_x not yet optimized. However, TEM identifies plentiful of about 1.0 nm $Pt/TaO_x/C$ clusters.

UHRTEM, and in particular STEM imaging in HAADF mode with aberration correction (FEI Titan 80-300), reveal grafted Pt nano-clusters on hybrid hypo-d-oxides supports, (Fig. 8.13a–d), with a rather uniform and evenly homogeneous distribution in average of about 2–2.4 nm in size on the best issues

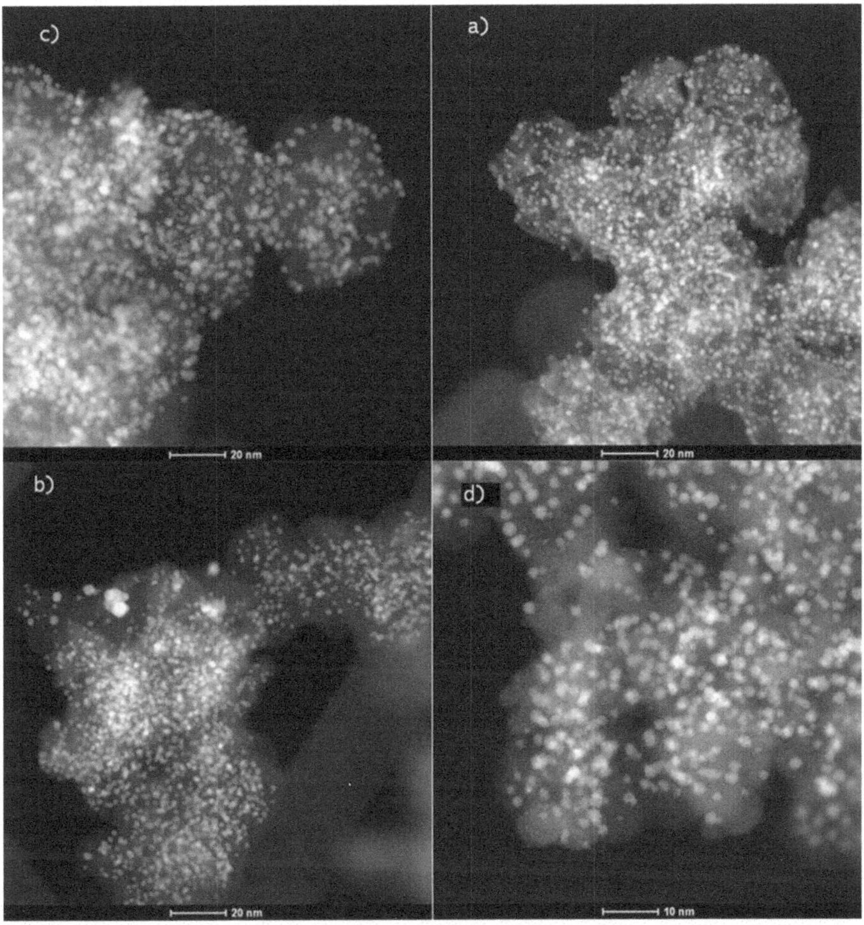

Fig. 8.13 Scanning transmission electron microscopy images in high-angle annular dark-field mode (HAADF-STEM) of TM hypo-d-oxide/carbon support material and nanostructured Pt electrocatalyst nanoparticles. (**a**) Composite simple basic Nb-oxide interactive (SMSI) supported electrocatalyst, {Pt (10 wt%)/20 wt% Nb_2O_5/C}; (**b**) Congenial mixed altervalent supported electrocatalyst, {Pt (10 wt%)/20 wt% Nb_2O_5,TiO_2/C}; (**c**) Composite interactive (SMSI) altervalent supported electrocatalyst of the same hypo-d-oxide structure as in (**b**), but of triple higher Pt weight percentage, {Pt (30 wt%)/20 wt% Nb_2O_5,TiO_2/C}; and (**d**) Congenial altervalent supported electrocatalyst based on mixed W-oxide and anatase titania interactive supported Pt clusters (Reprinted with permission from [7]. Copyright (2014) American Chemical Society)

{Pt (10 wt%)/20 wt% Nb_2O_5,TiO_2/C}, Fig. 8.13b), {Pt (10 wt%)/20 wt% Nb_2O_5/C}, Fig. 8.13a), and {Pt (10 wt%)/20 wt% WO_3,TiO_2/C}, Fig. 8.13d), obtained so far and in excellent agreement with size measurement with XRD. Even more so, with three times larger Pt amount {Pt (30 wt%)/20 wt% Nb_2O_5,TiO_2/C}, Fig. 8.13c), the homogeneous nanostructured Pt distribution keeps the same trend and in particular the average nano-size. Such a homogeneous, unusually narrow nano-size level of distribution has never so far been achieved (the interactive SMSI effect) and confirms the reliable basis for the rather pronounced spillover effect and the unique electrocatalytic achievements. No single Pt nanostructured cluster has been noticed on the prevailing carbon nano-particles percentage of their otherwise highly developed exposed surface area, no surface diffusion and no agglomeration, either. It would certainly be worthwhile noticing a rather allover homogeneous widespreading of the interactive hypo-d-oxide support structure and well distinct inter-d-d-bonded and interactive (SMSI) grafted fine Pt nano-sized clusters upon them, as a unique nanostructured Pt-bronze substantiation of advanced electrocatalytic properties, primarily and mostly extended by the pronounced interfering Pt-OH spillover effect for the cathodic ORR.

In addition, atomic-resolved HAADF-STEM observations strongly support the interacting (SMSI) and interfering spillover theory. The selective interactive hypo-hyper-d-d-interelectronic grafting distribution between Pt and Ta_2O_5 on the carbon support results in the interdependent Ta oxide network structure, symmetrically surrounding Pt cluster (Fig. 8.14). High-resolution HAADF images obtained with aberration-correctors of the electron probe (0.07 nm resolution) even resolve the individual Ta atoms of the oxide network supported on the carbon (Fig. 8.14b, c). This structure of Pt surrounded by the Ta oxide network would result, right at the interface Pt/Ta_2O_5, in the reinforcement of the overall SMSI effect at such a site, possibly with the $TaPt_3$ phase, already XPS confirmed for $TiPt_3$ (Fig. 8.15a) [11], and provide the highest electrocatalytic activity as the result.

The reinforced (SMSI) hypo-hyper-d-d-interelectronic bonded nanostructured synergistic electrocatalytic composite $Pt/TiPt_3/TiO_2$ has already been approved and tested, while symmetric monoatomic Ta distribution around Pt nano-cluster is the first step towards the inter-bonding and reinforcing $TaPt_3$ interphase at the interface Pt/Ta_2O_5, promising further electrocatalytic SMSI advances. In other words, the interactive Tauster hypo-hyper-d-d-(f)-supported electrocatalysts synergistically improve their monophase (Pt) activity, which after longer reduction proceeding with H-adatoms, or as the overall electrode reduction effect, finally emerge in the $Pt/TaPt3/Ta_2O_5,TiO_2/C$ state of maximal electrocatalytic activity.

Following the Pt loading on complex support (10 wt% Pt/20 wt% Nb_2O_5/C), HAADF, bright-field imaging and elemental mapping (EDXS) demonstrate that crystalline Pt catalyst nanostructured clusters are homogeneously dispersed on the hybrid mixed valence hypo-d-d-oxide support (Fig. 8.15b). Combining imaging and elemental analysis, it was found that the catalyst nanoparticles are embedded into (and/or properly symmetrically surrounded by) the oxide layer as pointed out using white arrows in Fig. 8.14. Higher loading on the same support, (30 wt% Pt/20 wt% Nb_2O_5/C), provides the evidence of maintained the same preceding uniform dispersion (Fig. 8.13). Based on elemental mapping with energy dispersive X-ray spectrometry, it was enabled to deduce that the Pt nanoparticles distribution is strongly correlated with the location of the transition metal oxide and rather avoiding any carbon (no d-d-interactive bonding). When present, the thickness of the disordered oxide network varies between 1 and 2 nm.

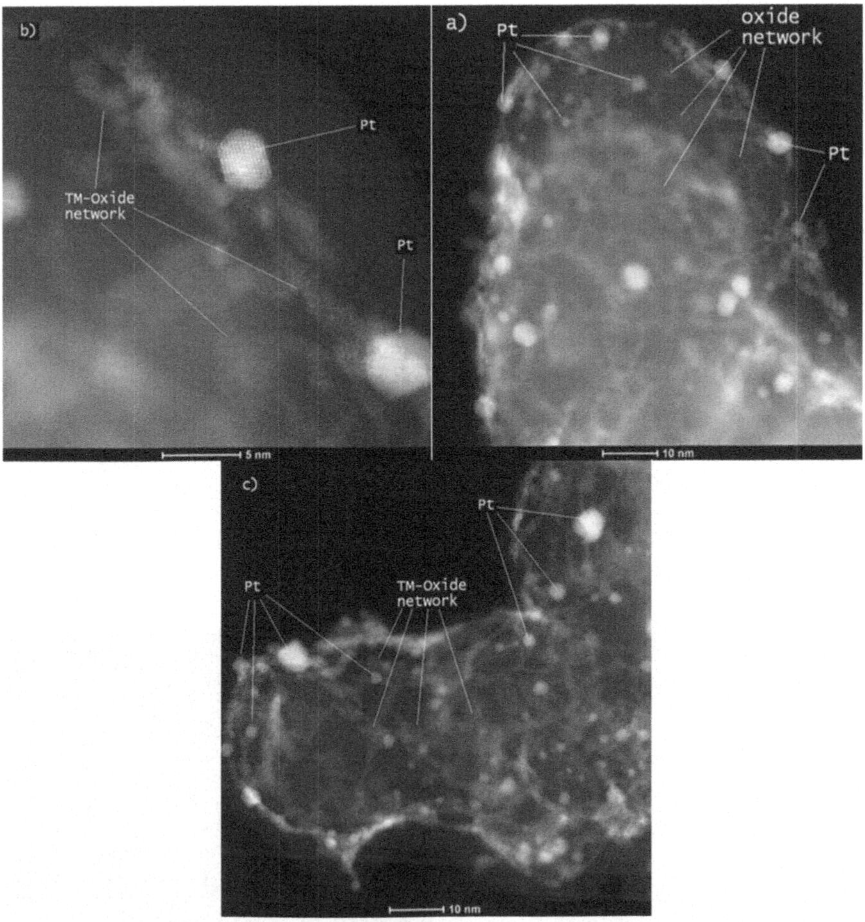

Fig. 8.14 Atomic-resolved HAADF STEM images analysis taken in different areas of the sample. Pt catalyst particles and transition metal Ta-oxide network are identified by labels and pointed out by relevant lines in images. (**a**) High resolution HAADF STEM general view of the carbon support, Pt and the Ta-oxide network; (**b**) detailed view of the Pt and Ta-oxide network showing the single Ta atoms of the oxide network (the oxygen atoms are not visible in this imaging condition), (**c**) overview of intricate Ta-Ox network on carbon support and strong interaction between the Pt catalysts only located on (or surrounded by) the Ta-oxide (Reprinted with permission from [7]. Copyright (2014) American Chemical Society)

8.12 Primary Oxide Spillover DL Charging and Discharging in the Light of the First Principle Theory

The overall summation (Σ) for the ion exchange properties of hypo-d-oxide hydrated membrane transferring matrix, under directional electric field effect or polarization, yields the reversible Pt-OH adsorptive and/or desorptive spillover

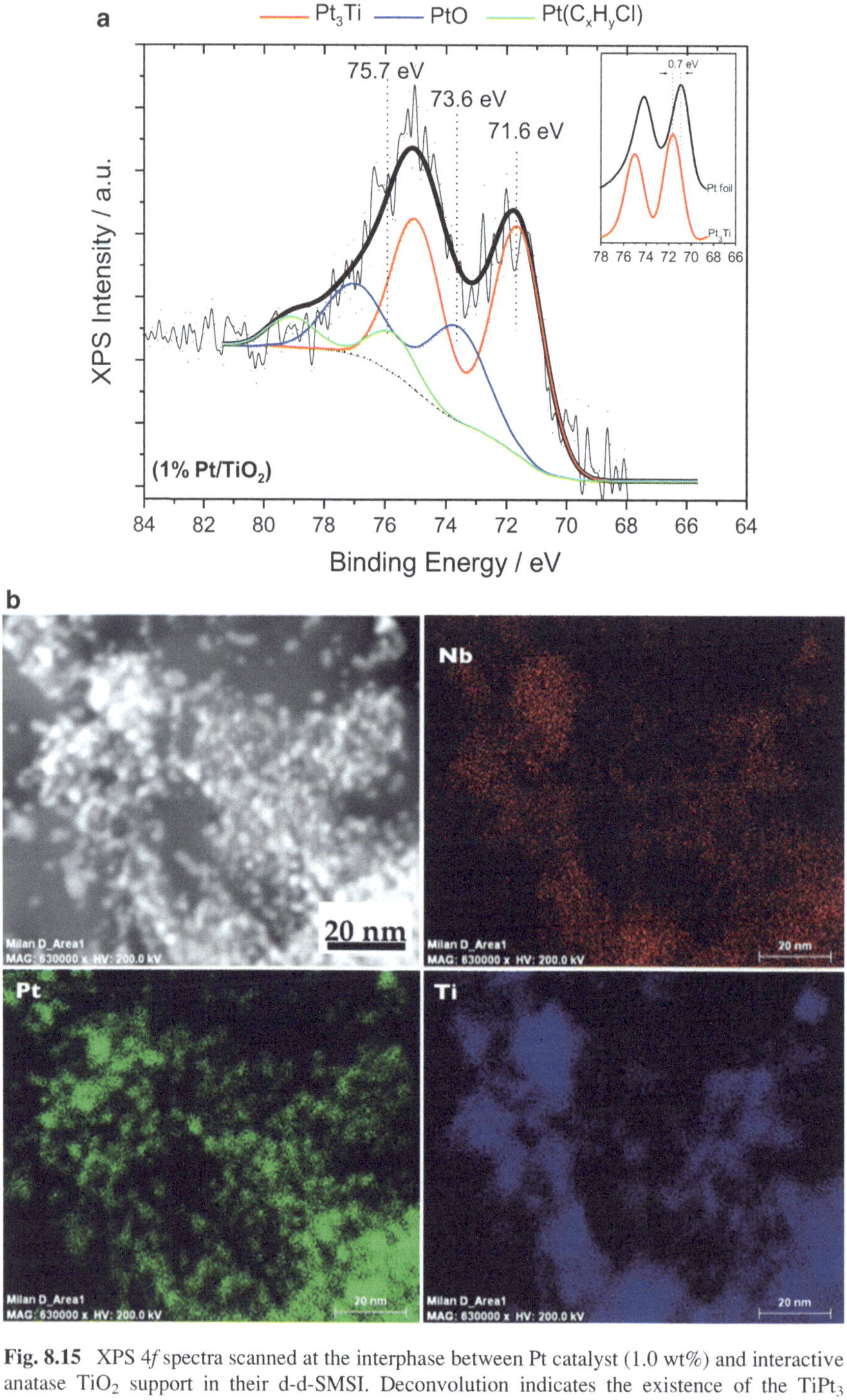

Fig. 8.15 XPS 4*f* spectra scanned at the interphase between Pt catalyst (1.0 wt%) and interactive anatase TiO_2 support in their d-d-SMSI. Deconvolution indicates the existence of the $TiPt_3$ intermetallic phase at the interphase Pt/TiO_2 with all individual spectral properties, as the prerequisite for the SMSI effect

upon, or from metal electrocatalyst reactive surface, resulting that way with the characteristic pronounced reversible effusion peaks of extremely fast electrode reaction [2–8], (Eq. (8.1-a), (8.1-b) and (8.1-c), yielding by such a summation the same former Eq. 8.1),

$$M + H_2O \Leftrightarrow M\text{-}OH + H^+ + e^- \tag{8.1}$$

The starting first principle thermodynamic relation for such a SMSI (exampled as M/TiO_2) system and just defined general reversible peak reaction, Eq. 8.1, arises therefrom in the following form,

$$\overline{\mu}_{H^+} = -\mu_{M-OH} - \overline{\mu}_e + \mu_{H_2O} + \mu_M \tag{8.7}$$

while from the main general self-explanatory relation for electrochemical potential definition,

$$\overline{\mu}_e = \mu_e - e\varphi = \mu_e - e\psi - e\chi \tag{8.8}$$

and particularly for hydronium ion (H_3O^+, abbreviated for simplification as H^+), there from Eq. 8.8 further follows an alternative relation

$$\overline{\mu}_{H^+} = \mu_{H^+} - e\varphi_{H^+} \tag{8.9}$$

To avoid and eliminate electrochemical potentials ($\overline{\mu}_{H^+}, \overline{\mu}_e$), by equalizing the right-hand sides of Eqs. 8.7 and 8.9, and replacing the corresponding value from Eq. 8.8, there straightforward yields the following relation,

$$\mu_{H^+} = -\mu_{M-OH} + \mu_M + \mu_{H_2O} + e\varphi - \mu_e + e\varphi_{H^+} \tag{8.10}$$

Meanwhile, since by definition, $\varphi = \varphi_{H^+} + \chi = \psi + \chi$, and thereby, $\varphi_{H^+} = \psi$, there further follows,

$$\mu_{H^+} = -\mu_{M-OH} - \mu_e + \mu_M + \mu_{H_2O} + 2e\psi + e\chi \tag{8.11}$$

Finally, by taking the derivative of Eq. 8.10, because $\Delta\mu_e = 0$, and since μ_M, μ_{H_2O}, and $\psi = \text{const.}$, while $\Delta\mu_{H^+} \ll e\,\Delta\chi$, there straightforward results,

$$\Delta\mu_{\text{M-OH}} = e\Delta\chi \tag{8.12}$$

exactly as cyclic voltammograms (Fig. 8.3) reveal: Every change in spillover of the M-OH, or of its chemical potential $(\Delta\mu_{M-OH})$, as the driving force of the effusion, or, simply its concentration variations, each corresponds to the increment of surface potential ($\Delta\chi$), or the change in capacity of DL because of M-OH dipole adsorption package, or desorption (as the spillover capacity), both of them being fast and substantially reversible, define DL charging or discharging (in its capacity, or pseudo-capacitance). Namely, since $\Delta\mu_{H^+}$ is logarithmic function of hydronium ions concentration, and according to the spillover electrode reaction (Eq. 8.1), primary oxide (M-OH) and H_3O^+ ions become simultaneously spent or produced in equivalent (1:1) amounts, while the contribution of surface energy is the linear function of surface potential (prevailing dipole effect), $e\Delta\chi$, then the inequality conclusion, $\Delta\mu_{H^+} \ll e\Delta\chi$, emerges quite logically.

Since the fundamental definition of work function (Φ) states [52, 66], that

$$e\Phi = -\mu_e + e\chi \tag{8.13}$$

and taking into account Eq. 8.12, by derivation of the latter (Eq. 8.13), there straight forward follows the broader fundamental definition for M-OH spillover adsorption and/or desorption, which is the same as the DL pseudo-capacitance charging and/or discharging,

$$e\Delta\Phi = e\Delta\chi = \Delta\mu_{M-OH} \tag{8.14}$$

Simple combination of Eq. 8.13 with Eq. 8.8, then yields the alternative first principle relation between work function and electrochemical potential, as the well known equation of general meaning and significance,

$$\bar{\mu}_e = -e\Phi - e\psi \tag{8.15}$$

which, for the considered issue of electrochemical cell (U_{WR}), assembled between working (W) and stable reference (R) electrode, can be straightforward written to read [52, 66],

$$\Delta\bar{\mu}_e = \Delta U_{WR} = -e\Delta\Phi - e\Delta\psi \tag{8.16}$$

The main contribution both to changes in work function ($\Delta\Phi$) and primary oxide dipole spillover ($\Delta\mu_{M-OH}$), comes from the surface potential increment ($\Delta\chi$) itself (Eq. 8.12), as the prevailing part of inner or Galvani potential ($\Delta\varphi$). Namely, Vayenas et al. [67, 68] have proved on plentiful systems in solid state electrolyte [69], aqueous media [70–72] and PEM Nafion 112 [73] the basic NEMCA or EPOC promotion relation in heterogeneous catalysis,

$$\Delta U_{WR} = e\Delta\Phi \tag{8.17}$$

that could also be considered as fundamental both in electrode kinetics [74–76] and spillover phenomena [66, 77]. Meanwhile, the latter (Eq. 8.17), when correlated with the preceding relation (Eq. 8.16), unambiguously and consequently yields that

$$\Delta\psi = 0, \psi_W = \psi_R = \text{const.} \tag{8.18}$$

which is also implied in above preceding equations (Eqs. 8.11 and 8.12) [78, 79]. Such conclusive and far-reaching statement is in particular reliably clear for the electrode steady state and with high concentration supporting electrolyte.

The whole thermodynamic equilibrium for the general issue of interactive (SMSI) metal/hypo-d-oxide supported catalyst, such as M/TiO_2, is now illustratively displayed in all interconnected energy details by Fig. 11 in [9]. As a consequence, any decrease of the M-OH chemical potential ($\Delta\mu_{M-OH}$), causes the work function to decrease correspondingly, and *vice versa*. In other words, any consumption of adsorbed M-OH upon metallic catalyst surface causes to decrease its chemical potential, and this way imposes imbalance within the established equilibrium. Thus, to keep constant both $\Delta\Phi$ and cell voltage between working (W) and reference (R) electrode (ΔU_{WR}), the titania phase, by its enriched latent storage of the primary oxide, automatically reacts by its membrane transferring mechanism and supplies hydroxide ions (Scheme 8.1), to emanate further as the primary oxide spillover adsorptive species on exposed metallic catalyst surface by corresponding electrocatalytic reaction (Eq. 8.1), both from hydrated anatase structure and by reacting of continuously supplied water molecules in their spontaneous adsorptive dissociation [21, 22].

The overall conclusion associated with otherwise broad experience in electrochemical promotion (NEMCA, EPOC) [66–68] of various chemical reactions in heterogeneous catalysis and particularly for Doebereiner (Ertl interpretation [30]) oxidation of hydrogen in water solutions, for catalysts interactive (SMSI) supported upon hypo-d-oxides, hydroxyl ions play the main membrane transferring and decisive promotion role, while M-OH, as defined dipole state, undertakes the spillover and features as the substantially mostly reacting species all over the metallic catalyst surface, enabling that way to keep established such thermodynamic equilibrium. In fact, there is no catalytic promotion for hydrogen oxidation in acidic aqueous media and apparently hydroxide ions impose such a substantial promoting efficiency, in particular within the interfering (600–900 mV vs. RHE) potential range, when NEMCA approaches promotion effects of two and even more orders of magnitude [68, 71–73]. Therefore, any disturbance of such an equilibrium, such as the consumption of spillover adsorbed M-OH species ($\Delta\mu_{M-OH}$), or simple imposition of an external polarization, as just analytically shown above, reflects as the driving force ($\Delta U_{WR} = e\Delta\Phi$), and instantaneously tends to reestablish the former equilibrium, or to keep its steady state.

Since the adsorption of the primary (M-OH) oxide plays a decisive and critical role in many important (electro)catalytic processes, in particular for CO tolerance in PEMFCs and the ORR, enthalpies of chemisorption might throw some light for *a priori* considerations and estimative assessments. Namely, from the measured isosteric heat of Pt-OH adsorption (ca 200 kJ/mol), the adsorptive Pt(1 1 1)-OH bond energy was estimated to be ca 136 kJ/mol, that is much less than the surface (Pt=O) oxide chemisorptive bonding energy (ca 350 kJ/mol), and this is taken as the argument for higher catalytic activity of the former in surface electrochemistry of CO [80]. In fact, a rough estimation of adsorptive energy for interactive supported Pt catalysts upon hypo-d-oxides tells that the above ratio might be even for more than an order higher, as a typical ratio between reversible *versus* polarizable properties, and consequently Pt-OH correspondingly much more active in heterogeneous catalysis and particularly for the ORR. Otherwise, since two Pt-oxides, Pt-OH and Pt=O, distinctly differ in their polarization of desorption and/or adsorption, and since the latter defines the enthalpy of adsorption, such pronounced overvoltage differences unavoidably should have to imply their corresponding adsorptive/desorptive enthalpy values and differences. In the same context, hydrogen atoms from bronze type interactive supported Pt electrocatalysts ($Pt/H_{0.3}WO_3$) are distinctly more electrocatalytically and catalytically active than H-adatoms from plain nanostructured (Pt/C), and feature much weaker adsorptive bonding strength.

8.13 XPS Evidence for the Primary Oxide Presence and Spillover

The link between model polycrystalline Pt (85 % in (1 1 1)) and the interactive supported nanostructured Pt electrocatalysts, on various individual, or mixed valence hypo-d-oxide compounds (Nb_2O_5, WO_3, MoO_3, Ta_2O_5), composed with anatase (1 0 1) titania (TiO_2) itself, imposed the need for advanced contemporary surface characterization, in particular as concerns the primary oxide (Pt-OH) appearance, latent storage and spillover [5, 6, 11, 65]. More specifically, in order to understand the origin of various oxidation states of Pt deposited on both individual (Nb_2O_5/C, Fig. 8.16) and/or mixed valence ($Pt/Nb_2O_5,TiO_2/C$, Fig. 8.17) hypo-d-oxide supports, *in situ* Pt $4f_{7/2}$ and O 1*s* XP spectra have been scanned and deconvoluted before and after the exposure in heated hydrogen stream, and by simple thermal annealing effect under UHV and/or relative to room temperature (RT) for *samples treatment avoiding any exposure to the atmosphere*. The shape Pt 4*f* spectra indicates the existence of more than one oxide type Pt species, so that in all issues, the photo-peaks can be analysed into three doublets [81]. The first component at binding energy 71.3 eV is attributed and identified to belong to metallic platinum, the peak at 72.7 eV is assigned to $Pt(OH)_x$ ($1 \leq x \leq 2$), due to combination of Pt with OH in the primary oxide (M–OH) type, or some other congenial O-species, while the third component at 74.9 eV is ascribed to PtO_x [81–83], finally ending-up under longer storage in air, as the stable surface oxide (PtO_2). The partial percentage contribution of each Pt-oxy-component to the total peak area is equivalently shown for both specimens in Table 8.1. As a strong XP spectral evidence for the primary oxide (Pt-OH) existence (Eq. 8.1) and interactive transference (Eq. 2), is quantitatively assessed that both the simple thermal, and in particular the annealing under hydrogen flow, causes the corresponding distinct reduction of the Pt oxy-specimens either as the thermal desorptive effect and even more pronounced by the reactive contribution of Pt-H itself. Similar experimental

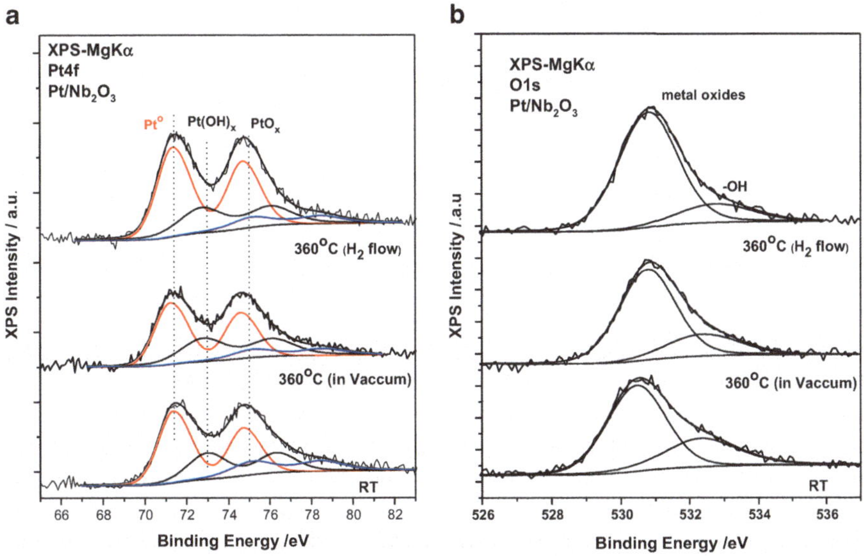

Fig. 8.16 (**a**, **b**) Pt 4*f* (**a**) and O1*s* (**b**) XPS core level spectra of the Pt catalyst on individual hypo-d-oxide, Nb_2O_5/C support at room temperature (RT), after annealing under UHV conditions at 360 °C and after reduction at 360 °C under 20 % H_2 in N_2 flow.

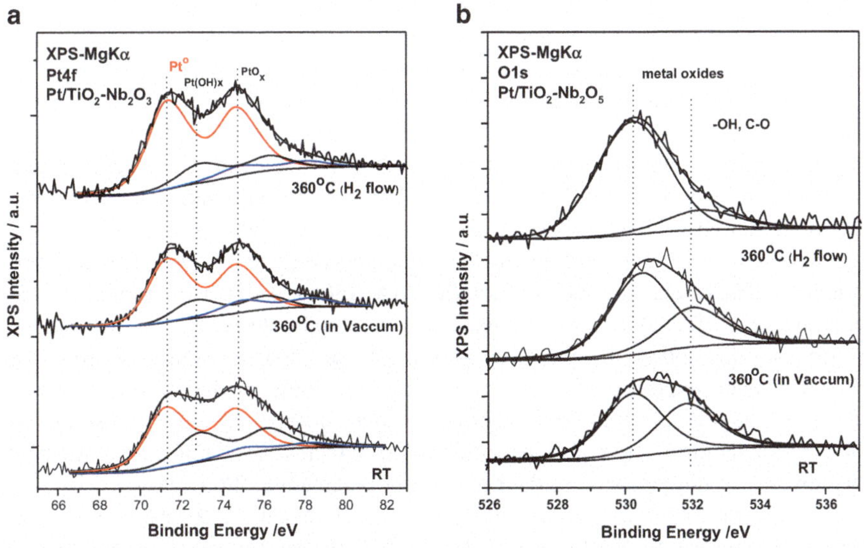

Fig. 8.17 (**a**, **b**) Pt 4*f* (**a**) and O1*s* (**b**) XPS core level spectra of the Pt catalyst on the mixed valence hypo-d-oxides, Nb_2O_5,TiO_2/C support at room temperature, after annealing under UHV conditions at 360 °C and after reduction at 360 °C under 20 % H_2 in N_2 flow

Table 8.1 Percentage contribution of each Pt4f component the total peak area

	Pt/Nb_2O_5-TiO_2			Pt/Nb_2O_5		
	Pt^0 (71.3 eV)	$Pt(OH)_x$ (72.7 eV)	PtO_2 (74.9 eV)	Pt^0 (71.3 eV)	$Pt(OH)_x$ (72.7 eV)	PtO_2 (74.9 eV)
RT	63.6	29.5	6.9	58.3	25.6	16.1
360 °C in vacuum	68.8	18.7	12.5	58.5	30.1	11.4
360 °C under H_2 flow	75.9	16	8	71	19.1	9.8

evidence has already been observed on the simple basic anatase titania (Pt/TiO_2, Fig. 8.4, [5]) interactive supported Pt electrocatalyst [5, 6, 11, 65]. Furthermore, a thorough parallel examination of the O 1*s* along with Pt 4*f* spectra, shows a pronounced decrease of the component attributed to the Pt-OH species by the former (BE ~ 533 eV), which apparently contradicts the enrichment in $Pt(OH)_x$ amounts, observed from the latter, Pt 4*f* spectrum. In other words, whilst the annealing causes a straight reduction of Pt oxi-species and desorption of the primary oxide (Pt-OH), the intensity discrepance in decrease of the O 1*s* component attributed to the same Pt-OH species might be compensated by the spillover from the enriched primary oxide latent storage of mixed valence hypo-d-oxide support. There is no other source for such remarkable difference and apparent discrepancies. In fact the same compensation (or, balance feeding spillover) effect enables anatase titania individually itself, but at correspondingly lower intensity [5, 6, 11, 65]. Meanwhile, all XP spectra distinctly indicate the existence, desorptive and/or Pt-H reactive removal and compensating latent storage spillover effect of the primary oxide [7]. In any case, the O 1*s* spectrum reveals a larger initial partial amounts of the primary oxide and other similar reactive oxi-species ($Pt(OH)_x$) for the issues of mixed valence interactive supported electrocatalysts. As a whole, the annealing in vacuum causes a distinct reduction of Pt oxi-species, in particular the decrease of the more active primary type $Pt(OH)_x$ oxides, and a slight increase of initially present, (long term of the storage nucleation aging effect) in relatively much lower, minor extents, or even initially absent, the PtO_x (PtO_2). At the same time, the O 1*s* component attributed to OH-species correspondingly decreases, indicating the significant desorption of hydroxyl groups in particular pronounced from the mixed valence hypo-d-oxide supports during annealing, as the strong clear evidence for the initial primary oxide enriched (latent storage) presence and spillover. In general, the most active primary oxide (Pt-OH) then keeps the low level, or even the absence of higher oxide in their amount or percentage, (PtO_x, PtO_2).

XP spectral *ex situ* measurements, Fig. 8.18a, b, reveal the corresponding Pt 4*f* deconvoluted spectra for the originally fresh prepared hypo-d-oxide interactive (SMSI) Pt electrocatalyst supports by Nb_2O_5, TiO_2-Nb_2O_5, TiO_2-WO_3 and TiO_2 itself, and for the same specimens after an annual storage. Three distinct oxo-components, just as above, participate now within the electrocatalytic Pt surface, appearing at three different binding photo-peaks, but the

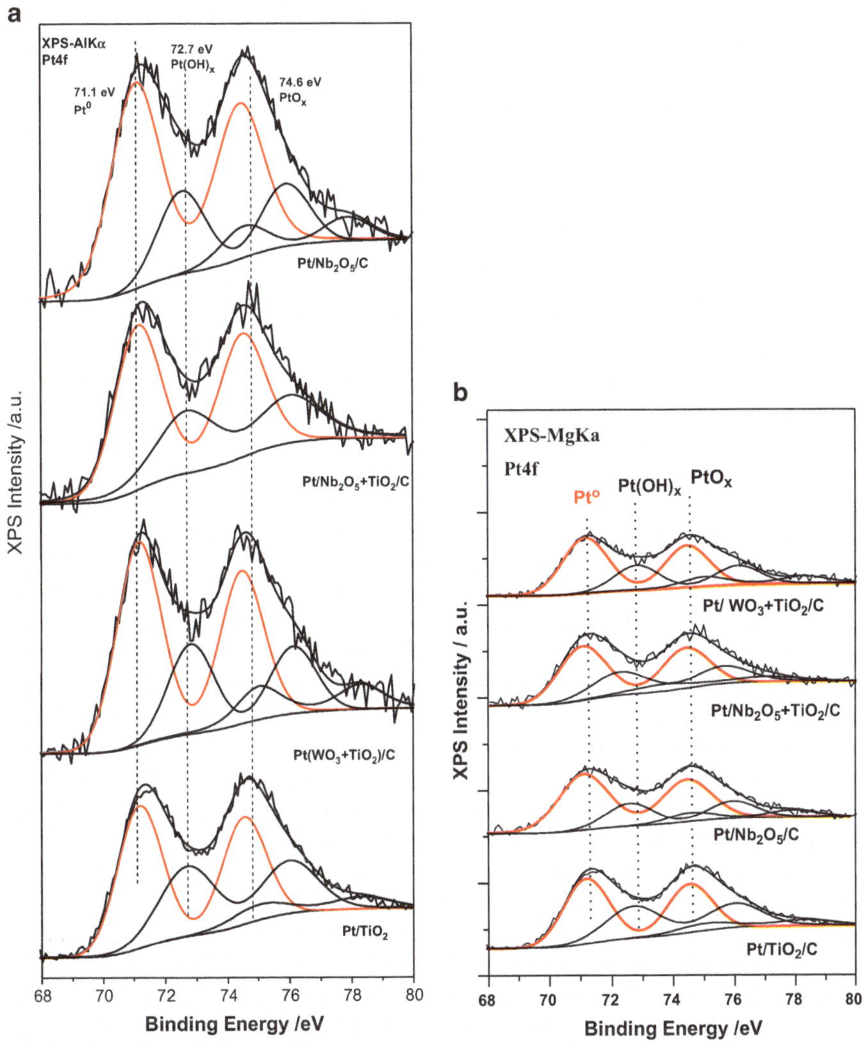

Fig. 8.18 (**a**, **b**) Pt 4*f* XPS core level spectra of Nb_2O_5, TiO_2-Nb_2O_5, TiO_2-WO_3, and TiO_2 supported Pt catalysts at room temperature: ((**a**) scanned fresh and (**b**) after aging of one year exposed in air).

non-stoichiometric PtO_x oxides ($1 \leq x \leq 2$), ending-up by the upper type surface oxide (PtO_2), as the final minor oxidation state [5, 6], (compare Fig. 8.10, [6]). The latter even does not appear at the Pt 4*f* spectrum of the fresh prepared Nb_2O_5-TiO_2 supported Pt catalyst, even after many repeating scans within a longer period (Fig. 8.18a). Such a decisive and conclusive remark is of substantial and fundamental significance for the present study: The primary oxide is proved initially available, in particular in the condensed wet state of the electrocatalysts. In other

Table 8.2 Percentage contribution of each Pt4f component to the total peak area for fresh prepared specimens and for an annual storage of the same specimens

	Percentage contribution of Pt4f components		
Catalyst support	Pt^0 (71.1 eV)	Pt(OH) (72.7 eV)	PtO_x (74.8 eV)
(a)			
Nb_2O_5	67.2 %	23.7 %	9.1 %
Nb_2O_5-TiO_2	65.3 %	34.7 %	–
WO_3-TiO_2	61.9 %	26.7 %	11.4 %
TiO_2	56.8 %	31.9 %	11.3 %
(b)			
Nb_2O_5	58.3 %	25.6 %	16.1 %
Nb_2O_5-TiO_2	63.6 %	29.5 %	6.9 %
WO_3-TiO_2	61.9 %	26.7 %	11.4 %
TiO_2	56.8 %	31.9 %	11.3 %

words, for such a state of experimental evidence, to complete the whole story, one should recall the self-catalytic effect of water molecules on the primary oxide *a priori* appearance and existence [30], and/or the Vittadni [21, 22] spontaneous dissociative water molecular adsorption, as the source of the latent storage for the Pt-OH spillover. Under such circumstances, highly reversible and electrocatalytically extremely active primary oxide, (Pt-OH), by the spontaneous interfering electrode reaction (Eq. 8.3), spends the exposed surface oxide (Pt=O), and prevents its upper level ($PtO_x \rightarrow PtO_2$) growth, so that the latter is either missing in fresh prepared specimens, or minor component after longer free air exposure.

The percentage contribution of each component to the total peak area is shown in Table 8.2a, which has some deeper theoretical significance. The roots and expectancies for the primary oxide spillover are indicated there on an almost quantitative scale basis. The decisive and pronounced cooperative effect of anatase titania is clearly indicated by its individual rather high initial primary oxide surface percentage, while the Pt/Nb_2O_5,TiO_2 features the most (dynamic latent storage) creative synergistic properties for the Pt-OH generation that further reflects on and definitely defines the overall distinctly advanced electrocatalytic activity for the ORR [7]. Shrinkage of the same XPS Pt-OH peak capacity both by the thermal and hydrogen reduction effect at higher temperature (Figs. 8.16 and 8.17a, b, see also Fig. 8.4, [5]), is observed as the additional experimental evidence for the existence of primary oxide initially available for further instantaneous spillover distribution under electrode polarization. In the same respect, the complementary DRIFT imprints (Fig. 8.9, [84]) on a similar manner reveal the same primary oxide thermal and Pt-H reduction features, but the desorption effect in the present XPS issues arises proportionally smaller since under UHV the examined specimens are deprived from water molecules. Meanwhile, the same specimens after unused open air storage show the nucleation of PtO_x, ending with PtO_2 (Fig. 8.17b), as the common transient in particular pronounced with more active the interactive supported Pt electrocatalysts is [5–11], and/or suppressed in the growth by more active and when accumulated in larger amounts the primary(Pt-OH) oxide.

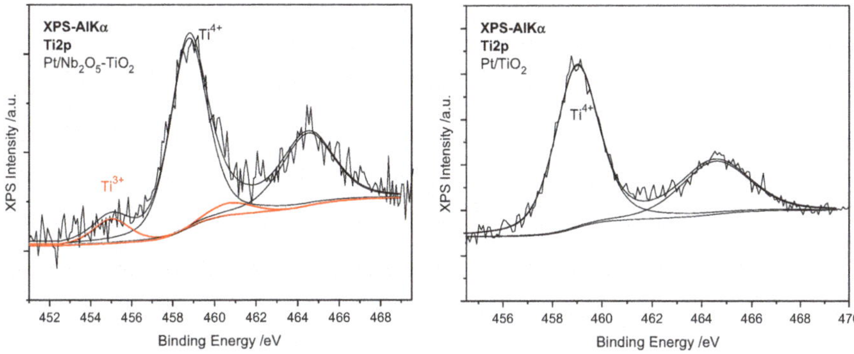

Fig. 8.19 Deconvoluted Ti 2*p* XPS core level spectra of TiO_2-Nb_2O_5 (*left*) and TiO_2 (*right*) supported Pt catalysts

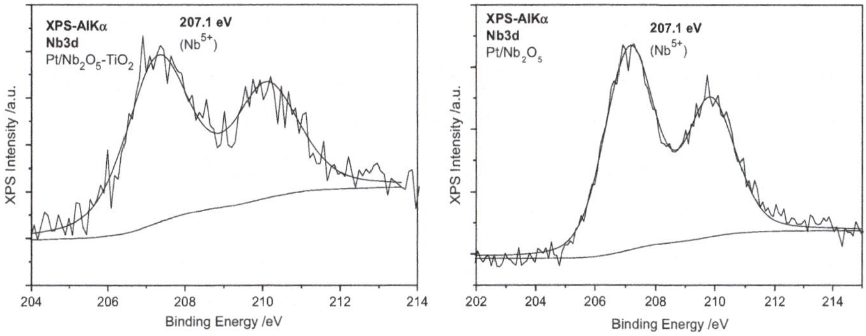

Fig. 8.20 Deconvoluted Nb 3*d* XPS core level spectra of TiO_2-Nb_2O_5 (*left*) and TiO_2 (*right*) supported Pt catalysts

In Fig. 8.19 the deconvoluted Ti 2*p* spectra of TiO_2-Nb_2O_5 and TiO_2 itself supported Pt catalysts are, in addition, comparatively presented. The main doublet in both spectra, at binding energy (Ti $2p_{3/2}$) 458.8 eV is attributed to Ti^{4+} species [81, 85, 86]. In the case of the TiO_2-Nb_2O_5 mixed valence hypo-d-oxide support, a second doublet at binding energy (Ti $2p_{3/2}$) 455 eV is attributed to the contribution of Ti atoms in the Ti^{3+} state. Ti atoms in the other mixed hypo-d-oxide support TiO_2-WO_3 are detected only in the Ti^{4+} state [87]. Such an effect might be the cause for the higher synergistic electrocatalytic activity of the interactive Nb_2O_5, and similarly Ta_2O_5, mixed valence compounds with anatase (1 0 1) titania (TiO_2), relative to other hypo-d-d-oxide composites.

In Fig. 8.20 the deconvoluted Nb 3*d* XPS core level spectra of TiO_2-Nb_2O_5 and Nb_2O_5 supported Pt catalysts are presented. In both cases only one doublet is apparent at binding energy (Nb $3d_{5/2}$) 207.1 eV that is characteristic for the Nb^{5+} state, which, when inserted into a titanium dioxide network, usually causes the effect in charge compensation [88]. The addition of such a charge can be compensated either by creating one vacancy of Ti per four introduced Nb ions, or by the reduction of Ti^{4+} to Ti^{3+} per each inserted Nb^{5+}-ion. Both of these effects can occur, with the latter being much more probable at relatively lower temperatures [88]. Quantitative analysis of the present results, using the Ti 2*p* and Nb 3*d* peak intensities (areas) corrected by the atomic sensitivity factors [89], shows that the appearance of each Ti^{3+}, corresponds indeed to the introduced one Nb^{5+}-ion, [90, 91]. In fact, EELS (Electron Energy Loss Spectroscopy) scans, in addition, clearly point to the valence variations of the oxide support, especially at the Ti edges (cf. [92, 93]).

Now, there arises a very interesting situation concerning the most promising electrocatalytic Nb-oxide structure, which at relatively low temperatures of calcination crystallizes as Nb_2O_5. The latter is well confirmed by the XPS analysis, while much more stable NbO_2 appears above 900 °C [94]. Such experimental evidence reveals the fifth hydroxide (OH^-) ion for the most easily transferable within the overall spillover mechanism,

$$Nb(OH)_5 + Pt \Leftrightarrow Nb(OH)_4{}^+ + Pt\text{-}OH + e^- \quad (8.19)$$

Meanwhile, XPS analysis has also revealed a further interrelating mechanism of similar exchanges with hydrated anatase titania,

$$Nb(OH)_4{}^+ + Ti(OH)_4 \Rightarrow Nb(OH)_5 + Ti(OH)_3{}^+ \quad (8.20)$$

or, when summed up,

$$Ti(OH)_4 + Pt \Leftrightarrow Ti(OH)_3{}^+ + Pt\text{-}OH + e^- \quad (8.21)$$

Then, the entire formalism clears up the equivalence and mutual facilitation between titania and niobia for the primary oxide spillover, as already concluded from Table 8.2a.

The conclusive observation from above XPS experimental evidence then has been that such membrane type ionic migration, both through the bulky network and over hypo-d-oxide surface exposure, ending up with the Pt-OH(Au-OH) spillover, is interdependent and can be further remarkably optimized by the common catalyst interactive support structure, in particular by hypo-f-oxide (CeO_2, GdO_2, HoO_2) ingredients with much larger, and thence, much more exposed (and thereby easier interactive and migrating) latent storage hydroxide structure. Let us then add the straight intimate and much more expanded interactive Pt catalyst hydrated interface (Pt/$Nb(OH)_5$, $Ti(OH)_4$, $Ce(OH)_4$) contact, and even direct interaction of NaOH from rich alkaline media [9],

$$OH^- + Pt \Leftrightarrow Pt\text{-}OH + e^- \quad (8.10)$$

which is the same and equivalent to Eq. 8.1 [9], to become aware of the multiple membrane type migration and even direct spillover Pt-OH generation. In such an electrocatalytic constellation surface and bulk migration represents now the main overall RDS and afforded open optimization area to provide already superior, broad range reversible electrocatalysts for oxygen electrode reactions (ORR, OER).

The reversible fast adsorptive interacting primary oxides (M-OH) lead up to adsorbed monolayers (Pt-OH $\rightarrow$ 1), and are distinctly marked by a specific potentiodynamic peaks, as the first step of water oxidation (Eqs. 8.1 and 8.2) [2–4], immediately after the initial classical double layer charging range of many metals, first of all on Ru, Au, Pt, Rh, Ag, Ni. Similar behavior is exhibited by some other transition metal primary type oxides of common formulas and structure, like MoO(OH), NiOOH, AuOOH, $WO_2(OH)$, and can be identified by corresponding peaks within their potentiodynamic and XPS spectra [53, 65]. All hypo-d-elements, in particular of high altervalent capacity, afford the reversible primary oxide type states, usually of pronounced catalytic activity and high electronic conductive properties (WO(OH), $MoO_2(OH)$), but unfortunately, in their oxidation sequences end-up with non-conductive and catalytically inactive higher valence oxide states (MoO_3, WO_3, Pt=O, PtO_2, Au=O, NiO_2), otherwise the whole aqueous electrochemistry would feature another and quite different physiognomy!

8.14 Striping Voltammetry Evidence for the Primary Oxide Spillover Effect

Striping voltammetry represents another indirect complementary experimental potentiodynamic evidence to scan and prove the primary oxide spillover effects in electrocatalysis. Interactive hypo-d-oxide supported and non-supported electrocatalysts (both Pt and RuPt) exhibit dramatically different activity for CO tolerance in LT PEMFC, and distinctly provide new additional cyclic voltammetry scanning spectra for the M-OH spillover effect [5, 6]. Ever since Watanabe [95] has shown that Ru even at submonolayer core-shell deposit, or while alloying with Pt, shifts the primary oxide growth to a much more negative potential range and enables CO tolerance, the primary oxide spillover became of substantial significance for PEMFCs [96]. Similarly, the hypo-*d*-oxide supported Pt and Ru (Pt/TiO_2/C, Ru/TiO_2/C) in their behavior *versus* these two plain and non-interactive supported metals (Pt/C, Ru/C) themselves, or even their interactive unsupported alloys, RuPt/C and RuPt/TiO_2/C, Fig. 8.21, have distinctly different catalytic properties, too, the interactive alternatives featuring an even more advanced and much more pronounced primary oxide spillover effect [5, 6]. Such distinct and clearly pronounced peak capacity difference between the interactive and non-interactive hypo-d-oxide supported Ru advanced electrocatalysts, belong to the best experimental evidence for the primary oxide spillover effect extensions by the latent storage. Since hypo-*d*-oxides, primarily anatase titania, zirconia and hafnia, and even more so tungstenia, niobia and tantalia, facilitate the spillover of M-OH, such facts clearly point to the advanced overall composite effect and advantages of membrane type OH^-—transferring within and over TiO_2, WO_3,

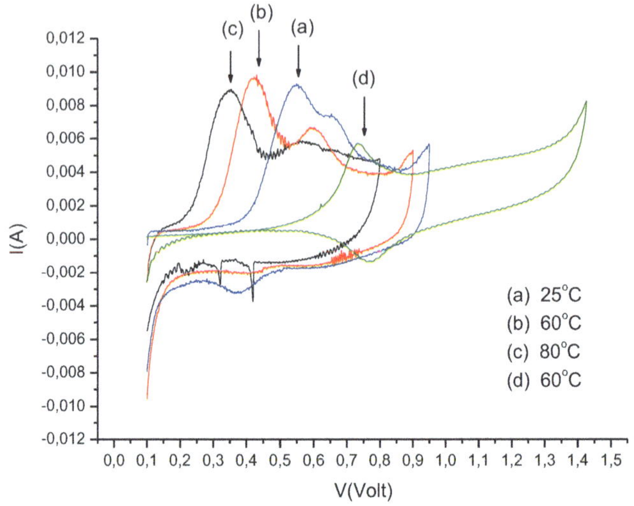

Fig. 8.21 The striping voltammograms for CO desorption from supported 10 wt% (0.4 mg cm^{-2}, 2 nm in average size, 1:1 atomic ratio Ru:Pt) $RuPt/TiO_2/C$ electrocatalyst CO-saturated at three different temperatures: 25 °C (**a**); 60 °C (**b**) and 80 °C (**c**), scanned at the scan rate of 2 mV s^{-1}; (**d**) the same stripping scans for CO desorption at 60 °C from unsupported 30 wt% (0.5 mg cm^{-2}) E-tek RuPt/C electrocatalyst of the same RuPt nano-size, atomic ratio and load, and sweep rate 10 mV·s^{-1}; its CO saturation at 55 °C

Nb_2O_5, TaO_2, and in particular within their substantially wet mixed networks of mixed altervalent catalyst supports, resulting in the speeded-up primary oxide effusion, relative to the plain non-interactive carbon (Pt/C).

In other words, while Ru metal itself facilitates both Pt-OH and Ru-OH spillover transfer even in the intermetallic non-interactive supported RuPt composite electrocatalyst [95, 96], the supporting effusion effect of titania advances the same effect for more than 300 mV relative to RuPt/C catalyst (Fig. 8.21). Anodic CO oxidation upon $Ru,Pt/TiO_2/C$ starts even within the potential range of UPD desorption of H-adatoms and becomes much more pronounced in the charge capacity relative to Ru, Pt/C. This important result is one of the most significant confirmations of the present interactive and dynamic spillover catalytic model, as implemented in electrocatalysis for hydrogen and oxygen electrode reactions. In fact, cyclic voltammogram spectra of plain Ru feature a unique type of shapes and in both scan directions stepwise reveal more than any other individual transition metal various interrelating oxide states and their peaks interdependances [97], and with such overall potentiodynamic sequences are closer to the reversible properties for oxygen electrode reactions than any other d-metal.

Mixed anatase (and even rutile) titania, and in particular tungstenia, niobia and tantalia form intermolecular solid oxide solutions of a high altervalent capacity (Scheme 8.2), compatible both in amorphous and crystalline forms of the edge sharing TiO_6 and the corner sharing WO_6 octahedrons, with pronouncedly

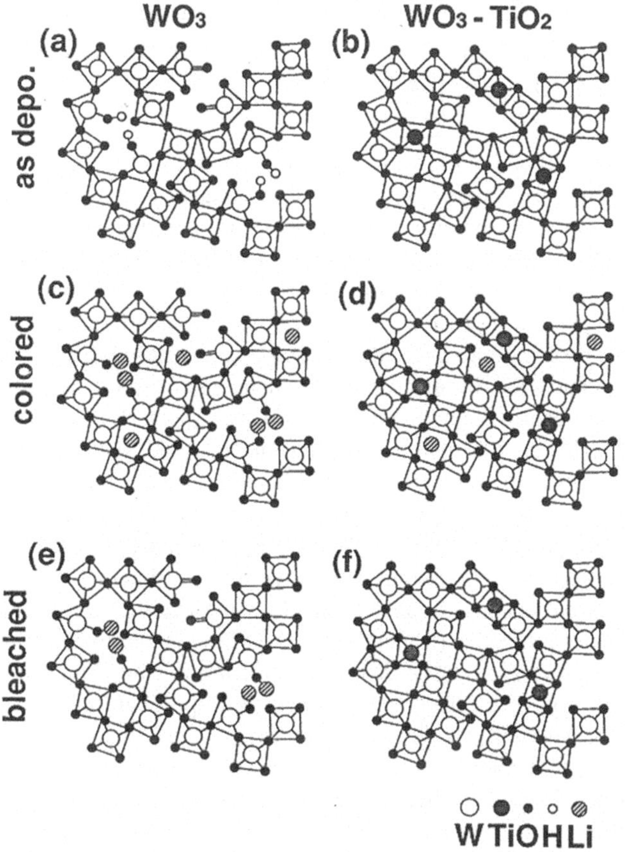

Scheme 8.2 Model presentation of electrochromic individual WO_3 layers and composite combinations of tungstenia and titania with corner and age sharing crystal units within the consistent mixed valence compounds. SEM and TEM confirmed for amorphous overall structure: (**a**, **b**) as deposited films, (**c**, **d**) as colored and (**e**, **f**) as bleached hypo d-oxides [92, 93]

increased electrochromic features even at high contents of the former [92, 93]. In fact, highly charged W^{6+} cations, like Nb^{5+}, additionally favor the reversible acidic dissociation of water molecules [21, 22], and thereby such electrochromic layers exhibit well defined ion exchange and electron conductive properties [23–25]. Thus, one of the fundamental contributions of the present paper is to show that prevailing anatase titania in the form of a composite mixed valence catalytic support with tungstenia (and/or even more so niobia, tantalia and molybdenia), behaves in a compatible way and regarding the ($Pt/H_{0.35}WO_3 \Leftrightarrow Pt/W(OH)_6$) bronze type equilibrium, so that the consequent reversibility features the same properties as pure tungsten bronze itself, all of them being further advanced by hypo-f-oxide ingredients (Ce, Gd, Ho, La, etc.)!

8.15 Superior Bronze Type Electrocatalysts for Reversible Alterpolar Interchanges in the Revertible (PEMFC vs WE) Cell Assembly

The striking target issue of the present paper has been to show the development and substantiation of corresponding spillover electrocatalysts for the superior revertible cell assembly for spontaneous reversible alterpolar interchanges between PEMFC and WE. In other words, L&MT PEMFCs, in particular large stationary systems, are unimaginable without revertible abilities with WE alternation, enabling to produce hydrogen during (usually overnight) excesses in electrical energy supply, and *vice versa*, to use stored amounts of the latter as a fuel in PEMFCs, in the course of (daily) deficiencies in electricity. In aqueous media classical plain Pt/C electrocatalysts feature catalytic surface properties of Pt-H and/or Pt=O, without any broader effusion of other interacting species, and such situation prevents reversible alterpolar interchanges, otherwise indispensable for the smooth revertible altering of PEMFC *versus* WE systems, when both alternatives are mutually equivalent and operate in their common overall harmony. In such a respect, the present paper introduces electron conductive and d-d-interactive individual and composite (mixed valence) hypo-d-(f)-oxide compounds of increased altervalent capacity, or their suboxides (Magneli phases), as the interactive catalytic supports and therefrom provides: (i) The SMSI interelectron-d-d-bonding and correspondingly improved catalytic effects, and (ii) Dynamic spillover interactive transfer of the M-OH, and alternatively the free fast effusion of H-adatoms for further electrode reactions, and thereby advance the overall electrocatalytic and alterpolar activities. Since hypo-d-oxides feature the exchange membrane properties, the higher the altervalent capacity, the higher the spillover effect. In fact, altervalent hypo-d-oxides impose spontaneous dissociative adsorption of water molecules and then spontaneously pronounced membrane spillover transferring properties instantaneously resulting with corresponding bronze type (Pt/H_xWO_3) under cathodic, and/or its hydrated state ($Pt/W(OH)_6$), responsible for Pt-OH effusion, under anodic polarization, this way establishing instantaneous reversibly revertible alterpolar bronze features ($Pt/H_{0.35}WO_3 \leftrightarrow Pt/W(OH)_6$), as the thermodynamic equilibrium, and thereby substantially advanced electrocatalytic properties of these composite interactive electrocatalysts, all along the potential axis between hydrogen and oxygen evolving limits, back and forth. Due to the dual spillover alterpolar interchanges, all four electrode reactions in aqueous media are reversibly altered (HER—Cathodic hydrogen evolution, HOR—Anodic hydrogen oxidation, ORR—Cathodic oxygen reduction, OER—Anodic oxygen evolution). In other words, the HER and OER are entirely equivalent in the operational sense to the couple of ORR and HOR, and in the revertible alterpolar interchanges between PEMFCs and WE that plain Pt by no means could enable itself, and this is the substance why we consider such electrocatalytic system for superior, and even ideal and unique. Such nanostructured type electrocatalysts, even of mixed valence hypo-d-oxide structures ($Pt/H_{0.35}WO_3,TiO_2/C$, or $Pt/H_xNbO_3,TiO_2/C$), have for the first time been synthesized by the sol-gel methods, including the super-critical drying with liquid CO_2, and shown to feature extra high stability, electron conductivity and non-exchanged initial pure mono-bronze spillover and catalytic properties. In other words, behaving as the unique, even ideal electrocatalytic system for substantiation of the revertible cell assembly for spontaneous reversible alterpolar interchanges between PEMFC and WE, though still being based on unavoidable Pt catalysts. Meanwhile, since nanostructured Pt clusters are grafted and inter-bonding fixed upon hypo-d-oxide support, just confirmed by UHRTEM nano-images, their life time has been dramatically increased, at least doubled, there is no agglomeration, nor hydrogen crossover through PEM, and there is now provided ability for prevailing Pt recovery. Namely, the substance consists in the reversible alterpolar interchangeable electrocatalytic operation with pronounced spillover (both Pt-H and Pt-OH) abilities; and then the bifunctional (PEMFC *versus* WE) cell construction is a simple routine engineering task.

8.16 Striking Electrocatalytic Conclusions

Classical potentiodynamic spectra in aqueous media, (Fig. 8.1), straight after the reversible adsorptive Pt-OH peak, reveal unusual, anomalous, longer and stronger anodic polarization even within the thermodynamic equilibrium potential values ($\Delta E = \Delta E_o + \eta_r + \eta_a$), and from the same cause and reasons, along the reverse cathodic sweep, too. Since the entire amount of reversibly adsorptive Pt-OH sooner, yet within the same reversible potential range (up to 1.29 V *vs.* RHE), undergoes complete disproportionation into highly polarizable and irreversibly deposited Pt=O monolayer, (Eq. 8.2), and thereby becomes unavailable for further reaction within such a longer (about 600 mVs) potential scans, as the result, there imposes the typical pronounced **reaction polarization**, (η_r), which suppresses further proceeding of the interfering Pt-OH/Pt=O reversible reaction (Eq. 8.3), within both, back and forth, scan directions. The latter extends further by the **activation polarization** (η_a), and all together they impose very strong and broad polarizable properties, (Fig. 8.1). On the contrary, interactive (SMSI) supports of mixed valence hypo-d-(f)-oxides, *a priori* behave as continuously renewable and recoverable enormous (Pt-OH) latent storage capacities and spillover sources (Scheme 8.1) all along the potential range between oxygen and hydrogen evolving limits. Thus, there is now no reaction polarization, ($\eta_r = 0$), and since the OER occurs straight from the enormous broad Pt-OH peak, and thereby exactly at the thermodynamic ROE value, ($\Delta E_o = 1.29$ V, $\eta_r = \eta_a = 0$), the reversible electrocatalysts of various interactive hypo-d-(f)-oxide support compositions for oxygen electrode reactions (ORR, OER) have now, finally been substantiated and tested on Pt and Au. The present conclusive discussion strongly asserts, implies and reveals why plain, non-modified Pt by no means can ever become reversible electrocatalyst for the oxygen electrode reactions, or at least not within the whole potential range between hydrogen and oxygen evolving limits (Figs. 8.1, 8.3 and 8.4).

The interplay of interconnected potentiodynamic spectra (Figs. 8.1, 8.2, 8.3, and 8.5), as the main electrochemical tools, have been thoroughly employed to investigate and define spillover phenomena and overall behavior of the primary (Pt-OH), in particular relative to the surface Pt=O (Au=O) oxide, and reveal therefrom the appearance of the strong reaction polarization range and barrier, as the prevailing effect of unavoidable broader irreversibility features of plain Pt(Au) along the potential axis between hydrogen and oxygen evolving limits. Various individual and/or better mixed valence of altervalent and strong interactive (SMSI) hypo-d-(f)-oxides, as the complementary catalytic supports have been broadly introduced as the counterbalancing composite species able to suppress and effectively eliminate the polarization impacts, by enriched latent storage for the Pt-OH (Au-OH) spillover, and achieve provided and long lasting the reversible electrocatalysts for the oxygen electrode (primarily the ORR) reactions. The entire approach has been based on intermetallic d-d-, d-f- and f-f-interelectronic bonding features and correlations, and consequently reflects in extra high stability, advanced electron conductivity and superior (electro)catalytic activity.

The properly conceived potentiodynamic spectral interplay of three complementary different cyclic voltammograms, as the main electrochemical tools, have been confirmed as the reliable system to investigate spillover phenomena and define compatible electrocatalysts for partially, but strongly and broadly polarizable electrode reaction, and then provide novel advanced electrocatalysts for the entire scanning and applied cycle. In such a respect the best hyper-d-electrocatalyst because of *a priori* advanced individual Ru-OH spillover features would be PtRu (AuRu not yet tested and optimized), in so far best optimized hypo-d-f-interactive supports (Nb_2O_5, Ta_2O_5, and WO_3, in combination with prevailing anatase (1 0 1) titania, TiO_2, and at least doped by CeO_2, GdO_2, HoO_2 mixed and altervalent composites). The primary oxide spillover interference between Ru, relative to Pt, the authors infer on, and hypo-d-d-oxides, mostly advanced by Nb_2O_5, Ta_2O_5 and WO_3, and promoted by hypo-f-oxide (CeO_2, GdO_2, HoO_2) ingredients, bring the overall electrocatalytic activity to remarkable higher values at higher current densities, and mark the general way and manner of further optimization.

Hypo-hyper-d-d-intermetallic bonding ($MoPt_3$, $HfPd_3$), while increasing the cohesive strength, correspondingly advances electrocatalytic activity for hydrogen electrode reactions, on the same way being further promoted by interactive (SMSI) hypo-d-oxide supports. Meanwhile, besides the same mixed valence hypo-hyper-d-d-(f)-interelectronic bonding effects for the oxygen electrode reactions, the primary oxide (PtRu-OH, Au-OH) latent storage and spillover imposes the more significant contribution of continuous renewable and recoverable, the reversible electrocatalysis for the oxygen electrode reactions (ORR, OER).

All interrelating steps considered, described and carried out in the course of development of interactive hypo-hyper-d-d-(f)-oxide supported electrocatalysts for the OER, have substantially been based on the d-d-(f)-ties and bonding strengths, and d-(f)-electronic density of states, since d-band is stated and proved for cohesive, adsorptive and catalytic orbital. When stated about their extra high stability in both acidic and alkaline media[5], this implies primarily metallic properties even of hypo-d-(f)-oxides, first of all as pronounced metallic type of high electron conductivity and the overall metallic strengths, features and behavior. This is the substance in the advanced reversible oxygen electrode electrocatalysis.

Acknowledgements The present chapter has been conceived and carried out at the Institute of Chemical Engineering Sciences, ICEHT/FORTH, Patras, Greece.

References

1. Grove WR (1842) On a gaseous voltaic battery. Philos Mag 21:417–420
2. Conway BE (1995) Electrochemical oxide film formation at noble metals as a surface-chemical process. Prog Surf Sci 49:331–452
3. Angerstein-Kozlowska H, Conway BE, Sharp WBA (1973) The real condition of electrochemically oxidized platinum surfaces: Part I. Resolution of component processes. J Electroanal Chem 43:9–36
4. Angerstein-Kozlovska H, Conway BE, Hamelin A, Stoicoviciu L (1986/1987) Elementary steps of electrochemical oxidation of single-crystal planes of Au. I. Chemical basis of processes involving geometry of anions and the electrode surfaces, Electrochim Acta 31:1051–1061; II. A chemical and structural basis of oxidation of the (111) plane, J Electroanal Chem 228:429–453
5. Jaksic JM, Krstajic NV, Vracar LJM, Neophytides SG, Labou D, Falaras P, Jaksic MM (2007) Spillover of primary oxides as a dynamic catalytic effect of interactive hypo-d-oxide supports. Electrochim Acta 53:349–361
6. Krstajic NV, Vracar LJM, Radmilovic VR, Neophytides SG, Labou D, Jaksic JM, Tunold R, Falaras P, Jaksic MM (2007) Advances in interactive supported electrocatalysts for hydrogen and oxygen electrode reactions. Surf Sci 601:1949–1966
7. Jaksic MM, Botton GA, Papakonstantinou GG, Nan F, Jaksic JM (2014) Primary oxide latent storage and spillover enabling electrocatalysts with reversible oxygen electrode properties and the alterpolar revertible (PEMFC *versus* WE) cell. J Phys Chem C 118:8723–8746
8. Papakonstantinou GD, Jaksic JM, Labou D, Siokou A, Jaksic MM (2011) Spillover phenomena and their striking impacts in electrocatalysis for hydrogen and oxygen electrode reactions. Adv Phys Chem 2011:1–22, Article ID 412165
9. Neophytides SG, Zafeiratos S, Jaksic MM (2003) Selective interactive grafting of composite bifunctional electrocatalysts for simultaneous anodic hydrogen and CO oxidation, I. Theoretical concepts and embodiment of novel type composite catalysts. J Electrochem Soc 150:E512–E526
10. Neophytides SG, Murase K, Zafeiratos S, Papakonstantinou GD, Paloukis FS, Krstajic NV, Jaksic MM (2006) Composite hypo-hyper-d-intermetallic phases as supported interactive electrocatalysts. J Phys Chem B 110:3030–3042
11. Jaksic JM, Labou D, Papakonstantinou GD, Siokou A, Jaksic MM (2010) Novel spillover interrelating reversible electrocatalysts for oxygen and hydrogen electrode reactions. J Phys Chem C 114:18298–18312
12. Ma Y, Balabuena PB (2007) Designing oxygen reduction catalysts: insights from metalloenzymes. Chem Phys Lett 440:130–133
13. Brewer L (1968) Bonding and structures of transition metals. Science 161:115–122
14. Jaksic MM (2000) Hypo-hyper-d-electronic interactive nature of synergism in catalysis and electrocatalysis for hydrogen reactions. Electrochim Acta 45:4085–4099
15. Tauster SJ, Fung SC (1978) Strong metal-support interactions: occurrence among the binary oxides of groups IIA–VB. J Catal 55:29–35

16. Tauster SJ, Fung SC, Baker RTK, Horsley JA (1981) Strong-interactions in supported metal catalysts. Science 211:1121–1125
17. Stevenson SA (1987) Metal-support interaction in catalysis, sintering and redispersion. Van Nostrand, New York
18. Haller GL, Resasco DE (1989) Metal–support interaction: Group VIII metals and reducible oxides. In: Eley DD, Pires H, Weisz PB (eds) Advances in catalysis, vol 36. Academic, San Diego, pp 173–235
19. Neophytides SG, Zafeiratos S, Papkonstantinou GD, Jaksic JM, Paloukis FE, Jaksic MM (2005) Extended Brewer hypo-hyper-d-interionic bonding theory, I. Theoretical considerations and examples for its experimental confirmation. Int J Hydrogen Energy 30:131–147
20. Neophytides SG, Zafeiratos S, Papakonstantinou GD, Jaksic JM, Paloukis FE, Jaksic MM (2005) Extended Brewer hypo-hyper-d-interionic bonding theory, II. Strong metal-support interaction grafting of composite electrocatalysts. Int J Hydrogen Energy 30:393–410
21. Vittadini A, Selloni A, Rotzinger FP, Gratzel M (1998) Structure and energetics of water adsorbed at TiO_2 anatase (101) and (001) surfaces. Phys Rev Lett 81:2954–2957
22. Lazzeri M, Vittadini A, Selloni A (2001) Structure and energetics of stoichiometric TiO_2 anatase surfaces. Phys Rev B 63. Article No. 155409
23. Livage J, Henry M, Sanchez C (1988) Sol-gel chemistry of transition metal oxides. Prog Solid State Chem 18:259–341
24. Judeinstein P, Livage J (1991) Sol-gel synthesis of WO_3 thin films. J Mater Chem 1:621–627
25. Livage J, Guzman G (1996) Aqueous precursors for electrochromic tungsten oxide hydrates. Solid State Ion 84:205–211
26. Koper MTM, Van Santen RA (1999) Interaction of H, O and OH with metal surfaces. J Electroanal Chem 472:126–136
27. Kohn HW, Boudart M (1964) Reaction of hydrogen with oxygen adsorbed on a platinum catalyst. Science 145:149–150
28. Benson JE, Kohn HW, Boudart M (1966) On the reduction of tungsten trioxide accelerated by platinum and water. J Catal 5:307–313
29. Boudart M, Vannice MA, Benson JE (1999) Adlineation, portholes and spillover. Z Phys Chem NF 64:171–177
30. Volkening S, Bedurftig K, Jacobi K, Wintterlin J, Ertl G (1999) Dual path mechanism for catalytic oxidation of hydrogen on platinum surface. Phys Rev Lett 83:2672–2675
31. Mavrikakis M, Stoltze P, Norskov JK (2000) Making gold less noble. Catal Lett 64:101–106
32. Awaludin Z, Moo JGS, Okajima T, Ohsaka T (2013) TaO_x-capped Pt nanoparticles as active and durable electrocatalysts for oxygen reduction. J Mater Chem A 1:14754–14765
33. Awaludin Z, Suzuki M, Masud J, Okajima T, Ohsaka T (2011) Enhanced electrocatalysis of oxygen reduction on Pt/TaO_x/GC. J Phys Chem C 115:25557–25567
34. Masuda T, Fukumitsu H, Fugane K, Togasaki H, Matsumura D, Tamura K, Nishihata Y, Yoshikawa H, Kobayashi K, Mori T, Uosaki K (2012) Role of cerium oxide in the enhancement of activity for the oxygen reduction reaction at Pt-CeO_x nanocomposite electrocatalyst—an *in situ* electrochemical X-ray absorption fine structure study. J Phys Chem C 116:10098–10102
35. Fugane K, Mori T, Ou DR, Yan P, Ye F, Yoshikawa H, Drennan J (2012) Improvement of cathode performance on Pt-CeO_x by optimization of electrochemical pretreatment conditions for PEMFC application. Langmuir 8:16692–16700
36. Ou DR, Mori T, Fugane K, Togasaki H, Ye F, Drennan J (2011) Stability of Ceria Supports in Pt–CeO_x/C catalysts. J Phys Chem C 115:19239–19245
37. Masud J, Alam MT, Okajima T, Ohsaka T (2011) Catalytic electrooxidation of formaldehyde at Ta_2O_5-modified Pt electrodes. Chem Lett Jpn 40:252–254
38. Masud J, Alam MT, Miah MR, Okajima T, Ohsaka T (2011) Enhanced electrooxidation of formic acid at Ta_2O_5-modified Pt electrode. Electrochem Commun 13:86–89
39. Hammer B, Norskov JK (1995) Why gold is the noblest of all the metals. Nature 376:238–240

40. Quaino P, Luque NB, Nazmutdinov R, Santos E, Schmickler W (2012) Why is gold such a good catalyst for oxygen reduction in alkaline media? Angew Chem Int Ed 51:1–5
41. Haruta M (2003) When gold is not noble: catalysis by nanoparticles. Chem Rec 3:75–87
42. Lina C, Song Y, Cao L, Chen S (2013) Oxygen reduction catalyzed by Au-TiO_2 nanocomposites in alkaline media. ACS Appl Mater Interfaces 5:13305–13311
43. Jaksic MM (1986) Advances in electrocatalysis for hydrogen evolution in the light of the Brewer-Engel valence-bond theory. J Mol Catal 38:161–202
44. Friedel J, Sayers CM (1977) On the role of d-d-electron correlations in the cohesion and ferromagnetism of transition metals, *J*. Physique 38:697–705
45. Gschneidner KA (1964) Physical properties and interrelations of metallic and semimetallic elements. In: Seitz F, Turnbull D (eds) Solid state physics, advances in research and applications, vol 16. Academic, New York, pp 275–427
46. Jaksic MM (2000) Volcano plots along the periodic table, their causes and consequences on electrocatalysis for hydrogen electrode reactions. J New Mater Electrochem Syst 3:153–168
47. Jaksic MM, Lacnjevac CM, Grgur BN, Krstajic NV (2000) Volcano plots along intermetallic hypo-hyper-d-electronic phase diagrams and electrocatalysis for hydrogen electrode reactions. J New Mater Electrochem Syst 3:169–182
48. Jaksic JM, Radmilovic VR, Krstajic NV, Lacnjevac CM, Jaksic MM (2011) Volcanic periodicity plots along transition series, Hypo-hyper-d-d-interelectronic correlations and electrocatalysis for hydrogen electrode reactions. Macedonian J Chem Chem Eng 30:3–18
49. Methfessel M, Hennig D, Schefler M (1992) Trends of the surface relaxations, surface energies, and work functions of the 4d transition metals. Phys Rev B 46:4816–4829
50. Kita H (1966) Periodic variation of exchange current density of hydrogen electrode reaction with atomic number and reaction mechanism. J Electrochem Soc 113:1095–1111
51. Miles MH (1975) Evaluation of electrocatalysts for water electrolysis in alkaline solutions. J Electroanal Chem 60:89–96
52. Trasatti S (1977) The work function in electrochemistry. In: Tobias CW, Goerischer H (eds) Advances in electrochemistry and electrochemical engineering, vol 10. Interscience, New York, pp 213–321
53. Jaksic JM, Vracar LJM, Neophytides SG, Zafeiratos S, Papakonstantinou GD, Krstajic NV, Jaksic MM (2005) Structural effects on kinetic properties for hydrogen electrode reactions and CO tolerance along Mo-Pt phase diagram. Surf Sci 598:156–173
54. Santos E, Schmickler W (2007) Electrocatalysis of hydrogen oxidation—theoretical foundations. Angew Chem Int Ed 46:8262–8265
55. Hammer B, Norskov JK (2000) Theoretical surface science and catalysis—calculations and concepts. Adv Catal 45:71–129
56. Christoffersen E, Liu P, Ruban A, Skriver HL, Norskov JK (2001) Anode materials for low-temperature fuel cells: a density functional theory study. J Catal 199:123–131
57. Peuckert M, Coenen FP, Bonzel HP (1984) XPS study of the electrochemical surface oxidation of platinum in 1N H_2SO_4 acid electrolyte. Electrochim Acta 29:1305–1314
58. Drawdy JE, Hoflund GB, Gardner SD, Yngvadottir E, Schryer DR (1990) Effect of pretreatment on a platinized tin oxide catalyst used for low-temperature CO oxidation. Surf Interface Anal 16:369–374
59. Akita T, Tanaka K, Tsubota S, Haruta M (2000) Analytical high-resolution TEM study of supported gold catalysts: orientation relationship between Au particles and TiO_2 supports. J Electron Microsc 49:657–662
60. Akita T, Lu P, Ichikawa S, Tanaka K, Haruta M (2001) Analytical TEM study on the dispersion of Au nanoparticles in Au/TiO_2 catalyst prepared under various temperatures. Surf Interface Anal 31:73–78
61. Haruta M (1997) Size- and support-dependency in the catalysis of gold. Catal Today 36:153–166
62. Date M, Haruta M (2001) Moisture effect on CO oxidation over Au/TiO_2 catalyst. J Catal 201:221–224

63. Boccuzzi F, Chiorino A, Tsubota S, Haruta M (1996) FTIR study of carbon monoxide oxidation and scrambling at room temperature over gold supported on ZnO and TiO_2. J Phys Chem 100:3625–3631
64. Boccuzzi F, Chiorino A, Manzoli M, Lu P, Akita T, Ichikawa S, Haruta M (2001) Au/TiO_2 nanosized samples: a catalytic, TEM, and FTIR study of the effect of calcination temperature on the CO oxidation. J Catal 202:256–267
65. Zafeiratos S, Papakonstantinou G, Jaksic MM, Neophytides SG (2005) The effect of Mo oxides and TiO_2 support on the chemisorption features of linearly adsorbed CO on Pt crystallites: an infrared and photoelectron spectroscopy study. J Catal 232:127–136
66. Riess I, Vayenas CG (2003) Fermi level and potential distribution in solid electrolyte cells with and without ion spillover. Solid State Ion 159:313–329
67. Vayenas CG, Bebelis S, Pliangos C, Brosda S, Tsiplakides D (2001) Electrochemical activation of catalysis: promotion, electrochemical promotion, and metal-support interactions. Kluwer, New York
68. Vayenas CG, Jaksic MM, Bebelis SI, Neophytides SG (1996) The electrochemical activation of catalytic reactions. In Bockris JO'M, Conway BE, White RE (eds), Modern aspects of electrochemistry, vol 29. Plenum Press, New York, pp 57–202
69. Vayenas CG, Bebelis S, Ladas L (1990) The dependence of catalytic activity on catalyst work function. Nature 343:625–627
70. Tsiplakides D, Nicole J, Vayenas CG, Comninellis C (1998) Work function and catalytic activity measurements of an IrO_2 film deposited on YSZ subjected to *in situ* electrochemical promotion. J Electrochem Soc 145:905–908
71. Neophytides S, Tsiplakides D, Stonehart P, Jaksic MM, Vayenas C (1994) Electrochemical enhancement of a catalytic reaction in aqueous solution. Nature 370:45–47
72. Neophytides SG, Tsiplakides D, Stonehart P, Jaksic MM, Vayenas CG (1996) Non-faradaic electrochemical modification of the catalytic activity of Pt for H_2 oxidation in aqueous alkaline media. J Phys Chem 100:14803–14814
73. Tsiplakides D, Neophytides SG, Enea O, Jaksic MM, Vayenas CG (1997) Non-faradaic electrochemical modification of the catalytic activity of Pt-black electrodes deposited on Nafion 117 solid polymer electrolyte. J Electrochem Soc 144:2072–2078
74. Tsiplakides D, Vayenas CG (2001) Electrode work function and absolute potential scale in solid-state electrochemistry. J Electrochem Soc 148:E189–E202
75. Tsiplakides D, Archonta D, Vayenas CG (2007) Absolute potential measurements in solid and aqueous electrochemistry using two Kelvin probes and their implications for the electrochemical promotion of catalysts. Top Catal 44:469–479
76. Trasatti S (1982) The concept of absolute electrode potential. An attempt at a calculation. J Electroanal Chem 139:1–13
77. Nicole J, Tsiplakides D, Pliangos C, Verykios XE, Comninellis C, Vayenas CG (2001) Electrochemical promotion and metal-support interactions. J Catal 204:23–34
78. Metcalfe IS (2001) Electrochemical promotion of catalysts, I. Thermodynamic considerations. J Catal 199:247–258
79. Metcalfe IS (2001) Electrochemical promotion of catalysts, II. The role of a stable spillover species and prediction of reaction rate modification. J Catal 199:259–272
80. Markovic NM, Ross PN (2002) Surface science studies of model fuel cell electrocatalysts. Surf Sci Rep 45:117–229
81. Siokou A, Ntais S (2003) Towards the preparation of realistic model Ziegler-Natta catalysts: XPS study of the $MgCl_2$/$TiCl_4$ interaction with flat SiO_2/Si(1 0 0). Surf Sci 540:379–388
82. Suh M, Bagus PS, Pak S, Rosynek MP, Lunsford JH (2000) Reactions of hydroxyl radicals on titania, silica, alumina, and gold surfaces. J Phys Chem B 104:2736–2742
83. Lang ND (1989) Theory of single-atom imaging in the scanning tunneling microscope. Comments Condens Matter Phys 14:253–257
84. Jaksic JM, Papakonstantinou GD, Labou D, Siokou A, Jaksic MM (2013) Spillover phenomena in electrocatalysis for oxygen and hydrogen electrode reactions. In: Suib SL (ed) New and

future developments in catalysis: hybrid materials, composites, and organocatalysts. Elsevier, Amsterdam, pp 175–212
85. Engelhard M, Baer D (2000) Third row transition metals by X-ray photoelectron spectroscopy. Surf Sci Spectra 7(1):1–68
86. Fuentes RE, Garcia BL, Weidner JW (2008) A Nb-doped TiO_2 electrocatalyst for use in direct methanol fuel cells. ECS Trans 12:239–248
87. Bokhimi MA, Novaro O, Lopez T, Sanchez E, Gomes R (1995) Effect of hydrolysis catalyst on the Ti deficiency and crystallite size of sol-gel-TiO_2 crystalline phases. J Mater Res 10:2788–2796
88. Arbiol J, Cerda J, Dezanneau G, Cirera A, Peiro F, Cornet A, Morante JR (2002) Effects of Nb doping on the TiO_2 anatase-to-rutile phase transition. J Appl Phys 92:853–861
89. Seah MP (1990) Quantification of AES and XPS. In: Briggs D, Seah MP (eds) Practical surface analysis, vol 1, 2nd edn. Wiley, New York, pp 201–256
90. Simões JAM, Beauchamp JL (1990) Transition metal-hydrogen and metal-carbon bond strengths: the keys to catalysis. Chem Rev 90:629–688
91. Anderson LC, Mooney CE, Lunsford JH (1992) Hydroxyl radical desorption from polycrystalline palladium: evidence for a surface phase transition. Chem Phys Lett 196:445–448
92. Hashimoto S, Matsuoka H (1992) Prolonged lifetime of electrochromism of amorphous WO_3–TiO_2 thin films. Surf Interface Anal 19:464–468
93. Hashimoto S, Matsuoka H (1991) Lifetime of electrochromism of amorphous WO_3-TiO_2 thin films. J Electrochem Soc 138:2403–2408
94. Sasaki K, Zhang L, Adzic RR (2008) Niobium oxide-supported platinum ultra-low amount electrocatalysts for oxygen reduction. Phys Chem Chem Phys 10:159–167
95. Watanabe M, Motoo S (1975) Electrocatalysis by ad-atoms: Part II. Enhancement of the oxidation of methanol on platinum by ruthenium ad-atoms. J Electroanal Chem 60:267–273
96. Davies JC, Hayden BE, Pegg DJ, Rendall ME (2002) The electro-oxidation of carbon monoxide on ruthenium modified Pt(1 1 1). Surf Sci 496:110–120
97. Hadzi-Jordanov S, Angerstein-Kozlowska HA, Vukovic M, Conway BE (1978) Reversibility and growth behavior of surface oxide films at ruthenium electrode. J Electrochem Soc 125:1471–1480

Chapter 9
Metal-Organic Frameworks as Materials for Fuel Cell Technologies

Henrietta W. Langmi, Jianwei Ren, and Nicholas M. Musyoka

9.1 Introduction

Depleting fossil fuel reserves and increasing environmental awareness have intensified research efforts to develop energy technologies that can meet the escalating demand for cheap, safe and sustainable clean energy. Among the alternative technologies, fuel cell technology has attracted considerable attention [1–3]. A fuel cell is an electrochemical device that directly converts the chemical energy of a fuel into electrical energy via electrochemical reactions. Fuel cells are attractive for a number of reasons such as high efficiency and reliability, minimum impact on environment and distinctive operating characteristics. Among the different types of fuel cells being developed, the polymer electrolyte membrane fuel cell (PEMFC) is considered the most appealing for a range of applications including portable power, transportation and stationary power systems, due to its unique features. The components of a PEMFC are an anode where hydrogen oxidation occurs, a cathode where oxygen reduction occurs, and an electrolyte membrane that allows protons to move from anode to cathode.

PEMFCs offer many advantages such as a low operating temperature, sustained operation at a high current density, low weight, compactness, rapid start-ups and suitability for discontinuous operation [4]. Despite the advantages of PEMFCs over other types of fuel cells several hurdles including cost, performance and durability, need to be overcome for PEMFCs to become commercially viable. There is high catalyst cost associated with the exclusive use of Pt and Pt-based catalysts in the fuel cell electrodes. In addition, there is inadequate performance durability as a result of cathode catalyst oxidation, catalyst migration, loss of electrode active

H.W. Langmi (✉) • J. Ren • N.M. Musyoka
HySA Infrastructure Centre of Competence, Materials Science and Manufacturing, Council for Scientific and Industrial Research (CSIR), PO Box 395, Pretoria 0001, South Africa
e-mail: hlangmi@csir.co.za

K.I. Ozoemena, S. Chen (eds.), *Nanomaterials for Fuel Cell Catalysis*,
Nanostructure Science and Technology, DOI 10.1007/978-3-319-29930-3_9

surface area, and corrosion of the carbon support [2, 5]. Furthermore, PEMFCs are prone to poisoning by carbon monoxide, which can be found in the fuel in trace amounts if it is obtained from fossil fuels. The sluggish kinetics of the oxygen reduction reaction (ORR) necessitates the use of catalysts at the anode, and this reaction accounts for the main source of voltage loss in low temperature PEMFCs, even in the presence of Pt-based catalysts [6]. Presently, the state-of-the-art catalyst for PEMFCs is supported Pt or supported Pt-based catalyst. In order to advance PEMFC technology research efforts are geared towards developing cheap non-Pt-based catalysts that are efficient and durable. A substantial amount of research has also focused on developing a better polymer electrolyte membrane (PEM) for fuel cells. The requirements for a superior PEM include high proton conductivity to enable efficient transport of protons, good thermal and chemical stability to retain electrolyte integrity under fuel cell operating conditions, serve as a gas tight separator to prevent fuel gases crossover, affordability, scalability to enable high volume production, and compatibility with other fuel cell components [7].

Fuel cells are a key component of the hydrogen energy value chain and therefore they are an enabling technology for the Hydrogen Economy. In the same light, other components of the hydrogen energy value chain such as hydrogen production and storage are critical enabling technologies for fuel cells. Hydrogen is widely considered to be an attractive candidate for the replacement of current carbon-based energy sources. It can be generated from clean and green sources and upon combustion only produces water as by-product (without any CO_2 emitted). In addition, hydrogen has an energy content that is about three times higher than that of other chemical fuels such as gasoline or diesel [8]. There are many technologies for hydrogen production with steam reforming from natural gas currently the dominant technology. Production of hydrogen from cheap and renewable sources is the key factor for the utilisation of hydrogen in fuel cells. However, technological and economic issues associated with the production and supply of hydrogen in this way will need to be overcome. Production of hydrogen from water using renewable energy sources, such as solar and wind, is an ideal method. In which case, the photocatalytic splitting of water is an important approach and efforts are being made towards developing low-cost and efficient photocatalysts for the production of hydrogen from water [9]. Currently, for the storage of hydrogen compressed and liquefied hydrogen are established technologies which are being employed in fuel cell vehicles [10]. However, both technologies have major drawbacks. As very high pressures are required to achieve high storage capacities this leads to heavy and bulky storage tanks. There are also serious safety risks associated with both storage technologies. In addition, hydrogen has a low boiling point (20 K) and therefore requires extensive cooling for liquefaction. This makes the storage system complex and results in boil off losses and a loss in some of the energy content of hydrogen [8]. Extensive research has been carried out on hydrogen storage in solid materials [11–14]. Solid materials-based hydrogen storage has an advantage over compressed and liquid hydrogen storage since hydrogen is stored in a safer and compact manner. In materials-based storage hydrogen may be stored through a physical adsorption process where molecular hydrogen binds weakly to the material, or through chemisorption where a stronger binding of atomic hydrogen occurs forming a new compound.

It is clearly evident that the development of effective, safe and affordable enabling fuel cell technologies as well as low-cost and efficient fuel cell materials present a serious materials challenge. Metal-organic frameworks (MOFs), also known as coordination polymers, are an extensive class of crystalline materials consisting of organic molecules that are bound to metals or metal clusters through coordination bonds to form one-, two- or three-dimensional structures. They possess large internal surface areas (experimental value of up to 7140 $m^2\ g^{-1}$, theoretical limit of 14600 $m^2\ g^{-1}$) [15] and ultra-high porosity (up to 90 % free volume) with pore sizes in the microporous and mesoporous ranges. The presence of both inorganic and organic building blocks creates almost a limitless number of possible structures and allows both the pore size and chemical environment to be tailored to attain specific properties. These unique characteristics of MOFs make them of tremendous interest for applications in a number of fields such as gas storage [16], separation [17], catalysis [18], sensors [19], drug delivery [20], magnetism [21], electrochemistry [22], clean energy [23] and fuel cells [24]. The huge interest in MOFs and their significance to the scientific community is reflected in the marked increasing trend in the number of MOF structures being prepared and investigated each year [25]. This chapter provides an overview of MOFs as promising materials for PEM fuel cell components and enabling fuel cell technologies. It first outlines the key structural characteristics of MOFs including a statement on their design. Then the use of MOFs as PEM in fuel cells is examined. This is followed by a review of MOFs as electrocatalysts in fuel cells. After that, the employment of MOFs in two key enabling technologies for fuel cells *i.e.* hydrogen production and storage is discussed. The chapter concludes with a summary and outlines the prospects for MOFs in fuel cell technologies.

9.2 Structural Features of MOFs

Figure 9.1 shows some examples of MOF structures [26]. Basically, a MOF is constructed by linking metal-containing units also known as secondary building units (SBUs) with organic linkers through coordination bonds. The self-assembly of these components in higher dimensions affords the desired MOF architecture. Although the complexity of the components can differ significantly from one assembly to another, in theory, they can usually be simplified. As Cook et al. [27] pointed out the important fundamental characteristics of a particular MOF include the number of binding sites available on the metal or metal clusters, the relative orientation of these binding sites, the number of Lewis-basic sites on a ligand, the modes of coordination and relative angularity of these sites. An appropriate choice of inorganic building units can afford diverse SBUs alongside precise control over predictable network topologies. Although organic building blocks possess predetermined shape and geometry, their flexibility generally affects the final

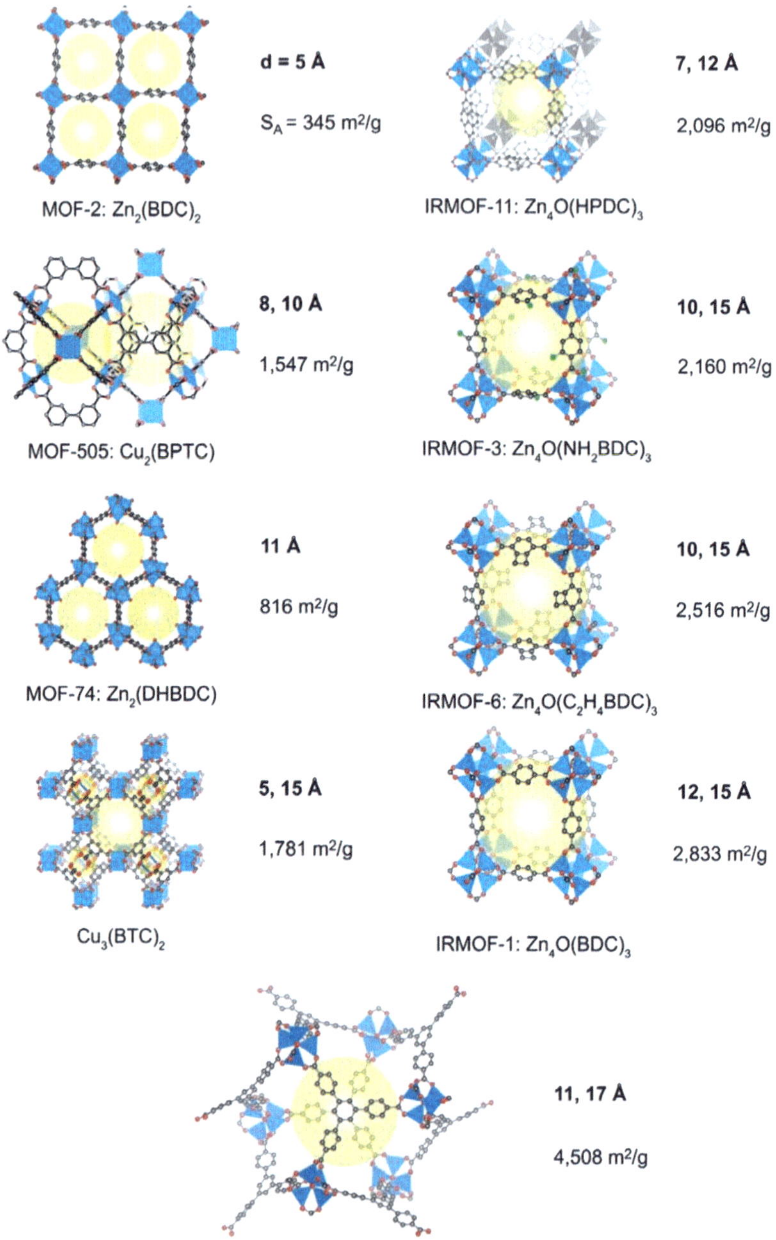

Fig. 9.1 Crystal structures of a variety of MOFs. The large *yellow spheres* represent the cavity space. For each MOF, the framework formula, pore size, and surface area are given. Reprinted with permission from [26]. Copyright (2005) American Chemical Society

architecture of the MOF. The organic linkers usually carry coordinating functional groups, which play a vital role in locking the metal ions tightly in place. Metal ions provide inadequate directionality which leads to mobility around the metal ion making it challenging to achieve such MOF structures as more than one structure is possible [28].

The design of MOFs has been explained through several models including the node-and-spacer [29], SBU [30, 31] and metal–organic polyhedral [32] models. However, it is challenging to begin from a particular model and achieve success in synthesising the corresponding material. In synthesising MOFs, by employing the principle that symmetrical SBUs attached to simple linkers generates only one high symmetry structure, enables control and prediction of the MOF structure and properties [28]. The design of an extended network, such as a MOF, can be attained by starting with well-defined and rigid molecular building blocks that will maintain their structural integrity during the construction process [31, 33]. Another promising approach to the design of these extended structures is the use of well-defined conditions that result in the formation of such building blocks *in situ* [31]. Yaghi et al. [28] highlighted that the practice of logical synthesis must start with knowing the target network 'blueprint' and identifying the building blocks necessary for its assembly; the implementation of such a process is referred to as reticular synthesis. Reticular synthesis is defined as the process of carefully assembling designed rigid molecular building blocks into predetermined ordered structures that are held together by strong bonding [28].

The SBU approach, which led to the design of rigid frameworks, resulted in the identification of a small number of preferred topologies that could be targeted in designed syntheses and was vital to attaining permanent porosity in MOFs [28]. The geometry of the SBU was shown to be dependent on the structure of the ligand and type of metal employed, as well as the metal-to-ligand ratio, solvent, and source of anions to balance the charge of the metal ion [34]. The principles of geometrical design and of tunability of the pore size by employing a known MOF as prototype have led to the development of isoreticular MOFs (IRMOFs) [35–37]. A typical example is the series of 16 highly crystalline IRMOFs based on the same geometry as that of MOF-5 (also referred to as IRMOF-1) [35]. MOF-5 consists of inorganic oxide-centred Zn_4O tetrahedral clusters coordinated with six carboxylate species of different terephthalate organic building blocks in an octahedral manner, forming a three-dimensional cubic porous framework. The IRMOF structures are based on the skeleton of MOF-5, wherein the pore functionality and size have been altered without changing the original cubic topology. Therefore, the IRMOFs differ in the nature of functional groups decorating the pores and in the metrics of their pore structure determined by the size of the organic building-block. MOFs are also structurally distinctive in that it is possible to construct frameworks according to the multivariable or mixed MOFs approach [38–40]. Here the MOF contains two or more isoreticular organic linkers, each with a different functionality that is dispersed within the framework in a random and homogeneous manner. The underlying topology of a MOF crystal structure is described by a net, designated by a three-letter symbol, and polyatomic groups serve as the vertices and edges of the net [41].

Another structural feature of MOFs is the pores, which are the empty spaces (cavities) generated when guest molecules are removed. It is possible for the structure to collapse during solvent removal and this is more likely to happen when the MOF contains large pores. Permanent porosity is attained when the framework remains intact and is easier to achieve this in microporous MOFs than in mesoporous MOFs. The removal of weakly coordinating solvent molecules also generates open metals sites, which can affect the properties of MOFs since open metal sites act as Lewis acids as well as serve as adsorption sites. The inclusion of specific functional groups in a MOF can change the chemical properties of the MOF's surface. Therefore, different functionalities of organic linkers can lead to MOFs with properties tailored for specific applications. The functionalisation of a MOF can be carried out before or after synthesis of the MOF [16]. In the former approach known as prefunctionalisation, the presence of a specific functional group on a ligand may sometimes prevent the formation of the targeted MOF. Postsynthetic modification in which functional groups are incorporated after formation of the desired MOF topology has emerged as a promising approach to alter the surface of MOFs [42, 43].

9.3 MOFs as Materials for Fuel Cell Components

9.3.1 Polymer Electrolyte Membranes

The development of PEMFC technology and its demonstration in transportation applications have grown rapidly over the past two decades. For the PEMFC to function the PEM must (i) conduct protons but not electrons, (ii) prevent fuel crossover and (iii) be resistant to the reducing environment at the cathode and the oxidative environment at the anode. Several types of polymers such as Nafion, polybenzimidazole (PBI) and sulfonated polyether-ether ketones (SPEEK) have been employed as electrolytes in PEMFCs. Nafion, a sulfonated fluoro-polymer, is the standard and most widely used with high proton conductivity of up to 10^{-1} S m^{-1} at 80 °C and 100 % relative humidity (RH) [22]. Proton conduction in these polymers occurs as a result of the presence of acidic components that create hydrophilic environments suitable for proton conduction. It is necessary to have such high conductivities for high power density PEMFCs. A drawback associated with these polymers arises from the humidification required which introduces complexity. In addition, the electrolyte may physically deteriorate upon hydration and dehydration leading to a loss in efficiency [7]. Moreover, the use of Nafion as membranes is limited to temperatures <100 °C. Efforts are being made to produce alternative lower cost, more robust electrolyte membranes with comparable conductivity to that of polymers such as Nafion. Furthermore, there is a search for novel proton-conducting membranes with higher conductivity at temperatures above 100 °C than currently available membranes. Fuel cells incorporating electrolytes with such higher operating temperature will be advantageous in terms of better CO tolerance, enhanced reaction kinetics and better water management.

Although MOFs are intrinsically porous crystalline materials they show great promise as materials for membranes in PEMFCs. The reason for this is because the guest species present in the pores of MOFs are capable of blocking the pores thereby preventing fuel crossover from anode to cathode and subsequent voltage loss [44]. Furthermore, the availability of counter ions in the pores of certain MOFs can concomitantly aid in proton conduction provided they contain water, acidic groups, or other simple functional groups containing hydrogen atoms. Moreover, incorporation of anhydrous guest molecules, such as heterocycles, in the pores and channels of MOFs can provide proton conduction pathways [45]. Therefore proton conduction in MOFs can occur through water-mediated hydrogen bonding networks, which provide pathways for the migration of protons or through non-water mediated proton-containing guest species that serve as proton conduction pathways. These proton-conduction pathways correspond to two temperature regimes; below 100 °C and above 100 °C, respectively. Besides exhibiting proton conductivity as a result of inclusion of protonic charge carriers in the MOF channels the framework itself can also display proton conductivity [45, 46]. Many research efforts have been directed towards investigating the proton conductivity of MOFs, and their potential for use as PEM is intensively being studied in MOFs research for fuel cells. The reader is referred to the comprehensive reviews by Ren et al. [24], Yoon et al. [45] and Ramaswamy et al. [7] for further reading.

9.3.1.1 Proton-Conducting MOFs Operating Below 100 °C

Although there are limited studies on proton conduction of one-dimensional (1D) coordination polymers it has been shown that there are some 1D coordination polymers that display proton conductivity [47–50]. A typical example is ferrous oxalate dihydrate, $Fe(ox)\cdot 2H_2O$, commonly referred to as Humboldtine [49]. Humboldtine consists of 1D chains formed by coordination of oxalate ligands to the equatorial positions of Fe^{2+} ions that act as metal nodes. Two water molecules coordinate axially to the Fe^{2+} nodes forming a 1D ordered arrangement of water molecules that interact with the framework via strong hydrogen bonds. It was suggested that such an arrangement of water molecules could provide a pathway for proton conduction. The proton conductivity was reported to be 1.3×10^{-3} S cm^{-1} at 25 °C and 98 % RH, which is high and comparable to Nafion. The electrical conductivity of Humboldtine was also measured and reported to be very low ($<10^{-10}$ S cm^{-1}), which will favour its application as electrolyte membrane in PEMFCs. In another study, Kanaizuka et al. [47] synthesised a coordination polymer, $[Zn(sbdc)(H_2O)_3]\cdot(H_2sbdc)\cdot H_2O$ (H_2sbdc: sulfone biphenyldicarboxylic acid) consisting of a 1D chain structure of alternating Zn^{2+} cations and sbdc anions. Non-coordinating uncharged acid (H_2sbdc) is present between the chains with hydrogen bonding networks. The proton conductivity of the coordination polymer was 2.5×10^{-7} S cm^{-1} at room temperature and 95 %

RH, and was strongly dependent on the RH. A 1D coordination polymer, M(dhbq)·$2H_2O$ (M = Mg, Mn, Co, Ni and Zn; H_2(dhbq) = 2,5-dihydroxy-1,4-benzoquinone) with axially coordinated water molecules was investigated [48, 51]. While the dihydrate showed a high proton conductivity of 4×10^{-5} S cm^{-1}, the anhydride did not display any proton conductivity, indicating that the water molecules were responsible for the proton conduction.

In 1979, the first investigation of proton conduction in MOFs was reported by Kanda et al. [52] for the two-dimensional (2D) MOF, [*N,N*'-bis(2-hydroxyethyl) dithiooxamido]copper(II), [$(HOC_2H_4)_2$(dtoa)Cu] (dtoa = dithiooxamide). Thereafter, very little attention was given to this subject of research until the early 2000s when Kitagawa and co-workers studied a series of 2D MOFs, which could incorporate different concentrations of guest water molecules between the 2D sheets. The MOFs were variations of the aforementioned 2D Cu-based MOF prepared by substituting the hydroxyl ethyl group [53–55]. Generally, the concentration of water molecules increases with increasing RH. Furthermore, proton conductivity also increases with increasing RH. Therefore, it has been suggested that water molecules, when present, play a role in proton conduction. At room temperature and 100 % RH, [$(HOC_2H_4)_2$(dtoa)Cu] exhibited a proton conductivity of 5×10^{-6} S cm^{-1}, with the MOF holding approximately 10 water molecules between the 2D sheets at such RH similarly to the widely used Nafion [56]. The derived series of MOFs displayed proton conductivity relatively lower than that of Nafion, ranging from 10^{-6} to 10^{-5} S cm^{-1} at room temperature and high RH (>75 %) [53–56]. Higher proton conductivity would have been expected for the MOFs with longer ligands since more water molecules can occupy the interplanar spacing of longer ligands [54]. However, it was revealed that [$(HO\text{-}H_4C_2)_2$dtoaCu] and [$(HO\text{-}H_6C_3)_2$dtoaCu] contain the same amount of water per dimeric unit under 100 % RH. A proton conductivity of 2.0×10^{-6} S cm^{-1} was observed for [$(HO\text{-}H_6C_3)_2$dtoaCu] at 100 % RH implying that the length of the hydrophobic moiety of the ligand does not significantly enhance proton conductivity.

To further investigate proton conductivity, Kitagawa and co-workers selected another 2D oxalate-bridged MOF, $(NH_4)_2$(adp)[$Zn_2(ox)_3$]·$3H_2O$ (adp = adipic acid) [57]. Adipic acid, NH_4^+ ions and water molecules occupy the pores of the MOF, with adipic acid and NH_4^+ providing additional protons which favour proton conductivity. The close proximity of the three species leads to the formation of 2D hydrogen bond network interactions involving all the guest molecules and the oxalate ions of the framework. Therefore, a high proton conductivity of 8×10^{-3} S cm^{-1} at 298 K and 98 % RH was obtained, which decreased with decreasing RH indicating that the water molecules present in the pores of the MOFs are critical for proton conduction. A series of oxalate-based MOFs, $NH(prol)_3$[$MCr(ox)_3$] (M = Mn(II), Fe(II), Co(II); $NH(prol)_3^+$ = tri(3-hydroxylpropyl)ammonium; ox = oxalate), was also investigated for proton conduction [58]. The MOFs consist of oxalate-bridged bimetallic layers that are intercalated by $NH(prol)_3^+$ ions. The MOFs showed a dependence of proton conductivity on RH and exhibited proton conductivity of approximately 1×10^{-4} S cm^{-1} at 75 % RH and 298 K, attributed to the 2D hydrophilic layers formed by the $NH(prol)_3^+$ ions. The study demonstrates

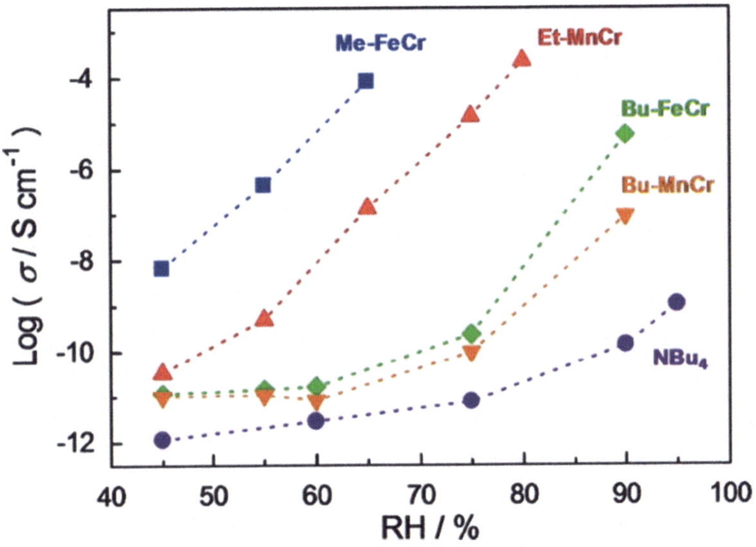

Fig. 9.2 Humidity dependence of the proton conductivity at 298 K of various MOFs. Reprinted with permission from [59]. Copyright (2012) American Chemical Society

that oxalate-bridged bimetallic complexes that contain hydrophilic ions are potentially good proton conductors. Furthermore, a study was carried out to investigate the influence of hydrophilicity of cations (containing different substituents) on the proton conductivity of the 2D oxalate-bridged bimetallic MOF, $[NR_3(CH_2COOH)][MCr(ox)_3)\cdot nH_2O^+]$ (R = Me (methyl), Et (ethyl), or Bu (n-butyl); and M = Mn or Fe) [59]. The RH dependence of the proton conductivity is presented in Fig. 9.2. The carboxyl groups in the MOFs act as proton carriers with the hydrophilicity of the cationic species controlled by NR_3. The proton conductivity of the MOFs increased with increasing adsorption of water and increasing hydrophilicity of the cationic species in the order $[NBu_3(CH_2COOH)]^+ < [NEt_3(CH_2COOH)]^+ < [NMe_3(CH_2COOH)]^+$. Therefore, the most hydrophilic MOF, $[N(Me_3)(CH_2COOH)][FeCr(ox)_3]$, adsorbed the largest amount of water and displayed the highest proton conductivity of 0.8×10^{-4} S cm^{-1} even at a low RH value of 65 % and ambient temperature.

Other 2D MOFs based on phosphonate ligands have been shown to have significant proton conductivity because phosphonates contain three oxygen atoms, two of which are available to act as proton acceptors following coordination of one to the metal node [60]. Taylor et al. [60] investigated the phosphonate-based proton-conducting MOF, PCMOF-3 ($Zn_3(L)H_2O\cdot 2H_2O$; L = [1,3,5-benzenetriphosphonate]$^{6-}$). The extensive hydrogen bonding network in this MOF between the water molecules and the phosphonate oxygen atoms gave a pathway for proton conduction. The proton conductivity was 3.5×10^{-5} S cm^{-1} at 25 °C and 98 % RH, and decreased to 4×10^{-8} S cm^{-1} at 44 % RH. Kundu et al. [61] attempted to influence proton conductivity by substituting the metal ions in MOFs, and as such constructed three 2D MOFs denoted as Ca-SBBA, Sr-SBBA and Ba-SBBA (SBBA = 4,4'-sulfobisbenzoic acid). No proton conductivity was observed in the as-synthesised MOFs. However, the MOFs displayed proton conductivity only after 24 h humidification. While Ca-SBBA and Sr-SBBA exhibited a proton conductivity of 8.6×10^{-6} and 4.4×10^{-5} S cm^{-1}, respectively at 25 °C under 98 % RH, Ba-SBBA did not display any proton conductivity under similar conditions. The higher proton conductivity observed for Sr-SBBA was as a result of a high carrier concentration, arising from the combination of metal clusters and solvent molecules. It was suggested that upon humidification, water molecules are absorbed within the crystal through hydrogen bond formation, which occurred with carboxylate bound metal clusters and also with sulfone groups. The lack of proton conductivity by Ba-SBBA was ascribed to phase change and loss of crystallinity upon humidification.

Besides the 1D and 2D MOFs, proton conduction studies have also been carried out on MOFs with 3D extended structures. An oxalate-bridged bimetallic 3D chiral quartz-like structured MOF, $(NH_4)_4[MnCr_2(ox)_6]\cdot 4H_2O$, was shown to have a high proton conductivity of 1.1×10^{-3} S cm^{-1} at room temperature and 96 % RH due to the water molecules in the functionalised channels of the MOFs being in close proximity with NH_4^+ ions and water molecules around the framework [62]. The effect of halogens and helical water chains on proton conductivity was evaluated for four homochiral MOF isomers; $[Zn(l\text{-}L_{Cl})(Cl)](H_2O)_2$, $[Zn(l\text{-}L_{Br})(Br)](H_2O)_2$, $[Zn(d\text{-}L_{Cl})(Cl)](H_2O)_2$, and $[Zn(d\text{-}L_{Br})(Br)](H_2O)_2$ (L = 3-methyl-2-(pyridin-4-ylmethylamino)butanoic acid). Both $[Zn(l\text{-}L_{Cl})(Cl)](H_2O)_2$ and $[Zn(d\text{-}L_{Cl})(Cl)](H_2O)_2$ consist of a 3D supramolecular network containing water molecules arranged in a continuous helical chain in 1D channels (Fig. 9.3). The arrangement of the water molecules in a helical manner is as a result of the repetition of one water molecule forming a weak hydrogen bond with the Cl atom and another water molecule forming a hydrogen bond with the first water molecule. Both $[Zn(l\text{-}L_{Cl})(Cl)](H_2O)_2$ and $[Zn(d\text{-}LD_{Cl})(Cl)](H_2O)_2$ showed proton conductivities of 4.45×10^{-5} and 4.42×10^{-5} S cm^{-1}, respectively at 30 °C and 98 % RH, which were higher than the values for the other two MOFs.

The influence of different functional groups on proton conductivity was demonstrated in a series of isostructural 3D MOFs denoted as MIL-53(M) (MIL = Matériaux de l'Institut Lavoisier; M = Al or Fe) [64]. The MOFs include $Al(OH)(BDC)(H_2O)$, $Al(OH)(BDC\text{–}NH_2)(H_2O)$, $Al(OH)(BDC\text{–}OH)(H_2O)_{1.5}$, and $Fe(OH)(BDC\text{–}(COOH)_2)(H_2O)$ (BDC = 1,4-benzenedicarboxylate). The proton conductivity was influenced by the acidity of the functional group. In fact, the order of proton conductivities for the MOFs with different substituents ($-NH_2 < -H < -OH < -COOH$) correlated well with the order of the pKa values of meta-substituted benzoic acids (R = $-NH_2$, –H, –OH, and –COOH and pKa = 4.74, 4.19, 4.08, and 3.62, respectively). This is the first example whereby proton conductivity has been readily controlled by incorporating different functional groups in a family of isostructural MOFs. Recently, the effect on proton conductivity of node-coordinated species and pore-filling solvents in MOFs was demonstrated for HKUST-1 (HKUST = Hong Kong University of Science and Technology), a 3D

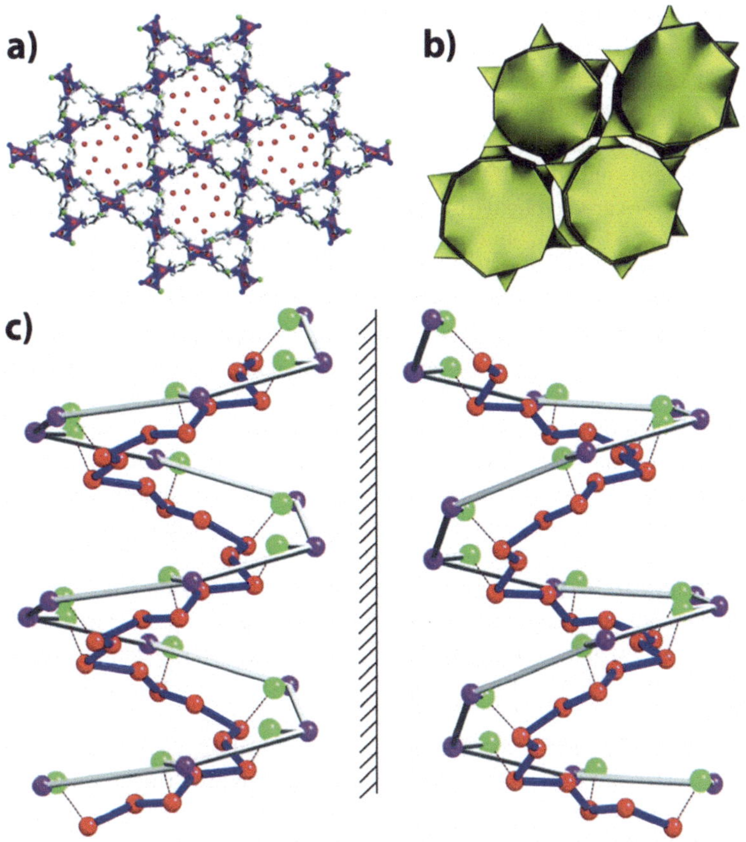

Fig. 9.3 (**a**) Polyhedral representation of the MOF, $[Zn(l\text{-}L_{Cl})(Cl)](H_2O)_2$ lattice viewed down the c-axis. (**b**) Tiling figure of $[Zn(l\text{-}L_{Cl})(Cl)](H_2O)_2$, showing zeolitic *unh*-topology along the c-axis. The tiling shows one kind of vertices, two kinds of edges, two kinds of faces, and one kind of tiles. (**c**) Mirror isomers of helical water chains surrounded by a molecular helix (outer helix). The molecular helix (outer helix) is shown as pink balls connected via gray bonds, and the helical water chain (inner helix) is shown as red balls connected via blue rods. Reprinted with permission from [63] Copyright (2011) American Chemical Society

MOF consisting of Cu (II) paddlewheel-type metal nodes with 1,3,5-benzene tricarboxylate linkers [65]. Postsynthetic modification of the MOF by coordination of H_2O at open Cu node sites resulted in a large (approximately 75-fold) enhancement in proton conductivity under methanol vapour relative to the MOF whose nodes were modified by acetonitrile. In fact, the conductivity of H_2O-HKUST-1 under methanol vapour was 1.5×10^{-5} S cm^{-1}. The results could be explained by the lack of dissociable protons by acetonitrile. As methanol vapour filled the pores, the coordinated' H_2O molecules served as proton donors, which led to an increase in the $CH_3OH_2^+$ carrier concentration and consequently an increase in the proton conductivity. Replacement of pore-filling solvent, methanol with *n*-hexane led to a massive drop in conductivity, which was more than five orders of magnitude lower than the value for the MOF containing nodes modified by H_2O.

Sen et al. [66] studied a MOF, $[\{(Zn_{0.25})_8(O)\}Zn_6(L)_{12}(H_2O)_{29}(DMF)_{69}(NO_3)_2]_n$ $\{H_2L = $ 1,3-bis(4-carboxyphenyl)imidazolium$\}$, based on a ligand containing both the carboxylate and imidazolium functionalities. The MOF has a 3D structure and possesses a unique Zn_8O cluster. Within the channels are solvent water and DMF as well as nitrate anions to balance the charge. The methylene groups of the imidazolium components are also aligned inside the channels. It was observed that proton conductivity of the MOF increased with humidity and reached a maximum of 2.3×10^{-3} S cm^{-1} at 25 °C and 95 % RH. The result was in agreement with water adsorption data, which revealed plateau regions at two, five and eight water molecules per formula unit. The MOF showed low proton conductivity up to hydration with four water molecules. Nevertheless, after hydration with the fifth water molecule the proton conductivity improved considerably. This demonstrates that the extent of hydration was a key parameter influencing proton conductivity of the MOF. More recently, Wei et al. [67] investigated composite MOFs containing both the carboxylate functionality and polyoxometalate (POM) anions. Two poly-POM-MOF composites, $\{H[Cu(Hbpdc)(H_2O)_2]_2[PM_{12}O_{40}]\cdot nH_2O\}_n$, ($H_2bpdc = $ 2,2′-bipyridyl-3,3′-dicarboxylic acid; M = W, Mo) were synthesised and evaluated. The 3D structures of the composites was built from $[Cu(Hbpdc)(H_2O)_2]_2^+$, $[PM_{12}O_{40}]^{3-}$, protons and water molecules. The 3D structures contain 1D hydrophilic channels wherein a continuous water phase is stabilised by hydrogen bonding. Both composites exhibited a low proton conductivity of approximately 3.0×10^{-7} S cm^{-1} at 25 °C and 98 % RH. However, the proton conductivity increased as the temperature increased reaching 1.25×10^{-3} and 1.56×10^{-3} S cm^{-1} at 100 °C and 98 % RH for M = Mo and W, respectively.

Costantino et al. [68] investigated two zirconium-MOFs based on phosphonate ligands designated as Na@Zr and NH_4@Zr, and containing two Na^+ and two NH_4^+ cations, respectively in the channels. The channels also contain five water molecules per formula unit, which interact strongly with each other (via hydrogen bonds) as well as with the P-O groups on the framework surface. The Na^+ and NH_4^+ cations could be substituted with protons and the MOFs displayed proton conductivity which increased with RH. Another example of proton conductivity in MOFs based on phosphonate ligands was demonstrated where tetraphosphonate ligands were used in the construction of La-based and Mg-based MOFs [68, 69]. Furthermore, Colodredo et al. [70] employed carboxyphosphonate to produce a series of La-MOFs, which displayed proton conductivity. Taylor et al. [60] reported proton conduction in a phosphonate-based 3D MOF, $[La(H_5L)(H_2O)_4]$ (L = 1,2,4,5-tetrakisphosphonomethylbenzene) designated as PCMOF-5. The flexible, tetrapodal ligand enabled the formation of a 3D framework with narrow 1D channels. The proton conductivity was 2.5×10^{-3} S cm^{-1} at 60 °C and 98 % RH. This was due to the formation of 1D acidic channels. Within these channels uncoordinated, diprotic phosphonic acid groups together with monodentate hydrogen phosphonate and uncoordinated water were present, which created facile proton conduction pathway. PCMOF-5 was also found to be very robust demonstrated by its high humidity stability.

9.3.1.2 Proton-Conducting MOFs Operating Above 100 °C

The low operating temperature of water-mediated proton conducting MOFs places a limitation on their use for fuel cells that operate at temperatures higher than 100 °C. As such some research efforts have been directed towards anhydrous proton-conducting MOFs. These MOFs incorporate anhydrous guest species in the channels that can provide proton conduction pathways above the dehydration temperature of the MOFs. In this regard, various guest species such as imidazole, triazole, histamine and ionic liquids have successfully been incorporated in MOFs and the MOFs investigated for proton conductivity [44, 71–74]. Unlike the case of water-mediated proton conducting MOFs the conductivity of the anhydrous proton-conducting MOFs may not be dependent on RH.

For instance, two Al-based MOFs, Al(μ_2-OH)(ndc) (ndc = 1,4-naphthalenedicarboxylate) and Al(μ_2-OH)(bdc) (bdc = 1,4-benzenedicarboxylate) containing imidazole molecules in their 1D channels were studied [71]. Imidazole is non-volatile and possesses a high boiling point. It exists as two tautomers with a proton that migrates between the two nitrogen atoms, giving a proton conduction pathway. The proton conductivity at room temperature was reported to be in the range of 10^{-8}–10^{-10} S cm^{-1} for both imidazole-containing Al-based MOFs. At 120 °C the proton conductivity increased considerably to 2.2×10^{-5} and 1.0×10^{-7} S cm^{-1} for Al(μ_2-OH)(ndc) and Al(μ_2-OH)(bdc), respectively. The difference in the proton conductivity of the two MOFs is substantial and this was attributed to the dynamic motion of imidazole molecules in the channels. For Al(μ_2-OH)(ndc), the hydrophobic channels interacted weakly with polar imidazole molecules allowing the imidazole molecules to move freely and conduct protons. Meanwhile, for Al(μ_2-OH)(bdc), since the channel is amphiphilic in nature, the hydrophilic sites interacted more strongly with some of the imidazole molecules thereby restricting the mobility of the imidazole molecules in the framework.

The effect of using histamine rather than imidazole as a proton conductor in Al(μ_2-OH)(ndc) was subsequently investigated [73]. Histamine possesses two proton-hopping sites on the imidazole ring and one on the amine group. The loading resulted in one histamine molecule per Al^{3+} ion. The histamine-loaded Al(OH)(ndc) displayed a high proton conductivity of 1.7×10^{-3} S cm^{-1} at 150 °C under anhydrous conditions, which was significantly higher than that of pure histamine. It was also observed that the proton conductivity dropped to 2.1×10^{-4} S cm^{-1} at 150 °C when the histamine content was halved. Furthermore, the proton conductivity of the histamine-loaded MOF was substantially higher than that of the corresponding aforementioned imidazole-loaded MOF. This massive difference was ascribed to the much higher concentration of histamine, which was twice as high as that of imidazole in the Al-based MOFs. Therefore, the concentration of the proton carrier is very important in influencing proton conductivity. The histamine-loaded MOF is viewed as a superionic conductor and the MOF hosts show promise in creating hybrid conducting materials. Another study on histamine-loaded Zn-MOF-74 revealed a low proton conductivity of 4.3×10^{-9} S cm^{-1} at 146 °C, which was attributed to the low mobility of histamine in the MOF, which limited the hopping of protons between imidazole rings [74].

A 3D proton-containing MOF (β-PCMOF2) based on a sulfonate ligand (2,4,6-trihydroxy-1,3,5-benzenetrisulfonate) with the channels occupied by triazole, which enhanced the proton conductivity and maximum operating temperature considerably, was reported [44]. Similarly to imidazole, triazole has a high boiling point and is non-volatile. The proton conductivity of the MOF could be controlled by varying the amount of triazole that was loaded in the channels. The conductivity was in the range 2–5 $\times 10^{-4}$ S cm^{-1} at 150 °C under anhydrous H_2 conditions, which was higher than the value for either pristine PCMOF2 or triazole. The high temperature phase MOF with a partial triazole loading (β-PCMOF2$(Tz)_{0.45}$) was used as a membrane in a H_2/air cell. The membrane was shown to be gas tight and gave an open circuit voltage of 1.18 V, stable for 72 h at 100 °C thus demonstrating the applicability of MOFs in a membrane electrode assembly of a fuel cell.

A different approach to non-water mediated proton conduction is to synthesise MOFs which possess intrinsic proton conductivity. This approach was pioneered by Kitagawa and co-workers, who synthesised a 2D MOF, $[Zn(H_2PO_4)2(TzH)_2]_n$ [46]. In this MOF, extended 2D layer structures are formed by octahedrally-coordinated Zn^{2+} ions connected to orthophosphates and bridging triazole molecules. Substantial interlayer hydrogen bonding occurs between triaozles and orthophosphates, yielding a proton conduction pathway. Proton conductivity measurements conducted on pellets gave a value of 1.2×10^{-4} S cm^{-1} at 150 °C, in the absence of guest proton conducting medium. Furthermore, proton conductivity measurements were also conducted on a single crystal and the results illustrated that the MOF conducted protons anisotropically. Kitagawa and co-workers also investigated another zinc phosphate MOF, $[Zn(HPO_4)(H_2PO_4)_2](ImH_2)_2$ [75]. The MOF comprises tetrahedrally coordinated Zn^{2+} ions and two types of orthophosphates, which form 1D chains, $[Zn(HPO_4)(H_2PO_4)_2]^{2-}$. Between the chains are two crystallographically independent protonated imidazole molecules, giving an ionic crystal system (Fig. 9.4). Considerable hydrogen bonding occurs between protonated imidazolium cations and the orthophosphates. Proton conductivity of the MOF was temperature dependent, increasing with increasing temperature reaching a value of 2.6×10^{-4} S cm^{-1} at 130 °C. The observed proton conductivity was ascribed to the local motion of the protonated imidazolium cations.

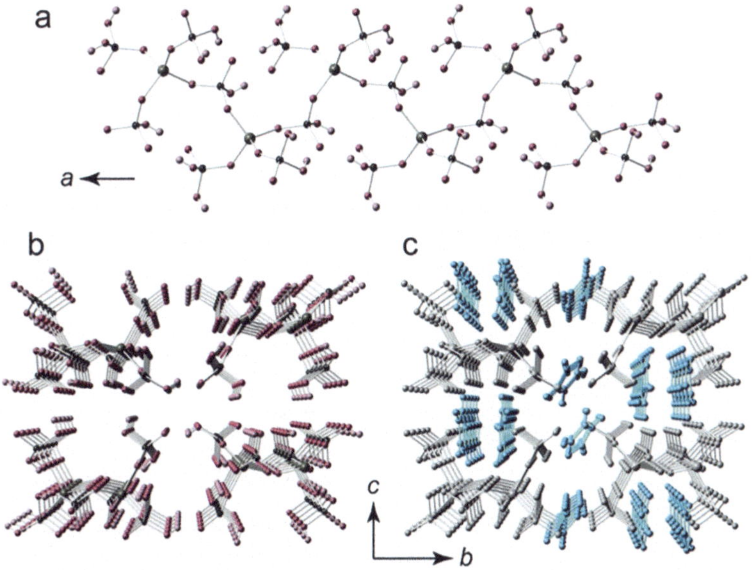

Fig. 9.4 (**a**) Crystal structure of negatively charged 1D coordination chains of $[Zn(HPO_4)(H_2PO_4)_2]^{2-}$ in $[Zn(HPO_4)(H_2PO_4)_2](ImH_2)_2$ and (**b**) packing structure of four 1D chains along the *a* axis. (**c**) Crystal structure of $[Zn(HPO_4)(H_2PO_4)_2](ImH_2)_2$. The ImH_2^+ ions are highlighted in blue and the networks are gray. H atoms in the networks have been omitted. Reprinted with permission from [75]. Copyright (2012) American Chemical Society

9.3.2 Catalysts or Catalyst Precursors for Oxygen Reduction Reaction

In PEMFCs oxygen reduction reaction (ORR) is the reaction that occurs at the cathode. It is a complex 4-electron reaction involving the breaking of a double bond and the formation of 4 OH-bonds via several stages. The kinetics of the ORR is sluggish necessitating the use of a catalyst. The sluggish kinetics of the ORR is one of the major factors limiting the performance and efficiency of fuel cells. Current fuel cell technologies employ Pt-based materials as cathode catalysts due to the high activity of Pt for the ORR. However, the high cost, limited reserves and poor durability of Pt-based catalysts hamper the large-scale commercialization of fuel cells. Over the past several decades extensive research has been devoted to developing alternative electrocatalysts, *i.e.* non-platinum group metal (PGM) catalysts for the ORR in fuel cells, so as to enhance their viability for practical applications. Although the non-PGM catalysts are much cheaper and highly abundant they are typically less efficient. In order to obtain decent efficiency higher catalyst content is required, which increases the thickness of the cathode layer leading to poor mass transport properties. Recently, MOFs have attracted attention as promising materials for ORR catalysts due to their excellent structural properties. The reader is referred to Morozan and Jaouen [22] and Ren et al. [24] for further reading on the subject.

Mao et al. [76] demonstrated the use of Cu-based MOFs as electrocatalysts for ORR. Cu-BTC (BTC = 1,3,5-tricarboxylate) was reported to be structurally unstable in aqueous media though it exhibited electrocatalytic activity. On the other hand, Cu-bipy-BTC (bipy = 2,2′-bipyridine) was structurally stable in water and displayed an excellent and stable electrocatalytic activity towards almost 4-electron reduction of O_2. In a phosphate buffer (pH 6.0), Cu-bipy-BTC showed a pair of well-defined redox peaks at ca. −0.15 V (Fig. 9.5). Upon bubbling of O_2 into the

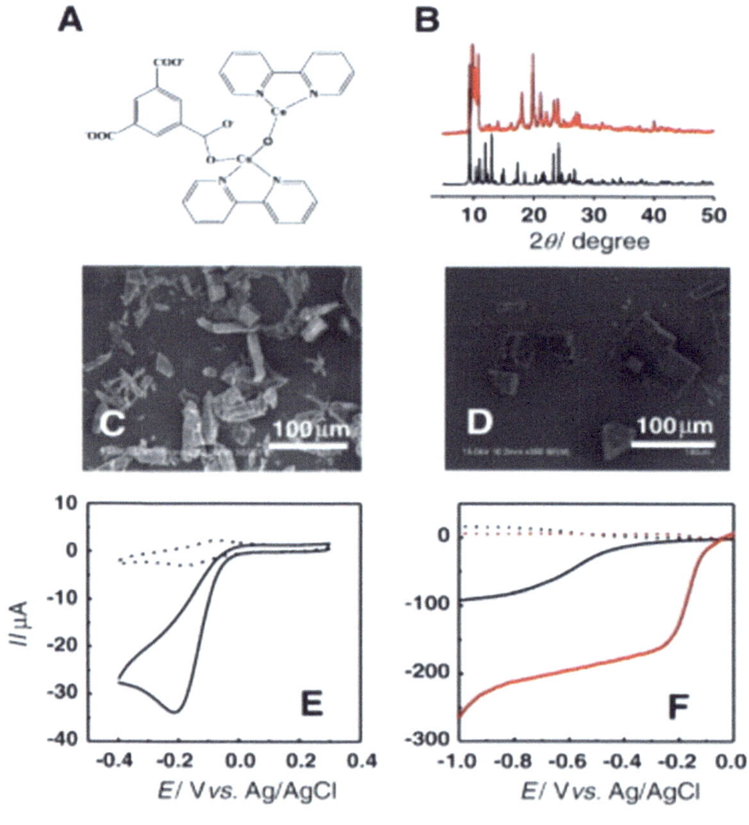

Fig. 9.5 (**a**) Coordination geometry of Cu atoms in Cu-bipy-BTC. (**b**) XRD patterns of the as-prepared Cu-bipy-BTC (*red curve*) and the simulated one (*black curve*). (**c**) Typical SEM image of fully crystallised Cu-bipy-BTC sample. (**d**) SEM image of Cu-bipy-BTC after being immersed in water for 24 h. (**e**) Typical CVs obtained at the Cu-bipy-BTC-modified GC electrodes in 0.10 M phosphate buffer (pH 6.0) saturated with N_2 (*dotted curve*) or O_2 (*solid curve*). Scan rate, 20 mV s^{-1}. (**f**) Typical RRDE voltammograms obtained with bare (*black curves*) and Cu-bipy-BTC-modified (*red curves*) GC electrodes as disk electrodes (*solid curves*) and platinum ring electrode (*dotted curves*) in 0.10 M phosphate buffer (pH 6.0) under air-saturated O_2. Electrode rotation rate, 400 rpm. Scan rate, 10 mV s^{-1}. Reprinted with permission from [76]. Copyright (2012) Elsevier B.V.

buffer the reduction peak current at −0.20 V increased while the reversed oxidation peak current of the redox wave decreased, indicating that ORR occurred under the electrocatalysis of Cu-bipy-BTC. A composite graphene–metalloporphyrin MOF (denoted as (G-dye–FeP)n) prepared from pyridine-functionalised graphene (denoted as G-dye) and an iron porphyrin (5,10,15,20-tetrakis(4-carboxyl)-iron-porphyrin; labelled FeP) was investigated for electrocatalytic activity [77]. G-dye served as a building block in the assembly of the MOF. The reason for using iron porphyrin was because metallo-porphyrins play a key role in biological reactions including oxygen transport and reduction reactions. It was reported that the contents of graphene and Fe porphyrin affected the crystallization process of the MOF. The composite displayed interesting catalytic activity towards the ORR in alkaline medium with an onset potential of ca. 0.93 V vs. reversible hydrogen electrode with a 4-electron ORR pathway. This result was a combined effect of the larger bond polarity arising from the nitrogen ligands in the G-dye, the catalytically active iron–porphyrin and the porosity of the MOF.

Several studies have reported the use of MOFs denoted as ZIFs (ZIF = zeolitic imidazolate framework) as sacrificial templates for the synthesis of ORR catalysts [78–82]. Ma et al. [78] first demonstrated the use of ZIFs-based ORR electrocatalyst by using Co-ZIF, $[Co(Im)_2{\cdot}0.5DMA]_\infty$ (Im = 3,5-imidazolate; DMA = *N*,*N*-dimethylaniline), as the template. After thermal activation above 600 °C the 3D MOF structure was not preserved. However, the structure enabled the creation of a high density of $Co–N_4$ catalytic sites which could be readily accessed by oxygen as sufficient porosity and surface area were retained. After pyrolysis at 750 °C in an inert atmosphere, the Co-ZIF exhibited optimal stability and activity for the ORR at pH 1, with an onset potential of 0.83 V vs. a reversible hydrogen electrode and ORR mechanism characterised by 3.2–3.5 electrons per O_2 molecule. A direct relationship existed between the surface area and the catalytic activity. The mesopores accounted for most of the surface area formed during heat treatment.

Proietti et al. [80] derived an electrocatalyst from a combination of iron acetate, phenanthroline and a Zn-based MOF, ZIF-8. The catalyst displayed increased volumetric activity and enhanced mass transport properties. ZIF-8 acted as a host for iron acetate and phenanthroline to produce a catalyst precursor that later underwent pyrolysis. ZIF-8 was selected as host due to its nitrogen-rich composition and large microporous surface area (1700 $m^2\ g^{-1}$), which are both crucial for ORR activity of the Fe/N/C catalysts. After the heat treatment processes ZIF-8 was transformed into a nitrogenated microporous carbon structure hosting $Fe–N_x$ active sites. A cathode made with the best Fe/N/C electrocatalyst in this study and tested in a H_2-O_2 fuel cell yielded a power density of 0.75 W cm^{-2} at 0.6 V. This value was comparable with that of a commercial Pt-based cathode that was tested under same conditions. The unprecedented catalytic activity was ascribed to enhanced mass transport properties due to the porous carbon nanostructure obtained following heat treatment. The microporous surface area of the optimised catalyst was approximately 1000 $m^2\ g^{-1}$. However, the results for durability and stability for the

MOF-derived Fe/C/N catalyst were relatively poor hindering its applicability on a commercial scale.

A polymetallic Fe-based ZIF was used as a precursor to prepare Fe-based electrocatalysts, which demonstrated excellent activity towards ORR in acidic medium [81]. Further enhancement of the catalytic activity was achieved by combining the electrocatalyst with ZIF-8 followed by pyrolysis. The newly derived catalyst as cathode showed an onset potential of 0.977 V and measured volumetric current density of 12 A cm^{-3} at 0.8 V in a single cell test. Very recently, Zhao et al. [82] demonstrated a facile synthesis approach for electrocatalysts involving pyrolysis of one-pot synthesised ZIF decorated with a uniformly dispersed iron complex. The electrocatalysts displayed excellent activity, which was influenced by the imidazole ligands, with the highest volumetric current density of 88.1 A cm^{-3} measured at 0.8 V in polymer electrolyte fuel cell tests. The cell performance at higher cell currents was affected by the surface area and porosity of the catalysts. This approach enables the feasibility of scale-up of ZIFs-based electrocatalysts and also offers the possibility of tailored synthesis of electrocatalysts in the burgeoning search for non-PGM catalysts.

A variation of the aforementioned study by Proietti et al. [80] was later carried out by the same research group where either the MOF was changed to Fe-BTC (specific surface area = 1500 m^2 g^{-1}; BTC = 1,3,5-tricarboxylate) or the Fe precursor changed to Fe–phthalocyanine or the organic precursor substituted with polyaniline [83]. It was observed that the changes in the electrochemical performance were small when ZIF-8 was present in the catalyst precursor mixture. However when ZIF-8 was substituted with Fe-BTC substantial adverse effects were reported. The reason for this is twofold. Firstly, ZIF-8 is a Zn-based MOF and Zn could easily evaporate at the high temperatures employed during pyrolysis. On the other hand, Fe present in Fe-BTC accumulated during thermal treatment and promoted graphitisation, which led to a decreased microporous surface area. In addition, the absence of nitrogen in Fe-BTC implies that the sole source of nitrogen in the catalyst precursor mixture for the active sites was the organic precursor. This study demonstrates the significance of MOF composition when used to derive non-PGM electrocatalysts by pyrolysis.

9.3.3 Anode Catalysts for Alcohol Electrooxidation

The use of alcohols is an alternative to the direct use of hydrogen in fuel cells. Alcohols, such as methanol and ethanol, possess high energy densities, and are stable liquids that can easily be transported and stored. Direct alcohol fuel cells (DAFCs) are being targeted for portable electronic devices and electric vehicles due to their high power density output and low pollutant emissions. PGM catalysts are highly active and stable particularly in acidic medium, which makes them attractive for use as anode catalysts for electrooxidation in fuel cells. However, PGMs are easily poisoned by intermediate products, such as CO, from alcohol oxidation.

Moreover, PGMs are expensive and limited in supply, which hampers the commercialisation of DAFCs. Consequently, there is need to find alternative low-cost, abundant catalysts with high efficiency and durability. Although very limited, mixed conduction of electrons and protons can be obtained in MOFs, which is desirable for their use as anode catalyst for alcohol oxidation.

Yang et al. [84] demonstrated the first example of the utilization of a noble-metal-free MOF material, $[(HOC_2H_4)_2dtoaCu]$ (H_2dtoa = dithiooxamide), as an electrocatalyst for ethanol electrooxidation reaction. The MOF was coated on a glassy-carbon electrode, and evaluated in 0.5 M H_2SO_4 solution. Cyclic voltammetry revealed one redox peak centred at 0.35 V *vs.* Ag/AgCl attributed to Cu^I/Cu^{II}. The ethanol electrooxidation on this MOF was dependent on the ethanol concentration. Upon increasing the ethanol concentration there was a continuous increase in current density. It was also noted that even when the ethanol concentration was as high as 2 M, no plateau was observed in the cyclic voltammograms, illustrating that the MOF catalyst was tolerant to the oxidation products of ethanol. Acetaldehyde was detected as the oxidation product of ethanol. The MOF catalyst performance in terms of oxidation potential and current density was comparable with Pt-based catalysts, even though ethanol was only partially oxidised to acetaldehyde under the conditions of investigation. Recently, the electrooxidation reaction mechanism of ethanol on the MOF electrocatalyst, $[(HOC_2H_4)_2dtoaCu]$, was elucidated theoretically for the first time by using the density functional theory method [85]. The indirect proton-transfer reaction from ethanol to the N atom via the hydroxyl group in HOC_2H_4 was suggested to be the key mechanism controlling the reactivity of the ethanol oxidation, since the direct proton-transfer reaction displayed approximately 44 kcal mol^{-1} higher activation energy. This novel ethanol electrooxidation reaction mechanism in the MOF is of fundamental importance to the application of noble-metal free MOF as electrocatalysts.

9.4 MOFs as Materials for Enabling Fuel Cell Technologies

9.4.1 Hydrogen Production

9.4.1.1 Photocatalysts for Hydrogen Production from Water

Much of the hydrogen produced in the world today is obtained from steam reforming of natural gas, which is composed predominantly of methane. In the future, hydrogen production from water is desirable to meet environmental and sustainability needs. Such a supply of hydrogen will be beneficial to the fuel cells industry. In particular, the reduction of water into hydrogen molecules in the presence of a photocatalyst is not dependent on fossil fuels, and is therefore ideally suited for clean energy generation. Some MOFs have been shown to display semiconducting behaviour whereby photon absorption generates a state of charge separation with a positive hole in the valence band and an electron in the conduction band [86, 87]. Semiconductors are generally applied in photocatalysis whereby light energy is converted into chemical energy [88]. Recently, MOFs have been explored as photocatalysts for generating hydrogen from water.

For instance, Kataoka et al. [89] reported the first example of MOFs (Ru-based MOFs) that function as an activity site for the reduction of water into hydrogen molecules in the presence of $Ru(bpy)_3^{2+}$ (bpy = 2,2′-bipyridine), MV^{2+} (methyl viologen; N,N′-dimethyl-4,4′-bipyridinium), and EDTA–2Na (EDTA = ethylenediaminetetraacetic acid) under visible light irradiation. The results showed that the highest turnover number was 8.16 based on Ru–MOFs and 81.6 based on $Ru(bpy)_3^{2+}$ after 4 h of light irradiation. The apparent quantum yield was 4.82 % at 450 nm. It was also shown that a porous MOF structure was valuable for improving the activity of the photochemical water reduction. A further study was conducted on the Ru-MOFs modified by changing the counter-ions within the framework to yield $[Ru_2(p\text{-bdc})_2X]_n$ (bdc = 1,4-benzenedicarboxylate; X = Cl^-,Br^- or BF_4^-) [90]. Hydrogen was generated when the Ru-MOFs were combined with $Ru(bpy)_3^{2+}$, MV^{2+} and EDTA. The observed trend in catalytic activity was as follows: $[Ru_2(p\text{-bdc})_2Br]_n > [Ru_2(_p\text{-bdc})_2BF_4]_n > [Ru_2(p\text{-bdc})_2Cl]_n$. The difference in the catalytic activity was ascribed to the surface modification, which was linked to the choice of counter ions. The MV^{2+} species were physically adsorbed on the Ru-MOFs surface and not within the cavities. The most active catalyst, $[Ru_2(p\text{-bdc})_2Br]_n$, displayed the highest catalytic activity generating 46.7 mmol H_2 with a turnover number of 18.7 after 4 h of irradiation. The study also revealed that the catalytic activity of MOFs with counter-ions changed significantly when the counter-ions were manipulated by molecular catalysts.

Two porphyrin-based MOFs, PCL-1 ($Ru_2(H_2tccp)BF_4$) and PCL-2 ($Ru_2(Zntccp)$ BF_4) (PCL = porphyrin coordination lattice, tccp = tetrakis(4-carboxyphenyl-porphyrin)), were studied by the same group [91]. It was also shown here that physical adsorption of MV^{2+} on the surface of the PCLs was necessary for photocatalysis with the adsorbed MV^{2+} acting as intermolecular electron relays from the porphyrin sites to the Ru metal centers. Irradiation of MV^{2+}, PCL and EDTA at 320 nm afforded turnover numbers of 20.8 and 29.9 for PCL-1 and PCL-2, respectively. The high turnover numbers is an indication of the positive effect porphyrin might have on MOF photocatalysts as porphyrin in this case functioned as a photosensitiser eliminating the need for $Ru(bpy)_3^{2+}$. In another report, water-stable porphyrin-based MOFs, Al-PMOF (H_2TCPP $[AlOH]_2(DMF_3(H_2O)_2)$ (H_2TCPP = *meso*-tetra(4-carboxyl-phenyl) porphyrin)) and Al/Zn-PMOF ($Zn_{0.986(12)}TCPP[AlOH]_2$) were investigated as photocatalysts for hydrogen generation from water [92]. Two different hydrogen generation systems, MOF/MV^{2+}/EDTA/Pt and MOF/EDTA/Pt were evaluated. For Al/Zn-PMOF and Al-PMOF hydrogen was generated at a rate of 100 and 200 mmol g^{-1} h^{-1}, respectively after an induction period of about 3 h.

Silva et al. [93] described the activity for the photocatalytic hydrogen generation of two Zr-based MOFs: UiO-66 ($[Zr_6O_4(OH)_4(bdc)_{12}]$, where bdc = 1,4-benzene dicarboxylate; UiO = University of Oslo) and UiO-66(NH_2) ([$Zr_6O_4(OH)_4(ata)_{12}$], where ata = 2-aminoterephthalate). The water-stable MOFs displayed photocatalytic activity for hydrogen generation in water/methanol upon irradiation at wavelength longer than 300 nm. In fact, the apparent quantum yield for H_2 generation using monochromatic light at 370 nm in a 3:1 water/methanol mixture was 3.5 % for UiO-66(NH_2). It was shown that the photocatalytic activity can be increased by rational design of the MOF and by addition of a hydrogen evolution co-catalyst. Laser flash photolysis enabled the photocatalytic activity for hydrogen generation to be correlated with the creation of charge separated states. The presence of the -NH_2 group in the bdc ligand did not affect the photochemistry but produced a bathochromic shift in the optical spectrum, showing the prospects of developing more efficient MOFs for generating hydrogen from water.

In a recent report, Wang et al. [94] loaded Pt nanoparticles in phosphorescent MOFs (Fig. 9.6) and demonstrated that the resulting Pt@MOF materials photocatalysed the generation of hydrogen by synergistic photoexcitation of the MOFs and electron injection into the Pt nanoparticles. In another study, MOF-253-Pt was synthesised by immobilising a platinum complex in 2,2′-bipyridine-based

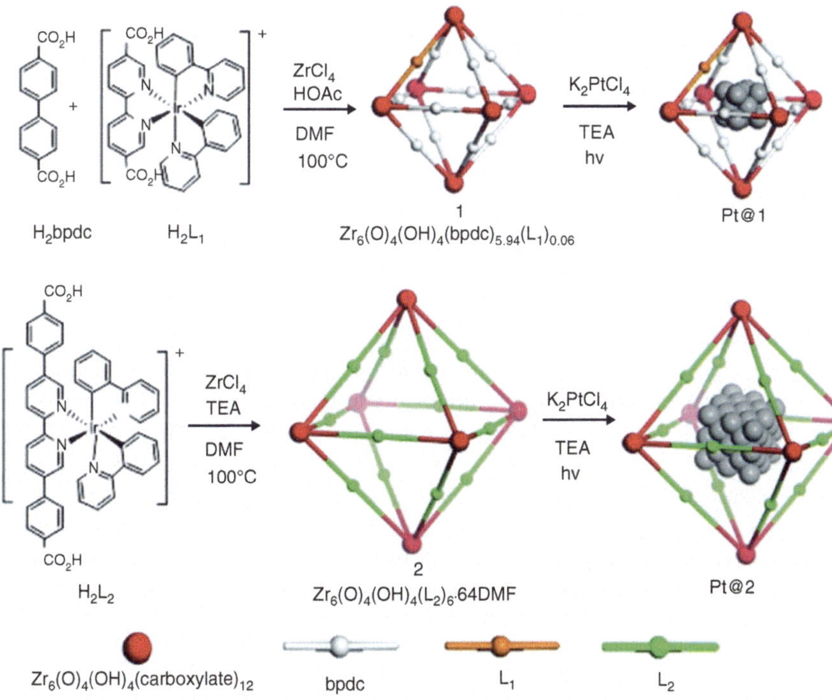

Fig. 9.6 Synthesis of phosphorescent Zr-carboxylate MOFs (**1** and **2**) of the *fcu* topology and subsequent loading of Pt nanoparticles inside MOF cavities via MOF-mediated photoreduction of K_2PtCl_4 to form the Pt@**1** and Pt@**2** assemblies. Reprinted with permission from [94]. Copyright (2012) American Chemical Society

microporous MOF (MOF-253) using a post-synthesis modification approach [95]. The functionalised MOF-253-Pt had a dual function; as a photosensitiser and a photocatalyst for hydrogen evolution under visible-light irradiation. The photocatalytic activity of MOF-253-Pt was about five times higher than that of the corresponding complex. The presence of the short Pt···Pt interactions played a key role in enhancing the photocatalytic activity of the resulting MOF. Even more recently, UiO-66 sensitised by adsorbed or directly added rhodamine B dye displayed photocatalytic activity for hydrogen evolution under visible-light illumination. In the presence of Pt as a co-catalyst, the adsorbed and directly added dye increased the photocatalytic activity remarkably to 30 and 26 times the value of neat Pt@UiO-66, respectively [96].

Besides serving as photocatalysts or hosts for photocatalysts MOFs have been employed as sacrificial templates or precursors of photocatalysts. For example, deKrafft et al. [97] developed a novel MOF-templated approach to prepare a mixed metal oxide nanocomposite material. Core-shell particles were first created and then calcined to decompose the core, producing crystalline octahedral nanoshells consisting of hematite Fe_2O_3 nanoparticles embedded in anatase TiO_2 with some Fe doping. The Fe_2O_3@TiO_2 material possessed interesting photophysical properties since it allowed photocatalytic hydrogen production from water using visible light, which could not be achieved by either Fe_2O_3 or TiO_2 alone, or a mixture of the two. More recently, He et al [98] demonstrated that embedding CdS in MOFs could considerably increase the photocatalytic efficiency of CdS for visible light-driven hydrogen production.

9.4.1.2 Catalysts or Catalyst Supports for Hydrogen Production from Hydrolysis of Chemical Hydrides

Hydrolysis of chemical hydrides such as NH_3BH_3 and $NaBH_4$ for the production of hydrogen for fuel cells is of great interest to many researchers because hydrolysis of these hydrides combines the best properties for storage and generation of hydrogen, such as solubility in water, fast controllable hydrolysis, overall stability and moderate exothermicity [99]. In the presence of an appropriate catalyst hydrogen can be generated from the catalytic hydrolysis of the chemical hydrides. Catalytic hydrolysis of chemical hydrides does not only generate hydrogen from the chemical hydride but also the hydrogen in water can be released, providing a promising method for on-site hydrogen supply [100]. Generally, the reactions for generation of hydrogen from hydrolysis of NH_3BH_3 and $NaBH_4$ are presented below, respectively:

$$NH_3BH_3 + 2H_2O \rightarrow NH_4BO_2 + 4H_2$$
$$NaBH_4 + 2H_2O \rightarrow NaBO_2 + 4H_2$$

In one report, a Co-based MOF, ZIF-9 (ZIF = zeolitic imidazolate framework), was synthesised via solvothermal method and the feasibility of the ZIF-9 as catalyst for

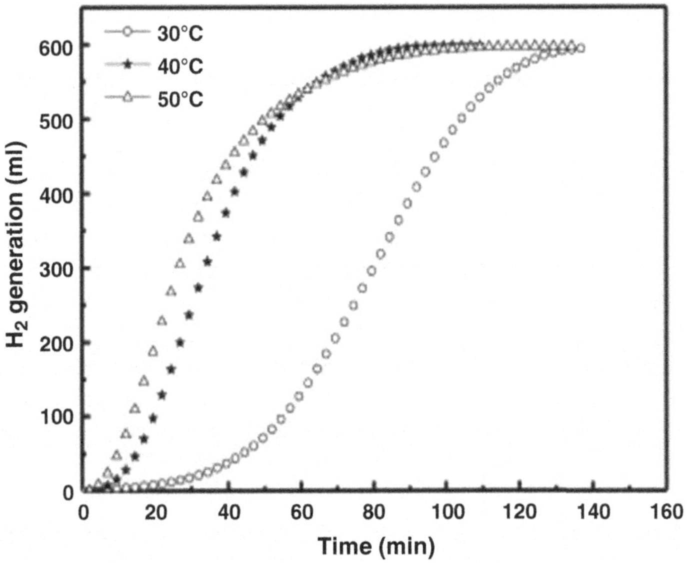

Fig. 9.7 The effect of reaction temperature on hydrogen production. $NaBH_4$: 0.5 wt%; NaOH: 5 wt.%; H_2O: 50 mL, catalyst: 0.025 g. Reprinted with permission from [99] Copyright (2012) Elsevier B.V.

hydrogen production from $NaBH_4$ hydrolysis was reported [99]. It was revealed that initial hydrogen generation rate of the ZIF-9 catalyst was relatively slow because of the gradual formation of CoB active centers. However, after the formation of CoB the hydrogen generation rate increased rapidly. Indeed, the hydrogen generation rate at 40 °C reached 3641.69 mL min^{-1} g^{-1}(Co) (Fig. 9.7). It was further demonstrated that the addition of NaOH increased the hydrolysis rate of $NaBH_4$ considerably. Upon cycling no obvious decrease of hydrogen production rate was observed implying ZIF-9 catalyst possessed relatively high stability. Furthermore, ZIF-9 maintained its basic crystal structure and crystallinity, but its long range order was altered to a certain degree during the operation.

Li et al. [101] successfully immobilised Ni nanoparticles in the MOF, ZIF-8, and demonstrated that the immobilised Ni nanoparticles acted as the catalytic sites. High catalytic activity and long durability for hydrogen generation from hydrolysis of aqueous NH_3BH_3 at room temperature was reported. It was suggested that the catalytic process occurred through the formation of an activated complex species originating from interaction between NH_3BH_3 molecules and the surface of Ni nanoparticles. For Ni@ZIF-8 prepared by chemical liquid deposition NH_3BH_3 hydrolysis could be completed with the release of hydrogen of $H_2/NH_3BH_3 = 3.0$ in 19 min ($Ni/NH_3BH_3 = 0.019$), corresponding to a turnover frequency of 8.4 min^{-1}. Meanwhile, Ni@ZIF-8 prepared using chemical vapour deposition displayed a higher activity for which the reaction could be completed in 13 min ($Ni/NH_3BH_3 = 0.016$), corresponding to a turnover frequency of 14.2 min^{-1}. In another study by the same group, ultrafine Pt nanoparticles were successfully immobilised inside the pores of the MOF, MIL-101, without aggregation of Pt nanoparticles on the external surfaces of the MOF [102]. The obtained Pt@MIL-101 was tested for the catalytic hydrolysis of NH_3BH_3 and the results showed that hydrogen was generated from aqueous NH_3BH_3 at room temperature in the presence of Pt@MIL-101 with different Pt loadings. All the Pt@MIL-101 catalysts were highly active for the hydrolysis of NH_3BH_3 with hydrogen release of $H_2/NH_3BH_3 = 3$. When 2 wt% Pt@MIL-101 (Pt/AB = 0.0029 in molar ratio) was employed, the hydrolysis of NH_3BH_3 was completed within 2.5 min, corresponding to $\sim 1.0 \times 10^4$ $L_{H2}\,mol_{Pt}{}^{-1}\,min^{-1}$, twice as high as the value of 2 wt% Pt/γ-Al_2O_3, which is the most active Pt catalyst for this reaction that has been reported.

MOFs have also been employed as precursors for hydrogen generation catalysts [103, 104]. For instance, an amorphous Co(0) catalyst was synthesised by the reduction of a MOF precursor, $Co_2(bdc)_2(dabco)$ (bdc = 1,4-benzenedicarboxylate; dabco = 1,4-diazabicyclo[2.2.2]octane) [104]. The active Co(0) sites were generated *in situ* with the reducing reaction through the micropores and channels of the MOF. The catalytic effective Co(0) sites were surrounded by the organic linkers and stabilised in the residue of the framework. The amorphous Co(0) catalyst displayed highly efficient activity in the hydrolysis of NH_3BH_3. Hydrogen generation of a 0.32 M aqueous NH_3BH_3 solution completed in 1.4 min at room temperature. The porous structure in the MOF was suggested to play a major role. This report demonstrated that MOFs exhibiting high surface area and sufficient pore structures, may ideally be suited as precursors for highly efficient heterogeneous catalysts.

9.4.2 *Hydrogen Storage*

9.4.2.1 Materials for Adsorptive Hydrogen Storage

Numerous studies have revealed that MOFs can store considerable amounts of hydrogen at liquid nitrogen temperature. At this temperature hydrogen adsorption occurs by a physical process involving weak van der Waals forces. Adsorptive hydrogen storage offers the benefits of fast kinetics of hydrogen uptake and release and complete reversibility of the adsorption process. There are many structural factors that determine hydrogen storage in MOFs such as surface area, pore size, pore volume‘, open metal sites and ligand functionalisation. Many research efforts are being directed towards enhancing hydrogen storage properties of MOFs so that they can become more viable for practical hydrogen storage. For a recent review of hydrogen storage in MOFs the reader is referred to Langmi et al. [105].

It has been reported that H_2 uptake in MOFs is influenced by pore size at low pressures, surface area at moderate pressures and free volume at high pressures [106]. It has also been established that at 77 K and a low pressure of 1 bar, hydrogen uptake can directly correlate with BET surface area in the range 100–2000 $m^2\,g^{-1}$

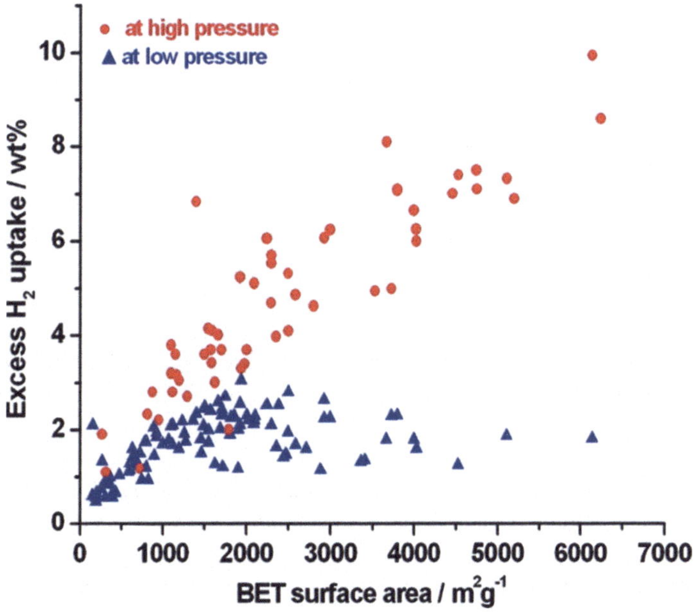

Fig. 9.8 A plot showing the relationship between H_2 uptake capacities at 77 K and BET surface areas of various MOFs. Low pressure is approximately 1 bar and high pressures are in the range of 10–90 bar. Reprinted with permission from [16]. Copyright (2012) American Chemical Society

(Fig. 9.8). Nevertheless, above 2000 $m^2\ g^{-1}$ the correlation breaks down since the MOF surface cannot be fully covered with hydrogen molecules [16]. A series of NbO-type MOFs (NOTT-101, PCN-46 and NOTT-102, NOTT = Nottingham; PCN = porous coordination network) was studied and a correlation between hydrogen uptake and length of ligand was reported [107, 108]. The longer the length of the ligand the higher the specific surface area, pore volume and high pressure hydrogen uptake. In fact, NOTT-101 possessed the shortest ligand (5.77 Å), lowest surface area (2316 $m^2\ g^{-1}$), pore volume (0.886 $cm^3\ g^{-1}$) and total H_2 uptake at 77 K and 60 bar (6.2 wt%). Meanwhile, NOTT-102 exhibited the highest values of 10.098 Å, 2942 $m^2\ g^{-1}$, 1.138 $cm^3\ g^{-1}$, 6.7 wt% respectively.

Generally, the interaction energy of hydrogen with adsorbent surfaces is low [109]. However, if the pores of the adsorbent are sufficiently small the potential fields from opposite walls can overlap, enhancing the interaction potential. Wang and Johnson [110] calculated the optimal pore size to be approximately 6 Å, which is roughly double the kinetic diameter of the hydrogen molecule (2.89 Å). This ideal pore size results in optimal interactions between hydrogen and the framework as it may allow the formation of a monolayer of hydrogen molecules on opposite walls thereby maximising the total van der Waals forces acting on the hydrogen molecules [111]. One strategy that has been employed to tailor pore size and enhance hydrogen interaction with the framework is catenation. Catenation is defined as the intergrowth of two or more identical frameworks. In the case where there is maximal displacement of frameworks from each other the catenation is known as interpenetration. Meanwhile, where there is minimal displacement of the frameworks it is regarded as interweaving [112, 113]. A typical example of the effect of catenation was illustrated for the isomer pair PCN-6′ (noncatenated) and PCN-6 (catenated) both with formula $Cu_3(TATB)_2$ (TATB = 4,4′,4″-*s*-triazine-2,4,6-triyltribenzoate) [114, 115]. At 1 bar and 77 K, hydrogen uptake for PCN-6' activated at 50 °C was 1.35 wt% while PCN-6 activated at the same temperature adsorbed 1.74 wt%, indicating that catenation can lead to enhancement in H_2 uptake [114]. Ma et al. [115] further illustrated from inelastic neutron scattering studies that for both isomers hydrogen initially bound to open Cu centers of the paddlewheel moieties with similar interaction energies. With increasing hydrogen pressure the H_2 molecules bound to or around the organic linkers. The latter interaction was substantially stronger in catenated PCN-6 than in non-catenated PCN-6′. Therefore, H_2 adsorption at 50 bar of the catenated framework PCN-6 was higher than that of non-catenated PCN-6' (*i.e.* 6.7 and 4.0 wt%, respectively). In non-catenated PCN-6′, the distance between opposite pore walls was 21.4 Å, which was long and did not favour overlap of potential fields, leaving an open structure with space in the middle of the pores that decreased the number of effective binding sites.

It has been demonstrated that open metal sites in MOFs have high affinities for hydrogen and as such play a key role in hydrogen storage [106, 116, 117]. The isosteric heat of hydrogen adsorption gives an indication of the strength of interaction between H_2 molecules and a MOF. Many of the reported MOFs generally display isosteric heat of adsorption in the range 4–12 kJ mol^{-1} [118, 119]. Although the inclusion of unsaturated metal sites in MOFs can enhance the binding of hydrogen to the MOFs [117, 120, 121], it has also been shown that the alignment of the unsaturated metal sites in relation to the H_2 molecules plays a part [122]. There have been numerous studies conducted on the NOTT-*nnn* series of MOFs [107, 118, 122, 123] (Fig. 9.9). For instance, Lin et al. [107] investigated isostructural MOFs constructed from binuclear Cu(II) paddlewheel nodes each bridged by four carboxylate. Total hydrogen adsorption of NOTT-103 reached 77.8 mg g^{-1} (7.22 wt%) at 77 K and 60 bar. Generally, the high hydrogen storage capacities of this series of MOFs were as a result of a vacant coordination site at each Cu(II) center and the large pore volumes of the MOFs. Although it was observed that the open Cu(II) sites were the strongest adsorption sites the isosteric heat of hydrogen adsorption for Cu(II) sites marginally exceeded those for other sites in the MOF [107]. Similarly, the high H_2 adsorption capacities for NOTT-112 and NOTT-116 were attributed to the presence of open metal sites in the cuboctahedral cages [123]. Another polyhedral MOF, NOTT-140, was reported [118]. The desolvated MOF with open Cu(II) sites had a total H_2 adsorption of 6.0 wt% at 20 bar and 77 K and an isosteric heat of 4.15 kJ mol^{-1} at zero coverage. This value is low in comparison with other Cu(II)-based polyhedral MOFs due to the different alignment of open Cu(II) sites [118]. A strategy was devised by Wang

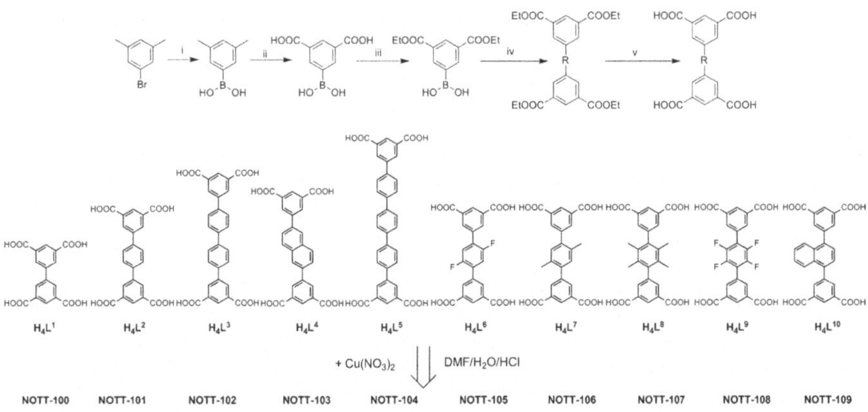

Fig. 9.9 Synthesis of ligands and their corresponding Cu complexes $[Cu(L)_2]$.˙ (*i*) Mg, Et_2O, B$(OMe)_3$; (*ii*) ^{t}BuOH/H_2O, $KMnO_4$; (*iii*) EtOH, H_2SO_4; (*iv*) BrRBr, $Pd(PPh_3)_4$, 80 °C, 3 days; (*v*) 2 M NaOH, HCl. Reprinted with permission from [107]. Copyright (2009) American Chemical Society

et al. [122] to increase the number of closest neighbouring open metal sites of each void that hosts hydrogen, and to align these sites toward the H_2 molecules, so as to make the interaction between hydrogen and the framework stronger.

Doping of MOFs with metal ions has attracted attention as several calculations have predicted enhancement in hydrogen uptake for MOFs doped with metal ions such as Li^+ ions. Such enhancement is associated with the stronger binding of hydrogen with the dopant cations [124–128]. Although these predictions have not been confirmed experimentally, a few experimental studies have been carried out to investigate the influence of chemical doping of MOFs on hydrogen storage [129–132]. For instance, the effect of doping MOFs with alkali metals (*i.e.* Li, Na, and K) was evaluated [130, 131]. In one study, a 2-fold interwoven MOF, $Zn_2(NDC)_2$(diPyNI) (NDC = 2,6-naphthalenedicarboxylate; diPyNI = *N*,*N*′-di-(4-pyridyl)-1,4,5,8-naphthalenetetracarboxydiimide) was first chemically reduced and then doped with Li^+, Na^+ and K^+ [131]. In another study, MOF-5 was doped with the same alkali metal ions [130]. There was a considerable increase in hydrogen uptake for the doped MOFs compared to the undoped MOFs [130, 131]. At 77 K and 1 bar, for the same amount of dopant, H_2 uptake followed the trend: $Li^+ < Na^+ < K^+$, in agreement with the trend of increasing size of dopant cation. However, the H_2 binding strength followed the reverse trend $Li^+ > Na^+ > K^+$. These observations were ascribed to the structural changes accompanying framework reduction, such as ligand polarizability and framework displacement, rather than the creation of particular metal-based adsorption sites [131, 133].

Apart from the metal sites, which are preferential sites for hydrogen adsorption, organic linkers have also been demonstrated to be favourable sites for hydrogen adsorption. An inelastic neutron scattering spectroscopy study of the rotational transitions of adsorbed H_2 molecules on MOF-5, $Zn_4O(BDC)_3$ (BDC = 1,4-benzenedicarboxylate) showed that there are two distinct binding sites attributed to hydrogen binding to Zn (II) metal centers and the BDC linkers [134]. It was reported that polarisable linkers promote H_2 uptake through improved interactions with H_2 molecules at higher surface coverage [135]. It was shown that the heat of H_2 adsorption decreased gradually from 7.1 kJ mol^{-1} at zero surface coverage to 4.5 kJ mol^{-1} at 7 mg g^{-1} (7 wt%) surface coverage for MOF-326, which has positively charged B and negatively charged N in the linkers. The polarised B-N bonds seemed to enhance the binding of H_2 to the MOF in comparison with simple aromatic units of other MOFs [135]. Nevertheless, aromatic rings in organic ligands can enhance the interactions between H_2 molecules and the framework. In particular, the presence of tetrazine rings can result in an even stronger interaction between H_2 molecules and the framework as tetrazine rings contain more electrons to generate an electron-rich conjugated π system for improved interaction with hydrogen [136]. Wang et al. [137] demonstrated that incorporation of aromatic moieties in IRMOF-3 (IR = isoreticular) positively influenced the H_2 binding capacity as a result of specific interactions between H_2 molecules and the added phenyl groups. Three isostructural MOFs denoted as NOTT-113, NOTT-114 and NOTT-115 (NOTT = Nottingham) with the same cuboctahedral cage structure constructed from 24 isophthalates from the ligands and 12 $\{Cu_2(RCOO)_4\}$ paddlewheel moieties were investigated [138]. The MOFs differed only in the functionality of the central core of the hexacarboxylate ligands with trimethylphenyl, phenylamine and triphenylamine moieties in NOTT-113, NOTT-114 and NOTT-115, respectively. Amongst the three MOFs, NOTT-115 displayed the highest heat of adsorption, indicating that functionalisation of the cage walls with more aromatic rings (*i.e.* triphenylamine) led to favourable interactions between H_2 and the framework. Conversely, amine functionalisation of the ligand core in NOTT-114 weakened the H_2–framework interaction.

One of the major challenges associated with MOFs is their low hydrogen storage capacity at ambient temperature. Although MOFs display high hydrogen storage capacities at cryogenic temperatures (for example, up to 9 wt% was reported experimentally at 56 bar and 77 K for NU-100; NU = Northwestern University [139]), at ambient temperature hydrogen storage is typically less than 1 wt% because of the weak binding (4–12 kJ mol^{-1} [118, 119] between hydrogen and the MOFs. This weak binding involves van der Waals forces that are responsible for physisorption leading to low hydrogen storage capacities at ambient temperature. In one theoretical study, it was predicted that the material should have an isosteric heat of adsorption of 15.1 kJ mol^{-1} in order to store hydrogen at ambient temperature and approximately 30 bar, and release at about 1.5 bar [140]. In another theoretical study on the effect of heat of adsorption on hydrogen storage and delivery between 1.5–120 bar, it was shown that the optimal isosteric heat of hydrogen adsorption is about 20 kJ mol^{-1} [141]. Therefore, it is crucial to increase the interactions between hydrogen and MOFs to foster their applicability in practical hydrogen storage systems. Even though theoretical calculations have predicted enhanced interactions of hydrogen with MOFs and high hydrogen storage capacity at room temperature when MOFs are doped with Li^+ ion, these predictions have not been confirmed by experiments [124–126].

An approach that has been debated in the literature for enhancement of hydrogen uptake at room temperature is 'spillover'. Hydrogen spillover is a well-documented phenomenon in catalysis. It involves the dissociation of hydrogen molecules into atoms on a supported metal surface followed by migration of the hydrogen atoms to the support also known as primary receptor. The hydrogen atoms may further migrate to a secondary receptor and this movement may be facilitated by constructing a bridge between the two receptors [142]. Hydrogen spillover has been investigated for improvement of hydrogen storage by incorporation of noble metals (such as Pt and Pd) in MOFs [143–147]. Despite earlier reports of encouraging hydrogen storage capacities for MOFs by spillover at room temperature ambiguity still exists about the mechanism involved in hydrogen storage enhancement from/by spillover. While a massive improvement in hydrogen uptake was reported in some studies no enhancement as a result of spillover was reported in other studies. As a result spillover for hydrogen storage in MOFs has been a subject of debate [148–151]. It is therefore important to exercise caution before attributing hydrogen storage in MOFs to spillover from a noble metal.

9.4.2.2 Hosts for Chemical Hydrogen Storage Materials

An alternative materials-based storage method to adsorptive hydrogen storage is chemical hydrogen storage. In this approach, hydrogen is stored via formation of chemical bonds such as in complex hydrides. The high hydrogen storage capacities of these materials make them particularly attractive for hydrogen storage (e.g. NH_3BH_3 and $LiBH_4$ have theoretical hydrogen storage capacities of 19.6 and 18.3 wt%, respectively). However, the poor kinetics of hydrogen uptake and release and/or unfavourable thermodynamics for these materials presents severe drawbacks [152, 153]. For instance, materials such as $LiBH_4$ and MgH_2 are too stable while others like AlH_3 are known to be too unstable for practical applications. In addition, materials such as MgH_2 are also known to possess sluggish kinetics. Nanoconfinement, a technique for compartmentalising nanoparticles within porous scaffolds, is a promising approach to enhance the kinetics and thermodynamic properties of chemical hydrogen storage materials [154]. Nanoconfinement is attractive due to the large atomic population at the particle surface that leads to an increase in surface energy. Such increase in the ratio between the surface and bulk atoms has been reported to afford higher surface area and also lead to enhanced surface reaction kinetics, particularly those involving diffusion within the particles [155]. MOFs have been employed as scaffolds for nanoconfining a range of chemical hydrogen storage materials. Their nano-sized pores, tunability and crystalline nature make them highly attractive for this purpose. The imposed nanostructural geometry of the hydrides in the scaffold matrix apparently generates many defects and specific interactions between the metal-hydrogen bond and the nanoporous walls may favour hydrogen release [156]. The inclusion of chemical hydrogen storage materials into the pores of MOFs may not only improve the kinetics, thermodynamics and reversibility of hydrogen desorption,

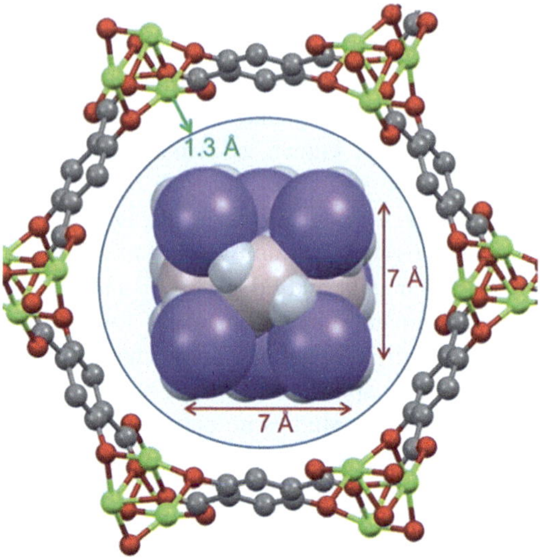

Fig. 9.10 Schematic representation of the $(NaAlH_4)_8$ clusters inside the MOF-74(Mg) pores. The space inside the sphere corresponds to the available space outside of magnesium van der Waals radius (1.3 Å). Reprinted with permission from [161]. Copyright (2012) American Chemical Society

but may also prevent the release of unwanted gases, making the materials more practical for hydrogen storage [157–159].

Recently, Sun et al. [160] reported that when a Cu-based MOF, HKUST-1, was used as the host for loading $LiBH_4$, redox reactions occurred between $LiBH_4$ and Cu–O units causing dehydrogenation to occur at a much lower temperature. In fact, dehydrogenation of $LiBH_4$@Cu-MOF began at about 60 °C, which is substantially lower than the temperature for pristine $LiBH_4$ (380 °C). Remarkably, during the loading of $LiBH_4$ into the hydrated HKUST-1, the coordinated water molecules present in HKUST-1 could react with $LiBH_4$ to release H_2 at room temperature. In another report, Bhakta et al. [158] incorporated nanoclusters of $NaAlH_4$ (4 wt%) in HKUST-1. Nanoconfinement in this manner led to faster H_2 desorption kinetics, as such, decomposition occurred at about 100 °C lower for $NaAlH_4$@HKUST-1 than bulk $NaAlH_4$. Bulk $NaAlH_4$ released 70 % of its hydrogen at about 250 °C while the nanoconfined $NaAlH_4$ began to release hydrogen at 70 °C, with 80 % of its hydrogen released at 155 °C. Therefore, $NaAlH_4$ was destabilised in terms of H_2 desorption by nanoscaling through confinement in the MOF material. Reversible hydrogen storage by $NaAlH_4$ confined within the pores of Ti-functionalised MOF-74(Mg) was later reported [161]. Nanoconfinement of $NaAlH_4$ within the MOF pores (Fig. 9.10) led to a modification of the decomposition pathway of $NaAlH_4$. Both Ti-doped and undoped nano-$NaAlH_4$@MOF-74(Mg) started to release hydrogen at a much lower temperature of about 50 °C, which was roughly 100 °C lower than the onset desorption temperature for bulk $NaAlH_4$. The presence of Ti catalyst enabled an almost total reversibility of rehydrogenation.

Li et al. [159] used an yttrium-based MOF to nanoconfine NH_3BH_3. Enhanced H_2 release kinetics and lower H release temperature were reported. Nanoconfined NH_3BH_3 started to desorb hydrogen at 50 °C reaching a peak at 85 °C, which was 30 °C lower than that of neat NH_3BH_3. At 85 °C nanoconfined NH_3BH_3 desorbed

8.0 wt% hydrogen within 10 min and 11 wt% within 3 h. Meanwhile, no hydrogen desorption occurred for neat NH_3BH_3 at this temperature. At 95 °C, H_2 desorption from nanoconfined NH_3BH_3 even reached 10.2 wt% in only 10 min and 13 wt% in 3 h. Moreover, the release of NH_3 alongside H_2 was eliminated due to the interaction between open Y^{3+} sites and NH_3BH_3. In another report, Wahab et al. [162] also pointed out that the formation of ammonia in nanoconfined NH_3BH_3 could be eliminated if there are metallic sites in the host material such as MOF. Clean H_2 generation and enhanced kinetics has further been demonstrated for NH_3BH_3 nanoconfined in Fe-MIL-53 [163]. Gadipelli et al. [164] examined nanoconfinement of NH_3BH_3 within the pores of MOF-74(Mg). It was also revealed that nanoconfinement of NH_3BH_3 within the pores of MOF-74(Mg) led to significant enhancement of the dehydrogenation kinetics at temperatures below 100 °C. The release of unwanted by-products (*i.e.* ammonia, borazine and diborane) was also prevented. In a recent investigation, Lim et al. [157] reported the incorporation of magnesium nanocrystals (Mg NCs) in the pores of SNU-90', [Zn_4O $(atb)_2$] (SNU = Seoul National University; atb = aniline-2,4,6-tribenzoate). The study showed that the hybrid hydrogen storage material possessed both physisorption and chemisorption properties, displaying synergistic behaviour by increasing the isosteric heat of H_2 physisorption for the MOF, and decreasing the temperatures for H_2 chemisorption and desorption for magnesium. The thermal stability of MOFs is important for their application as hosts for chemical hydrogen storage materials. In this regard, the thermal stabilities of HKUST-1(Cu), MIL-53 (Al) and ZIF-8(Zn) in a hydrogen environment have been evaluated [165, 166].

9.5 Conclusions and Future Prospects

Fuel cells are an integral component in the hydrogen energy value chain serving as conversion devices to convert the chemical energy of hydrogen into electrical energy via electrochemical reactions. MOFs, a class of inorganic-organic hybrid porous crystalline materials, are attractive for a variety of applications due to their many unique features such as tunable porosity and functionality, extraordinarily high surface area and huge structural diversity as a result of an almost limitless possibility of combining metal centres and organic linkers. This chapter has reviewed MOFs as materials for fuel cell technologies including their application in fuel cell components (*i.e.* polymer electrolyte membrane and electrocatalysts) and enabling technologies for fuel cells (*i.e.* hydrogen production and storage).

Several properties of polymeric membranes are required to meet certain standards for application in fuel cells, such as high proton conduction, good strength (*i.e.* mechanical, chemical and thermal strength) and low gas permeability. Development of a robust inexpensive substitute to state-of-the-art membrane like Nafion is desirable. MOFs exhibiting significant proton conduction below 100 °C mediated by water where the hydrogen-bonding networks provide a proton-conduction pathway, have been explored in many studies. Proton conductivity in these MOFs relates directly to relative humidity. In order for MOFs to act as proton conductors for practical applications, it is desirable that the MOFs function at relatively high operating temperatures (>100 °C) and under anhydrous conditions. In this light, guest molecules such as imidazole, triazole, histamine and ionic liquids have been incorporated in MOFs and these molecules play a key role in attaining high proton conductivity in the MOFs. Significant progress has been made in proton-conducting MOFs but there are still challenges that need to be overcome and these provide opportunities for future research. Proton-conducting MOFs with proton conductivities matching that of Nafion have been reported. Further research on the long-term performance of proton-conducting MOFs is warranted for their practical application in fuel cells. The flexibility in design, and tunability of pore size, functionality, chemical and thermal stability, will all play a role in the development of MOFs for use as proton-conducting polymer electrolyte membranes.

One of the barriers to overcome before PEMFC can be properly commercialised is associated with the electrocatalyst for the ORR. For operation with pure hydrogen and air, Pt is the most active material that is currently employed. The use of non-PGMs as catalysts for the ORR will reduce the cost but the cell performance will need to be improved. Although there have been limited reports on the applicability of MOFs for the ORR a few studies have examined the direct use of MOFs or MOF composites as electrocatalysts for the ORR. These studies reported interesting catalytic activity of these materials for the ORR. However, there is need to investigate a wider range of MOFs for this role before any concrete conclusions can be made about their direct applicability as electrocatalysts for the ORR. It seems that MOFs will have more potential as precursors or sacrificial templates for electrocatalysts for the ORR. The studies that have been carried out using this approach show promise. In this regard, the surface area and porosity of the MOF will play a role in determining the properties of the derived catalysts. A wide variety of MOFs with pores in the microporous to mesoporous range should be exploited for this purpose, and the activity, stability and durability of the derived catalyst as well as potential for scalability are all issues that will need to be addressed.

Hydrogen production and storage are known to be key enabling technologies for fuel cells. Without these technologies the transition to the Hydrogen Economy cannot be fully realised. For hydrogen generation the modifiable pore size, surface area, metal centres and organic linkers of MOFs are major contributing factors. The role of MOFs in hydrogen production has been demonstrated through photocatalytic splitting of water and hydrolysis of chemical hydrides such as $NaBH_4$ and NH_3BH_3. For photocatalytic hydrogen generation, when catalysis occurs at the inorganic cluster the reported catalytic activities are low, but when the MOF is used as a scaffold or for light harvesting higher catalytic performances are realised. Hydrolysis of chemical hydrides is advantageous over pyrolysis as production of unwanted gases alongside hydrogen is eliminated. While significant hydrogen generation has been achieved by both photocatalytic water splitting and hydrolysis of chemical hydrides, it still remains a challenge to improve the catalytic activity of MOFs for hydrogen production. By appropriate design of active sites and corresponding functionalities MOFs are sure to make their mark in hydrogen generation.

In the context of hydrogen storage, MOFs have been employed for adsorptive hydrogen storage and as hosts for chemical hydrides. For adsorptive hydrogen storage, the storage capacities of MOFs at cryogenic temperatures are quite promising especially at high pressures. However, as the temperature increases the capacities decrease. Various structural and compositional factors influence hydrogen storage in MOFs such as surface area, pore volume, pore size, catenation, open metal sites, dopant cations and ligand functionalisation. Despite the encouraging storage capacities of MOFs at cryogenic temperatures the main challenge facing MOFs for practical applications is how to enhance hydrogen adsorption at ambient temperature. Further research should focus on optimising existing MOFs through various modifications and functionalisation or even formation of MOF composites to enhance storage capacities at ambient temperatures. The potential of MOFs to act as host for chemical hydrides offers the benefits of reduced dehydrogenation temperature and enhanced hydrogen release kinetics for the hydride, as well as suppression of unwanted gaseous products. The design of suitable MOFs with high thermal stability for this purpose is one hurdle that will need to be overcome.

In summary, MOFs are receiving increasing interest in fuel cells and their enabling technologies. In this regard, significant progress has been made but many challenges are yet to be overcome. The key to using MOFs in real fuel cell applications may lie in modification of existing MOFs to obtain desirable properties or synthesis of novel MOFs with tailored properties. With the many attributes of MOFs such as the vast structural diversity, tunability of pore size and surface functionality, extraordinary porosity and surface area, and their rich chemistry, it can be said that the future of MOFs in fuel cell technologies is certainly promising.

References

1. Steele BCH, Heinzel A (2001) Materials for fuel-cell technologies. Nature 414:345–352
2. Bashyam R, Zelenay P (2006) A class of non-precious metal composite catalysts for fuel cells. Nature 443:63–66
3. Kamarudin MZF, Kamarudin SK, Masdar MS, Daud WRW (2013) Review: direct ethanol fuel cells. Int J Hydrogen Energy 38:9438–9453
4. Wee J (2007) Applications of proton exchange membrane fuel cell systems. Renew Sustain Energy Rev 11:1720–1738
5. Xu H, Kunz R, Fenton JM (2007) Investigation of platinum oxidation in PEM fuel cells at various relative humidities. Electrochem Solid State Lett 10:B1–B5
6. Gasteiger HA, Kocha SS, Sompalli B, Wagner FT (2005) Activity benchmarks and requirements for Pt, Pt-alloy, and non-Pt oxygen reduction catalysts for PEMFCs. Appl Catal B Environ 56:9–35
7. Ramaswamy P, Wong NE, Shimizu GKH (2014) MOFs as proton conductors-challenges and opportunities. Chem Soc Rev 43:5913–5932
8. Schlapbach L, Züttel A (2001) Hydrogen-storage materials for mobile applications. Nature 414:353–358

9. Ni M, Leung MKH, Leung DYC, Sumathy K (2007) A review and recent developments in photocatalytic water-splitting using TiO_2 for hydrogen production. Renew Sustain Energy Rev 11:401–425
10. Eberle U, Müller B, Von Helmolt R (2012) Fuel cell electric vehicles and hydrogen infrastructure: status 2012. Energy Environ Sci 5:8780–8798
11. Mandal TK, Gregory DH (2009) Hydrogen storage materials: present scenarios and future directions. Annu Rep Prog Chem Sect A 105:21–54
12. McWhorter S, Read C, Ordaz G, Stetson N (2011) Materials-based hydrogen storage: attributes for near-term, early market PEM fuel cells. Curr Opin Solid State Mater Sci 15:29–38
13. Langmi HW, McGrady GS (2008) Ternary nitrides for hydrogen storage: Li-B-N, Li-Al-N and Li-Ga-N systems. J Alloys Compd 466:287–292
14. Ramirez-Cuesta AJ, Mitchell PCH, Ross DK, Georgiev PA, Anderson PA, Langmi HW, Book D (2007) Dihydrogen in zeolite CaX-An inelastic neutron scattering study. J Alloys Compd 446–447:393–396
15. Farha OK, Eryazici I, Jeong NC, Hauser BG, Wilmer CE, Sarjeant AA, Snurr RQ, Nguyen ST, Yazaydin AO, Hupp JT (2012) Metal-organic framework materials with ultrahigh surface areas: is the sky the limit? J Am Chem Soc 134:15016–15021
16. Suh MP, Park HJ, Prasad TK, Lim D (2012) Hydrogen storage in metal-organic frameworks. Chem Rev 112:782–835
17. Li J, Sculley J, Zhou H (2012) Metal-organic frameworks for separations. Chem Rev 112:869–932
18. Yoon M, Srirambalaji R, Kim K (2012) Homochiral metal-organic frameworks for asymmetric heterogeneous catalysis. Chem Rev 112:1196–1231
19. Kreno LE, Leong K, Farha OK, Allendorf M, Van Duyne RP, Hupp JT (2012) Metal-organic framework materials as chemical sensors. Chem Rev 112:1105–1125
20. Horcajada P, Chalati T, Serre C, Gillet B, Sebrie C, Baati T, Eubank JF, Heurtaux D, Clayette P, Kreuz C, Chang J, Hwang YK, Marsaud V, Bories P, Cynober L, Gil S, Férey G, Couvreur P, Gref R (2010) Porous metal-organic-framework nanoscale carriers as a potential platform for drug delivery and imaging. Nat Mater 9:172–178
21. Kurmoo M (2009) Magnetic metal-organic frameworks. Chem Soc Rev 38:1353–1379
22. Morozan A, Jaouen F (2012) Metal organic frameworks for electrochemical applications. Energy Environ Sci 5:9269–9290
23. Li S, Xu Q (2013) Metal-organic frameworks as platforms for clean energy. Energy Environ Sci 6:1656–1683
24. Ren Y, Chia GH, Gao Z (2013) Metal-organic frameworks in fuel cell technologies. Nano Today 8:577–597
25. Furukawa H, Cordova KE, O'Keeffe M, Yaghi OM (2013) The chemistry and applications of metal-organic frameworks. Science 341:1230444-1–1230444-1
26. Millward AR, Yaghi OM (2005) Metal-organic frameworks with exceptionally high capacity for storage of carbon dioxide at room temperature. J Am Chem Soc 127:17998–17999
27. Cook TR, Zheng Y, Stang PJ (2013) Metal-organic frameworks and self-assembled supramolecular coordination complexes: comparing and contrasting the design, synthesis, and functionality of metal-organic materials. Chem Rev 113:734–777
28. Yaghi OM, O'Keeffe M, Ockwig NW, Chae HK, Eddaoudi M, Kim J (2003) Reticular synthesis and the design of new materials. Nature 423:705–714
29. Moulton B, Zaworotko MJ (2001) From molecules to crystal engineering: supramolecular isomerism and polymorphism in network solids. Chem Rev 101:1629–1658
30. O'Keeffe M, Eddaoudi M, Li H, Reineke T, Yaghi OM (2000) Frameworks for extended solids: geometrical design principles. J Solid State Chem 152:3–20
31. Eddaoudi M, Moler DB, Li H, Chen B, Reineke TM, O'Keeffe M, Yaghi OM (2001) Modular chemistry: secondary building units as a basis for the design of highly porous and robust metal-organic carboxylate frameworks. Acc Chem Res 34:319–330

32. Perry VIJJ, Perman JA, Zaworotko MJ (2009) Design and synthesis of metal-organic frameworks using metal-organic polyhedra as supermolecular building blocks. Chem Soc Rev 38:1400–1417
33. Yaghi OM, Li H, Davis C, Richardson D, Groy TL (1998) Synthetic strategies, structure patterns, and emerging properties in the chemistry of modular porous solids. Acc Chem Res 31:474–484
34. Collins CS, Sun D, Liu W, Zuo J, Zhou H (2008) Reaction-condition-controlled formation of secondary-building-units in three cadmium metal-organic frameworks with an orthogonal tetrakis(tetrazolate) ligand. J Mol Struct 890:163–169
35. Eddaoudi M, Kim J, Rosi N, Vodak D, Wachter J, O'Keeffe M, Yaghi OM (2002) Systematic design of pore size and functionality in isoreticular MOFs and their application in methane storage. Science 295:469–472
36. Cavka JH, Jakobsen S, Olsbye U, Guillou N, Lamberti C, Bordiga S, Lillerud KP (2008) A new zirconium inorganic building brick forming metal organic frameworks with exceptional stability. J Am Chem Soc 130:13850–13851
37. Surblé S, Serre C, Mellot-Draznieks C, Millange F, Férey G (2006) A new isoreticular class of metal-organic-frameworks with the MIL-88 topology. Chem Commun 3:284–286
38. Burrows AD, Frost CG, Mahon MF, Richardson C (2008) Post-synthetic modification of tagged metal-organic frameworks. Angew Chem Int Ed 47:8482–8486
39. Kleist W, Jutz F, Maciejewski M, Baiker A (2009) Mixed-linker metal-organic frameworks as catalysts for the synthesis of propylene carbonate from propylene oxide and CO_2. Eur J Inorg Chem 24:3552–3561
40. Deng H, Doonan CJ, Furukawa H, Ferreira RB, Towne J, Knobler CB, Wang B, Yaghi OM (2010) Multiple functional groups of varying ratios in metal-organic frameworks. Science 327:846–850
41. O'Keeffe M, Peskov MA, Ramsden SJ, Yaghi OM (2008) The Reticular Chemistry Structure Resource (RCSR) database of, and symbols for, crystal nets. Acc Chem Res 41:1782–1789
42. Tanabe KK, Cohen SM (2011) Postsynthetic modification of metal-organic frameworks—a progress report. Chem Soc Rev 40:498–519
43. Wang Z, Cohen SM (2007) Postsynthetic covalent modification of a neutral metal-organic framework. J Am Chem Soc 129:12368–12369
44. Hurd JA, Vaidhyanathan R, Thangadurai V, Ratcliffe CI, Moudrakovski IL, Shimizu GKH (2009) Anhydrous proton conduction at 150°C in a crystalline metal-organic framework. Nat Chem 1:705–710
45. Yoon M, Suh K, Natarajan S, Kim K (2013) Proton conduction in metal-organic frameworks and related modularly built porous solids. Angew Chem Int Ed 52:2688–2700
46. Umeyama D, Horike S, Inukai M, Itakura T, Kitagawa S (2012) Inherent proton conduction in a 2D coordination framework. J Am Chem Soc 134:12780–12785
47. Kanaizuka K, Iwakiri S, Yamada T, Kitagawa H (2010) Design and characterization of a polarized coordination polymer of a zinc(II) biphenyldicarboxylate bearing a sulfone group. Chem Lett 39:28–29
48. Morikawa S, Yamada T, Kitagawa H (2009) Crystal structure and proton conductivity of a one-dimensional coordination polymer, {Mn(DHBQ)(H2O)2}. Chem Lett 38:654–655
49. Yamada T, Sadakiyo M, Kitagawa H (2009) High proton conductivity of one-dimensional ferrous oxalate dihydrate. J Am Chem Soc 131:3144–3145
50. Dey C, Kundu T, Banerjee R (2012) Reversible phase transformation in proton conducting Strandberg-type POM based metal organic material. Chem Commun 48:266–268
51. Yamada T, Morikawa S, Kitagawa H (2010) Structures and proton conductivity of one-dimensional $M(dhbq)\cdot nH_2O$ ($M = Mg$, Mn, Co, Ni, and Zn, H2(dhbq) = 2,5-dihydroxy-1,4-benzoquinone) promoted by connected hydrogen-bond networks with absorbed water. Bull Chem Soc Jpn 83:42–48
52. Kanda S, Yamashita K, Ohkawa K (1979) A proton conductive coordination polymer. I. [N, N′-Bis(2 hydroxyethyl)dithiooxamido]copper(II). Bull Chem Soc Jpn 52:3296–3301

53. Nagao Y, Ikeda R, Kanda S, Kubozono Y, Kitagawa H (2002) Complex-plane impedance study on a hydrogen-doped copper coordination polymer: N,N′-bis-(2-hydroxyethyl) dithiooxamidato-copper(II). Mol Cryst Liq Cryst Sci Technol Sect A Mol Cryst Liq Cryst 379:89–94
54. Nagao Y, Ikeda R, Iijima K, Kubo T, Nakasuji K, Kitagawa H (2003) A new proton-conductive copper coordination polymer, (HOC3H6)2dtoaCu (dtoa = dithiooxamide). Synth Met 135–136:283–284
55. Nagao Y, Fujishima M, Ikeda R, Kanda S, Kitagawa H (2003) Highly proton-conductive copper coordination polymers. Synth Met 133–134:431–432
56. Kitagawa H, Nagao Y, Fujishima M, Ikeda R, Kanda S (2003) Highly proton-conductive copper coordination polymer, H2dtoaCu (H2dtoa = dithiooxamide anion). Inorg Chem Commun 6:346–348
57. Sadakiyo M, Yamada T, Kitagawa H (2009) Rational designs for highly proton-conductive metal-organic frameworks. J Am Chem Soc 131:9906–9907
58. Okawa H, Shigematsu A, Sadakiyo M, Miyagawa T, Yoneda K, Ohba M, Kitagawa H (2009) Oxalate-bridged bimetallic complexes {NH(prol)3}[MCr(ox)3] (M = MnII, FeII, CoII; NH (prol)3^{+} = tri(3-hydroxypropyl)ammonium) exhibiting coexistent ferromagnetism and proton conduction. J Am Chem Soc 131:13516–13522
59. Sadakiyo M, Okawa H, Shigematsu A, Ohba M, Yamada T, Kitagawa H (2012) Promotion of low-humidity proton conduction by controlling hydrophilicity in layered metal-organic frameworks. J Am Chem Soc 134:5472–5475
60. Taylor JM, Mah RK, Moudrakovski IL, Ratcliffe CI, Vaidhyanathan R, Shimizu GKH (2010) Facile proton conduction via ordered water molecules in a phosphonate metal-organic framework. J Am Chem Soc 132:14055–14057
61. Kundu T, Sahoo SC, Banerjee R (2012) Alkali earth metal (Ca, Sr, Ba) based thermostable metal-organic frameworks (MOFs) for proton conduction. Chem Commun 48:4998–5000
62. Pardo E, Train C, Gontard G, Boubekeur K, Fabelo O, Liu H, Dkhil B, Lloret F, Nakagawa K, Tokoro H, Ohkoshi S, Verdaguer M (2011) High proton conduction in a chiral ferromagnetic metal-organic quartz-like framework. J Am Chem Soc 133:15328–15331
63. Sahoo SC, Kundu T, Banerjee R (2011) Helical water chain mediated proton conductivity in homochiral metal-organic frameworks with unprecedented zeolitic unh -topology. J Am Chem Soc 133:17950–17958
64. Shigematsu A, Yamada T, Kitagawa H (2011) Wide control of proton conductivity in porous coordination polymers. J Am Chem Soc 133:2034–2036
65. Jeong NC, Samanta B, Lee CY, Farha OK, Hupp JT (2012) Coordination-chemistry control of proton conductivity in the iconic metal-organic framework material HKUST-1. J Am Chem Soc 134:51–54
66. Sen S, Nair NN, Yamada T, Kitagawa H, Bharadwaj PK (2012) High proton conductivity by a metal-organic framework incorporating Zn 8O clusters with aligned imidazolium groups decorating the channels. J Am Chem Soc 134:19432–19437
67. Wei M, Wang X, Duan X (2013) Crystal structures and proton conductivities of a MOF and Two POM-MOF composites based on CuII ions and 2,2′-bipyridyl-3,3′-dicarboxylic acid. Chem Eur J 19:1607–1616
68. Costantino F, Donnadio A, Casciola M (2012) Survey on the phase transitions and their effect on the ion-exchange and on the proton-conduction properties of a flexible and robust Zr phosphonate coordination polymer. Inorg Chem 51:6992–7000
69. Colodrero RMP, Olivera-Pastor P, Losilla ER, Hernández-Alonso D, Aranda MAG, Leon-Reina L, Rius J, Demadis KD, Moreau B, Villemin D, Palomino M, Rey F, Cabeza A (2012) High proton conductivity in a flexible, cross-linked, ultramicroporous magnesium tetraphosphonate hybrid framework. Inorg Chem 51:7689–7698
70. Colodrero RMP, Papathanasiou KE, Stavgianoudaki N, Olivera-Pastor P, Losilla ER, Aranda MAG, León-Reina L, Sanz J, Sobrados I, Choquesillo-Lazarte D, García-Ruiz JM, Atienzar P, Rey F, Demadis KD, Cabeza A (2012) Multifunctional luminescent and

proton-conducting lanthanide carboxyphosphonate open-framework hybrids exhibiting crystalline-to-amorphous- to-crystalline transformations. Chem Mater 24:3780–3792
71. Bureekaew S, Horike S, Higuchi M, Mizuno M, Kawamura T, Tanaka D, Yanai N, Kitagawa S (2009) One-dimensional imidazole aggregate in aluminium porous coordination polymers with high proton conductivity. Nat Mater 8:831–836
72. Chen W, Xu H, Zhuang G, Long L, Huang R, Zheng L (2011) Temperature-dependent conductivity of $Emim^+$ ($Emim^+ =$ 1-ethyl-3-methyl imidazolium) confined in channels of a metal-organic framework. Chem Commun 47:11933–11935
73. Umeyama D, Horike S, Inukai M, Hijikata Y, Kitagawa S (2011) Confinement of mobile histamine in coordination nanochannels for fast proton transfer. Angew Chem Int Ed 50:11706–11709
74. Inukai M, Horike S, Umeyama D, Hijikata Y, Kitagawa S (2012) Investigation of post-grafted groups of a porous coordination polymer and its proton conduction behavior. Dalton Trans 41:13261–13263
75. Horike S, Umeyama D, Inukai M, Itakura T, Kitagawa S (2012) Coordination-network-based ionic plastic crystal for anhydrous proton conductivity. J Am Chem Soc 134:7612–7615
76. Mao J, Yang L, Yu P, Wei X, Mao L (2012) Electrocatalytic four-electron reduction of oxygen with copper(II)-based metal-organic frameworks. Electrochem Commun 19:29–31
77. Jahan M, Bao Q, Loh KP (2012) Electrocatalytically active graphene-porphyrin MOF composite for oxygen reduction reaction. J Am Chem Soc 134:6707–6713
78. Ma S, Goenaga GA, Call AV, Liu D (2011) Cobalt imidazolate framework as precursor for oxygen reduction reaction electrocatalysts. Chem Eur J 17:2063–2067
79. Palaniselvam T, Biswal BP, Banerjee R, Kurungot S (2013) Zeolitic imidazolate framework (ZIF)-derived, hollow-core, nitrogen-doped carbon nanostructures for oxygen-reduction reactions in PEFCs. Chem Eur J 19:9335–9342
80. Proietti E, Jaouen F, Lefèvre M, Larouche N, Tian J, Herranz J, Dodelet J (2011) Iron-based cathode catalyst with enhanced power density in polymer electrolyte membrane fuel cells. Nat Commun 2:416
81. Zhao D, Shui J, Chen C, Chen X, Reprogle BM, Wang D, Liu D (2012) Iron imidazolate framework as precursor for electrocatalysts in polymer electrolyte membrane fuel cells. Chem Sci 3:3200–3205
82. Zhao D, Shui J, Grabstanowicz LR, Chen C, Commet SM, Xu T, Lu J, Liu D (2014) Highly efficient non-precious metal electrocatalysts prepared from one-pot synthesized zeolitic imidazolate frameworks. Adv Mater 26:1093–1097
83. Lefèvre M, Dodelet JP (2012) Recent advances in non-precious metal electrocatalysts for oxygen reduction in PEM fuel cells. ECS Trans 45:35–44
84. Yang L, Kinoshita S, Yamada T, Kanda S, Kitagawa H, Tokunaga M, Ishimoto T, Ogura T, Nagumo R, Miyamoto A, Koyama M (2010) A metal-organic framework as an electrocatalyst for ethanol oxidation. Angew Chem Int Ed 49:5348–5351
85. Ishimoto T, Ogura T, Koyama M, Yang L, Kinoshita S, Yamada T, Tokunaga M, Kitagawa H (2013) A key mechanism of ethanol electrooxidation reaction in a noble-metal-free metal-organic framework. J Phys Chem C 117:10607–10614
86. Llabrés i Xamena FX, Corma A, Garcia H (2007) Applications for Metal-Organic Frameworks (MOFs) as quantum dot semiconductors. J Phys Chem C 111:80–85
87. Alvaro M, Carbonell E, Ferrer B, Llabrés I, Xamena FX, Garcia H (2007) Semiconductor behavior of a metal-organic framework (MOF). Chem Eur J 13:5106–5112
88. Fox MA, Dulay MT (1993) Heterogeneous photocatalysis. Chem Rev 93:341–357
89. Kataoka Y, Sato K, Miyazaki Y, Masuda K, Tanaka H, Naito S, Mori W (2009) Photocatalytic hydrogen production from water using porous material [Ru2(p-BDC)2]n. Energy Environ Sci 2:397–400
90. Kataoka Y, Miyazaki Y, Sato K, Saito T, Nakanishi Y, Kiatagwa Y, Kawakami T, Okumura M, Yamaguchi K, Mori W (2011) Modification of MOF catalysts by manipulation of counter-ions: experimental and theoretical studies of photochemical hydrogen production

from water over microporous diruthenium (II, III) coordination polymers. Supramol Chem 23:287–296
91. Sato K, Kataoka Y, Mori W (2012) Photochemical production of hydrogen from water using microporous porphyrin coordination lattices. J Nanosci Nanotechnol 12:585–590
92. Fateeva A, Chater PA, Ireland CP, Tahir AA, Khimyak YZ, Wiper PV, Darwent JR, Rosseinsky MJ (2012) A water-stable porphyrin-based metal-organic framework active for visible-light photocatalysis. Angew Chem Int Ed 51:7440–7444
93. Silva CG, Luz I, Llabrés I, Xamena FX, Corma A, García H (2010) Water stable Zr-benzenedicarboxylate metal-organic frameworks as photocatalysts for hydrogen generation. Chem Eur J 16:11133–11138
94. Wang C, Dekrafft KE, Lin W (2012) Pt nanoparticles@photoactive metal-organic frameworks: efficient hydrogen evolution via synergistic photoexcitation and electron injection. J Am Chem Soc 134:7211–7214
95. Zhou T, Du Y, Borgna A, Hong J, Wang Y, Han J, Zhang W, Xu R (2013) Post-synthesis modification of a metal-organic framework to construct a bifunctional photocatalyst for hydrogen production. Energy Environ Sci 6:3229–3234
96. He J, Wang J, Chen Y, Zhang J, Duan D, Wang Y, Yan Z (2014) A dye-sensitized Pt@UiO-66(Zr) metal-organic framework for visible-light photocatalytic hydrogen production. Chem Commun 50:7063–7066
97. Dekrafft KE, Wang C, Lin W (2012) Metal-organic framework templated synthesis of Fe_2O_3/TiO_2 nanocomposite for hydrogen production. Adv Mater 24:2014–2018
98. He J, Yan Z, Wang J, Xie J, Jiang L, Shi Y, Yuan F, Yu F, Sun Y (2013) Significantly enhanced photocatalytic hydrogen evolution under visible light over CdS embedded on metal-organic frameworks. Chem Commun 49:6761–6763
99. Li Q, Kim H (2012) Hydrogen production from $NaBH_4$ hydrolysis via Co-ZIF-9 catalyst. Fuel Process Technol 100:43–48
100. Bai Y, Wu C, Wu F, Yi B (2006) Carbon-supported platinum catalysts for on-site hydrogen generation from NaBH4 solution. Mater Lett 60:2236–2239
101. Li P, Aranishi K, Xu Q (2012) ZIF-8 immobilized nickel nanoparticles: highly effective catalysts for hydrogen generation from hydrolysis of ammonia borane. Chem Commun 48:3173–3175
102. Aijaz A, Karkamkar A, Choi YJ, Tsumori N, Rönnebro E, Autrey T, Shioyama H, Xu Q (2012) Immobilizing highly catalytically active Pt nanoparticles inside the pores of metal-organic framework: a double solvents approach. J Am Chem Soc 134:13926–13929
103. Li Y, Song P, Zheng J, Li X (2010) Promoted H_2 generation from NH_3BH_3 thermal dehydrogenation catalyzed by metal-organic framework based catalysts. Chem Eur J 16:10887–10892
104. Song P, Li Y, Li W, He B, Yang J, Li X (2011) A highly efficient Co(0) catalyst derived from metal-organic framework for the hydrolysis of ammonia borane. Int J Hydrogen Energy 36:10468–10473
105. Langmi HW, Ren J, North B, Mathe M, Bessarabov D (2014) Hydrogen storage in metal-organic frameworks: a review. Electrochim Acta 128:368–392
106. Lin X, Jia J, Zhao X, Thomas KM, Blake AJ, Walker GS, Champness NR, Hubberstey P, Schröder M (2006) High H_2 adsorption by coordination-framework materials. Angew Chem Int Ed 45:7358–7364
107. Lin X, Telepeni I, Blake AJ, Dailly A, Brown CM, Simmons JM, Zoppi M, Walker GS, Thomas KM, Mays TJ, Hubberstey P, Champness NR, Schröder M (2009) High capacity hydrogen adsorption in Cu(II) tetracarboxylate framework materials: the role of pore size, ligand functionalization, and exposed metal sites. J Am Chem Soc 131:2159–2171
108. Zhao D, Yuan D, Yakovenko A, Zhou H (2010) A NbO-type metal-organic framework derived from a polyyne-coupled di-isophthalate linker formed in situ. Chem Commun 46:4196–4198

109. Zhao X, Xiao B, Fletcher AJ, Thomas KM, Bradshaw D, Rosseinsky MJ (2004) Hysteretic adsorption and desorption of hydrogen by nanoporous metal-organic frameworks. Science 306:1012–1015
110. Wang Q, Johnson JK (1999) Molecular simulation of hydrogen adsorption in single-walled carbon nanotubes and idealized carbon slit pores. J Chem Phys 110:577–586
111. De La Casa-Lillo MA, Lamari-Darkrim F, Cazorla-Amorós D, Linares-Solano A (2002) Hydrogen storage in activated carbons and activated carbon fibers. J Phys Chem B 106:10930–10934
112. Batten SR, Robson R (1998) Interpenetrating nets: ordered, periodic entanglement. Angew Chem Int Ed 37:1460–1494
113. Chen B, Eddaoudi M, Hyde ST, O'Keeffe M, Yaghi OM (2001) Interwoven metal-organic framework on a periodic minimal surface with extra-large pores. Science 291:1021–1023
114. Ma S, Sun D, Ambrogio M, Fillinger JA, Parkin S, Zhou H (2007) Framework-catenation isomerism in metal-organic frameworks and its impact on hydrogen uptake. J Am Chem Soc 129:1858–1859
115. Ma S, Eckert J, Forster PM, Ji WY, Young KH, Chang J, Collier CD, Parise JB, Zhou H (2008) Further investigation of the effect of framework catenation on hydrogen uptake in metal-organic frameworks. J Am Chem Soc 130:15896–15902
116. Dinca M, Long JR (2007) High-enthalpy hydrogen adsorption in cation-exchanged variants of the microporous metal-organic framework Mn3[(Mn4CI) 3(BTT)8(CH3OH)10]2. J Am Chem Soc 129:11172–11176
117. Liu Y, Eubank JF, Cairns AJ, Eckert J, Kravtsov VC, Luebke R, Eddaoudi M (2007) Assembly of Metal-Organic Frameworks (MOFs) based on indium-trimer building blocks: a porous MOF with soc topology and high hydrogen storage. Angew Chem Int Ed 46:3278–3283
118. Tan C, Yang S, Champness NR, Lin X, Blake AJ, Lewis W, Schröder M (2011) High capacity gas storage by a 4,8-connected metal-organic polyhedral framework. Chem Commun 47:4487–4489
119. Chen B, Zhao X, Putkham A, Hong K, Lobkovsky EB, Hurtado EJ, Fletcher AJ, Thomas KM (2008) Surface interactions and quantum kinetic molecular sieving for H_2 and D_2 adsorption on a mixed metal-organic framework material. J Am Chem Soc 130:6411–6423
120. Dinca M, Dailly A, Liu Y, Brown CM, Neumann DA, Long JR (2006) Hydrogen storage in a microporous metal-organic framework with exposed Mn^{2+} coordination sites. J Am Chem Soc 128:16876–16883
121. Latroche M, Surblé S, Serre C, Mellot-Draznieks C, Llewellyn PL, Lee J, Chang J, Sung HJ, Férey G (2006) Hydrogen storage in the giant-pore metal-organic frameworks MIL-100 and MIL-101. Angew Chem Int Ed 45:8227–8231
122. Wang X, Shengqian M, Rauch K, Simmons JM, Yuan D, Wang X, Yildirim T, Cole WC, López JJ, De Meijere A, Zhou H (2008) Metal-organic frameworks based on double-bond-coupled Di-isophthalate linkers with high hydrogen and methane uptakes. Chem Mater 20:3145–3152
123. Yan Y, Telepeni I, Yang S, Lin X, Kockelmann W, Dailly A, Blake AJ, Lewis W, Walker GS, Allan DR, Barnett SA, Champness NR, Schröder M (2010) Metal-organic polyhedral frameworks: high H_2 adsorption capacities and neutron powder diffraction studies. J Am Chem Soc 132:4092–4094
124. Blomqvist A, Araújo CM, Srepusharawoot P, Ahuja R (2007) Li-decorated metal-organic framework 5: a route to achieving a suitable hydrogen storage medium. Proc Natl Acad Sci U S A 104:20173–20176
125. Sang SH, Goddard WA III (2007) Lithium-doped metal-organic frameworks for reversible H_2 storage at ambient temperature. J Am Chem Soc 129:8422–8423
126. Mavrandonakis A, Tylianakis E, Stubos AK, Froudakis GE (2008) Why Li doping in MOFs enhances H_2 storage capacity? A multi-scale theoretical study. J Phys Chem C 112:7290–7294

127. Mavrandonakis A, Klontzas E, Tylianakis E, Froudakis GE (2009) Enhancement of hydrogen adsorption in metal-organic frameworks by the incorporation of the sulfonate group and Li cations. A multiscale computational study. J Am Chem Soc 131:13410–13414
128. Klontzas E, Mavrandonakis A, Tylianakis E, Froudakis GE (2008) Improving hydrogen storage capacity of MOF by functionalization of the organic linker with lithium atoms. Nano Lett 8:1572–1576
129. Mulfort KL, Farha OK, Stern CL, Sarjeant AA, Hupp JT (2009) Post-synthesis alkoxide formation within metal-organic framework materials: a strategy for incorporating highly coordinatively unsaturated metal ions. J Am Chem Soc 131:3866–3868
130. Chu C, Chen J, Lee T (2012) Enhancement of hydrogen adsorption by alkali-metal cation doping of metal-organic framework-5. Int J Hydrogen Energy 37:6721–6726
131. Mulfort KL, Hupp JT (2008) Alkali metal cation effects on hydrogen uptake and binding in metal-organic frameworks. Inorg Chem 47:7936–7938
132. Himsl D, Wallacher D, Hartmann M (2009) Improving the hydrogen-adsorption properties of a hydroxy-modified MIL-53(A1) structural analogue by lithium doping. Angew Chem Int Ed 48:4639–4642
133. Mulfort KL, Hupp JT (2007) Chemical reduction of metal-organic framework materials as a method to enhance gas uptake and binding. J Am Chem Soc 129:9604–9605
134. Rosi NL, Eckert J, Eddaoudi M, Vodak DT, Kim J, O'Keeffe M, Yaghi OM (2003) Hydrogen storage in microporous metal-organic frameworks. Science 300:1127–1129
135. Tranchemontagne DJ, Park KS, Furukawa H, Eckert J, Knobler CB, Yaghi OM (2012) Hydrogen storage in new metal-organic frameworks. J Phys Chem C 116:13143–13151
136. Chang Z, Zhang D, Chen Q, Li R, Hu T, Bu X (2011) Rational construction of 3D pillared metal-organic frameworks: synthesis, structures, and hydrogen adsorption properties. Inorg Chem 50:7555–7562
137. Wang Z, Tanabe KK, Cohen SM (2010) Tuning hydrogen sorption properties of metal-organic frameworks by postsynthetic covalent modification. Chem Eur J 16:212–217
138. Yan Y, Blake AJ, Lewis W, Barnett SA, Dailly A, Champness NR, Schröder M (2011) Modifying cage structures in metal-organic polyhedral frameworks for H_2 storage. Chem Eur J 17:11162–11170
139. Farha OK, Yazaydin AÖ, Eryazici I, Malliakas CD, Hauser BG, Kanatzidis MG, Nguyen ST, Snurr RQ, Hupp JT (2010) De novo synthesis of a metal-organic framework material featuring ultrahigh surface area and gas storage capacities. Nat Chem 2:944–948
140. Bhatia SK, Myers AL (2006) Optimum conditions for adsorptive storage. Langmuir 22:1688–1700
141. Bae Y, Snurr RQ (2010) Optimal isosteric heat of adsorption for hydrogen storage and delivery using metal-organic frameworks. Micropor Mesopor Mater 132:300–303
142. Lachawiec AJ Jr, Qi G, Yang RT (2005) Hydrogen storage in nanostructured carbons by spillover: bridge-building enhancement. Langmuir 21:11418–11424
143. Cheon YE, Suh MP (2009) Enhanced hydrogen storage by palladium nanoparticles fabricated in a redox-active metal-organic framework. Angew Chem Int Ed 48:2899–2903
144. Li Y, Yang RT (2006) Hydrogen storage in metal-organic frameworks by bridged hydrogen spillover. J Am Chem Soc 128:8136–8137
145. Li Y, Yang RT (2006) Significantly enhanced hydrogen storage in metal-organic frameworks via spillover. J Am Chem Soc 128:726–727
146. Li Y, Yang RT (2008) Hydrogen storage in metal-organic and covalent-organic frameworks by spillover. AIChE J 54:269–279
147. Liu Y, Zeng J, Zhang J, Xu F, Sun LX (2007) Improved hydrogen storage in the modified metal-organic frameworks by hydrogen spillover effect. Int J Hydrogen Energy 32:4005–4010
148. Hirscher M (2010) Remarks about spillover and hydrogen adsorption—comments on the contributions of A.V. Talyzin and R.T. Yang. Micropor Mesopor Mater 135:209–210

149. Li YW, Wang L, Yang RT (2010) Response to "hydrogen adsorption in Pt catalyst/MOF-5 materials" by Luzan and Talyzin. Micropor Mesopor Mater 135:206–208
150. Luzan SM, Talyzin AV (2011) Comment to the "response to 'hydrogen adsorption in Pt catalyst/MOF-5 materials " by Li et al. Micropor Mesopor Mater 139:216–218
151. Szilágyi PA, Callini E, Anastasopol A, Kwakernaak C, Sachdeva S, Van De Krol R, Geerlings H, Borgschulte A, Züttel A, Dam B (2014) Probing hydrogen spillover in Pd@MIL-101(Cr) with a focus on hydrogen chemisorption. Phys Chem Chem Phys 16:5803–5809
152. Bérubé V, Radtke G, Dresselhaus M, Chen G (2007) Size effects on the hydrogen storage properties of nanostructured metal hydrides: a review. Int J Energy Res 31:637–663
153. Fichtner M (2011) Nanoconfinement effects in energy storage materials. Phys Chem Chem Phys 13:21186–21195
154. Rude LH, Nielsen TK, Ravnsbæk DB, Bösenberg U, Ley MB, Richter B, Arnbjerg LM, Dornheim M, Filinchuk Y, Besenbacher F, Jensen TR (2011) Tailoring properties of borohydrides for hydrogen storage: A review. Phys Status Solidi A Appl Mater Sci 208:1754–1773
155. Zlotea C, Latroche M (2013) Role of nanoconfinement on hydrogen sorption properties of metal nanoparticles hybrids. Colloids Surf A Physicochem Eng Asp 439:117–130
156. Wu H (2008) Strategies for the improvement of the hydrogen storage properties of metal hydride materials. ChemPhysChem 9:2157–2162
157. Lim D, Yoon JW, Ryu KY, Suh MP (2012) Magnesium nanocrystals embedded in a metal-organic framework: hybrid hydrogen storage with synergistic effect on physi- and chemisorption. Angew Chem Int Ed 51:9814–9817
158. Bhakta RK, Herberg JL, Jacobs B, Highley A, Behrens R Jr, Ockwig NW, Greathouse JA, Allendorf MD (2009) Metal-organic frameworks as templates for nanoscale NaAIH4. J Am Chem Soc 131:13198–13199
159. Li Z, Zhu G, Lu G, Qiu S, Yao X (2010) Ammonia borane confined by a metal-organic framework for chemical hydrogen storage: Enhancing kinetics and eliminating ammonia. J Am Chem Soc 132:1490–1491
160. Sun W, Li S, Mao J, Guo Z, Liu H, Dou S, Yu X (2011) Nanoconfinement of lithium borohydride in Cu-MOFs towards low temperature dehydrogenation. Dalton Trans 40:5673–5676
161. Stavila V, Bhakta RK, Alam TM, Majzoub EH, Allendorf MD (2012) Reversible hydrogen storage by NaAlH4 confined within a titanium-functionalized MOF-74(Mg) nanoreactor. ACS Nano 6:9807–9817
162. Wahab MA, Zhao H, Yao XD (2012) Nano-confined ammonia borane for chemical hydrogen storage. Front Chem Sci Eng 6:27–33
163. Srinivas G, Ford J, Zhou W, Yildirim T (2012) Zn-MOF assisted dehydrogenation of ammonia borane: enhanced kinetics and clean hydrogen generation. Int J Hydrogen Energy 37:3633–3638
164. Gadipelli S, Ford J, Zhou W, Wu H, Udovic TJ, Yildirim T (2011) Nanoconfinement and catalytic dehydrogenation of ammonia borane by magnesium-metal-organic-framework-74. Chem Eur J 17:6043–6047
165. Grzech A, Yang J, Dingemans TJ, Srinivasan S, Magusin PCMM, Mulder FM (2011) Irreversible high-temperature hydrogen interaction with the metal organic framework Cu3 (BTC)2. J Phys Chem C 115:21521–21525
166. Chen H, Wang L, Yang J, Yang RT (2013) Investigation on hydrogenation of metal-organic frameworks HKUST-1, MIL-53, and ZIF-8 by hydrogen spillover. J Phys Chem C 117:7565–7576

Chapter 10
Sonoelectrochemical Production of Fuel Cell Nanomaterials

Bruno G. Pollet and Petros M. Sakkas

10.1 Introduction

When oil, one of the most important energy sources in the history of mankind, was first discovered in Pennsylvania (USA) almost 150 years ago, the fuel cell had already been known for 20 years in 1839 by Sir William Grove who developed the first fuel cell based on reversing the electrolysis of water by accident [1]. However, the first operational commercial fuel cell was developed in 1950 by Francis Bacon at Cambridge University who demonstrated a 5 kW alkaline fuel cell (AFC) and in the 1970s International Fuel Cells developed a 12 kW AFC for NASA's space shuttle orbiter which provided continuous and reliable power.

Since the 1960's until today, USA, Canada and Japan have received large governmental R&D funding programmes (unlike the UK and Europe who have so far received little funding in comparison) to develop further fuel cell systems for stationary and transport applications [2]. Today, fuel cells are widely considered to be an efficient (offering a much higher energy density and energy efficiency compared to any other current energy storage devices), a non-polluting power source and therefore a promising energy device for the transport, mobile and stationary sectors [3]. However, the deployment of fuel cells has, so far, not been very successful due to its high cost and limited durability.

A fuel cell is an 'electrochemical' device operating at various temperatures (up to 1000 °C) that transforms the chemical energy of a fuel (hydrogen, methanol or natural gas etc) and an oxidant (air or pure oxygen) in the presence of a catalyst

B.G. Pollet (✉)
Eau2Energy, Nottingham NG14 6DX, England, UK
e-mail: info@eau2energy.com

P.M. Sakkas
School of Chemical Engineering, National Technical University of Athens,
Zografou Campus 15780, Greece

K.I. Ozoemena, S. Chen (eds.), *Nanomaterials for Fuel Cell Catalysis*,
Nanostructure Science and Technology, DOI 10.1007/978-3-319-29930-3_10

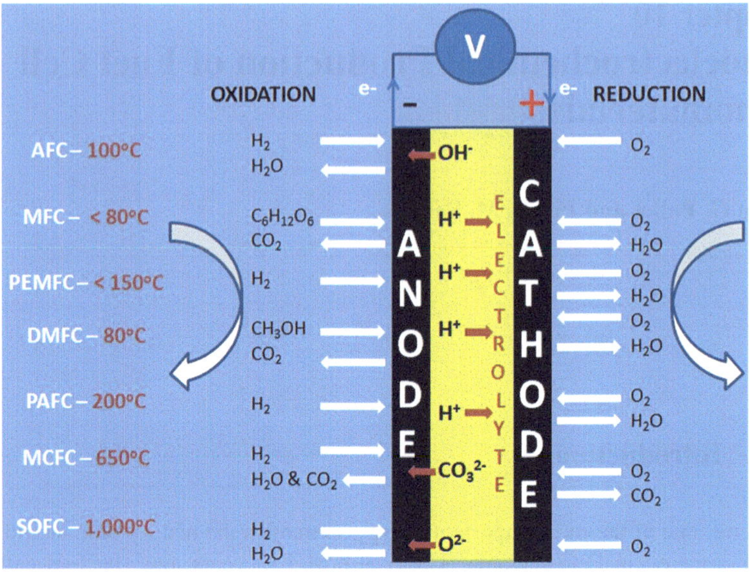

Fig. 10.1 Schematic of various fuel cells

into water, heat and electricity. Furthermore, the power generated by a fuel cell depends largely upon catalytic electrodes and materials used [1–4].

There are currently six main groupings of fuel cell available namely [5] (Fig. 10.1): (i) Proton Exchange Membrane Fuel Cell (PEMFC) including Direct Methanol Fuel Cell (DMFC), (ii) Alkaline Fuel Cell (AFC), (iii) Phosphoric Acid Fuel Cell (PAFC), (iv) Molten Carbonate Fuel Cell (MCFC), (v) Solid Oxide Fuel Cell (SOFC) and (vi) Microbial Fuel Cell (MFC). PEMFC, AFC, PAFC, and MFC operate at low temperatures in the range of [50–200 °C] while SOFC and MCFC at high operating temperatures in the range of [650–1000 °C].

The heart of fuel cells consists of a non-conductive electrolyte material sandwiched between two electrodes; the anode and cathode (Fig. 10.1). The fuel and the oxidant are fed continuously to the anode and the cathode sides respectively. At the anode side the fuel is decomposed into ions and electrons. The insulator electrolyte material allows only ions to flow from both the anode and cathode sides. Free electrons generated at the anode flow to the cathode side through an external electrical circuit. The recombination of the ions with oxidant occurs at the cathode to form pure water. Contrarily to water electrolysis, the polarity of a fuel cell on the anode is negative and on the cathode is positive [5].

Currently, the focus in Fuel Cell technologies is to offer significant quantitative improvements in: (a) lowering costs (by a factor of 10–100) and significantly improving reliability, durability and performance for the stationary, transport and portable markets; (b) addressing the challenges related to hydrogen generation, storage and utilisation; and (c) the acceleration of fuel cell technologies deployment to various markets such as cars and buildings [1].

As all fuel cells are normally distinguished by the materials used, e.g., the electrolyte and catalyst materials, although the manufacturing of the fuel cell electrodes is also different in each case, the main objective in fuel cell technologies is to develop low-cost, high-performance and durable materials. Currently, fuel cell systems are too expensive and not durable. For example, taking the case of PEMFCs, the platinum (Pt) catalyst accounts for more than half [*the average price of platinum has risen more than four times in 10 years between 2000 and 2010*] of the total stack cost with Pt tending to aggregate or/and dissolve when used in long operating conditions [4].

There are several ways to reduce the cost and increasing the performance of a fuel cell by [1–5]:

(i) decreasing the catalyst loading in fuel cell electrodes,
(ii) decreasing the catalyst nanoparticle size,
(iii) developing Pt-free electrocatalysts,
(iv) using novel fabrication methods to synthesize catalysts and producing better catalyst dispersion on fuel cell electrodes yielding better utilisation,
(v) using new techniques to increase mass-transport at the fuel cell electrode surface.

One of the most promising method and technique for producing performing and efficient fuel cell catalysts, electrodes and electrolyte materials is to employ ultrasound.

10.2 Ultrasound

Many of us are familiar with the use of ultrasound [6, 7], for example in biochemistry ultrasound is commonly used for the disruption of cells and living tissues in order to extract effectively important constituents. Ultrasound is also used as a medical imaging tool (e.g. pre-natal image scanning) and as a diagnostic tool (e.g. non-destructive testing of materials) in the frequency range of 2–10 MHz, but recently there has been an upsurge of interest in the application of low-frequency high-energy power ultrasound (20 kHz–2 MHz). First observed in the nineteenth century with the discovery of the piezoelectric effect by Curie and the ultrasonic whistle by Galton, ultrasound is usually defined as a sound wave with a frequency above 16 kHz (16,000 Hz or 16,000 cycles per second) with the upper limit usually taken to be 5 MHz for gases and 500 MHz for liquids and solids (Fig. 10.2).

The application of ultrasound in chemical, physical and biological sciences can be divided into two main groups: (i) low frequency or power ultrasound (20–100 kHz), also known in the 1950s as "Macrosonics" and (ii) high frequency or diagnostic ultrasound (2–10 MHz).

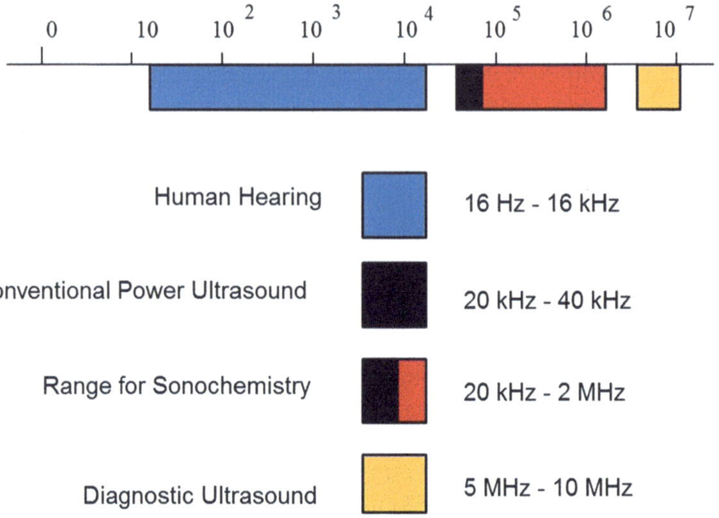

Fig. 10.2 The frequency ranges of (ultra)sound

Diagnostic ultrasound is often used in chemical analysis, medical scanning and in the study of relaxation phenomena [6, 7]. Low amplitude waves are used to determine the velocity and absorption coefficient of the sound wave by the medium, i.e. the effect of the medium on the ultrasonic wave. However, power ultrasound can be regarded as the effect of the sound wave on the medium. Power ultrasound used in liquid systems causes (i) a zone of extreme 'mixing' close to the ultrasonic source (i.e. the ultrasonic transducer), (ii) degassing, (iii) surface cleaning and pitting (erosion) and, (iv) an increase in bulk temperature (Fig. 10.3). It is for these reasons that low frequency and high energy waves are used in ultrasonic cleaning, drilling, soldering, chemical processes and emulsification [1].

Ultrasonic cleaning is probably the most common and known application of ultrasound. However, there are several areas where power ultrasound has been successfully employed (Fig. 10.4) such as: water and soil remediation (e.g. destruction of bacteria and organics, heavy metals removal, etc.), manufacturing of food ingredients and products (e.g. emulsification of oil/water based fluids, flavourings and vitamins extraction, etc.), impregnation of various materials, crystallisation and precipitation of organic and inorganic compounds, nano-sized materials production, polymerisation, drilling, soldering, cutting, plastic welding, surface treatment and preparation (for activation and modification) prior to plating and electroplating, metal finishing and precision engineering (e.g. the aerospace industry) [6, 7].

More recently, power ultrasound has been successfully demonstrated as an effective industrial process intensification technique to treat water effluents and produce pharmaceutical materials. The systems are used as non-chemical processes to either (i) control and eradicate microbes and bacteria (microbial activity) in

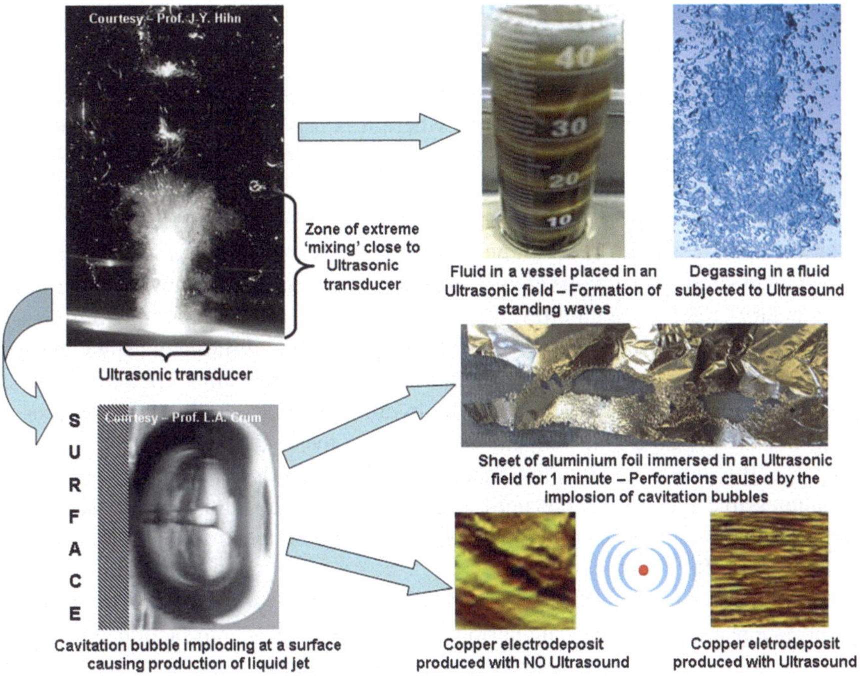

Fig. 10.3 The effect of ultrasound in a liquid and near a surface

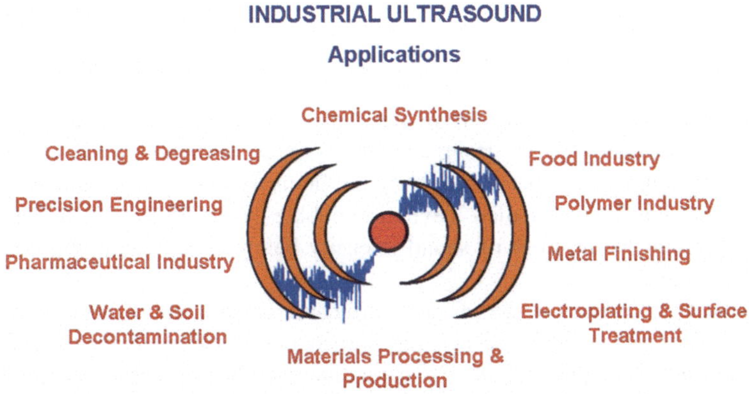

Fig. 10.4 The use of ultrasound in industrial applications

contaminated soils and waters (in other words, power ultrasound acting as a powerful bactericide) mainly in the water treatment industry, or (ii) produce micro- and nano-sized pharmaceutical ingredients (process known as *Sonocrystallisation*) in the fine chemicals and pharmaceutical industries. The systems are safe, robust and proven technologies.

10.3 What Is Sonochemistry?

Over the past few years the use of power ultrasound has found wide applications in the chemical and processing industries where it is used to enhance both synthetic and catalytic processes and to generate new products. This area of research has been termed *sonochemistry*, which mainly concerns reactions involving a liquid leading to an increase in reaction rates, product yields and erosion of surfaces [6, 7]. However, the main reason for most of the observed effects of ultrasound on surfaces and chemical reactions is recognised as being due to 'cavitation' effect which occurs as a secondary effect when an ultrasonic wave passes through a liquid medium.

Cavitation was first reported in 1895 by Thornycroft and Barnaby [8] when they observed that the propeller of a submarine eroded over short operating times, caused by collapsing bubbles induced by hydrodynamic cavitation in turns generating intense pressure and temperature gradients locally. In the late 1920s, Lord Rayleigh [9] published the first mathematical model describing 'cavitation' in incompressible fluids. It is not until 1927 that the use of ultrasound on chemical and biological systems was first observed and recognised as a useful tool by Richard and Loomis [10].

As an ultrasonic wave passes through a liquid; fluctuating pressures are rapidly set up as a result of the alternate periods of compression and rarefaction associated with the wave [6–10]. During the compression cycle, the liquid is subjected to a sufficiently positive pressure which pushes the molecules of the liquid together, whilst during the following rarefaction cycle, the liquid is subjected to an equal but negative pressure which pulls the molecules of the liquid away from each other. Increasing the amplitude of the wave leads to an increase in the magnitude of the positive and negative pressures. If the liquid is subjected to a sufficiently large negative pressure during the rarefaction cycle, the molecules are torn away from each other producing very small cavities, called 'microbubbles'. In other words these negatives pressures are strong enough to overcome the intermolecular forces binding the liquid. The process of tearing the liquid apart is known as 'cavitation' and the microbubbles are called 'cavitation bubbles' (Fig. 10.5) [6–10]. The cavitation threshold is the limit of sound intensity below which cavitation does not occur in a liquid.

Cavitation phenomenon is well known to cause erosion, emulsification, molecular degradation, sonoluminescence and sonochemical enhancements of reactivity solely attributed to the collapse of cavitation bubbles [6–10]. It is now well accepted in the field that the cavitation bubble collapse leads to near adiabatic heating of the vapour that is inside the bubble, creating the so-called "hot-spot" in the fluid, where:

(1) High temperatures (ca. 5000 K) and high pressures (ca. 2000 atm/200 MPa) are generated with a collision density of 1.5 kg cm^{-2} and pressure gradients of 2 TPa cm^{-1}, with lifetimes shorter than 0.1 μs and cooling rates above 10^{9-10} K s^{-1} during the collapsing of cavitation bubbles are observed. Here, water vapour is 'pyrolyzed' into hydrogen radicals ($H^•$) and hydroxyl radicals ($OH^•$),

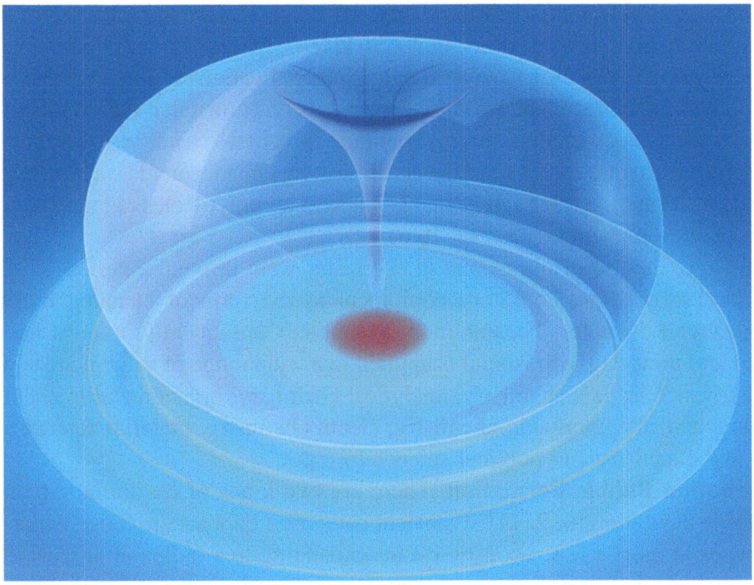

Fig. 10.5 Representation of a cavitation bubble imploding

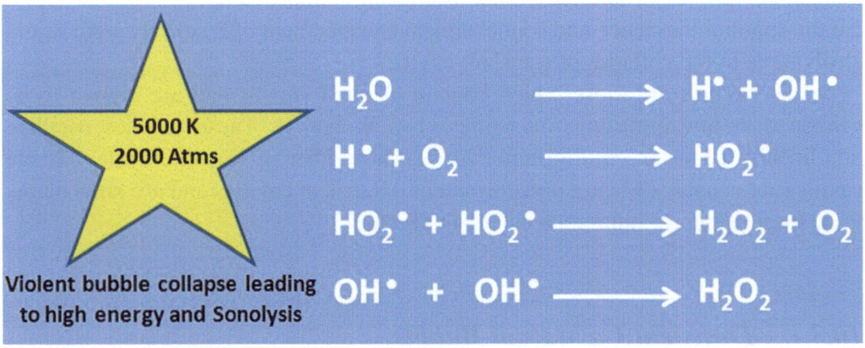

Fig. 10.6 Schematic of water sonolysis

known as water sonolysis (Fig. 10.6). Note that the exact temperatures and pressures generated during cavity implosion are difficult both to calculate theoretically and to determine experimentally.

(2) The interfacial region between the cavitation bubbles and the bulk solution is paramount. The temperature is lower in the interior of the bubbles than the exterior but high enough for thermal decomposition of the solutes to take place with greater local hydroxyl radical concentrations in this region.
(3) The reactions of solute molecules with hydrogen atoms and hydroxyl radicals occur in the bulk solution at ambient temperature.

Because of ultrasound's 'extraordinary' effects, extensive work has been carried out in which high power ultrasound (20 kHz–2 MHz) was applied to various chemical processes leading to several industrial applications and many publications over a wide range of subject areas [6, 7]. It has been shown that the effects of high intensity ultrasonic irradiation on chemical processes lead to both chemical and physical effects, for example, mass-transport enhancement, surface cleaning and radical formation ($HO^•$, $HO_2^•$ and $H^•$) due to homolytic cleavage *via* sonolysis due to cavitation phenomena (Fig. 10.6) [6, 7].

Although, the use of ultrasound in chemistry went through a period of neglect until the 1980s when laboratory equipment in the form of cleaning baths and biological cell disrupters became more available. Consequently, interest has been revived and nowadays ultrasound is applied to a wide range of subject areas within chemistry such as nanochemistry, analytical chemistry, organometallic chemistry, organic chemistry, inorganic chemistry, electrochemistry, material science, environmental chemistry and polymer science [6, 7].

Cavitation Bubble Formation. There are two kinds of cavitation—**transient** and **stable** cavitation [6, 7, 11]. Transient cavitation bubbles are those which exist for one, or at most a few, acoustic cycles expanding to a radius of at least double their initial size before collapsing violently into several smaller bubbles. These tiny bubbles may act as nuclei for other bubbles. Transient bubbles occur mainly in liquids subjected to ultrasound intensities which are greater than 10 W cm^{-2}. It was thought initially that the effects such as erosion, emulsification, molecular degradation, sonoluminescence and sonochemical enhancement of reactivity were solely attributable to the collapse of transient cavities.

Stable cavities are those which contain gas and vapour and are known to be generated at low ultrasonic intensities (1–2 W cm^{-2}). These cavities oscillate non-linearly about an equilibrium size over a number of acoustic cycles. Stable cavities are capable of being transformed into transient cavities and are known now to be responsible for numerous chemical effects.

10.3.1 Transient Cavitation Bubble

As the name implies, transient cavitation bubbles have a short life-time (ca. 10^{-5} s in a 20 kHz ultrasonic field) before they collapse violently on compression and disintegrate into smaller bubbles. These smaller bubbles may act as nuclei for further bubbles, or if they are of sufficiently small radius, they can dissolve into the bulk of the solution. During the short lifetime of the transient bubble, it is assumed that little or no diffusion of dissolved gas can take place from the solution bulk into the cavity, whereas condensation and evaporation of liquid is assumed to take place freely. Since there is no gas present in the bubble to act as a cushion, the implosion leads to a very violent collapse.

It was assumed by many authors, such as Noltingk and Neppiras [12], that adiabatic collapse of the bubbles would allow for a calculation of the temperature or pressure within the bubble (Eq. 10.1 and 10.2). For example, if it is assumed that the vibrations of the bubbles occur so very fast that little heat exchange can occur with the surrounding liquid environment, then the vapour inside the bubble is heated during the compression cycle and one may deduce, under these adiabatic conditions, that the maximum pressure, P_{max} and temperature, T_{max}, can be given as follows:

$$P_{max} = P\left[\frac{P_m(\gamma - 1)}{P}\right]^{\frac{\gamma}{\gamma-1}} \tag{10.1}$$

$$T_{max} = T_o\left[P_m\frac{(\gamma - 1)}{P}\right] \tag{10.2}$$

where T_o is the ambient temperature,
γ is the ratio of the specific heat capacities of the gas (or gas vapour) mixture,
P is the pressure in the bubble at its maximum size (usually assumed to be equal to the vapour pressure, P_v, of the liquid), and
P_m is the pressure in the liquid at the moment of transient collapse.

At ambient temperature and pressure, Noltingk and Neppiras [12] were able to deduce that extremely high temperatures (T_{max}) and pressures (P_{max}) were produced in the final phase of implosion equivalent to 5000 K and 2000 atm. The release of the pressure as a shock wave is a factor which has been used to explain increased chemical reactivity, due to increased molecular collision, and polymer degradation; the high temperatures within the bubble have formed the basis for the explanation of radical production and sonoluminescence [6, 7, 11].

10.3.2 Stable Cavitation Bubble

A stable bubble is one of which is thought to exist for many cycles. Stable bubbles contain mainly gas and vapour. They are produced at low intensities (1–3 W cm^{-2}) and oscillate about an equilibrium size for several acoustic cycles [6, 7, 11]. The time-scale over which they exist is long enough for both mass transfer and thermal diffusion to occur freely. This process occurs as follows:

In the rarefaction phase of the sound wave gas diffuses from the liquid into the bubble causing the bubble to expand, whilst in the compression phase gas diffuses out of the bubble into the liquid. The increased surface area increases the rate of gas and vapour diffusion into the bubble. As a succeeding compression wave passes through the liquid, the bubble is compressed. Therefore, gas and vapour diffuse out. As a result, the rate of inward diffusion will become greater than the rate of diffusion back to the liquid and this will lead to an overall growth of the bubble. As the bubble grows, the bubble will become more compressible due to changes in the acoustical and environmental conditions of the medium.

The stable bubble may be transformed to a transient bubble and undergo collapse, the violence of the collapse being less than that of a vapour-filled transient bubble since the gas present cushions the implosion. There are two possible fates for stable cavitation bubbles. They may either grow sufficiently large to be capable of rising to the surface of the liquid—this is the process of ultrasonic degassing—they can become unstable due to differences between the resonant frequency and the driving frequency (i.e. the frequency from the transducer) and be transformed into transient bubbles.

It was also found that the maximum temperatures and pressures of these cavitation bubbles generated upon collapse are less than for transient cavitation bubbles due to the 'cushioning' effect [6, 7, 11]. However, it was shown that contracted bubbles cause temperature rise within the bubble and changes in local hydrostatic pressure [6, 7, 11]. Calculations of the local pressures due to these resonance vibrations have resulted in values which exceed the hydrostatic pressure by a factor of 150,000. There is no doubt that the intense local strains in the vicinity of the resonating bubble are the cause of the many disruptive mechanical effects of ultrasound.

In summary, cavitation bubble formation is a three step process consisting of (1) nucleation, (2) bubble growth and (3) collapse of gas vapour filled bubbles in a liquid phase. These bubbles transform the low energy density of a sound field into a high energy density by absorbing energy from the sound waves over one or several cycles and releasing it during very short intervals. During rarefaction cycles, negative pressures developed by the high power ultrasound are strong enough to overcome the intermolecular forces binding the fluid. The succeeding compression cycles can cause the microbubles to collapse almost instantaneously with the release of a large amount of energy. When the cavitation bubble collapses close to or on a solid surface it can only do so asymmetrically due to the surface which leads to a microjet of liquid being directed towards the surface of the material at speeds of up to 200 m s^{-1} [6, 11, 13].

10.4 Sonoelectrochemistry: The Use of Ultrasound in Electrochemistry

The use of ultrasound on electrochemical systems or Sonoelectrochemistry was first observed by Moriguchi as early as 1934 [14] and since then, continues to be an active and exciting research area. In the 1930s, Schmid and Ehert [15] also studied the ultrasonic effects on the passivity of metals and the generation of gases by electrolysis [16]. Nearly 30 years later, in 1956, Nyborg et al. [17] demonstrated that the movement of liquid induced by ultrasound, known as acoustic streaming and in the same year Penn et al. [18] studied the effect of ultrasound on concentration gradient in the electrolyte and at the electrode surface. In 1965, for the first time, Bard [19] showed that ultrasound caused an increase in mass transport of electroactive species from the bulk solution to the electrode surface in controlled potential coulometry experiments. Since then, extensive work has been carried out in which high power ultrasound (20–100 kHz) was applied to various electrochemical processes leading to several industrial applications and many publications over a wide range of subject areas such as electrodeposition, electroplating, electrochemical dissolution, corrosion testing and nanotechnology.

For over 80 years, nearly a thousand papers have been written on the subject with many original work, general reviews [6] in sonoelectrochemistry (with the first one from Mason et al. [20] in 1990, including the effects of power ultrasound on special media), organic sonoelectrosynthesis, sonoelectroanalysis, sonoelectrochemical production of nanomaterials, and recently the sonoelectrochemical production of fuel cell materials [6].

In all these papers and reviews, it was clearly shown that the effects of high intensity ultrasonic irradiation on electrochemical processes lead to both chemical and physical effects, for example, mass-transport enhancement, surface cleaning and radical formation. Many workers have also investigated the distribution of ultrasonic waves or energy in various electrochemical reactors operating in the lower ultrasonic frequency range (20–100 kHz) and at high ultrasonic powers. Several methods for such determination have been proposed e.g. aluminium foil erosion, sonoluminescence, calorimetric methods, chemical dosimetry [7] and laser-sheet visualization [6].

In sonoelectrochemistry, ultrasound is known for its capacity to promote especially heterogeneous reactions mainly through intense mass-transport, interfacial cleaning and thermal effects. In addition, homogeneous chemical reactions have been reported to be affected and the generation of highly radical species [6, 7] [e.g. the production of $H^{\bullet}$ and $OH^{\bullet}$ radicals by sonolysis in intense sound fields is an important aspect for the use of ultrasound. For example, in the detoxification of environmentally harmful wastes [containing heavy metals, hydrocarbons (PAH's) and chlorinated compounds (PCB's)], ultrasound has been found to be extremely beneficial.

The vast variety of ultrasonically induced effects observed in sonoelectrochemistry may be ascribed to the generation, pulsation and collapse of cavitation bubbles in the electrolyte medium near the electrode surface. This ultrasonic cavitation occurs at low to high ultrasonic intensities (ultrasonic power per tip or transducer or cell base area). A pulsating cavitation bubble close to the electrode surface generates microstreaming. When the cavitation bubble reaches a resonant size, it collapses asymmetrically leading to the formation of high velocity jet of liquid toward the surface. This phenomenon leads to a thinning of the diffusion layer and can improve the overall mass-transfer and hence reaction rates [6, 7].

Since most of the observed effects of ultrasound in electrochemical processes are thought to be due to the cavitation effect together with micro-streaming, the application of ultrasound is known to be very beneficial in the electrochemical industry. This has led to investigations into mass-transport, electron-transfer processes and electrode surface adsorption [6, 7]. The reader is invited to consult the following book 'Power Ultrasound in Electrochemistry: From Versatile Laboratory

Tool to Engineering Solution' by BG Pollet (ed) [6]. This book deals with the aspect of electrochemistry combined with ultrasound and explains the various electrochemical phenomena occurring at the electrode surface when a potential is applied across it. For this purpose, electrode kinetic and mass-transport parameters are defined. The book also outlines the theory, principles and applications of sonoelectrochemistry in various branches of chemistry. Finally it is shown how coupling ultrasound with electrochemistry could be used to improve electrochemical processes, enhance detection limits in the electroanalysis of toxic samples and produce nanomaterials.

In 1998, Pollet et al. [6, 13] showed, with the aid of mathematical models based on mass-balance equations and using the *quasi-reversible* redox couple $Fe(CN)_6^{3-}$/ $Fe(CN)_6^{4-}$ as an electrochemical model, that a *Levich-like* equation relating the limiting current, the inverse square root of the electrode radius and the inverse square root of the electrode-horn distance, the transmitted ultrasonic intensity (and thus the transmitted ultrasonic power) may be generated for ultrasonic frequencies of 20 and 40 kHz (probe systems only and at 298 K) using Eq. (10.3) also known as the *Pollet* equation [6, 13]:

$$I_{\mathrm{lim}} = 0.84nFAD_{\mathrm{o}}{}^{2/3}n^{-1/6}r_{\mathrm{e}}{}^{-1/2}A_{\mathrm{uht}}{}^{-1/2}d^{-1/2}C^{*}P_{\mathrm{T}}{}^{1/2} \tag{10.3}$$

where I_{lim} is the limiting current (A), n is the number of electrons transferred during the electrochemical process, F is the Faraday constant (C mol^{-1}), A is the electrode area (cm^2), D_{o} is the diffusion coefficient ($\mathrm{cm}^2\ \mathrm{s}^{-1}$) of the electroactive species, d is the ultrasonic horn-electrode distance (cm), ν is the kinematic viscosity ($\mathrm{cm}^2\ \mathrm{s}^{-1}$), r_{e} is the working electrode radius (cm), A_{uht} is the ultrasonic horn tip area (cm^2), C^* is the bulk concentration of the electroactive species (mol cm^{-3}) and P_{T} is the ultrasonic power transmitted (W).

From Eq. 10.3, experimental limiting current values and assuming that all the above parameters are known, the transmitted ultrasonic power (P_{T}) can be calculated.

10.5 Sonoelectrochemical Production of Noble Metals and Fuel Cell Electrocatalysts

The sonoelectrochemical method is the use of power ultrasound applied to various electrochemical processes leading to both chemical and physical effects, such as increase in mass-transport, improved surface cleaning, increase in deposit hardness, deposit thickness, deposition rates and efficiencies as well as improved deposit adhesion and radical formation [6].

There are a few studies reporting the use of the sonoelectrochemical method to produce noble metals and fuel cell electrocatalysts. Here, the sonoelectrochemical technique involves either depositing a metal under continuous electrical current and ultrasound or producing nanosize metals at various currents and ultrasonic pulses (a few ms) at a vibrating electrode (Fig. 10.7).

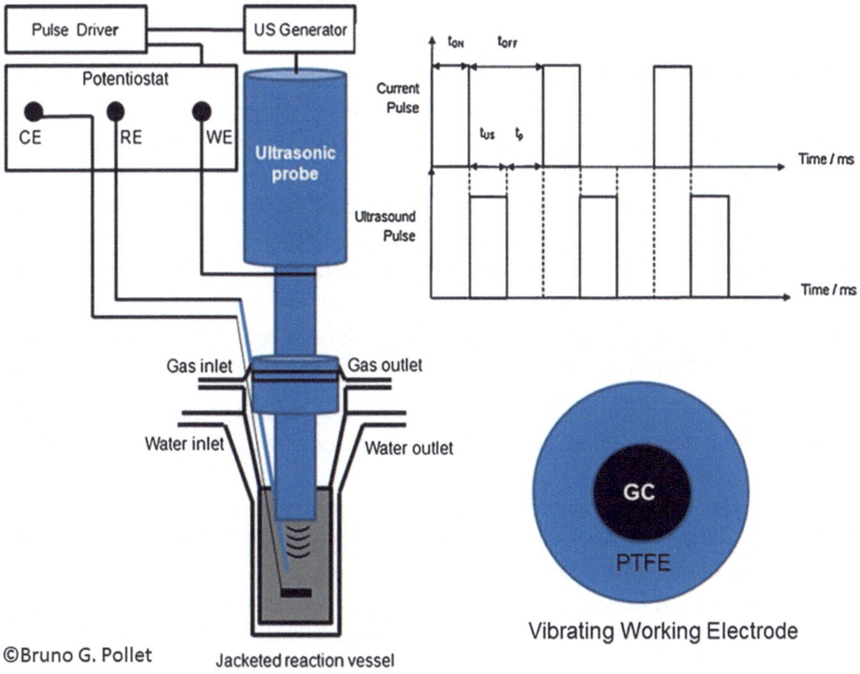

Fig. 10.7 Sonoelectrochemical experimental set up

10.5.1 For PEM Fuel Cells

Shen et al. [21] produced uniform spherical three-dimensional dendritic Pt nanostructures (DPNs) with an average dimension of 2.5 ± 0.5 nm, at room temperature sonoelectrochemically (20 kHz; 20 W; pulse width of the current = 1.0 s, resting time of the current pulse = 0.5 s; duration of the ultrasonic pulse = 0.3 s) by using mixtures of hexachloroplatinum(IV) acid hydrate in various surfactants [PVP, poly(ethylene glycol)$_{20}$-poly(propylene glycol)$_{20}$-poly(ethyleneglycol)$_{20}$ (P123), SDS and poly-diallyl-dimethyl ammonium chloride (PDDA)] at current densities up to 40 mA cm^{-2}. They showed that at low current densities the reduction rate of Pt is slower, the number of nuclei is small and the rate of growth is faster than that of nucleation, which lead to large nanoparticles (5–7 nm). As the current density increases the reduction rate of Pt^{4+} increases and the nucleation rate is faster than that of growth leading to the generation of more nuclei and the formation of smaller primary Pt nanoparticles. They also showed that stabilized Pt nanoparticles do not aggregate in PVP due to stronger bonds between the Pt precursors with the C=O groups of the PVP in other words PVP adheres to the nanoparticles through a charge-transfer interaction between the pyrrolidone rings and Pt atoms. The DPNs showed improved electrocatalytic activity towards MOR due to monodisperse Pt nanoparticles, and improved porosity structure leading to large effective surface area. They proposed a mechanism whereby Pt ions are reduced by the electrical current and formed Pt primary nanoparticles which are subsequently dislodged by the vibrating electrode. The primary nanoparticles then in solution spontaneously assemble together and formed small spherical DPNs. They showed that ultrasound leads to small primary nanoparticles in favour of the crystallite reorganization and growth of a stable near-single crystal.

Zin et al. [22] produced platinum nanoparticles from aqueous chloroplatinic solutions by the pulse sonoelectrochemical (20 kHz, up to 118 W cm^{-2}) method on titanium alloy electrodes in the absence of any surfactants, alcohols and polymers by producing short applied current pulses triggered and followed immediately by ultrasonic pulses at the working electrode (in this case the cathode). The time management sequence employed was as follows [Fig. 10.7]:

1. A short current pulse of $|i| = 50$ mA cm^{-2} was applied to the *sonoelectrode*, whereby the titanium horn acted as an electrode only (t_{ON}); the time of this phase typically varied between 0.3 and 0.5 s.
2. Immediately after the electrochemical pulse was turned off, an ultrasonic pulse was sent to the *sonoelectrode* and here it acted only as a vibrating ultrasonic horn (t_{US}); this second phase lasted no more than 0.5 s.
3. A rest time, t_p, followed the two previous phases (this was useful to restore the initial electrolyte conditions close to the *sonoelectrode*).

They showed that Pt mean grain size ranging from 11 to 15 nm was produced and globular clusters had a mean size ranging between 100 and 200 nm which in turn aggregated and built complex structures.

Shen et al. [23] showed that by using the sonoelectrochemical (20 kHz; 20 W; pulse on time of the current = 0.5 s; pulse off time of the current = 0.5 s, duration of the ultrasonic pulse = 0.3 s) method, it is possible to realize the morphology-controlled synthesis of palladium nanostructures [spherical (SNP), multitwinned (MTP) and spongelike (SSP)] at room temperature in the presence of various surfactants and polymers [cethyltrimethylammoniumbromide (CTAB), PVP and PDDA]. They showed that the size and shape of the Pd nanostructures may be controlled by varying the current density and the precursor solution pH value. Furthermore, the electrocatalytic activities of the produced spongelike Pd nanostructures for direct alcohol oxidation in alkaline media showed higher electrochemical active surface than other Pd nanostructures. Qui et al. [24] also synthesized highly dispersed spherical Pd nanoparticles of a dendritic superstructure in the presence of CTAB by the pulse sonoelectrochemical (20 kHz) method at room temperature and a reaction time above 2.5 h. They explained that the dendritic-structured Pd nanoparticles had a treelike structure and agreed with the diffusion-limited aggregation (DLA) model, involving cluster formation by the adhesion of particles to a selected seed on contact and allowing the particle to diffuse and stick to the growing structure. They stipulated that it is possible that Pd particles reach the anode and grow into a dendritic structure. They concluded that the shape and size of spherical nanocrystalline Pd may be controlled by varying the current density (8–13 mA cm^{-2}), the interval between continuous ultrasonic pulses,

Table 10.1 A summary of the sonoelectrochemical production of noble mono-metallics at various ultrasonic frequencies and powers in several surfactants

Noble metals	Ultrasonic frequency (kHz)	Ultrasonic power	Surfactant	Solvent	Particle size (nm)	Authors
Pt	20	20 W	PVP, P123, SDS, PDDA	–	2.5 ± 0.5	Shen et al. [21]
Pt	20	Up to 118 W cm^{-2}	–	–	<200	Zin et al. [22]
Pd	20	–	PVP, PDDA, CTAB	–	<200	Shen et al. [23]
Pd	20	20–120 W cm^{-2}	CTAB	–	5–10 (depending on current densities and CTAB concentrations)	Qui et al. [24]

ultrasonic intensity (20–120 W cm^{-2}) and the CTAB concentration. For example, they observed that the shape of the nanoparticles appeared irregular and agglomerated below 20 and 120 W cm^{-2}.

For all the sonoelectrochemical production of nano-size metals, a mechanism has been proposed where metallic ions are reduced by a short current pulse to produce metallic nanoparticles on the *sonoelectrode* surface, which are then dislodged by the ultrasonic pulse. The metallic nanoparticles in solution tend to spontaneously assemble together whereby under insonation, the Ostwald ripening process is accelerated, leading to smaller primary nanoparticles [6].

Table 10.1 shows a summary of the sonoelectrochemical production of noble mono- metallics at various ultrasonic frequencies and powers in several surfactants.

10.5.2 For Solid Oxide Fuel Cells (SOFC)

10.5.2.1 Basic Components

SOFC technology has the most remarkable property of running using a wide variety of gas fuels, from pure H_2, CO and H_2/CO rich reformate gases, to gaseous or gasified liquid hydrocarbons such as natural gas [25]. The electrode exposed to fuel is anode and plays a multi-functional role. On the side where anode is attached to solid electrolyte (typically yttria-stabilized zirconia, YSZ) the former receives oxygen ions delivered by the later at temperatures beyond 600 °C. On the side where anode is exposed to fuel gaseous phase it receives reactants (typically H_2) and yields products (typically H_2O). Finally the solid electrode phase takes up electrons, establishing one of the electric current flow poles. This is the concept of *three-phase boundary* (TPB), where transport pathways of ions, gaseous chemical species and electrons intercept. Electrochemistry can only take place in locations where all three components are spatially close to each other and thus be enhanced when a high volumetric TPB length is achieved [26].

Such a composite material, consisting of ceramic and metal species, is referred to as *cermet* electrode. The latter has to retain a porous structure with the metal particles well dispersed in its bulk, so as to deliver high electrochemical performance, i.e. let O^{2-} oxidize the fuel that resides at the porous anode side. So far, Ni/YSZ fulfils most of the requirements for an SOFC anode. This composite acts as an electronic conductor, due to Ni, as well as an ionic conductor, due to YSZ [27]. An alternative to YSZ ionic conductor is gadolinium-doped ceria, GDC, as it exhibits high activity for hydrocarbon oxidation and high ionic conductivity at the same time [28]. Since the first attempts to use ceria for the SOFC anodes in the '60s, introduction of $CeO_{2\text{-}\delta}$ based additives, e.g. Ni/GDC, is widely considered among the most promising directions in anodes development [29].

As far as the electrons conductive part of an anode is concerned, Pt electrodes were originally used due to their stability, high conductivity, and a substantial catalytic activity, while later it was substituted with less expensive components, primarily Ni-based cermets for the anodes [30]. Although Ni has been the standard primary metallic component in anode cermets for decades, Ni/YSZ is thermally unstable and results easily to Ni agglomeration and island formation. Nevertheless, Ni/YSZ systems demonstrate a tendency towards accumulation of impurities and formation of secondary phases at two- and three-phase boundaries [26]. As a result, continuous efforts are being made since the 70's in order to develop optimized anode formulations, particularly via the use of transition metal components [29].

10.5.2.2 Importance of Nanomaterials Addition in SOFC

One of the most complex parts of an SOFC is the anode, as it has to (a) let as many O^{2-} as possible migrate from the electrolyte towards its porous surface, (b) fulfill fuel oxidation at intermediate (typically 700–850 °C) or high temperatures (beyond 900 °C) and (c) let current be conducted through its bulk and be collected at its surface. At the same time, it should ideally (d) remain porous for as long as possible and (e) not experience any considerable sintering of its catalysis aiding metallic components. To date, the superior catalytic activity of Ni for electrochemical oxidation of hydrocarbon fuels renders cermets containing Ni/YSZ or Ni/GDC as the preferred anode materials. As in all catalytic applications, the smaller and the more dispersed these components are, the better the fuel utilization is expected to be.

Moreover, several combinations of metal components are utilized on the anode side particularly for the case of direct non-pure hydrogen oxidation [31, 32]. A good selection of materials combinations may be the key for achieving the best performance characteristics for a SOFC. For instance, it has been shown that part of the anode overpotential is enhancing the Ni particles sintering and that this process can be significantly delayed by increasing the work of adhesion between Ni and YSZ particles. The latter can be achieved by trace amounts of additives, such as CeO_2 or TiO_2 at the Ni/YSZ interface [33]. Other studies have also shown that composites of Ni/CeO_2 have additional significance because of the possible application in high temperature fuel cells [34]. Moreover, the important problem of SOFC cermet anodes degradation has been studied via a theoretical model from a fundamental viewpoint. The model describes the gradual degradation of the anode due to nickel particle sintering and the concomitant loss of TPB length [35].

Among recent developments many studies concern the direct feeding of SOFCs with hydrocarbons from fossil or renewable sources. Even though the above strategy appears more attractive compared to the state-of-the-art technology of using hydrogen as fuel, there still remain some major challenges to be addressed for extended fuel application in SOFCs. One of them is anode poisoning caused by unfavorable reactions of catalytic anode materials with C and/or S species, which are present in readily available hydrocarbon fuels such as natural gas, biogas or coal syngas [36]. In recent years there have been several reviews that tackle the development challenges of SOFC anode materials [37–40].

Carbon deposition in SOFCs may originate from the internal steam or dry reforming of hydrocarbons or biofuels. Many factors that influence the likelihood of such formation have been identified. The catalyst nature is one of the most important [41]. For example, carbon formation is more likely over Ni than, for instance, over Au, Ag, Cu, Ru or LSCM [27, 41–43]. In a more recent study Sn/Ni surface alloy has been identified as much more carbon tolerant than monometallic Ni under conditions of steam reforming of methane, propane and isooctane at moderate steam to carbon ratios [44]. As a result, a state of the art Ni-based cermet enhanced with a second or more metal compounds could operate under steam or dry reforming conditions with high carbon tolerance in the intermediate temperature range.

Sulfur poisoning of SOFC systems operating at intermediate temperatures can be irreversible while at high temperatures in tends to be reversible [45, 46]. Inhibition of sulfur adsorption leading to sulfides formation has been reported in reviews that investigate different anodes and conditions [36, 47]. Several studies prove that by decorating Ni surface with elements such as Mo, Re, W, Pt, Ag or their oxycarbides, can modify the surface electronic properties [48–53].

10.5.2.3 Preparation of SOFC Materials

Sonoelectrochemical synthetic route has an extended nanoparticles preparation showcase of many metallic, metal oxides or alloy species. All these compounds can be used for cermets preparation either by (a) a *step-by-step* approach where the sole nanoparticles synthesis is followed by impregnation or decoration on the fuel cells ceramic materials or (b) a *synchronous* approach where the nanoparticles are co-deposited on the working electrode along with the ceramic substrates.

Sole Synthesis of Metal Species

Numerous research groups worldwide have extensively investigated metal, metal oxides or metallic alloys sonoelectrochemical synthesis. As a result, there have been published some very comprehensive reviews that compile and compare several methods [54–57]. From all species reviewed Ag, Ni, Cu, Pt and Au have been multiple times studied.

A typical method to categorize the various synthesis routes is according to the position of the ultrasound emitter according to the working electrode. Three types are reported: (a) the conventional *face-on* where the sonohorn is opposite to the electrode, (b) the *side-on* where the electrode is placed perpendicular to the ultrasonic waves, and finally (c) the *sonotrode* where the actual ultrasonic horn is at the same time the electrode [58]. Use of the sonotrode is suited for nanoparticles synthesis that will be dispersed in the electrochemical bath usually with the help of surfactandts also dissolved in the bath. Furthermore, the sonoelectrochemical technique may follow either (a) the typical sequence of electrochemical (potentio-/galvano-static) and ultrasound pulses or (b) electrochemical pulses under continuous sonication.

Nickel is the typical element used in SOFC anodes because it enhances catalytic activity and offers good electrical conductivity being a much more economic alternative to Pt. Originally, Delplancke et al. prepared crystalline Ni powders with a sharp diameter distribution around 100 nm combining pulsed current and pulsed ultrasound on the same electrode surface [59]. Later on, Davis et al. studied the advantageous effects that the acoustic streaming and cavitation brings in Ni sonoelectroanalysis [60]. Jia et al. synthesized for the first time first time nanoporous Ni particles in dimethyl sulphoxide (DMSO) [61]. It was found that these nanopores were produced by the aggregation and/or melting of primary nanoparticles of about 4–5 nm.

Gold nanoparticles are of the largest interest for SOFC anode modification after Ni. Liu et al. synthesized Au nanoparticles from 2 to 15 nm under cathodic overpotentials and ultrasonication [62]. These experiments revealed that adjusting the sonoelectrochemical reduction time could control the ratio of Au nanoparticles to Au-containing nanocomplexes in solutions. On the other hand, Aqil et al. made a study of Au nanoparticles synthesized under potentiostatic conditions and by experimenting with several stabilizing agents [63]. They claimed that the mean size could be easily tuned between 5 and 35 nm by simple control of the electrochemical parameters. Shen et al. managed to prepare Au nanoparticles with no use of stabilizers. Size and shape was controlled by adjusting current density, reaction time and pH of bath. Particles sizes of 10–50 nm were synthesized with an obvious tendency to form agglomerates [64]. Sakkas et al. conducted a series of experiments for Au nanoparticles synthesis at various potentiostatic pulses and continuous sonication by using PVP as surfactant [65]. The average size distributions ranged from 10 to 35 nm according to the electrochemical pulse. No crystallinity was evident under HR-TEM study. This confirms previous studies, which proved that sonoelectrodeposition under continuous ultrasonication leads to the formation of amorphous materials [66].

Copper or its oxides have been investigated as an alternative to the more expensive Au or Ag. Haas et al. used galvanostatic conditions and PVP as stabilizer. They experimented with current duration, experiment duration, PVP concentration, temperature, and ultrasonic power to obtain particles ranging from 10 to 55 nm. A suggested possible bonding framework for the interaction between metallic Cu and PVP is also presented [67]. On the other hand, Sáez et al. present a study of Cu electrocrystallization for obtaining a narrow size distribution. They used both galvanostatic and potentiostatic conditions that yielded mean particles sizes of 200 nm and 81 nm respectively. In some of the experiments PVP was also used [68]. Furthermore, Sakkas et al. prepared Cu nanoparticles using galvanostatic conditions in presence of PVP. Their study yielded that for brief electrodeposition pulses the mean particle size distribution is 8 nm, smaller than for longer pulses where they obtained 50 nm Cu nanoparticles [65]. Finally, Schneider et al. conducted, among others, a series of synthesis experiments at room temperature using a copper sulfate based electrolyte and following the face-on route [69]. They grew a fine Cu film on a typical electronic quartz crystal microbalance (EQCM) using continuous ultrasonic irradiation at different intensities. Their attempts revealed that the surface of Cu deposition gets significantly coarser compared to silent conditions, which in turn leads to higher area of the film grown [69].

Silver has also been investigated by many research schemes. Zhu et al. worked with a typical Ag electrochemical bath adding nitrilotriacetate (NTA). They reported that varying the electrosonic time, the Ag precursor and NTA concentration affects the nanoparticles shape [70]. Liu et al. from the same group attempted Ag nanoparticles synthesis from a different electrochemical solution and this time in the presence of gelatin. They prepared the electrochemical bath following a quite intricate route and keeping as low light exposure as possible. The reduction was done with galvanostatic conditions. The powder that yielded was amorphous and the size distribution from 12.2 nm to more than 60 nm [71]. Liu et al. followed a potentiostatic approach of Ag nanoparticles synthesis, and by increasing the cathodic overpotential they managed to control the size distribution from as low as 2–20 nm [72]. Jiang et al. followed galvanostatic conditions with the presence of PVP as surfactant. They were able to achieve Ag nanoparticles of 5–35 nm as well as dendritic formations by varying the current density or the PVP presence [73]. Finally, Vu et al. managed to synthesize not only spherical but also rod-like Ag nanoparticles in presence of SDS as surfactant. In their studies they used galvanostatic conditions and a stainless steel foil as anode. Ultrasonic horn was immersed in the bath in the space between the opposite facing electrodes. By varying the precursor as well as surfactant concentration they ended up with fine particles of 5–14 nm or rods of 100–500 nm [74].

Platinum has been so far synthesized in two different nanoformations, i.e. dendritic Pt nanostructures (DPNs) and monodisperse. Shen et al. managed to synthesize DNPs that exhibited remarkable porosity structure and enhanced effective surface area, thus adding to the electrocatalytic activity of this compound

[21]. On the other hand, Zin et al. prepared Pt nanoagregates with the Pt nanoparticles ranging from 11 to 15 nm. Their approach involved galvanostatic synthesis with use of no surfactant [22].

Tungsten nanoparticles has been rarely synthesized via sonoelectrochemistry. In their work, Lei et al. managed to prepare body center cubic W nanoparticles galvanostatically by varying the current density as well as the ultrasound pulse period and intensity. Spherical nanoparticles of about 30 nm yielded [75].

Synchronous Deposition of Ceramic and Metal Species

A combined action of electrochemical metal deposition and ultrasonication for SOFC cermet composites preparation was utilized for the first time by Argirusis et al. [76]. In their experiments they used state of the art doped oxide nanoparticles (GDC) suspended in a typical metal precursor-bearing electrolyte for Ni/GDC synthesis. They worked with a Ni sulfamate electrolyte at 50 °C and 5 g/L of GDC in suspension. GDC crystallite size was 5–10 nm while the pre ultrasonication agglomerates reached up to 149 μm. Ultrasound was used prior and during electrodeposition, continuously, for achieving deaglomeration and full dispersion respectively. An ultrasonic horn tip of 13 mm diameter was placed at a distance of 22 mm face on [58] to the Au working electrode and operated at 20 kHz and 25 W cm^{-2}. With the application of ultrasound, currents increased and became noisier compared to sole Ni depositions conducted as reference. Co-deposited Ce amount was reported to have reached 2.5 wt% according to SEM/EDX data.

On the other hand, Ni/CeO_2 cermets were synthesized by Xue et al. [77]. They conducted, among others, a study on Ni electrochemical pulsed current deposition and synchronous CeO_2 nanoparticles deposition under continuous ultrasonication. In their experiments they used a typical sulfamate electrolyte at 45 °C and 40 g/L of suspended CeO_2 with particles average size estimated at 40 nm. The two electrodes were placed vertically and the bath was immersed in an ultrasonic bath with the aid of magnetic stirring at 1000 rpm. Current density was kept at 4 A dm^{-2} in all samples. SEM analysis revealed that grain size of cermets prepare via this method is reduced significantly, with the CeO_2 nanoparticles uniformly distributed.

An attempt of Cu/CeO_2 nanocomposite cermet was reported by Lee et al. [78]. They used a mixture of copper(II) sulfate pentahydrate with sulfuric acid as electrolyte at 50 °C and 20 g/L ceria with approximate mean particles diameter of 20 nm. Anode and cathode were dipped in bath in parallel with a distance of 20 mm. A 3 mm diameter ultrasonic tip was also immersed in the suspension at a position where it was out of the inter-electrode area and used synchronously to the electrodeposition. At an ultrasonic intensity of 120 W cm^{-2} and diluted electrolyte concentration, a higher volume fraction of CeO_2 snanoparticles in Cu film of about 20.5 vol.% was observed compared to the silent deposition.

Decoration of Ceramic Species with Nanoparticles

Most of the times, nanoparticles prepared via the sonoelectrochemical route remain in a colloidal solution state that is prone to slow or rapid agglomeration depending on whether a suitable surfactant was used. In order to take advantage of the synthesized species minuscule size and a good dispersion of them on the ceramic powders one has to proceed to their *decoration* on ceramic substrate material. This has been achieved in the past via *impregnation*. According to it, substrate powders are dispersed in the colloidal solution and accordingly agitated until successful residing of nanoparticles in nanopores or cavities of the substrate powders. However the use of ultrasonication has demonstrated remarkable results in decoration. Gedanken has reported a study of decoration on several substrates via ultrasonication [79]. Especially in case of Au nanoparticles decoration on state-of-the-art GDC powders, Sakkas et al. have used direct high power ultrasonication with successful results [65, 80].

10.5.3 For Other Fuel Cells

Molten carbonate fuel cells experience cathode material dissolution under extreme conditions. Nickel is used widely as cathode material. Addition of $LiCoO_2$ on the surface of the cathode is known to reduce the precipitation of Ni in the bath [81]. It is also reported that CeO_2 layers and CoO/CeO_2 layers reduce NiO dissolution, and catalyze the Li incorporation into the NiO cathode [82]. As a result, there is strong interest in Co incorporation in CeO_2 composites for MCFCs cathodes protection. Sonoelectrochemical synthetic route has been used many times for preparation of Co, Ni or their alloys. Argirusis et al. achieved a co-deposition of Co and GDC with CeO_2 content of up to 8 wt% [76]. Nevertheless, Reisse et al. reported the formation of a new alloy that consists of 75 %Co-25 %Ni [54]. Delplancke et al. have also reported Co nanoparticles synthesis on average size distribution of 100 nm [59]. Finally, Dabalà et al. reported the synthesis of Co-Fe alloys [83].

10.6 Conclusions

The sonoelectrochemical methods for the preparation of efficient *mono-* and *bi-*metallic nanoparticle electrocatalysts have an advantage over many other methods due to the unusual experimental conditions caused by cavitation, water sonolysis and enhanced mass-transport phenomenon. These methods are expected to be promising in fuel cell technologies as they are cost-effective, easy to use and less time-consuming than any other conventional methods.

References

1. Sammes N (ed) (2006) Fuel cell technology: reaching towards commercialization. Springer, London
2. Blomen LJMJ, Mugerwa MN (1993) Fuel cell systems. Plenum Press, New York
3. Mench MM (2008) Fuel cell engines. Wiley, London
4. O'Hayre R, Colella W, Cha S-W, Prinz FB (2009) Fuel cell fundamentals. Wiley, London
5. Srinivasan S (2006) Fuel cells: from fundamentals to applications. Springer, London
6. Pollet BG (ed) (2012) Power ultrasound in electrochemistry: from versatile laboratory tool to engineering solution. Wiley, Chichester
7. Mason TJ, Lorimer JP (1998) Sonochemistry, theory, applications and uses of ultrasound in chemistry. Ellis Horwood, Chichester
8. Thorneycroft J, Barnaby SW (1895) Torpedo-boat destroyers. Inst Civ Eng 122
9. Rayleigh L (1917) On the pressure developed in a liquid during the collapse of a spherical cavity. Philos Mag 34(199-04):94–98
10. Richards WT, Loomis AL (1927) Chemical effects of high frequency sound waves I. A preliminary survey. J Am Chem Soc 49:3086–3100
11. Pollet BG (1998) The effect of ultrasound upon electrochemical processes. Dissertation, Coventry University, England, UK
12. Noltingk BE, Neppiras EA (1950) Cavitation produced by ultrasonics. Proc Phys Soc B 63:674
13. Pollet BG, Hihn J-Y, Doche M-L, Lorimer JP, Mandroyan A, Mason TJ (2007) Transport limited currents close to an ultrasonic horn equivalent flow velocity determination. J Electrochem Soc 154:E131–E138
14. Moriguchi N (1934) The influence of supersonic waves on chemical phenomena. III. The influence on the concentration polarisation. J Chem Soc Jpn 55:749–750
15. Schmid G, Ehret L (1937) Beeinflussung der Metallpassivität durch Ultraschall. Z Elektrochem 43:408–415
16. Schmid G, Ehret L (1937) Beeinflussung der Elektrolytischen Abscheidungspotentiale von Gasen durch Ultraschall. Z Elektrochem 43:597–608
17. Kolb J, Nyborg W (1956) Small-scale acoustic streaming in liquids. J Acoust Soc Am 28:1237–1242
18. Penn R, Yager E, Hovorka F (1959) Effect of ultrasonic waves on concentration gradients. J Acoust Soc Am 31:1372
19. Bard A (1965) High speed controlled potential coulometry. Anal Chem 35:1125–1128
20. Mason TJ, Lorimer JP, Walton DJ (1990) Sonoelectrochemistry. Ultrasonics 28:333–337
21. Shen Q, Jiang L, Zhang H, Min Q, Hou W, Zhu J-J (2008) Three-dimensional dendritic Pt nanostructures: sonoelectrochemical synthesis and electrochemical applications. J Phys Chem C 112:16385–16392
22. Zin V, Pollet BG, Dabalá M (2009) Sonoelectrochemical (20 kHz) production of platinum nanoparticles from aqueous solutions. Electrochim Acta 54:7201–7206
23. Shen Q, Min Q, Shi J, Jiang L, Zhang J-R, Hou W, Zhu J-J (2009) Morphology-controlled synthesis of palladium nanostructures by sonoelectrochemical method and their application in direct alcohol oxidation. J Phys Chem C 113:1267–1273
24. Qiu X-F, Xu J-Z, Zhu J-M, Zhu J-J, Xu S, Chen HY (2003) Controllable synthesis of palladium nanoparticles via a simple sonoelectrochemical method. J Mater Res 18:1399–1404
25. Steele BCH (1999) Fuel cell technology: running on natural gas. Nature 400:619–620
26. Bessler WG, Vogler M, Störmer H, Gerthsen D, Utz A, Weber A, Ivers-Tiffée E (2010) Model anodes and anode models for understanding the mechanism of hydrogen oxidation in solid oxide fuel cells. Phys Chem Chem Phys 12:13888–13903
27. Park S, Vohs JM, Gorte RJ (2000) Direct oxidation of hydrocarbons in a solid-oxide fuel cell. Nature 404:265–267
28. Trovarelli A (1996) Catalytic properties of ceria and ceria-containing materials. Catal Rev Sci Eng 38:439–520

29. Tsipis EV, Kharton VV (2008) Electrode materials and reaction mechanisms in solid oxide fuel cells: a brief review—I.Performance-determining factors. J Solid State Electrochem 12:1039–1060
30. Minh NQ, Takahashi T (1995) Science and technology of ceramic fuel cells. Elsevier, Amsterdam
31. Ford DC, Nilekar AU, Xu Y, Mavrikakis M (2010) Partial and complete reduction of O_2 by hydrogen on transition metal surfaces. Surf Sci 604:1565
32. Peng G, Mavrikakis M (2015) Adsorbate diffusion on transition metal nanoparticles. Nano Lett 15:629
33. Presvytes D, Vayenas CG (2007) Mathematical modeling of the operation of SOFC Nickel-cermet anodes. Ionics 13:9–18
34. Qu NS, Zhu D, Chan KC (2006) Fabrication of Ni–CeO_2 nanocomposite by electrodeposition. Scr Mater 54:1421–1425
35. Faes A, Hessler-Wyser A, Presvytes D, Vayenas CG, Van herle J (2009) Nickel-zirconia anode degradation and triple phase boundary quantification from microstructural analysis. Fuel Cells 9:841–851
36. Gong M, Liu X, Trembly J, Johnson C (2007) Sulfur-tolerant anode materials for solid oxide fuel cell application. J Power Sources 168:289–298
37. Mark Ormerod R (2003) Solid oxide fuel cells. Chem Soc Rev 32:17–28
38. Brandon NP, Skinner S, Steele BCH (2003) Recent advances in materials for fuel cells. Annu Rev Mater Res 33:183–213
39. Tao S, John Irvine JTS (2004) Catalytic properties of the perovskite oxide $La_{0.75}Sr_{0.25}Cr_{0.5}Fe_{0.5}O_{3-\delta}$ in relation to its potential as a solid oxide fuel cell anode material. Chem Mater 16:4116–4121
40. Myung J-H, Ko H-J, Lee J-J, Lee J-H, Hyun S-H (2012) Synthesis and characterization of NiO/GDC-GDC dual nano-composite powders for high performance methane fueled solid oxide fuel cells. Int J Hydrogen Energy 37:11351–11359
41. Gavrielatos I, Drakopoulos V, Neophytides SG (2008) Carbon tolerant Ni-Au SOFC electrodes operating under internal steam reforming conditions. J Catal 259:75–84
42. Bebelis S, Neophytides SG, Kotsionopoulos N, Triantafyllopoulos N, Colomer MT, Jurado J (2006) Methane oxidation on composite ruthenium electrodes in YSZ cells. Solid State Ion 177:2087–2091
43. Tao S, Irvine JTS (2004) Synthesis and Characterization of (La0.75Sr0.25)Cr0.5Mn0.5O3-δ, a Redox-Stable, Efficient Perovskite Anode for SOFCs. J Electrochem Soc 151:A252
44. Nikolla E, Schwank J, Linic S (2009) Comparative study of the kinetics of methane steam reforming on supported Ni and Sn/Ni alloy catalysts: the impact of the formation of Ni alloy on chemistry. J Catal 263:220–227
45. Sasaki K, Susuki K, Iyoshi A, Uchimura M, Imamura N, Kusaba H, Teraoka Y, Fuchino H, Tsujimoto K, Uchida Y, Jingo N (2006) H_2S poisoning of solid oxide fuel cells. J Electrochem Soc 153:A2023–A2029
46. Matsuzaki Y, Yasuda I (2000) The poisoning effect of sulfur-containing impurity gas on a SOFC anode: Part I. Dependence on temperature, time, and impurity concentration. Solid State Ion 132:261–269
47. Trembly JP, Marquez AI, Ohrn TR, Bayless DJ (2006) Effects of coal syngas and H_2S on the performance of solid oxide fuel cells: single-cell tests. J Power Sources 158:263–273
48. Liu M, Wei G, Luo J, Sanger AR, Chuang KT (2003) Use of metal sulfides as anode catalysts in H_2S-Air SOFCs. J Electrochem Soc 150:A1025–A1029
49. Hahn K, Mavrikakis M (2014) Atomic and molecular adsorption on Re(0001). Top Catal 57:54
50. McEvoy AJ, Smith MJ (2007) Regeneration of anodes exposed to sulfur. ECS Trans 7:373–380
51. Yentekakis IV, Vayenas CG (1989) Chemical cogeneration in solid electrolyte cells. The oxidation of formula to formula. J Electrochem Soc 136:996–1002

52. Nilekar AU, Sasaki K, Farberow CA, Adzic RR, Mavrikakis M (2011) Mixed-metal Pt monolayer electrocatalysts with improved CO tolerance. J Am Chem Soc 133:18574
53. Wei GL, Liu M, Luo JL, Sanger AR, Chuang KT (2003) Influence of gas flow rate on performance of H_2S/air solid oxide fuel cells with MoS_2-NiS-Ag anode. J Electrochem Soc 150:A463–A469
54. Reisse J, Caulier T, Deckerkheer C, Fabre O, Vandercamrnen J, Delplancke JL, Winand R (1996) Quantitative sonochemistry. Ultrason Sonochem 3:147–151
55. Sáez V, Mason TJ (2009) Review—sonoelectrochemical synthesis of nanoparticles. Molecules 14:4284–4299
56. González-García J, Esclapez MD, Bonete P, Hernández YV, Garretón LG, Sáez V (2010) Current topics on sonoelectrochemistry. Ultrasonics 50:318–322
57. Pollet BG (2010) The use of ultrasound for the fabrication of fuel cell materials. Int J Hydrogen Energy 35:11986–12004
58. Compton RG, Eklund JC, Marken F, Rebbitt TO, Akkermans RP, Waller DN (1997) Dual activation: coupling ultrasound to electrochemistry—an overview. Electrochim Acta 42:2919–2927
59. Delplancke J-L, Di Bella V, Reisse J, Winand R (1994) Production of metal nanopowders by sonoelectrochemistry. MRS Proc 372:75
60. Davis J, Vaughan DH, Stirling D, Nei L, Compton RG (2002) Cathodic stripping voltammetry of nickel: sonoelectrochemical exploitation of the Ni(III)/Ni(II) couple. Talanta 57:1045–1051
61. Jia F, Hu Y, Tang Y, Zhang L (2007) A general nonaqueous sonoelectrochemical approach to nanoporous Zn and Ni particles. Powder Technol 176:130–136
62. Liu Y-C, Lin L-H, Chiu W-H (2004) Size-controlled synthesis of gold nanoparticles from bulk gold substrates by sonoelectrochemical methods. J Phys Chem B 108:19237–19240
63. Aqil A, Serwas H, Delplancke JL, Jérôme R, Jérôme C, Canet L (2008) Preparation of stable suspensions of gold nanoparticles in water by sonoelectrochemistry. Ultrason Chem 15:1055–1061
64. Shen Q, Min Q, Shi J, Jiang L, Hou W, Zhu J-J (2011) Synthesis of stabilizer-free gold nanoparticles by pulse sonoelectrochemical method. Ultrason Sonochem 18:231–237
65. Sakkas P, Schneider O, Martens S, Thanou P, Sourkouni G, Argirusis C (2012) Fundamental studies of sonoelectrochemical nanomaterials preparation. J Appl Electrochem 42:763–777
66. Rao CNR, Muller A, Cheetan AK (2008) The chemistry of nanomaterials synthesis, properties and applications, vol 1. Wiley-VCH Verlag GmbH & Co., Weinheim, p 151
67. Haas I, Shanmugam S, Gedanken A (2006) Pulsed sonoelectrochemical synthesis of size-controlled copper nanoparticles stabilized by poly(N-vinylpyrrolidone). J Phys Chem B 110:16947–16952
68. Sáez V, Graves J, Paniwnyk L, Mason TJ (2010) Copper electrocrystallization on titanium electrodes: controlled growth of copper nuclei using a potential step technique. Phys Procedia 3:111–115
69. Schneider O, Matić S, Argirusis C (2008) Application of the electrochemical quartz crystal microbalance technique to copper sonoelectrochemistry Part 1. Sulfate-based electrolytes. Electrochim Acta 53:5485–5495
70. Zhu J, Liu S, Palchik O, Koltypin Y, Gedanken A (2000) Shape-controlled synthesis of silver nanoparticles by pulse sonoelectrochemical methods tools. Langmuir 16:6396–6399
71. Liu S, Huang W, Chen S, Avivi S, Gedanken A (2001) Synthesis of X-ray amorphous silver nanoparticles by the pulse sonoelectrochemical method. J Non Cryst Solids 283:231–236
72. Liu YC, Lin L-H (2004) New pathway for the synthesis of ultrafine silver nanoparticles form bulk silver substrates in aqueous solutions by sonoelectrochemical methods. Electrochem Commun 6:1163–1168
73. Jiang L-P, Wang A-N, Zhao Y, Zhang J-R, Zhu J-J (2004) Novel route for the preparation of monodisperse silver nanoparticles via a pulsed sonoelectrochemical technique. Inorg Chem Commun 7:506–509

74. Vu LV, Long NN, Doanh SC, Trung BQ (2009) Preparation of silver nanoparticles by pulsed sonoelectrochemical method and studying their characteristics. J Phys Conf Ser 187:012077
75. Lei H, Tang Y-J, Wei J-J, Li J, Li X-B, Shi H-L (2007) Synthesis of tungsten nanoparticles by sonoelectrochemistry. Ultrason Sonochem 14:81–83
76. Argirusis C, Matić S, Schneider O (2008) An EQCM study of ultrasonically assisted electrodeposition of Co/CeO_2 and Ni/CeO_2 composites for fuel cell applications. Phys Status Solidi A 205:2400–2404
77. Xue Y-J, Liu H-B, Lan M-M, Li J-S, Li H (2010) Effect of different electrodeposition methods on oxidation resistance of $Ni\text{-}CeO_2$ nanocomposite coating. Surf Coat Technol 204:3539–3545
78. Lee D, Gan YX, Chen X, Kysar JW (2007) Influence of ultrasonic irradiation on the microstructure of Cu/Al_2O_3, CeO_2 nanocomposite thin films during electrocodeposition. Mater Sci Eng A 447:209–216
79. Gedanken A (2007) Doping nanoparticles into polymers and ceramics using ultrasound radiation. Ultrason Sonochem 14:418–430
80. Sakkas PM, Schneider O, Sourkouni G, Argirusis C (2014) Sonochemistry in the service of SOFC research. Ultrason Sonochem 21:1939–1947
81. Brenscheidt T, Nitschke F, Söllner O, Wendt H (2001) Molten carbonate fuel cell research II. Comparing the solubility and the in-cell mobility of the nickel oxide cathode material in lithium: potassium and lithium: sodium carbonate melts. Electrochim Acta 46:783
82. Kim MH, Hong MZ, Kim Y-S, Park E, Lee H, Ha H-W, Kim K (2006) Cobalt and cerium coated Ni powder as a new candidate cathode material for MCFC. Electrochim Acta 51:6145
83. Dabalà M, Pollet BG, Zin V, Campadello E, Mason TJ (2008) Sonoelectrochemical (20 kHz) production of $Co_{65}Fe_{35}$ alloy nanoparticles from Aotani solutions. J Appl Electrochem 38:395–402

Chapter 11
Direct Ethanol Fuel Cell on Carbon Supported Pt Based Nanocatalysts

T.S. Almeida, N.E. Sahin, P. Olivi, T.W. Napporn, G. Tremiliosi-Filho, A.R. de Andrade, and K.B. Kokoh

11.1 Introduction

One of the biggest problems nowadays comes from the progressive increase in the concentration of toxic gases due to combustion of fossil fuels, which represents a large part of the energy consumed by the society. The greenhouse gas emissions, especially in large urban centers, reach every year increasingly levels. Besides the large industries, urban transport is another main factor responsible for pollution.

The need for cleaner energy sources and converter systems that combine high efficiency and reduction of environmental impacts has attracted increasing interest from the scientific community. In this context, fuel cells, system capable of converting chemical energy into electrical energy, are receiving special attention mainly due to its ability to generate electricity efficiently and not aggressive to the environment. The versatility of such systems is the feasible use of various fuels, efficient energy conversion, silent operation, and high applicability. These make this system an attractive technology for a "clean" future [1].

T.S. Almeida
Universidade Federal do Triângulo Mineiro—Campus Iturama,
Iturama 382800-000, MG, Brazil

N.E. Sahin • T.W. Napporn • K.B. Kokoh (✉)
Université de Poitiers, IC2MP UMR CNRS 7285, "Equipe SAMCat",
4 rue Michel Brunet B27, TSA 51106, Poitiers Cedex 09 86073, France
e-mail: boniface.kokoh@univ-poitiers.fr

P. Olivi • A.R. de Andrade
Departamento de Química, Faculdade de Filosofia Ciências e Letras de Ribeirão Preto,
Universidade de São Paulo, Ribeirão Preto 14040-901, SP, Brazil

G. Tremiliosi-Filho
Instituto de Quimica de São Carlos, Universidade de São Paulo,
São Carlos 13560-970, SP, Brazil

K.I. Ozoemena, S. Chen (eds.), *Nanomaterials for Fuel Cell Catalysis*,
Nanostructure Science and Technology, DOI 10.1007/978-3-319-29930-3_11

Table 11.1 Types of fuel cells classified according to the electrolyte used

Type	Electrolyte	Working temperature (°C)	Efficiency (%)
Alkaline (AFC)	KOH	60–90	50–60
Polymeric electrolyte (PEMFC)	Polymer: Nafion® (H^+)	50–90	50–60
Phosphoric acid (PAFC)	H_3PO_4 (H^+)	160–200	55
Molten Carbonate (MCFC)	Molten carbonate (CO_3^-)	650–700	60–65
Solid oxide (SOFC)	ZrO_2 (O_2^-)	800–900	55–65

Fuel cell systems have a higher thermodynamic efficiency than internal combustion engines or turbines, which convert chemical energy into heat to perform useful work. Electrical devices produce energy without combustion, and thus there is no need of moving parts. Additionally, fuel cells also reduce the emission of pollutants. There are several types of fuel cells which are classified according to the electrolytes employed [2]. Table 11.1 shows the main types of fuel cells.

Among the fuel cells cited above the proton exchange membrane fuel cell (PEMFC) is one of the most promising converter systems for portable devices [3, 4]. This system operates at low temperatures (50–90 °C) using as electrolyte a proton exchange membrane, with operational simplicity, high efficiency and low gas pollutant emissions which makes its use even more attractive [2]. Nowadays the best performance of PEMFCs is observed for those that use hydrogen as the fuel. Hydrogen can be supplied to the cell either directly or indirectly by the reform of liquid substances [5, 6]. However, the problems in the storage, handling, distribution, and high cost of the equipment used for the hydrogen generation represent a major drawback to its implementation [7, 8]. For this reason, liquid fuels such as methanol and ethanol are being investigated as a convenient alternative to the replacement of hydrogen.

Ethanol becomes particularly attractive as an alternative fuel for use in direct ethanol fuel cell (DEFC) because it can be easily produced from the biomass such as sugar cane, corn and others, it is not harmful and presents high theoretical energy density (8.6 kWh kg^{-1}) compared to methanol (6.1 kWh kg^{-1}) [9]. However, this fuel presents challenging oxidation kinetics, requiring the development of catalysts capable of breaking down the C-C bond [10, 11].

The principle of PEM-DEFC operation is illustrated in Fig. 11.1. The anode is fed by an ethanol solution, while the cathode is composed by humidified oxygen, so that good conductivity is maintained in the proton exchange membrane (PEM).

Ethanol is oxidized to carbon dioxide, and protons and electrons are produced on the anode. Protons are then transferred through the polymeric electrolyte to the cathode side, where they react with oxygen and electrons to produce water. The equations are shown below:

$$\text{Anode: } CH_3CH_2OH + 3H_2O \rightarrow 2CO_2 + 12H^+ + 12e^- \quad E^\circ = -0.084 \text{ V vs. SHE} \tag{11.1}$$

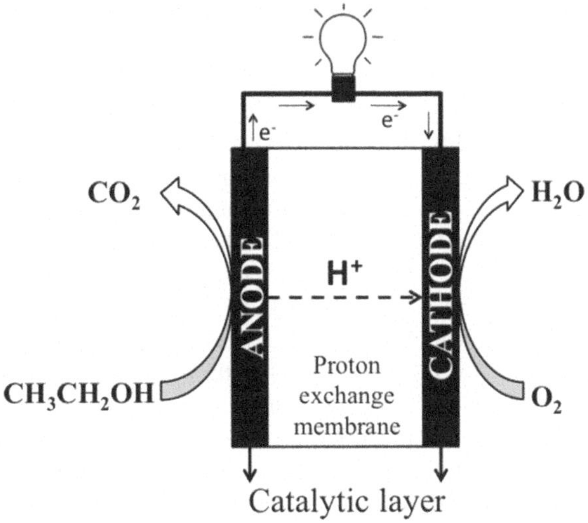

Fig. 11.1 Schematic diagram of a DEFC

$$\text{Cathode: } 3O_2 + 12H^+ + 12e^- \rightarrow 6H_2O \qquad E^\circ = 1.229\,\text{V vs. SHE} \tag{11.2}$$

$$\text{Global Equation: } CH_3CH_2OH + 3O_2 \rightarrow 2CO_2 + 3H_2O \quad E^\circ = 1.313\,\text{V vs. SHE} \tag{11.3}$$

There are two challenge that must be overcome to implement this technology: (i) on the anode side, one must find a way to achieve the complete oxidation of ethanol to CO_2, to release the theoretical 12 electrons and also beat the strong adsorption of the intermediates of ethanol oxidation onto the active sites of the catalyst, which leads to a loss of efficient energy [10, 12]; (ii) on the cathode side, enhance the slow kinetics of oxygen reduction at the temperature in which the membrane is thermically stable (below 120 °C).

11.2 Bi and Tri-Metallic Materials: Effect of the Co-catalyst on the Pt-Based Anode

Pt-based catalysts are considered as the main material for ethanol electro-oxidation in acidic medium. However, pure Pt does not display good catalytic activity due to poisoning of its active sites with intermediates generated during ethanol electrooxidation [13–15]. Bimetallic and trimetallic Pt-based catalysts can enhance the oxidation of CO and small organic molecules due to a bifunctional mechanism [16, 17] or to electronic effects [18]. For instance, metals such as Sn, Ni, Rh, Ru, and W are largely investigated as co-catalysts in the composition of the electrode materials in order to minimize the poisoning of Pt sites [13, 14, 19–29].

Of these, Pt-Sn composites display excellent activity for the oxidation of ethanol, as reported by Purgato et al. [30, 31], Neto et al. [13], and Almeida et al. [22, 32]. This improvement is attributed to the bifunctional mechanism exerted by the Sn and its oxides form which are responsible for the H_2O activation at their surfaces at low overpotentials as compared to pure Pt, which leads to a reduced onset potential and enhanced catalytic activity of ethanol oxidation [20, 33–35]. Nevertheless, even if PtSn has shown better results than pure Pt catalyst, most of the products formed at this bimetallic composite still contain C-C bonds, which suggests that the activity of PtSn/C, Pt and PtSn must be further enhanced by the addition of a third element to improve the dehydrogenation reaction and the cleavage of the C-C bond.

Ru is another metal largely investigated as a co-catalyst and plays a similar role as Sn in ethanol electrooxidation. The adsorption and decomposition of ethanol and its intermediate reaction products occur on Pt active sites, while the dissociative adsorption of water occurs over Sn or Ru sites to form oxygen-containing surface species. As reported by Emerson et al. [36] the addition of Ru to PtSn-based catalysts showed good performance in the electrochemical oxidation of ethanol. Its presence lowered the onset potential of ethanol oxidation (near 0.2 V), but acetaldehyde was the major byproduct of the reaction. Antolini et al. [14] also found that Ru addiction to PtSn catalysts enhances the catalytic activity and this effect is related to the Ru/Sn ratio in the alloy and the synergistic effect of Ru and Sn oxides.

More recently Ni has been studied as the second and/or third element added to Pt, PtSn or PtRu catalysts. The main advantage of the introduction of this metal is the shift of the oxidation potential of ethanol to lower values, coupled with an increase in current density. The literature reveals that Sn and Ni decrease the energy of the chemisorption of ethanol and its oxidation intermediates, such as CO on Pt-sites [37, 38]. The addition of a third metal improves the oxophilic character of the surface, thus raising the strength of the Sn-O bond and contributing to increased acidity of the surface of the Sn-OH sites. This accentuates the bifunctional mechanism facilitating the ethanol electrooxidation [39].

The beneficial effect of Ni addition has been reported by Ribadeneira and Hoyos [40] who evaluated the catalytic effect of Ni on binary PtSn/C and PtRu/C and concluded that the addition of Ni on PtRu/C and PtSn/C catalysts increases significantly their catalytic activities to ethanol electro-oxidation. Bonesi et al. [41] also studied the Ni effect on PtSn/C and PtRu/C and the chronoamperometry characterization showed a higher current density for the ternary composition PtSnNi/C, again demonstrating the beneficial effect of Ni addition.

Rh has received great attention as a co-catalyst in ethanol electrooxidation due to its ability to break the C-C bond more efficiently than other metals studied so far. Souza et al. [42] carried out DEMS and *in-situ* FTIR studies and monitored the catalytic activity of Pt, Rh and PtRh. They showed that Pt-Rh combination decreased significantly the concentration of acetaldehyde and increased CO_2

concentration during the ethanol electrooxidation as compared pure platinum. They attributed this behavior to the formation of Rh-O species that act as oxygen supply promoting the formation of CO-O bonds and finally CO_2, therefore, releasing 12 electrons from the ethanol molecule. Sagar et al. [43] investigated Pt:Rh (90:10, 70:30 and 35:65) catalysts and have shown an activity strongly dependent on the Rh amount. Its presence above 30 % weight has a positive contribution to the ethanol oxidation because it assists in breaking the C-C bond. Lima and Gonzalez [19] investigated Pt-Ru, Pt-Rh, Pt-Ru-Rh activities and observed by XANES characterization a pronounced electronic modification of the Pt 5d-band in the presence of Rh and Ru. This consequently reduces the strong adsorption of ethanol intermediates on Pt, increasing their rates of oxidation and decreases its oxidation onset potential. The presence of Rh forms a larger quantity of acetic acid and CO_2. Rh has presented promising results for the C-C bond cleavage [44, 45]; its high cost represents a barrier since it means more expensive catalysts.

Less investigated than other metals described so far, W has been reported in some studies towards ethanol electrooxidation. Maillard [46] investigated the tolerance to CO on Pt-WO_x composites supported on carbon, where the characterization by X-ray absorption showed that WO_x promotes slight modification of the electronic structures of the platinum surface. The presence of WO_x species assist the electrooxidation of ethanol acting as a Bronstëd acid in the process of dehydrogenation and water molecules dissociation. Ribeiro et al. [25] and dos Anjos et al. [26] observed a higher catalytic activity for the oxidation of ethanol at PtSnW than at PtSn/C. They attributed the improvement of the catalytic activity to the presence of Sn and W that decreased the overpotential for the intermediate species removal.

A considerable number of different metals were evaluated as possible Pt structure modifiers. However, the metals described above have shown more significant effects. With the large amount of materials already studied for ethanol electrooxidation since the beginning of the development of fuel cells to date, it can be found in the literature a large library of materials with different compositions and metal ratios.

11.3 Catalyst Synthesis Methods

The synthesis methods are an essential part to obtain efficient catalysts for applying in DEFC. It is known that the catalyst activity is dependent on their composition, morphology, and particle size [47, 48] and the methodology applied to their synthesis can lead to better control of these parameters [41, 49, 50]. In this context, many research groups have dedicated efforts to develop efficient methods for the synthesis of nanostructured catalysts, aiming at obtain materials with homogeneous metallic distribution, small particle size, and high catalytic activity [49, 51–53]. Below we describe some of the methods employed in the catalysts synthesis.

Microemulsion method: This method is convenient to synthesize nanoparticles by using a reducing agent, normally $NaBH_4$, in combination with different strategies for better control of nanoparticle size and composition. More precisely, the microemulsion method is based on the use of two immiscible solvents in the presence of a surfactant forming stable microscopic structures called micelles or reverse micelles. On a macroscopic scale, the resulting solution seems to be homogeneous but in fact, on a nanometer scale, the solution is heterogeneous and it is composed of nanodroplets dispersed in a continuous organic phase. The micellar medium is responsible to control the chemical reduction rate of metallic cations and particles size. The solvent and the surfactant used must be chosen carefully to allow a good control of the micelle volume which is directly related to the particle size, and the reducing agent should be stable in the medium (solvent) avoiding parallel reaction with the micelle components [54].

Pt-based catalysts can be easily synthesized by preparing a microemulsion by dissolving the metal precursors in ultrapure water then adding an organic solvent such as n-heptane following by the addition of polyethylene glycol dodecyl ether (Brij® 30, surfactant). After mixing the solutions to form reverse micelles the carbon support is added following by the reducing agent sodium borohydride ($NaBH_4$). Finally after nanoparticle formation, the material is filtered, washed and ground [54, 55]. For example, Godoy et al. [21] have prepared PtSn/C catalysts with sizes around 2.5 nm that were well dispersed on the carbon support. The authors claimed that the reduction of metals trapped in the micelles enabled the achievement of well-distributed particles.

Alcohol-reduction method: Developed by Toshima and Yonezawa [56, 57] this method provides colloidal dispersions of nanoparticles with uniform particle size and homogeneous distribution. The method consists in refluxing an alcohol solution that acts as a solvent and reducing agent [52], usually ethylene glycol, containing the metallic ions and a stabilizing agent (polymer or surfactant), providing homogeneous colloidal dispersions of the corresponding metal nanoparticles. This method offers some advantages such as good reproducibility, satisfactory distribution, and small particle size.

Neto et al. [13] have investigated PtRu/C, PtSn/C, and PtSnRu/C catalysts prepared by the alcohol-reduction method using an ethylene glycol/water solution. Particle of the order of 2.7 nm were achieved for the catalysts. Spinacé et al. [23] have used this method for the production of PtSn/C, PtRh and PtSnRh/C catalysts. They also obtained catalysts with small particle size (2.0 nm), which culminated in significant catalytic activity for ethanol electrooxidation.

Formic acid method: This method was developed with the purpose of obtaining dispersed platinum catalysts supported on carbon for PEMFC fuel cells [58]. Basically, the catalyst is obtained by the addition of a high surface area carbon support (Vulcan XC-72, carbon nanotube, etc.) in a formic acid solution, which is used as a reducing agent. The resulting mixture is heated to 80 °C followed by addition of aliquots of a solution containing platinum and other metal salts to the reaction vessel. After complete metal reduction, the nanocatalyst is filtered, dried, and grounded. Antolini et al. [14] have looked into the effects of introducing ruthenium into PtSn/C catalysts prepared by the formic acid method. They obtained the required catalytic composition with particles of small size (3.5 nm) homogeneously distributed onto the carbon support.

Bönnemann method [55, 59]: This method consists in the production of a stable colloid in an inert and dry atmosphere using non-hydrated metallic chlorides and solvents. This approach can be successfully applied to obtain Pt-based catalysts containing different metals and metal oxides, in particular Sn, V, W and Mo, and other transition-metal elements such as Cu, Fe, Co and Ni [60].

The anhydrous salts are dissolved in anhydrous tetrahydrofuran (THF) with an appropriate amount of tetraoctylammonium bromide [$N(oct)_4Br$]. The reducing agent is produced by dissolving [$N(oct)_4Br$] in THF and adding [$KHBr(Et)_3$], resulting in the formation of a stark reducing agent, triethylhydroborate of tetraoctylammonium [$N(oct)_4HBr(Et)_3$] (Eq. 11.4). The reducing agent will also act as a surfactant after metal reduction, preventing any agglomeration of the metallic particles.

$$N\,(oct)_4Br + KHB(et)_3 \rightarrow N(oct)_4HB(et)_3 + KBr \tag{11.4}$$

For the reduction of metallic ions, a 50 % excess over the stoichiometric amount of [$N(oct)_4HBr(Et)_3$] is added to the solution of metal salts and heated under stirring. The reduction occurs with hydrogen evolution, as shown in Eq. 11.5, and the solution darkening.

$$MeX_n + N(oct)_4HBr(et)_3 \rightarrow M^*\left[N(oct)_4\right]^+ + nB(et)_3 + {}^{n}\!/_{2}H_2 + nX^- \tag{11.5}$$

After the synthesis of the stable colloid, a suspension of carbon black in THF is prepared and the colloid solution is added in a dropwise fashion to the carbon black dispersion, finishing the preparation of the powder catalyst with filtration and several washes using THF and ethanol. The catalyst is calcined at 300 °C for 1 h under air atmosphere to remove the organic surfactant.

Thermal decomposition of polymeric precursors (DPP) method: The methodology based on the Pechini method [61] was initially developed for the preparation of ceramic materials and later adapted for the preparation of electrode materials. Firstly, citric acid (CA) is dissolved in a polyhydroxylic alcohol (ethylene glycol, EG) at 60 °C. Metallic salt (Pt or another metal) is added and the temperature is raised to 90 °C, forming a polymeric resin containing the metallic cation [22, 25, 30, 31]. To obtain the catalyst, the resins are stoichiometrically mixed with carbon Vulcan XC-72 or any other support, and fired at high temperatures; e.g., 300–400 °C. The major advantage of this route is the attainment of robust catalysts with experimental composition close to the nominal one. This method has been employed in the synthesis of a series of Pt-based materials [25, 26, 30, 62].

It is important to know that the degree of esterification, hardness and porosity of the resin can be controlled by taking in account the molar ratio between the complexing agent [CA], the metal cation [METAL], and the ethylene glycol

[EG]. Thus, two molar ratios: [CA]/[METAL] = *CM* and [CA]/[EG] = *CE* can be defined. The microstructure of the resulting powder is related with these two ratios. *CM* describes the degree of the chelation process of the metal in the organic product. For solutions with low *CM* ratios ($CM < 1$), there may be precipitation of the metals salts in the solution, and therefore, no uniformity in the chelation of the metals [63], which is due to an insufficient amount of organic molecules to chelate metallic cations. On the other hand, for relatively high *CM* ratios, it is easy to dissolve the metal ions but the amount of organics to be removed is excessive. *CE* describes the degree of esterification between the chelating agent and the ethylene glycol and affects powder morphologies. Equimolar ratio leads to porous resin [64] generating a rigid polyester net that reduces any segregation of metals during the polymer decomposition process at high temperatures [65].

Microwave-assisted heating method: This method has gained importance due to its operational simplicity, efficiency, and reduced preparation time [23, 53, 66, 67]. It promotes rapid reduction of the metallic precursors and is responsible for the formation of nanometric particles size [68]. However, its application must meet some criteria; the solvent acts as a reducing agent and must show high reduction conversion of metallic ions, high capacity for conversion of electromagnetic energy into thermal energy, and suitable temperature profile for reduction for all metallic ions. Figure 11.2 depicts a voltammogram of a PtSn/C electrode material obtained from this synthetic method.

The efficiency of reducing agents in stabilizing the nanoparticles and preventing formation of metal clusters is related to the oxidizing species that is generated during the heating process [53]. Oxidation of small-chain alcohols occurs through interaction between the hydroxyl group(s) present in these molecules and the metallic ions, which are subsequently reduced to metallic particles forming carbonyl and/or carboxyl species which adsorbs onto the particle surface promoting their stabilization and preventing their agglomeration, which culminates in formation of small particles [69].

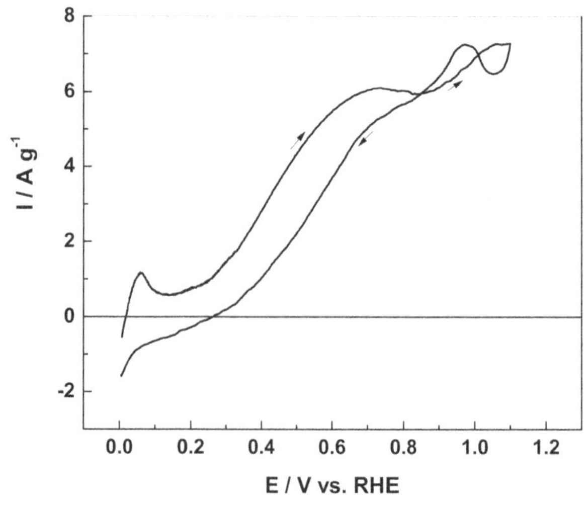

Fig. 11.2 Cyclic voltammogram of the $Pt_{70}Sn_{30}/C$ (20 wt.% metal loading) electrode material recorded in supporting electrolyte 0.05 mol L^{-1} H_2SO_4 and in the presence of 1.0 mol L^{-1} ethanol at 10 mV s^{-1}. Reprinted with permission from [82]. Copyright (2012), Elsevier

Ethanol may be used as a reducing agent and its oxidation under microwave-assisted heating leads to formation of acetaldehyde and acetic acid, which are the stabilizing species. When ethylene glycol is applied as the reducing agent, there is also the formation of acetaldehyde, which acts also as a reducing agent, and acetic acid as stabilizing agent [70, 71]. In another proposal, Bock et al. [69] claimed that the stabilizing agent from microwave-assisted heating of ethylene glycol should be oxalic and glycolic acid. The use of propylene glycol differs from ethylene glycol only in the final obtained stabilizer that are lactic and pyruvic acids [72, 73].

Besides the formation of stabilizing species during the heating processes, some stabilizing compounds can be used in the nanoparticle synthesis that improve the size control and the particle distribution [66, 74, 75]. These stabilizing agents can act in two different ways depending on their chemical structure. When the stabilizing species is a surfactant it can hinder the nanoparticle growth due to steric hindrance, where the presence of a large carbon chain in the structure of these molecules prevents the formation of metal clusters during the reduction process [76]. In another approach, the use of anions such as citrate and acetate assists in electrostatic stabilization. These molecules have the ability to form chelates that electrostatically repels species with similar charge preventing the formation of metallic clusters [70, 77]. Electrostatic stabilizing agents act in two ways: (i) in the formation of stabilizing agent-metal ion pairs; and (ii) adsorbing on fresh-formed nanoparticles surface to prevent further reduction and cluster formation. Thus, the more efficient the stabilizing agents the smaller the particle size and more homogeneous will be its distribution onto de carbon support [78].

Table 11.2 summarizes different catalyst compositions synthesized by various methodologies, highlighting the particle and crystallite sizes. As can be seen in

Table 11.2 Average particle size and crystallite size for different catalysts prepared by different literature methods

Methodology	Composition	Particle size (TEM) (nm)	Crystallite size (DRX) (nm)	Ref.
Microemulsion	$Pt_{75}Sn_{25}$	3.0	–	[21]
Alcohol reduction	$Pt_{52}Sn_{48}$	3.0	2.5	[23]
	$Pt_{52}Sn_{42}Ni_{06}$	2.4	–	[79]
	$Pt_{67}Sn_{11}Rh_{22}$	2.0	3.0	[23]
	$Pt_{49}Sn_{29}Ru_{22}$	3.3	–	[13]
Formic acid	$Pt_{77}Sn_{33}$	4.5	3.9	[52]
	$Pt_{40}Sn_{45}Pd_{15}$	2.5	3.1	[58]
	$Pt_{30}Sn_{37}Rh_{32}$	1.8	3.0	[80]
DPP	$Pt_{78}Sn_{22}$	5.0	7.4	[25]
	$Pt_{72}Sn_{28}$	8.2	6.5	[22]
	$Pt_3Sn_1Ni_1$	5.1	3.7	[81]
	$Pt_{83}Sn_{09}Ni_{06}$	4.3	3.3	[22]
	$Pt_{78}Sn_{13}Ru_{09}$	5.7	4.8	[36]
MW	$Pt_{35}Sn_{65}$	4.2	3.9	[62]
	$Pt_{73}Sn_{27}$	2.8	2.0	[82]

Table 11.2, the particle and crystallite size is virtually the same for all the methods listed. However it is important to notice that each method has specific advantages that may be useful depending on the desired catalyst structure and morphology. Here we listed only usual methods of synthesis of nanoparticles dispersed on carbon. However, for specific structures, such as core-shell, single crystals and others more specific methods must be applied.

11.4 Oxygen Reduction Reaction on Carbon Supported Pt Catalyst

The electrocatalytic reduction of molecular oxygen has been intensively researched for many practical systems because this reaction plays a key role for achieving the overall reaction in fuel cells. A multi-electron transfer electrode for molecular oxygen reduction in acid solution is of primary concern because of its possible utilization as cathode in fuel cells. The mechanism of the electrochemical oxygen reduction reaction is quite complicated and involves many intermediates, primarily depending on the nature of the electrode material, catalyst and electrolyte. Stonehart and Ross [83] proposed the use of rotating thin layer electrode (RTLE) for the direct measurement of electrochemical reaction rate constant for the first time. To elucidate the oxygen reduction mechanism, the rotating disk electrode (RDE) and rotating ring disk electrode (RRDE) have been widely used since RDE and RRDE enable the determination of kinetics and the quantitative detection of the intermediate reaction species [84–90]. In these methods the oxygen reduction reaction (ORR) occurs on the disk electrode and the produced hydrogen peroxide is detected at the ring.

11.4.1 Oxygen Reduction Reaction Pathways

As the ORR involves four electrons, four protons and a O-O bond cleavage, it has a complex mechanism in acidic [91–93], alkaline [88, 94, 95] and neutral [90] electrolytes at low and moderate temperatures. Depending on the electrode material where the electrochemical oxygen reduction reaction takes place, H_2O_2 may be generated during the ORR as an intermediate or a final product [96–99].

Since the first reaction scheme proposed for the oxygen reduction reaction by Damjanovic et al. [100], several reaction schemes have been proposed in order to identify the mechanistic reaction pathways. Besides the oxygen reduction reaction schemes proposed by Damjanovic et al. [100], Wroblowa et al. [101], Appleby et al. [102] Anastasijevic et al. [88], and Zurilla et al. [103], also suggested modification on the ORR mechanism. Figure 11.3 shows a simplified scheme that gives a diagnostic criterion to discriminate between the different pathways.

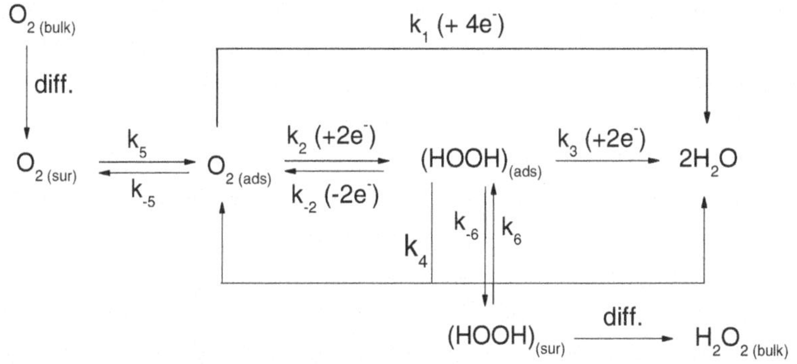

Fig. 11.3 Scheme showing the oxygen reduction reaction pathways proposed by Wroblowa et al. [101] in acidic media. Reprinted with permission from [101]. Copyright (1976), Elsevier

According to the scheme shown in Fig. 11.3, one O_2 molecule is adsorbed on a catalytic site at the electrode surface, and it may either follow the direct four-electron reduction pathway to form water by means of Eq. 11.6 and k_1, or two-electron pathway to generate hydrogen peroxide by means of Eq. 11.7 and k_2. Hydrogen peroxide either desorbs from the surface (k_{-6}), is further reduced to water by means of Eq. 11.8 and k_3, moreover it can desorb from the surface or may chemically decompose to form water and oxygen by means of Eq. 11.9 and k_4. For the hydrogen peroxide decomposition, two pathways (electrochemical and radical) were envisaged. In the case of platinum, hydrogen peroxide decomposition is a combination of electrochemical reduction and oxidation. Hydrogen peroxide decomposition can also be defined as a radical reaction [97, 104, 105]. Molecular oxygen may be converted into oxygenated (O*), hydroxyl (OH*) and superhydroxyl (OOH*) intermediate species, which are not detectable experimentally. Adsorption and coverage of O_2 on the platinum active surface may affect the reaction pathway depending on the dissociative and associative mechanism [106, 107]. According to the density functional theory (DFT) calculations [106], oxygen reduction tends to follow an associative mechanism where superhydroxyl (OOH*) is first formed and then O-O bond is cleaved in the case of two-electron pathway. Additionally, in the direct four-electron reduction pathway, dissociative mechanism is dominant at low oxygen coverage where O-O bond is cleaved before superhydroxyl (OOH*) formation.

$$O_2 + 4H^+ + 4e^- \rightarrow 2H_2O \quad E^\circ = 1.229\ V\ vs.\ SHE \tag{11.6}$$

$$O_2 + 2H^+ + 2e^- \leftrightarrows H_2O_2 \quad E^\circ = 0.682\ V\ vs.\ SHE \tag{11.7}$$

$$H_2O_2 + 2H^+ + 2e^- \rightarrow 2H_2O \quad E^\circ = 1.776\ V\ vs.\ SHE \tag{11.8}$$

$$2H_2O_2 \rightarrow 2H_2O + O_2 \tag{11.9}$$

From the thermodynamic data based on Nernst equation for the electrochemical reactions (in Eqs. 11.6–11.8), the corresponding equilibrium potentials are 1.229, 0.695, 1.763 V *versus* the standard hydrogen electrode (SHE), respectively (Eqs. 11.10–11.12).

$$E_{O_2/H_2O} = E^o_{O_2/H_2O} - \frac{RT}{4F} ln \frac{a^2_{H_2O}}{a_{O_2} a^4_{H^+}} \quad \left(E^o_{O_2/H_2O} = +1.229\ V\right) \tag{11.10}$$

$$E_{O_2/H_2O_2} = E^o_{O_2/H_2O_2} - \frac{RT}{2F} ln \frac{a_{H_2O_2}}{a_{O_2} a^2_{H^+}} \quad \left(E^o_{O_2/H_2O_2} = +0.682\ V\right) \tag{11.11}$$

$$E_{H_2O_2/H_2O} = E^o_{H_2O_2/H_2O} - \frac{RT}{2F} ln \frac{a^2_{H_2O}}{a_{H_2O_2} a^2_{H^+}} \quad \left(E^o_{H_2O_2/H_2O} = +1.776\ V\right) \tag{11.12}$$

where the E^o_n is defined as standard potential of electrochemical reaction n (V), R is the ideal gas constant (8.314 J mol^{-1} K^{-1}), T is the absolute temperature (K), F is the Faraday constant (96,485 C mol^{-1}) and a_i is the activity of the compound i [108, 109].

In the case of alkaline media, while direct four-electron pathway takes place according to Eq. 11.13, peroxide pathway is carried out via Eq. 11.14, and followed by either the reduction of peroxide ion through Eq. 11.15 or decomposition of peroxide *via* Eq. 11.16.

The peroxide pathways related to acidic medium (Eqs. 11.7 and 11.8) and alkaline medium (Eqs. 11.14 and 11.15) may contain adsorbed-states of superoxide $\left(O_2^{*-}\right)$, superhydroxide (HO_2^*) and peroxide $\left(O_2H^-\right)$, (H_2O_2). As the adsorbed peroxide species tend to desorb from the electrode surface, the mechanism can be distinguished [97, 110, 111].

$$O_2 + 2H_2O + 4e^- \rightarrow 4OH^- \quad E^\circ = 0.401\ V \tag{11.13}$$

$$O_2 + H_2O + 2e^- \rightarrow HO_2^- + OH^- \quad E^\circ = -0.076\ V \tag{11.14}$$

$$HO_2^- + H_2O + 2e^- \rightarrow 3OH^- \quad E^\circ = 0.867\ V \tag{11.15}$$

$$2HO_2^- \rightarrow 2OH^- + O_2 \tag{11.16}$$

Thanks to its high energy conversion, avoiding of corrosion of carbon supports, and the elimination of possible degradation of electrode material because of the peroxide formation and free radicals, the most efficient pathway is the direct four-electron reduction of adsorbed oxygen to water without peroxide formation. Moreover, peroxide formation tends to attack the membrane and accelerate the dissolution of platinum, which causes a deterioration of the durability. Therefore, oxygen reduction with high selectivity for the direct four-electron reduction pathway is desired. Great effort has been put into the development of selectively catalyzing the four-electron reduction of oxygen to produce water with low overpotential and low

H_2O_2-production. That's why, it is important to understand the intrinsic characteristics of the surface of the platinum catalyst by quantifying the yield of H_2O_2, that is the molar ratio of produced H_2O_2 to the total of all reduced oxygen [99, 106, 112].

11.4.2 RDE Study on Oxygen Reduction Kinetic

Metal nanoparticles have attracted a great deal of interest in scientific research and industrial applications due to their unique properties based on large surface to volume ratio and quantum size effects [113–115]. It is clear that geometric (Pt-Pt bond distance) [116] and electronic (Pt d-band vacancy) [117] effects are of great importance in the electrocatalytic activity of the Pt/C electrocatalyst. The most efficient catalyst for oxygen reduction reaction is platinum, as its surface reactivity supply catalytic process mainly depending on O and OH bonding energy on the platinum surface [118]. With sluggish reaction kinetics the oxygen reduction reaction includes a multi-electron transfer process involving different reaction intermediates, thus the oxygen reduction requires high overpotentials. To overcome this electrochemical voltage loss, catalysis of oxygen reduction requires a high platinum loading (typically 0.1–0.5 mg cm^{-2}) in fuel cell cathodes [119].

The oxygen reduction reaction has been analyzed on carbon supported platinum (Pt/C) electrocatalysts in 0.1 mol L^{-1} $HClO_4$ through cyclic voltammetry (CV) [120, 121] and rotating disk electrode (RDE) [60, 84, 122–124] techniques. Electrochemical rotating disk electrode (RDE) experiments were conducted in a conventional three-electrode electrochemical cell in 0.1 mol L^{-1} $HClO_4$. The working electrode was composed of 25 μg of catalyt powders by dropping a catalytic ink onto the freshly polished glassy carbon disk. A saturated calomel electrode (0.244 V vs. RHE) and the glassy carbon plate (5 cm^2 geometric surface area) were the reference electrode and the counter electrode, respectively. To avoid the electrolyte contamination by chlorides the reference electrode was separated from the working electrode compartment by a closed Luggin-Haber ionic bridge. The working electrode potential is controlled by an applied driving potential difference between the reference and working electrodes. Owing to the very short distance between the Luggin-Haber capillary tip and the working electrode, in most cases ohmic (iR) corrections are not required. All potentials used were converted to reversible hydrogen electrode (RHE) scale.

11.4.3 Levich and Koutecký-Levich Plots

Rotating disk electrode (RDE) measurements allow analysis of the electrocatalytic activity of the oxygen reduction reaction. Under the steady-state conditions and considering the electrode process is described in Eq. 11.17, which gives a sigmoidal polarization curves with a diffusion-limited current density (j_d).

$$O + n_{e^-} \rightarrow R \tag{11.17}$$

Figure 11.4a displays the polarization curves plotted by current density versus disk potential for the Pt/C in 0.1 mol L^{-1} $HClO_4$ recorded at 5 mV s^{-1} and 20 °C as a function of the rotation rates. It is clearly seen that the oxygen reduction reaction takes place under a mixed kinetic and diffusion control region in the potential range 0.94 and 0.70 V vs. RHE followed by a purely diffusion-limited region. By using Levich plots (j versus $\omega^{1/2}$), the diffusion-limited current density (j_d) without any film formation can be calculated using Eq. 11.18 from the Levich law [125].

$$j_d = 0.20\, nFD_{O_2}^{2/3} C_{O_2} v^{-1/6} \omega^{1/2} = B\omega^{1/2} \tag{11.18}$$

where B is the Levich constant, n is the number of transferred electrons per oxygen molecule, F is the Faraday constant (96,485 C mol^{-1}); D_{O_2}, is the coefficient diffusion of oxygen in 0.1 mol L^{-1} $HClO_4$ (1.7×10^{-5} cm^2 s^{-1}), C_{O_2} is the oxygen

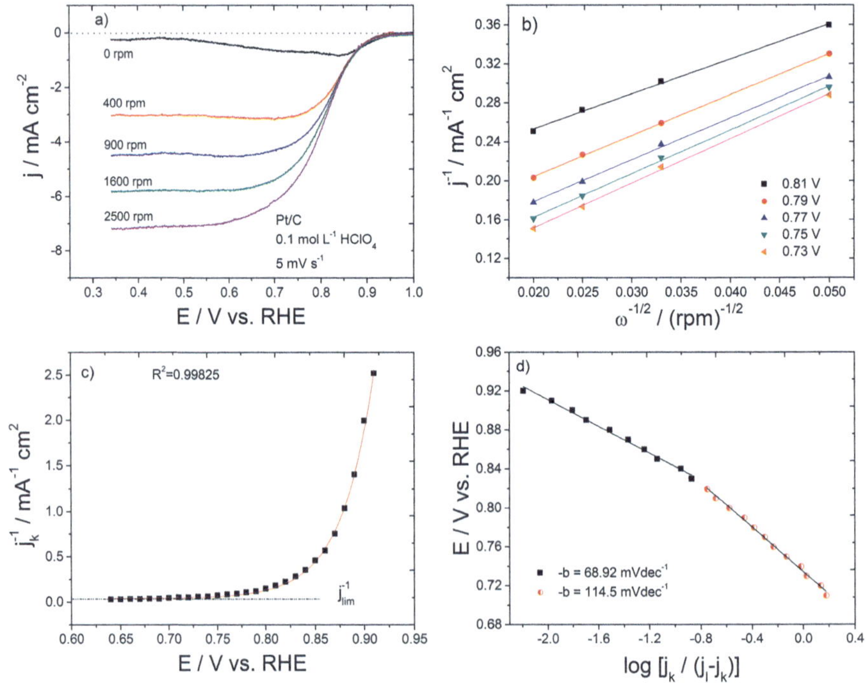

Fig. 11.4 RDE measurement results performed with O_2 saturated 0.1 mol L^{-1} $HClO_4$ recorded at 5 mV s^{-1} and 20 °C. (**a**) polarization curves on Pt/C electrocatalyst at various rotation rates (0, 400, 900, 1600, 2500 rpm), (**b**) Koutecký-Levich plots at different potentials, in which the reaction is controlled by both diffusion and kinetic effects (0.81, 0.79, 0.77, 0.75, 0.73 V vs. RHE), (**c**) exponential plot of limiting current density, and (**d**) Tafel plots for low current density (l.c.d) and high current density (h.c.d) regions

concentration in the considered electrolyte (1.3×10^{-3} mol dm^{-3}) in the electrolyte, v is the kinetic viscosity of the electrolyte (1.0×10^{-2} cm^2 s^{-1}), ω is the rotation rate of the disk electrode, and *0.20* is the coefficient used when the ω is expressed in revolution per minute (rpm) [122]. The calculated j_d value from the Levich law according to the Eq. (11.18) is 7.14 mA cm^{-2} for four transferred electrons. As can be seen in Fig. 11.4a, the current density plateau increases proportionally by increasing the rotation rate in diffusion controlled area and the current density at 2500 rpm is nearly 7.2 mA cm^{-2}, corresponding to an oxygen reduction pathway involving mainly direct four-electrons pathway. Additionally, from the slopes of the Levich plots, the number of transferred electrons involved in oxygen reduction can be calculated by using Eq. 11.18 [126, 127].

For the reactions controlled by both diffusion and kinetics through the hydrodynamic voltammetry, the measured current density (j) can be written as dependent on the diffusion-limited current density (j_d) and the kinetic current density (j_k) in Eq. 11.19 as well-known Koutecký-Levich (K-L) approach [124, 128–130], which provides the kinetic parameters as regards the catalyst surface. Jahn and Vielstich illustrated the Koutecký-Levich approach on the rates of the Fe^{2+}/Fe^{3+} redox system [131]. A typical Koutecký-Levich plot is a linear function between inverse of the current densities versus inverse of the square root of the rotation rates ($1/j$ versus $1/\omega^{1/2}$) for the Pt/C catalyst as shown in Fig. 11.4b. The plots of $1/j$ versus $1/\omega^{1/2}$ are anticipated to yield straight lines with the intercept corresponding to j_k in Eq. 11.20 and the slopes reflecting the so-called B constant value.

$$\frac{1}{|j|} = \frac{1}{|j_d|} + \frac{1}{|j_k|} \tag{11.19}$$

In the case of a catalytic film, the kinetic current density (j_k) can be represented by Eq. 11.20.

$$\frac{1}{|j|} = \frac{1}{|j_d|} + \frac{1}{|j_f|} + \frac{1}{|j_{ads}|} + \frac{1}{|j_o(\theta/\theta_e) \times e^{(\eta/b)}|} \tag{11.20}$$

where j_f and j_{ads} correspond to diffusion-limited current density in the catalytic film and to the adsorption-limited current density of molecular oxygen, respectively. Additionally, $\eta = E\text{-}E_{eq}$ is the overpotential, b is the Tafel slope, j_o is the exchange current density, θ and θ_e are the coverage of platinum surface by species are caused oxygen adsorption at potential E and at the equilibrium potential E_{eq} [124, 132]. If assuming that the electron transfer is the rate determining step, then the adsorption step is faster than the electron transfer reaction, so $\theta \approx \theta_e$ for all electrode potentials [111, 128]. As the film diffusion-limited current density (j_f) and the adsorption-limited current density (j_{ads}) do not depend on the disk electrode rotation rate and applied potential (E), the kinetic current density can be expressed with the limiting current density (j_{lim}) in Eq. 11.21.

Table 11.3 RDE and RRDE results: number of electrons (n_e), limiting current density (j_{lim}), exchange current density (j_o), Tafel slopes (b), kinetic current density from intercept value at 0.90 V (j_k), and real kinetic current density normalized with the active surface area (j_k) for the Pt/C catalyst

Pt/C	n_e	j_{lim} (mA cm^{-2})	b (mV dec^{-1})	j_o (mA cm^{-2})	b (mV dec^{-1})	j_o (mA cm^{-2})	j_k @0.90 V (mA cm^{-2})
RDE	3.91	33.25	68.92	1.58×10^{-5}	114.50	2.59×10^{-3}	0.012
RRDE	3.93	38.50	66.74	1.65×10^{-5}	116.44	3.84×10^{-3}	0.011

$$\frac{1}{|j_k|} = \frac{1}{|j_{lim}|} + \frac{1}{|j_o(\theta/\theta_e) \times e^{(\eta/b)}|} \tag{11.21}$$

By extrapolating the intercepts of the K-L plots as a function of the electrode potential, the limiting current density (j_{lim}) can be estimated as shown in Fig. 11.4c. Then, in order to access Tafel slope, Eq. 11.21 is written in Eq. 11.22. Tafel plots of the kinetic current densities giving a straight line are drawn between $log\left|\frac{j_k}{j_l - j_k}\right|$ versus electrode potential in Fig. 11.4d. From the Tafel slope and the intercept at the equilibrium electrode potential, one could estimate $[-log(j_{lim}/j_o)]$ value. Then the exchange current density is evaluated with the calculated (j_{lim}) value by using Eq. 11.22. The kinetic parameters (n_e, j_{lim}, b, j_o) obtained from K-L analysis were listed in Table 11.3. The RDE measurements indicated that the Pt/C catalyst is active for molecular oxygen reduction in acid media with multi-electron charge transfer ($n = 4$), to form water.

$$\eta = -b\left[\log\left(\frac{j_{lim}}{j_o}\right) + \log\left(\frac{j_k}{j_{lim} - j_k}\right)\right] \tag{11.22}$$

Tafel slope is expressed in unit of mV/dec. The unit "dec" refers to decades of current density. Two distinct Tafel slopes are obtained with the same values, indicating that mechanism of oxygen adsorption and the first electron transfer are similar for RDE and RRDE experiments. Typically, the Tafel slopes are 60 mV/dec and 120 mV/dec for the oxygen reduction reaction at low current density (l.c.d.) and high current density (h.c.d.) regions, respectively [133–135]. Tafel slopes were close to 67.0 ± 1.0 mV/dec at low current density regions and 115.0 ± 1.0 mV/dec at high current density regions for the Pt/C electrocatalyst on RDE and RRDE experiments, indicating that Tafel slope regions change from Temkin to Langmuir conditions for changing coverage of surface by adsorbed oxygen species.

The intrinsic catalytic activity of an electrochemical reaction is defined by the exchange current density, which is an important kinetic parameter representing the electrochemical reaction rate at equilibrium. For an electrochemical reaction, both forward and backward reactions can occur. At equilibrium, the net current density of the reaction is zero. The current density of the forward reaction and equals that of the backward reaction. This current density is called exchange current density.

The magnitude of the exchange current density determines how rapidly the electrochemical reaction can occur. The catalytic activity of the Pt/C catalyst can be usually evaluated in terms of the current density per geometric area *j*, and the kinetic current density (j_k), which is free from the mass transport, can be evaluated by active surface area under identical operating conditions [136, 137]. The calculated values of the kinetic current density on RDE and RRDE listed in Table 11.3.

11.4.4 Oxygen Reduction Reaction Order

Different potentials in the region of mass transport control can be selected to determine the reaction order with respect to O_2, and the limiting current at each rotation rate. The plots of log(*i*) *versus* log[(i_d – *i*)/i_d] at measured potentials where the ORR is expected to be under the mixed kinetic/diffusion control should yield straight and parallel lines, confirming that the oxygen reduction on Pt/C catalyst surface is the first order reaction [85, 93, 103].

$$log(i) = \mathrm{m}\log\left(1 - \frac{i}{i_d}\right) + \mathrm{log}\mathrm{i}_\mathrm{k} \tag{11.23}$$

where *i* is the measured current, i_d is the diffusion-limiting current, i_k is the kinetic current and m is the reaction order. The slope of the log(*i*) *versus* log[(i_d – *i*)/i_d] plot in Fig. 11.5 found about 0.99 for Pt/C electrocatalyst, confirming that kinetics of oxygen reduction in $HClO_4$ electrolyte is the first order reaction.

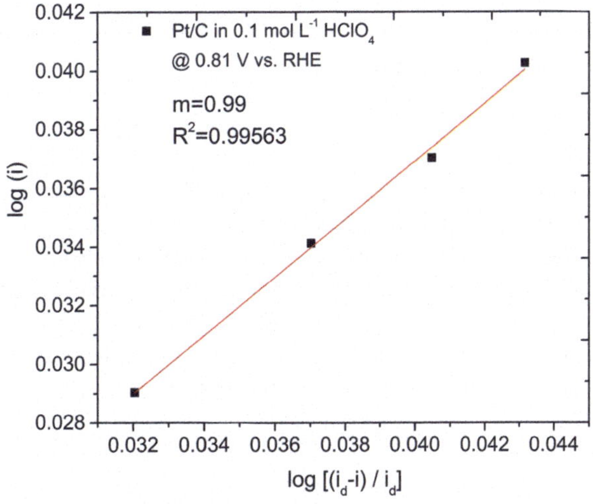

Fig. 11.5 Reaction order (*m*) plots of log (i) versus log [(i_d – i)/i_d] at various rotation rates at 0.81 V vs. RHE for oxygen reduction reaction on the Pt/C electrocatalyst in 0.1 mol L^{-1} $HClO_4$

11.4.5 RRDE Study on Oxygen Reduction Pathways

Rotating ring disk electrode (RRDE) consisted of a glassy carbon disk (0.196 cm^2 geometric area) and a Pt ring (0.11 cm^2 geometric area) sealed in a polytetrafluoroethylene holder. The electrochemical measurements were performed with a bipotentiostat and rotation control. The RRDE technique provides further information than RDE during the oxygen reduction reaction. The Pt ring electrode potential was kept at 1.2 V vs. RHE, at which all H_2O_2 molecules that reach at the ring were oxidized to H_2O as a function of the applied potential. In addition to the oxidation of H_2O_2 at the ring, which produces an anodic current, decomposition of the H_2O_2 without current flow may occur according to reaction in Eq. 11.9.

11.4.5.1 Collection Efficiency (*N*)

A typical ferrocyanide/ferricyanide redox reversible couple was used for determination of the collection efficiency in 0.1 mol L^{-1} NaOH supporting electrolyte, containing 0.01 mol L^{-1} $K_3[Fe(CN)_6]$. The constant ring potential of 1.20 V vs. RHE enables the oxidation of the $[Fe(CN)_6]^{4-}$ at the disk electrode, to $[Fe(CN)_6]^{3-}$ in 0.1 mol L^{-1} NaOH electrolyte. That is, while the reduction of ferric ions to ferrous ions on the disk electrode according to Eq. 11.24, the oxidation of ferrous ions occurs *via* mass transport limited rate at the ring electrode according to Eq. 11.25.

$$Fe^{3+} + e^- \rightarrow Fe^{2+} \tag{11.24}$$

$$Fe^{2+} \rightarrow Fe^{3+} + e^- \tag{11.25}$$

The collection efficiency was determined by averaging the ratio of the disk current between 0.4 and 1.0 V vs. RHE and the corresponding ring current according to Eq. 11.26 [112, 138].

$$N = -i_r/i_d \tag{11.26}$$

Figure 11.6 shows (a) ring and (b) disk currents regarding the collection efficiency related to RRDE at various rotation rates. The experimental collection efficiency value decreases with increasing the rotation rates because of the combined effects of the disk current that increases more rapidly than ring electrode response, depending on the laminar flow. A plot of the ring to disk currents (i_r/i_d) versus disk potential E is shown in Fig. 11.6c. From the six independent measurements, the collection efficiency was evaluated as $N = 0.20$.

Fig. 11.6 Linear scan voltammograms obtained on RRDE in 0.1 mol L^{-1} NaOH containing 0.01 mol L^{-1} $K_3[Fe(CN)_6]$ recorded at (**a**) ring, (**b**) disk electrode with 100, 400, 900, 1600 and 2500 rpm at a constant ring potential of 1.20 V vs. RHE and 20 mV s^{-1}, and (**c**) the collection efficiency obtained by the ring to disk current ratio

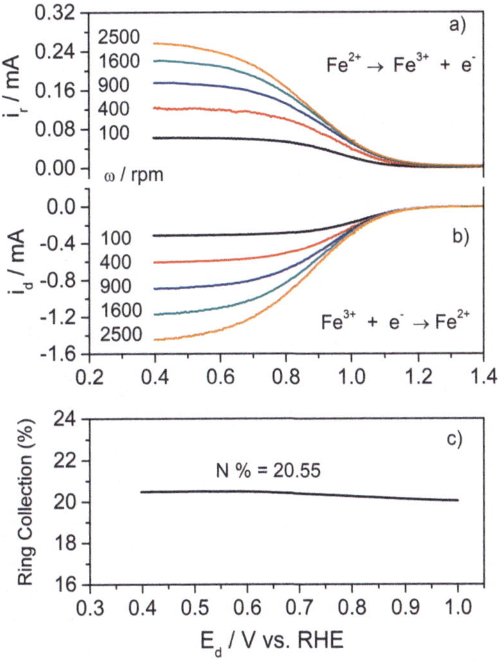

11.4.5.2 H_2O_2 Detection and Kinetic Parameters by RRDE

From rotating ring disk electrode measurements as shown in Fig. 11.7, it is possible to assume that oxygen reduction reaction on Pt/C catalyst involves the four-electron transfer per oxygen molecule by producing mainly H_2O. The calculations based on Eq. 11.27 [87] demonstrate that H_2O proportion efficiency $p(H_2O)$ is 99.71 % listed in Table 11.4. N represents the collection efficiency, j_d represents the disk current density and $j_{r,l}$ represent the limiting current density for the disk potentials E_d lower than 0.9 V vs. RHE and $j^o_{r,l}$ represents the limiting current density for disk potentials E_d higher than 0.9 V vs. RHE.

$$p(H_2O) = \frac{N\left(-\frac{j_d}{j_{r,l}} - j^o_{r,l}\right) - 1}{N\left(\frac{-j_d}{j_{r,l}} - j^o_{r,l}\right) + 1} \qquad (11.27)$$

From the calculated data using Eq. 11.27 in Fig. 11.7, a very small amount of H_2O_2 was detected at high overpotentials, *i.e.* below 0.8 V vs. RHE for the cathodic scan direction, corresponding to $n = 3.93$ as shown in Fig. 11.8. During the oxygen reduction reaction on the disk, two-electron and four-electron reduction take place forming peroxide as intermediate, which is detected on the ring electrode. The current corresponds to the two-electron reduction on the disk is equal to $|i_r/N|$,

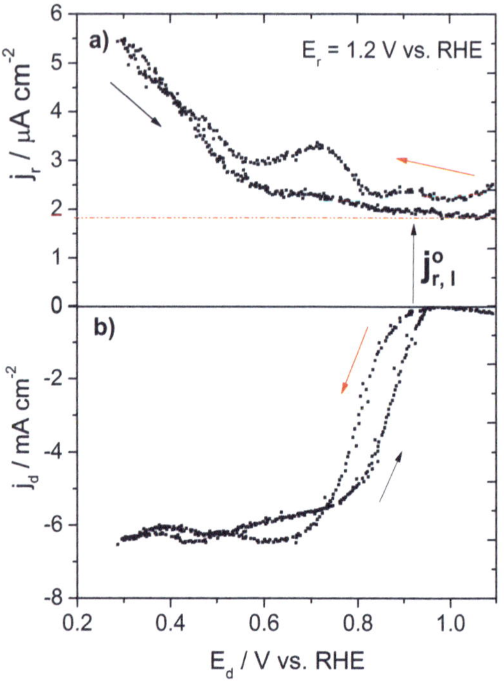

Fig. 11.7 Polarization curves (**a**) ring, (**b**) disk electrode oxygen reduction reaction on Pt/C electrocatalyst recorded at 1600 rpm in O_2 saturated 0.1 mol L^{-1} $HClO_4$ and 5 mV s^{-1}

Table 11.4 For RRDE—collection efficiency (N), the number of exchanged electrons (n_e), H_2O proportion efficiency $p(H_2O)$, the percentage of peroxide formation (% H_2O_2) at 1600 rpm and the total flux of consumed oxygen (F_d) at disk electrode for Pt/C catalyst

	N %	n_e	% pH_2O	% H_2O_2	$F_d \times 10^{-5}$
RRDE	20.55	3.93	99.71	0.29	2.84

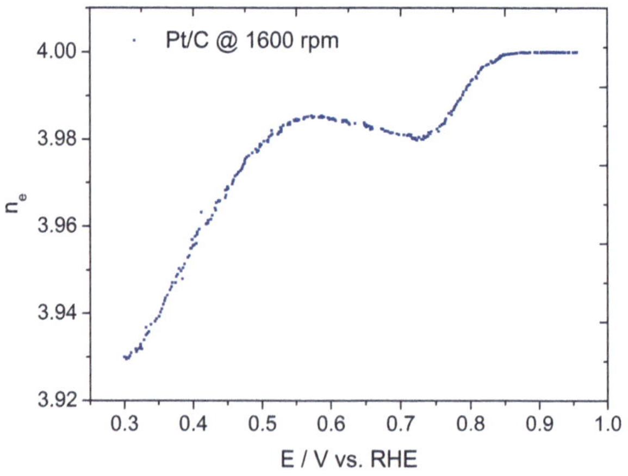

Fig. 11.8 Number of transferred electrons determined from ring and disk currents by using Eq. 11.29 for the oxygen reduction reaction on Pt/C electrocatalyst recorded at 1600 rpm in O_2 saturated 0.1 mol L^{-1} $HClO_4$ and 5 mV s^{-1}

while the current corresponds to the four-electron reduction is the difference between total measured disk current $|i_d|$ *and* $|i_r/N|$. Therefore, the total flux of oxygen consumed at the disk electrode (F_d) is determined using Eq. 11.28 [93, 112, 139, 140].

$$F_d = \frac{|i_r/N|}{2FA} + \frac{i_d - |i_r/N|}{4FA} = \frac{i_d}{nFA} \tag{11.28}$$

Figure 11.8 represents the average number of electron transfer (n_e), which is evaluated by using Eq. 11.29 as a function of the disk potential as regards the Pt/C electrode catalyst. There is a decrease around 0.75 V vs. RHE and a further decrease towards lower potentials, where the peroxide formation occurs. Equation 11.29 is used to calculate the average number of transferred electrons involved in oxygen reduction reaction. The results show that the value of n_e is very close to 4, indicating that the oxygen reduction reaction on Pt/C surface takes place through direct four-electron reaction to form water.

$$n_e = \frac{4}{1 + |i_r/N \times i_d|} \tag{11.29}$$

According to the polarization curves of the Pt/C in Fig. 11.9, the region of the mixed kinetic and diffusion control (0.95 V < E < 0.70 V) is followed by diffusion limited current region (0.70 V < E < 0.30 V). The kinetic data regarding the rotating ring disk electrode experiments are summarized in Table 11.4. Moreover, the behavior of the ring electrode clearly denotes the amount of H_2O_2 formed, which increases when the disk potential decreases [98]. When the potential range decrease from 0.55 to 0.30 V, a decrease of ring current and an increase of disk current are observed with increasing the rotation rate, indicating that the intermediate H_2O_2 continues to reduce to water. The H_2O_2 produced at the disk electrode is oxidized at the ring by the reverse reaction in Eq. 11.7. The percentage of the current to generate water is calculated by using the collection efficiency, ring and disk currents in Eq. 11.30. The amount of peroxides was 0.29 %, which is in good agreement with the data obtained from Eq. 11.27 as summarized in Table 11.4.

$$p(H_2O) = \frac{N|I_d| - I_r}{N|I_d| + I_r} \tag{11.30}$$

By using the polarization curves of the Pt/C obtained from RRDE results in Fig. 11.10, Koutecký-Levich plots (Fig. 11.10a), limiting current density (Fig. 11.10b), Tafel plots (Fig. 11.10c) and peroxide amount (Fig. 11.10d) formed during the oxygen reduction reaction obtained from Eq. 11.30 were confirmed by rotating disk electrode results in Table 11.3.

We can conclude from the data related to Tables 11.3 and 11.4 that O_2 reduction proceeds principally through the four-electron reduction pathway with 99.71 % percentage of H_2O and 0.29 % H_2O_2 formation, as shown in Fig. 11.10d.

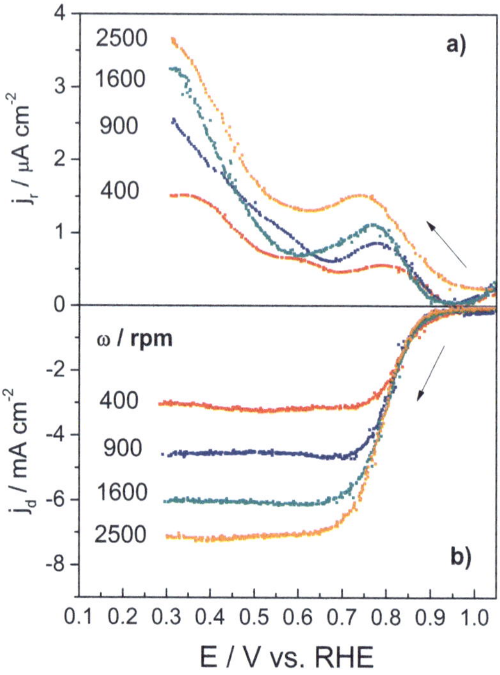

Fig. 11.9 Polarization curves (**a**) ring, (**b**) disk electrode response for the oxygen reduction reaction on Pt/C electrocatalyst recorded at different rotation rates in O_2 saturated 0.1 mol L^{-1} $HClO_4$ and 5 mV s^{-1}

11.5 Analytical Determination of Intermediates and Reaction Products Resulted from Ethanol Oxidation

In order to achieve the development of suitable electrocatalysts for the electrooxidation of organic molecules it is important to know what are the products and intermediates formed in that kind of reaction. These parameters are necessary to determine the reaction mechanism and the role of the distinct catalytic materials. Different techniques may be employed in order to identify intermediates and reactions products. *In situ* techniques must be used in order to determine intermediates that can be formed and consumed during the reaction and that exist only into the double layer region. This kind of determination is difficult to be done in the fuel cell and they are applied normally using electrochemical cells with special configuration. The formed products can be determined *ex situ* using both fuel cells and electrochemical cells by the application of different techniques.

Several studies on the electrooxidation of ethanol have been devoted to the identification and quantification of the adsorbed intermediates and by-products onto the electrode by means of various techniques, such as Differential Electrochemical Mass Spectrometry (DEMS) [141–146], *in situ* Fourier-Transform Infrared Spectroscopy (FTIR) [24, 147, 148], and High Performance Liquid Chromatography (HPLC) [22, 62, 149, 150].

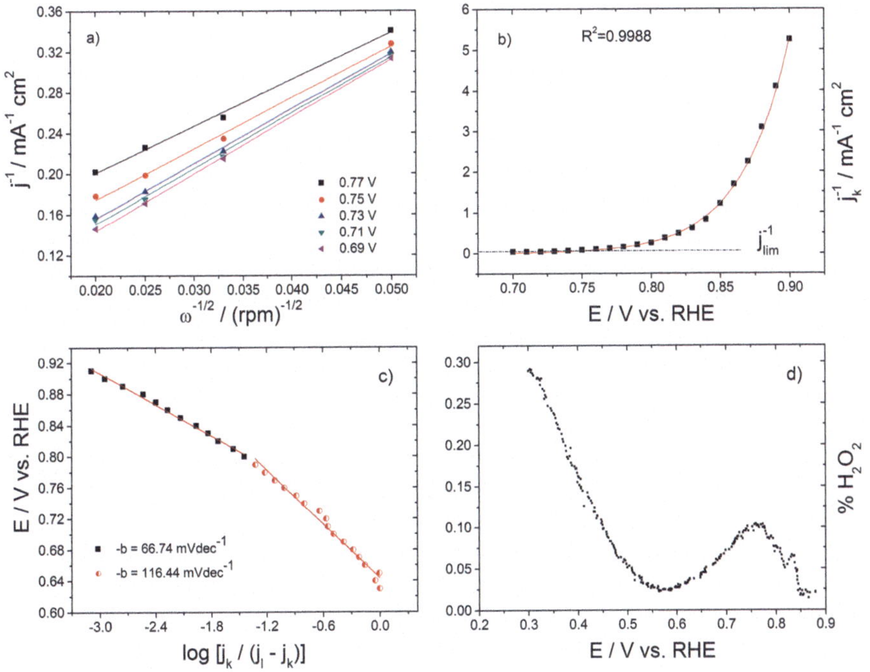

Fig. 11.10 Polarization curves (**a**) ring, (**b**) disk electrode responses for the oxygen reduction reaction on Pt/C electrocatalyst recorded at different rotation rates in O_2 saturated 0.1 mol L^{-1} $HClO_4$ and 5 mV s^{-1}

11.5.1 In Situ *Infrared Spectroscopy*

In situ FTIR technique allows tracking of the formation and evolution of the products during ethanol oxidation. The technique also provides information about the reaction intermediates adsorbed on the surface of the catalyst or the reaction products present in the immediate vicinity of the electrode.

The ethanol oxidation involves a large number of intermediates and reaction products such as acetaldehyde, acetic acid, carbon monoxide and carbon dioxide. The use of FTIR techniques is also useful to compare the selectivity of the various catalysts. FTIR studies can be accomplished on the adsorption and oxidation of key reaction intermediates such as CO, acetaldehyde or acetic acid in relation to the applied potential to refine the understanding of the reaction mechanism.

In these techniques, the cell geometry is adapted to the spectroscopic apparatus in order to allow the external reflection and the recording of the reflected beam [151]. The working electrode is typically a glassy carbon or gold disk in which surface the catalytic material is dispersed. The electrode is inserted against a window that is transparent to incident radiation in the spectral range concerned. In the investigation of ethanol electrooxidation, a calcium fluoride (CaF_2) window which is transparent in the 1000–3000 cm^{-1} wavenumber range [152], is usually employed.

There are two main *in situ* FTIR reflection techniques generally used in order to investigate ethanol oxidation or any organic molecule oxidation. The SPAIRS (Single Potential Alteration Infrared Reflectance Spectroscopy) technique is used to monitor the presence of reaction intermediates *in situ*, and the formation of reaction products, such as CO_2, on and in the vicinity of the electrode. On other hand, the SNIFTIRS (Subtractively Normalized Interfacial Fourier Transform Infrared Reflectance Spectroscopy) technique is more suitable to the *in situ* detection of adsorbed species from dissociative chemisorption of ethanol. These two techniques can give us complementary information.

The SPAIRS method is based on the acquisition of reflectivity spectra at different potentials recorded during a slow variation of the potential (1 mV s^{-1}) in a linear voltammetry or first voltammetric cycle. The spectra are recorded every 50 mV and the potential range depends on the catalyst composition considering that some metals such as Ru, Mo, Rh and others are soluble at high potential values. Normally the studies are done in between 0.05 and 1.4 V vs. RHE. The results recorded consist in a spectrum obtained from the accumulation of *n* interferograms processed by Fourier transform subtracting recorded reference potential spectrum as showed by the expression below (Eq. 11.31):

$$\Delta absorbance = \frac{\Delta R}{R_{Eref}} = \frac{R_{Ei} - R_{Eref}}{R_{Eref}} \tag{11.31}$$

where R_{Ei} represents the reflectivity on the potential where the spectra accumulation were recorded and R_{Eref} is the spectrum of reference taken on the beginning or on the end of the positive potential variation.

When the calculation of ΔR/R is done with $E_i > E_{ref}$, each negative band observed on spectrum corresponds to the formation of a species while a positive band corresponds to the consumption of a reagent.

Figure 11.11 shows representative SPAIRS spectra during ethanol oxidation.

The interpretation of SPAIRS can be easily made knowing the respective wavenumber of the all intermediates and byproducts from ethanol oxidation. Table 11.5 summarizes the species formed during the oxidation of ethanol and their respective bands observed in the infrared spectrum.

The intense band at 1640 cm^{-1} is attributed to interfacial water [154] and the strong band at 1200 cm^{-1} is characteristic of HSO_4^- ions in the thin layer of solution between the working electrode and the CaF_2 window. The anion is formed due to the reaction of SO_4^{2-} with H^+ released during the oxidation of ethanol [30, 155].

The band at 1100–1120 cm^{-1} is assigned to the vibration of acetaldehyde C-H bonds, δ-(C-H) [24]. The formation of acetic acid can be seen by the appearance of two band at 1280 cm^{-1} and 1380 cm^{-1} [156–158] arising from the C-O stretch and –OH deformation, respectively. The presence of this molecule is also evidenced by the broad band at 2615-2650 cm^{-1} that is assigned to the asymmetric stretching

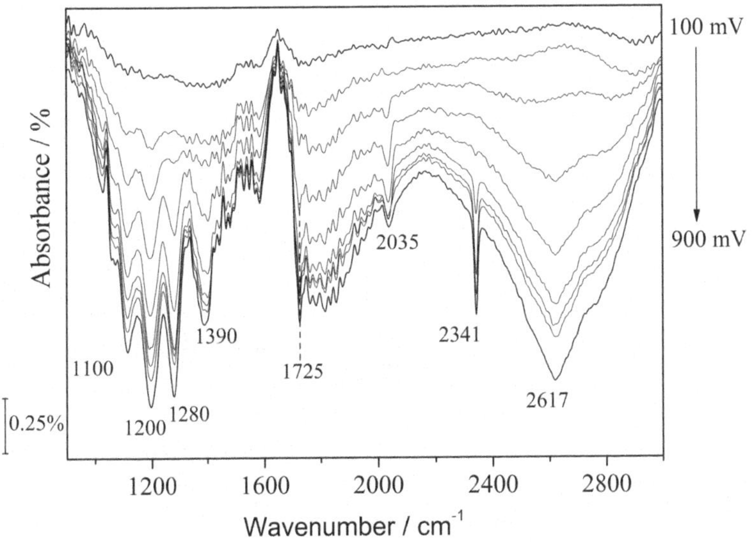

Fig. 11.11 Representative SPAIRS spectra of the species coming from ethanol oxidation on PtSn/C. 0.1 mol L^{-1} Ethanol in 0.5 mol L^{-1} H_2SO_4, T = 25 °C

Table 11.5 Different species and their respective bands in the infrared spectrum [24, 153]

Wavenumber (cm^{-1})	Functional group	Assignment
2983, 2906	CH_3, CH_2	C-H stretch
2632	$-CH_3$ (acetic acid)	C-H stretch
2341	CO_2	C-O asymmetric stretch
2055–2060	–CO	Linearly bonded
1725	COOH or CHO	C=O stretch
1402	$-CH_3COO^-$	C-O asymmetric stretch
1370/1280	COOH	C-O stretch + OH deformation
1600	Interfacial water	H-O-H stretch

ν(C-H) of the $-CH_3$ group [24]. The band at 1725 cm^{-1} corresponds to the stretching of the carbonyl group ν(C=O) in both acetaldehyde and acetic acid, indicating their formation [141, 156, 159]. The band between 2030 and 2050 cm^{-1} corresponds to CO adsorbed linearly on the Pt sites [160–162] and can be observed appearing weakly in 2035 cm^{-1} at 300 mV vs. RHE. Above 400 V, CO_L begins to be oxidized to CO_2, as evidenced by the appearance of the band at 2343 cm^{-1}, attributed to the asymmetrical stretching (O=C=O).

The use of SPAIRS techniques provides access to the potential in which different species appear and disappear and allow comparison of different catalyst materials. The example below (Fig. 11.12) compares different PtSn-based catalyst towards ethanol oxidation by monitoring the evolution of the CO_L and CO_2 bands.

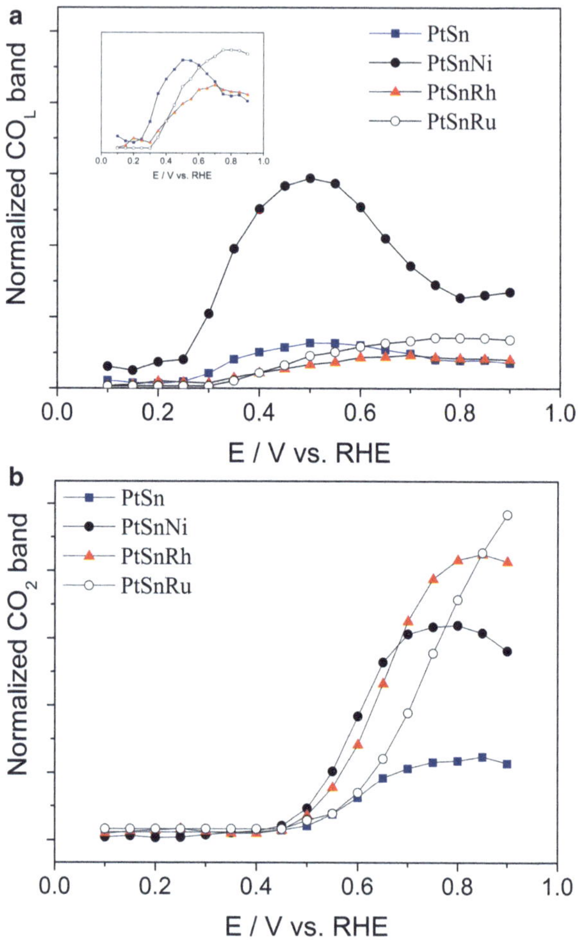

Fig. 11.12 Evolution of CO (**a**) and CO_2 (**b**) band intensity against the potential for different catalyst composition. Inset corresponds to the band intensities of the catalysts PtSn, PtSnRh, and PtSnRu

It can be seen in Fig. 11.12 that the catalyst containing Ni exhibits the most intense CO_L band, justifying their liability to break the C-C bond. CO_L appears around 300 mV and has a maximum at 500 mV vs. RHE, and then there is a drop in intensity due to its oxidation to CO_2. The formation of CO_2 as evidenced by the band at 2340 cm^{-1} was observed for all compositions and this molecule begins to be formed at 500–600 mV vs. RHE. As a conclusion, one can say that the catalysts containing Ni, Rh and Ru enhanced the CO_2 formation in comparison with binary catalysts which showed a low amount of CO_2 indicating that these catalysts are less active for breaking the C-C bond.

11.5.2 High Performance Liquid Chromatography

The quantification of ethanol oxidation products can be obtained using HPLC technique with two different approaches. First, a conventional electrochemical cell is used in a semi-cell condition in which the anodic reaction is the ethanol electrooxidation. This approach allows investigation of different conditions such as temperature, ethanol concentration, catalyst composition, and so on, even in the potentiostatic or galvanostatic modes. The products are analyzed after the electrolysis or small aliquots can be withdrawn in order to determine the product concentration and evolution. Volatile compounds can be determined by trapping them in appropriate solutions. In this experiment, a proton exchange membrane separates the electrolytic cell containing two compartments. The anode side must remain sealed to avoid losses of volatile compounds like CO_2, acetaldehyde and even ethanol (Fig. 11.13).

The second method is to analyze the solution from the anode side of a direct ethanol fuel cell according to the scheme shown in Fig. 11.14a. In both systems, the products are separated with an HPX-87H column under isocratic conditions using 3.33 mmol L^{-1} H_2SO_4 and a flow rate of 0.6 mL min^{-1} (Fig. 11.14b).

At the end of the long-term electrolysis, N_2 is bubbled through the solution, to quantify the volatile compounds produced during the process in separate trap compartments. Acetaldehyde is trapped in a 0.2 wt% of 2,4-dinitrophenyl-hydrazine solution in 2.0 mol L^{-1} HCl. Subsequently, the concentration of the solid

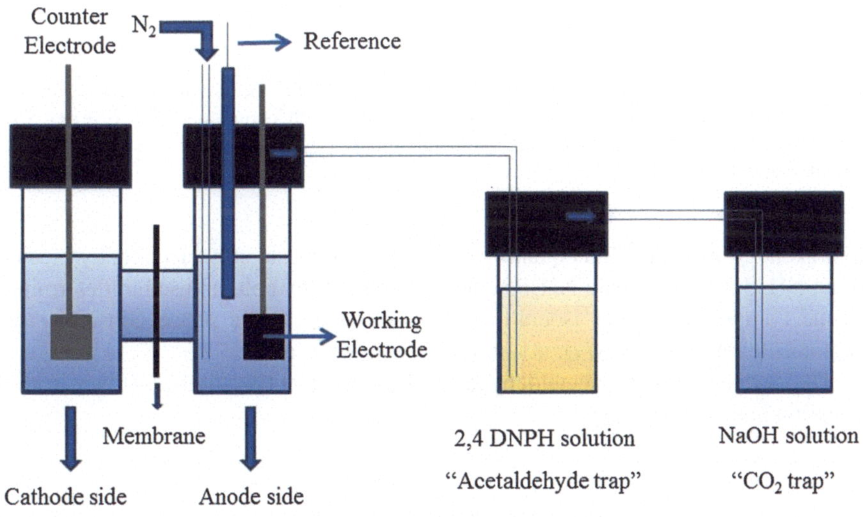

Fig. 11.13 Electrolytic cell scheme used to study ethanol oxidation and determine its products by chromatography analysis. Reprinted with Permission from [149]

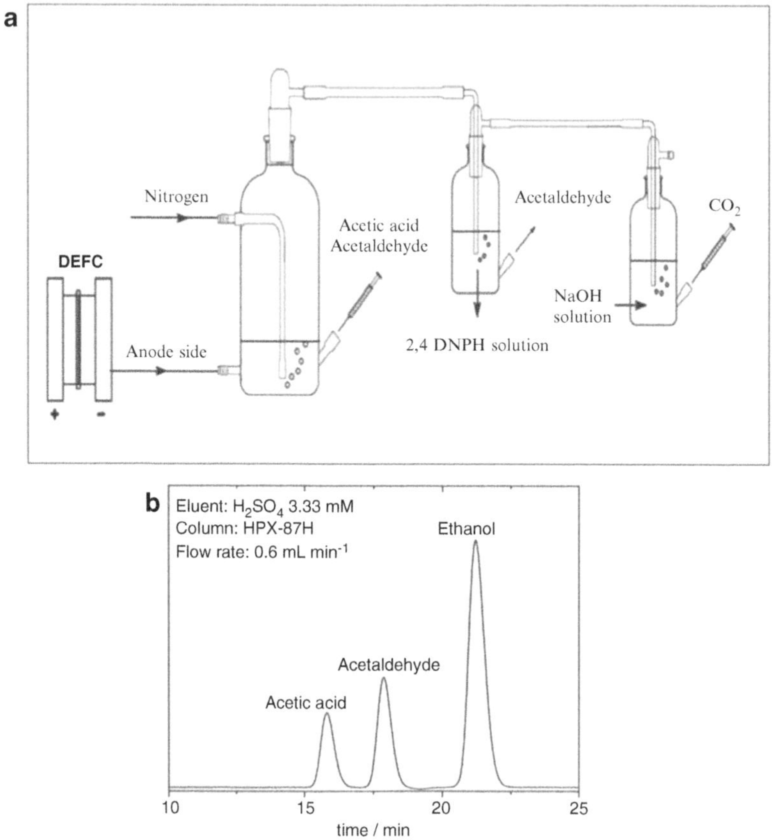

Fig. 11.14 (**a**) Set-up for recovering reaction products of DEFC for chromatography analysis; Reprinted with permission from [149]. Copyright (2006), with permission from Elsevier. (**b**) separation of the reaction products in the bulk solution. Reprinted from [30]. Copyright (2009), with permission from Elsevier

hydrazone formed therein is quantified after dissolution in ethyl acetate, by means of an NH_2P-50 (Asanhipak NH_2P series) column or any column used to determine amine compounds, under isocratic elution using a mobile phase consisting of acetonitrile: H_2O (40:60 v/v), at a 0.6 mL min^{-1} flow rate.

The derivatization of acetaldehyde to allow its quantification is shown in the equation below (Eq. 11.32):

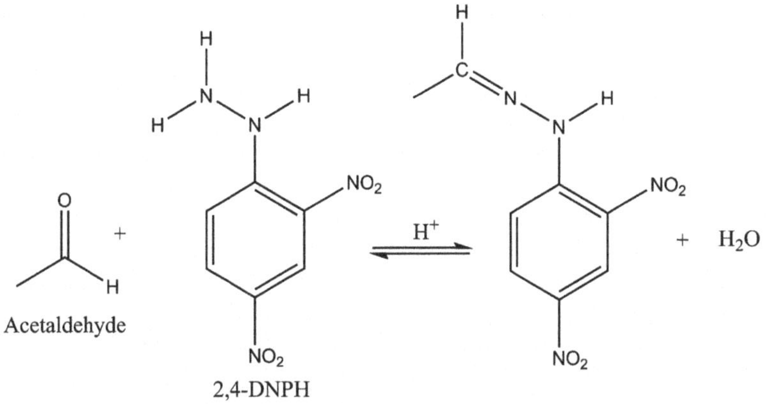

(11.32)

The released CO_2 was trapped in 0.1 mol L^{-1} NaOH. The possibly formed carbonate is quantitatively analyzed by comparison with a Na_2CO_3 reference prepared under the same conditions.

Based on the results of the techniques described above the global mechanism of ethanol oxidation may be summarized according to the following scheme [10]:

$$CH_3CH_2OH \rightarrow [CH_3CH_2OH]_{ads} \rightarrow CO_2(\text{total oxidation}) \quad (11.33)$$

$$CH_3CH_2OH \rightarrow [CH_3CH_2OH]_{ads} \rightarrow CH_3CHO \rightarrow CH_3COOH\ (\text{partial oxidation}) \quad (11.34)$$

From this general equation, highlights of the detailed mechanisms have been proposed by different groups [34, 141, 159, 161] and can be summarized as described below [25, 30]:

$$Pt + CH_3CH_2OH \rightarrow Pt - (OCH_2CH_3)_{ads} + e^- + H^+ \quad (11.35)$$

$$Pt - (OCH_2CH_3)_{ads} \rightarrow Pt + CH_3CHO + e^- + H^+ \quad (11.36)$$

or,

$$Pt + CH_3CH_2OH \rightarrow Pt - (CHOH - CH_3)_{ads} + e^- + H^+ \quad (11.37)$$

$$Pt - (CHOH - CH_3)_{ads} \rightarrow Pt + CH_3CHO + e^- + H^+ \quad (11.38)$$

$$Pt + CH_3CHO \rightarrow Pt - (CO - CH_3)_{ads} + e^- + H^+ \quad (11.39)$$

$$Pt + H_2O \rightarrow Pt - OH_{ads} + e^- + H^+ \quad (11.40)$$

$$Pt - (CO - CH_3)_{ads} + Pt - OH_{ads} \rightarrow 2Pt + CH_3COOH \quad (11.41)$$

$$Pt + CH_3CH_2OH \rightarrow Pt - CO_{ads} + Pt - CH_x \quad (11.42)$$

$$Pt - CO_{ads} + Pt - OH_{ads} \rightarrow CO_2 + e^- + H^+ \quad (11.43)$$

11.6 Electrical Performances of a Single DEFC

Finding optimum operation condition of DEFC must be analyzed in order to have realistic information about the practical use of this device. Heysiattalab et al. [163] have reported a systematic study of key parameters that affect the DEFC performance. Among them, they investigated temperature, ethanol concentration and oxygen pressure effects. These authors have shown that the increase of temperature not only contributes to the kinetic process but also influences the ethanol oxidation and oxygen reduction rates, because the increase of the diffusion will increase concentration of ethanol and oxygen in the catalyst layer, leading to a rise of ethanol oxidation and oxygen reduction. Also, temperature increase enhances the membrane conductivity, which will increase the current density and fuel cell performance. The increase of ethanol concentration leads to an enhanced fuel cell performance, because it improves diffusion and consequently the ethanol concentration in the catalyst layer. However, they noticed that high ethanol concentration leads to a decrease in the cell performance, mainly because of ethanol crossover through the membrane that causes a decrease in the oxygen reduction rate in the cathode side. In their study, they have found that 0.5 mol L^{-1} of ethanol was the best concentration to reach the maximum current density and power. The pressure in the cathode side has the same effect of ethanol concentration; by increasing O_2 pressure the cell performance improves because more O_2 would be introduced in the catalyst layer that facilitates its reduction.

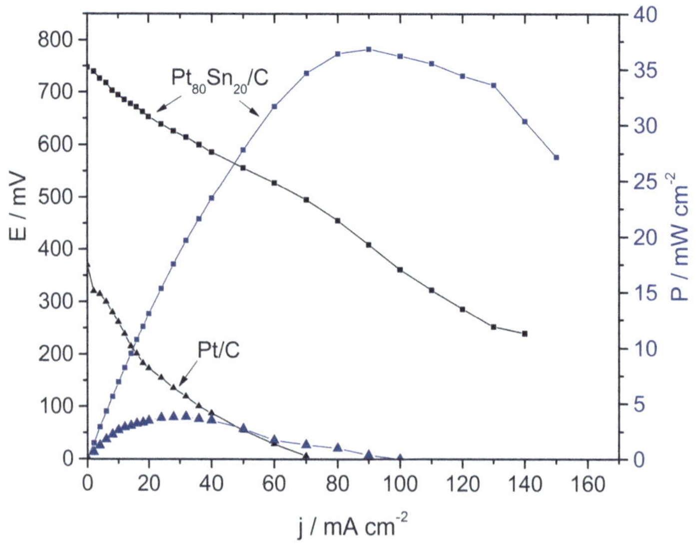

Fig. 11.15 Electrical performances of a 5 cm^2 single DEFC recorded at 90 °C using tin modified platinum catalyst. (2 mg cm^{-2} catalyst loading, 40 % catalyst on Vulcan XC-72 carbon); cathode material: 2 mg cm^{-2} (Pt/C); membrane: Nafion® 117; [Ethanol] = 2 mol L^{-1}; $P_{ethanol}$ = 1 bar; P_{O2} = 3 bar. Reproduced and adapted with permission from [31]. Copyright (2012), Elsevier

It is important to notice that the catalyst composition and the deposited amount on the diffuse layer also significantly affect the performance of the fuel cell (Fig. 11.15). There is still no consensus between the different groups that investigate DEFC performances related to the control of the parameters that influence the performance of the cell. Table 11.6 summarizes some catalyst compositions, operating conditions of the cell and the observed power to give us an overview of results obtained so far.

Table 11.6 Summary of PEM-DEFC performances

Catalyst	Operating conditions	Power (mW cm^{-2})	Ref.
Pt_3Sn_1	$P_{O2}=2$ bar; [EtOH] = 1.5 mol L^{-1}; T = 90 °C; anode (2.0 mg_{Pt} cm^{-2})	82	[164]
Pt_1Sn_1	$P_{O2}=2$ bar; [EtOH] = 2.0 mol L^{-1}; T = 100 °C; anode (1.0 mg_{Pt} cm^{-2})	60	[165]
Pt_3Sn_1	$P_{O2}=2$ bar; [EtOH] = 1.5 mol L^{-1}; T = 90 °C; anode (1.5 mg_{Pt} cm^{-2})	55	[166]
Pt_3Sn_1	$P_{O2}=2$ bar; [EtOH] = 1.0 mol L^{-1}; T = 90 °C; anode (1.5 mg_{Pt} cm^{-2})	80	[167]
Pt_3Sn_2	$P_{O2}=2$ bar; [EtOH] = 1.0 mol L^{-1}; T = 75 °C; anode (1.3 mg_{Pt} cm^{-2})	40	[168]
Pt_1Sn_1	$P_{O2}=2$ bar; [EtOH] = 1.0 mol L^{-1}; T = 90 °C; anode (1.0 mg_{Pt} cm^{-2})	55	[20]
$Pt_{89}Sn_{11}$	$P_{O2}=3$ bar; [EtOH] = 2.0 mol L^{-1}; T = 90 °C; anode (2.0 mg_{Pt} cm^{-2})	35	[169]
$Pt_{90}Sn_{10}$	$P_{O2}=3$ bar; [EtOH] = 2.0 mol L^{-1}; T = 110 °C; anode (2.0 mg_{Pt} cm^{-2})	72	[62]
$Pt_{80}Sn_{20}$	$P_{O2}=3$ bar; [EtOH] = 2.0 mol L^{-1}; T = 90 °C; anode (2.0 mg_{Pt} cm^{-2})	37	[31]
$Pt_{75}Sn_{25}$	$P_{O2}=3$ bar; [EtOH] = 1.0 mol L^{-1}; T = 90 °C; anode (1.0 mg_{Pt} cm^{-2})	20	[52]
$Pt_{50}Ru_{50}$	$P_{O2}=2$ bar; [EtOH] = 1.0 mol L^{-1}; T = 90 °C; anode (2.0 mg_{Pt} cm^{-2})	45	[20]
$Pt_{52}Ru_{48}$	$P_{O2}=1$ bar; [EtOH] = 2.0 mol L^{-1}; T = 80 °C; anode (3.0 mg_{Pt} cm^{-2})	60	[39]
$Pt_{50}Ru_{50}$	$P_{O2}=2$ bar; [EtOH] = 2.0 mol L^{-1}; T = 90 °C; anode (1.0 mg_{Pt} cm^{-2})	20	[75]
$Pt_{86}Sn_{10}Ru_{04}$	$P_{O2}=3$ bar; [EtOH] = 2.0 mol L^{-1}; T = 80 °C; anode (3.0 mg_{Pt} cm^{-2})	50	[149]
$Pt_{65}Sn_{19}Ru_{15}$	$P_{O2}=3$ bar; [EtOH] = 2.0 mol L^{-1}; T = 80 °C; anode (2.0 mg_{Pt} cm^{-2})	10	[170]
$Pt_{51}Sn_{38}Rh_{11}$	$P_{O2}=2$ bar; [EtOH] = 2.0 mol L^{-1}; T = 100 °C; anode (1.0 mg_{Pt} cm^{-2})	42	[23]

(continued)

Table 11.6 (continued)

Catalyst	Operating conditions	Power (mW cm^{-2})	Ref.
$Pt_{86}Sn_{07}Rh_{07}$	$P_{O2} = 3$ bar; [EtOH] = 1.0 mol L^{-1}; T = 80 °C; anode (2.0 mg_{Pt} cm^{-2})	10	[171]
$Pt_{67}Sn_{20}Rh_{13}$	$P_{O2} = 3$ bar; [EtOH] = 2.0 mol L^{-1}; T = 80 °C; anode (2.0 mg_{Pt} cm^{-2})	7	[170]
$Pt_{75}Sn_{14}Pd_{11}$	$P_{O2} = 3$ bar; [EtOH] = 1.0 mol L^{-1}; T = 90 °C; anode (2.0 mg_{Pt} cm^{-2})	8	[171]
$Pt_{40}Sn_{45}Pd_{15}$	$P_{O2} = 3$ bar; [EtOH] = 2.0 mol L^{-1}; T = 90 °C; anode (2.0 mg_{Pt} cm^{-2})	14	[170]
$Pt_{81}Sn_{11}Ni_{08}$	$P_{O2} = 3$ bar; [EtOH] = 1,0 mol L^{-1}; T = 80 °C; anode (2.0 mg_{Pt} cm^{-2})	34	[171]
$Pt_{81}Sn_{11}Ni_{08}$	$P_{O2} = 3$ bar; [EtOH] = 2.0 mol L^{-1}; T = 80 °C; anode (2.0 mg_{Pt} cm^{-2})	45	[172]
$Pt_{82}Sn_{09}Ni_{09}$	$P_{O2} = 3$ bar; [EtOH] = 2.0 mol L^{-1}; T = 80 °C; anode (2.0 mg_{Pt} cm^{-2})	45	[170]

11.7 Conclusion

Based on fundamental aspects, investigations were carried out on both the ethanol oxidation reaction at various Pt-based catalysts and the oxygen reduction reaction on Pt/C. The helpful results obtained gave useful information on the electrochemical behaviors of the fuel (ethanol) and the oxidant (O_2) in acid medium for the construction of a DEFC device. Indeed, ethanol is a sustainable molecule; its oxidation at the surface of mono, bi and trimetallic Pt-based anodes showed that a number of transition metals must be used as co-catalysts to decrease the content of the precious noble metal, to enhance the current densities and to shift the onset potential toward lower values. Otherwise, the kinetics of the ORR have been introduced with details on the main pathway of this reduction process at the Pt/C surface. Note that it proceeds with a nearly 4-electron process, indicating the formation of hydrogen peroxide in low amount. For the moment and in the better case, the delivered power density in the single DEFC performed was inferior to 100 mW cm^{-2} whatever the catalyst composition, which is mainly due to a weak C-C bond cleavage. It is hoped that the oxophilics transition metals such as Sn, Ru, Rh were active to remove CO to CO_2, leading to notice that the main reaction products of ethanol oxidation were acetaldehyde and acetic acid. Anyway the findings collected in this chapter open an excellent window of many studies on various renewable and friendly environmentally fuels that will pave the way for further developments in direct alcohol fuel cell technology.

References

1. Wu J, Yuan XZ, Martin JJ, Wang H, Zhang J, Shen J, Wu S, Merida W (2008) A review of PEM fuel cell durability: degradation mechanisms and mitigation strategies. J Power Sources 184:104–119
2. Wendt H, Götz M, Linardi M (2000) Tecnologia de células a combustível. Quím Nova 23:538–546
3. Wang CY (2004) Fundamental models for fuel cell engineering. Chem Rev 104:4727–4765
4. Song SQ, Tsiakaras P (2006) Recent progress in direct ethanol proton exchange membrane fuel cells (DE-PEMFCs). Appl Catal B Environ 63:187–193
5. Perna A (2007) Hydrogen from ethanol: theoretical optimization of a PEMFC system integrated with a steam reforming processor. Int J Hydrogen Energy 32:1811–1819
6. Chutichai B, Authayanun S, Assabumrungrat S, Arpornwichanop A (2013) Performance analysis of an integrated biomass gasification and PEMFC (proton exchange membrane fuel cell) system: hydrogen and power generation. Energy 55:98–106
7. Ströbel R, Oszcipok M, Fasil M, Rohland B, Jörissen L, Garche J (2002) The compression of hydrogen in an electrochemical cell based on a PE fuel cell design. J Power Sources 105:208–215
8. Takeichi N, Senoh H, Yokota T, Tsuruta H, Hamada K, Takeshita HT, Tanaka H, Kiyobayashi T, Takano T, Kuriyama N (2003) "Hybrid hydrogen storage vessel", a novel high-pressure hydrogen storage vessel combined with hydrogen storage material. Int J Hydrogen Energy 28:1121–1129
9. Lamy C, Lima A, LeRhun V, Delime F, Coutanceau C, Léger J-M (2002) Recent advances in the development of direct alcohol fuel cells (DAFC). J Power Sources 105:283–296
10. Antolini E (2007) Catalysts for direct ethanol fuel cells. J Power Sources 170:1–12
11. Calegaro ML, Suffredini HB, Machado SAS, Avaca LA (2006) Preparation, characterization and utilization of a new electrocatalyst for ethanol oxidation obtained by the sol-gel method. J Power Sources 156:300–305
12. Lamy C, Rousseau S, Belgsir EM, Coutanceau C, Léger J-M (2004) Recent progress in the direct ethanol fuel cell: development of new platinum–tin electrocatalysts. Electrochim Acta 49:3901–3908
13. Neto AO, Dias RR, Tusi MM, Linardi M, Spinacé EV (2007) Electro-oxidation of methanol and ethanol using PtRu/C, PtSn/C and PtSnRu/C electrocatalysts prepared by an alcohol-reduction process. J Power Sources 166:87–91
14. Antolini E, Colmati F, Gonzalez ER (2007) Effect of Ru addition on the structural characteristics and the electrochemical activity for ethanol oxidation of carbon supported Pt-Sn alloy catalysts. Electrochem Commun 9:398–404
15. Tripkovic AV, Lovic JD, Popovic KD (2010) Comparative study of ethanol oxidation at Pt-based nanoalloys and UPD-modified Pt nanoparticles. J Serb Chem Soc 75:1559–1574
16. Watanabe M, Motoo S (1975) Electrocatalysis by ad-atoms. 2. Enhancement of oxidation of methanol on platinum by ruthenium ad-atoms. J Electroanal Chem 60:267–273
17. Watanabe M, Motoo S (1975) Electrocatalysis by ad-atoms. 3. Enhancement of oxidation of carbon-monoxide on platinum by ruthenium ad-atoms. J Electroanal Chem 60:275–283
18. Grant JL, Fryberger TB, Stair PC (1985) Charge-transfer electronic effects on chemically modified Mo(100) surfaces. Surf Sci 159:333–352
19. Lima FHB, Gonzalez ER (2008) Ethanol electro-oxidation on carbon-supported Pt-Ru, Pt-Rh and Pt-Ru-Rh nanoparticles. Electrochim Acta 53:2963–2971
20. Li H, Sun G, Cao L, Jiang L, Xin Q (2007) Comparison of different promotion effect of PtRu/C and PtSn/C electrocatalysts for ethanol electro-oxidation. Electrochim Acta 52:6622–6629
21. Godoi DRM, Perez J, Villullas HM (2010) Alloys and oxides on carbon-supported Pt-Sn electrocatalysts for ethanol oxidation. J Power Sources 195:3394–3401
22. Almeida TS, Kokoh KB, De Andrade AR (2011) Effect of Ni on Pt/C and PtSn/C prepared by the Pechini method. Int J Hydrogen Energy 36:3803–3810

23. Spinacé EV, Dias RR, Brandalise M, Linardi M, Neto AO (2010) Electro-oxidation of ethanol using PtSnRh/C electrocatalysts prepared by an alcohol-reduction process. Ionics 16:91–95
24. Garcia-Rodriguez S, Rojas S, Pena MA, Fierro JL, Baranton S, Léger J-M (2011) An FTIR study of Rh-PtSn/C catalysts for ethanol electrooxidation: effect of surface composition. Appl Catal B Environ 106:520–528
25. Ribeiro J, dos Anjos DM, Léger J-M, Hahn F, Olivi P, de Andrade AR, Tremiliosi-Filho G, Kokoh KB (2008) Effect of W on PtSn/C catalysts for ethanol electrooxidation. J Appl Electrochem 38:653–662
26. Dos Anjos DM, Hahn F, Léger J-M, Kokoh KB, Tremiliosi-Filho G (2008) Ethanol electrooxidation on Pt-Sn and Pt-Sn-W bulk alloys. J Braz Chem Soc 19:795–802
27. Shen PK, Xu C (2006) Alcohol oxidation on nanocrystalline oxide Pd/C promoted electrocatalysts. Electrochem Commun 8:184–188
28. Xu C, Shen PK, Liu Y (2007) Ethanol electrooxidation on Pt/C and Pd/C catalysts promoted with oxide. J Power Sources 164:527–531
29. Xu C, Tian Z, Shen P, Jiang SP (2008) Oxide (CeO2, NiO, Co(3)O(4) and Mn3O4)-promoted Pd/C electrocatalysts for alcohol electrooxidation in alkaline media. Electrochim Acta 53:2610–2618
30. Purgato FLS, Olivi P, Léger J-M, de Andrade AR, Tremiliosi-Filho G, Gonzalez ER, Lamy C, Kokoh KB (2009) Activity of platinum-tin catalysts prepared by the Pechini-Adams method for the electrooxidation of ethanol. J Electroanal Chem 628:81–89
31. Purgato FLS, Pronier S, Olivi P, de Andrade AR, Léger J-M, Tremiliosi-Filho G, Kokoh KB (2012) Direct ethanol fuel cell: electrochemical performance at 90 °C on Pt and PtSn/C electrocatalysts. J Power Sources 198:95–99
32. Palma LM, Almeida TS, De Andrade AR (2013) high catalytic activity for glycerol electrooxidation by binary Pd-based nanoparticles in alkaline media. ECS Trans 58:651–661
33. Su BJ, Wang KWE, Tseng CJE, Wang CHA, Hsueh YU (2012) Synthesis and catalytic property of PtSn/C toward the ethanol oxidation reaction. Int J Electrochem Sci 7:5246–5255
34. Arenz M, Stamenkovic V, Blizanac BB, Mayrhofer KJ, Markovic NM, Ross PN (2005) Carbon-supported Pt–Sn electrocatalysts for the anodic oxidation of H_2, CO, and H_2/CO mixtures: Part II: The structure–activity relationship. J Catal 232:402–410
35. Koper MTM (2004) Electrocatalysis on bimetallic and alloy surfaces. Surf Sci 548:1–3
36. Cunha EM, Ribeiro J, Kokoh KB, de Andrade AR (2011) Preparation, characterization and application of Pt–Ru–Sn/C trimetallic electrocatalysts for ethanol oxidation in direct fuel cell. Int J Hydrogen Energy 36:11034–11042
37. Moghaddam RB, Pickup PG (2012) Support effects on the oxidation of ethanol at Pt nanoparticles. Electrochim Acta 65:210–215
38. Balakrishnan K, Schwank J (1992) FTIR study of bimetallic Pt-Sn/Al_2O_3 catalysts. J Catal 138:491–499
39. Liu ZL, Ling XY, Su XD, Lee JY, Gan LM (2005) Preparation and characterization of Pt/C and Pt-Ru/C electrocatalysts for direct ethanol fuel cells. J Power Sources 149:1–7
40. Ribadeneira E, Hoyos BA (2008) Evaluation of Pt–Ru–Ni and Pt–Sn–Ni catalysts as anodes in direct ethanol fuel cells. J Power Sources 180:238–242
41. Bonesi A, Garaventa G, Triaca WE, Castro Luna AM (2008) Synthesis and characterization of new electrocatalysts for ethanol oxidation. Int J Hydrogen Energy 33:3499–3501
42. De Souza JPI, Queiroz SL, Bergamaski K, Gonzalez ER, Nart FC (2002) Electro-oxidation of ethanol on Pt, Rh, and PtRh electrodes. A study using DEMS and in-situ FTIR techniques. J Phys Chem B 106:9825–9830
43. Sen Gupta S, Datta J (2006) A comparative study on ethanol oxidation behavior at Pt and PtRh electrodeposits. J Electroanal Chem 594:65–72
44. Kowal A, Li M, Shao M, Sasaki K, Vukmirovic MB, Zhang J, Marinkovic NS, Liu P, Frenkel AI, Adzic RR (2009) Ternary Pt/Rh/SnO_2 electrocatalysts for oxidizing ethanol to CO_2. Nat Mater 8:325–330

45. Kowal A, Gojkovic SL, Lee KS, Olszewski P, Sung YE (2009) Synthesis, characterization and electrocatalytic activity for ethanol oxidation of carbon supported Pt, Pt-Rh, Pt-SnO_2 and Pt-Rh-SnO_2 nanoclusters. Electrochem Commun 11:724–727
46. Maillard F, Peyrelade E, Soldo-Olivier Y, Chatenet M, Chainet E, Faure R (2007) Is carbon-supported Pt-WOx composite a CO-tolerant material? Electrochim Acta 52:1958–1967
47. Nores-Pondal FJ, Vilella IMJ, Troiani H, Granada M, De Miguel SR, Scelza OA, Corti HR (2009) Catalytic activity vs. size correlation in platinum catalysts of PEM fuel cells prepared on carbon black by different methods. Int J Hydrogen Energy 34:8193–8203
48. Liu ZL, Ling XY, Su XD, Lee JY (2004) Carbon-supported Pt and PtRu nanoparticles as catalysts for a direct methanol fuel cell. J Phys Chem B 108:8234–8240
49. Lim D-H, Choi D-H, Lee W-D, Lee H-I (2009) A new synthesis of a highly dispersed and CO tolerant PtSn/C electrocatalyst for low-temperature fuel cell; its electrocatalytic activity and long-term durability. Appl Catal B Environ 89:484–493
50. Spinacé EV, Farias LA, Linardi M, Neto AO (2008) Preparation of PtSn/C and PtSnNi/C electrocatalysts using the alcohol-reduction process. Mater Lett 62:2099–2102
51. Vidal-Iglesias FJ, Al-Akl A, Watson DJ, Attard GA (2006) A new method for the preparation of PtPd alloy single crystal surfaces. Electrochem Commun 8:1147–1150
52. Colmati F, Antolini E, Gonzalez ER (2007) Ethanol oxidation on a carbon-supported Pt75Sn25 electrocatalyst prepared by reduction with formic acid: effect of thermal treatment. Appl Catal B Environ 73:106–115
53. Zhao J, Chen W, Zheng Y, Li X, Xu Z (2006) Microwave polyol synthesis of Pt/C catalysts with size-controlled Pt particles for methanol electrocatalytic oxidation. J Mater Sci 41:5514–5518
54. Cushing BL, Kolesnichenko VL, O'Connor CJ (2004) Recent advances in the liquid-phase syntheses of inorganic nanoparticles. Chem Rev 104:3893–3946
55. Coutanceau C, Brimaud S, Lamy C, Léger J-M, Dubau L, Rousseau S, Vigier F (2008) Review of different methods for developing nanoelectrocatalysts for the oxidation of organic compounds. Electrochim Acta 53:6865–6880
56. Wang Z-B, Yin G-P, Zhang J, Sun Y-C, Shi P-F (2006) Investigation of ethanol electrooxidation on a Pt-Ru-Ni/C catalyst for a direct ethanol fuel cell. J Power Sources 160:37–43
57. Toshima N, Yonezawa T (1998) Bimetallic nanoparticles-novel materials for chemical and physical applications. New J Chem 22:1179–1201
58. Antolini E, Colmati F, Gonzalez ER (2009) Ethanol oxidation on carbon supported $(PtSn)_{alloy}/SnO_2$ and $(PtSnPd)_{alloy}/SnO_2$ catalysts with a fixed Pt/SnO_2 atomic ratio: effect of the alloy phase characteristics. J Power Sources 193:555–561
59. Franco EG, Neto AO, Linardi M, Aricó E (2002) Synthesis of electrocatalysts by the Bönnemann method for the oxidation of methanol and the mixture H_2/CO in a proton exchange membrane fuel cell. J Braz Chem Soc 13:516–521
60. Jeon MK, Zhang Y, McGinn PJ (2010) A comparative study of PtCo, PtCr, and PtCoCr catalysts for oxygen electro-reduction reaction. Electrochim Acta 55:5318–5325
61. Pechini MP (1967) USA Patent 3,330,697
62. Simoes FC, dos Anjos DM, Vigier F, Léger J-M, Hahn F, Coutanceau C, Gonzalez ER, Tremiliosi-Filho G, de Andrade AR, Olivi P, Kokoh KB (2007) Electroactivity of tin modified platinum electrodes for ethanol electrooxidation. J Power Sources 167:1–10
63. Galceran M, Pujol MC, Aguiló M, Díaz F (2007) Sol-gel modified Pechini method for obtaining nanocrystalline $KRE(WO_4)_2$ (RE = Gd and Yb). J Sol-Gel Sci Technol 42:79–88
64. Laberty-Robert C, Ansart F, Deloget C, Gaudon M, Rousset A (2001) Powder synthesis of nanocrystalline ZrO_2–8%Y_2O_3 via a polymerization route. Mater Res Bull 36:2083–2101
65. Kwon SW, Park SB, Seo G, Hwang ST (1998) Preparation of lithium aluminate via polymeric precursor routes. J Nucl Mater 257:172–179

66. Li X, Chen WX, Zhao J, Xing W, Xu ZD (2005) Microwave polyol synthesis of Pt/CNTs catalysts: effects of pH on particle size and electrocatalytic activity for methanol electrooxidization. Carbon 43:2168–2174
67. Kadirgan F, Beyhan S, Atilan T (2009) Preparation and characterization of nano-sized Pt-Pd/C catalysts and comparison of their electro-activity toward methanol and ethanol oxidation. Int J Hydrogen Energy 34:4312–4320
68. Tsuji M, Hashimoto M, Nishizawa Y, Kubokawa M, Tsuji T (2005) Microwave-assisted synthesis of metallic nanostructures in solution. Chem Eur J 11:440–452
69. Bock C, Paquet C, Couillard M, Botton GA, MacDougall BR (2004) Size-selected synthesis of PtRu nano-catalysts: reaction and size control mechanism. J Am Chem Soc 126:8028–8037
70. Yang J, Deivaraj TC, Too HP, Lee JY (2004) Acetate stabilization of metal nanoparticles and its role in the preparation of metal nanoparticles in ethylene glycol. Langmuir 20:4241–4245
71. Yu WY, Tu WX, Liu HF (1999) Synthesis of nanoscale platinum colloids by microwave dielectric heating. Langmuir 15:6–9
72. Tsujino T, Ohigashi S, Sugiyama S, Kawashiro K, Hayashi H (1992) Oxidation of propylene glycol and lactic acid to pyruvic acid in aqueous phase catalyzed by lead-modified palladium-on-carbon and related systems. J Mol Catal 71:25–35
73. Pinxt H, Kuster BFM, Marin GB (2000) Promoter effects in the Pt-catalysed oxidation of propylene glycol. Appl Catal A Gen 191:45–54
74. Zhao J, Wang P, Chen W, Liu R, Li X, Nie Q (2006) Microwave synthesis and characterization of acetate-stabilized Pt nanoparticles supported on carbon for methanol electro-oxidation. J Power Sources 160:563–569
75. Neto AO, Verjulio-Silva RWR, Linardi M, Spinacé EV (2009) Preparation of PtRu/C electrocatalysts using citric acid as reducing agent and OH^- ions as stabilizing agent for direct alcohol fuel cell (DAFC). Int J Electrochem Sci 4:954–961
76. Tu WX, Liu HF (2000) Rapid synthesis of nanoscale colloidal metal clusters by microwave irradiation. J Mater Chem 10:2207–2211
77. Viau G, Brayner R, Poul L, Chakroune N, Lacaze E, Fievet-Vincent F, Fievet F (2003) Ruthenium nanoparticles: size, shape, and self-assemblies. Chem Mater 15:486–494
78. Ozkar S, Finke RG (2002) Nanocluster formation and stabilization fundamental studies: ranking commonly employed anionic stabilizers via the development, then application, of five comparative criteria. J Am Chem Soc 124:5796–5810
79. Spinacé EV, Linardi M, Neto AO (2005) Co-catalytic effect of nickel in the electro-oxidation of ethanol on binary Pt–Sn electrocatalysts. Electrochem Commun 7:365–369
80. Colmati F, Antolini E, Gonzalez ER (2008) Preparation, structural characterization and activity for ethanol oxidation of carbon supported ternary Pt–Sn–Rh catalysts. J Alloys Compd 456:264–270
81. Parreira LS, da Silva JCM, D'Villa -Silva M, Simões FC, Garcia S, Gaubeur I, Cordeiro MAL, Leite ER, dosSantos MC (2013) PtSnNi/C nanoparticle electrocatalysts for the ethanol oxidation reaction: Ni stability study. Electrochim Acta 96:243–252
82. Almeida TS, Palma LM, Leonello PH, Morais C, Kokoh KB, De Andrade AR (2012) An optimization study of PtSn/C catalysts applied to direct ethanol fuel cell: effect of the preparation method on the electrocatalytic activity of the catalysts. J Power Sources 215:53–62
83. Stonehart P, Ross PN Jr (1976) The use of porous electrodes to obtain kinetic rate constants for rapid reactions and adsorption isotherms of poisons. Electrochim Acta 21:441–445
84. Schmidt TJ, Gasteiger HA, Stäb GD, Urban PM, Kolb DM, Behm RJ (1998) Characterization of high surface area electrocatalysts using a rotating disk electrode configuration. J Electrochem Soc 145:2354–2358
85. Vesovic V, Anastasijevic N, Adzic RR (1987) Rotating disk electrode: a re-examination of some kinetic criteria with a special reference to oxygen reduction. J Electroanal Chem Interfacial Electrochem 218:53–63

86. Hsueh KL, Chin DT, Srinivasan S (1983) Electrode kinetics of oxygen reduction: a theoretical and experimental analysis of the rotating ring-disc electrode method. J Electroanal Chem Interfacial Electrochem 153:79–95
87. Jakobs RCM, Janssen LJJ, Barendrecht E (1985) Oxygen reduction at polypyrrole electrodes—I. Theory and evaluation of the rrde experiments. Electrochim Acta 30:1085–1091
88. Anastasijević NA, Vesović V, Adžić RR (1987) Determination of the kinetic parameters of the oxygen reduction reaction using the rotating ring-disk electrode: Part I. Theory. J Electroanal Chem Interfacial Electrochem 229:305–316
89. Anastasijević NA, Vesović V, Adžić RR (1987) Determination of the kinetic parameters of the oxygen reduction reaction using the rotating ring-disk electrode: Part II. Applications. J Electroanal Chem Interfacial Electrochem 229:317–325
90. Zečević S, Dražić DM, Gojković S (1989) Oxygen reduction on iron: Part III. An analysis of the rotating disk-ring electrode measurements in near neutral solutions. J Electroanal Chem Interfacial Electrochem 265:179–193
91. Maruyama J, Inaba M, Ogumi Z (1998) Rotating ring-disk electrode study on the cathodic oxygen reduction at Nafion®-coated gold electrodes. J Electroanal Chem 458:175–182
92. Anastasijević NA, Dimitrijević ZM, Adžić RR (1986) Oxygen reduction on a ruthenium electrode in acid electrolytes. Electrochim Acta 31:1125–1130
93. Jiang T, Brisard GM (2007) Determination of the kinetic parameters of oxygen reduction on copper using a rotating ring single crystal disk assembly ($RRD_{Cu(hkl)}E$). Electrochim Acta 52:4487–4496
94. Damjanovic A, Genshaw MA, Bockris JOM (1967) The mechanism of oxygen reduction at platinum in alkaline solutions with special reference to H_2O_2. J Electrochem Soc 114:1107–1112
95. Appel M, Appleby AJ (1978) A ring-disk electrode study of the reduction of oxygen on active carbon in alkaline solution. Electrochim Acta 23:1243–1246
96. Markovic N (2002) Surface science studies of model fuel cell electrocatalysts. Surf Sci Rep 45:117–229
97. Yeager E (1984) Electrocatalysts for O_2 reduction. Electrochim Acta 29:1527–1537
98. Antoine O, Durand R (2000) RRDE study of oxygen reduction on Pt nanoparticles inside Nafion®: H_2O_2 production in PEMFC cathode conditions. J Appl Electrochem 30:839–844
99. Ke K, Hatanaka T, Morimoto Y (2011) Reconsideration of the quantitative characterization of the reaction intermediate on electrocatalysts by a rotating ring-disk electrode: The intrinsic yield of H_2O_2 on Pt/C. Electrochim Acta 56:2098–2104
100. Damjanovic A, Genshaw MA, Bockris JOM (1966) Distinction between intermediates produced in main and side electrodic reactions. J Chem Phys 45:4057–4059
101. Wroblowa HS, Yen Chi P, Razumney G (1976) Electroreduction of oxygen: a new mechanistic criterion. J Electroanal Chem Interfacial Electrochem 69:195–201
102. Appleby AJ, Savy M (1978) Kinetics of oxygen reduction reactions involving catalytic decomposition of hydrogen peroxide: application to porous and rotating ring-disk electrodes. J Electroanal Chem Interfacial Electrochem 92:15–30
103. Zurilla RW, Sen RK, Yeager E (1978) The kinetics of the oxygen reduction reaction on gold in alkaline solution. J Electrochem Soc 125:1103–1109
104. Sánchez-Sánchez CM, Bard AJ (2009) Hydrogen peroxide production in the oxygen reduction reaction at different electrocatalysts as quantified by scanning electrochemical microscopy. Anal Chem 81:8094–8100
105. Olson TS, Pylypenko S, Fulghum JE, Atanassov P (2010) Bifunctional oxygen reduction reaction mechanism on non-platinum catalysts derived from pyrolyzed porphyrins. J Electrochem Soc 157:B54–B63
106. Nørskov JK, Rossmeisl J, Logadottir A, Lindqvist L, Kitchin JR, Bligaard T, Jónsson H (2004) Origin of the overpotential for oxygen reduction at a fuel-cell cathode. J Phys Chem B 108:17886–17892
107. Guo S, Zhang S, Sun S (2013) Tuning nanoparticle catalysis for the oxygen reduction reaction. Angew Chem Int Ed Engl 52:8526–8544

108. Katsounaros I, Schneider WB, Meier JC, Benedikt U, Biedermann PU, Auer AA, Mayrhofer KJ (2012) Hydrogen peroxide electrochemistry on platinum: towards understanding the oxygen reduction reaction mechanism. Phys Chem Chem Phys 14:7384–7391
109. Katsounaros I, Cherevko S, Zeradjanin AR, Mayrhofer KJ (2014) Oxygen electrochemistry as a cornerstone for sustainable energy conversion. Angew Chem Int Ed Engl 53:102–121
110. Jiang R, Dong S (1990) Rotating ring disk electrode (RRDE) theory dealing with non stationary electrocatalysis: study of the electrocatalytic reduction of dioxygen at cobalt protoporphrin modified electrode. J Phys Chem 94:7471–7476
111. Demarconnay L, Coutanceau C, Léger J-M (2004) Electroreduction of dioxygen (ORR) in alkaline medium on Ag/C and Pt/C nanostructured catalysts—effect of the presence of methanol. Electrochim Acta 49:4513–4521
112. Paulus UA, Schmidt TJ, Gasteiger HA, Behm RJ (2001) Oxygen reduction on a high-surface area Pt/Vulcan carbon catalyst: a thin-film rotating ring-disk electrode study. J Electroanal Chem 495:134–145
113. Kawabata A (1996) Electronic properties of metallic fine particles. Surf Rev Lett 03:9–12
114. Volokitin Y, Sinzig J, de Jongh LJ, Schmid G, Vargaftik MN, Moiseevi II (1996) Quantum-size effects in the thermodynamic properties of metallic nanoparticles. Nature 384:621–623
115. Halperin WP (1986) Quantum size effects in metal particles. Rev Mod Phys 58:533–606
116. Jalan V, Taylor EJ (1983) Importance of interatomic spacing in catalytic reduction of oxygen in phosphoric acid. J Electrochem Soc 130:2299–2302
117. Toda T, Igarashi H, Uchida H, Watanabe M (1999) Enhancement of the electroreduction of oxygen on Pt alloys with Fe, Ni, and Co. J Electrochem Soc 146:3750–3756
118. Lim D-H, Wilcox J (2012) Mechanisms of the oxygen reduction reaction on defective graphene-supported Pt nanoparticles from first-principles. J Phys Chem C 116:3653–3660
119. Rabis A, Rodriguez P, Schmidt TJ (2012) Electrocatalysis for polymer electrolyte fuel cells: recent achievements and future challenges. ACS Catal 2:864–890
120. Angerstein-Kozlowska H, Conway BE, Sharp WBA (1973) The real condition of electro-chemically oxidized platinum surfaces: Part I. Resolution of component processes. J Electroanal Chem Interfacial Electrochem 43:9–36
121. Pozio A, De Francesco M, Cemmi A, Cardellini F, Giorgi L (2002) Comparison of high surface Pt/C catalysts by cyclic voltammetry. J Power Sources 105:13–19
122. Grolleau C, Coutanceau C, Pierre F, Léger J-M (2010) Optimization of a surfactant free polyol method for the synthesis of platinum–cobalt electrocatalysts using Taguchi design of experiments. J Power Sources 195:1569–1576
123. Chen W, Kim J, Sun S, Chen S (2008) Electrocatalytic reduction of oxygen by FePt alloy nanoparticles. J Phys Chem C 112:3891–3898
124. Grolleau C, Coutanceau C, Pierre F, Léger J-M (2008) Effect of potential cycling on structure and activity of Pt nanoparticles dispersed on different carbon supports. Electrochim Acta 53:7157–7165
125. Bard AJ, Faulkner LR (2001) Electrochemical methods: fundamentals and applications. Wiley, New York
126. Choi SI, Lee SU, Kim WY, Choi R, Hong K, Nam KM, Han SW, Park JT (2012) Composition-controlled PtCo alloy nanocubes with tuned electrocatalytic activity for oxygen reduction. ACS Appl Mater Interfaces 4:6228–6234
127. Lima FHB, Ticianelli EA (2004) Oxygen electrocatalysis on ultra-thin porous coating rotating ring/disk platinum and platinum–cobalt electrodes in alkaline media. Electrochim Acta 49:4091–4099
128. Coutanceau C, Croissant MJ, Napporn T, Lamy C (2000) Electrocatalytic reduction of dioxygen at platinum particles dispersed in a polyaniline film. Electrochim Acta 46:579–588
129. Wang JX, Markovic NM, Adzic RR (2004) Kinetic analysis of oxygen reduction on Pt(111) in acid solutions: intrinsic kinetic parameters and anion adsorption effects. J Phys Chem B 108:4127–4133

130. Wang JX, Brankovic SR, Zhu Y, Hanson JC, Adžić RR (2003) Kinetic characterization of PtRu fuel cell anode catalysts made by spontaneous Pt deposition on Ru nanoparticles. J Electrochem Soc 150:A1108–A1117
131. Jahn D, Vielstich W (1962) Rates of electrode processes by the rotating disk method. J Electrochem Soc 109:849–852
132. Lebègue E, Baranton S, Coutanceau C (2011) Polyol synthesis of nanosized Pt/C electrocatalysts assisted by pulse microwave activation. J Power Sources 196:920–927
133. Prakash J, Tryk DA, Yeager EB (1999) Kinetic investigations of oxygen reduction and evolution reactions on lead ruthenate catalysts. J Electrochem Soc 146:4145–4151
134. Murthi VS, Urian RC, Mukerjee S (2004) Oxygen reduction kinetics in low and medium temperature acid environment: correlation of water activation and surface properties in supported Pt and Pt alloy electrocatalysts. J Phys Chem B 108:11011–11023
135. Bakir ÇC, Şahin N, Polat R, Dursun Z (2011) Electrocatalytic reduction of oxygen on bimetallic copper–gold nanoparticles–multiwalled carbon nanotube modified glassy carbon electrode in alkaline solution. J Electroanal Chem 662:275–280
136. Diabaté D, Napporn TW, Servat K, Habrioux A, Arrii-Clacens S, Trokourey A, Kokoh KB (2013) Kinetic study of oxygen reduction reaction on carbon supported Pd-based nanomaterials in alkaline medium. J Electrochem Soc 160:H302–H308
137. Mayrhofer KJJ, Strmcnik D, Blizanac BB, Stamenkovic V, Arenz M, Markovic NM (2008) Measurement of oxygen reduction activities via the rotating disc electrode method: from Pt model surfaces to carbon-supported high surface area catalysts. Electrochim Acta 53:3181–3188
138. Gojković SL, Gupta S, Savinell RF (1999) Heat-treated iron(III) tetramethoxyphenyl porphyrin chloride supported on high-area carbon as an electrocatalyst for oxygen reduction: Part III. Detection of hydrogen-peroxide during oxygen reduction. Electrochim Acta 45:889–897
139. Maruyama J, Inaba M, Morita T, Ogumi Z (2001) Effects of the molecular structure of fluorinated additives on the kinetics of cathodic oxygen reduction. J Electroanal Chem 504:208–216
140. Vogel W, Lundquist L, Ross P, Stonehart P (1975) Reaction pathways and poisons—II: The rate controlling step for electrochemical oxidation of hydrogen on Pt in acid and poisoning of the reaction by CO. Electrochim Acta 20:79–93
141. Iwasita T, Pastor E (1994) A dems and FTir spectroscopic investigation of adsorbed ethanol on polycrystalline platinum. Electrochim Acta 39:531–537
142. García G, Tsiouvaras N, Pastor E, Peña MA, Fierro JLG, Martínez-Huerta MV (2012) Ethanol oxidation on PtRuMo/C catalysts: in situ FTIR spectroscopy and DEMS studies. Int J Hydrogen Energy 37:7131–7140
143. Schmiemann U, Müller U, Baltruschat H (1995) The influence of the surface structure on the adsorption of ethene, ethanol and cyclohexene as studied by DEMS. Electrochim Acta 40:99–107
144. Ianniello R, Schmidt VM, Rodríguez JL, Pastor E (1999) Electrochemical surface reactions of intermediates formed in the oxidative ethanol adsorption on porous Pt and PtRu. J Electroanal Chem 471:167–179
145. Wang H, Jusys Z, Behm RJ (2006) Ethanol electro-oxidation on carbon-supported Pt, PtRu and Pt_3Sn catalysts: a quantitative DEMS study. J Power Sources 154:351–359
146. Sun S, Halseid MC, Heinen M, Jusys Z, Behm RJ (2009) Ethanol electrooxidation on a carbon-supported Pt catalyst at elevated temperature and pressure: a high-temperature/high-pressure DEMS study. J Power Sources 190:2–13
147. Pastor E, Iwasita T (1994) D/H exchange of ethanol at platinum electrodes. Electrochim Acta 39:547–551
148. Camara GA, de Lima RB, Iwasita T (2005) The influence of PtRu atomic composition on the yields of ethanol oxidation: a study by in situ FTIR spectroscopy. J Electroanal Chem 585:128–131

149. Rousseau S, Coutanceau C, Lamy C, Léger J-M (2006) Direct ethanol fuel cell (DEFC): electrical performances and reaction products distribution under operating conditions with different platinum-based anodes. J Power Sources 158:18–24
150. Palma LM, Almeida TS, de Andrade AR (2012) Development of plurimetallic electrocatalysts prepared by decomposition of polymeric precursors for $EtOH/O_2$ fuel cell. J Braz Chem Soc 23:555–564
151. Eneau-Innocent B, Pasquier D, Ropital F, Léger J-M, Kokoh KB (2010) Electroreduction of carbon dioxide at a lead electrode in propylene carbonate: a spectroscopic study. Appl Catal B Environ 98:65–71
152. Bewick A, Kunimatsu K, Pons BS, Russell JW (1984) Electrochemically modulated infrared spectroscopy (EMIRS): experimental details. J Electroanal Chem Interfacial Electrochem 160:47–61
153. Xia XH, Liess HD, Iwasita T (1997) Early stages in the oxidation of ethanol at low index single crystal platinum electrodes. J Electroanal Chem 437:233–240
154. Rodes A, Pastor E, Iwasita T (1994) An FTIR study on the adsorption of acetate at the basal planes of platinum single-crystal electrodes. J Electroanal Chem 376:109–118
155. Batista EA, Malpass GRP, Motheo AJ, Iwasita T (2004) New mechanistic aspects of methanol oxidation. J Electroanal Chem 571:273–282
156. Vigier F, Coutanceau C, Hahn F, Belgsir EM, Lamy C (2004) On the mechanism of ethanol electro-oxidation on Pt and PtSn catalysts: electrochemical and in situ IR reflectance spectroscopy studies. J Electroanal Chem 563:81–89
157. Corrigan DS, Krauskopf EK, Rice LM, Wieckowski A, Weaver MJ (1988) Adsorption of acetic acid at platinum and gold electrodes: a combined infrared spectroscopic and radiotracer study. J Phys Chem 92:1596–1601
158. Leung LWH, Weaver MJ (1988) Real-time FTIR spectroscopy as a quantitative kinetic probe of competing electrooxidation pathways for small organic molecules. J Phys Chem 92:4019–4022
159. Iwasita T, Rasch B, Cattaneo E, Vielstich W (1989) A sniftirs study of ethanol oxidation on platinum. Electrochim Acta 34:1073–1079
160. Beden B, Lamy C, Bewick A, Kunimatsu K (1981) Electrosorption of methanol on a platinum electrode. IR spectroscopic evidence for adsorbed CO species. J Electroanal Chem 121:343–347
161. Perez JM, Beden B, Hahn F, Aldaz A, Lamy C (1989) "In situ" infrared reflectance spectroscopic study of the early stages of ethanol adsorption at a platinum electrode in acid medium. J Electroanal Chem Interfacial Electrochem 262:251–261
162. Léger J-M, Beden B, Lamy C, Bilmes S (1984) Carbon monoxide electrosorption on low index platinum single crystal electrodes. J Electroanal Chem Interfacial Electrochem 170:305–317
163. Heysiattalab S, Shakeri M, Safari M, Keikha MM (2011) Investigation of key parameters influence on performance of direct ethanol fuel cell (DEFC). J Ind Eng Chem 17:727–729
164. Zhu M, Sun G, Xin Q (2009) Effect of alloying degree in PtSn catalyst on the catalytic behavior for ethanol electro-oxidation. Electrochim Acta 54:1511–1518
165. Crisafulli R, Antoniassi RM, Neto AO, Spinacé EV (2014) Acid-treated PtSn/C and PtSnCu/C electrocatalysts for ethanol electro-oxidation. Int J Hydrogen Energy 39:5671–5677
166. Zhu M, Sun G, Li H, Cao L, Xin Q (2008) Effect of the Sn(II)/Sn(IV) redox couple on the activity of PtSn/C for ethanol electro-oxidation. Chin J Catal 29:765–770
167. Jiang L, Sun G, Sun S, Liu J, Tang S, Li H, Zhou B, Xin Q (2005) Structure and chemical composition of supported Pt–Sn electrocatalysts for ethanol oxidation. Electrochim Acta 50:5384–5389
168. Tsiakaras PE (2007) PtM/C (M = Sn, Ru, Pd, W) based anode direct ethanol–PEMFCs: structural characteristics and cell performance. J Power Sources 171:107–112

169. Ribeiro J, dos Anjos DM, Kokoh KB, Coutanceau C, Léger J-M, Olivi P, de Andrade AR, Tremiliosi-Filho G (2007) Carbon-supported ternary PtSnIr catalysts for direct ethanol fuel cell. Electrochim Acta 52:6997–7006
170. Almeida TS, Palma LM, Morais C, Kokoh KB, De Andrade AR (2013) Effect of adding a third metal to carbon-supported PtSn-based nanocatalysts for direct ethanol fuel cell in acidic medium. J Electrochem Soc 160:F965–F971
171. Beyhan S, Coutanceau C, Léger J-M, Napporn TW, Kadırgan F (2013) Promising anode candidates for direct ethanol fuel cell: carbon supported PtSn-based trimetallic catalysts prepared by Bönnemann method. Int J Hydrogen Energy 38:6830–6841
172. Beyhan S, Léger J-M, Kadırgan F (2013) Pronounced synergetic effect of the nano-sized PtSnNi/C catalyst for ethanol oxidation in direct ethanol fuel cell. Appl Catal B Environ 130–131:305–313

Chapter 12
Direct Alcohol Fuel Cells: Nanostructured Materials for the Electrooxidation of Alcohols in Alkaline Media

Hamish Andrew Miller, Francesco Vizza, and Alessandro Lavacchi

12.1 Introduction: The Potential of Biomass Derived Alcohols in Direct Fuel Cells

Renewable biomass derived alcohols are attractive fuels for application in direct fuel cells due to their low toxicity and their potential as part of carbon neutral energy transformation processes [1, 2]. Among these ethanol has been the focus of most interest, due to the possibility of production from the fermentation of biomass or the steam explosion of lignocelluloses with reasonably low energy cost [3, 4]. Alcohols with a higher molecular weight are also attracting attention as they offer high energy densities, low vapor pressure and some of them can be obtained from renewable resources (Table 12.1). Included in this group are ethylene glycol [5, 6] (EG) and glycerol [7, 8] (G). Ethylene glycol has a volumetric energy density of 5.9 kWh L^{-1} and can be produced by the heterogeneous hydrogenation of cellulose derivatives [9, 10]. Glycerol has a volumetric energy density of 6.3 kWh L^{-1}, and is a by-product of biodiesel production and, as such, is inexpensive (0.3 US$ kg^{-1}) and readily available (2.4 million tonnes produced per year) [11].

Biomass production is a virtuous cycle in terms of greenhouse emissions. Indeed CO_2 emitted during the production process of ethanol and the CO_2 eventually released by ethanol oxidation in direct ethanol fuel cells (DEFCs) can be reverted into biomasses by the plants from which the fuel is produced. Hence, ethanol is potentially a *sustainable* energy source. On the other hand we should consider that a *life cycle* analysis of bioethanol renders an energy payback ratio of the fuel in the range of 1 or lower (*i.e.* a negative return on energy invested) if the production is from corn for example [12]. Hence, it is not at all convenient to use corn for producing ethanol. Alternatively, sugar-rich plants such as rapeseeds and sugarcane

H.A. Miller (✉) • F. Vizza • A. Lavacchi
ICCOM-CNR, Via Madonna del Piano 10, Sesto Fiorentino (Firenze) 50019, Italy
e-mail: hamish.miller@iccom.cnr.it

K.I. Ozoemena, S. Chen (eds.), *Nanomaterials for Fuel Cell Catalysis*,
Nanostructure Science and Technology, DOI 10.1007/978-3-319-29930-3_12

Table 12.1 Physical properties of various alcohol fuels

Fuel	Specific energy density (kWh kg^{-1})	Density at 20 °C g cm^{-3}	Volumetric energy density (kWh dm^{-3})	E_{cell} (V)	n
Methanol	6.1	0.79	4.8	1.21	6e$^-$
Ethanol	8.0	0.79	6.3	1.15	12e$^-$
Ethylene glycol	5.2	1.11	5.8	1.22	10e$^-$
Glycerol	5.0	1.26	6.4	1.09	14e$^-$
Propan-1-ol	9.1	0.81	7.4	1.13	18e-
Propan-2-ol	9.0	0.79	7.1	1.12	18e$^-$

can produce an energy return ratio for derived bioethanol that ranges from 4 to 8. More importantly the production of bioethanol from unused nonagricultural lands from spontaneously grown plants such as *arundo donax* also known as giant cane, has also been reported [13]. In such cases it should be possible to get an energy payback ratio even larger than ten. Other alcohols such as glycerol and ethylene glycol can also be derived by a variety of processes from biomass. These two alcohols have received much attention recently as they can potentially be oxidized in direct alcohol fuel cells to render not only electrical energy but also valuable chemicals from the formation of partially oxidized intermediates. This will form an important part of the discussion in this chapter.

Anion exchange membrane (AEM) direct alcohol fuel cells (DAFCs) represent devices in which the exploitation of these bio-alcohols can be achieved [2, 14–17]. In alkaline media the kinetics of the oxygen reduction reaction (ORR) are greatly enhanced due to improved charge transfer [1]. The high pH also offers a much less corrosive environment and hence non-precious metal catalysts can be employed especially at the cathode [18–20]. Polarization losses due to alcohol crossover from the anode to the cathode are also alleviated by using these catalysts, which are highly selective for the ORR.

To date for alcohols like EtOH, G and EG, complete oxidation to CO_2 is yet to be achieved and remains a very challenging task [21, 22]. To the best of our knowledge, no known transition metal based electrocatalyst promotes complete oxidation in either acidic or alkaline media. The major reaction pathways involve the formation of various carboxylic acids or carboxylates while C-C scission and formation of CO_2 or carbonate is only a minor route. Indeed, the C-C bond cleavage of EtOH has only been observed on Pd at pHs less than 13 [23]. For EG complete oxidation is also difficult with the major pathways involving the formation of oxalate and glycolate [8, 24]. In the case of G the electrooxidation mechanism is much more complex and significant amounts of carbonate can be formed. On nanostructured Pd electrocatalysts this occurs through the oxidation of the glycolate intermediate [8]. Partial and selective oxidation of poly-alcohols to important fine chemicals is potentially a route to the realization of the so-called bio-refinery. One of the main goals in such processes is increasing the selectivity towards a given partial oxidation product. Tartronate, dihydroxyacetone, hydroxyperuvate, mexoxylate and lactate are all value added products of industrial interest. In order to realize such processes the anode catalyst employed in the fuel cell must guarantee two aspects: (i) selectivity in the partial oxidation and (ii) high activity. The recent development of *molecular* anode electrocatalysts employed in Organo Metallic Fuel Cells (OMFCs) with tunable single catalytic sites has enabled this to be achieved for a number of alcohols of industrial interest. In fact for the first time the oxidation of substrates such as 1,3-propandiol and 1,4-butandiol have been exploited in fuel cells.

The challenge facing researchers in this field is to use nanotechnology to tune anode materials to reach such performance targets. In this chapter, we discuss the most recent strategies in nanostructuring of anode materials for DAFCs. After looking at the mechanisms of alcohol electrooxidation in alkaline media we examine the most recently developed materials on the basis of the active metal present. This is followed by examples of nanostructured metal catalysts with shape and structure control where the formation of high index faceted nanoparticles has led to activity enhancements. Next we take a look at a molecular approach to alcohol electrooxidation in particular the development of organometallic complexes as anode catalysts which show remarkable selectivity in the formation of partially oxidized valuable chemicals (e.g. lactate from 1,2-propandiol). The emphasis in the discussion here is to highlight materials that have been employed in DAFCs with anion exchange membranes, thus showing the state of the art in alkaline DAFC performance.

12.2 Alcohol Electrooxidation in Alkaline Media

In this overview of alcohol electrooxidation we have excluded methanol which is more than sufficiently reviewed elsewhere. Under acidic conditions platinum-ruthenium alloys exhibit the lowest overpotentials and resistance to poisoning due to the synergistic effect of the two metals. Platinum is extremely active for the dissociative chemisorption of methanol, while the oxidation of carbonaceous adsorbates (*e.g.* CO) is favored by the presence of ruthenium oxides [25]. Indeed, carbon supported PtRu alloys are presently the state of the art for methanol electrooxidation in acidic conditions like those found in DMFCs. We will discuss in detail higher alcohols of which selected physical properties are detailed in Table 12.1.

12.2.1 Ethanol Electrooxidation

Ethanol is the most obvious candidate to replace methanol in DAFCs. Scheme 12.1 shows the variety of oxidation products attainable in principle through electrooxidation. Complete oxidation to CO_2 renders $12e^-$ but as mentioned

Scheme 12.1 Ethanol electrooxidation reaction products

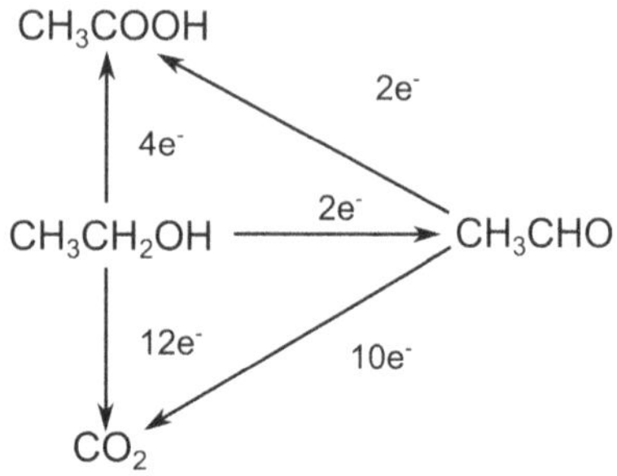

previously is difficult to obtain as this requires a breakage of the C-C bond. Usually, partial oxidation is obtained with the major products being acetaldehyde and acetic acid which deliver $2e^-$ and $4e^-$ respectively [26].

Detailed studies of the reaction products of ethanol electrooxidation in acidic media on Pt electrodes by chromatographic techniques and Differential Electrochemical Mass Spectroscopy (DEMS) has allowed a detailed description of the complex mechanisms involved. Reaction (Eq. 12.1) occurs at higher electrode potentials ($E < 0.6$ V) whereas reaction (Eq. 12.2) is favored at lower potentials ($E < 0.6$ V).

$$CH_3CH_2OH + H_2O \rightarrow CH_3COOH + 4H^+ + 4e^- \tag{12.1}$$

$$CH_3CH_2OH \rightarrow CH_3CHO + 2H^+ + 2e^- \tag{12.2}$$

At intermediate potentials ($0.6\ V \leq E \leq 0.8$ V), the dissociative adsorption of water occurs (Eq. 12.3):

$$Pt + H_2O \rightarrow Pt - OH_{ads} + H^+ + e^- \tag{12.3}$$

The oxidation of adsorbed CH_3CHO may then produce acetic acid as follows (Eq. 12.4):

$$\begin{aligned} &CH_3CHO + Pt \rightarrow Pt - (CH_3CHO)_{ads} \quad \text{and} \\ &Pt - (CH_3CHO)_{ads} + Pt - (OH)_{ads} \rightarrow CH_3COOH + H^+ + e^- \end{aligned} \tag{12.4}$$

It is difficult to obtain further oxidation to CO_2 at low temperatures but both CO (as poisoning species), CO_2 and methane have been observed as reaction products. The mechanisms are as follows:

$$Pt + Pt(CO - CH_3) \rightarrow Pt - (CO)_{ads} + Pt - (CH_3)_{ads} \tag{12.5}$$

Equation 12.5 illustrates the dissociation of the adsorbed aldehyde into adsorbed CO and CH_3. Hence, the adsorbed methyl may recombine with adsorbed hydrogen to produce methane and refresh catalytic metal sites (Eqs. 12.6 and 12.7).

$$2Pt + H_2O \rightarrow Pt - H_{ads} + Pt - (OH)_{ads} \tag{12.6}$$

$$Pt - (CH_3)_{ads} + Pt - H_{ads} \rightarrow CH_4 + 2Pt \tag{12.7}$$

In addition, CO may react with the hydroxyl adsorbed at the platinum surface to produce CO_2.

$$Pt - (CO)_{ads} + Pt - (OH)_{ads} \rightarrow CO_2 + H^+ + e^- + 2Pt \tag{12.8}$$

The formation of strongly adsorbing intermediates like CO that block the Pt catalyst surface sites together with the difficulty in C-C bond breaking are the two key obstacles to overcome for exploiting as much as possible of the energy density of ethanol as fuel. For this the role that the adsorbed hydroxyl species play in the oxidation of ethanol is fundamental. The presence of adsorbed CO species at the platinum surface hampers catalytic activity. Nevertheless, its occurrence is essential to produce a full oxidation to CO_2. Indeed it is the presence of adsorbed hydroxyl which allows CO to be oxidized to CO_2 (Eq. 12.8). Coupling materials capable of increasing the rate of formation of the adsorbed hydroxyl at the platinum surface is indeed a key for increasing the effectiveness of ethanol electrooxidation.

Ethanol electrooxidation in alkaline media has become increasingly investigated as Pt can be effectively substituted by other transition metals in particular Pd and Au. Nevertheless, C-C bond cleavage in alkaline environments has been shown to *not* occur for ethanol at pHs larger than 13 [23], acetate being the only oxidation product according to Eq. 12.9.

$$C_2H_5OH + 5OH^- \rightarrow CH_3COO^- + 4H_2O + 4e^- \tag{12.9}$$

which proceeds according to the mechanism illustrated in Scheme 12.2.

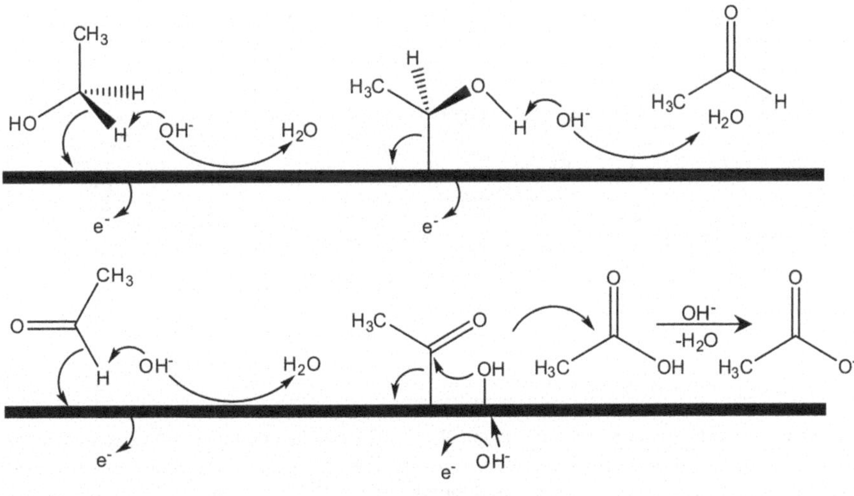

Scheme 12.2 Ethanol electrooxidation in alkaline media

In order to produce full oxidation to acetate the adsorption of the hydroxyl anion is essential. Indeed, it has been demonstrated that at low over potentials on Pd catalysts, hydroxyl adsorption is the rate-determining step [27] (Eq. 12.10).

$$Pd + OH^{-} \rightarrow Pd - OH_{ads} + e^{-} \tag{12.10}$$

The nanostructuring of electrocatalytic materials which increases the hydroxyl adsorption rate of palladium materials has been shown to be effective in enhancing ethanol electrooxidation. In one prominent example, the addition of CeO_2 to a Vulcan-XC72 carbon supported Pd catalyst improved significantly the kinetics of ethanol electrooxidation. The enhanced activity was attributed to the phenomenon known as primary oxide spillover [28].

The importance of OH_{ads} on the ethanol electrooxidation mechanism on Pd in alkaline media has also been supported by *ab initio* molecular dynamics simulations [29]. These studies have shown that within the *inner* Helmholtz layer (IHL), OH_{ads} is more active for α-CH bond breakage than Pd surface sites indicating that OH_{ads} is the active centre for dehydrogenation. After the formation of OH_{ads}, the predominant channel for the generation of electric current occurs through water dissociation, with new OH_{ads} formed for further dehydrogenation. Ethanol decomposition and water dissociation hence form a complete electrocatalytic cycle, transferring H to the aqueous solvent on a continuous basis. Consequently, the concentration of OH_{ads} (OH^{-} anion) on or near the surface should have a great impact on the electroactivity of ethanol oxidation. Compared to the typical multistep mechanisms, the mechanism proposed by these researchers (shown in Fig. 12.1) involves limited intermediates that concur with experimental observations.

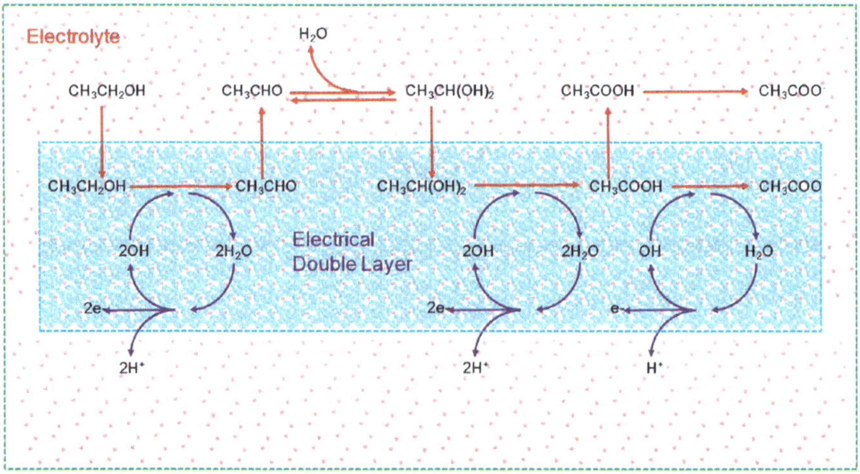

Fig. 12.1 General scheme describing ethanol electrooxidation in a DEFC on a Pd electrode in the presence of the electrical double layer. The *red arrows* indicate ethanol dehydrogenation while the *purple arrows* indicate water dissociation. Reprinted with permission from [29] Copyright (2014) American Chemical Society

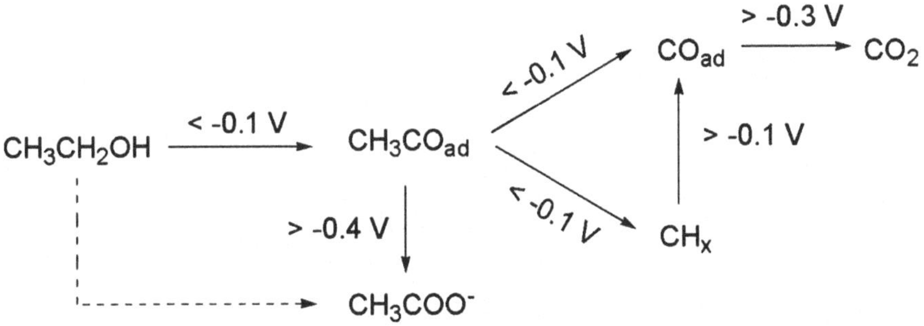

Scheme 12.3 Reaction pathways for interfacial ethanol oxidation on a Pd electrode in alkaline media (The solid-line arrows denote the findings described in [30]). Reprinted with permission from [30] Copyright (2014) American Chemical Society

Our understanding of the ethanol electrooxidation mechanism on Pd in alkaline media is still much less complete as compared to Pt in acidic media despite the number of recent *in-situ* investigations. In an important example, Yang and co-workers used *in-situ attenuated total reflection surface enhanced infrared adsorption spectroscopy* (ATR-SEIRAS) to study ethanol electrooxidation on Pd in alkaline media and proposed a duel pathway (Scheme 12.3) [30]. This study showed that at OCV adsorbed acyl species (CH_3CO_{ads}) form *via* a non electrochemical pathway. The adsorbed acyl species thus formed can be further electrooxidized either through a C2 pathway to form acetate (at high anode potentials) or a C1 pathway to CO_{ad} and CH_x and eventually CO_2 (at low potentials). The real-time spectral measurements with and without isotope labeling indicated that ethanol may undergo dehydrogenation at α-C to form adsorbed acyl rather than acetaldehyde, followed by a successive decomposition to C1 species, including CO_{ad} and CH_x at open circuit potential or lower potentials. As the potential becomes more positive, the adsorbed acetyl may be converted to either acetate in the C2 pathway or CO_2 via the C1 species described above; adsorbed CH_x may also be converted to CO_{ad} species at high potentials. Figure 12.2 clearly shows that the intensity of the vs(OCO) acetate band observed in the ATR-SEIRA spectra (D) largely traces the CV current density curve (C), suggesting a predominant share of the C2 pathway in the overall EOR current density.

It is worthwhile to note that both the C1 and C2 pathways are often used to account for EOR on both Pt and Pd electrodes, and in the majority of cases the C2 pathway remains the predominant one. One of the interesting conclusions of this study was the suggestion that inhibiting direct oxidation of the adsorbed acyl should be an effective way of enhancing the C1 pathway of EOR.

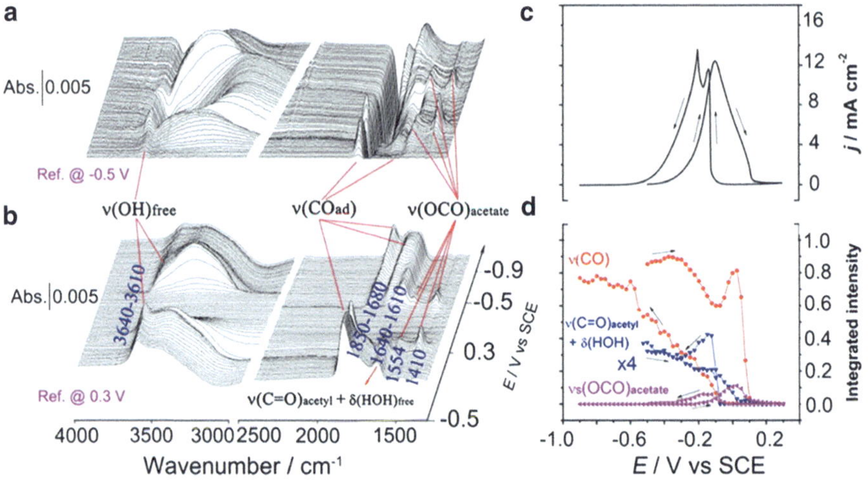

Fig. 12.2 *In situ* ATR-SEIRA spectra obtained with a time resolution of 5 s on a Pd electrode in 0.1 M NaOH/0.5 M EtOH solution, taking the single-beam spectrum at −0.5 V (**a**) and 0.3 V (**b**) as the reference spectrum. (**c**) Corresponding CV of the Pd electrode in a 0.1 M NaOH/0.5 M EtOH solution at 5 mV/s, together with (**d**) potential-dependent band intensities for $\nu(CO_{ad})$ (red), νs (OCO) of adsorbed acetate (pink), and ν(C=O)acetyl + δ(HOH)free (blue). Reprinted with permission from [30] Copyright (2014) American Chemical Society

12.2.2 Ethylene Glycol and Glycerol

The number of polyalcohols which can be potentially employed in alkaline DAFCs is quite extensive with recent extension to 1,2-propandiol, 1,3-propandiol and 1,4-butandiol good examples [31]. Here we limit the discussion of the mechanisms to the electrooxidation of EG and G as there are many more reports of alkaline fuel cells employing these fuels in the recent literature [32]. As with ethanol complete oxidation of these species to CO_2 has not yet been achieved although for EG or G some C-C bond scission may occur in alkaline media with eventual formation of carbonate although this remains a minor reaction pathway as compared to the formation of various carboxylates. Such reactions are, in turn, not completely selective with usually a mixture of products formed. As shown in Scheme 12.4, in the case of EG on Pd-based anodes in alkaline media, carbonate can be produced by the oxidation of glycolate (route c), whereas the major product, oxalate, is produced by the direct oxidation of EG (path b).

The oxidation of G is not surprisingly more complex than that of EG. Scheme 12.5 illustrates the possible reaction mechanisms proposed for G oxidation on Pd, Pt and Au based anode electrocatalysts. On Pt the main product is glycerate produced via glyceraldehyde. Glycerate is the primary oxidation product on Au electrodes. At higher potentials (>0.8 V), glyceric acid can be further oxidized to glycolate, formate and CO_2 [33].

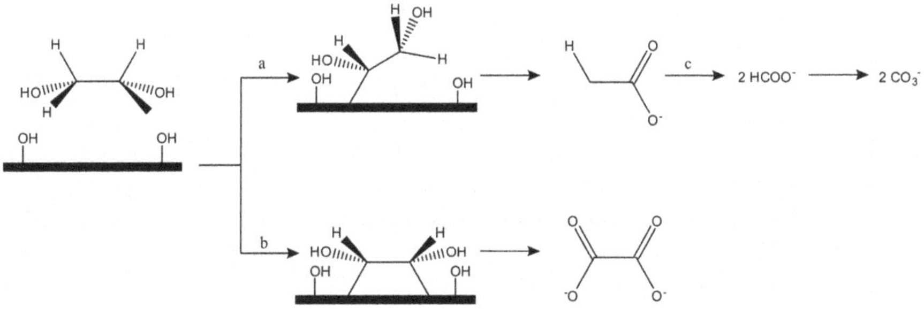

Scheme 12.4 Direct and sequential routes for the electrooxidation of EG on Pd in alkaline media. Reproduced from [32] with permission of John Wiley and Sons

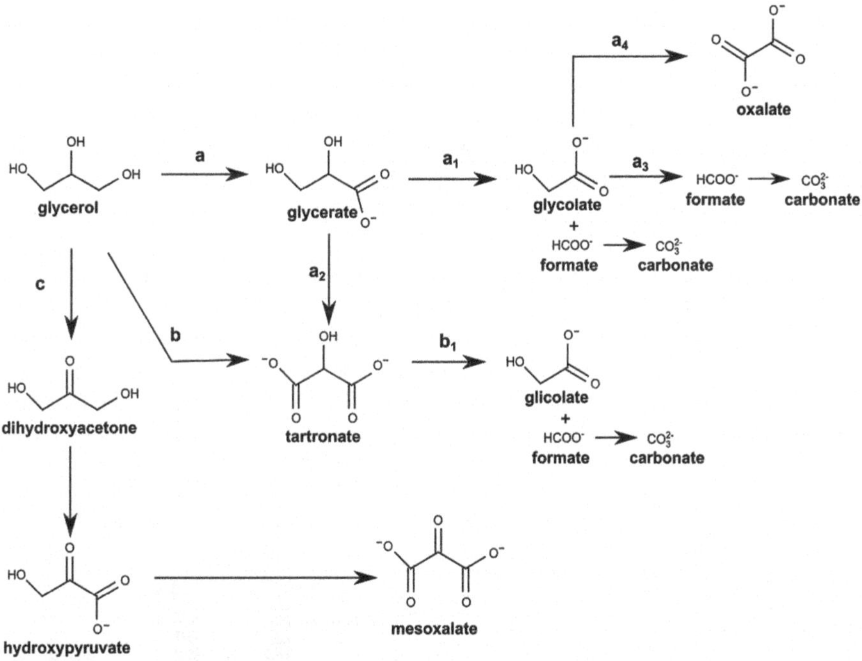

Scheme 12.5 Main oxidation pathways of glycerol oxidation in alkaline media

Glycerol on Pd-based anodes is first oxidized to the aldehyde (path a), which in turn is quickly oxidized to glycerate in a subsequent two-electron transfer step. Glycerate is further oxidized to tartronate (path a2) and, by cleavage of C-C bond, into glycolate and formate (path a2 and b1). The latter species leads to carbonate and glycolate is oxidized to oxalate (path a4) [33–36]. G is also oxidized directly to tartronate (path b) by chelating adsorption on catalytic surfaces. The oxidation of G yields significant amounts of carbonate. Since oxalate is very slowly oxidized on

Pd-based electrodes [34, 35, 37, 38] CO_2 is prevalently a by-product of the oxidation of either glycerate or glycolate (paths a1 and a3).

Longer chain alcohols have also been proposed for use as fuels in DAFCs. Isopropanol oxidation on Pt and Pd in alkaline media has been shown to produce acetone as the only soluble product [39]. Other more complex poly-alcohols that have up till now not been reported in DAFCs, such as 1,2-propandiol, 1,3-propandiol and 1,4-butandiol have only been reported using organometallic anode electrocatalysts in OMFCs [31]. This will be discussed in the section devoted to molecular electrocatalysts.

12.3 Direct Alkaline Alcohol Fuel Cells

Alcohols can be oxidized both in acidic and alkaline environments. However, the employment of higher alcohols (*e.g.* EtOH, G and EG) in proton exchange membrane fuels cells (despite intensive investigation) is limited by a number of factors including; (i) extremely sluggish anode electrode kinetics requiring high noble metal loadings, (ii) incomplete oxidation [24] and (iii) severe poisoning of Pt based catalysts by carbonaceous intermediates. Consequently, acidic DAFC devices have been realized that produce power densities in the range of only a few tens of mW cm^{-2} and with very low fuel efficiencies.

At present the best performing DAFCs [40] are those operating in alkaline environments [14] employing anion exchange polymeric membranes (AEMs) combined with a liquid electrolyte in the fuel solution (*e.g.* KOH) (Fig. 12.3). Both the kinetics of alcohol oxidation reactions and the ORR are faster at high pH. It has also been shown that when an acid electrolyte is changed to alkaline media, the fuel cell efficiency increases [41–43].

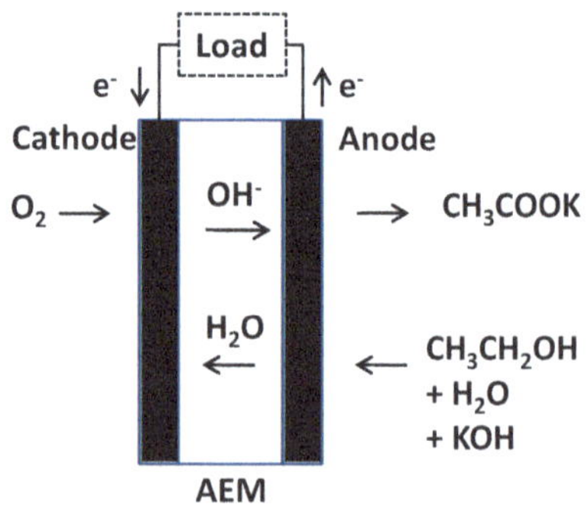

Fig. 12.3 Schematic representation of a DAFC fueled with ethanol operating with an anion exchange membrane (AEM) and KOH electrolyte

12.3.1 Cathode Catalysts and Anion Exchange Membranes

Regarding the cathode electrocatalyst in alkaline environments, platinum can be readily replaced by non noble metal materials [20, 44–47]. This is an important advantage as these catalysts are often not active toward alcohol oxidation hence limiting the consequences of fuel cross-over to the cathode. The most commonly utilized non PGM cathode catalysts employed in alkaline DAFCs have been transition metal (Fe, Co, Ni, Mn, Cr) phthalocyanine based macrocycles [48, 49] as well as Fe/C [50], Cu/C [51], FeCu/C [52–54] FeAg/C [19] and FeCo/C [18] based catalysts produced from carbon supported heat treated (600–1000 °C) metal phthalocyanine precursors. A number of other non PGM cathodes have now been reported derived from separate metal and nitrogen precursors brought together with a carbonaceous support in a heat treatment step under inert atmosphere (600–1000 °C) [55, 56]. Nitrogen doped carbon materials that may or may not contain transition metals have also attracted much attention as alternatives to Pt [57, 58]. For example nitrogen containing carbon nanotube arrays (NCNTs) have been shown by Gong et al. to be highly stable and have an onset potential for the ORR comparable to Pt/C in 0.1 M KOH [59]. Unlike Pt/C the NCNTs were shown to be immune to the presence of methanol, H_2 and CO.

The range of anion exchange membranes that have been investigated for electrochemical systems including DAFCs has been recently nicely reviewed by Varcoe et al. [17]. However, for the most part commercially available AEMs have been employed to obtain the best performance including those produced by Tokuyama, Solvay and Fumatech.

12.3.2 Fuel Cell Performance

In Tables 12.2, 12.3, and 12.4 a selection of the best recent DAFC performance data reported in the literature is shown for the three fuels EtOH, EG and G. Apart from isolated examples the best performance has been obtained using FeCo/C or FeCu/C cathode catalysts combined with Tokuyama AEMs (A201 or A901). Fuel concentrations ranging from 1 to 3 M alcohol and up to 6 M KOH liquid electrolyte have been used. The importance of the KOH electrolyte to performance is vital not only to maintain ionic conductivity in the electrode layer and membrane but also provides sufficient OH^- for the alcohol oxidation reaction as the AOR is highly pH dependent. Keeping the pH above 14 favors the direct $4e^-$ oxidation to carboxylates avoiding poisoning intermediates that can be formed at lower pHs (*e.g.* CO).

Ethanol has been by far the most investigated fuel. As can be seen from Table 12.2 the most successful anode catalysts are based primarily on Pd. Enhancements in performance with respect to simple Pd/C catalysts have been achieved through both alloy formation (*e.g.* PdIrNi/C and Pd_3Ru/C) and by exploiting strong metal-support interactions using metal oxide supports (*e.g.* CeO_2,

Table 12.2 Selected DAFC performance with ethanol

Anode	Cathode	Membrane	P_{max} (mW cm^{-2})	T_{cell} (°C)	OCV (mV)	Fuel	Ref.
Pd/C-CeO_2	FeCo/C	A201	140	80	870	10 % EtOH, 2 M KOH	[60]
1 mg cm^{-2}	2 mg cm^{-2}						
Pd/TNTA-web	FeCo/C	A201	210	80	600	10 % EtOH, 2 M KOH	[61]
1.5 mg cm^{-2}	2 mg cm^{-2}						
Pd/TNTA-web	FeCo/C	A201	335	80	900	10 % EtOH, 2 M KOH	[61]
6 mg cm^{-2}	2 mg cm^{-2}						
Pd/Ni-foam	FeCu/C	A201	164	60	870	EtOH 3 M, 3 M KOH	[62]
3 mg cm^{-2}	2 mg cm^{-2}						
PdNi/C	FeCo/C	A201	130	80	900	EtOH 3 M, 5 M KOH	[63]
2 mg cm^{-2}	2 mg cm^{-2}						
PdIrNi/C	FeCo/C	A201	92	60	900	EtOH 3 M, 5 M KOH	[64]
1 mg cm^{-2}	2 mg cm^{-2}						
Pd_3Ru/C	MnO_2 nanotube	A201	176	80	800	EtOH 3 M, 3 M KOH	[65]
1 mg cm^{-2}	2 mg cm^{-2}						

Table 12.3 Selected DAFC performance with ethylene glycol

Anode	Cathode	Membrane	P_{max} (mW cm^{-2})	T_{cell} (°C)	OCV (mV)	Fuel	Ref.
Pt/C	FeCu/C	A201	71	50	868	1 M EG, 2 M KOH	[66]
1 mg cm^{-2}	1 mg cm^{-2}						
Au/C	FeCu/C	A201	7.3	50	478	1 M EG, 2 M KOH	[66]
1 mg cm^{-2}	1 mg cm^{-2}						
Pd/TNTA-web	FeCo/C	A201	170	80	920	10 % EG, 2 M KOH	[61]
1.5 mg cm^{-2}	2 mg cm^{-2}						
PdNi	FeCo/C	PBI	112	90		1 M EG, 7 M KOH	[67]
1 mg cm^{-2}	1 mg cm^{-2}						

Table 12.4 Selected DAFC performance with glycerol fuel

Anode	Cathode	Membrane	P_{max} (mW cm^{-2})	T_{cell} (°C)	OCV (mV)	Fuel	Ref.
SD-PtCo/CNT	FeCu/C	A901	268.5	80	860	3 M G, 6 M KOH	[68]
0.5 mg	3 mg cm^{-2}						
Pt/C	FeCu/C	A901	169	80	790	3 M G, 6 M KOH	[68]
0.5 mg	3 mg cm^{-2}						
Pt/CNT	FeCu/C	A901	221	80	840	3 M G, 6 M KOH	[68]
	3 mg cm^{-2}						
Pd/TNTA-web	FeCo/C	A201	160	80	940	10 % G, 2 M KOH	[61]
1.5 mg cm^{-2}	2 mg cm^{-2}						

TiO_2 and Ni/NiO). Ma and coworkers using a carbon supported Pd_3Ru anode and MnO_2 nanotube cathode achieved a peak power with ethanol at 80 °C of 176 mW cm^{-2} while Bambagioni et al. obtained 140 mW cm^{-2} peak power also at 80 °C using a Pd nanoparticle catalyst supported on a mixed carbon-CeO_2 support. Even higher power densities have been obtained using Pd nanoparticles supported on titania nanotube arrays (TNTAs). Chen et al. was able to reach 335 mW cm^{-2} peak power by using a 6 mg_{Pd} cm^{-2} loaded TNTA-web material at 80 °C.

Reports of complete cells using EG and G are much less prolific. In most cases simple mono-metal nanoparticle anodes have been used. Unlike with EtOH the oxidation kinetics on Pd are more sluggish and as a consequence good performance has been obtained also with Pt/C catalysts. Xin et al. compared Pt/C and Au/C anodes with EG and was able to obtain 7.3 mW cm^{-2} with Au while using Pt/C produced ten times better power output. With glycerol as fuel higher power outputs have been reported. Qi and coworkers using CNT supported Pt catalysts obtained >200 mW cm^{-2} at 80 °C with a very impressive 268.5 mW cm^{-2} obtained using surface dealloyed PtCo nanoparticles supported on CNTs (SD-PtCo/CNT).

Nickel has been cited often as a possible low cost alternative to precious metals for alcohol electrooxidation [69–74]. Despite many research efforts the overpotential for alcohol oxidation on nickel catalysts is too high for practical application in fuel cells. In the next section we will discuss how nanostructuring metal catalysts have been used in improving the kinetics and overpotential of alcohol electrooxidation in alkaline media. Emphasis will be given to materials actually employed in DAFCs and the selectivity for the preparation of partially oxidized intermediates will also be discussed.

12.4 Nanostructured Materials for Alcohol Electrooxidation in DAFCs

12.4.1 Palladium Based Nanocatalysts

A recent direct comparison of the ethanol electrooxidation behaviors of Pt/C and Pd/C in alkaline media demonstrated that Pd is superior for two main reasons; (i) the higher oxophilic nature of Pd that promotes the adsorption of OH groups and (ii) the relatively inert nature of Pd to C-C bond cleavage [75]. Pd is consequently more stable and less susceptible to poisoning in the direct oxidation of ethanol to acetate. Indeed, the number and diversity of Pd based compounds that have been developed for alcohol electrooxiation under alkaline conditions is large. As we have seen in the previous section DAFCs equipped with Pd anodes can produce peak power densities of over 100 mW cm^{-2} [1, 60, 76].

The catalytic activity and selectivity of palladium electrocatalysts can be improved by the addition of a second metal or metal oxide with a promotional effect. The activity of Pd/C catalysts for ethanol electrooxidation (EOR) has been enhanced by a number of researchers by alloy formation with non noble metals. Shen and co-workers obtained a peak power density of 90 mW cm^{-2} using a Pd_2Ni_3/C alloy as anode catalyst in a DAFC operating at 60 °C. X-ray photoelectron spectroscopy (XPS) analysis of the anode catalyst revealed the chemical states of Ni, including metallic Ni, NiO, $Ni(OH)_2$ and NiOOH [77]. In another recent report Pt-Pd alloy nanostructures exhibited enhanced activity and stability for ethanol electrooxidation (EOR) through an electronic, synergistic alloy effect between Pt and Pd atoms, and the presence of unique interconnected nanostructures [78]. The direct oxidation pathway from ethanol to acetate was found to be favored by this catalyst. The addition of Ni based compounds to Pd catalysts generally leads to an enhancement in the EOR. Jiang and coworkers prepared a Pd–Ni–P compound for ethanol oxidation. It was found that the presence of Ni and P atoms modify the crystal structure, and charge transfer to the nearby Pd atoms thus enhancing the EOR. Compared to a Pd black catalyst a 110 mV decrease in overpotential was observed for the EOR on the Pd-Ni-P catalyst [79]. The four electron direct oxidation to acetate was also confirmed for this material.

The combination of Pd and Au in alloyed and core-shell materials either carbon supported or without support has been the subject of a number of recent reports regarding enhancements in EOR activity [80–84]. The addition of gold to palladium nanostructures has an important ameliorative effect on the EOR. Firstly, because the co-metal Au favors the adsorption of OH_{ads} onto the Pd surface of the catalyst. Secondly, as the lattice constant of Pd is smaller than that of Au, the addition of Au to Pd causes a tensile strain in the structure of surface Pd and an up shift of the d-band center of Pd [85]. According to d band theory of Nørskov and coworkers the trend of reactivity follows the trend in d-band center values of overlayer and impurity atoms [86–90]. When metals with small lattice constants are overlayed or alloyed on metals with larger lattice constants, the d-band center shifts up and vice versa, which subsequently affects the reaction rate. Since the lattice parameter for Au (4.08 Å) is larger than that for Pd (3.89 Å), the lattice mismatch between Au and Pd leads to an expansive strain of Pd and an up-shift of d-band center, which weakens the interaction between the adsorbed intermediate species CH_3CO_{ads} and the surface of Pd and thus results in an enhanced activity toward EOR [82, 91]. This effect has been supported by XPS results that show the binding energies of Pd shift to higher values as the Au/Pd ratios increased [80]. Two recent examples of bimetallic PdAu non supported catalysts in the form of nanoporous solids [83] and nanowire networks [80] (NNW) have been shown to have in addition to enhanced activity also greater stability. Such nano-networks can possess the advantages of the one-dimensional structure such as improved electron transport characteristics during catalysis as a result of the path-directing effects of structural anisotropy [92, 93] combined with the fact that self-supported materials avoid the severe corrosion and oxidation of the carbon supports [94, 95].

Researchers incorporating Sn or SnO_2 into Pd based catalysts have obtained remarkable improvements in ethanol oxidation activity in alkaline media

[96–99]. In Pd-Sn binary alloys prepared by Du et al. using a polyol method, the Sn was found to be partially oxidized in the otherwise homogeneous alloy [96]. Using DFT calculations the optimum Sn content of 14 wt% was determined. DFT calculations also confirmed that Pd–Sn alloy structures would result in lower reaction energies for the dehydrogenation of ethanol, compared to the pure Pd crystal. All of the PdSn/C catalysts produced at least twice the peak current density for ethanol oxidation in CVs with respect to Pd/C. The results also confirmed the formation of acetate as major product of ethanol oxidation. Atanassov and co-workers recently studied the mechanism of ethanol electrooxidation on SnO_2 supported Pd nanoparticles. *In situ* IR reflection absorbance spectroscopy (IRRAS) studies showed that in 1 M KOH the SnO_2 support acted as a co-catalyst providing hydroxide ions to the interface layer, increase the turnover rate, and limiting the final product to acetate whereas in 0.1 M KOH some CO_2 was formed at high potentials [100].

The nature of the catalyst support material can have a dramatic effect on performance. Increasingly, researchers are investigating novel materials both carbon and non-carbon based. An important recent example is reduced graphene oxide (RGO). Catalysts prepared with this material have ultra-high dispersion of nanoparticles combined with strong SMSIs (Strong Metal Support Interactions). Zhang et al. used a one-pot solvothermal synthesis to prepare Pd-on-Cu nanoparticles evenly distributed on reduced graphene oxide (Pd-on-Cu/RGO) [101]. The as-prepared nanocomposites exhibited a large EASA of 64.98 $m^2\ g_{Pd}^{-1}$ and as a consequence improved electrocatalytic activity, and better stability for ethanol oxidation in alkaline media. RGO modified with various organic materials such as Dimethyldiallylammonium chloride (DMDAAC) [102] has been used as a support for a number of highly active Pd based EOR catalysts including Pd [103], PdCo [104], PdCu [105], Ni_xPd_{100-x} [106] and Ni@Pd [107]. All showed enhanced activity and stability for the EOR. A number of other exotic supports have been used to support Pd nanoparticles for EOR. An anode consisting of Pd supported on hydroxyapatite [108] combined with a Pt/C cathode and anion exchange membrane produced a peak power density of 50 mW cm^{-2} at room temperature with 1 M ethanol and 1 M KOH (Pd loading of anode was 0.5 mg cm^{-2}). MWCNT supported Pd nanoparticles have been investigated in DAFCs. Together with a FeCo/C cathode and anion exchange membrane (Tokuyama A201) peak power densities of 95, 75 and 80 were obtained at 80 °C with methanol, ethanol and glycerol respectively [109]. The same authors also supported Pd on a composite FLG-CNT material where the CNTs were shown to improve Pd dispersion by inhibiting aggregation of the graphene sheets [110].

Another strategy to prepare Pd catalysts without carbon supports has been to prepare self supported materials which combine active phase and high surface area support. A number of self supported bimetallic Pd catalysts have been developed for the electrooxidation of glycerol and ethylene glycol such as Pd-In [111] and Pd_xBi [112]. These catalysts were prepared using the Sacrificial Support Method (SSM) developed at the University of New Mexico. This synthetic method produces self-supported, porous nanostructured materials. The reaction products of glycerol oxidation were studied using the Pd_xBi material. This important study showed how the catalyst composition and morphology such as the formation of pores that act as nanoreactors influence the product selectivity between the formation of aldehydes and ketones at low potentials as opposed to hydroxypyruate and CO_2 at high potentials. In another recent example, the synergistic effect of Au-Pd bimetallic surfaces in Au-covered Pd and Pt nanowires for ethanol electrooxidation in alkaline media has been reported [113, 114]. Cherevko and co-workers prepared highly ordered Pt, Pd and Au nanowire arrays using a home-made AAO (Anodic Aluminum Oxide) template by electrodeposition [114]. After which the AAO template was removed. Decoration of Pt and Pd on the as prepared Au nanowire array was achieved using chemical reduction with ascorbic acid. The ethanol electrooxidation activities of the decorated materials were several times greater than the Pd or Pt only arrays although the synergistic effect of the two metals was not determined but could be due to an up-shift in the energy of the d-states.

12.4.1.1 Product Selectivity on Palladium

As described previously ethanol electrooxidation at high pHs on Pd proceeds through the direct formation of acetate. The product selectivity of electrooxidation of G and EG in alkaline fuel cells is not the same and can be largely influenced by the nanostructure of Pd electrodes. This was shown by Vizza and co-workers comparing a simple Pd/C catalyst with a nanostructured Pd/(Ni-Zn)-C anode in DAFCs operating at room temperature [32]. Both the conversion and the product selectivity varied between the two catalysts. The improvement in conversion with the Pd/(Ni-Zn)-C anode was assigned to the ability of the (Ni-Zn)-C support to increase the amount of OH_{ads} groups on the Pd catalyst surface. On the other hand the Pd/C catalyst was much more selective in EG oxidation, yielding 89.5 % of glycolate whereas Pd/(Ni-Zn)-C yielded in addition to glycolate both oxalate and carbonate. With G there was little difference in the product selectivities between the two catalysts with a mixture of glycolate, glycerate, tartronate, oxalate, formate, and carbonate formed. No secondary alcohol oxidation products were detected with either catalyst (dihydroxyacetone, hydroxypiruvate, or mesoxalate).

12.4.2 Nanostructured Gold Electrocatalysts

Gold electrodes in acidic media typically exhibit very low activity for alcohol oxidation. In alkaline media however Au is substantially more active for alcohol oxidation and the electrooxidation of a wide range of substrates have been studied on Au electrodes (*e.g.* D-glucose, xylose, *etc.*) [115]. The improved oxidation of alcohols on gold electrodes in alkaline media has been attributed to the deprotonation of the R–OH group that occurs at sufficiently high pHs, and leads to the formation of the alkoxide species (RO^-), that are much more active in oxidation reactions [115]. Varella and de Lima studied the catalytic oxidation of ethanol on a polycrystalline gold electrode in alkaline media [116]. Cyclic voltammetry combined with *in-situ* FTIR, showed that adsorbed CO formed at very low overpotentials (circa 600 mV *vs.* RHE), indicating that a small amount of C-C bond breakage occurs. Acetate ions were however identified as the major product of electrooxidation.

The electrooxidation of ethanol on Au(1 1 1), Au(1 0 0), Au(1 1 0) and Au(2 1 0) single crystal electrodes was studied by Beyhan and co-workers as well as on a Au polycrystalline electrode [117]. The activity for the electrooxidation of ethanol in the low potential region was ranked as follows: Au(1 1 0) > Au(2 1 0) > Au (1 0 0) > Au(poly) > Au(1 1 1), suggesting that defect sites on the Au surface dominate. Adsorbed OH was shown to play an important role in the ethanol oxidation reaction and FTIR spectroscopy measurements showed that the main reaction product is acetate. Ethanol electrooxidation has been widely investigated on carbon supported gold nanoparticle catalysts; including gold-conducting polymer nanocomposites (Polyaniline, polypyrrole, polythiophene and poly (3,4-ethylenedioxythiophene) [118]. In another recent example Kumar et al. prepared a nanostructured Au decorated graphene catalyst. The onset of ethanol electrooxidation was shown to shift favorably [119].

Despite being extensively studied, the overpotential for ethanol electrooxidation on nanostructured Au electrodes is too high to have practical use in DAFCs. By contrast gold based catalysts have been more readily applied as anodes in fuel cells and electrolysis cells fed with glycerol [120] as the onset potential for glycerol oxidation has been shown to be similar to palladium. On gold the glycerol electrooxidation reaction has been shown to be structure sensitive (Au(1 1 0) > Au(1 0 0) and (Au(1 1 1)) and the rate determining step in alkaline media is the adsorption of hydroxyl ions with partial charge transfer (Eq. 12.11).

$$Au + OH^- \rightarrow AuOH^{\lambda -} + (1 - \lambda)e^- \tag{12.11}$$

The research group of Li at Michigan Technological University has extensively applied gold and gold alloy catalysts in DAFCs using both glycerol [7, 121–126] and ethylene glycol [66] as fuels. A nanoparticle Au/C (3.5 nm) catalyst was prepared and studied for EG and G electrooxidation in alkaline media. The Electrochemical Active Surface Area (EASA) of the Au/C catalyst was calculated as 26.8 $m^2\ g^{-1}$. An anode prepared with this catalyst (1 $mg_{Au}\ cm^{-2}$) and combined with a Tokuyama A201 membrane and FeCu/C cathode was studied in a DAFC fueled with 1 M alcohol and 2 M KOH. The behavior of the cell was investigated with G, EG and methanol under the same test conditions at 80 °C. When fed with 2.0 M KOH + 1.0 M methanol, the cell yielded an OCV of 0.29 V and an peak power density of 0.8 mW cm^{-2} (at 8 mA cm^{-2}). Much better performance was achieved with EG; OCV 0.58 V PP 20.3 mW cm^{-2} and G OCV 0.67 V and PP 57.9 mW cm^{-2}.

12.4.2.1 Product Selectivity on Gold

Li's group also studied the product distribution of working DAFCs using the Au/C anode. Product analysis showed that on Au/C high selectivity of EG electrooxidation to glycolic acid (>98 %) was obtained while the electrooxidation of glycerol favored the production of deeper-oxidized chemicals: *i.e.* tartronate, mesoxalate and oxalate. The same authors showed that the product distribution could be changed by tuning the anode potential; by moving from 0.35 to 0.65 V, the selectivity to tartronate dropped from 79 to 26 %, while that to mesoxalate increased from 0 to 57 % [123]. Further tuning of the electrode structure and reaction conditions helped to produce >60 % yield of tartronate from glycerol. A gold nanoparticle catalyst supported on CNTs was also used in an electrolyzer set at 1.6 V (SHE). Under these conditions 85 % selectivity for the formation of glycolate was achieved avoiding the tartronate route confirming that the anode potential has a large effect on selectivity for electrooxidation on Au [121].

12.4.3 Fine Tuning the Surface Structure of Nanocatalysts at the Atomic Level

The electrooxidation of alcohols like ethanol, EG and G in DAFCs is only partial and this is the case in both acid and alkaline media. Further oxidation requires the cleavage of the C–C bond and as we have seen for ethanol the main products are acetaldehyde and acetic acid, while CO_2 production contributes to less than 2 % of the total current even on the most effective electrocatalysts (*e.g.* Pt – Sn alloys) [127]. Consequently, the fuel utilization efficiency of an ethanol fed DAFC is actually very low. Higher conversion to CO_2 has been demonstrated only working with very dilute ethanol solutions in acidic DEFCs (Faradaic efficiency 64 % for 0.1 M ethanol at 80 °C) [128].

A number of studies on the reactivity of Pt single-crystal planes with hydrocarbons (*e.g.* hydrogenolysis and isomerization) have revealed that the product selectivity depends highly on surface atomic arrangements and the presence of low coordinated step atoms, especially kink atoms, which can promote the breaking of C–C bonds. Surfaces with a high number of low coordinated surface atoms have *high-index* facets. High index planes have at least one Miller index larger than 1. The (1 1 1) and (1 0 0) planes are atomically flat with closely packed highly coordinated surface atoms. Pt high-index planes exhibit higher reactivity than these low index planes. This knowledge has been very helpful when designing catalysts used for the electrooxidation of small organic molecules that contain C–C bonds. For example Tarnowski et al. [129] studied the effects of step atoms on the selectivity of ethanol electrooxidation by using Pt(5 3 3), Pt(7 5 5), and Pt(1 1 1) single crystal electrodes. The yield of CO_2 was increased by introducing the (1 0 0) step on a (1 1 1) terrace. At the same time the yield of acetic acid on the Pt(5 3 3) electrode was only approx. 25–30 % of that on the Pt(1 1 1). Sun and co-workers [130] in a similar way studied the electrooxidation of isopropanol on platinum and found that Pt(6 1 0) was the most active surface for yielding CO_2. The order of activity for producing CO_2 was Pt(6 1 0) > Pt(1 1 1) > Pt(1 0 0) > Pt(2 1 1) > Pt (1 1 0). These results clearly demonstrate that high-index surfaces with a high density of step and kink atoms do significantly promote the complete electrooxidation of fuels containing C–C bonds.

In some cases this strategy has been applied to the EOR in alkaline media. For example Zhang and coworkers successfully prepared concave palladium polyhedra (size >50 nm) by an *in situ* facet-selective etching growth route. Representative electron microscopic images of the as-obtained Pd nanoparticles are shown in Fig. 12.4 [131]. Figure 12.4c depicts model images of ideal concave polyhedra in different orientations, and the corresponding nanocrystals (NCs) are indicated in Fig. 12.4b. Each particle is comprised of 32 facets in total, including 24 identical five-edge facets and 8 identical triangular facets. Aberration-corrected high-resolution (HR)TEM was used to characterize the surface structure (Fig. 12.4d). It can be seen that the concave polyhedral Pd NCs possess a high density of atomic steps on their surface (see the border atoms highlighted in 12.4e, f). Due to the presence of a high density of atomic steps, these nanoparticles exhibited an enhanced specific electrocatalytic activity towards ethanol oxidation in alkaline media.

Sang Woo Han and co-workers have recently reported the synthesis of Au-Pd alloyed nanoparticles enclosed exclusively by high-index facets {5 4 1} [132]. These nanocrystals were realized using a simple one-pot aqueous synthesis that did not require seed or additional metal ions as structure regulating agent. The nanocrystals with an average size of 114 nm displayed an hexaoctahedral structure (i.e. a polyhedron bound by 48 triangular high index facets) and exhibited higher catalytic performance toward the electro-oxidation of ethanol in alkaline media than equivalent Au-Pd nanocrystals bound by low-index facets.

The majority of bottom-up methods for shape and structure control of bimetallic core-shell or faceted metal nanoparticles such as "seeded growth" or "kinetic controlled growth" tend to lead to particles sizes larger than 50 nm. At the same time we know that electrocatalytic activity is optimized on nanoparticles in the sub 10 nm size range [133, 134]. Such size control has been achieved by some researchers using extremely slow growth. For example Tilley and coworkers prepared Pd shell on Au core-shell nanocrystals with precise layer by layer control of the shell thickness [135]. More facile methods to obtain particle sizes in the sub 10 nm range involve top-down milling. Top-down nanostructuring with precious metal decoration of core-shell particle formation has been achieved by Ozoemena et al. A precursor material FeCo@Fe/C (Fe shell on FeCo core) with large particle sizes (>210 nm), subjected to a rapid solvothermal microwave reaction in the presence of a Pd salt, lead to the downsizing of the core particles and decoration of Pd (particle sizes 3–7 nm) [136]. The resulting FeCo@Fe@Pd/C core-shell particles exhibited enhanced activity for EtOH, EG and G electrooxidation in alkaline media when compared to a Pd/C catalyst with the same metal loading [137].

Fig. 12.4 (**a**) TEM and (**b**) SEM images of concave polyhedral Pd NCs. (**c**) Model images in different orientations corresponding to those of the nanoparticles marked with the same number in (**b**). (**d–f**) Aberration-corrected HRTEM images of concave polyhedral Pd NCs oriented along the ‹110› zone axis. Panel (**e**) and (**f**) is the magnified HRTEM image taken from the selected area marked by the *red* and *blue line* in panel (**d**), respectively. The inset in (**e**) and (**f**) is the corresponding FFT pattern. Reproduced from [131] with permission of John Wiley and Sons

Sun and co-workers in 2007 announced the development of a novel electrochemical method for the synthesis of small (2–10 nm) Pt nanocrystals enclosed with high-index facets [138, 139]. The process involves an electrochemical square-wave potential deposition process of Pt onto a glassy carbon (GC) electrode in a solution containing 0.1 M H_2SO_4 and 30 mM ascorbic acid. This electrochemical shape control method was also used to obtain high index faceted Pt nano-crystals supported on carbon black with a size (2–10 nm) comparable to that of standard commercial Pt/C catalysts [140]. A much higher density of low coordinate atomic steps was obtained and this led to a doubling of the cleavage of the C-C bond in the electrooxidation of ethanol as evidenced by *in situ* FTIR spectroscopy. Other researchers have followed suit using square wave potential treatments to obtain shape transformations. For example Zhou et al. transformed Pt nanocubes to tetrahexahedra with a size of around 10 nm [141]. During the process the surface structure was changed from {1 0 0} low-index facets to {3 1 0} high-index facets. As a consequence the electrocatalytic activity for ethanol oxidation was also greatly enhanced.

The electrochemical square-wave potential method has also been used, with some modifications, for the preparation of Palladium nanocrystals with high-index facets [142]. In this work Pd nanocrystals were electrodeposited (from solution) onto a GC electrode from a 0.2 mM $PdCl_2 + 0.1$ M $HClO_4$ solution by programmed electrodeposition. A SEM image of the Pd nanoparticles is shown in Fig. 12.5 (mean particle diameter 61 nm).

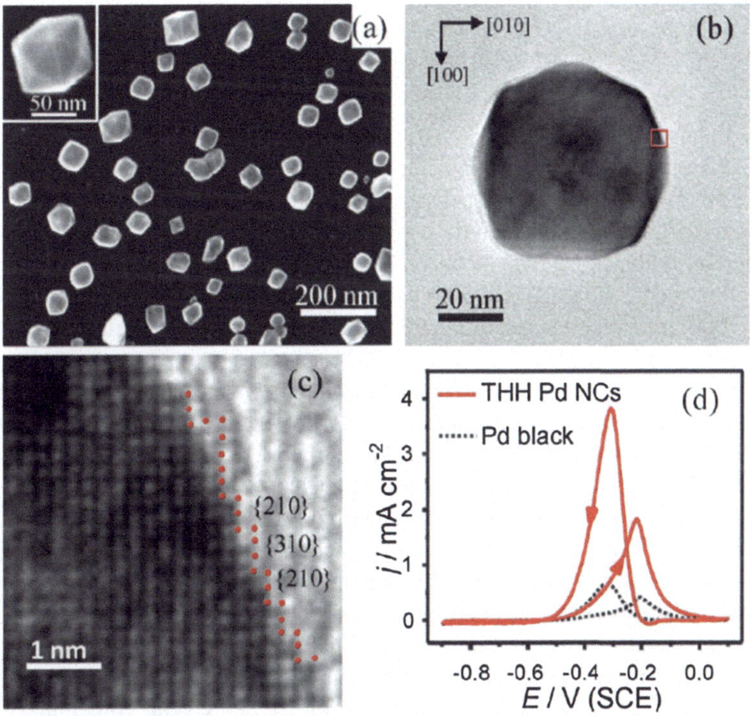

Fig. 12.5 (**a**) SEM image of Pd nano-crystals. The inset is a high magnification SEM image. (**b**) TEM image of a Pd NC recorded along the [0 0 1] direction. (**c**) HRTEM image recorded from the boxed area in (**b**), showing some {2 1 0} and {3 1 0} steps that have been marked by *red dots*. (**d**) Cyclic voltammograms of Pd NCs (*solid line*) and Pd black catalyst (*dashed line*) at 10 mV s^{-1} in 0.1 M ethanol + 0.1 M NaOH. Reprinted with permission from [142]. Copyright (2011) American Chemical Society

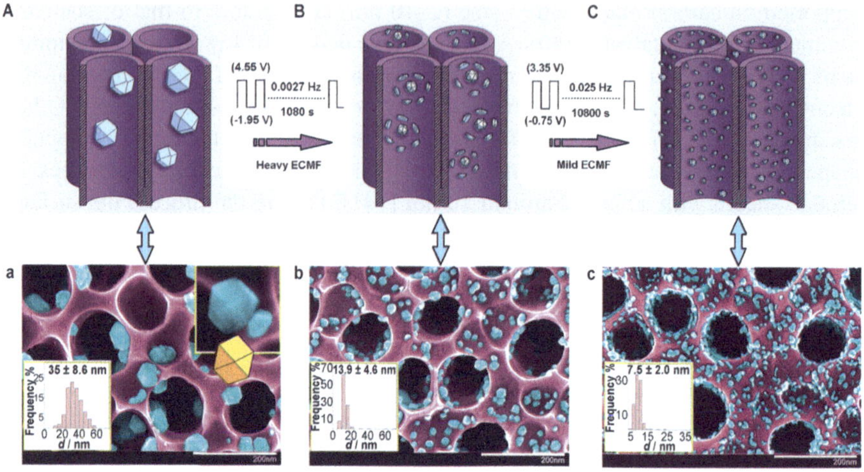

Fig. 12.6 (**A**) TNTAs with as-deposited Pd and (**a**) the corresponding SEM image. (**B**) TNTAs with Pd after heavy ECMF and (**b**) the corresponding SEM image. (**C**) TNTAs with Pd after heavy and mild ECFM and (**c**) the corresponding SEM image. False coloring of the SEM images shows Pd NPs (*light blue*) and TNTA support (*violet*). The white scale bars in (**a**–**c**) are 200 nm. Reprinted from [143] with permission from John Wiley and Sons

The exposed facets (mainly {7 3 0}) were determined by HRTEM and SAED. The {2 1 0} and {3 1 0} steps are visible on the border atoms in the HRTEM image Fig. 12.5b, c. The presence of such a high density of surface active sites, yielded approx 4–6 times higher catalytic activity per unit surface area compared to a commercial Pd black catalyst (Johnson Matthey, Inc.) for ethanol electrooxidation in alkaline media (Fig. 12.5d). This Pd catalyst also showed enhanced stability. After 1000 potential cycles, 75.0–95.5 % of the initial catalytic activity was maintained.

An adaption of the square wave method has been recently reported with the name *Electrochemical Milling and Faceting* (ECMF) [143]. Large Pd nanoparticles (35 nm) with low-index facets were first supported on TiO_2 nanotube arrays (TNTAs) (Fig. 12.6). A two-step square wave electrochemical treatment was then applied. In the first "heavy" step, a palladium oxidation was applied at 4.55 V (vs. RHE) for 180 s, followed by the reduction of the Pd oxides at −1.95 V (vs. RHE) for 180 s (Fig. 12.6b). This was followed by a milder treatment with a frequency of 0.025 Hz for 3 h between +3.35 and −0.75 V (vs. RHE) (Fig. 12.6). The overall treatment resulted in not only a net reduction in mean particle size to 7 nm particles (milling) but also the formation of high index facets, multiple twins and a high density of step atoms (Faceting) (Fig. 12.7). Cyclic voltammetry with ethanol under alkaline conditions showed a peak current density of 201 mA cm^{-2} (Fig. 12.8, curve 3), corresponding to a normalized mass-specific activity for Pd of 8965 Ag^{-1}. The onset potential for the oxidation of ethanol shifted 170 mV more negative than the potential obtained for the as-deposited sample.

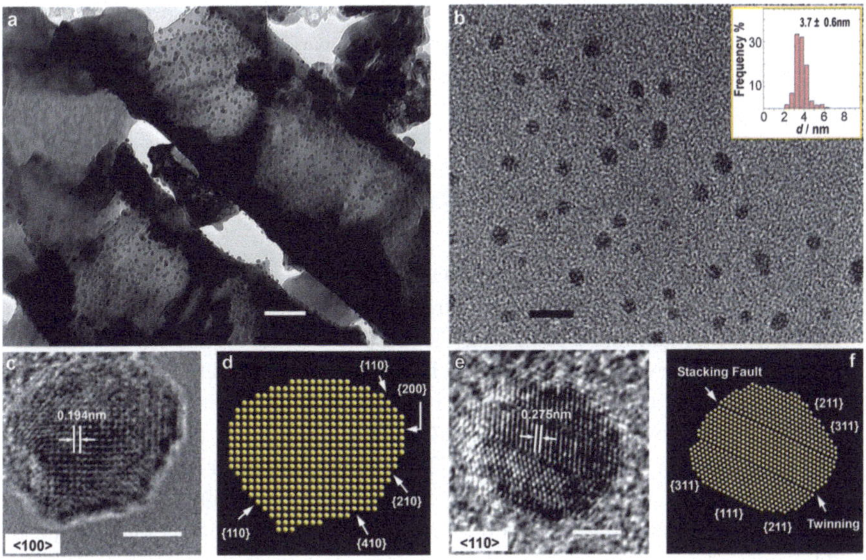

Fig. 12.7 (**a**) TEM image of the Pd-loaded TNTA electrode after heavy and mild ECMF (scale bar = 50 nm). (**b**) Pd nanoparticles found in the electrolyte after heavy and mild ECMF (scale bar = 35 nm). (**c**) HRTEM image (scale bar = 2 nm) and (**d**) atomic models with face assignment of the TNTA-supported Pd nanoparticle along the ‹100› direction. (**e**) HRTEM image (scale bar = 2 nm) and (**f**) face assignment of the TNTA-supported Pd nanoparticles along the ‹110› direction. Reprinted from [143] with permission from John Wiley and Sons

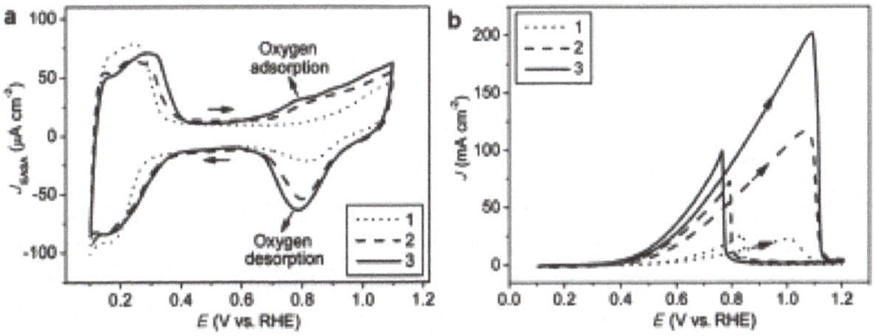

Fig. 12.8 Cyclic voltammograms of TNTAs with deposited Pd recorded in (**a**) 0.1 M $HClO_4$ and (**b**) 2 M KOH with 10 wt% EtOH. Scan rate: 50 mV s^{-1}. *Curve 1*: TNTA-Pd as deposited. *Curve 2*: TNTA-Pd after heavy ECMF. *Curve 3*: TNTA-Pd after heavy and mild ECMF. Reprinted from [143] with permission from John Wiley and Sons

The same authors have applied analogous square wave methods to treat polycrystalline Pd and Pt electrodes [144, 145]. Constant potential oxidation reduction cycles applied to both electrodes resulted in an increase in activity for alcohol electrooxidation, which was attributed to a combination of increased electro active surface area (EASA) and an increased concentration of low coordination surface atoms. Electrochemical in-situ FTIR spectroscopy showed that on the treated Pt electrode C-C bond breaking was actually limited when compared to the pristine electrode thus leading to more stable performance as poisoning intermediates such as CO are avoided [145].

12.4.4 Nanostructured Metal Oxide Supports

Up till now we have described a wide range of Pd and Pd alloyed catalysts that show excellent initial activity for alcohol electrooxidation under alkaline conditions both in half cell investigations and in DAFCs. Despite the large quantity of materials reported to date very few investigations have dealt with the actual fuel utilization efficiency of these catalysts employed in alkaline DAFCs. In one rare example Bambagioni and coworkers using a Pd/C catalyst in a room temperature DEFC found a fuel energy efficiency of circa 2.6 % [60]. A number of factors have an effect on the fuel efficiency including incomplete conversion to CO_2, catalyst poisoning, fuel consumption as well as consumption of supporting electrolyte. In addition palladium based electrocatalysts also suffer from the fact that at anode potentials higher than 0.6 V (RHE) in half cells and 0.15 in monoplanar DAFCs, the surface adsorbed Pd-OH_{ads} species (Eq. 12.12), active for the oxidation of ethanol, start converting into inactive Pd-O according to Eqs. 12.13 and 12.14. As a result, the number of active sites on the electrode surface decreases, and fuel conversion is slowed down and ultimately inhibited.

$$Pd + OH^- \rightarrow Pd - OH_{ads} + e^- \tag{12.12}$$

$$Pd - OH_{ads} + OH^- \rightarrow Pd - O + \mathrm{H_2O} + \mathrm{e}^- \tag{12.13}$$

$$Pd - OH_{ads} + Pd - OH_{ads} \rightarrow Pd - O + \mathrm{H_2O} \tag{12.14}$$

In one strategy developed to diminish the extent of Pd oxidation to Pd-O on DAFC anodes, a small amount of $NaBH_4$ was introduced to the ethanol fuel solution. This reagent was able to reduce any Pd-O formed back to Pd metal [146]. An alternative approach involves anticipating the oxidation of Pd^0 to Pd^I-OH_{ads} through the use of co-supports that promote the transfer of adsorbed OH groups to the Pd catalyst surface. One of the most effective materials used so far is ceria. CeO_2 is a mixed conductor, showing both electronic and ionic conduction, with many applications in catalysis in conjunction with transition metals. The first investigation of DAFCs containing anode electrocatalysts made of Pd nanoparticles supported on a mixture of carbon and ceria showed that, at the same metal loading and experimental conditions, the energy efficiency of a DEFC assembled with the Pd–CeO_2/C anode electrocatalyst was twice as much as that supplied by the cells with an analogous Pd/C electrocatalyst [60]. It was proposed that ceria promotes the formation at low potentials of Pd^I-OH_{ads}, the species responsible for ethanol oxidation. In DEFCs using the Pd-CeO_2/C anode power densities as high as 66 mW cm^{-2} at 25 °C and 140 mW cm^{-2} at 80 °C were obtained.

Carbon supported nanoparticle electrocatalysts suffer from limited long term durability [147, 148]. Dissolution, and metal particle growth (aggregation and Ostwald ripening), and carbon corrosion account for most of the loss of the EASA during fuel cell operation [94, 149, 150]. Carbon materials are susceptible to corrosion/oxidation at high potentials (>0.7 V *vs.* NHE) under fuel cell operating conditions, processes which accelerate further the above degradation mechanisms [151]. The hydrophobicity of carbon also limits the mass transport of electro-active species inside the catalyst layer which impacts negatively on the kinetics of electrochemical reactions in the liquid phase.

Alternative catalyst supports have indeed shown improved electrode stability compared to carbon supported nanoparticles during ethanol electrooxidation in alkaline media. For example a PdAg alloy supported on a Nb-doped-TiO_2 support exhibited a remarkable improvement in stability compared to Pd/C under fixed potential and cycling tests [152]. In another example, Pd-PANI-Pd sandwich nanotubes used for ethanol electrooxidation showed the advantage of an open electrode structure that favors fast mass transport in the liquid phase [153]. Chen and coworkers have recently reported a novel electrode structure comprised of a 2 μm thick layer of TiO_2 nanotube arrays (TNTA) that cover the surface of a titanium fiber non-woven web electrode (Fig. 12.9) [154]. Onto this material Pd nanoparticles are supported by deposition and chemical reduction with $NaBH_4$. This material was applied as an anode electrode in two electrochemical devices (i) in electrochemical reforming of alcohols for hydrogen production [154] and (ii) as anode electrode in DAFCs [61]. The fibrous open structure of the web electrode, provides in the liquid phase, rapid mass transport and short diffusion paths for electroactive species (Fig. 12.9). This material, employed as anode electrode in alkaline DAFCs, combined with an anion exchange membrane (A201) and a FeCo/C cathode produced peak power densities at 80 °C of 210, 170 and 160 mW cm^{-2} with EtOH, EG and G respectively (Fig. 12.10). The flexibility of this material allowed an increase in Pd loading (from 1.5 to 6 mg cm^{-2}) by combining four single electrodes on the anode side of the DAFC. With ethanol at 80 °C the cell produced a peak power of 335 mW cm^{-2} and a maximum current density of 2.25 A cm^{-2} at 0.1 V cell potential (Fig. 12.10).

12.5 Molecular Anode Catalysts for DAFCs

Increasingly, researchers in this field are striving to use DAFCs to combine the combination of the production of renewable energy (with no CO_2 emissions) together with the production of industrially relevant feedstocks (such as aldehydes, ketones and carboxylic acids) [7, 31, 34, 66, 76, 121, 123, 126]. In this way the free energy of an alcohol is converted into electrical energy and the alcohol itself is transformed into an oxidation product or products which are different from CO_2. Two types of DAFCs have been developed for this purpose: (1) traditional devices, where the anode and cathode are separated by an anion-exchange membrane and

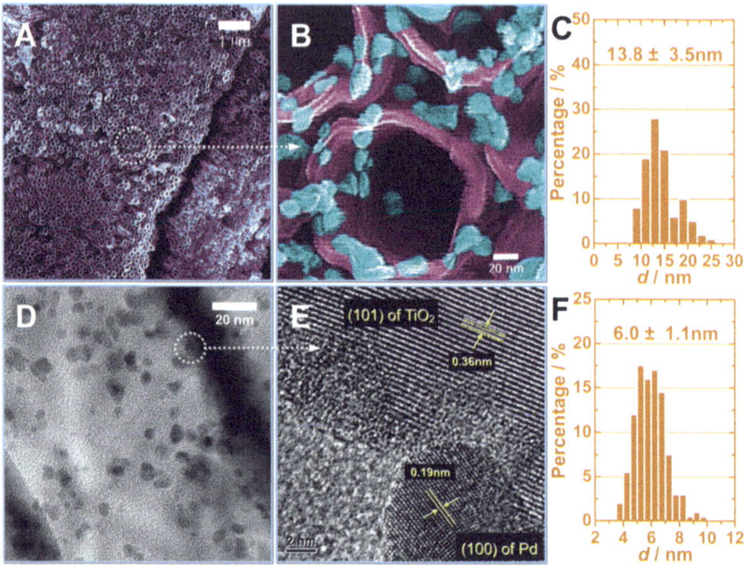

Fig. 12.9 Representative FESEM images of Pd/TNTA-web (**a**, **b**). TEM (**d**) and HRTEM (**e**) analysis of Pd/TNTA-web. Particle size distribution (**c**) from FESEM analysis of Pd particles around the opening of the nanotubes and (**f**) particle size distribution from TEM analysis of particles situated along the tube walls. Reprinted from [61] with permission from John Wiley and Sons

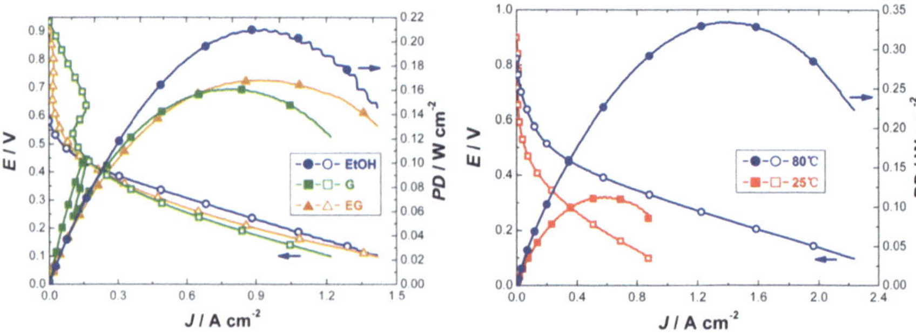

Fig. 12.10 (*Left*) Potentiodynamic and power density curves for DAFCs fuelled with 2.0 M KOH and alcohol (10 wt%) solutions Pd/TNTA-web (1.5 $mg_{Pd}\,cm^{-2}$) and (*right*) potentiodynamic and power density curves for combined Pd/TNTA-web anodes (6 $mg_{Pd}\,cm^{-2}$) ethanol 10 wt% in 2 M KOH. Reprinted from [61] with permission from John Wiley and Sons

are coated with an electrocatalyst, generally a nanostructured noble metal, supported on a conductive material [36, 76, 109, 155–157] and (2) enzymatic biofuel cells (EBFC) utilizing oxidation enzymes such as dehydrogenases in conjunction with an electron transfer mediator [158, 159]. Efficient devices of this type

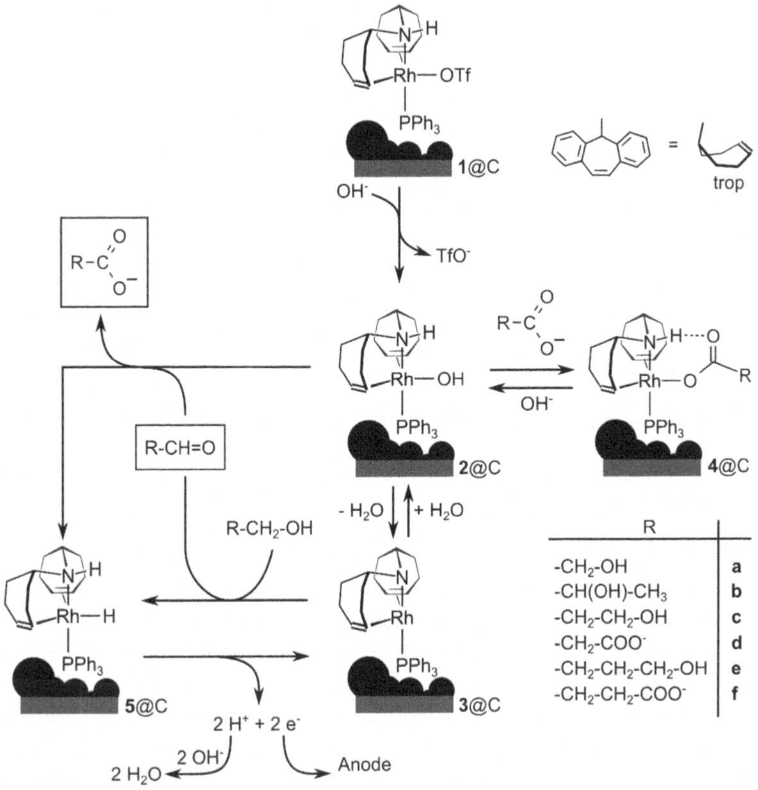

Scheme 12.6 Proposed mechanism for the reactions occurring on the anode of the OMFC. Reprinted from [31] with permission from John Wiley and Sons

have been recently developed for a variety of renewable alcohols and polyalcohols such as ethylene glycol, glycerol and 1,2-propandiol [36, 156].

Recently a third type of fuel cell has been developed, known as an OrganoMetallic Fuel Cell (OMFC), operating in alkaline media where the anode catalyst is a molecular metal complex. Several research groups have demonstrated this concept showing that electrocatalysts based on specific metal complexes can oxidize alcohols [160–162] and glucose [163, 164] at low overpotentials. In the most successful example, an organometallic complex [Rh(OTf)(trop$_2$NH)(PPh$_3$)] ((12.1) in Scheme 12.6) has been shown to selectively catalyze the oxidation of alcohols to carboxylic compounds [28, 31, 165].

The metal complex evolves through fast chemical equilibria to form specific catalysts for ethanol dehydrogenation, aldehyde dehydrogenation and H^+/electron transfer. The [Rh(OTf)(trop$_2$NH)(PPh$_3$)] complex (12.1) can be deposited intact onto a conductive carbon support such as Vulcan XC-72 or Ketjenblack EC 600 JD to form the anode electrode. The mechanism for alcohol oxidation is shown in Scheme 12.6.

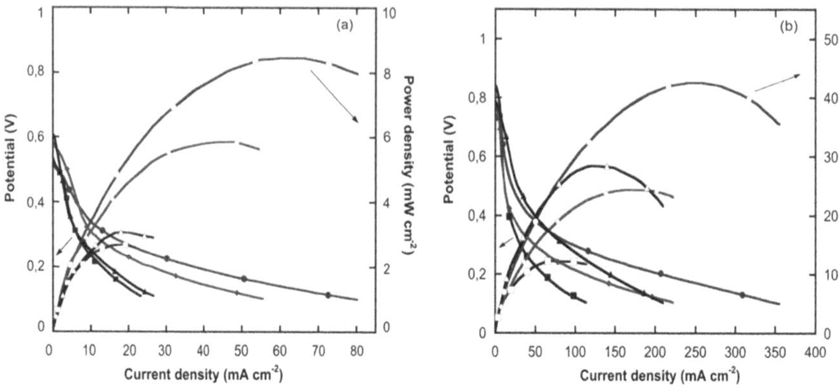

Fig. 12.11 Polarization (*filled circle*) and power density (*circle*) curves of OMFCs fueled with: (*filled circle*) 1,4-B; (*filled diamond*) 1,3-P; (*filled triangle*) 1,2-P; (*filled square*) EG aqueous solutions 5 wt% in 2 M KOH. (**a**) Passive air-breathing OMFC at 22 °C and (**b**) active OMFC at 80 °C. Reprinted from [31] with permission from John Wiley and Sons

On the electrode surface, the precursor 1@C is rapidly converted into the hydroxo complex, [Rh(OH)(trop2NH)(PPh3)]@C (2@C), which is in a rapid equilibrium with the amide [Rh(trop2N)(PPh3)]@C (3@C) and water. The amide 3@C dehydrogenates diols to aldehydes. Under the basic reaction conditions, the aldehydes are rapidly further converted promoted by catalyst 2@C to form carboxylate ions and 5@C. The latter complex is oxidized at the electrode, releasing two H+ (neutralized to give water under basic conditions) and two electrons with regeneration of the amide 3@C. There is some resemblance with the enzymatic biofuel cell, but the main characteristic of this system is that one molecular rhodium complex is capable of evolving through fast chemical equilibria in the course of the catalytic cycle to form a specific catalyst for alcohol dehydrogenation (the amide 3@C), a specific catalyst for aldehyde dehydrogenation (the hydroxo complex 2@C), and a specific catalyst for the H^+/ electron transfer (the hydride 5@C).

MEAs were fabricated comprising of a nickel foam anode coated with 1@C (ca. 1 mg cm^{-2} Rh), a carbon paper cathode coated with a Fe-Co/C electrocatalyst and a Tokuyama A-201 anion-exchange membrane. Figure 12.11a shows typical polarization and power density curves for passive cells fueled with aqueous solutions of ethylene glycol (EG), 1,2-propanediol (1,2-P), 1,3-propanediol (1,3-P), and 1,4-butanediol (1,4-B) in 2 M KOH at room temperature. The cell containing 1,4-B exhibits the highest peak power density (8.5 mW cm^{-2} at 0.14 V) compared to the other diols. The power density supplied by these OMFCs increases with cell temperature (Fig. 12.11b). Indeed, 12.4 mW cm^{-2}, 26.9 mW cm^{-2}, 24.3 mW cm^{-2}, and 42.2 mWcm$^{-\ 2}$, were obtained with EG, 1,2- P, 1,3-P, and 1,4-B, respectively at 80 °C. To date this is the first report of fuel cells that are fed with 1,2-P, 1,3-P, and 1,4-B.

Room temperature OMFCs were subjected to galvanostatic experiments (20 mA cm^{-2}) and the fuel exhausts were quantitatively and qualitatively analyzed

Table 12.5 Electrooxidation selectivity data for OMFCs at 22 °C and 80 °C with various alcohol fuels

Fuel	Temperature (°C)	Max power density (mW cm^{-2})	TON	Conversion % (mmol)	Galvanostatic time (20 mA cm^{-2})	Products
EG	22	2.7	86	38 % (3.2)	17 h 14 m	100 % glycolate
1,2-P	22	3.1	92	50 % (3.4)	18 h 17 m	100 % lactate
1,3-P	22	5.9	173	83 % (6.1)	34 h 16 m	92 % 3-hydroxy-propanate
						8 % malonate
1,4-B	22	8.5	114	78 % (114)	28 h 38 m	79 % 4-hydroxy-butanoate
						21 % succinate
EG	80	12.4	329	28 % (12.1)	12 h 45 m	100 % glycolate
1,2-P	80	26.9	318	34 % (12.0)	12 h 39 m	100 % lactate
1,3-P	80	24.3	445	45 % (16.3)	17 h 29 m	96 % 3-hydroxy-propanate
						4 % malonate
1,4-B	80	42.2	486	79 % (18.4)	22 h 11 m	74 % 4-hydroxy-butanoate
						26 % succinate

Reprinted from [31] with permission from John Wiley and Sons

by $^{13}C\{^{1}H\}$ NMR spectroscopy and HPLC (Table 12.5 summarizes the selectivity data). The OMFCs fed with EG and 1,2-P, yielded selectively glycolate and lactate. The galvanostatic experiment with 1,3-P gave 3-hydroxy-propanoate (92 %) and malonate (8 %) with a total conversion of 83 %. Using 1,4-B, 4-hydroxy-butanoate (79 %) and succinate (21 %) were obtained. In general this molecular anode electrocatalyst was shown to convert short chain diols (e.g. EG and 1,2-P) into the mono-carboxylate whereas diols with longer spacer chains produced also a small amount of the di-carboxylate (e.g. 1,3-P and 1,4-B).

12.6 Conclusions and Future Perspectives

The recent flurry of activity in this field attests to this fact that the application of biomass derived alcohols in DAFCs is an attractive target for researchers working towards a system of sustainable energy production and distribution. The move to study devices that work in alkaline media rather than under corrosive acidic conditions is advantageous as many more materials can be investigated and the prohibitive costs associated with the use of large amounts of platinum based catalysts can be avoided. The availability of commercial anion exchange membranes together with highly active non precious metal cathode electrocatalysts means that work can be focused on improving the anode electrocatalyst. This chapter has shown that the most active metals for the alcohol oxidation reaction are Pt, Pd and Au. Palladium stands out as the most active material. The addition of a second transition metal (e.g. Ni, Pt, Cu, Co, Ru, In, Bi) or oxide (e.g. CeO_2, TiO_2, SnO_2) to Pd catalysts leads to dramatic improvements in activity. Researchers have also manipulated the surface structure of Pd nanoparticles on an atomic level. A high concentration of low coordinated surface atoms on high index faceted nanoparticles have been prepared by both chemical and electrochemical means. These faceted particles exhibit enhanced activity for alcohol oxidation. Important advances in stability have been achieved by replacing carbon supports with inorganic conductive oxides. The most remarkable results have been obtained by employing Pd doped titania nanotube arrays (TNTA) where peak power densities in DAFCs of up to 335 mW cm^{-2} have been obtained with ethanol.

As the alcohol used as fuel becomes more complex (*e.g.* 1,2-propandiol and 1,4-butandiol) the electrooxidation pathways also become complicated with a wide range of possible pathways and intermediates. Indeed, the most challenging aspect of developing DAFCs as power sources is to obtain complete oxidation of the fuel molecule to CO_2, a feat yet to be achieved even for the simplest, ethanol. Taking advantage of this aspect some investigators have worked at tuning selectivity of the oxidation process toward the formation of a valuable partially oxidized intermediate (e.g. lactate from 1,2-propandiol). For this the most successful anode catalysts have been organometallic complexes which are single-site molecular catalysts that have been shown to selectively oxidize a range of polyalcohols to their monocarboxylates.

In summary, as a power source both catalyst stability and complete oxidation of alcohol fuels to CO_2 are essential in order for DAFCs to be exploited commercially. As part of the bio-refinery concept *i.e.* cogeneration of electrical power and industrially relevant intermediates from renewable bio-alcohols, DAFCs will have an important role to play. However, in such devices better selectivity and higher conversion is required from the anode electrocatalyst.

References

1. Brouzgou A, Podias A, Tsiakaras P (2013) PEMFCs and AEMFCs directly fed with ethanol: a current status comparative review. J Appl Electrochem 43(2):119–136. doi:10.1007/s10800-012-0513-2
2. Yu EH, Krewer U, Scott K (2010) Principles and materials aspects of direct alkaline alcohol fuel cells. Energies 3(8):1499–1528. doi:10.3390/En3081499

3. Scordia D, Cosentino SL, Jeffries TW (2013) Effectiveness of dilute oxalic acid pretreatment of Miscanthus × giganteus biomass for ethanol production. Biomass Bioenerg 59:540–548. doi:10.1016/j.biombioe.2013.09.011
4. Shi AM, Du ZY, Ma XC, Cheng YL, Min M, Deng SB, Chen P, Li D, Ruan R (2013) Production and evaluation of biodiesel and bioethanol from high oil corn using three processing routes. Bioresour Technol 128:100–106. doi:10.1016/j.biortech.2012.10.007
5. Livshits V, Philosoph A, Peled E (2008) Direct ethylene glycol fuel-cell stack—study of oxidation intermediate products. J Power Sources 178(2):687–691. doi:10.1016/j.jpowsour.2007.07.054
6. Kaplan D, Burstein L, Rosenberg Y, Peled E (2011) Comparison of methanol and ethylene glycol oxidation by alloy and core-shell platinum based catalysts. J Power Sources 196 (20):8286–8292. doi:10.1016/j.jpowsour.2011.06.023
7. Xin L, Zhang ZY, Wang ZC, Li WZ (2012) Simultaneous generation of mesoxalic acid and electricity from glycerol on a gold anode catalyst in anion-exchange membrane fuel cells. Chemcatchem 4(8):1105–1114. doi:10.1002/cctc.201200017
8. Marchionni A, Bevilacqua M, Bianchini C, Chen YX, Filippi J, Fornasiero P, Lavacchi A, Miller H, Wang LQ, Vizza F (2013) Electrooxidation in alkaline media of ethylene glycol and glycerol on Pd-(Ni-Zn)/C anodes in direct alcohol fuel cells. Chemsuschem 6 (3):390–390. doi:10.1002/cssc.201300154
9. Ji N, Zhang T, Zheng MY, Wang AQ, Wang H, Wang XD, Chen JGG (2008) Direct catalytic conversion of cellulose into ethylene glycol using nickel-promoted tungsten carbide catalysts. Angew Chem Int Ed 47(44):8510–8513. doi:10.1002/anie.200803233
10. Zheng MY, Wang AQ, Ji N, Pang JF, Wang XD, Zhang T (2010) Transition metal-tungsten bimetallic catalysts for the conversion of cellulose into ethylene glycol. Chemsuschem 3 (1):63–66. doi:10.1002/cssc.200900197
11. Van Gerpen J (2005) Biodiesel processing and production. Fuel Process Technol 86 (10):1097–1107. doi:10.1016/j.fuproc.2004.11.005
12. Pimentel D, Patzek T (2005) Ethanol production using corn, switchgrass, and wood; biodiesel production using soybean and sunflower. Nat Resour Res 14(1):65–76. doi:10.1007/s11053-005-4679-8
13. Scordia D, Cosentino SL, Lee J-W, Jeffries TW (2011) Dilute oxalic acid pretreatment for biorefining giant reed (Arundo donax L.). Biomass Bioenergy 35(7):3018–3024. http://dx.doi.org/10.1016/j.biombioe.2011.03.046
14. Antolini E, Gonzalez ER (2010) Alkaline direct alcohol fuel cells. J Power Sources 195 (11):3431–3450. doi:10.1016/j.jpowsour.2009.11.145
15. Wang YJ, Qiao JL, Baker R, Zhang JJ (2013) Alkaline polymer electrolyte membranes for fuel cell applications. Chem Soc Rev 42(13):5768–5787. doi:10.1039/C3cs60053j
16. Tarasevich MR, Korchagin OV, Kuzov AV (2013) Electrocatalysis of anodic oxidation of ethanol. Russ Chem Rev 82(11):1047–1065. doi:10.1070/Rc2013v082n11abeh004276
17. Varcoe JR, Atanassov P, Dekel DR, Herring AM, Hickner MA, Kohl PA, Kucernak AR, Mustain WE, Nijmeijer K, Scott K, Xu TW, Zhuang L (2014) Anion-exchange membranes in electrochemical energy systems. Energy Environ Sci 7(10):3135–3191. doi:10.1039/C4ee01303d
18. Bambagioni V, Bianchini C, Filippi J, Lavacchi A, Oberhauser W, Marchionni A, Moneti S, Vizza F, Psaro R, Dal Santo V, Gallo A, Recchia S, Sordelli L (2011) Single-site and nanosized Fe-Co electrocatalysts for oxygen reduction: synthesis, characterization and catalytic performance. J Power Sources 196(5):2519–2529. doi:10.1016/j.jpowsour.2010.11.030
19. Miller HA, Bevilacqua M, Filippi J, Lavacchi A, Marchionni A, Marelli M, Moneti S, Oberhauser W, Vesselli E, Innocenti M, Vizza F (2013) Nanostructured Fe-Ag electrocatalysts for the oxygen reduction reaction in alkaline media. J Mater Chem A 1 (42):13337–13347. doi:10.1039/C3ta12757e
20. Kruusenberg I, Matisen L, Shah Q, Kannan AM, Tammeveski K (2012) Non-platinum cathode catalysts for alkaline membrane fuel cells. Int J Hydrogen Energy 37 (5):4406–4412. doi:10.1016/j.ijhydene.2011.11.143

21. Shen SY, Zhao TS, Wu QX (2012) Product analysis of the ethanol oxidation reaction on palladium-based catalysts in an anion-exchange membrane fuel cell environment. Int J Hydrogen Energy 37(1):575–582. doi:10.1016/j.ijhydene.2011.09.077
22. He QG, Shyam B, Macounova K, Krtil P, Ramaker D, Mukerjee S (2012) Dramatically enhanced cleavage of the C-C bond using an electrocatalytically coupled reaction. J Am Chem Soc 134(20):8655–8661. doi:10.1021/Ja301992h
23. Fang X, Wang LQ, Shen PK, Cui GF, Bianchini C (2010) An in situ Fourier transform infrared spectroelectrochemical study on ethanol electrooxidation on Pd in alkaline solution. J Power Sources 195(5):1375–1378. doi:10.1016/j.jpowsour.2009.09.025
24. Lamy C, Lima A, LeRhun V, Delime F, Coutanceau C, Leger JM (2002) Recent advances in the development of direct alcohol fuel cells (DAFC). J Power Sources 105(2):283–296. doi:10.1016/S0378-7753(01)00954-5, Pii: S0378-7753(01)00954-5
25. Gasteiger HA, Markovic N, Ross PN, Cairns EJ (1993) Methanol electrooxidation on well-characterized Pt-Rn alloys. J Phys Chem 97(46):12020–12029. doi:10.1021/J100148a030
26. Lamy C, Lima A, LeRhun V, Delime F, Coutanceau C, Léger J-M (2002) Recent advances in the development of direct alcohol fuel cells (DAFC). J Power Sources 105(2):283–296. doi:10.1016/S0378-7753(01)00954-5
27. Liang ZX, Zhao TS, Xu JB, Zhu LD (2009) Mechanism study of the ethanol oxidation reaction on palladium in alkaline media. Electrochim Acta 54(8):2203–2208. doi:10.1016/j.electacta.2008.10.034
28. Bevilacqua M, Bianchini C, Marchionni A, Filippi J, Lavacchi A, Miller H, Oberhauser W, Vizza F, Granozzi G, Artiglia L, Annen SP, Krumeich F, Grutzmacher H (2012) Improvement in the efficiency of an organometallic fuel cell by tuning the molecular architecture of the anode electrocatalyst and the nature of the carbon support. Energy Environ Sci 5 (9):8608–8620. doi:10.1039/C2ee22055e
29. Sheng T, Lin WF, Hardacre C, Hu P (2014) Role of water and adsorbed hydroxyls on ethanol electrochemistry on Pd: new mechanism, active centers, and energetics for direct ethanol fuel cell running in alkaline medium. J Phys Chem C 118(11):5762–5772. doi:10.1021/Jp407978h
30. Yang YY, Ren J, Li QX, Zhou ZY, Sun SG, Cai WB (2014) Electrocatalysis of ethanol on a Pd electrode in alkaline media: an in situ attenuated total reflection surface-enhanced infrared absorption spectroscopy study. ACS Catal 4(3):798–803. doi:10.1021/Cs401198t
31. Bellini M, Bevilacqua M, Filippi J, Lavacchi A, Marchionni A, Miller HA, Oberhauser W, Vizza F, Annen SP, Grutzmacher H (2014) Energy and chemicals from the selective electrooxidation of renewable diols by organometallic fuel cells. Chemsuschem 7 (9):2432–2435. doi:10.1002/cssc.201402316
32. Marchionni A, Bevilacqua M, Bianchini C, Chen Y-X, Filippi J, Fornasiero P, Lavacchi A, Miller H, Wang L, Vizza F (2013) Electrooxidation of ethylene glycol and glycerol on Pd-(Ni-Zn)/C anodes in direct alcohol fuel cells. Chemsuschem 6(3):518–528. doi:10.1002/cssc.201200866
33. Kwon Y, Schouten KJP, Koper MTM (2011) Mechanism of the catalytic oxidation of glycerol on polycrystalline gold and platinum electrodes. Chemcatchem 3(7):1176–1185. doi:10.1002/cctc.201100023
34. Bambagioni V, Bevilacqua M, Bianchini C, Filippi J, Marchionni A, Vizza F, Wang LQ, Shen PK (2010) Ethylene glycol electrooxidation on smooth and nanostructured Pd electrodes in alkaline media. Fuel Cells 10(4):582–590. doi:10.1002/fuce.200900120
35. Bambagioni V, Bianchini C, Filippi J, Oberhauser W, Marchionni A, Vizza F, Psaro R, Sordelli L, Foresti ML, Innocenti M (2009) Ethanol oxidation on electrocatalysts obtained by spontaneous deposition of palladium onto nickel-zinc materials. Chemsuschem 2(1):99–112. doi:10.1002/cssc.200800188
36. Bianchini C, Shen PK (2009) Palladium-based electrocatalysts for alcohol oxidation in half cells and in direct alcohol fuel cells. Chem Rev 109(9):4183–4206. doi:10.1021/cr9000995

37. Christensen PA, Hamnett A (1989) The oxidation of ethylene glycol at a platinum electrode in acid and base: an in situ FTIR study. J Electroanal Chem Interfacial Electrochem 260 (2):347–359. http://dx.doi.org/10.1016/0022-0728(89)87149-9
38. Hahn F, Beden B, Kadirgan F, Lamy C (1987) Electrocatalytic oxidation of ethylene glycol: Part III. In-situ infrared reflectance spectroscopic study of the strongly bound species resulting from its chemisorption at a platinum electrode in aqueous medium. J Electroanal Chem Interfacial Electrochem 216(1–2):169–180. http://dx.doi.org/10.1016/0022-0728(87) 80205-X
39. Santasalo-Aarnio A, Kwon Y, Ahlberg E, Kontturi K, Kallio T, Koper MTM (2011) Comparison of methanol, ethanol and iso-propanol oxidation on Pt and Pd electrodes in alkaline media studied by HPLC. Electrochem Commun 13(5):466–469. doi:10.1016/j.elecom.2011. 02.022
40. An L, Zhao TS, Wu QX, Zeng L (2012) Comparison of different types of membrane in alkaline direct ethanol fuel cells. Int J Hydrogen Energy 37(19):14536–14542. http://dx.doi. org/10.1016/j.ijhydene.2012.06.105
41. An L, Zhao TS, Shen SY, Wu QX, Chen R (2010) Performance of a direct ethylene glycol fuel cell with an anion-exchange membrane. Int J Hydrogen Energy 35(9):4329–4335. http:// dx.doi.org/10.1016/j.ijhydene.2010.02.009
42. An L, Zhao TS, Chen R, Wu QX (2011) A novel direct ethanol fuel cell with high power density. J Power Sources 196(15):6219–6222. http://dx.doi.org/10.1016/j.jpowsour.2011.03.040
43. Li YS, Zhao TS, Liang ZX (2009) Performance of alkaline electrolyte-membrane-based direct ethanol fuel cells. J Power Sources 187(2):387–392. http://dx.doi.org/10.1016/j. jpowsour.2008.10.132
44. Lee J, Jeong B, Ocon JD (2013) Oxygen electrocatalysis in chemical energy conversion and storage technologies. Curr Appl Phys 13(2):309–321. doi:10.1016/j.cap.2012.08.008
45. Othman R, Dicks AL, Zhu ZH (2012) Non precious metal catalysts for the PEM fuel cell cathode. Int J Hydrogen Energ 37(1):357–372. doi:10.1016/j.ijhydene.2011.08.095
46. Brouzgou A, Song SQ, Tsiakaras P (2012) Low and non-platinum electrocatalysts for PEMFCs: current status, challenges and prospects. Appl Catal B Environ 127:371–388. doi:10.1016/j.apcatb.2012.08.031
47. Jaouen F, Proietti E, Lefevre M, Chenitz R, Dodelet JP, Wu G, Chung HT, Johnston CM, Zelenay P (2011) Recent advances in non-precious metal catalysis for oxygen-reduction reaction in polymer electrolyte fuel cells. Energy Environ Sci 4(1):114–130. doi:10.1039/ C0ee00011f
48. Guo JS, Zhou J, Chu D, Chen RR (2013) Tuning the electrochemical interface of Ag/C electrodes in alkaline media with metallophthalocyanine molecules. J Phys Chem C 117 (8):4006–4017. doi:10.1021/Jp310655y
49. Ding L, Xin Q, Zhou XJ, Qiao JL, Li H, Wang HJ (2013) Electrochemical behavior of nanostructured nickel phthalocyanine (NiPc/C) for oxygen reduction reaction in alkaline media. J Appl Electrochem 43(1):43–51. doi:10.1007/s10800-012-0503-4
50. Lalande G, Faubert G, Cote R, Guay D, Dodelet JP, Weng LT, Bertrand P (1996) Catalytic activity and stability of heat-treated iron phthalocyanines for the electroreduction of oxygen in polymer electrolyte fuel cells. J Power Sources 61(1–2):227–237. doi:10.1016/S0378-7753 (96)02356-7
51. Ding L, Qiao JL, Dai XF, Zhang J, Zhang JJ, Tian BL (2012) Highly active electrocatalysts for oxygen reduction from carbon-supported copper-phthalocyanine synthesized by high temperature treatment. Int J Hydrogen Energy 37(19):14103–14113. doi:10.1016/j. ijhydene.2012.07.046
52. He QG, Yang XF, Ren XM, Koel BE, Ramaswamy N, Mukerjee S, Kostecki R (2011) A novel CuFe-based catalyst for the oxygen reduction reaction in alkaline media. J Power Sources 196(18):7404–7410. doi:10.1016/j.jpowsour.2011.04.016
53. He QG, Yang XF, He RH, Bueno-Lopez A, Miller H, Ren XM, Yang WL, Koel BE (2012) Electrochemical and spectroscopic study of novel Cu and Fe-based catalysts for oxygen

reduction in alkaline media. J Power Sources 213:169–179. doi:10.1016/j.jpowsour.2012.04.029
54. Piana M, Boccia M, Filpi A, Flammia E, Miller HA, Orsini M, Salusti F, Santiccioli S, Ciardelli F, Pucci A (2010) H-2/air alkaline membrane fuel cell performance and durability, using novel ionomer and non-platinum group metal cathode catalyst. J Power Sources 195 (18):5875–5881. doi:10.1016/j.jpowsour.2009.12.085
55. Lefevre M, Proietti E, Jaouen F, Dodelet JP (2009) Iron-based catalysts with improved oxygen reduction activity in polymer electrolyte fuel cells. Science 324(5923):71–74. doi:10.1126/science.1170051
56. Proietti E, Jaouen F, Lefevre M, Larouche N, Tian J, Herranz J, Dodelet JP (2011) Iron-based cathode catalyst with enhanced power density in polymer electrolyte membrane fuel cells. Nat Commun 2:416. doi:10.1038/Ncomms1427
57. Trogadas P, Fuller TF, Strasser P (2014) Carbon as catalyst and support for electrochemical energy conversion. Carbon 75:5–42. doi:10.1016/j.carbon.2014.04.005
58. Yang Z, Nie HG, Chen X, Chen XH, Huang SM (2013) Recent progress in doped carbon nanomaterials as effective cathode catalysts for fuel cell oxygen reduction reaction. J Power Sources 236:238–249. doi:10.1016/j.jpowsour.2013.02.057
59. Gong KP, Du F, Xia ZH, Durstock M, Dai LM (2009) Nitrogen-doped carbon nanotube arrays with high electrocatalytic activity for oxygen reduction. Science 323(5915):760–764. doi:10.1126/science.1168049
60. Bambagioni V, Bianchini C, Chen YX, Filippi J, Fornasiero P, Innocenti M, Lavacchi A, Marchionni A, Oberhauser W, Vizza F (2012) Energy efficiency enhancement of ethanol electrooxidation on Pd-CeO2/C in passive and active polymer electrolyte-membrane fuel cells. Chemsuschem 5(7):1266–1273. doi:10.1002/cssc.201100738
61. Chen Y, Bellini M, Bevilacqua M, Fornasiero P, Lavacchi A, Miller HA, Wang L, Vizza F (2014) Direct alcohol fuel cells: toward the power densities of hydrogen-fed proton exchange membrane fuel cells. Chemsuschem 8(3):524–533. doi:10.1002/cssc.201402999
62. Li YS, He YL (2014) Layer reduction method for fabricating Pd-coated Ni foams as high-performance ethanol electrode for anion-exchange membrane fuel cells. RSC Adv 4 (32):16879–16884. doi:10.1039/C4ra01399a
63. Li YS, Zhao TS (2011) A high-performance integrated electrode for anion-exchange membrane direct ethanol fuel cells. Int J Hydrogen Energy 36(13):7707–7713. doi:10.1016/j.ijhydene.2011.03.090
64. Shen SY, Zhao TS, Xu JB, Li YS (2011) High performance of a carbon supported ternary PdIrNi catalyst for ethanol electro-oxidation in anion-exchange membrane direct ethanol fuel cells. Energy Environ Sci 4(4):1428–1433. doi:10.1039/C0ee00579g
65. Ma L, He H, Hsu A, Chen RR (2013) PdRu/C catalysts for ethanol oxidation in anion-exchange membrane direct ethanol fuel cells. J Power Sources 241:696–702. doi:10.1016/j.jpowsour.2013.04.051
66. Xin L, Zhang ZY, Qi J, Chadderdon D, Li WZ (2012) Electrocatalytic oxidation of ethylene glycol (EG) on supported Pt and Au catalysts in alkaline media: reaction pathway investigation in three-electrode cell and fuel cell reactors. Appl Catal B Environ 125:85–94. doi:10.1016/j.apcatb.2012.05.024
67. An L, Zeng L, Zhao TS (2013) An alkaline direct ethylene glycol fuel cell with an alkali-doped polybenzimidazole membrane. Int J Hydrogen Energy 38(25):10602–10606. doi:10.1016/j.ijhydene.2013.06.042
68. Qi J, Xin L, Zhang ZY, Sun K, He HY, Wang F, Chadderdon D, Qiu Y, Liang CH, Li WZ (2013) Surface dealloyed PtCo nanoparticles supported on carbon nanotube: facile synthesis and promising application for anion exchange membrane direct crude glycerol fuel cell. Green Chem 15(5):1133–1137. doi:10.1039/C3gc36955b
69. Suleimanov NM, Khantimerov SM, Kukovitsky EF, Matukhin VL (2008) Electrooxidation of ethanol on carbon nanotubes-nickel nanoparticles composites in alkaline media. J Solid State Electrochem 12(7–8):1021–1023. doi:10.1007/s10008-008-0519-1

70. Xu CW, Hu YH, Rong JH, Jiang SP, Liu YL (2007) Ni hollow spheres as catalysts for methanol and ethanol electrooxidation. Electrochem Commun 9(8):2009–2012. doi:10.1016/j.elecom.2007.05.028
71. Jin GP, Baron R, Xiao L, Compton RG (2009) Ultrasonic synthesis of nickel nanostructures on glassy carbon microspheres and their application for ethanol electrooxidation. J Nanosci Nanotechnol 9(4):2719–2725. doi:10.1166/Jnn.2009.462
72. Yi Y, Uhm S, Lee J (2010) Electrocatalytic oxidation of ethanol on nanoporous Ni electrode in alkaline media. Electrocatalysis 1(2–3):104–107. doi:10.1007/s12678-010-0015-0
73. Wang ZH, Du YL, Zhang FY, Zheng ZX, Zhang YZ, Wang CM (2013) High electrocatalytic activity of non-noble Ni-Co/graphene catalyst for direct ethanol fuel cells. J Solid State Electrochem 17(1):99–107. doi:10.1007/s10008-012-1855-8
74. Muench F, Oezaslan M, Rauber M, Kaserer S, Fuchs A, Mankel E, Brotz J, Strasser P, Roth C, Ensinger W (2013) Electroless synthesis of nanostructured nickel and nickel-boron tubes and their performance as unsupported ethanol electrooxidation catalysts. J Power Sources 222:243–252. doi:10.1016/j.jpowsour.2012.08.067
75. Ma L, Chu D, Chen RR (2012) Comparison of ethanol electro-oxidation on Pt/C and Pd/C catalysts in alkaline media. Int J Hydrogen Energy 37(15):11185–11194. doi:10.1016/j.ijhydene.2012.04.132
76. Bianchini C, Bambagioni V, Filippi J, Marchionni A, Vizza F, Bert P, Tampucci A (2009) Selective oxidation of ethanol to acetic acid in highly efficient polymer electrolyte membrane-direct ethanol fuel cells. Electrochem Commun 11(5):1077–1080. doi:10.1016/j.elecom.2009.03.022
77. Shen SY, Zhao TS, Xu JB, Li YS (2010) Synthesis of PdNi catalysts for the oxidation of ethanol in alkaline direct ethanol fuel cells. J Power Sources 195(4):1001–1006. doi:10.1016/j.jpowsour.2009.08.079
78. Liu XY, Zhang Y, Gong MX, Tang YW, Lu TH, Chen Y, Lee JM (2014) Facile synthesis of corallite-like Pt-Pd alloy nanostructures and their enhanced catalytic activity and stability for ethanol oxidation. J Mater Chem A 2(34):13840–13844. doi:10.1039/C4ta02522a
79. Jiang RZ, Tran DT, McClure JP, Chu D (2014) A class of (Pd-Ni-P) electrocatalysts for the ethanol oxidation reaction in alkaline media. ACS Catal 4(8):2577–2586. doi:10.1021/Cs500462z
80. Hong W, Wang J, Wang EK (2014) Facile synthesis of highly active PdAu nanowire networks as self-supported electrocatalyst for ethanol electrooxidation. ACS Appl Mater Interfaces 6(12):9481–9487. doi:10.1021/Am501859k
81. Geraldes AN, da Silva DF, Pino ES, da Silva JCM, de Souza RFB, Hammer P, Spinace EV, Neto AO, Linardi M, dos Santos MC (2013) Ethanol electro-oxidation in an alkaline medium using Pd/C, Au/C and PdAu/C electrocatalysts prepared by electron beam irradiation. Electrochim Acta 111:455–465. doi:10.1016/j.electacta.2013.08.021
82. Feng YY, Liu ZH, Xu Y, Wang P, Wang WH, Kong DS (2013) Highly active PdAu alloy catalysts for ethanol electro-oxidation. J Power Sources 232:99–105. doi:10.1016/j.jpowsour.2013.01.013
83. Chen LY, Chen N, Hou Y, Wang ZC, Lv SH, Fujita T, Jiang JH, Hirata A, Chen MW (2013) Geometrically controlled nanoporous PdAu bimetallic catalysts with tunable Pd/Au ratio for direct ethanol fuel cells. ACS Catal 3(6):1220–1230. doi:10.1021/Cs400135k
84. Huang ZY, Zhou HH, Li CH, Zeng FY, Fu CP, Kuang YF (2012) Preparation of well-dispersed PdAu bimetallic nanoparticles on reduced graphene oxide sheets with excellent electrochemical activity for ethanol oxidation in alkaline media. J Mater Chem 22(5):1781–1785. doi:10.1039/C1jm13024b
85. Nguyen ST, Law HM, Nguyen HT, Kristian N, Wang SY, Chan SH, Wang X (2009) Enhancement effect of Ag for Pd/C towards the ethanol electro-oxidation in alkaline media. Appl Catal B Environ 91(1-2):507–515. doi:10.1016/j.apcatb.2009.06.021
86. Hammer B, Norskov JK (2000) Theoretical surface science and catalysis—calculations and concepts. Adv Catal 45:71–129

87. Greeley J, Norskov JK (2005) A general scheme for the estimation of oxygen binding energies on binary transition metal surface alloys. Surf Sci 592(1-3):104–111. doi:10.1016/j.susc.2005.07.018
88. Greeley J, Norskov JK, Mavrikakis M (2002) Electronic structure and catalysis on metal surfaces. Annu Rev Phys Chem 53:319–348. doi:10.1146/annurev.physchem.53.100301.131630
89. Stamenkovic V, Mun BS, Mayrhofer KJJ, Ross PN, Markovic NM, Rossmeisl J, Greeley J, Norskov JK (2006) Changing the activity of electrocatalysts for oxygen reduction by tuning the surface electronic structure. Angew Chem Int Ed 45(18):2897–2901. doi:10.1002/anie.200504386
90. Ruban A, Hammer B, Stoltze P, Skriver HL, Norskov JK (1997) Surface electronic structure and reactivity of transition and noble metals. J Mol Catal A Chem 115(3):421–429. doi:10.1016/S1381-1169(96)00348-2
91. He QG, Chen W, Mukerjee S, Chen SW, Laufek F (2009) Carbon-supported PdM (M = Au and Sn) nanocatalysts for the electrooxidation of ethanol in high pH media. J Power Sources 187(2):298–304. doi:10.1016/j.jpowsour.2008.11.065
92. Xia YN, Yang PD, Sun YG, Wu YY, Mayers B, Gates B, Yin YD, Kim F, Yan YQ (2003) One-dimensional nanostructures: synthesis, characterization, and applications. Adv Mater 15 (5):353–389. doi:10.1002/adma.200390087
93. Zhang LG, Li N, Gao FM, Hou L, Xu ZM (2012) Insulin amyloid fibrils: an excellent platform for controlled synthesis of ultrathin superlong platinum nanowires with high electrocatalytic activity. J Am Chem Soc 134(28):11326–11329. doi:10.1021/Ja302959e
94. Wang YJ, Wilkinson DP, Zhang JJ (2011) Noncarbon support materials for polymer electrolyte membrane fuel cell electrocatalysts. Chem Rev 111(12):7625–7651. doi:10.1021/Cr100060r
95. Lavacchi A, Miller H, Vizza F (2013) Other support nanomaterials. Nanotechnol Electrocatal Energy 170:145–187. doi:10.1007/978-1-4899-8059-5_6
96. Du WX, Mackenzie KE, Milano DF, Deskins NA, Su D, Teng XW (2012) Palladium-tin alloyed catalysts for the ethanol oxidation reaction in an alkaline medium. ACS Catal 2 (2):287–297. doi:10.1021/Cs2005955
97. Wang H, Liu ZY, Ji S, Wang KL, Zhou TB, Wang RF (2013) Ethanol oxidation activity and structure of carbon-supported Pt-modified PdSn-SnO2 influenced by different stabilizers. Electrochim Acta 108:833–840. doi:10.1016/j.electacta.2013.07.061
98. Mao HM, Wang LL, Zhu PP, Xu QJ, Li QX (2014) Carbon-supported PdSn SnO2 catalyst for ethanol electro-oxidation in alkaline media. Int J Hydrogen Energy 39(31):17583–17588. doi:10.1016/j.ijhydene.2014.08.079
99. Wang RF, Liu ZY, Ma YJ, Wang H, Linkov V, Ji S (2013) Heterostructure core PdSn-SnO2 decorated by Pt as efficient electrocatalysts for ethanol oxidation. Int J Hydrogen Energy 38 (31):13604–13610. doi:10.1016/j.lihydene.2013.08.044
100. Martinez U, Serov A, Padilla M, Atanassov P (2014) Mechanistic insight into oxide-promoted palladium catalysts for the electro-oxidation of ethanol. Chemsuschem 7 (8):2351–2357. doi:10.1002/cssc.201402062
101. Zhang QL, Zheng JN, Xu TQ, Wang AJ, Wei J, Chen JR, Feng JJ (2014) Simple one-pot preparation of Pd-on-Cu nanocrystals supported on reduced graphene oxide for enhanced ethanol electrooxidation. Electrochim Acta 132:551–560. doi:10.1016/j.electacta.2014.03.159
102. Zhang MM, Xie JM, Sun Q, Yan ZX, Chen M, Jing JJ, Hossain AMS (2013) In situ synthesis of palladium nanoparticle on functionalized graphene sheets at improved performance for ethanol oxidation in alkaline media. Electrochim Acta 111:855–861. doi:10.1016/j.electacta.2013.08.135
103. Gao LN, Yue WB, Tao SS, Fan LZ (2013) Novel strategy for preparation of graphene-Pd, Pt composite, and its enhanced electrocatalytic activity for alcohol oxidation. Langmuir 29 (3):957–964. doi:10.1021/La303663x

104. Wang Y, Zhao Y, Yin J, Liu MC, Dong Q, Su YQ (2014) Synthesis and electrocatalytic alcohol oxidation performance of Pd-Co bimetallic nanoparticles supported on graphene. Int J Hydrogen Energy 39(3):1325–1335. doi:10.1016/j.ijhydene.2013.11.002
105. Dong Q, Zhao Y, Han X, Wang Y, Liu MC, Li Y (2014) Pd/Cu bimetallic nanoparticles supported on graphene nanosheets: facile synthesis and application as novel electrocatalyst for ethanol oxidation in alkaline media. Int J Hydrogen Energy 39(27):14669–14679. doi:10.1016/j.ijhydene.2014.06.139
106. Ahmed MS, Jeon S (2014) Highly active graphene-supported NixPd100-x binary alloyed catalysts for electro-oxidation of ethanol in an alkaline media. ACS Catal 4(6):1830–1837. doi:10.1021/Cs500103a
107. Zhang MM, Yan ZX, Sun Q, Xie JM, Jing JJ (2012) Synthetic core-shell Ni@Pd nanoparticles supported on graphene and used as an advanced nanoelectrocatalyst for methanol oxidation. New J Chem 36(12):2533–2540. doi:10.1039/C2nj40651a
108. Cui Q, Chao SJ, Bai ZY, Yan HY, Wang K, Yang L (2014) Based on a new support for synthesis of highly efficient palladium/hydroxyapatite catalyst for ethanol electrooxidation. Electrochim Acta 132:31–36. doi:10.1016/j.electacta.2014.03.129
109. Bambagioni V, Bianchini C, Marchionni A, Filippi J, Vizza F, Teddy J, Serp P, Zhiani M (2009) Pd and Pt-Ru anode electrocatalysts supported on multi-walled carbon nanotubes and their use in passive and active direct alcohol fuel cells with an anion-exchange membrane (alcohol = methanol, ethanol, glycerol). J Power Sources 190(2):241–251. doi:10.1016/j.jpowsour.2009.01.044
110. Machado BF, Marchionni A, Bacsa RR, Bellini M, Beausoleil J, Oberhauser W, Vizza F, Serp P (2013) Synergistic effect between few layer graphene and carbon nanotube supports for palladium catalyzing electrochemical oxidation of alcohols. J Energy Chem 22 (2):296–304
111. Serov A, Martinez U, Atanassov P (2013) Novel Pd-In catalysts for alcohols electrooxidation in alkaline media. Electrochem Commun 34:185–188. doi:10.1016/j.elecom.2013.06.003
112. Zalineeva A, Serov A, Padilla M, Martinez U, Artyushkova K, Baranton S, Coutanceau C, Atanassov PB (2014) Self-supported PdxBi Catalysts for the electrooxidation of glycerol in alkaline media. J Am Chem Soc 136(10):3937–3945. doi:10.1021/Ja412429f
113. Cheng FL, Dai XC, Wang H, Jiang SP, Zhang M, Xu CW (2010) Synergistic effect of Pd-Au bimetallic surfaces in Au-covered Pd nanowires studied for ethanol oxidation. Electrochim Acta 55(7):2295–2298. doi:10.1016/j.electacta.2009.11.076
114. Cherevko S, Xing XL, Chung CH (2011) Pt and Pd decorated Au nanowires: extremely high activity of ethanol oxidation in alkaline media. Electrochim Acta 56(16):5771–5775. doi:10.1016/j.electacta.2011.04.052
115. Rodriguez P, Koper MTM (2014) Electrocatalysis on gold. Phys Chem Chem Phys 16 (27):13583–13594. doi:10.1039/C4cp00394b
116. de Lima RB, Varela H (2008) Catalytic oxidation of ethanol on gold electrode in alkaline media. Gold Bull 41(1):15–22
117. Beyhan S, Uosaki K, Feliu JM, Herrero E (2013) Electrochemical and in situ FTIR studies of ethanol adsorption and oxidation on gold single crystal electrodes in alkaline. J Electroanal Chem 707:89–94. doi:10.1016/j.jelechem.2013.08.034
118. Pandey RK, Lakshminarayanan V (2012) Ethanol electrocatalysis on gold and conducting polymer nanocomposites: a study of the kinetic parameters. Appl Catal B Environ 125:271–281. doi:10.1016/j.apcatb.2012.06.002
119. Kumar VL, Siddhardha RSS, Kaniyoor A, Podila R, Molli M, Kumar VSM, Venkataramaniah K, Ramaprabhu S, Rao AM, Ramamurthy SS (2014) Gold decorated graphene by laser ablation for efficient electrocatalytic oxidation of methanol and ethanol. Electroanalysis 26(8):1850–1857. doi:10.1002/elan.201400244
120. Simoes M, Baranton S, Coutanceau C (2012) Electrochemical valorisation of glycerol. Chemsuschem 5(11):2106–2124. doi:10.1002/cssc.201200335
121. Zhang ZY, Xin L, Qi J, Wang ZC, Li WZ (2012) Selective electro-conversion of glycerol to glycolate on carbon nanotube supported gold catalyst. Green Chem 14(8):2150–2152. doi:10.1039/C2gc35505a

122. Zhang ZY, Xin L, Qi J, Wang ZC, Chadderdon D, Li WZ (2012) Simultaneous generation of valuable chemicals and electricity from selective electrocatalytic oxidation of glycerol in anion exchange membrane fuel cell reactors. Abstr Pap Am Chem S 244
123. Zhang ZY, Xin L, Qi J, Chadderdon DJ, Sun K, Warsko KM, Li WZ (2014) Selective electro-oxidation of glycerol to tartronate or mesoxalate on Au nanoparticle catalyst via electrode potential tuning in anion-exchange membrane electro-catalytic flow reactor. Appl Catal B Environ 147:871–878. doi:10.1016/j.apcatb.2013.10.018
124. Zhang ZY, Xin L, Qi J, Chadderdon DJ, Li WZ (2013) Supported Pt, Pd and Au nanoparticle anode catalysts for anion-exchange membrane fuel cells with glycerol and crude glycerol fuels. Appl Catal B Environ 136:29–39. doi:10.1016/j.apcatb.2013.01.045
125. Zhang ZY, Xin L, Li WZ (2012) Supported gold nanoparticles as anode catalyst for anion-exchange membrane-direct glycerol fuel cell (AEM-DGFC). Int J Hydrogen Energy 37 (11):9393–9401. doi:10.1016/j.ijhydene.2012.03.019
126. Qi J, Xin L, Chadderdon DJ, Qiu Y, Jiang YB, Benipal N, Liang CH, Li WZ (2014) Electrocatalytic selective oxidation of glycerol to tartronate on Au/C anode catalysts in anion exchange membrane fuel cells with electricity cogeneration. Appl Catal B Environ 154:360–368. doi:10.1016/j.apcatb.2014.02.040
127. Wang Q, Sun GQ, Jiang LH, Xin Q, Sun SG, Jiang YX, Chen SP, Jusys Z, Behm RJ (2007) Adsorption and oxidation of ethanol on colloid-based Pt/C, PtRu/C and Pt3Sn/C catalysts: In situ FTIR spectroscopy and on-line DEMS studies. Phys Chem Chem Phys 9(21):2686–2696. doi:10.1039/B700676b
128. James DD, Pickup PG (2012) Measurement of carbon dioxide yields for ethanol oxidation by operation of a direct ethanol fuel cell in crossover mode. Electrochim Acta 78:274–278. doi:10.1016/j.electacta.2012.05.120
129. Tarnowski DJ, Korzeniewski C (1997) Effects of surface step density on the electrochemical oxidation of ethanol to acetic acid. J Phys Chem B 101(2):253–258. doi:10.1021/Jp962450c
130. Sun SG, Lin Y (1998) Kinetics of isopropanol oxidation on Pt(111), Pt(110), Pt(100), Pt(610) and Pt(211) single crystal electrodes—studies of in situ time-resolved FTIR spectroscopy. Electrochim Acta 44(6–7):1153–1162. doi:10.1016/S0013-4686(98)00218-7
131. Zhang ZC, Nosheen F, Zhang JC, Yang Y, Wang PP, Zhuang J, Wang X (2013) Growth of concave polyhedral Pd nanocrystals with 32 facets through in situ facet-selective etching. Chemsuschem 6(10):1893–1897. doi:10.1002/cssc.201300346
132. Lee YW, Kim D, Hong JW, Kang SW, Lee SB, Han SW (2013) Kinetically controlled growth of polyhedral bimetallic alloy nanocrystals exclusively bound by high-index facets: AuPd hexoctahedra. Small 9(5):660–665. doi:10.1002/smll.201201813
133. Bell AT (2003) The impact of nanoscience on heterogeneous catalysis. Science 299 (5613):1688–1691. doi:10.1126/science.1083671
134. Shao MH, Peles A, Shoemaker K (2011) Electrocatalysis on platinum nanoparticles: particle size effect on oxygen reduction reaction activity. Nano Lett 11(9):3714–3719. doi:10.1021/Nl2017459
135. Henning AM, Watt J, Miedziak PJ, Cheong S, Santonastaso M, Song MH, Takeda Y, Kirkland AI, Taylor SH, Tilley RD (2013) Gold-palladium core-shell nanocrystals with size and shape control optimized for catalytic performance. Angew Chem Int Ed 52 (5):1477–1480. doi:10.1002/anie.201207824
136. Fashedemi OO, Julies B, Ozoemena KI (2013) Synthesis of Pd-coated FeCo@Fe/C core–shell nanoparticles: microwave-induced ‘top-down’ nanostructuring and decoration. Chem Commun 49:2034–2036
137. Fashedemi OO, Ozoemena KI (2014) Comparative electrocatalytic oxidation of ethanol, ethylene glycol and glycerol in alkaline medium at Pd-decorated FeCo@Fe/C core-shell nanocatalysts. Electrochim Acta 128:279–286. doi:10.1016/j.electacta.2013.10.194
138. Tian N, Zhou ZY, Sun SG, Ding Y, Wang ZL (2007) Synthesis of tetrahexahedral platinum nanocrystals with high-index facets and high electro-oxidation activity. Science 316 (5825):732–735. doi:10.1126/science.1140484

139. Ding Y, Gao Y, Wanga ZL, Tian N, Zhou ZY, Sun SG (2007) Facets and surface relaxation of tetrahexahedral platinum nanocrystals. Appl Phys Lett 91(12):121901. doi:10.1063/1.2785953
140. Zhou ZY, Huang ZZ, Chen DJ, Wang Q, Tian N, Sun SG (2010) High-index faceted platinum nanocrystals supported on carbon black as highly efficient catalysts for ethanol electrooxidation. Angew Chem Int Ed 49(2):411–414. doi:10.1002/anie.200905413
141. Zhou ZY, Shang SJ, Tian N, Wu BH, Zheng NF, Xu BB, Chen C, Wang HH, Xiang DM, Sun SG (2012) Shape transformation from Pt nanocubes to tetrahexahedra with size near 10 nm. Electrochem Commun 22:61–64. doi:10.1016/j.elecom.2012.05.023
142. Tian N, Zhou ZY, Yu NF, Wang LY, Sun SG (2010) Direct electrodeposition of tetrahexahedral Pd nanocrystals with high-index facets and high catalytic activity for ethanol electrooxidation. J Am Chem Soc 132(22):7580–7581. doi:10.1021/Ja102177r
143. Chen YX, Lavacchi A, Chen SP, di Benedetto F, Bevilacqua M, Bianchini C, Fornasiero P, Innocenti M, Marelli M, Oberhauser W, Sun SG, Vizza F (2012) Electrochemical milling and faceting: size reduction and catalytic activation of palladium nanoparticles. Angew Chem Int Ed 51(34):8500–8504. doi:10.1002/anie.201203589
144. Wang L, Bevilacqua M, Chen Y-X, Filippi J, Innocenti M, Lavacchi A, Marchionni A, Miller H, Vizza F (2013) Enhanced electro-oxidation of alcohols at electrochemically treated polycrystalline palladium surface. J Power Sources 242:872–876
145. Wang LQ, Bevilacqua M, Filippi J, Fornasiero P, Innocenti M, Lavacchi A, Marchionni A, Miller HA, Vizza F (2015) Electrochemical growth of platinum nanostructures for enhanced ethanol oxidation. Appl Catal Environ 165:185–191. doi:10.1016/j.apcatb.2014.10.009
146. Wang L, Bambagioni V, Bevilacqua M, Bianchini C, Filippi J, Lavacchi A, Marchionni A, Vizza F, Fang X, Shen PK (2010) Sodium borohydride as an additive to enhance the performance of direct ethanol fuel cells. J Power Sources 195(24):8036–8043. doi:10.1016/j.jpowsour.2010.06.101
147. Kocha SS (2012) Electrochemical degradation: electrocatalyst and support durability. In: Polymer electrolyte fuel cell degradation. Academic, Boston, pp 89–214
148. Chen ZW, Waje M, Li WZ, Yan YS (2007) Supportless Pt and PtPd nanotubes as electrocatalysts for oxygen-reduction reactions. Angew Chem Int Ed 46(22):4060–4063. doi:10.1002/anie.200700894
149. Borup R, Meyers J, Pivovar B, Kim YS, Mukundan R, Garland N, Myers D, Wilson M, Garzon F, Wood D, Zelenay P, More K, Stroh K, Zawodzinski T, Boncella J, McGrath JE, Inaba M, Miyatake K, Hori M, Ota K, Ogumi Z, Miyata S, Nishikata A, Siroma Z, Uchimoto Y, Yasuda K, Kimijima KI, Iwashita N (2007) Scientific aspects of polymer electrolyte fuel cell durability and degradation. Chem Rev 107(10):3904–3951. doi:10.1021/Cr0501821
150. Shao-Horn Y, Sheng WC, Chen S, Ferreira PJ, Holby EF, Morgan D (2007) Instability of supported platinum nanoparticles in low-temperature fuel cells. Top Catal 46(3-4):285–305. doi:10.1007/s11244-007-9000-0
151. Cao MN, Wu DS, Cao R (2014) Recent advances in the stabilization of platinum electrocatalysts for fuel-cell reactions. Chemcatchem 6(1):26–45. doi:10.1002/cctc.201300647
152. Nguyen ST, Yang YH, Wang X (2012) Ethanol electro-oxidation activity of Nb-doped-TiO_2 supported PdAg catalysts in alkaline media. Appl Catal B Environ 113:261–270. doi:10.1016/j.apcatb.2011.11.046
153. Wang AL, Xu H, Feng JX, Ding LX, Tong YX, Li GR (2013) Design of Pd/PANI/Pd sandwich-structured nanotube array catalysts with special shape effects and synergistic effects for ethanol electrooxidation. J Am Chem Soc 135(29):10703–10709. doi:10.1021/Ja403101r
154. Chen YX, Lavacchi A, Miller HA, Bevilacqua M, Filippi J, Innocenti M, Marchionni A, Oberhauser W, Wang L, Vizza F (2014) Nanotechnology makes biomass electrolysis more energy efficient than water electrolysis. Nat Commun 5:4036. doi:10.1038/ncomms5036

155. Antolini E (2007) Catalysts for direct ethanol fuel cells. J Power Sources 170(1):1–12. doi:10.1016/j.jpowsour.2007.04.009
156. Matsuoka K, Iriyama Y, Abe T, Matsuoka M, Ogumi Z (2005) Alkaline direct alcohol fuel cells using an anion exchange membrane. J Power Sources 150:27–31. doi:10.1016/j.jpowsour.2005.02.020
157. Vigier F, Rousseau S, Coutanceau C, Leger JM, Lamy C (2006) Electrocatalysis for the direct alcohol fuel cell. Top Catal 40(1–4):111–121. doi:10.1007/s11244-006-0113-7
158. Cracknell JA, Vincent KA, Armstrong FA (2008) Enzymes as working or inspirational electrocatalysts for fuel cells and electrolysis. Chem Rev 108(7):2439–2461. doi:10.1021/cr0680639
159. Vincent KA, Barton SC, Canters GW, Heering HA (2009) Electrocatalysis for fuel cell at enzyme-modified electrodes. In: Koper MTM (ed) Fuel cell catalysis. Wiley, New York. doi:10.1002/9780470463772.ch17
160. Yamazaki S, Yao M, Fujiwara N, Siroma Z, Yasuda K, Ioroi T (2012) Electrocatalytic oxidation of alcohols by a carbon-supported Rh porphyrin. Chem Commun 48 (36):4353–4355. doi:10.1039/C2cc30888f
161. Brownell KR, McCrory CCL, Chidsey CED, Perry RH, Zare RN, Waymouth RM (2013) Electrooxidation of alcohols catalyzed by amino alcohol ligated ruthenium complexes. J Am Chem Soc 135(38):14299–14305. doi:10.1021/Ja4055564
162. Weiss CJ, Das P, Miller DL, Helm ML, Appel AM (2014) Catalytic oxidation of alcohol via nickel phosphine complexes with pendant amines. ACS Catal 4(9):2951–2958. doi:10.1021/Cs500853f
163. Elouarzaki K, Haddad R, Holzinger M, Le Goff A, Thery J, Cosnier S (2014) MWCNT-supported phthalocyanine cobalt as air-breathing cathodic catalyst in glucose/O-2 fuel cells. J Power Sources 255:24–28. doi:10.1016/j.jpowsour.2013.12.109
164. Elouarzaki K, Le Goff A, Holzinger M, Thery J, Cosnier S (2012) Electrocatalytic oxidation of glucose by rhodium porphyrin-functionalized MWCNT electrodes: application to a fully molecular catalyst-based glucose/O-2 fuel cell. J Am Chem Soc 134(34):14078–14085. doi:10.1021/Ja304589m
165. Annen SP, Bambagioni V, Bevilacqua M, Filippi J, Marchionni A, Oberhauser W, Schonberg H, Vizza F, Bianchini C, Grutzmacher H (2010) A biologically inspired organometallic fuel cell (OMFC) that converts renewable alcohols into energy and chemicals. Angew Chem Int Ed 49(40):7229–7233. doi:10.1002/anie.201002234

Chapter 13
Effects of Catalyst-Support Materials on the Performance of Fuel Cells

Paul M. Ejikeme, Katlego Makgopa, and Kenneth I. Ozoemena

13.1 Introduction

In the twenty-first century, the energy crisis and environmental pollution are both stark challenges faced by human society because of speedily rising energy demand and enormous combustion of fossil derived fuels. It is generally believed that the conventional notions of energy generation and conversion are not suitable for the world's sustainable development [1, 2]. Renewable energy technologies therefore hold the promise to meet the increasing energy demands of over a seven billion people planet without depleting our natural resources and compromising the quality of our environment [3]. Recently, the emerging need for clean energy and high speed electronic has motivated researchers across the globe to discover, develop, and assemble new classes of nanomaterials in unconventional device architectures. A great deal of research has thus been carried out to explore clean and renewable energy sources as well as their devices [4, 5]. Among the different energy systems, hydrogen and DAFCs involving mainly methanol and ethanol have been

P.M. Ejikeme (✉)
Department of Chemistry, University of Pretoria, Pretoria 0002, South Africa

Department of Pure and Industrial Chemistry, University of Nigeria, Nsukka 410001, Nigeria
e-mail: paul.ejikeme@unn.edu.ng

K. Makgopa
Department of Chemistry, University of Pretoria, Pretoria 0002, South Africa

K.I. Ozoemena (✉)
Council for Scientific and Industrial Research, Pretoria, South Africa

Energy Materials Unit, Materials Science and Manufacturing, Council for Scientific and Industrial Research (CSIR), Pretoria 0001, South Africa

Molecular Sciences Institute, School of Chemistry, University of the Witwatersrand, Johannesburg 2050, South Africa
e-mail: kozoemena@csir.co.za

K.I. Ozoemena, S. Chen (eds.), *Nanomaterials for Fuel Cell Catalysis*,
Nanostructure Science and Technology, DOI 10.1007/978-3-319-29930-3_13

extensively studied as ideal energy converters that convert chemical energy of alcohols and an oxidant directly to electrochemical energy [6, 7].

Fuel cells (FCs) are 'electrochemical' devices that convert directly the chemical energy / heat of combustion of a fuel (e.g., hydrogen, methanol, ethanol, natural gas, hydrocarbons, etc.) and an oxidant (i.e., air or pure oxygen) in the presence of a catalyst into electricity, heat and water [8–10], at an efficiency that is typically 2–3 times that of an internal combustion engine [9]. A German scientist, Christian Friedrich Schoenbein (1799–1868), first discovered the fuel cell effect in January 1839, and reported observing electric current induced by the combination of hydrogen and oxygen, (i.e., the inverse electrolysis process). The best known early fuel cell experiments were however performed in 1842 by the British physicist and lawyer, Sir William R. Grove (1811–1896), who produced an electric current by connecting a hydrogen anode and an oxygen cathode [11, 12]. Fuel cell technologies are only classed as renewable energy systems if the fuel used is renewable (i.e., biofuels, hydrogen from wind or solar conversion of water) [3]. They show promise as high-efficient energy conversion technology for both mobile and stationary power applications, and have shown even greater promise as a substitute to the conventional combustion engine for transportation applications [13]. The FC process, just as any other redox process, involves electron transfer during the oxidation and reduction reactions with an essentially invariant electrode-electrolyte system.

13.1.1 Classification of Fuel Cells

Fuel cells are generally classified into six (6) broad groups based on the choice of operating fuel and electrolyte [14] as shown in Fig. 13.1 and described elsewhere in this book. Alkaline fuel cells (AFC) utilise an aqueous potassium hydroxide (KOH) solution as the alkaline electrolyte while proton exchange membrane fuel cells (PEMFC) and direct methanol fuel cells (DMFC) use polymer membranes as the electrolyte for conducting protons. Furthermore, phosphoric acid fuel cells (PAFC) use pure phosphoric acid as the electrolyte, molten carbonate fuel cells (MCFC) use a molten mixture of lithium, sodium, and potassium carbonates for conducting carbonate ions while solid oxide fuel cells (SOFC) make use of ceramic (solid oxide) material as electrolytes, to conduct negative oxygen ions from the cathode to the anode. It is also common to classify fuel cells according to their operating temperature regimes (Fig. 13.1). Thus, low temperature (i.e., AFC, PEMFC, DMFC); medium temperature (PAFC); and high-temperature (i.e., MCFC, DCFC, and SOFC) fuel cells are also known.

Today, the low-temperature fuel cells (PEMFC and AFC) shown in Fig. 13.2 are quite popular. While the PEMFC operates in acidic conditions and uses proton-exchange membranes (PEM) and platinum catalysts, the AFC operates in alkaline media with anion-exchange membrane (AEM) and uses low-Pt or Pt-free catalysts.

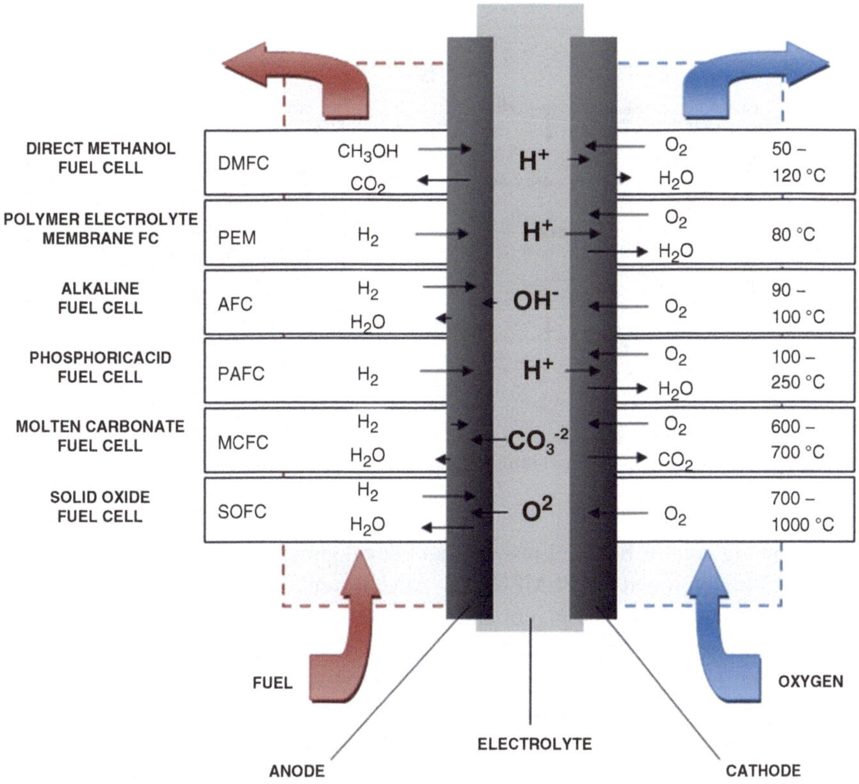

Fig. 13.1 Schematic representation of different types of fuel cells, electrode reactions and products, and operating temperatures (*Source*: http://www.fuelcells.org/uploads/FuelCell Types.jpg, Accessed on 13th June, 2015)

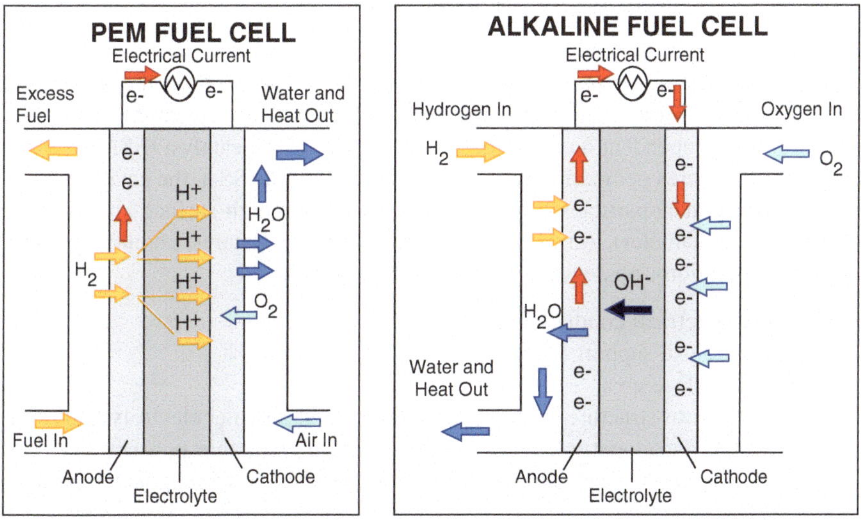

Fig. 13.2 Schematic representation of PEMFC and AFC (*Source*: Green Car Congress, Accessed on 13 June 2015, http://www.greencarcongress.com/2005/05/zap_and_apollo_.html)

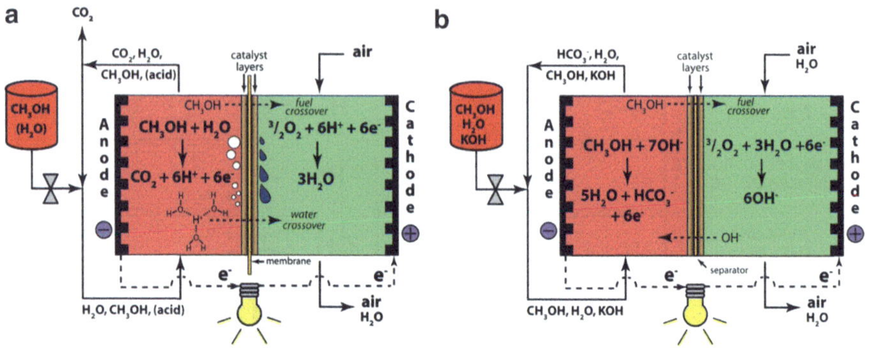

Fig. 13.3 Schematic representation of a direct methanol fuel cell (DMFC) operating in (**a**) acid and (**b**) alkaline conditions (*Source*: http://clipart-finder.com/oxygen-clipart.html, Accessed 13th June 2015)

A growing area of fuel cell research and development is the direct alcohol fuel cell wherein hydrogen in PEMFC and AFC is replaced with alcohol (such as methanol, ethanol, ethylene glycol and glycerol). Alcohols are considered alternatives to hydrogen since the power density of the alcohols (liquids with no pressure requirement during storage) in terms of energy by volume of fuel is much higher than that of hydrogen at standard conditions [15]. Figure 13.3 shows a generalised DAFC scheme using methanol as fuel. DAFCs (especially alkaline-based systems involving higher alcohols) yield products that are valuable intermediates for the fine chemical, pharmaceutical, and agrochemical sectors.

13.2 Catalyst-Supports for Fuel Cells

Electrocatalysts are the main drivers of the reactions that occur at the anode and cathode sides of a fuel cell. The performance of each of the electrode reaction is predominantly dependent on the specific surface area of the catalyst (SSA, i.e., real catalyst surface area per gram). To increase the value of the SSA, the catalysts must be dispersed as nanoparticles (2–5 nm in diameter) on a high-surface-area support (10–50 nm in diameter). An ideal support for the electrocatalyst should, amongst other considerations, possess the following properties [6, 14]:

(i) a good electrical conductivity,
(ii) good catalyst–support interaction,
(iii) a large surface area,
(iv) a mesoporous structure enabling the ionomer and polymer electrolyte to bring the catalyst nanoparticles close to the reactants (suitable porosity to allow good reactant and product flux), i.e. to maximise the triple-phase boundary (TPB),
(v) a good water handling capability to avoid flooding,

(vi) good corrosion resistance/high stability in fuel cell environments, and
(vii) ease of catalyst recovery.

It has been recognized [14] that a good interaction between the catalyst and the support not only improves catalysts' effectiveness while decreasing catalyst loss, but also governs charge transfer, assist in sufficiently enhancing the catalyst performance and robustness by reducing catalyst deactivation, and in some cases affects the catalyst particle size. From the foregoing, the choice of support material is vital in shaping the behaviour, performance, durability and cost effectiveness of the catalyst in particular, and the overall fuel cell performance in general [15].

13.2.1 Carbon-Based Supports

The reputation of active carbon as a precious metal support in industrial chemistry is well documented. The existence of carbon in various allotropes [16] (Fig. 13.4), coupled with the established advantages of graphitized carbon as support material,

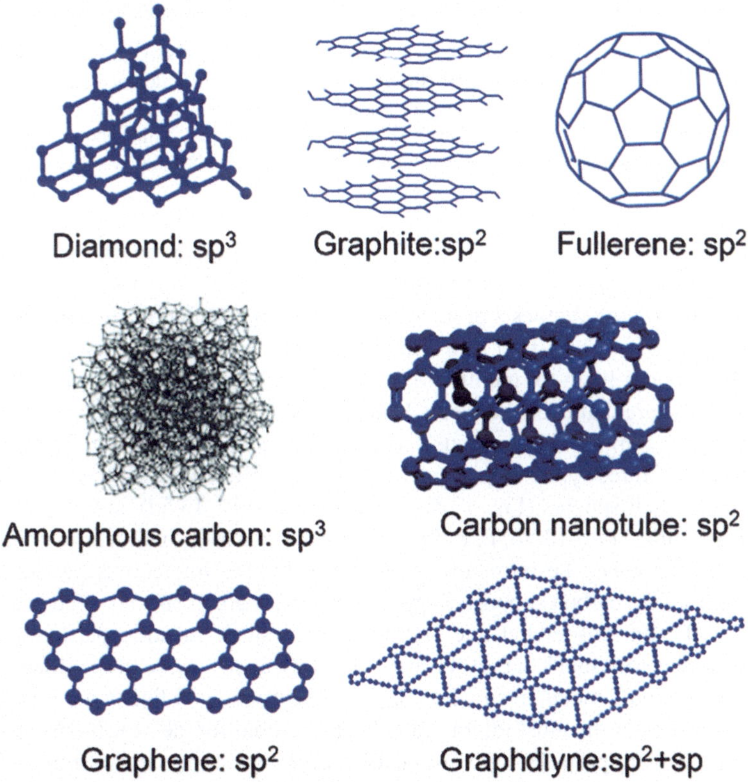

Fig. 13.4 Naturally existing and man-made carbon allotropes (*Source*: Li et al., Chem. Soc. Rev., 2014, 43, 2572–2586, Reproduced with permission from [16], copyright 2014 Royal Society of Chemistry)

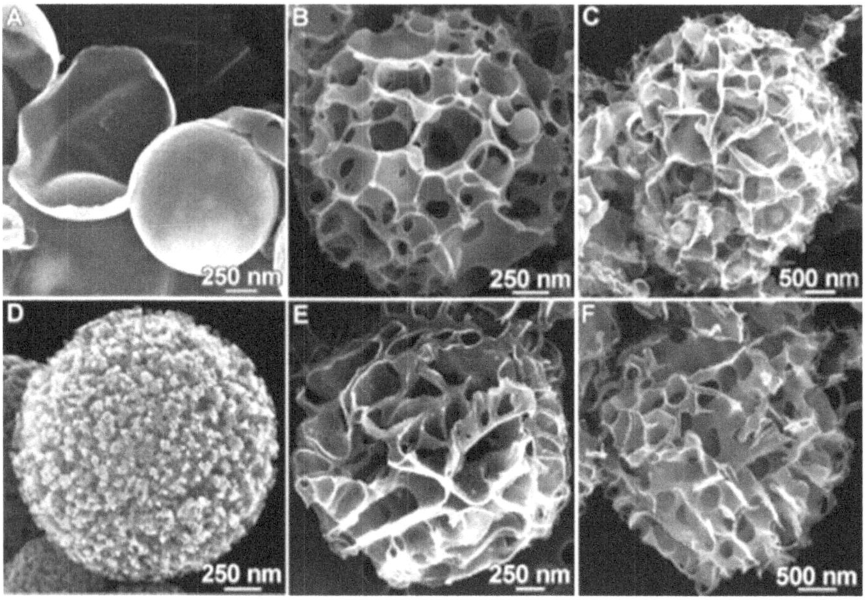

Fig. 13.5 SEM images of USP porous carbons. Reaction conditions: 1.5 M solutions, 700 °C, Ar at 1.0 standard litre per minute (slpm). Product from (**a**) lithium chloroacetate (**b**) sodium chloroacetate (**c**) potassium chloroacetate (**d**) lithium dichloroacetate (**e**) sodium dichloroacetate and (**f**) potassium dichloroacetate (*Source*: Skrabalak and Suslick, J. Am. Chem. Soc., 2006, 128, 12642–12643. Reproduced with permission from [17], copyright 2006 American Chemical Society)

with different modifications available, has however continued to make it a choice material for the support of FC catalysts.

Depending on the synthesis protocol adopted, carbon supports come in different morphologies (Fig. 13.5a–f) [17] ranging from core-shells, spheres, hollow spheres, nanotubes, onion-like, to mention but a few. Generally, they can either be mesoporous (pore diameters between 2 and 50 nm) or microporous (pore diameters of less than 2 nm) or macroporous (pore diameters of greater than 50 nm) or a mixture of the three types of pore sizes. Figure 13.5 exemplifies the presence of hollow core-shell spheres (Fig. 13.5a), mesoporous (Fig. 13.5d) and macroporous carbon networks (Fig. 13.5b, c, e, f). Carbons are stable in both acidic and basic media and can easily be burnt off; a simple technique for recovering the metal catalysts from the carbon support. Supported catalysts are of special interest as they allow for the fine dispersion and stabilisation of small metallic particles. They provide access to a much larger number of catalytically active atoms than in the corresponding bulk metal, even when the latter is ground to a fine powder [18, 19].

Carbons remain the most important support material for electrocatalysts used in fuel cells. The importance of carbon as support material stems from its high availability and low-cost. However, carbon suffers shortcomings that limit its rate performance as catalyst-support, which include:

(i) Severe corrosion/oxidation with use under normal operating conditions (during fuel cell start-up and shut-down) at high potentials (>0.7 V vs. NHE), resulting in poor durability of the electrocatalyst because of dissolution, Ostwald ripening, and aggregation [14, 20, 21];
(ii) Presence of high amount of micropores (<1 nm), which can impede fuel supply to the surface, and present low accessible surface area for the deposition of metal particles;
(iii) Low polarity and high hydrophobicity, which impede permeability of gases and liquids;
(iv) Poor stability at temperatures higher than 373 K and lack of proton conductivity [22, 23].

The degradation mechanism of carbon-supported platinum catalyst in ORR was investigated in a very recent study by Zhang et al. [24]. From a combination of methods based on accelerated stress test (AST) protocols, the authors observed activity loss in electrochemical active surface area (ECSA) and ORR which they attributed to the Pt dissolution/redeposition, agglomeration, detachment and carbon corrosion.

The corrosion reaction of carbon material in aqueous acid electrolytes including proton exchange membranes is generalized as follows [25, 26]:

$$C + 2H_2O \rightarrow CO_2 + 4H^+ + 4e^- \quad E^\circ = 0.207\ \text{V (vs SHE)} \tag{13.1}$$

Equation 13.1 indicates that carbon can be thermodynamically oxidized at potentials above 0.207 V. Although this reaction is thermodynamically feasible at the potentials which the fuel cell cathode operates, it is believed to be almost non-existent or negligible in low-temperature fuel cells (60–90 °C). However, despite the sluggishness of the reaction, long-term operations can cause a decrease in carbon content in the catalyst layers [27]. Furthermore, under certain circumstances when electrode potentials are raised extremely high, there is a rapid degradation of carbon supports as well as the catalyst. Electrochemical corrosion of carbon materials as catalyst-supports of PEMFCs leads to two major effects: (i) electrical isolation of the catalyst particles as they are separated from the support and (ii) aggregation of catalyst small particles (Ostwald ripening) and consequently to the evolution of an inhomogenous structure over time (Fig. 13.6) [28].

This stems from the fact that molecules on the surface of a particle are energetically less stable than the ones already well ordered and packed in the interior. These dual effects result in a decrease in the electrochemical active surface area (EASA) of the catalyst and an increase in the hydrophilicity of the surface. This can, in turn, result in a decrease in gas permeability as the pores become more likely to be filled with liquid water films that can hinder gas transport [25].

This problem associated with catalyst-supports has been extensively studied by several workers [29–31]. For example, Mathias et al. [29] subjected MEA cathodes to accelerated carbon support corrosion at 1.2 V and 80 °C at various times and demonstrated that carbon weight losses between 5 and 10 % occurred (Fig. 13.7),

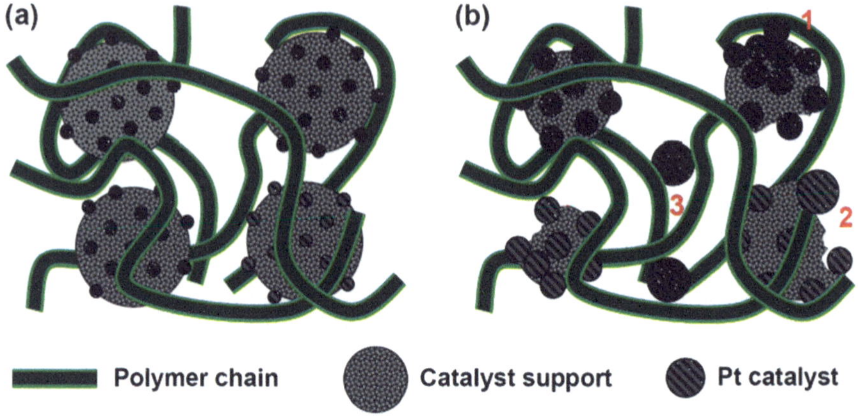

Fig. 13.6 Schematic representation of the effect of corrosion of carbon on (*1*) agglomeration, (*2*) coalescence, and (*3*) loss of catalyst particles in the membrane electrode assembly (MEA) during operation of PEMFCs: (**a**) normal (corrosion-resistant) electrode and (**b**) corroded electrode (*Source*: Huang et al., J. Am. Chem. Soc., 2009, 131, 13898–13899. Reproduced with permission from [28], copyright 2009 American Chemical Society)

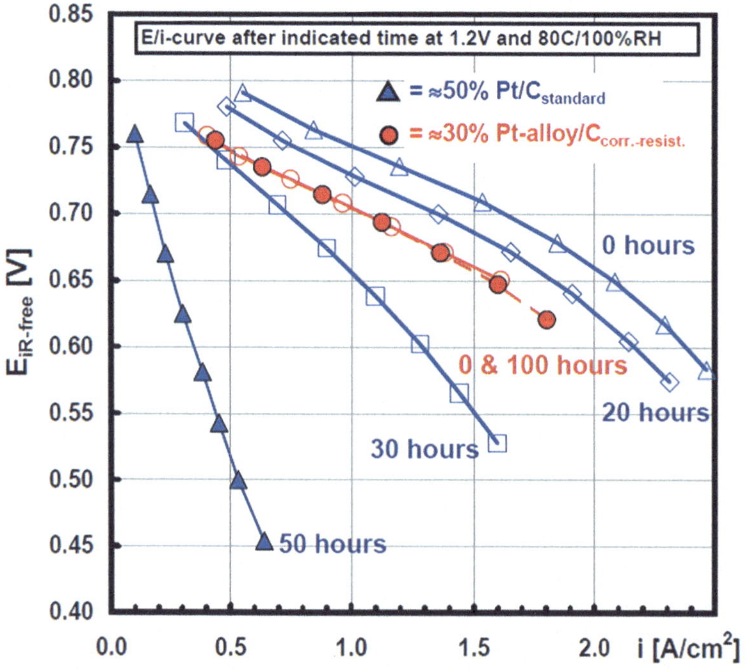

Fig. 13.7 Impact of accelerated carbon support corrosion on the E/I performance curves of 50 cm^2 MEAs with both a ~50 % Pt/C standard and a ≈ 30 wt% Pt alloy/$C_{corr.-resist.}$ cathode catalyst (*Source*: Matthias et al., The Electrochemical Society Interface, 2005, 14(3), 24–35. Reproduced with permission from [29], copyright 2005 *The Electrochemical Society*)

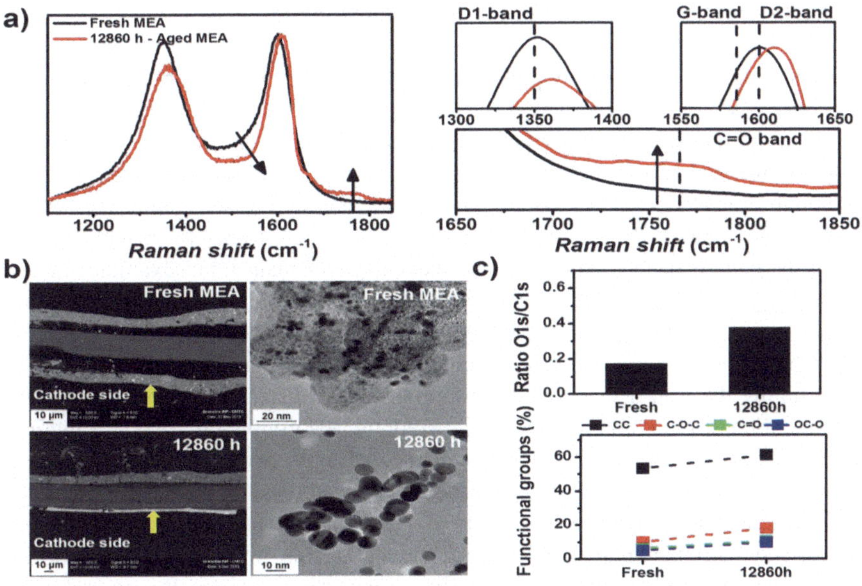

Fig. 13.8 Characterization of MEAs operated under "real-life" PEMFC conditions: (**a**) Raman spectra, (**b**) scanning electron microscopy (SEM) and transmission electron microscopy (TEM) images, and (**c**) XPS-derived data. PEMFC operating conditions: $j = 0.25$ A cm^{-2}, $0.65 < E_{cell} < 0.77$ V, $333 < T < 338$ K, t = 12,860 h, 250 start/stop events (*Source*: Castanheira et al., ACS Catal., 2014, 4, 2258–2267. Reproduced with permission from [31], copyright 2014 American Chemical Society)

leading to unacceptably large performance deterioration under studied operating conditions; implying that standard carbon supports (Vulcan-XC72 and Ketjen black) do not satisfy automotive requirements with respect to either start-up / shut-down cycles or prolonged idle conditions. Similar findings were reported by Wang et al. [30] in their study with graphene. In a related study [31], it was demonstrated that the disordered domains of the carbon supports were preferentially oxidized at voltages related to the PEMFC cathode ($0.40 < E < 1.0$ V).

Experiments at electrode potentials greater than 1.0 V (obtained during start-up, fuel starvation events and shut-down of PEMFC systems) were reported [31] to have accelerated this degradation process (Fig. 13.8). The decrease of the D1-band intensity and the thinning of the G- and D2-bands at 1610 cm^{-1} (Fig. 13.8a) confirmed the preferential degradation of the disordered graphite domains of the supports while the emergence of the carbonyl band at 1760 cm^{-1} showed that the ordered graphite crystallites were also corroded. The SEM images of the MEA (Fig. 13.8b) show that the cathode thickness estimated from SEM images decreased (Fig. 13.8b, yellow arrows) from 13.1 μm at the fresh state to 3.8 μm after 12,860 h of operation while the TEM images revealed an increased fraction of agglomerated Pt nanoparticles, as well as changes of the structure of the carbon particles. The larger value of the O1s/C1s ratio indicates that, on the average, the surface was more oxidized after 12,860 h of operation than at the fresh state.

13.2.1.1 Physico-Chemical Properties of Carbon Supports

The performance of carbon supports is dependent on their unique physico-chemical properties such as size, shape, pore structure/distribution, surface defects and chemical properties. Several types of carbons have been studied as catalyst-supports for fuel cell and they include carbon blacks, carbon nanotubes (CNTs), carbon microspheres, carbon nanofibers and graphenes. The shape and structure of carbons are some of the factors that can impact on their ability to strongly adsorb catalysts on their surfaces for enhanced fuel cell performance. For example, Cuong et al. [32] used density functional theory to unravel the interplay between the carbon supports (i.e., graphene sheet, a metallic single wall carbon nanotube (SWCNT) and a series of semiconducting SWCNTs) and Pt nano-clusters in determining the stability and electronic properties of Pt nanoclusters. They found that the Pt cluster could be best stabilized by adsorption on carbon nanotube supports than on graphene support, and the optimal curvature of SWCNTs for adsorption can be found. Considering that the dissolution and/or loss of electrocatalyst's surface area over a period of time due to sintering remains one of the most fundamental problems in fuel cells, this finding may explain the superiority in catalytic activity of carbon nanotube supported-Pt catalyst. The increased adsorption energy of Pt13 cluster on the carbon nanotubes, 4.0 eV SWCNTs vs 2.21 eV for graphene, which is an indication of an enhanced stability was interpreted in term of geometric factors. According to the authors, "*due to the curvature-induced pyramidalization and misalignment of p-orbitals of carbon atoms in SWCNT, carbon atoms on the outside wall of the SWCNT have more* sp^3 *nature than those on a flat graphene sheet*". Also, considering that the Pt-C bonding is due to hybridization between d-states of Pt atoms and p-states of adjacent C atoms, it means that the Pt atoms are able to strongly interact with carbon atoms in the case of carbon nanotubes, leading to an increase in the adsorption energy. The high curvature of the carbon nanotube can promote the overlap between p and d wave functions, thus increasing the adsorption energy of Pt on SWCNTs.

Importantly, the physico-chemical properties of carbon materials are intricately linked with the methods used in their production or surface modification. Generally, surface functionalization of carbons is aimed at improving the hydrophilicity (e.g., treatment with acids and bases) or hydrophobicity (e.g., treatment with organic compounds such as benzene) of the carbon supports. The functional groups, such as the oxygen-containing groups, provide the binding sites for the growth of the metal catalyst ions. The most popular surface modification of carbon support is the functionalization with acids or bases, a simple treatment that allows for the introduction of oxygen-containing groups such as carboxyl, hydroxyl or phenolic groups on the carbon surfaces. In their study, Kim and Park [33] interrogated the effects of chemical treatment of carbon supports on electrochemical behaviours for platinum catalysts of fuel cells. Three different carbon supports were studied with Pt catalysts; virgin carbon blacks (CBs), neutral-treated carbon blacks (NCBs obtained by treating CBs with 0.2 M benzene), base-treated carbon blacks (BCBs, obtained by treating CBs with 0.2 M KOH), and acid-treated carbon blacks (ACBs, obtained by treating CBs with 0.2 M H_3PO_4). The authors found that the size and the Pt-loading were strongly dependent on surface characteristics of the carbons blacks: BCBs-supported Pt gave the smallest particle size (2.65 nm) and the highest loading (97 %) compared to other carbons investigated. The electroactivity of the Pt catalysts was enhanced with BCBs and NCBs, but decayed with ACBs.

CNTs are increasingly becoming attractive supports for electrocatalysts for fuel cells due to their excellent mechanical strength, high conductivity, and high surface areas. However, for the CNTs to be effectively used as support materials, they need to be functionalized to enhance their dispersibility in solvents, and to permit anchorage of the metal. Some notable methods for surface-functionalization of CNTs include acid-treatments [34, 35] and sulfonation processes [36, 37]. The sulfonation process was introduced by Peng et al. [38]. To improve the sulfonation process, Ozoemena and co-workers [39] first functionalised the pristine CNTs to increase the concentration of the oxo-groups (mainly the –COOH) by the three-step acid-treatment process: (i) refluxing in 2.6 M HNO_3 for 24 h; (ii) sonicating in H_2SO_4/HNO_3 mixture (3:1 ratio) for 24 h; and (iii) stirring in H_2SO_4/H_2O_2 mixture (4:1 ratio) at 70 °C. The acid-functionalised CNTs (MWCNT-COOH) were then sulfonated using a mixture of H_2SO_4 and acetic anhydride at 70 °C for 2 h using the procedure reported by Sun et al. [36]. The catalysts were supported onto the sulfonated-CNTs (MWCNT-SO_3H) using the fast microwave-assisted polyol synthesis, as shown in Fig. 13.9.

Recently, Ozoemena and co-workers introduced a new technique called "*microwave-induced top-down nanostructuring and decoration* (*MITNAD*)" to synthesize Pd-based ternary core–shell-shell (FeCo@Fe@Pd) nanocatalyst [40–43], Figs. 13.10, 13.11, and 13.12. To understand the effect of chemical treatments on the CNTs, they studied the catalytic properties of the FeCo@Fe@Pd nanocatalyst supported on MWCNT-COOH and MWCNT-SO_3H.

It was clearly proved that the surface functional groups (mainly –COOH and –SO_3H) on the MWCNT support played a critical role on the physico-chemical properties of the FeCo@Fe@Pd nanocatalyst towards the electrocatalytic oxidation of ethylene glycol and glycerol. The FeCo@Fe@Pd/MWCNT-COOH gave smaller particle size (ca. 7.4 nm), more uniform dispersion or loading of the catalyst on the support, higher electrochemically-active surface area (ECSA, *ca.* 75 $m^2\ g^{-1}$) and more enhanced electrocatalytic activity than the FeCo@Fe@Pd/MWCNT-SO_3H (~11 nm particle size, with ECSA of *ca.* 42 $m^2\ g^{-1}$). The result is in agreement with the findings of Kim and Park [33] as already discussed above, where carbon black treated with acid showed poor catalyst loading compared to the base-treated or neutral carbon blacks.

The MWCNT-COOH support (a weak acid) might have resulted in complete reduction of the Pd, thus allowing a more uniform dispersion and higher loading of catalysts than the MWCNT-SO_3H support (strong acid). In a nutshell, the high catalytic performance of the FeCo@Fe@Pd/MWCNT-COOH may be related to the improved electronic properties of the catalysts coupled with the high affinity of its –COOH surface with the catalyst species.

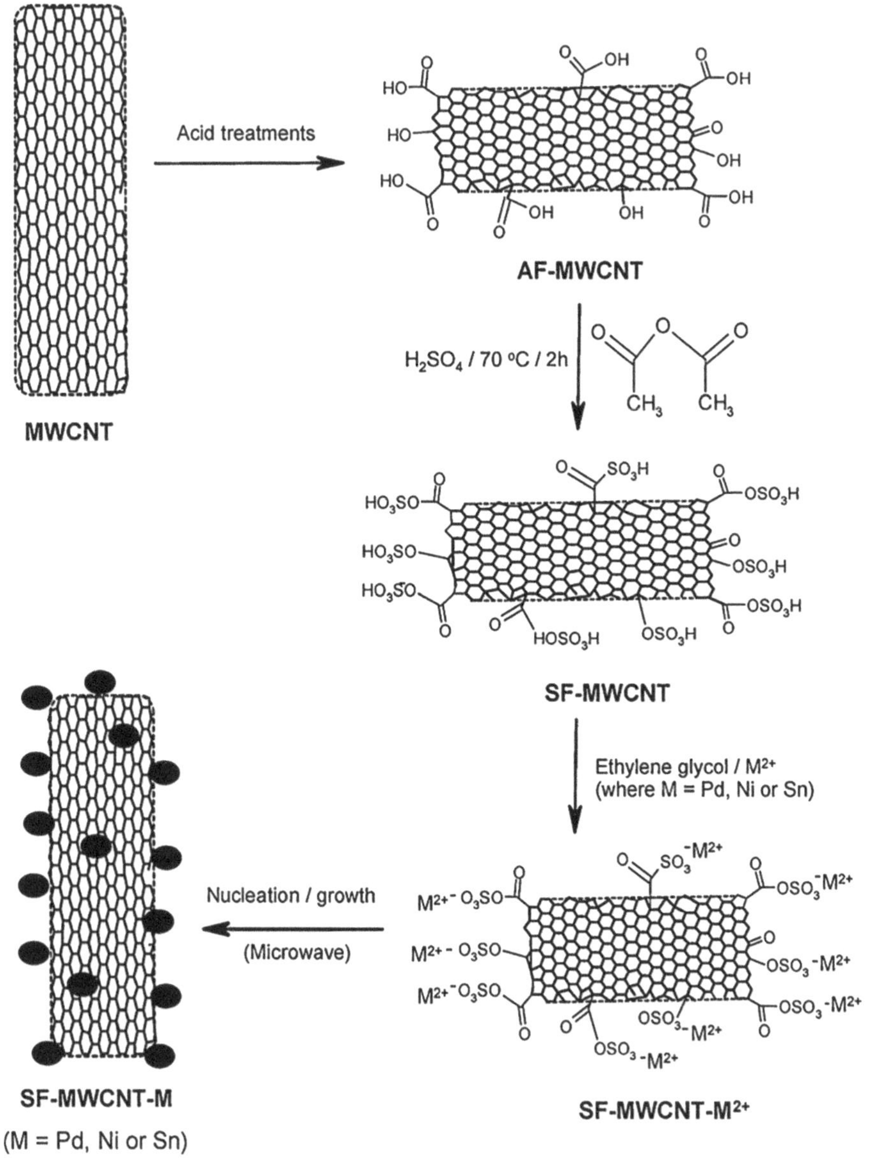

Fig. 13.9 Schematic of the preparation SF-MWCNT-M (M = Pd, Ni or Sn) nanocomposites using microwave-assisted strategy (*Source*: Ramulipho et al., Electrochimica Acta 2012, 59, 310–320. Reproduced with permission from [39], copyright 2012 Elsevier LTD)

Chemical oxidation of carbon supports provides oxygen-containing functional groups as the binding sites for the growth of metal ions. However, the oxidised sites of the carbon supports may accelerate the degradation process of the support material, thereby inducing significant aggregation of the catalyst nanoparticles.

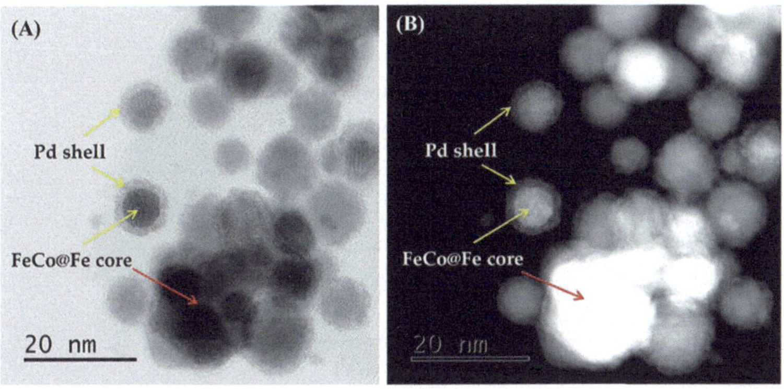

Fig. 13.10 (**a**) High-angle annular dark-field (HAADF)- and (**b**) bright-field (BF)-STEM images simultaneously acquired for the FeCo@Fe@Pd/MWCNT-COOH (*Source*: Fashedemi et al., *J. Mater. Chem. A*, 2015, 3, 7145–7156. Reproduced from [43], copyright 2015 Royal Society of Chemistry)

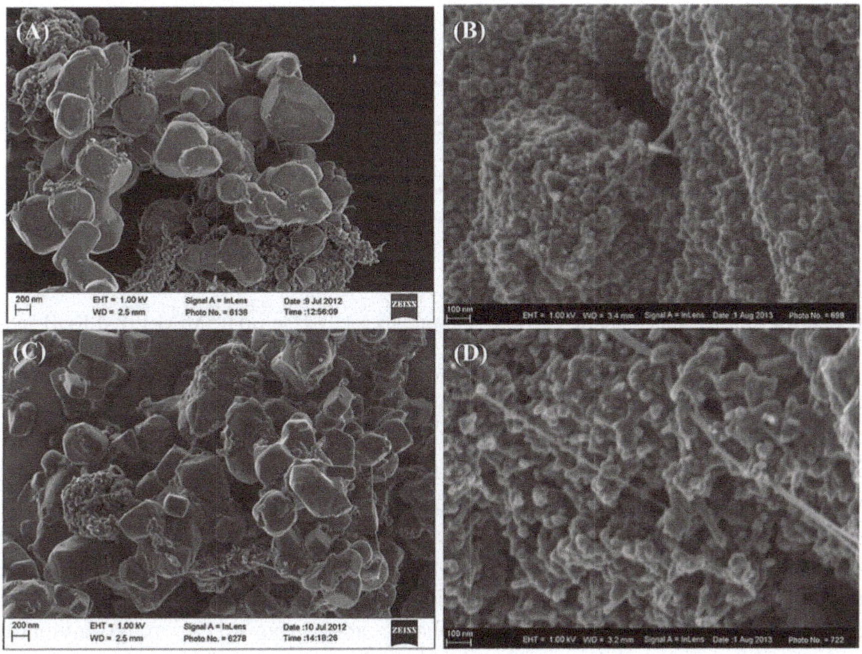

Fig. 13.11 FESEM images of (**a**) FeCo@Fe/MWCNT-COOH, (**b**) FeCo@Fe@Pd/ MWCNT-COOH, (**c**) FeCo@Fe/MWCNT-SO3H and (**d**) FeCo@Fe@Pd/MWCNT-SO3H (*Source*: Fashedemi et al., *J. Mater. Chem. A*, 2015, 3, 7145–7156. Reproduced from [43], copyright 2015 Royal Society of Chemistry)

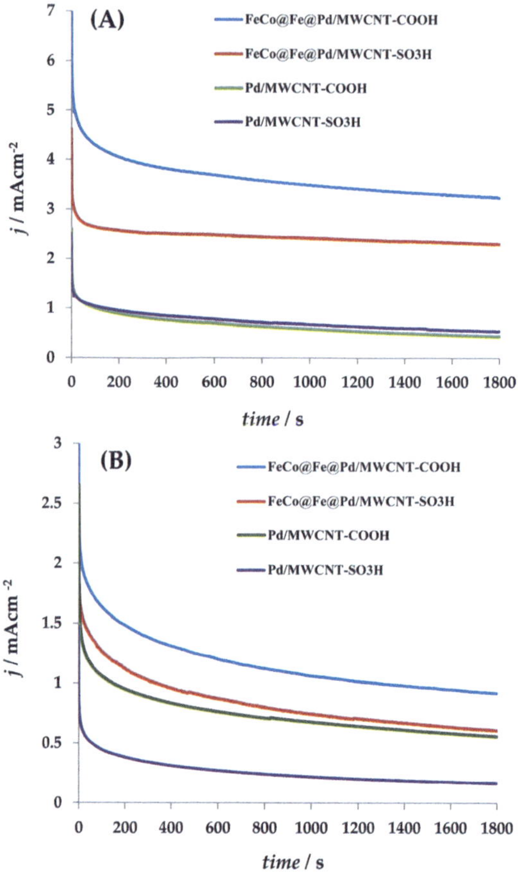

Fig. 13.12 Chronoamperometric curves of the various catalysts in (**a**) 0.5MEG/1.0 M KOH and (**b**) 0.5MGLY/1MKOH at a fixed potential of 0.16 V vs. Ag|AgCl (3 M KCl) (*Source*: Fashedemi et al., *J. Mater. Chem. A*, 2015, 3, 7145–7156. Reproduced from [43], copyright 2015 Royal Society of Chemistry)

To curb these potential oxidation problems, it is possible to wrap the carbon support with polymeric materials that contain appropriate functional groups for the binding of the catalyst metal ions. For example, Fujigaya et al. [44–50] have elegantly wrapped CNTs with polymeric materials that contain nitrogen atoms (i.e., pyridine-doped polybenzimidazole [50]), sulfonic acid (i.e., sulfonated polysulfone and sulfonated polyimide [46]) and phosphonic acid groups (i.e., poly(vinylphosphonicacid)-doped polybenzimidazole [44]). In a recent study [48], the group wrapped CNTs with poly[2,29-(2,6-pyridine)-5,59-bibenzimidazole] (PyPBI; Fig. 13.13), a polymer with N-atoms as the binding sites for the growth of Pt ions.

The wrapping was made possible through the π-π interactions between the CNTs and PyPBI. A series of CNT/PyPBI/Pt composite were prepared by changing the initial amount of Pt salt used in each reaction with the view of controlling the

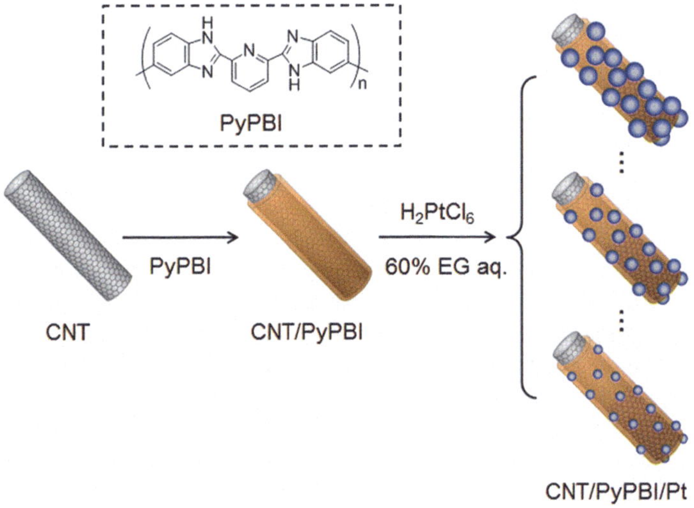

Fig. 13.13 Schematic drawing describing the preparation of series of CNT/PyPBI/Pt composites (*Source*: Hafez et al., Scientific Reports 2014, 4, 6295. doi:10.1038/srep06295. Reproduced with permission from [48], copyright 2014 Nature Publishing Group)

diameter of the Pt catalysts. The composite with the smallest Pt nanoparticle size of 2.3 nm diameter gave about 8 times higher mass activity than the fuel cell containing Pt with a 3.7 nm diameter. This represents the first example of the ability to control the diameter of Pt on polymer-wrapped carbon supporting materials. The catalysts displayed excellent stability with durability of 10 times higher than that of the commercial carbon black/Pt.

13.2.2 Non-carbon Supports

As already discussed, the major drawback associated with the use of carbon electrodes is corrosion. The corrosion problem of carbon-based support can be ameliorated by the use of non-carbon support materials that are corrosion-resistant and still maintain the desirable properties of carbon such as high-surface-area and high electrical conductivity. To this end, porous metals and metal alloys, inorganic metal oxides and mixed oxides, metal carbides and nitrides amongst others are being studied and used as briefly described hereunder.

13.2.2.1 Conducting Polymers

The use of heterocyclic conducting polymers (Fig. 13.14) as support materials has gained wide acceptance in fuel cells. Conducting polymers (CPs) such as

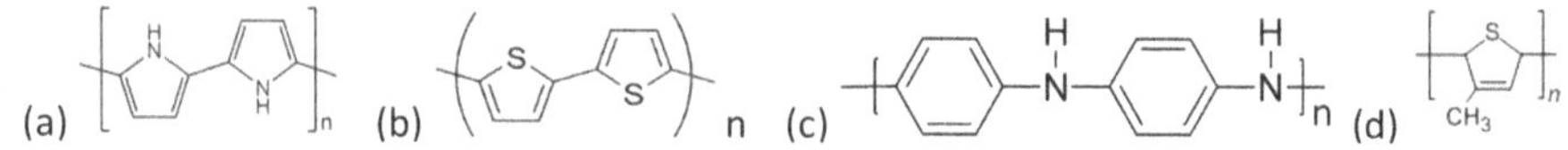

Fig. 13.14 Chemical structures of some common heterocyclic conducting polymers ((**a**) polypyrrole, (**b**) polythiophene, (**c**) polyaniline, (**d**) poly-(3-methylthiophene))

polypyrrole (PPY), polythiophene, polyaniline (PANI), and poly-(3-methylthiophene) possess conjugated backbones and are, therefore, convenient for proton transport [51], since the repeating cyclic groups are weak Lewis base sites favourable for proton hopping. In addition, the extensive conjugation-induced structural co-planarity reduces the steric hindrance in proton transport.

These common conducting polymers and their derivatives used as supports for catalysts are not only conjugated but most of them also possess heteroatoms which could easily serve as the site for catalyst attachments, thus avoiding the necessity of additional functionalization. In an a recent review by Dutta et al. [52], the authors elegantly discussed the use of CPs as catalyst-supporting matrices, especially their properties that have permitted them to provide better dispersion, distribution and anchoring of catalysts for enhanced electrochemical performance. A number of reports exist in the literature on various conducting polymers that have been used as support materials [51, 53–58]. Antolini and Gonzalez [59] established that porous polymeric matrixes not only increase the specific area of catalysts but, most importantly, present higher tolerance to poisoning due to the adsorption of CO species compared to carbon supports. CPs are highly electron-conducting (in the range 10^{-6} to 10^{3} S cm^{-1}, which is about two orders of magnitude higher than that of conventionally used carbon supports [60, 61]) and also proton-conducting materials. The inherent proton- and electron-conducting properties of CPs make them good candidates for replacing Nafion ionomers in the catalyst layer of fuel cell electrodes providing enhanced performance in the process. The morphology of the CPs also play a role in their performance as support materials. For example, the electrocatalytic oxidation of methanol on nanofibular polyaniline (nPANI) and granular polyaniline (gPANI) electrodes modified with Pt microparticles showed that nPANI had better conductivity and higher specific surface area, exhibiting a considerably higher electrocatalytic activity on the methanol oxidation than the gPANI electrode [53].

Huang et al. [62] investigated polypyrrole (PPy) samples, obtained by *in-situ* chemical oxidative polymerization, as an alternate cathode catalyst support material (i.e., for oxygen reduction reaction, ORR) in PEMFC. The electrochemical stabilities of the PPy samples and commercial carbon supports (Vulcan XC-72) were examined by cyclic voltammetry (CV). In contrast to the carbon supports (Vulcan XC-72), the PPy only showed a small anodic current up to 1.8 V, indicating resistance towards oxidation under high positive potentials. Pt/PPy was electrochemically more stable than the Pt/C electrocatalyst. Tafel plots for the ORR activity showed two-fold higher activity for Pt/PPy catalyst than that of Pt black catalyst at 0.9 V. The Pt/PPy catalyst demonstrated good ORR kinetics and comparable fuel cell performance with that of the commercial E-TEK Pt/C catalyst.

Some of the advantages of the use of the CPs as support materials in fuel cells (e.g., DMFC) include: (i) ability to significantly reduce CH_3OH crossover, with a slight compromise on proton conductivity; (ii) ability to allow the diffusion of gases out of the MEA due to the porous nature of CPs; (iii) replacement of the costly Nafion, thus reducing the overall cost of the fuel cells, and (iv) CPs are thermally stable in the temperature range within which DMFC operates, thus rendering stability to the overall system [51]. Other major advantages of CPs (especially the aromatic CPs) as support constituent include the prospect of (i) adequate catalyst binding on the inherently present heteroatoms in common ACPs, and its derivatives, thus avoiding the necessity for additional functionalization steps [54, 58, 61], and (ii) fine dispersion and uniform distribution of expensive catalysts, resulting in their exhibiting the maximum active surface area for catalytic reactions leading to their reduced consumption and maximum utilization [54, 63].

Another important CP is the poly(3,4-ethylenedioxythiophene)-polystyrene sulphonic acid (PEDOT-PSSA) which has been shown to perform better than the traditional Vulcan XC-72R carbon, as shown in a solid-polymer-electrolyte direct methanol fuel cell (SPE-DMFC using Pt-Ru catalyst) [61]. PEDOT-PSSA performed better than Vulcan XC-72 carbon in terms of electrical conductivity and current response. The high performance of the PEDOT-PSSA was attributed to its mixed-conducting nature that enhanced proton and electron transport within the anode catalyst, improving the activity of the Pt-Ru catalyst.

Recently, Wang et al. [30] investigated the gold nanoparticles supported on various supports for glycerol electro-oxidation. The authors found that Au nanocatalysts supported on extended poly(4-vinylpyridine) functionalized graphene (Au-P4P/G) exhibited much higher activity and selectivity for three carbon products than those on carbon black, P4P functionalized reduced graphene oxide (Au-P4P/rGO) and poly(*m*-aminophenol) (PmAP) wrapped graphene (Au-PmAP/G).

13.2.2.2 Metal Oxides and Hydroxides

Metal oxides and hydroxides are among the non-carbon materials used as support for direct alcohol fuel cell electrocatalysis in a bid to circumvent the perennial problem of carbon corrosion. The following metal oxides have been incorporated as co-catalysts for fuel cells (Table 13.1): NbO, NbO_2 and Nb_2O_5 [64–75], niobium-doped titanium dioxide [76–79], ZrO_2 [80–85], CeO_2 used alone [86–90] or in conjunction with other oxides or carbon [91–96]. Lei et al. recently showed that the excellent durability of Pt-CeO_2 nanocubes-graphene oxide catalyst was attributable to the free radical scavenging activity of CeO_2 that significantly slowed down the chemical degradation of Nafion binder in catalytic layers, consequently alleviating the decay of Pt catalysts, resulting in catalyst's excellent cycle life [97]. Other metallic oxides which have been found useful include MoO_x [98–103], RuO_2 and Ru-mixed oxides [104–110], titanium oxides (TiO_2) and their composites [28, 106, 111–118], MnO_2, Mn_2O_3, Mn_3O_4 and MnO [89, 90, 119–122], cobalt oxides and

Table 13.1 Other oxide support materials employed in DAFCs

Active catalyst	Oxide type	Process involved	Reference
Pt	ZnO	MeOH oxidation	[144]
PdCo	tGO#	ORR	[145]
Pd-Zn	Al_2O_3	MeOH reforming	[146]
Pd	Al_2O_3	MeOH reforming	[146]
Au	γFe_2O_3	CO oxidation	[147]
Pt	γFe_2O_3	CO oxidation	[148]
Pt	RuO_2-TiO_2	ROH oxidation	[149]
Pt	RuO_2/xH_2O-TiO_2	ROH oxidation	[149]
Pt	SnO_2	ORR	[150]
Pt	SnO_2/Ketjen black	ORR	[151]
Pt	SnO_2-Nb	I-V performance	[152]
Pt	MgO/C	EtOH oxidation	[153]
Pt	MgO/C	EtOH oxidation	[154]
Pd	NiO/MgO@C	EtOH oxidation	[155]
Pd	In_2O_3/CNTs	EtOH oxidation	[156]
Pd	TiO2-C	MeOH oxidation	[157]
Pt	MgO/CNT	EtOH oxidation	[158]
Mn	$CaMn_3O_6$	ORR	[159]
Mn	$CaMnO_3$	ORR	[159]
Pt	$Mn_3(PO_4)_2$@BSA	MeOH oxidation	[119]

Key: tGO# = thiolated grapheme oxide

hydroxides [89, 90, 123–132] as well as silica, a unique oxide that forms hydroxyl species even in acidic pH and their hybrids [133–143]. The vast number of transition metal oxides arising from the variability in the oxidation states of the transition metals has made possible the design of numerous transition metal-based oxides for FC catalyst support. Other promising alternative metal oxide support materials employed in the search for ways of alleviating the problem of corrosion of the carbon support of electrocatalysts under cathode conditions in FCs are summarized in Table 13.1. Bimetallic catalysts combined with some oxides, like TiO_2 and exhibiting definite morphologies (Fig. 13.15) have shown promising results in methanol oxidation.

Metal oxides are considered to be more appealing as they spontaneously form surface hydroxyl groups that helps to circumvent adsorption of catalyst poisons, CO, albeit most of them in an alkaline medium.

13.2.2.3 Transition Metal Carbides

Transition metal carbides (TMCs) are produced by incorporating C atoms into the interstitial sites of transition metal atoms, including all 3d elements, as well as 4d and 5d elements of group's 4–6 early transition metals. It was reported [160–162]

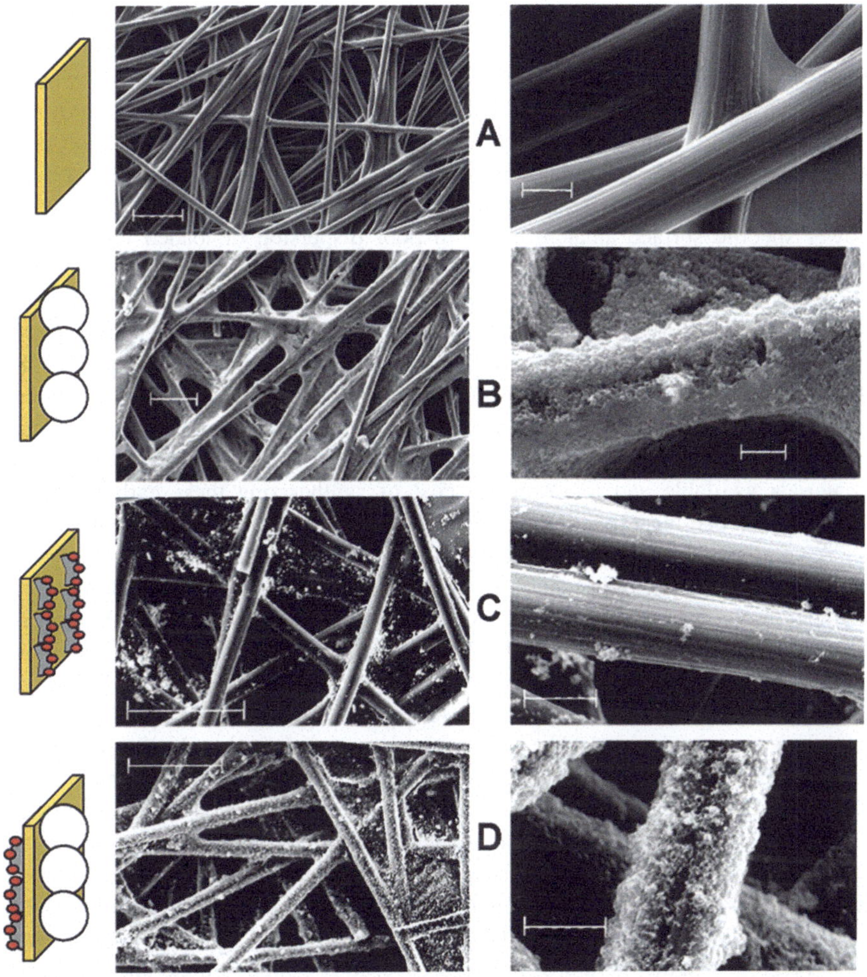

Fig. 13.15 Scanning electron micrograph showing the carbon fiber paper and its modification with TiO_2 and Pt catalyst: (**a**) CFE, (**b**) CFE/TiO_2, (**c**) Pt-Ru/CFE, and (**d**) TiO_2/CFE/Pt-Ru. (Magnification: Bar in left column represents 50 μm, bar in right column represents 6 μm). (*Source*: Drew et al., J. Phys. Chem. B 2005, 109, 11851–11857. Reproduced with permission from [111], copyright 2005 American Chemical Society)

that the groups 4–6 TMCs typically display properties characteristic of three different classes of materials that include the extreme hardness and brittleness of covalent solids, the ordered bulk structure and high melting temperatures of ionic solids, as well as electric conductivity and heat capacity of metals. Since the electrochemical properties of metal carbides can vary greatly depending on operating conditions (i.e., electrochemical potential, current density) as well as pH environment (i.e., acidic, neutral, basic) [163] it is therefore of particular concern to recognize the electrochemical stability of transition metal carbide catalysts over a wide range of chemical environments to determine their usefulness for different electrochemical applications. The use of these metal carbides either as catalysts or catalyst supports in a variety of electrochemical devices including fuel cells, photoelectrochemical cells, etc. is expected to lead to reduction in the amount of noble metals used in electrochemical applications. The literatures are replete with conclusions that the interaction between carbides and noble metals increases with the decrease of carbide particle size, [164, 165] (i.e., the larger carbide particles having lower specific surface area, resulting in poor dispersion of loaded noble metal nanoparticles) [166]. The ability of the transition metals of roughly groups 4–6 to form stable carbides, including multiple stoichiometries and polymorphs, that are resistant to electrochemical corrosion in addition to being more abundant and orders of magnitude less expensive than the Pt-group metal catalysts [167, 168] have ultimately lead to increased interest in these materials for electrochemical applications. The establishment of syntheses methods such as salt flux synthesis and temperature programmed reduction developed to increase the range of carbide materials available and to lower the cost of synthesizing these carbides [167] have been added incentive in the electrochemical application of these materials.

A comparison of the valence band X-ray photoelectron spectrum of tungsten carbide with the spectra of tungsten and platinum shows that, near the Fermi level, the electronic density of states of tungsten carbide more nearly resembles that of platinum than that of tungsten [169, 170]. This overall resemblance in the electronic as well as in catalytic properties of Pt and tungsten monocarbide (WC) [171], has made tungsten carbide one of the most studied [172] amongst these transition metals, either as electrocatalyst or as support material for low loadings of Pt. It should be noted though that contrary to the hypothesis that the platinum-like catalytic activity of tungsten carbide results from the contribution of carbon valence electrons to the 5d band of tungsten, the width of the unfilled portion of the d band increases continuously from tungsten to tungsten carbide, according to Houston et al. [173]. From the foregoing, a number of reports on the use of transition metal carbides as electrocatalysts exists [163, 167, 171, 172, 174, 175] while even more could be found in the literature on their use as catalyst support materials [164, 166, 171, 176–179]. For further reading on this subject matter, reference should be made to some extensive and excellent reviews in the literature [13, 160, 174, 175].

13.2.2.4 Transition Metal Nitrides

The use of transition metal nitrides as catalyst support in direct alcohol fuel cells electrocatalysis has become a subject of great interest due to their inherent good catalyst support interaction, thermal stability, high corrosion resistance and high conductivity compared to carbon supports. Some of the nitrides investigated include the mesoporous binary nitrides (e.g., CrN, TiN, Fig. 13.16 [3, 180]) and ternary nitride (e.g., $Ti_{0.5}Nb_{0.5}N$ [181]) complexes. In all cases, Yang et al. [180–183] reported that these nitride supports for Pt showed excellent electronic conductivity, high surface area, and better performance towards

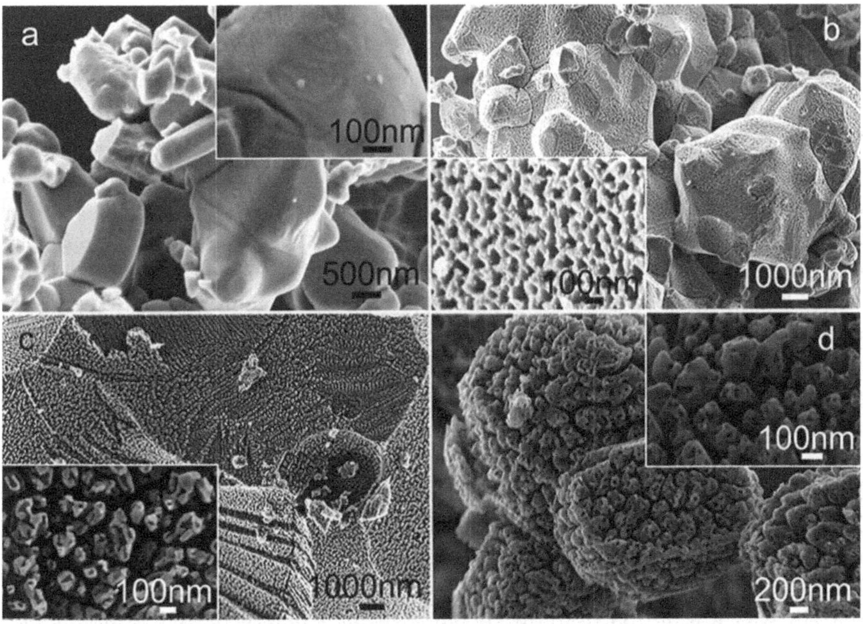

Fig. 13.16 SEM images of TiN samples from ammonolysis of Zn_2TiO_4 at (**a**) 500 1C, (**b**) 600 1C, (**c**) 700 1C and (**d**) 800 1C for 8 h (*Source*: Yang et al., Phys. Chem. Chem. Phys., 2013, 15, 1088–1092. Reproduced with permission from [183], copyright 2013 Royal Society of Chemistry)

methanol electro-oxidation in both acid and alkaline media than commercial Pt/C. In addition, they showed better electrochemical stability and higher tolerance to corrosion than the commercial Pt/C. Unlike other nitrides investigated, Mo_2N was found to be electrochemically unstable even at a relative low potential (0.5 V, vs. SCE) [184]. This finding suggests that Mo_2N may not be able to promote ORR, especially if one considers that ORR occurs at reasonable rates between ~0.45 and 0.65 V (RHE) and its standard equilibrium potentials is ~0.80 V (RHE).

Thotiyl and co-workers [185] indicated that titanium nitride nanoparticles (TiN NPs) are nitrides of choice because it does not only exhibit excellent corrosion resistance and electrochemical stability, but also acts as a co-catalyst adsorbing OH groups and facilitating the oxidation of the poisonous intermediates (CO) adhered on adjacent Pt surface, enhancing the methanol oxidation reaction (MOR) activity in the process (especially for high temperature PEM fuel cells, to replace carbon materials [165]). The authors observed that the onset of ethanol oxidation was negatively shifted by at least 100 mV on Pt-TiN as compared to Pt-C and that the peak currents were more than doubled when Pt was loaded onto TiN showing the positive effect of TiN support for Pt towards ethanol oxidation reaction (EOR). A positive chemical shift of Pd (3d) binding energy was used by the authors to confirm the existence of metal-support interaction between Pd and TiN, which in turn helped to weaken the Pd-CO synergetic bonding interaction during the electro-oxidation process [186].

A number of other researchers have also probed into the possibilities of using metallic nitrides as electro-oxidation catalysts, with varying degrees of success [165, 187–192] and recommendations [193–198]. The possibility of whether the introduction of molybdenum element into TiN might improve the electrocatalytic activity of supported Pt catalyst on one hand, and preserve the intrinsic electrochemical stability of the TiN nanostructures on the other towards methanol oxidation was the subject of a recent investigation [199]. The titanium molybdenum nitride ($Ti_{0.8}Mo_{0.2}N$) supported Pt catalyst did not only exhibit a much higher mass activity and durability than that of the conventional Pt/C (E-TEK) electrocatalysts for MOR but also led to a significant decrease of Pt density on carbon support after the accelerated durability test (ADT) test, demonstrating that the major cause for ECSA loss of Pt/C was Pt detaching, ripening and the corrosion of the carbon support. DAFC support for Pt nanoparticles with high surface area, hollow and interior porous structure, titanium nitride nanotubes (TiN NTs), was obtained by Xiao et al. [200] via a combination of solvothermal alcoholysis and post-nitriding method. The support reportedly displayed enhanced activity and durability during MOR compared with the commercial Pt/C (E-TEK) catalyst. The performance of the TiN NTs was attributed to the smaller size and better dispersion of the Pt NPs (providing high EASA), the unique hollow and porous structure (Fig. 13.17) of the TiN NTs (speeding up the diffusion rate and oxidation of the carbonaceous species), as well as co-catalyst effect, and strong interaction with Pt nanoparticles.

13.2.2.5 Dual-Purpose Supports

Some materials play dual role as support as well as current collector. Some of the materials that have served this dual purpose include glassy carbon rods [201], nickel foam and carbon cloths [202]. For example, Modibedi et al. [202] employed

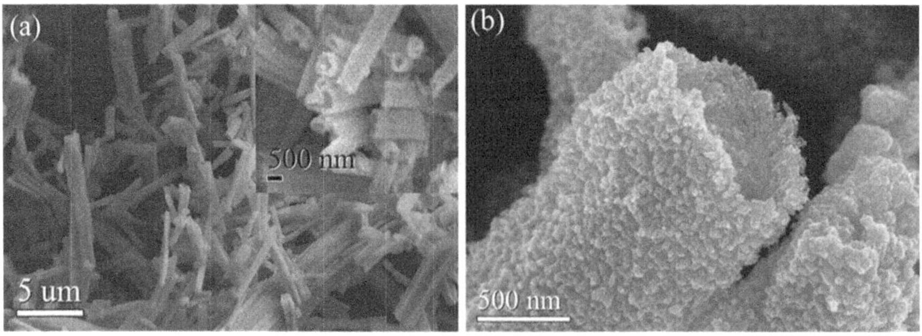

Fig. 13.17 (**a**) SEM image of TiN NTs and (insert) magnified SEM image of the vertical section of some nanotubes. (**b**) A typical enlarged SEM image of a TiN NT. [Xiao et al., Electrochim. Acta, 2014, 141, 279–285. Reproduced with permission from [200], copyright 2014 Elsevier Ltd]

Ni foam and carbon cloth as both supports and current collectors for Pt and Pd catalysts obtained by electrochemical atomic layer deposition (EC-ALD). They observed that methanol oxidation proceeded better on carbon paper than on Ni foam; while the reverse was the case with ethanol oxidation and that the two supports produced good quality deposits.

13.3 Summary and Future Prospects

In this book chapter, we have reviewed the concept of fuel cell generally and particularly various types of catalyst supports that have been developed over time for use in fuel cell eletrocatalysis. The direct conversion of the chemical energy/heat of combustion of fuels and oxidants in the presence of a catalyst into electricity, heat and water at an efficiency higher than that of internal combustion engine have revolutionised the search for environmentally benign fuel. The challenges of the use of carbon supports in fuel cells are gradually being addressed by the modification of conventional carbon supports as well as the use of non-carbon supports. Significant progress has been made in the commercialization of fuel cells for various applications to major corporations while fuel cells for aircraft auxiliary power units (APUs) that provide energy for functions other than propulsion as well as fuel cells for fire prevention are very high potential outlets.

Due to the high price premiums associated with fuel cells, added value features need to be exploited in order to make them more appealing and increase unit sales and market penetration. With the limited availability and high cost of platinum and platinum group metals, research should be sustained in the search for low cost materials that may uniquely exhibit catalytic and stability properties equivalent to, or surpassing that of platinum. Irrespective of the effort already made in developing modern fuel cell technologies, there is need for newer technologies that shall alleviate the prevailing economic challenges and exceed the advantages of the existing technologies to be fully acceptable for mass production. It is also recommended that research effort should be channelled towards the development of fuel cell materials that will address most of the challenges highlighted in this chapter and elsewhere in this book as well as full optimization of fuel cell operating conditions. It is equally important that the new catalysts being developed keep their electrochemical performance for ORR and their alcohol tolerance ability over a long usage time.

References

1. Huang H, Wang X (2014) Recent progress on carbon-based support materials for electrocatalysts of direct methanol fuel cells. J Mater Chem A 2:6249–6670. doi:10.1039/c3ta14754a

2. Wang H, Dai H (2013) Strongly coupled inorganic-nano-carbon hybrid materials for energy storage. Chem Soc Rev 42:3088–3113. doi:10.1039/c2cs35307e
3. Linares N, Silvestre-Albero AM, Serrano E, Silvestre-Albero J, Garcia-Martinez J (2014) Mesoporous materials for clean energy technologies. Chem Soc Rev 43:7681–7717. doi:10.1039/c3cs60435g
4. Guo YG, Hu JS, Wan LJ (2008) Nanostructured materials for electrochemical energy conversion and storage devices. Adv Mater 20:2878–2887. doi:10.1002/adma.200800627
5. Choi NS, Chen Z, Freunberger SA, Ji X, Sun YK, Amine K et al (2012) Challenges facing lithium batteries and electrical double-layer capacitors. Angew Chem Int Ed 51:9994–10024. doi:10.1002/anie.201201429
6. Zhang Z, Shimizu T, Senz S, Gösele U (2009) Ordered high-density Si [100] nanowire arrays epitaxially grown by bottom imprint method. Adv Mater 21:2824–2828. doi:10.1002/adma.200802156
7. Jariwala D, Sangwan VK, Lauhon LJ, Marks TJ, Hersam MC (2013) Carbon nanomaterials for electronics, optoelectronics, photovoltaics, and sensing. Chem Soc Rev 42:2824–2860. doi:10.1039/c2cs35335k
8. Aiyejina A, Sastry MKS (2012) PEMFC flow channel geometry optimization: a review. J Fuel Cell Sci Technol 9:011011. doi:10.1115/1.4005393
9. Grujicic M, Chittajallu KM (2004) Optimization of the cathode geometry in polymer electrolyte membrane (PEM) fuel cells. Chem Eng Sci 59:5883–5895. doi:10.1016/j.ces.2004.07.045
10. Carcadea E, Stefanescu I, Ionete RE, Ene H, Ingham DB, Ma L (2008) PEM fuel cell geometry optimisation using mathematical modelling. Int J Multiphys 2:313–326. doi:10.1260/175095408786927462
11. Grove WR (1839) On a small voltaic battery of great energy; some observations on voltaic combinations and forms of arrangement; and on the inactivity of a copper positive electrode in nitro-sulphuric acid. Philos Mag 15:287–293. doi:10.1080/14786443908649881
12. Grove WR (1839) On voltaic series and the combination of gases by platinum. Philos Mag 14:127–130. doi:10.1080/14786443908649684
13. Li L, Hu L, Li J, Wei Z (2015) Enhanced stability of Pt nanoparticle electrocatalysts for fuel cells. Nano Res 8:418–440. doi:10.1007/s12274-014-0695-5
14. Sharma S, Pollet BG (2012) Support materials for PEMFC and DMFC electrocatalysts—a review. J Power Sources 208:96–119. doi:10.1016/j.jpowsour.2012.02.011
15. Martínez-Huerta M, Tsiouvaras N, García G, Peña M, Pastor E, Rodriguez J et al (2013) Carbon-supported ptrumo electrocatalysts for direct alcohol fuel cells. Catalysts 3:811–838. doi:10.3390/catal3040811
16. Li Y, Xu L, Liu H, Li Y (2014) Graphdiyne and graphyne: from theoretical predictions to practical construction. Chem Soc Rev 43:2572–2586. doi:10.1039/c3cs60388a
17. Skrabalak SE, Suslick KS (2006) Porous carbon powders prepared by ultrasonic spray pyrolysis. J Am Chem Soc 128:12642–12643. doi:10.1021/ja064899h
18. Bianchi CL, Biella S, Gervasini A, Prati L, Rossi M (2003) Gold on carbon: influence of support properties on catalyst activity in liquid-phase oxidation. Catal Lett 85:91–96. doi:10.1023/A:1022176909660
19. Auer E, Freund A, Pietsch J, Tacke T (1998) Carbons as supports for industrial precious metal catalysts. Appl Catal A Gen 173:259–271. doi:10.1016/S0926-860X(98)00184-7
20. Qiu Z, Huang H, Du J, Tao X, Xia Y, Feng T et al (2014) Biotemplated synthesis of bark-structured TiC nanowires as Pt catalyst supports with enhanced electrocatalytic activity and durability for methanol oxidation. J Mater Chem A 2:8003. doi:10.1039/c4ta00277f
21. Wilson MS (1993) Surface area loss of supported platinum in polymer electrolyte fuel cells. J Electrochem Soc 140:2872. doi:10.1149/1.2220925
22. Philippot K, Serp P (2012) Concepts in nanocatalysis. In: Nanomaterials in catalysis, 1st edn. Wiley-VCH, Weinheim, pp 1–54

23. Yuan X, Ding X-L, Wang C-Y, Ma Z-F (2013) Use of polypyrrole in catalysts for low temperature fuel cells. Energy Environ Sci 6:1105–1124. doi:10.1039/C3EE23520C
24. Zhang Y, Chen S, Wang Y, Ding W, Wu R, Li L et al (2015) Study of the degradation mechanisms of carbon-supported platinum fuel cells catalyst via different accelerated stress test. J Power Sources 273:62–69. doi:10.1016/j.jpowsour.2014.09.012
25. Borup R, Meyers J, Pivovar B, Kim YS, Mukundan R, Garland N et al (2007) Scientific aspects of polymer electrolyte fuel cell durability and degradation. Chem Rev 107:3904–3951
26. Kinoshita K (1988) Carbon: electrochemical and physicochemical properties. Wiley, New York, ISBN: 978-0471848028
27. Sasaki K, Shao M, Adzic R (2009) Dissolution and stabilization of platinum in oxygen cathodes. In: Proton exchange membrane fuel cell durability. Springer, New York, pp 7–28
28. Huang S, Ganesan P, Park S, Popov BN (2009) Development of a titanium dioxide supported platinum catalyst with ultrahigh stability for polymer electrolyte membrane fuel cell applications. J Am Chem Soc 131:13898–13899. doi:10.1021/ja904810h
29. Mathias MF, Makharia R, Gasteiger HA, Conley JJ, Fuller TJ, Gittleman CJ et al (2005) Two fuel cell cars in every garage? Electrochem Soc Interface 14:24–35
30. Wang H, Thia L, Li N, Ge X, Liu Z, Wang X (2015) Selective electro-oxidation of glycerol over Au supported on extended poly(4-vinylpyridine) functionalized graphene. Appl Catal B Environ 166–167:25–31. doi:10.1016/j.apcatb.2014.11.009
31. Castanheira L, Dubau L, Mermoux M, Berthomé G, Caqué N, Rossinot E et al (2014) Carbon corrosion in proton-exchange membrane fuel cells: from model experiments to real-life operation in membrane electrode assemblies. ACS Catal 4:2258–2267. doi:10.1021/cs500449q
32. Cuong NT, Fujiwara A, Mitani T, Chi DH (2008) Effects of carbon supports on Pt nano-cluster catalyst. Comput Mater Sci 44:163–166. doi:10.1016/j.commatsci.2008.01.061
33. Kim S, Park SJ (2006) Effects of chemical treatment of carbon supports on electrochemical behaviors for platinum catalysts of fuel cells. J Power Sources 159:42–45. doi:10.1016/j.jpowsour.2006.04.041
34. Siswana MP, Ozoemena KI, Nyokong T (2006) Electrocatalysis of asulam on cobalt phthalocyanine modified multi-walled carbon nanotubes immobilized on a basal plane pyrolytic graphite electrode. Electrochim Acta 52:114–122. doi:10.1016/j.electacta.2006.03.090
35. Liu J, Rinzler AG, Dai H, Hafner JH, Bradley RK, Boul PJ et al (1998) Fullerene pipes. Science 280:1253–1256. doi:10.1126/science.280.5367.1253
36. Sun ZP, Zhang XG, Liang YY, Li HL (2009) A facile approach towards sulfonate functionalization of multi-walled carbon nanotubes as Pd catalyst support for ethylene glycol electro-oxidation. J Power Sources 191:366–370. doi:10.1016/j.jpowsour.2009.01.093
37. Du CY, Zhao TS, Liang ZX (2008) Sulfonation of carbon-nanotube supported platinum catalysts for polymer electrolyte fuel cells. J Power Sources 176:9–15. doi:10.1016/j.jpowsour.2007.10.016
38. Peng F, Zhang L, Wang H, Lv P, Yu H (2005) Sulfonated carbon nanotubes as a strong protonic acid catalyst. Carbon N Y 43:2405–2408. doi:10.1016/j.carbon.2005.04.004
39. Ramulifho T, Ozoemena KI, Modibedi RM, Jafta CJ, Mathe MK (2012) Fast microwave-assisted solvothermal synthesis of metal nanoparticles (Pd, Ni, Sn) supported on sulfonated MWCNTs: Pd-based bimetallic catalysts for ethanol oxidation in alkaline medium. Electrochim Acta 59:310–320. doi:10.1016/j.electacta.2011.10.071
40. Fashedemi OO, Julies B, Ozoemena KI (2013) Synthesis of Pd-coated FeCo@Fe/C core-shell nanoparticles: microwave-induced "top-down" nanostructuring and decoration. Chem Commun (Camb) 49:2034–2036. doi:10.1039/c3cc38672d
41. Fashedemi OO, Ozoemena KI (2013) Enhanced methanol oxidation and oxygen reduction reactions on palladium-decorated FeCo@Fe/C core-shell nanocatalysts in alkaline medium. Phys Chem Chem Phys 15:20982–20991. doi:10.1039/c3cp52601a

42. Fashedemi OO, Ozoemena KI (2014) Comparative electrocatalytic oxidation of ethanol, ethylene glycol and glycerol in alkaline medium at Pd-decorated FeCo@Fe/C core-shell nanocatalysts. Electrochim Acta 128:279–286. doi:10.1016/j.electacta.2013.10.194
43. Fashedemi OO, Miller HA, Marchionni A, Vizza F, Ozoemena KI (2015) Electro-oxidation of ethylene glycol and glycerol at palladium-decorated FeCo@Fe core–shell nanocatalysts for alkaline direct alcohol fuel cells: functionalized MWCNT supports and impact on product selectivity. J Mater Chem A 3(2015):7145–7156. doi:10.1039/C5TA00076A
44. Berber MR, Fujigaya T, Sasaki K, Nakashima N (2013) Remarkably durable high temperature polymer electrolyte fuel cell based on poly(vinylphosphonic acid)-doped polybenzimidazole. Sci Rep 3:1764. doi:10.1038/srep01764
45. ChemCatChem (2013) Interfacial engineering of Pt catalysts for fuel cells—MeOH oxidation is dramatically improved by polymer coating on a Pt catalyst. 7:1701–1704. doi:10.1002/cctc.201300157
46. Fujigaya T, Kim CR, Matsumoto K, Nakashima N (2013) Effective anchoring of Pt-nanoparticles onto sulfonated polyelectrolyte-wrapped carbon nanotubes for use as a fuel cell electrocatalyst. Polym J 45:326–330. doi:10.1038/pj.2012.145
47. Fujigaya T, Nakashima N (2013) Fuel cell electrocatalyst using polybenzimidazole-modified carbon nanotubes as support materials. Adv Mater 25:1666–1681. doi:10.1002/adma.201204461
48. Hafez IH, Berber MR, Fujigaya T, Nakashima N (2014) Enhancement of platinum mass activity on the surface of polymer-wrapped carbon nanotube-based fuel cell electrocatalysts. Sci Rep 4:6295. doi:10.1038/srep06295
49. Matsumoto K, Fujigaya T, Sasaki K, Nakashima N (2011) Bottom-up design of carbon nanotube-based electrocatalysts and their application in high temperature operating polymer electrolyte fuel cells. J Mater Chem 21:1187. doi:10.1039/c0jm02744h
50. Fujigaya T, Okamoto M, Nakashima N (2009) Design of an assembly of pyridine-containing polybenzimidazole, carbon nanotubes and Pt nanoparticles for a fuel cell electrocatalyst with a high electrochemically active surface area. Carbon N Y 47:3227–3232. doi:10.1016/j.carbon.2009.07.038
51. Dutta K, Kumar P, Das S, Kundu PP (2014) Utilization of conducting polymers in fabricating polymer electrolyte membranes for application in direct methanol fuel cells. Polym Rev 54:1–32. doi:10.1080/15583724.2013.839566
52. Dutta K, Das S, Rana D, Kundu PP (2015) Enhancements of catalyst distribution and functioning upon utilization of conducting polymers as supporting matrices in DMFCs: a review. Polym Rev 55:1–56. doi:10.1080/15583724.2014.958771
53. Zhou HH, Jiao SQ, Chen JH, Wei WZ, Kuang YF (2004) Effects of conductive polyaniline (PANI) preparation and platinum electrodeposition on electroactivity of methanol oxidation. J Appl Electrochem 34:455–459. doi:10.1023/B:JACH.0000016635.35555.04
54. Dutta K, Kundu PP (2014) A review on aromatic conducting polymers-based catalyst supporting matrices for application in microbial fuel cells. Polym Rev 54:401–435. doi:10.1080/15583724.2014.881372
55. Rajesh B, Thampi KR, Bonard J-M, Mathieu HJ, Xanthopoulos N, Viswanathan B (2003) Conducting polymeric nanotubules as high performance methanol oxidation catalyst support. Chem Commun (Camb) 21:2022–2023. doi:10.1039/b305591d
56. Maiyalagan T (2008) Electrochemical synthesis, characterization and electro-oxidation of methanol on platinum nanoparticles supported poly(o-phenylenediamine) nanotubes. J Power Sources 179:443–450. doi:10.1016/j.jpowsour.2008.01.048
57. Chen Z, Xu L, Li W, Waje M, Yan Y (2006) Polyaniline nanofibre supported platinum nanoelectrocatalysts for direct methanol fuel cells. Nanotechnology 17:5254–5259. doi:10.1088/0957-4484/17/20/035
58. Ma Y, Jiang S, Jian G, Tao H, Yu L, Wang X et al (2009) CNx nanofibers converted from polypyrrole nanowires as platinum support for methanol oxidation. Energy Environ Sci 2:224. doi:10.1039/b807213m

59. Antolini E, Gonzalez ER (2009) Polymer supports for low-temperature fuel cell catalysts. Appl Catal A Gen 365:1–19. doi:10.1016/j.apcata.2009.05.045
60. Qi Z, Pickup PG (1998) High performance conducting polymer supported oxygen reduction catalysts. Chem Commun 1998:2299–2300. doi:10.1039/a805322g
61. Tintula KK, Pitchumani S, Sridhar P, Shukla AK (2010) A solid-polymer-electrolyte direct methanol fuel cell (DMFC) with Pt-Ru nanoparticles supported onto poly (3,4-ethylenedioxythiophene) and polystyrene sulphonic acid polymer composite as anode. J Chem Sci 122:381–389. doi:10.1007/s12039-010-0043-6
62. Huang SY, Ganesan P, Popov BN (2009) Development of conducting polypyrrole as corrosion-resistant catalyst support for polymer electrolyte membrane fuel cell (PEMFC) application. Appl Catal B Environ 93:75–81. doi:10.1016/j.apcatb.2009.09.014
63. Liu FJ, Huang LM, Wen TC, Li CF, Huang SL, Gopalan A (2008) Effect of deposition sequence of platinum and ruthenium particles into nanofibrous network of polyaniline-poly (styrene sulfonic acid) on electrocatalytic oxidation of methanol. Synth Met 158:603–609. doi:10.1016/j.synthmet.2008.04.002
64. Kremliakova N (2013) Stable, durable carbon supported catalyst composition for fuel cell. US patent 2013/0164655 A1
65. Campbell SA, Kremliakova N (2014) Synthesis of stable and durable catalyst composition for fuel cell. US patent 8722284 B2
66. Hardman S, Chandan A, Steinberger-Wilckens R (2014) Fuel cell added value for early market applications. J Power Sources 287:297–306. http://dx.doi.org/10.1016/j.jpowsour.2015.04.056
67. Brosha EL, Blackmore KJ, Burrell AK, Henson NJ, Phillips J (2010) Engineered nano-scale ceramic supports for PEM fuel cells. In: 2010 fuel cell seminar and exposition, San Antonio, TX
68. Merzougui B, Shao M, Protsailo LV (2011) Fuel cell catalyst support with boron carbide-coated metal oxides/phosphates and method of manufacturing same. US patent 20110136047 A1
69. Zhang L, Wang L, Holt CMB, Zahiri B, Li Z, Malek K et al (2012) Highly corrosion resistant platinum–niobium oxide–carbon nanotube electrodes for the oxygen reduction in PEM fuel cells. Energy Environ Sci 5:6156. doi:10.1039/c2ee02689a
70. Morozan A, Jousselme B, Palacin S (2011) Low-platinum and platinum-free catalysts for the oxygen reduction reaction at fuel cell cathodes. Energy Environ Sci 4:1238. doi:10.1039/c0ee00601g
71. Tripković V, Abild-Pedersen F, Studt F, Cerri I, Nagami T, Bligaard T et al (2012) Metal oxide-supported platinum overlayers as proton-exchange membrane fuel cell cathodes. ChemCatChem 4:228–235. doi:10.1002/cctc.201100308
72. Bonakdarpour A, Tucker RT, Fleischauer MD, Beckers NA, Brett MJ, Wilkinson DP (2012) Nanopillar niobium oxides as support structures for oxygen reduction electrocatalysts. Electrochim Acta 85:492–500. doi:10.1016/j.electacta.2012.08.005
73. Huang K, Li Y, Yan L, Xing Y (2014) Nanoscale conductive niobium oxides made through low temperature phase transformation for electrocatalyst support. RSC Adv 4:9701. doi:10.1039/c3ra47091a
74. Chris MB, Michael H, Zhang L, Wang L, Holt CMB, Navessin T et al (2010) NRC Publications Archive (NPArC) Archives des publications du CNRC (NPArC) oxygen reduction reaction activity and electrochemical stability of thin-film bilayer systems of platinum on niobium oxide. J Phys Chem C 114:16463–16474
75. Xu C, Pietrasz P, Yang J, Soltis R, Sun K, Sulek M, et al (2013) Pt-based ORR catalyst on carbon-supported amorphous niobium oxide support. In: 224th ECS meeting. The Electrochemical Society
76. Huang S-Y, Ganesan P, Popov BN (2010) Electrocatalytic activity and stability of niobium-doped titanium oxide supported platinum catalyst for polymer electrolyte membrane fuel cells. Appl Catal B Environ 96:224–231. doi:10.1016/j.apcatb.2010.02.025

77. Do TB, Cai M, Ruthkosky MS, Moylan TE (2010) Niobium-doped titanium oxide for fuel cell application. Electrochim Acta 55:8013–8017. doi:10.1016/j.electacta.2010.03.027
78. Rosenfeld D, Schmid P, Széles S, Lévy F, Demarne V, Grisel A (1996) Electrical transport properties of thin-film metal-oxide-metal Nb2O5 oxygen sensors. Sensors Actuators B Chem 37:83–89. doi:10.1016/S0925-4005(96)01991-0
79. Morris D, Dou Y, Rebane J, Mitchell C, Egdell R, Law D et al (2000) Photoemission and STM study of the electronic structure of Nb-doped TiO_2. Phys Rev B 61:13445–13457. doi:10.1103/PhysRevB.61.13445
80. Iordache C, Blair S, Lycke D, Huff S (2008) Catalysts including metal oxide for organic fuel cells. US 2008/0014494 A1
81. Suzuki Y, Ishihara A, Mitsushima S, Kamiya N, Ota K-I (2007) Sulfated-zirconia as a support of Pt catalyst for polymer electrolyte fuel cells. Solid-State Lett 10:B105–B107. doi:10.1149/1.2730625
82. Subban C, Zhou Q, Leonard B, Ranjan C, Edvenson HM, Disalvo FJ et al (2010) Catalyst supports for polymer electrolyte fuel cells. Philos Trans A Math Phys Eng Sci 368:3243–3253. doi:10.1098/rsta.2010.0116
83. Lv H, Mu S (2014) Nano-ceramic support materials for low temperature fuel cell catalysts. Nanoscale 6:5063–5074. doi:10.1039/c4nr00402g
84. Ji S, Cho GY, Yu W, Su P-C, Lee MH, Cha SW (2015) Plasma-enhanced atomic layer deposition of nanoscale yttria-stabilized zirconia electrolyte for solid oxide fuel cells with porous substrate. ACS Appl Mater Interfaces 7:2998–3002. doi:10.1021/am508710s
85. Kim J-W (1999) Polarization effects in intermediate temperature, anode-supported solid oxide fuel cells. J Electrochem Soc 146:69. doi:10.1149/1.1391566
86. Sun C, Xie Z, Xia C, Li H, Chen L (2006) Investigations of mesoporous CeO_2-Ru as a reforming catalyst layer for solid oxide fuel cells. Electrochem Commun 8:833–838. doi:10.1016/j.elecom.2006.03.018
87. Tada M, Bal R, Mu X, Coquet R, Namba S, Iwasawa Y (2007) Low-temperature PROX (preferential oxidation) on novel CeO(2)-supported Cu-cluster catalysts under fuel-cell operating conditions. Chem Commun (Camb) 2:4689–4691. doi:10.1039/b709176a
88. Sun C, Stimming U (2007) Recent anode advances in solid oxide fuel cells. J Power Sources 171:247–260. doi:10.1016/j.jpowsour.2007.06.086
89. Xu C, Tian Z, Shen P, Jiang SP (2008) Oxide (CeO_2, NiO, Co_3O_4 and Mn_3O_4)-promoted Pd/C electrocatalysts for alcohol electrooxidation in alkaline media. Electrochim Acta 53:2610–2618. doi:10.1016/j.electacta.2007.10.036
90. Ye KH, Zhou SA, Zhu XC, Xu CW, Shen PK (2013) Stability analysis of oxide (CeO_2, NiO, Co_3O_4 and Mn_3O_4) effect on Pd/C for methanol oxidation in alkaline medium. Electrochim Acta 90:108–111. doi:10.1016/j.electacta.2012.12.012
91. Scibioh MA, Kim S-K, Cho EA, Lim T-H, Hong S-A, Ha HY (2008) Pt-CeO_2/C anode catalyst for direct methanol fuel cells. Appl Catal B Environ 84:773–782. doi:10.1016/j.apcatb.2008.06.017
92. Avgouropoulos G, Ioannides T, Papadopoulou C (2002) A comparative study of Pt/γ-Al_2O_3, Au/α-Fe_2O_3 and CuO-CeO_2 catalysts for the selective oxidation of carbon monoxide in excess hydrogen. Catal Today 75:157–167
93. Yu HB, Kim J-H, Lee H-I, Scibioh MA, Lee J, Han J et al (2005) Development of nanophase CeO_2-Pt/C cathode catalyst for direct methanol fuel cell. J Power Sources 140:59–65. doi:10.1016/j.jpowsour.2004.08.015
94. Guo JW, Zhao TS, Prabhuram J, Chen R, Wong CW (2006) Development of PtRu-CeO_2/C anode electrocatalyst for direct methanol fuel cells. J Power Sources 156:345–354. doi:10.1016/j.jpowsour.2005.05.093
95. Ou DR, Mori T, Togasaki H, Takahashi M, Ye F, Drennan J (2011) Microstructural and metal-support interactions of the Pt-CeO_2/C catalysts for direct methanol fuel cell application. Langmuir 27:3859–3866. doi:10.1021/la1032898

96. Liu Y, Fu Q, Stephanopoulos MF (2004) Preferential oxidation of CO in H_2 over CuO-CeO_2 catalysts. Catal Today 93–95:241–246. doi:10.1016/j.cattod.2004.06.049
97. Lei M, Wang ZB, Li JS, Tang HL, Liu WJ, Wang YG (2014) CeO2 nanocubes-graphene oxide as durable and highly active catalyst support for proton exchange membrane fuel cell. Sci Rep 4:7415. doi:10.1038/srep07415
98. Ioroi T, Akita T, Yamazaki SI, Siroma Z, Fujiwara N, Yasuda K (2006) Comparative study of carbon-supported Pt/Mo-oxide and PtRu for use as CO-tolerant anode catalysts. Electrochim Acta 52:491–498. doi:10.1016/j.electacta.2006.05.030
99. Vellacheri R, Unni SM, Nahire S, Kharul UK, Kurungot S (2010) Pt-MoOx-carbon nanotube redox couple based electrocatalyst as a potential partner with polybenzimidazole membrane for high temperature polymer electrolyte membrane fuel cell applications. Electrochim Acta 55:2878–2887. doi:10.1016/j.electacta.2010.01.012
100. Ugalde-Reyes O, Hernandez-Maya R, Ocampo-Flores AL, Ramirez FA, Sosa-Hernandez E, Angeles-Chavez C et al (2014) Study of the electrochemical activities of Mo-modified Pt catalysts, for application as anodes in direct methanol fuel cells: effect of the aggregation route. J Electrochem Soc 162:H132–H141. doi:10.1149/2.0521503jes
101. Yan Z, Xie J, Jing J, Zhang M, Wei W, Yin S (2012) MoO_2 nanocrystals down to 5 nm as Pt electrocatalyst promoter for stable oxygen reduction reaction. Int J Hydrogen Energy 37:15948–15955. doi:10.1016/j.ijhydene.2012.08.033
102. Zhang Z, Liu J, Gu J, Cheng L (2014) An overview of metal oxide materials as electrocatalysts and supports for polymer electrolyte fuel cells. Energy Environ Sci 7:2535–2558. doi:10.1039/c3ee43886d
103. Ma L, Zhao X, Si F, Liu C, Liao J, Liang L et al (2010) A comparative study of Pt/C and Pt-MoOx/C catalysts with various compositions for methanol electro-oxidation. Electrochim Acta 55:9105–9112. doi:10.1016/j.electacta.2010.08.034
104. Antolini E, Gonzalez ER (2009) Ceramic materials as supports for low-temperature fuel cell catalysts. Solid State Ion 180:746–763. doi:10.1016/j.ssi.2009.03.007
105. Parrondo J, Han T, Niangar E, Wang C, Dale N, Adjemian K et al (2014) Platinum supported on titanium-ruthenium oxide is a remarkably stable electrocatayst for hydrogen fuel cell vehicles. Proc Natl Acad Sci U S A 111:45–50. doi:10.1073/pnas.1319663111
106. Lo CP, Wang G, Kumar A, Ramani V (2013) TiO_2-RuO_2 electrocatalyst supports exhibit exceptional electrochemical stability. Appl Catal B Environ 140–141:133–140. doi:10.1016/j.apcatb.2013.03.039
107. Shinde VM, Madras G (2013) Synthesis of nanosized Ce0.85 M0.1Ru0.05O2-delta (M = Si, Fe) solid solution exhibiting high CO oxidation and water gas shift activity. Appl Catal B Environ 138–139:51–61. doi:10.1016/j.apcatb.2013.02.021
108. Kumar A, Ramani VK (2013) RuO_2-SiO_2 mixed oxides as corrosion-resistant catalyst supports for polymer electrolyte fuel cells. Appl Catal B Environ 138–139:43–50. doi:10.1016/j.apcatb.2013.02.015
109. Thanh Ho VT, Pillai KC, Chou H-L, Pan C-J, Rick J, Su W-N, et al (2011) Robust non-carbon Ti0.7Ru0.3O2 support with co-catalytic functionality for Pt: enhances catalytic activity and durability for fuel cells. Energy Environ Sci 4:4194. doi:10.1039/c1ee01522b
110. Ho VTT, Nguyen NG, Pan CJ, Cheng JH, Rick J, Su WN et al (2012) Advanced nanoelectrocatalyst for methanol oxidation and oxygen reduction reaction, fabricated as one-dimensional pt nanowires on nanostructured robust Ti0.7Ru0.3O2 support. Nano Energy 1:687–695. doi:10.1016/j.nanoen.2012.07.007
111. Drew K, Girishkumar G, Vinodgopal K, Kamat P (2005) Boosting fuel cell performance wit a semiconductor photocatalyst: TiO_2/Pt-Ru hybrid catalyst for methanol oxidation. J Phys Chem B 109:11851–11857
112. Gojković SL, Babić BM, Radmilović VR, Krstajić NV (2010) Nb-doped TiO_2 as a support of Pt and Pt-Ru anode catalyst for PEMFCs. J Electroanal Chem 639:161–166. doi:10.1016/j.jelechem.2009.12.004

113. Naeem R, Ahmed R, Ansari MS (2014) TiO_2 and Al_2O_3 promoted Pt/C nanocomposites as low temperature fuel cell catalysts for electro oxidation of methanol in acidic media. IOP Conf Ser Mater Sci Eng 60:012031. doi:10.1088/1757-899X/60/1/012031
114. Park KW, Han SB, Lee JM (2007) Photo(UV)-enhanced performance of Pt-TiO_2 nanostructure electrode for methanol oxidation. Electrochem Commun 9:1578–1581. doi:10.1016/j.elecom.2007.02.020
115. Park K-W, Seol K-S (2007) Nb-TiO_2 supported Pt cathode catalyst for polymer electrolyte membrane fuel cells. Electrochem Commun 9:2256–2260. doi:10.1016/j.elecom.2007.06.027
116. Sun S, Zhang G, Sun X, Cai M, Ruthkosky M (2012) Highly stable and active Pt/Nb-TiO_2 carbon-free electrocatalyst for proton exchange membrane fuel cells. J Nanotechnol 2012:13–15. doi:10.1155/2012/389505
117. Zhao G, Zhao TS, Yan XH, Zeng L (2015) A high catalyst-utilization electrode for direct methanol fuel cells. Electrochim Acta 164:337–343. doi:10.1016/j.electacta.2015.02.181
118. Zhao L, Wang Z-B, Liu J, Zhang J-J, Sui X-L, Zhang L-M et al (2015) Facile one-pot synthesis of Pt/graphene-TiO_2 hybrid catalyst with enhanced methanol electrooxidation performance. J Power Sources 279:210–217. doi:10.1016/j.jpowsour.2015.01.023
119. Zhang Z, Zhang Y, He L, Yang Y, Liu S, Wang M et al (2015) A feasible synthesis of $Mn_3(PO_4)_2$@BSA nanoflowers and its application as the support nanomaterial for Pt catalyst. J Power Sources 284:170–177. doi:10.1016/j.jpowsour.2015.03.011
120. Deleebeeck L, Ippolito D, Hansen KK (2015) Catalytic enhancement of carbon black and coal-fueled hybrid direct carbon fuel cells. J Electrochem Soc 162:F327–F339. doi:10.1149/2.0761503jes
121. Dong X, Takahashi M, Nagao M, Hibino T (2011) Compact bipolar plate-free direct methanol fuel cell stacks. Chem Commun (Camb) 47:5292–5294. doi:10.1039/c1cc10493d
122. Dong H-Q, Chen Y-Y, Han M, Li S-L, Zhang J, Li J-S et al (2014) Synergistic effect of mesoporous Mn_2O_3-supported Pd nanoparticle catalysts for electrocatalytic oxygen reduction reaction with enhanced performance in alkaline medium. J Mater Chem A 2:1272. doi:10.1039/c3ta13585c
123. Menezes PW, Indra A, Sahraie NR, Bergmann A, Strasser P, Driess M (2015) Cobalt-manganese-based spinels as multifunctional materials that unify catalytic water oxidation and oxygen reduction reactions. ChemSusChem 8:164–171. doi:10.1002/cssc.201402699
124. Selcuk S, Selloni A (2015) DFT + U study of the surface structure and stability of Co_3O_4 (110): dependence on U. J Phys Chem C 4:150427132340009. doi:10.1021/acs.jpcc.5b02298
125. Zou X, Goswami A, Asefa T (2013) Efficient noble metal-free (electro)catalysis of water and alcohol oxidations by zinc-cobalt layered double hydroxide. J Am Chem Soc 135:17242–17245. doi:10.1021/ja407174u
126. Yeo BS, Bell AT (2011) Enhanced activity of gold-supported cobalt oxide for the electrochemical evolution of oxygen. J Am Chem Soc 133:5587–5593. doi:10.1021/ja200559j
127. Frydendal R, Busch M, Halck NB, Paoli EA, Krtil P, Chorkendorff I et al (2015) Enhancing activity for the oxygen evolution reaction: the beneficial interaction of gold with manganese and cobalt oxides. ChemCatChem 7:149–154. doi:10.1002/cctc.201402756
128. Walton AS, Fester J, Bajdich M, Arman MA, Osiecki J, Knudsen J et al (2015) Interface controlled oxidation states in layered cobalt oxide nanoislands on gold. ACS Nano 9:2445–2453
129. Jiao F, Frei H (2010) Nanostructured cobalt and manganese oxide clusters as efficient water oxidation catalysts. Energy Environ Sci 3:1018. doi:10.1039/c002074e
130. Wang HF, Kavanagh R, Guo YL, Guo Y, Lu G, Hu P (2012) Origin of extraordinarily high catalytic activity of Co_3O_4 and its morphological chemistry for CO oxidation at low temperature. J Catal 296:110–119. doi:10.1016/j.jcat.2012.09.005
131. Larmier K, Chizallet C, Raybaud P (2015) Tuning the metal-support interaction by structural recognition of cobalt-based catalyst precursors. Angew Chem Int Ed 54(23):6824–6827. doi:10.1002/anie.201502069

132. Jiao F, Frei H (2009) Nanostructured cobalt oxide clusters in mesoporous silica as efficient oxygen-evolving catalysts. Angew Chem Int Ed 48:1841–1844. doi:10.1002/anie.200805534
133. Melvin AA, Joshi VS, Poudyal DC, Khushalani D, Haram SK (2015) Electrocatalyst on insulating support?: hollow silica spheres loaded with Pt nanoparticles for methanol oxidation. ACS Appl Mater Interfaces 7:6590–6595. doi:10.1021/am508830h
134. Duval Y, Mielczarski JA, Pokrovsky OS, Mielczarski E, Ehrhardt JJ (2002) Evidence of the existence of three types of species at the quartz-aqueous solution interface at pH0-10: XPS surface group quantification and surface complexation modeling. J Phys Chem B 106:2937–2945. doi:10.1021/jp012818s
135. Seger B, Kongkanand A, Vinodgopal K, Kamat PV (2008) Platinum dispersed on silica nanoparticle as electrocatalyst for PEM fuel cell. J Electroanal Chem 621:198–204. doi:10.1016/j.jelechem.2007.09.037
136. Liu B, Chen JH, Zhong XX, Cui KZ, Zhou HH, Kuang YF (2007) Preparation and electrocatalytic properties of Pt-SiO2 nanocatalysts for ethanol electrooxidation. J Colloid Interface Sci 307:139–144. doi:10.1016/j.jcis.2006.11.027
137. Lin C, Hsu S, Ho W (2015) Using SiO_2 nanopowders in anode catalyst layer to improve the performance of a proton exchange membrane fuel cell at low humidity. J Mater Sci Chem Eng 3:72–79
138. Vu THT, Tran TTT, Le HNT, Tran LT, Nguyen PHT, Nguyen HT et al (2015) Solvothermal synthesis of Pt -SiO2/graphene nanocomposites as efficient electrocatalyst for methanol oxidation. Electrochim Acta 161:335–342. doi:10.1016/j.electacta.2015.02.100
139. Uzunoglu A, Dundar F, Ata A (2015) Modification of vulcan XC-72 for enhanced durability of PEMFC catalyst layer. Int JRenew Energy Res 5:111–120
140. Pinchuk OA, Dundar F, Ata A, Wynne KJ (2012) Improved thermal stability, properties, and electrocatalytic activity of sol-gel silica modified carbon supported Pt catalysts. Int J Hydrogen Energy 37:2111–2120. doi:10.1016/j.ijhydene.2011.10.093
141. Pinchuk OA, Dundar F, Ata A, Wynne KJ (2012) Improved thermal stability, properties, and electrocatalytic activity of sol-gel silica modified carbon supported Pt catalysts. Int J Hydrogen Energy 37:2111–2120. doi:10.1016/j.ijhydene.2011.10.093
142. Inaba M, Suzuki T, Hatanaka T, Morimoto Y (2015) Fabrication and cell analysis of a Pt/SiO_2 platinum thin film electrode. J Electrochem Soc 162:F634–F638. doi:10.1149/2.0201507jes
143. Dundar F, Uzunoglu A, Ata A, Wynne KJ (2012) Durability of carbon-silica supported catalysts for proton exchange membrane fuel cells. J Power Sources 202:184–189. doi:10.1016/j.jpowsour.2011.12.010
144. Su C, Hsueh Y, Kei C, Lin C, Perng T (2013) Fabrication of high-activity hybrid Pt@ZnO catalyst on carbon cloth by atomic layer deposition for photoassisted electro-oxidation of methanol. J Phys Chem C
145. Yun M, Ahmed MS, Jeon S (2015) Thiolated graphene oxide-supported palladium cobalt alloyed nanoparticles as high performance electrocatalyst for oxygen reduction reaction. J Power Sources 293:380–387. doi:10.1016/j.jpowsour.2015.05.094
146. Ilinich O, Liu Y, Castellano C, Koermer G, Moini A, Farrauto R (2008) A new palladium-based catalyst for methanol steam reforming in a miniature fuel cell power source. Platin Met Rev 52:134–143. doi:10.1595/147106708X324403
147. Chen LY, Chang BK, Lu Y, Yang WG, Tatarchuk BJ (2002) Selective catalytic oxidation of CO in H_2: fuel cell applications. Abstr Pap Am Chem Soc 224:249–254. doi:10.1016/S0920-5861(00)00426-0
148. Korotkikh O, Farrauto R (2000) Selective catalytic oxidation of CO in H-2: fuel cell applications. Catal Today 62:249–254
149. Lo C-P, Wang G, Kumar A, Ramani V (2011) RuO_2/xH_2O-TiO_2 as catalyst support for PEM fuel cells. ECS Trans 41:1249–1255. doi:10.1149/1.3635656
150. Horiguchi D, Tsukatsune T, Noda Z, Hayashi A, Sasaki K (2014) Pt/SnO_2 electrocatalysts on conductive fillers. ECS Trans 64:215–220. doi:10.1149/06403.0215ecst

151. Kinumoto T, Kitayama S, Matsuoka M, Tsumura T, Toyoda M (2014) Correlation between preparation condition and performance of Pt/SnO_2/KB for cathode catalyst of PEMFC. ECS Trans 64:199–205
152. Nagamatsu Y, Kanda K, Noda Z, Hayashi A, Sasaki K (2014) Electrochemical performance of MEAs with Pt/SnO_2 mixed with conductive fillers. ECS Trans 64:207–213
153. Xu C, Shen PK, Ji X, Zeng R, Liu Y (2005) Enhanced activity for ethanol electrooxidation on Pt-MgO/C catalysts. Electrochem Commun 7:1305–1308. doi:10.1016/j.elecom.2005.09.015
154. Li N, Zeng Y, Chen S, Xu C, Shen P (2014) Ethanol oxidation on Pd/C enhanced by MgO in alkaline medium. Int J Hydrogen Energy 39:1–5. doi:10.1016/j.ijhydene.2013.12.122
155. Mahendiran C, Maiyalagan T, Scott K, Gedanken A (2011) Synthesis of a carbon-coated NiO/MgO core/shell nanocomposite as a Pd electro-catalyst support for ethanol oxidation. Mater Chem Phys 128:341–347. doi:10.1016/j.matchemphys.2011.02.067
156. Chu D, Wang J, Wang S, Zha L, He J, Hou Y et al (2009) High activity of Pd-In_2O_3/CNTs electrocatalyst for electro-oxidation of ethanol. Catal Commun 10:955–958. doi:10.1016/j.catcom.2008.12.041
157. Liang R, Hu A, Persic J, Zhou YN (2013) Palladium nanoparticles loaded on carbon modified TiO_2 nanobelts for enhanced methanol electrooxidation. Nano Micro Lett 5:202–212. doi:10.5101/nml.v5i3.p202-212
158. Liu B, Chen JH, Xiao CH, Cui KZ, Yang L, Pang HL et al (2007) Preparation of Pt/MgO/CNT hybrid catalysts and their electrocatalytic properties for ethanol electrooxidation. Energy Fuels 21:1365–1369. doi:10.1021/ef060452i
159. Han X, Zhang T, Du J, Cheng F, Chen J (2013) Porous calcium–manganese oxide microspheres for electrocatalytic oxygen reduction with high activity. Chem Sci 4:368–376. doi:10.1039/c2sc21475j
160. Stottlemyer AL, Kelly TG, Meng Q, Chen JG (2012) Reactions of oxygen-containing molecules on transition metal carbides: surface science insight into potential applications in catalysis and electrocatalysis. Surf Sci Rep 67:201–232. doi:10.1016/j.surfrep.2012.07.001
161. Chen JG (1996) Carbide and nitride overlayers on early transition metal surfaces: preparation, characterization, and reactivities. Chem Rev 96:1477–1498. doi:10.1021/cr950232u
162. Oyama ST (1992) Preparation and catalytic properties of transition metal carbides and nitrides. Catal Today 15:179–200. doi:10.1016/0920-5861(92)80175-M
163. Weidman MC, Esposito DV, Hsu Y-C, Chen JG (2012) Comparison of electrochemical stability of transition metal carbides (WC, W2C, Mo2C) over a wide pH range. J Power Sources 202:11–17. doi:10.1016/j.jpowsour.2011.10.093
164. Yan Z, He G, Cai M, Meng H, Shen PK (2013) Formation of tungsten carbide nanoparticles on graphitized carbon to facilitate the oxygen reduction reaction. J Power Sources 242:817–823. doi:10.1016/j.jpowsour.2013.05.161
165. Wang YJ, Wilkinson DP, Zhang JJ (2011) Noncarbon support materials for polymer electrolyte membrane fuel cell electrocatalysts. Chem Rev 111:7625–7651
166. Yan Z, Gao L, Zhang M, Xie J, Chen M (2015) Angstrom-scale vanadium carbide rods as Pt electrocatalyst support for efficient methanol oxidation reaction. RSC Adv 5:9561–9564. doi:10.1039/C4RA11798K
167. Regmi YN, Waetzig GR, Duffee KD, Schmuecker SM, Thode JM, Leonard BM (2015) Carbides of group IVA, VA and VIA transition metals as alternative HER and ORR catalysts and support materials. J Mater Chem A 3:10085–10091. doi:10.1039/C5TA01296A
168. Liu X, Fechler N, Antonietti M (2013) Salt melt synthesis of ceramics, semiconductors and carbon nanostructures. Chem Soc Rev 42:8237–8265. doi:10.1039/c3cs60159e
169. Bennett LH, Cuthill JR, McAlister AJ, Erickson NE, Watson RE (1974) Electronic structure and catalytic behavior of tungsten carbide. Science 181:563–565. doi:10.1126/science.184.4136.563
170. Levy RB, Boudart M (1973) Platinum-like behavior of tungsten carbide in surface catalysis. Science 181(1973):547–549. doi:10.1126/science.181.4099.547

171. Kimmel YC, Esposito DV, Birkmire RW, Chen JG (2012) Effect of surface carbon on the hydrogen evolution reactivity of tungsten carbide (WC) and Pt-modified WC electrocatalysts. Int J Hydrogen Energy 37:3019–3024. doi:10.1016/j.ijhydene.2011.11.079
172. Gong XB, You SJ, Wang XH, Gan Y, Zhang RN, Ren NQ (2013) Silver-tungsten carbide nanohybrid for efficient electrocatalysis of oxygen reduction reaction in microbial fuel cell. J Power Sources 225:330–337. doi:10.1016/j.jpowsour.2012.10.047
173. Houston JE, Laramore GE, Park RL (1974) Surface electronic properties of tungsten, tungsten carbide, and platinum. Science 185:258–260
174. Chen W-F, Muckerman JT, Fujita E (2013) Recent developments in transition metal carbides and nitrides as hydrogen evolution electrocatalysts. Chem Commun (Camb) 49:8896–8909. doi:10.1039/c3cc44076a
175. Ham DJ, Lee JS (2009) Transition metal carbides and nitrides as electrode materials for low temperature fuel cells. Energies 2:873–899. doi:10.3390/en20400873
176. Esposito DV, Chen JG (2011) Monolayer platinum supported on tungsten carbides as low-cost electrocatalysts: opportunities and limitations. Energy Environ Sci 4:3900. doi:10.1039/c1ee01851e
177. Hsu IJ, Kimmel YC, Dai Y, Chen S, Chen JG (2012) Rotating disk electrode measurements of activity and stability of monolayer Pt on tungsten carbide disks for oxygen reduction reaction. J Power Sources 199:46–52. doi:10.1016/j.jpowsour.2011.10.024
178. Lu JL, Li ZH, Jiang SP, Shen PK, Li L (2012) Nanostructured tungsten carbide/carbon composites synthesized by a microwave heating method as supports of platinum catalysts for methanol oxidation. J Power Sources 202:56–62. doi:10.1016/j.jpowsour.2011.11.018
179. Yan Z, Cai M, Shen PK (2013) Nanosized tungsten carbide synthesized by a novel route at low temperature for high performance electrocatalysis. Sci Rep 3:1646. doi:10.1038/srep01646
180. Yang M, Cui Z, DiSalvo FJ (2013) Mesoporous chromium nitride as a high performance non-carbon support for the oxygen reduction reaction. Phys Chem Chem Phys 15:7041–7044. doi:10.1039/c3cp51109j
181. Cui Z, Burns RG, DiSalvo FJ (2013) Mesoporous Ti0.5Nb0.5N ternary nitride as a novel non carbon support for oxygen reduction reaction in acid and alkaline electrolytes. Chem Mater 25:3782–3784. doi:10.1021/cm4027545
182. Yang M, Guarecuco R, Disalvo FJ (2013) Mesoporous chromium nitride as high performance catalyst support for methanol electrooxidation. Chem Mater 25:1783–1787. doi:10.1021/cm400304q
183. Yang M, Cui Z, DiSalvo FJ (2013) Mesoporous titanium nitride supported Pt nanoparticles as high performance catalysts for methanol electrooxidation. Phys Chem Chem Phys 15:1088–1092. doi:10.1039/C2CP44215A
184. Cui Z, Yang M, Di Salvo FJ (2013) Mo2N/C hybrid material as a promising support for the electro-oxidation of methanol and formic acid. Electrochem Commun 33:63–67. doi:10.1016/j.elecom.2013.04.017
185. Thotiyl MMO, Sampath S (2011) Electrochemical oxidation of ethanol in acid media on titanium nitride supported fuel cell catalysts. Electrochim Acta 56:3549–3554. doi:10.1016/j.electacta.2010.12.091
186. Thotiyl MMO, Kumar TR, Sampath S (2010) Pd supported on titanium nitride for efficient ethanol oxidation. J Phys Chem C 114:17934–17941. doi:10.1021/jp1038514
187. Cesiulis H, Ziomek-Moroz M (2000) Electrocrystallization and electrodeposition of silver on titanium nitride. J Appl Electrochem 30:1261–1268. doi:10.1023/A:1026553712521
188. Evans SAG, Terry JG, Plank NOV, Walton AJ, Keane LM, Campbell CJ et al (2005) Electrodeposition of platinum metal on TiN thin films. Electrochem Commun 7:125–129. doi:10.1016/j.elecom.2004.11.014
189. Musthafa OTM, Sampath S (2008) High performance platinized titanium nitride catalyst for methanol oxidation. Chem Commun (Camb) 7:67–69. doi:10.1039/b715859a

190. Kakinuma K, Wakasugi Y, Uchida M, Kamino T, Uchida H, Deki S et al (2012) Preparation of titanium nitride-supported platinum catalysts with well controlled morphology and their properties relevant to polymer electrolyte fuel cells. Electrochim Acta 77:279–284. doi:10.1016/j.electacta.2012.06.001
191. Avasarala B, Murray T, Li W, Haldar P (2009) Titanium nitride nanoparticles based electrocatalysts for proton exchange membrane fuel cells. J Mater Chem 19:1803. doi:10.1039/b819006b
192. Higgins DC, Choi J-Y, Wu J, Lopez A, Chen Z (2012) Titanium nitride–carbon nanotube core–shell composites as effective electrocatalyst supports for low temperature fuel cells. J Mater Chem 22:3727. doi:10.1039/c2jm15014j
193. Jia Y, Wang Y, Dong L, Huang J, Zhang Y, Su J et al (2015) A hybrid of titanium nitride and nitrogen-doped amorphous carbon supported on SiC as a noble metal-free electrocatalyst for oxygen reduction reaction. Chem Commun 51:2625–2628. doi:10.1039/C4CC08007F
194. Pan Z, Xiao Y, Fu Z, Zhan G, Wu S, Xiao C et al (2014) Hollow and porous titanium nitride nanotubes as high performance catalyst support for oxygen reduction reaction. J Mater Chem A 2:13966–13975. doi:10.1039/c4ta02402h
195. Chen J, Takanabe K, Ohnishi R, Lu D, Okada S, Hatasawa H et al (2010) Nano-sized TiN on carbon black as an efficient electrocatalyst for the oxygen reduction reaction prepared using an mpg-C3N4 template. Chem Commun (Camb) 46:7492–7494. doi:10.1039/c0cc02048f
196. Alhajri NS, Yoshida H, Anjum DH, Garcia-Esparza AT, Kubota J, Domen K et al (2013) Synthesis of tantalum carbide and nitride nanoparticles using a reactive mesoporous template for electrochemical hydrogen evolution. J Mater Chem A 1:12606. doi:10.1039/c3ta12984e
197. Dong Y, Wu Y, Liu M, Li J (2013) Electrocatalysis on shape-controlled titanium nitride nanocrystals for the oxygen reduction reaction. ChemSusChem 6:2016–2021
198. Xiao Y, Zhan G, Fu Z, Pan Z, Xiao C, Wu S et al (2015) Titanium cobalt nitride supported platinum catalyst with high activity and stability for oxygen reduction reaction. J Power Sources 284:296–304. doi:10.1016/j.jpowsour.2015.03.001
199. Xiao Y, Fu Z, Zhan G, Pan Z, Xiao C, Wu S et al (2015) Increasing Pt methanol oxidation reaction activity and durability with a titanium molybdenum nitride catalyst support. J Power Sources 273:33–40. doi:10.1016/j.jpowsour.2014.09.057
200. Xiao Y, Zhan G, Fu Z, Pan Z, Xiao C, Wu S et al (2014) Robust non-carbon titanium nitride nanotubes supported Pt catalyst with enhanced catalytic activity and durability for methanol oxidation reaction. Electrochim Acta 141:279–285. doi:10.1016/j.electacta.2014.07.070
201. Mkwizu TS, Mathe MK, Cukrowski I (2010) Electrodeposition of multilayered bimetallic nanoclusters of ruthenium and platinum via surface-limited redox-replacement reactions for electrocatalytic applications. Langmuir 26:570–580. doi:10.1021/la902219t
202. Modibedi RM, Louw EK, Mathe MK, Ozoemena KI (2013) The electrochemical atomic layer deposition of Pt and Pd nanoparticles on Ni foam for the electro oxidation of alcohols. ECS Trans 50:9–18

Chapter 14
Applications of Nanomaterials in Microbial Fuel Cells

R. Fogel and J. L. Limson

14.1 Principles of Microbial Fuel Cells

Microbial fuel cells are biotechnological devices that transform the chemical energy present in biodegradable organic compounds into electrical energy, primarily through the use of microorganisms as catalysts for oxidation/reduction reactions [1, 2]. As is the convention in electrochemical cells, a fuel cell can be broadly categorised into possessing an anodic compartment i.e. the site of the electron source reactions and a cathodic compartment, where reducing reactions accept the electron. The most commonly-reported configurations of MFCs make use of biologically modified anodes (bioanodes), but research into the use of biocathodes are increasingly common [1, 2]. Anodes and cathodes can be either physical separation of the electrode types, or conceptual divisions of the reactions that occur in MFCs. A physical separation is considered to be useful in that it provides separate environmental conditions required for efficient anodic and cathodic reactions to occur. Typically, the anode and the cathode are separated from one another by a cation-exchange membrane, or CEM, in order to facilitate the transport of biogenerated protons from the anode to the cathode (Fig. 14.1).

The primary purpose of microbial fuel cells has traditionally been perceived to be the recovery of energy, either through direct electrical energy generation, or to energistically support the production of energy-rich compounds e.g. H_2 from waste [2, 3]. Due to the inclusion of metabolising bacteria within this technology, they bear the additional promise of combining bioremediation processes with power generation [2, 3]. Increasing emphasis has been placed on the utilisation of different wastewaters and the potential that MFCs hold for bioremediation of specific wastes, either as a standalone technology or coupled to other wastewater treatment

R. Fogel • J.L. Limson (✉)
Biotechnology Innovation Centre, Rhodes University, Grahamstown 6140, South Africa
e-mail: r.fogel@ru.ac.za; J.Limson@ru.ac.za

K.I. Ozoemena, S. Chen (eds.), *Nanomaterials for Fuel Cell Catalysis*,
Nanostructure Science and Technology, DOI 10.1007/978-3-319-29930-3_14

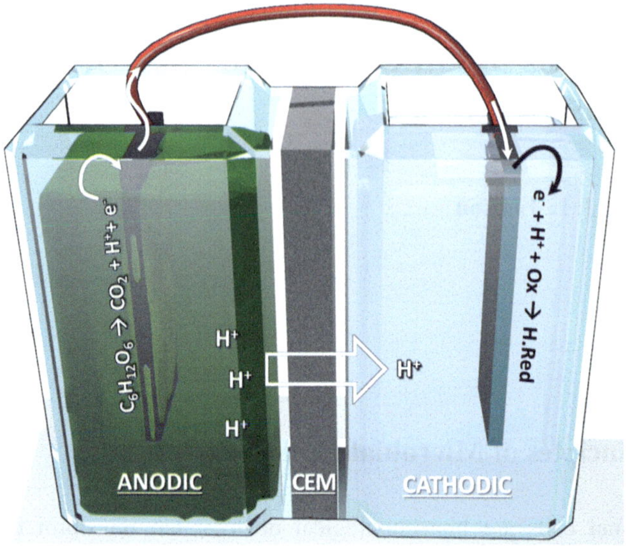

Fig. 14.1 Schematic diagram of general components of a conventional bioanodic microbial fuel cells fuelled with oxygen, one of the most common examples of microbial fuel cell configurations available in current literature. Electrons generated through catabolic microbial processes enter the anode and travel via the external circuit to the cathode. Protons generated through microbial processes diffuse through the Cation Exchange Membrane (CEM) to the cathode to react in reduction reactions

processes such as anaerobic digesters. As such, a very large list of dissolved organic compounds, ranging from complex lignocellulose fractions [4], abundant and nutrient-rich domestic [2] and industrial [4] wastewaters, toxic phenols [5], as well as inorganic nitrates, sulphates, and heavy metals [3] have been studied as to their ability to support waste bioremediation technologies. Given this broad scope, research into the application of nanomaterials to address enduring concerns in MFC design and architecture takes this duality of purpose into consideration, with several such studies conducted using specific wastewaters in research aimed at real applications.

Considering the core purpose of microbial fuel cells, research is primarily addressed at improving both power output and to a lesser extent, enhancing bioremediation. Here issues of the nature of the bacteria, viability and lifespan, nature of the electrode materials with respect to cost, stability and biocompatibility are frequently considered along with foci aimed at MFC architecture around issues of scalability in particular.

Currently, as a direct means of generating electrical power, microbial fuel cells are too uneconomic to manufacture or operate at the scale required to process the volumes of wastewater to make this technology viable. Several key aspects limiting this technology have been identified across several different publications and can be summarised as: the reliance on expensive catalysts/prohibitive amounts of catalysts, especially platinum [6, 7]; ineffective use of catalysts e.g. poor dispersal [7]; the reliance on expensive reagents/materials for optimum function, especially the addition of electron transfer mediators and electrode materials [6]; toxicity of electron transfer mediators [8]; low power generation [8]; poor durability of electrode materials [6, 7]; poor scaleability; poor mass transport across the system and within compartments [7] and the requirement for external power input for optimum function e.g. agitation/heat.

Electrode materials for any low-cost industrial process, such as MFCs, should be mainly comprised of low-cost materials that remain chemically stable during operating lifetimes, have a high surface area and be easily scaled-up [2]. Due to the biological nature of microbial fuel cells, electrodes interacting with the biocatalysts should, ideally, promote cellular adhesion (or, at least, not be inimical to bacteria in the anode) and possess a limited tendency towards chemo- and bio-fouling, to improve their long-term operational times. For these outlined reasons, carbonaceous electrodes are often included as electrode materials.

Given the cost-effectiveness and ease-of-production, combined with their chemical stability, good biocompatibility and good conductivity of many of the carbon-based materials, there is an obvious research interest in incorporating them as macroscale electrode materials for fuel cells. However, compared to their metal counterparts, the carbonaceous electrodes suffer from relatively poor conductivity and high electrochemical overpotentials.

Considering the above, the application of nanomaterials and research within this area thus encompasses several aspects, most notably, electrode modification and the potential thereof for impacting (1) anodic electron transferpower processes, (2) mass transport within the system and (3) cathodic electron transfer processes. As such, in this chapter, we overview how nanomaterials may have a positive impact on key limitations in terms of electrode materials and surface area, as catalysts in the oxygen reduction reaction, and as mediators for enhancing bacterial electron transfer efficiency.

14.2 Nanostructured Electrode Materials in Microbial Fuel Cells

Table 14.1 provides a brief list of samples of the available microbial fuel cells reported in recent literature, emphasising those where nanomaterial modifications were investigated. Where available, power density values prior to and after nanomaterial inclusion at the anode or cathode (as indicated) of the MFC configurations are listed, with associated percentage increases noted.

Given the diversity of electrode materials used for MFC fabrication, microbial catalysts selected and organic substrates investigated as well as nanomaterials utilised in modifying one or both of the electrodes, it is difficult to create any but the most general comparisons from data presented in Table 14.1. Indeed, this lack of standardisation was identified previously [23], but is still limiting the ability of

Table 14.1 outputs of various reported MFCs present in the current literature (2014 and previous)

Anolyte (bacteria) / Catholyte	Anode modifiers	Cathode modifiers	P.D. (mW m^{-2})[a]	% increase[b]	Ref.
Anaerobic sludge consortium	Carbon cloth	Carbon cloth	468	189	[9]
50 mM $Fe(CN)_6^{3-/4-}$ in PBS	GO/PANI NFs	Carbon cloth	1390		
Palm oil mill effluent(consort.)	Carbon paper	Carbon paper	38		[10]
	–	Copper Pc	118	210	
	–	Pt	121	218	
PB with NH_4Cl	–	Ni NPs	94	147	
Unreported	Carbon	Carbon	32	272	[11]
Air cathode	–	MnO_2	119		
Urban wastewater (consort.) PB	Graphite	$Pd/TiO_2/Ti$	200		[12]
Domestic wastewater (consort.)	Carbon	Carbon	306	18	[13]
PB ± $Fe(CN)_6^{3-/4-}$	Au NPs	–	363		
(*Escherichia coli* K12)	Polymer/Pt	FePc NPs + carbon	2957		[14]
PB, pH 3.5					
Synthetic nutrient medium (consort.)	Carbon paper	Pt carbon paper Carageenan	80	−15	[15]
PBS with NH_4Cl	–	NPs + polypyrrole	72		
(*Clostridium butyricum*)	PTFE + Platinised Pt.	Graphitised carbon	260,000		[16]
$Fe(CN)_6^{3-/4-}$ in PB, pH 7.0					
1 g/L glucose PBS (*E. coli* K12)	Toray carbon paper	Pt. carbon	468	32	[17]
	–	Co/Fe/N/graphite		161	
Air cathode	–	Co/Fe/N/CNTs	751		
2 g/L glucose medium (*E. coli*)	Carbon felt	Pt/carbon paper	~80		[18]
	MoO_2C	–	~170	113	
	CNTs	–	~180	125	
Air cathode	MoO_2C/CNTs	–	~1050	1215	

Synthetic medium (*Enterobacter cloacae*)	Carbon paper	Carbon paper	4		[19]
	–	Carbon Black/FePc	17	323	
PB, pH 7	Carbon Black	Carbon Black/FePc	43	975	
Synthetic medium (*E. cloacae*) PB	Carbon paper	Carbon paper	2.3		[20]
	CNTs/FePc	CNTs/FePc	44	1913	
Synthetic medium (sludge consortium)	Carbon cloth	Carbon cloth	430	330	[21]
Air cathode	–	MnO_2 NT/GO	1850		
Glucose (*E. coli*)	Glassy carbon	Pt	457		[22]
	MWCNT	–	699	53	
PB ± $Fe(CN_6)^{3-/4-}$	MWCNT/SnO_2	–	1421	211	

Underlined text in column 1 indicates the catholyte

Abbreviations: *AuNPs* gold nanoparticles, *CNTs* carbon nanotubes, *consort* consortium, *FePc* Iron(II) Phthalocyanine, *GO* graphene oxide, *NFs* nanofibers, *NPs* nanoparticles, *NT* nanotubes, *PB/S* phosphate-buffered/saline, *PANI* polyaniline

[a]Maximum power density (P.D.) reported during the study

[b]Percentage increase of maximum power density, as compared to electrode configuration lacking nanomaterial modification reported in that study

the MFC research community to unify findings and provide an underlying sense of progress. As demonstrated in Table 14.1, microbial fuel cells are generally characterised by having quite low reported power densities than other contemporary electrical energy generating devices. Reported peak power densities obtained at microbial fuel cells using conventional electrodes typically range from the mW m^{-2}, and rarely, to the low W m^{-2} of reported electrode surface area (Table 14.1) and is an often-cited limitation to the broader application of MFC [24]. This range of power generation is cited variously as 2–3 orders of magnitude lower [25] than other energy-generating technologies, including their next counterparts—chemical fuel cells. Compared to reported power density of catalysed direct methanol fuel cells operating under STP (~14,000 mW m^{-2} in [26]), the power generating ability of microbial fuel cells are currently very low. The proposed target of recoverable energy of ~1 kWhr kg^{-1} of organic pollutant removed and generation capacity of 1 kW m^{-3} [3] can be achieved with very few of these systems.

This is as a result of multiple intersecting factors that lower the overall obtainable power densities (Fig. 14.1). Major discrepancies are notable in the literature, mainly attributed to the selection of the bacteria and overall differences to the configuration and design of the fuel cells. This uniformly low power response however remains the single greatest challenge to the field. Arguably, this is the area of MFC research to which nanotechnology can yield the greatest results and has already demonstrated great promise. Overall, in the literature and as indicated with specific examples in Table 14.1, the rational inclusion of nanomaterials as electrode materials/modifiers appears to have significantly improved the power generation of the MFC systems. From the data indicated in Table 14.1, for example, the utilisation of carbon nanotubes and FePc at the cathode serves to synergistically enhance power density by 1913 % relative to the use of unmodified carbon paper electrodes (e.g. [20] in Table 14.1). Similarly, increases in power density were observed for modifications of the anode, with values of 125 % increase at metal oxide modified electrodes with the inclusion of carbon nanotubes by Wang et al. in 2014 ([18] in Table 14.1).

Considering this, lowering the cost of catalysts is an area where nanomaterials also play a particular role, platinum-based catalysts being a case in point. Indeed, the cost-benefit of including nanomaterials based catalysts are frequently touted as a reason for their study. Given the role of platinum in catalysing the oxygen reduction reaction, several papers compare the application of metallophthalocyanine-based catalysts (in the presence and absence of nanomaterials) as alternatives to platinum. Reported power densities obtained at MFCs operated with cathodes modified with copper phthalocyanines [10] and iron phthalocyanines [20] were comparable to those operating using platinum-modified cathodes. The cost of nanomaterials is also an important consideration in any scale up operation. For this purpose, Mshoperi et al., 2014 [19] examined untreated carbon black as an alternative support for iron phthalocyanine to carbon nanotubes, obtaining near identical power density values to studies using acid functionalised carbon nanotubes. Such lowering in cost and complexity of pre-treatment holds obvious utility in the scale up and real application of MFC technology.

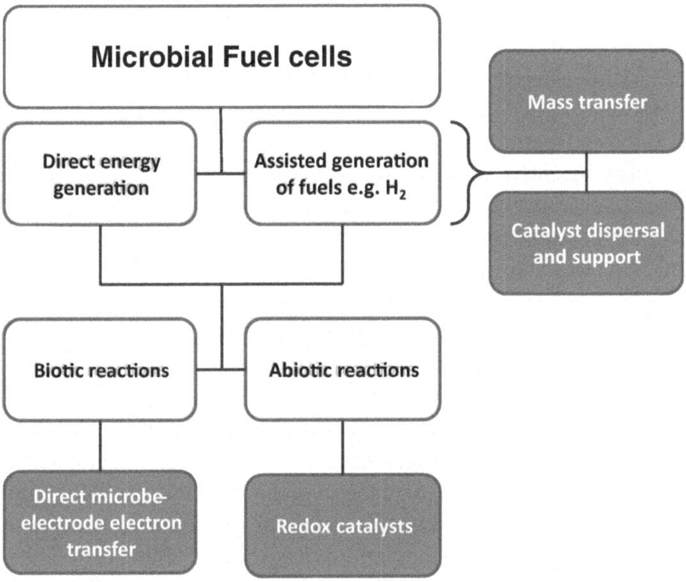

Fig. 14.2 Overview of the current limitations of the major divisions of fuel cell functions and processes (*black text, white background*) and the limitations in fuel cell technology addressable by nanotechnology *(white text, grey background)*

In order to further contextualise the role that nanomaterials may play, it is necessary to consider the inherent limitations of MFC technology, summarised briefly in Fig. 14.2.

14.3 Inherent Limitations of Microbial Fuel Cells

14.3.1 Thermodynamic Limitations of Using Biological Systems

Part of the power limitations reported at bioanodic microbial fuel cells and noted in Table 14.1 arises due to inherent constraints of the technology, but may be addressable by futuristic technologies. This limitation arises because living organisms are responsible for the initial oxidation reactions, liberating electrons from carbonaceous molecules. The final processes of energy generation in biological systems is mainly electrochemical—the chemical energy present in nutrient compounds is converted to reducing equivalents (e^- and H^+) and stored in the reduced forms of the high-energy redox intermediates, NAD(H) and FAD(H_2). A series of biochemical electron-transfer reactions serves to convert this electrical energy into a proton imbalance across membranes, the relaxation of which drives ATP synthesis [23, 27].

A portion, typically the majority, of the harvestable energy arising from catabolism is retained and used by the biocatalyst to maintain cellular processes, rather than entering the external circuit via the anode. Despite billions of years of evolution, the energy yield from the metabolism of as simple a compound as glucose is only between 55 and 65 % energy efficient [27]. When one considers that the majority of electron-transport reactions occur on the intracellular face of cell membranes and are (currently) unavailable for transfer to the external circuit, the very nature of cellular metabolism imposes a severe restriction on the amount of energy that is harvestable from the system. Conversely, chemical and enzymatic fuel cells do not suffer from similar restrictions, with the additional benefit in chemical fuel cells that temperature and pressure of the reactants can be varied for optimal performances without degrading the catalysts.

The potential energy of electrons exiting the bacteria is low. Comparison of the formal reduction potentials between thc penultimate cellular electron acceptors (members of the cytochrome protein family, cytochromes a and cyotochrome c, have formal reduction potentials of +0.633 and +0.703 V vs. SHE, respectively) and the most common terminal electron acceptor molecule, O_2 (formal reduction potential of +1.229 V [27]), sets the limit of maximum potential difference reported to be between 0.5 and 0.8 V at OCV, under physiological conditions [6]. While there might indeed be technological interventions in the future that might address this limitation, these are currently relegated to conjectural stages of research.

Efforts at improving the economic sustainability and the cost-effectiveness of MFCs is rather a more achievable and immediate manner of improving the implementation and deployment of MFC technology [8]. Despite this, engineering and design interventions have demonstrated power densities of >2.4 W/m^2 electrode surface area [8, 23], effectively demonstrating the amount of harvestable energy that is available from these systems without complex electrode engineering and materials sciences.

Alterations to the rate and efficiency of electron transfer have been investigated and optimised under several conditions: electrode selection and configuration [2], microbial fuel cell design and configuration, composition and concentration of electrolytes [23], presence, composition and type of permiselective membranes (if any).

14.3.2 Mass Transfer Limitations

As with any heterogeneous chemical process, microbial fuel cells suffer from mass transfer limitations. Two separate classes of reactions of typical MFCs have been determined to limit power generation: mass transfer supplying substrates to catalysts and removing products within an electrode's compartment [28] and the transfer of charge carriers e.g. H^+ between compartments, usually separated through a permiselective membrane, e.g., Nafion [29, 30]. Mass transfer limitations affect both cell voltage (as described by the classical Nernst equation) and the overall power obtainable, by decreasing the obtainable current [28].

An important criterion in microbial fuel cell technology is the surface- and system-engineering of energy-efficient ways of improving desirable mass-transfer reactions, such as those mentioned above, while minimising undesirable mass transfer reactions, e.g., the diffusion of oxygen from aerobic compartments to anaerobic compartments [29, 30], the diffusion of microbial cells to abiotic compartments, etc. Solution agitation, while noted to produce significant improvements on the power generation of fuel cells e.g. [19], requires too much energy input to produce a sustainable solution.

One of the most-often cited methods of improving the effectiveness of MFC power generation overall is the improvement of effective electrode surface areas. This not only increases the catalyst loading of electrodes, but also improves the mass transfer rates between the electrode and the compartment space, through enhanced surface area to volume ratios. It is in this area that the use of conductive nanoparticles and nanomaterials play significant roles as non-catalytic electrode modifiers. The translation of materials from the macro- and microscale to the nanoscale results in improvements of surface areas of several orders of magnitude. Table 14.2 shows that carbon materials can range from a macropscopic size of $0.005\ m^2\ g^{-1}$ (carbon cloth, Table 14.2, [32]) to $1662\ m^2\ g^{-1}$ (reported for activated carbon powder [33]), an impressive span of five orders of magnitude. Indeed it is this increase in effective surface area that is most commonly attributed with the increases in power density with the nanomodification of electrodes recorded in Table 14.1, particularly when studying carbonaceous nanomaterials.

Table 14.2 Reported and calculated surface areas of common electrode materials and nanoscaled electrode modifiers

Material	Surface area $m^2\ g^{-1}$(method)	Reference
Micro- and macroscopic materials		
Activated carbon granules	0.125 (BET, N_2)	[31]
Carbon cloth	0.005 (cited)	[32]
Carbon fibers	7.11 (cited)	[32]
Graphite rods	0.007 (cited)	[32]
Graphite felt	0.020 (cited)	[32]
Nanoscopic materials		
Graphene	264	[32]
Activated carbon powder, SPC-01	1662 (BET)	[33]
Graphitic carbon nanostructures	112 (BET)	[34]
Graphene oxide—AuNP (φ 20–30 nm)	984 (BET)	[33]
Pristine carbon nanotubes	34 (BET)	[35]
Acid-functionalised CNTs	85 (BET)	[35]
Hydrothermally-synthesised MnO_2 nanoparticles	112 (BET)	[11]
20 % Pt dispersed on Vulcan XC-72	65 (RRDE Voltammetry)	[36]

A combination of improved mass transport conditions, and the rational design of electrodes supporting more facile electron-transfer reactions and nanotechnological enhancements have all yielded significant overall improvements on the efficiency and output capacity of microbial fuel cells as evidenced in Table 14.1.

In the rest of this chapter we detail further how these limitations impact on the MFC system for both the biotic and abiotic processes commonly reported in the literature, highlighting in each section how nanomaterials can be harnessed to address these.

14.4 Nanomaterials Supporting Biotic Processes in Microbial Fuel Cells

Microbial fuel cells can make use of microbial catalysts in both the anodic and cathodic sections. The most commonly-encountered microbial fuel cells make use of microbes as anodic catalysts, primarily used in order to biodegrade organic matter and act as the proton and electron sources which participate in the rest of the MFC process [2, 3]. Considered as cathodic catalysts, microbes are capable of acting as intermediate electron acceptors, transferring the incoming electrons to a variety of researched pollutants and acceptors, such as nitrates, sulphates, perchlorates, metal ions and dioxygen [2, 3, 28, 37]. For either purpose, the method by which microbial-electrode electron transfer occurs are much the same. As such, they are discussed simultaneously.

Due to their integral function within the MFC system, the selection of the microbial catalyst has a profound influence on the nature and the extent of the performance of the microbial fuel cell. This is due to, firstly, the ability of a biocatalyst to thrive in the conditions present in the anodic/cathodic chambers e.g. nutrient availability and the absence/presence of oxygen, pH, pollutant presence, etc. Secondly, the selection of microbial catalyst influences the manner by which electrons transfer to the anode and enter the external circuit [3]. Cells owe much of their metabolic function to electrochemical reactions, through the use of oxidorectase enzymes. This class of enzymes catalyses the transfer of electron between reactants, via redox centres embedded in their protein structures, and participate greatly in the electrochemical reactions that generate cellular energy [38]. As mentioned previously, the means by which energy is gathered during catabolism reactions is by a series of electrochemical reactions.

In eukaryote cells, the majority of the catabolic electron-transport metabolic processes are conducted within the specialised organelles of the mitochondria. This limits the accessibility of the electrodes to interact with electrons arising from these processes. Hence, the majority of the microbes studied for microbial fuel cell technologies are primarily bacterial or archaeal organisms. Nonetheless, research on some facultative anaerobic eukaryotes, primarily yeasts, are proceeding. The majority of researched microbes are either obligate anaerobes and facultative

Table 14.3 Common microbial catalysts investigated in current research and pertinent characteristics—2014

Organism name	Cell wall	Cited #[a]	Respiration	Electron transfer [Ref]
Shewanella spp.	Gram-negative	277	Facultative	Mixed [23, 40, 41]
Pseudomonas spp.	Gram-negative	159	Facultative	Self-mediated [42,43]
Geobacter sulfurreducens	Gram-negative	131	Anaerobic	Direct [23]
Escherichia coli	Gram-negative	127	Facultative	Mixed [18, 44]
Clostridium spp.	Gram-positive	>100	Anaerobic	Mixed [23, 45]
Saccharomyces cereviseae	Fungal (yeast)	87	Facultative	Not reported
Archae family	Archaea	39	Varied	Not reported
Enterobacter cloacae	Gram-negative	17	Facultative	Direct [46]

[a]Cited # established via keyword search in Scopus.com in December, 2014

anaerobes i.e. use terminal electron acceptors other than O_2—the rationale behind this is that the reliance on other, less abundant and cell-permeable, electron acceptors acts as a selective pressure for extracellular electron transfer reactions [23].

Gram-negative bacteria, compared to the Gram-positive bacteria, maintain a thinner outer cell membrane, a point which has been implicated in providing greater extracellular electron transfer efficiencies, both when direct and indirect bacteria-anode electron transfer is attempted, e.g. [39]. They therefore are more commonly researched for MFC purposes (e.g. Table 14.3). However, the *Clostridium* genus is often included in microbial fuel studies, due to its historical role in biofuel fermentation and have been shown to successfully produce power e.g. [23]. Other studies have also found no influence on the outer cell wall structure with power generation e.g. [47], but no concerted attempt to distinguish between direct electron transfer and mediated electron transfer has yet been made between the two cell wall classes. When cultured in a consortium, Gram-positive bacteria have been exhibited to allow electron-transfer reactions when co-cultured with Gram-negative bacteria that secrete electron transport mediator compounds [42].

Exoelectrogenesis, the processes by which electrons are transported from outside their liberating cells, can occur via two pathways: direct and indirect. A schematic of this appears in Fig. 14.3. Direct cell-electrode electron transfer is possible and is discussed in greater detail below.

The electrical connection between the cells and electrode by indirect transport occurs with the deliberate addition, or microbial secretion, of electroactive mediator compounds that diffusively shuttle electrons between the microbial catalyst and the anode. Some examples of mediators produced by cultures operating under mediated electron transport include flavins, quinones and phenazines [39, 42, 48]. Depending on the system under examination, the dissolved electron transport mediators interface with the outer portion of the cell wall, i.e., cytochromes, or can penetrate through the cell wall to react with the inner metabolic machinery of the microbe, in order to abstract electrons from the system. A much wider range of tested microbes can make use of mediator compounds as electron acceptors than have been observed to produce them [3].

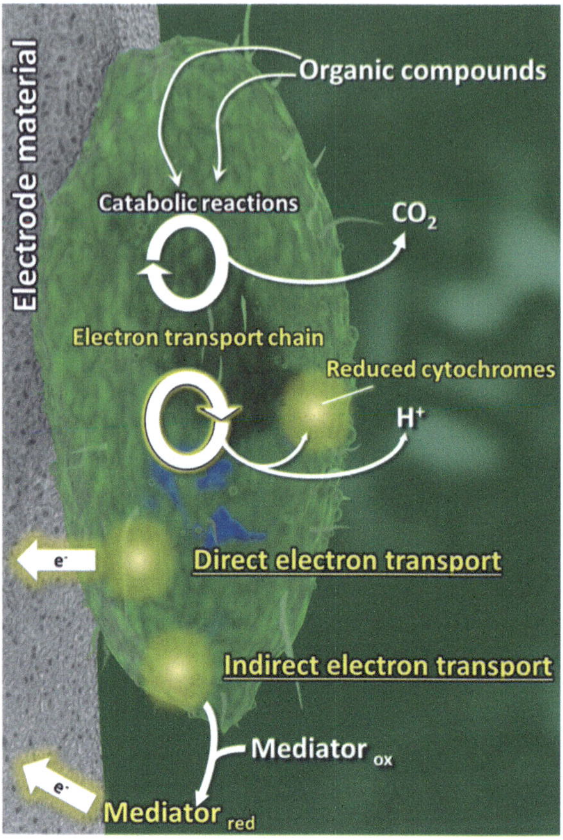

Fig. 14.3 Overview of general processes occurring at most bioanodes reported in literature. In the case of biocathodes, electron transport originates at the electrode and the cell, or produced metabolites, act as electron acceptor

Overall, the means by which exoelectrogenesis occurs appears to be more dependant on the system present in the anode, rather than be a key feature of the microbe selected e.g. [18]. It is quite likely that microbes use a variety of means in order to transfer electrons to their metabolic electron acceptor, depending on local environmental conditions, in order to maintain metabolic flexibility.

14.4.1 Nanomaterials Supporting Direct Electron Transport

The transfer of electrons from the bacteria to electrode via direct, or "mediatorless", mechanisms is highly desirable for electrochemical fuel cells: the absence of added mediator improves scaleability of the technology, decreases mass transfer losses, lowers the operational costs and circumvents the toxicity considerations of including certain electron transport mediators e.g. ferricyanide [49].

Since both Gram-positive and Gram-negative bacteria cell walls possess a net negative charge at physiological pHs [50], both direct and indirect electron transfer has been reported to be passively improved by enhancing cell adhesion to the electrode material through alteration of the electrostatic charge of the electrode e.g. ammonia treatment [49]. Two separate catalyst classes are, therefore, distinguishable when considering improving anodic processes—those that improve direct electron transfer between the immobilised microbe and the electrode and those that catalyse oxidation reactions of reduced microbial mediators or metabolites.

Direct electron transport between microbe and solid electrode appears to rely primarily on the interaction of the electrode materials with cellular c-type cytochrome proteins bound to the outer cell membrane. These heme-containing proteins are widely distributed among bacteria and archaea [48]. c-Type cytochromes have been noted to be expressed at high levels under certain growth conditions with certain microbial species e.g. *Shewanella putrefasciens* and *Geobacter sulferreducens* [23], which make use of these proteins to transfer electrons between cytochromes and extracellular insoluble metals, e.g., Fe^{3+} and Mn^{4+} oxides, as terminal electron transfers during metabolism. This phenomenon can be further aided by the microbial extrusion of pili containing a high surface density of these proteins, resulting in a conductive bacterial "nanowire", which maintains electrochemical activity via the cytochromes (Fig. 14.4 serves as an example) Naturally, this phenomenon is of great interest to researchers seeking to achieve better direct electrical connectivity between microbial catalysts and proximal solid electrodes to lower mass transfer losses [2]; more so when pili have been noted to connect microbial cells to one-another, both intra- and inter-species [51], allowing for the

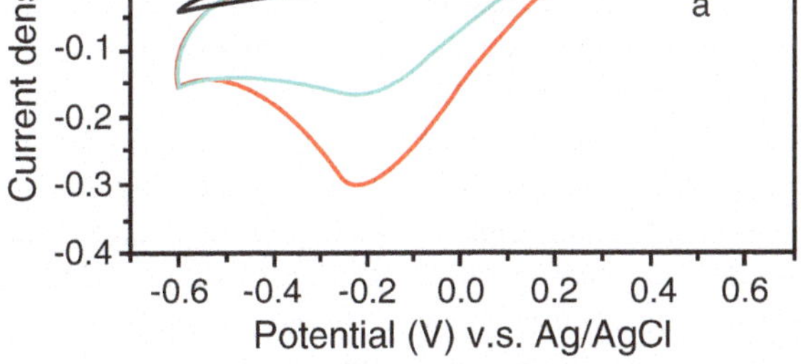

Fig. 14.4 Cyclic voltammograms of carbon felt bianodes modified with carbon nanotubes/ MoO_2C nanoparticles/ PTFE nanocomposite. Cyclic voltammograms display electrodes following immersion in growth medium containing *E. coli* for: (**a**) 0 h, (**b**) 48 h and (**c**) 120 h. Peaks observed are attributed to the reduction/oxidation of c-type cytochromes, demonstrating successful colonising of the electrode and direct bacteria-anode electron transfer. Reproduced from [18] with permission from Elsevier

possibility of electrically connecting an entire biofilm to solid electrodes via this mechanism, to improve power generation [6].

Electrode engineering research aimed at improving electrical connectivity between the biocatalyst and the anode, as well as improving the surface area of the electrode system has resulted in several physical and chemical modifications proposed to enhance microbe-electrode attachment. Polypyrrole- and polyaniline-based nanomaterials also demonstrate significant improvement of cellular adhesion, by reasons of increased surface roughness, coupled with the presence of cationic nitrogenous groups to improve cellular adhesion electrostatically [24]. A more targeted approach, through the covalent attachment of electron transfer mediators to the surfaces of electrodes, in order to stimulate direct electron transfer between membrane-bound electroactive cellular proteins and the electrode have also met with some success [49].

Nanoparticles, through a combination of enhanced electrode surface area, alteration of surface chemistry and the presentation of electroactive moieties to the microbial cells, have been reported to play a role in the improvement of power generation at microbial fuel cells [3, 25, 49]. To this effect, Ni, Pd, Au and Fe_2O_3 nanoparticles have been employed to enhance direct electron transfer [49]. In the case of carbon-based nanoparticles, the additional improvements in conductivity between nanoscopic and macroscopic materials is also cited as a contributing factor [25]. The inherent biocompatibility of carbonaceous nanomaterials has been cited as a major influence on cell adherence to electrode materials [49]: for this reason, numerous research articles study the inclusion of carbon nanotubes [49], carbon blacks [19] and graphite/graphene nanomaterials [49] to improve microbe-electrode interactions. Metal nanoparticles have been researched for their ability to improve direct electron transfer, chiefly by mimicking the natural electron acceptors used by pilli-forming microbes.

The improvements in current response in electrodes modified with nanoparticles are often attributed to an increase in electrode surface area. Related benefits of improving surface area using nanomaterials includes increasing the mass-transport of solutes around the electrode and improving the surface area for cell adhesion at anodes [52]. Certainly, anodic modifications with nanoparticles (Table 14.1) (Fig. 14.5) are responsible for significant improvements in power generation at MFCs, when compared to control MFC configurations with unmodified anode e.g. [9, 18–20]. In most cases, improved biocompatibility, resulting in higher levels of microbial colonisation is cited as a major contributing factor behind power generation increases, as well as improved cell-electron electron transfer efficiency. Figure 14.6 provides a typical example of improved direct electron transfer efficiencies, and subsequent improvements in power generation from nanomodified bioanodes in the manner commonly reported in the literature.

In addition to improving electron transfer efficiency, recent research has uncovered the direct influence of nanoparticles to the microbial catalysts' behaviour. Iron nanoparticle-decorated graphite anodes were noted to produce a significant increase in bioanodic compartment power generation [54] and improve direct electron transfer (Fig. 14.6) [53]. Examination of the gene expression differences between

Fig. 14.5 Scanning electron micrograph of carbon black: FePc nanocomposite coated carbon microfiber paper electrode

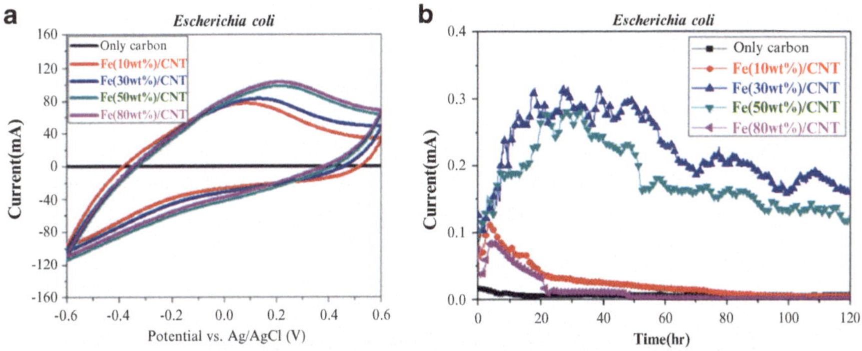

Fig. 14.6 Electrochemical characteristics of carbon paper anodes modified with bionanocomposite mixture of carbon nanotubes, Fe_3O_4 nanoparticles and *Escherichia coli* cells: (**a**) cyclic voltammetric analysis of anode of bionanocomposite with varying Fe_3O_4 content, with peak formation attributed to microbial-electrode electron transfer. (**b**) MFC performance using nanomodified bioanodes, with varying Fe_3O_4 content. Reproduced from [53], copyright 2014 with permission from Elsevier

populations of *Shewanella oneidasis,* grown in the presence or absence of the nanoparticles, indicated increased expression of biofilm-generating genes, notably those implicated in organism-electrode transfer, such as those regulating flavin production and cytochrome c proteins [54] for the reasons discussed above. In a similar study, a mixed consortium of microbial species grown either on unmodified carbon paper bioanodes or on ones modified with gold nanoparticles showed significant difference between colonising species present in either biofilms, concomitant with an increase in power generation [13]. Both phenomenons were demonstrated to alter with alteration in surface density of the nanoparticles.

14.5 Nanomaterials Supporting Abiotic Processes

As means of improving power generation at bioanodes, the use of catalysts to improve mediated electron transfer between cells and solid electrodes has previously been investigated, but given the uncertainty regarding the mechanism by which electrons are transferred to the external circuit, power improvements and electrochemical behaviour cannot be conclusively attributed to an improvement of direct, or mediated electron transfer reactions. Certainly, the ability of nanoparticles to act as catalysts and mediators for the oxidation of NADH [55], flavins [56] and other molecules implicated in microbe-electrode mediated transport e.g. phenolic compounds [57] are fairly well documented in the literature and have been explored for these roles in enzymatic fuel cells (e.g. [56] and biosensors [55].

As part of the ongoing integration of this technology towards bioremediation, several water pollutants have been studied as potential electron sinks. Nitrates, sulphates and perchlorates are among the pollutants studied, but diooxygen reduction is still the most commonly-researched electron acceptor in cathodes for MFC purposes. The addition of microbial and nanoscale catalysts and modifiers have been consistently studied for their application in improving cathodic reactions and to increase the overall economic viability of microbial fuel cells. The remainder of this section is devoted to an examination of the role of nanomaterials in improving the oxygen reduction reaction, for microbial fuel cell purposes

14.5.1 Overview of the Oxygen Reduction Reaction

Diooxygen (O_2) is already well established as an effective electron acceptor. The overall reaction rate for oxygen reduction to water is well known to proceed via several different mechanisms—either through 1-electron, 2-electron, or 4-electron processes, as outlined in Table 14.4.

Table 14.4 Standard reduction potentials of oxygen reduction reactions

Half-reaction	E_o' (V vs. SHE)
$\frac{1}{2}O_2 + 2H^+ + 2e^- \rightarrow H_2O$	1.229
$O_2 + 2H^+ + 2e^- \rightarrow H_2O_2$	0.703
$O_2 + 2H_2O + 4\,e^- \rightarrow 4OH^-$	0.633
$H_2O_2 + 2H^+ + 2e^- \rightarrow 2H_2O$	0.473

Reduction potentials drawn from [58]

Given the strong oxidising potentials of diooxygen and its associated reducible hydrides, from +1.3 to +0.6 V vs. SHE (Table 14.4), as well as its relative abundance in the environment, it is unsurprising that a large number of research publications have been generated with a focus on its successful integration into conventional MFC technologies. Indeed, the majority of microbial fuel cell research publications make use of oxygen as the terminal electron acceptor in their reported systems.

It is tacitly accepted that there are three major obstacles that must be overcome when considering oxygen as an electron acceptor—the relatively poor solubility of diooxygen in water (~8.2 mg/L maximum solubility in pure water, at STP using atmospheric air) imposes a large mass transfer limitation on the system overall; the large overpotentials of oxygen reduction reactions observed at unmodified carbonaceous electrodes [59] and the tendency of carbon electrodes to produce peroxide intermediates under overpotential conditions. These are undesirable for a further number of reasons:

The two-electron reduction of oxygen to form hydrogen peroxide has a lower formal reduction potential than the four-electron process due to the lowered number of electrons transferred. This decreases the overall emf of the MFC.

The four-electron and two-electron reduction pathways are distinguished by the O-O bond in diooxygen. In the case of the four-electron reduction pathway, this bond is broken as the initial reaction of the pathway, forming two adsorping oxygen atoms on the electrode surface which then undergo two separate $2H^{+}/2e^{-}$ reductions [59] to form H_2O as in Table 14.4. The O-O bond in the diooxygen molecule remains whole while undergoing a $2H^{+}/2e^{-}$ reduction to produce H_2O_2 [59]; this occurs more readily at carbonaceous electrodes than metal electrodes. Due to the more negative formal potential of the subsequent reduction reaction to generate water from H_2O_2 (Table 14.4), this allows hydrogen peroxide to accumulate at appreciable levels in these systems [59]. Peroxides, and their reaction progenitors, are corrosive and may damage the electrode and vessel materials.

Hydrogen peroxides undergo spontaneous oxidation to form water and O_2 i.e. this side-reaction abstracting protons without electrogenerated electron transfer reactions and lowers the overall efficiency of MFCs.

Significant effort has been expended in order to prevent the diffusion of oxygen into the anodic compartment, primarily through engineering research into the MFC cation exchange membrane [60]. Even if the selection of microbial catalysts is not limited to the strict anaerobes (who would die if exposed to aerobic conditions), oxygen, with its high oxidation potential, would compete with the anode as the electron acceptor.

Regardless of the exact reduction mechanism employed by the system, oxygen reduction is generally agreed to initiate by a rate-determining adsorption step at both metallic and non-metallic surfaces [61]. Enhancing the rate at which this occurs, largely through increasing the surface area of electrodes and the dispersion of catalyst, is as important as thermodynamic improvements to the system through further catalytic steps.

14.5.2 Nanomaterials Supporting Oxygen Reduction in Microbial Fuel Cells

The platinum group metals, iridium, osmium, palladium, platinum, rhodium and ruthenium, are well known as hydrogenation reaction catalysts [62] and it is unsurprising to note their common inclusion as catalysts for lowering the oxygen reduction reaction overpotential in the literature. Indeed, these form amongst the best electrocatalysts for both hydrogen formation from H^+ and the reduction of water [63]. There is good evidence in the literature that oxygen reduction via platinides proceeds along a four-electron transfer route described in Table 14.4, validated both for the bulk metal and for Pt nanoparticles dispersed along carbon electrodes, measured via RDE studies [64]. Due to the typically large separation of overpotential between platinum-catalysed and carbon electro-reduction of oxygen, almost complete separation of reactions has been noted to be able to occur, substantially decreasing the amount of hydroxyl anion radicals and hydrogen peroxide generated near the cathode surface.

The porphyrin-like prosthetic groups in proteins, e.g., heme, have long been known to possess both oxygen-binding capabilities and strong redox catalyst behaviour. It is well known that the synthetic porphyrin macro-class not only exhibit strong chelating properties towards oxygen, but also strong catalytic activity towards oxygen reduction processes, as part of a wider diversity of reduction/oxidation catalysis reactions [65]. Porphyrins and phthalocyanines e.g. Fig. 14.7, both metal-free [66] and metal-containing [20] are known to have strong catalytic activity towards oxygen reduction reactions—their investigation as cost-effective

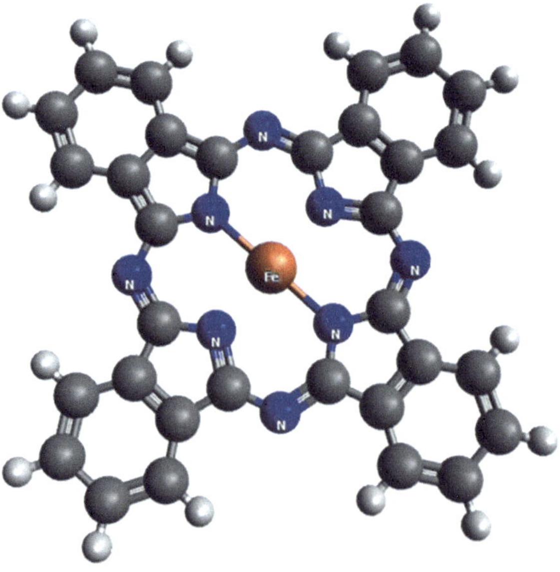

Fig. 14.7 Iron(II) phthalocyanine, a nitrogen-carbon macrocyclic organometallic compound widely researched as a low-cost catalyst for the oxygen reduction reaction

alternatives to the platinum-group metal catalysts is still a major focus of ongoing investigation [14, 19].

Depending on the metal centres chelated within these rings, four-electron, or two-electron oxygen reduction reactions have been noted, with the nitrogen periphery at the metal substitution moiety appearing to be responsible for ORR catalysis [61, 67]. As nanoscale functional analogues to porphyrins, nitrogen- and nitrogen-metal-doped carbon nanomaterials demonstrate application as metal-free ORR catalysts [61, 68]. Several examples of metal-based and carbon-based nanocomposite cathodes for ORR are presented in Fig. 14.8.

Apart from acting as catalysts themselves, carbon nanoparticles serve a separate, but equally important, role as low-cost methods of catalyst dispersal at a nanoscale.

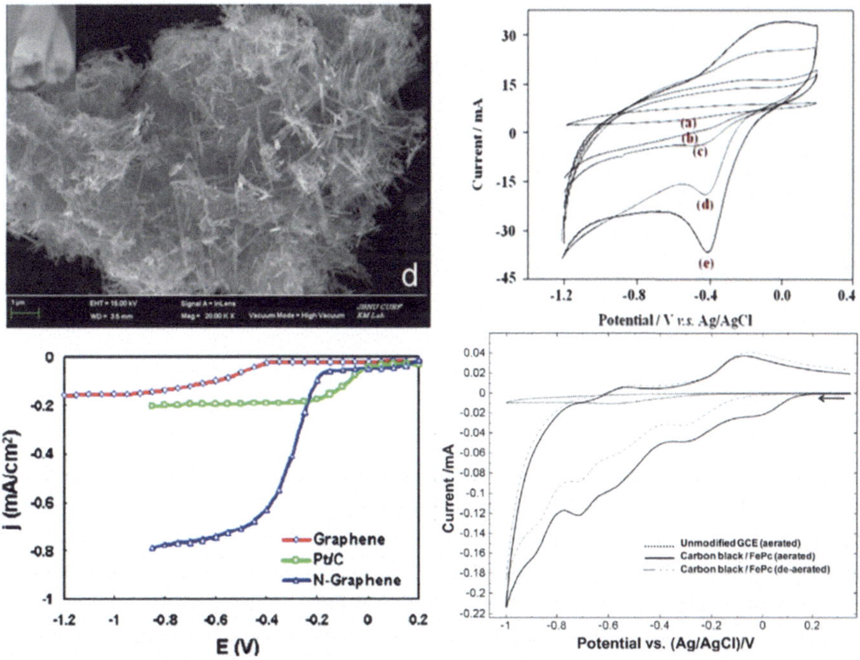

Fig. 14.8 Nanocomposite carbon-based oxygen reduction reaction catalysts found in literature. (**i**) Fe-SEM micrograph of MnO_2/Graphene oxide nanocomposite. Reproduced from [21], copyright 2014, with permission from Elsevier; (**ii**) cyclic voltammetry demonstrating ORR obtained at (*a*) unmodified carbon cloth working electrode Oxygen reduction and (*e*) the MnO_2/ Graphene oxide nanocomposite. Reproduced from [21], copyright 2014, with permission from Elsevier; (**iii**) rotating ring-disk electrode voltammograms for the ORR reaction obtained at glassy carbon working electrodes modified with either graphene (Graphene), platinised Vulcan XC-72R carbon (Pt/C) and nitrogen-containing graphene (N-graphene). Reprinted with permission from [68], copyright 2010, American Chemical Society; (**iv**) Carbon black/Iron(II)phthalocyanine nanocomposite, demonstrating the ORR profile obtained at unmodified GCE working electrodes (*asterisk, right*) and at the nanocomposite-modified electrode (*asterisk, left*). Reproduced from [19], copyright 2014, with permission from Elsevier. For all presented voltammograms, note the movement of the ORR to more anodic potentials

As mentioned earlier, adequate support and dispersion of catalysts is essential to achieve uniform deposition of catalysts along electrode surfaces in order to maximise mass transfer to the electrode [17]. In this area, carbon nanomaterials in particular, have been employed to great effect. Metal nanoparticles have been dispersed using carbon nanotubes to produce catalyst-modified conductive nanomaterial cathodes for improved ORR [10]. Given the high degree of aromaticity along pristine stretches of their walls and potential for π-π electron stacking, multi-walled carbon nanotubes have been demonstrated to evenly disperse binuclear phthalocyanines [17, 69], resulting in superior electrocatalytic performance of the nanocomposite towards ORR, in turn improving the obtainable power from the MFC [20].

14.5.3 Other Electron Acceptors

To mimic the heterogeneity of the overall denitrification process making use of nitrate anions as electron acceptors, a multi-metal heterogenous electrocatalyst, consisting of gold nanoparticles, heme proteins and cadmium ions deposited onto glassy carbon electrodes has shown some weak denitrification [70] catalytic activity, albeit to a lesser extent than previously reported.

14.6 Conclusions

The ultimate goal of research in MFC technology is the realisation of large scale [8], cost-effective systems that generate appreciable levels of power. Given the nature of the system towards remediation or the utilisation of waste materials as an energy source, such large-scale efforts aim for power generated to be sufficient to power the operations of the treatment facility, at least. That MFC technology will not match the power density outputs of chemical fuel cell technology without futuristic approaches identified, owing to the thermodynamic limitations of the biological systems engaged, is clear. Notwithstanding other significant areas of progress in recent years, nanomaterials offer substantial opportunities to enhance baseline power density output through improved electron transfer and mass transfer processes that limit the power generation of MFC, both considering the biotic and abiotic processes present in this technology. Their inclusion resulting in increased surface area and as a dispersal agent or immobilisation matrix for catalysts, as has been shown. Nanomaterial catalysts used show substantial benefits for lowering the operating cost of MFC technology, an important curtailment in scale up and widespread adoption of such technology. Fifty percent of the capital costs of MFC systems are estimated to arise from catalyst inclusion [24] and the ongoing research into effective, low-cost catalysts and catalyst dispersants to replace expensive platinum-group metal catalysts can only improve the cost-effectiveness of this technology.

Several research groups have expended efforts on the study of different wastewaters, and it is here where the biggest leaps in terms of the future practical adoption and large scale utilisation of MFC technology may yet come from. This is stated in light of increasing legislation globally on industrial, municipal and domestic wastewater as well as tenuous access to potable water, and increasingly so in so-called first world nations. Here beneficiation of wastewater to generate a usable product, e.g. potable or agricultural grade water is an attractive reality, in addition to the system generating both direct electricity and achieving remediation of a wastewater. Such opportunities are easily represented in areas with food or domestic wastewater rather than industrial wastewater, given the complexity of the latter. However, industrial wastewater, given legislative requirements may represent financially viable opportunities given the premium on remediation. The wide variety of manufacturing industry however means that the realisation of such a usable system would need to be developed with consideration of chemical and biological complexity of the waste, and on a waste-by-waste basis. Curtailments in efficiency of MFC systems are obvious, given differences in environmental parameters and chemical composition. In particular the nature of the biomaterials used in the anode and cathode are particularly vulnerable to fluctuating conditions. In order to limit such vulnerability, for example of bacteria to toxins in industrial waste streams, efforts at bacterial encapsulation to isolate and protect these from the immediate environment are needed, for example, sol-gel encapsulation [71].

Nanomaterials and nanotechnology approaches may yet have further roles to play in enhancing aspects of remediation and is an area currently open for further exploration, particularly as this applies to the realisation of future large-scale MFC systems.

References

1. Logan BE, Rabaey K (2012) Review: conversion of wastes into bioelectricity and chemicals by using microbial electrochemical technologies. Science 337:686–690. doi:10.1126/science.1217412
2. Mohan SV, Velvizhi G, Modestra JA, Srikanth S (2014) Microbial fuel cell: critical factors regulating bio-catalyzed electrochemical process and recent advancements. Renew Sustain Energy Rev 40:779–797. doi:10.1016/j.rser.2014.07.109
3. Rabaey K, Rodríguez J, Blackall LL, Keller J, Gross P, Batstone D, Verstraete W, Nealson KH (2007) Microbial ecology meets electrochemistry: electricity-driven and driving communities. ISME J 1:9–18. doi:10.1038/ismej.2007.4
4. Pant D, Van Bogaert G, Diels L, Vanbroekhoven K (2010) A review of the substrates used in microbial fuel cells (MFCs) for sustainable energy production. Bioresour Technol 101:1533–1541. doi:10.1016/j.biortech.2009.10.017
5. Friman H, Schechter A, Ioffe Y, Nitzan Y, Cahan R (2013) Current production in a microbial fuel cell using a pure culture of *Cupriavidus basilensis* growing in acetate or phenol as a carbon source. J Microbial Biotechnol 6:425–434. doi:10.1111/1751-7915.12026

6. Logan BE, Regan JM (2006) Microbial fuel cells—challenges and applications. Environ Sci Technol 40:5172–5180. doi:10.1021/es0627592
7. Pollet BG (2010) The use of ultrasound for the fabrication of fuel cell materials. Int J Hydrogen Energy 35:11986–12004. doi:10.1016/j.ijhydene.2010.08.021
8. Logan BE (2010) Scaling up microbial fuel cells and other bioelectrochemical systems. Appl Microbiol Biotechnol 85:1665–1671. doi:10.1007/s00253-009-2378-9
9. Hou J, Liu Z, Zhang P (2013) A new method for fabrication of graphene/polyaniline nanocomplex modified microbial fuel cell anodes. J Power Sources 224:139–144. doi:10.1016/j.jpowsour.2012.09.091
10. Ghasemi M, Daud WRW, Rahimnejad M, Rezayid M, Fatemi A, Jafari Y, Somalu MR, Manzour A (2013) Copper-phthalocyanine and nickel nanoparticles as novel cathode catalysts in microbial fuel cells. Int J Hydrogen Energy 38:9533–9540. doi:10.1016/j.ijhydene.2013.01.177
11. Haoran Y, Lifang D, Tao L, Yong C (2014) Hydrothermal synthesis of nanostructured manganese oxide as cathodic catalyst in a microbial fuel cell fed with leachate. Sci World J. ID: 791672. doi:10.1155/2014/791672
12. Hosseini MG, Ahadzadeh I (2012) A dual-chambered microbial fuel cell with Ti/nano-TiO_2/Pd nano-structure cathode. J Power Sources 220:292–297. doi:10.1016/j.jpowsour.2012.07.096
13. Alatraktchi FA, Zhang Y, Angelidaki I (2014) Nanomodification of the electrodes in microbial fuel cell: impact of nanoparticle density on electricity production and microbial community. Appl Energy 116:216–222. doi:10.1016/j.apenergy.2013.11.058
14. Zhao F, Harnischa F, Schröder U, Scholz F, Bogdanoff P, Hermann I (2005) Application of pyrolysed iron(II) phthalocyanine and CoTMPP based oxygen reduction catalysts as cathode materials in microbial fuel cells. Electrochem Commun 7:1405–1410. doi:10.1016/j.elecom.2005.09.032
15. Esmaeili C, Ghasemi M, Heng LY, Hassan SHA, Abda MM, Daud WRW, Ilbeygi H, Ismail AF (2014) Synthesis and application of polypyrrole/carrageenan nano-biocomposite as a cathode catalyst in microbial fuel cells. Carbohydr Polym 114:253–259. doi:10.1016/j.carbpol.2014.07.072
16. Niessen J, Schröder U, Scholz F (2004) Exploiting complex carbohydrates for microbial electricity generation—a bacterial fuel cell operating on starch. Electrochem Commun 6:955–958. doi:10.1016/j.elecom.2004.07.010
17. Deng L, Zhou M, Liu C, Liu L, Liu C, Dong S (2010) Development of high performance of Co/Fe/N/CNT nanocatalyst for oxygen reduction in microbial fuel cells. Talanta 81:444–448. doi:10.1016/j.talanta.2009.12.022
18. Wang Y, Li B, Cui D, Xiang X, Li W (2014) Nano-molybdenum carbide/carbon nanotubes composite as bifunctional anode catalyst for high-performance *Escherichia coli*-based microbial fuel cell. Biosensors Bioelectron 51:349–355. doi:10.1016/j.bios.2013.07.069
19. Mshoperi E, Fogel R, Limson J (2014) Application of carbon black and iron phthalocyanine composites in bioelectricity production at a brewery wastewater fed microbial fuel cell. Electrochim Acta 128:311–317. doi:10.1016/j.electacta.2013.11.016
20. Edwards SL, Fogel R, Mtambanengwe K, Togo C, Laubscher R, Limson JL (2012) Metallophthalocyanine/carbon nanotube hybrids: extending applications to microbial fuel cells. J Porphyrins Phthalocyanines 16:917–926. doi:10.1142/S1088424612501027
21. Kumar GG, Awan Z, Nahm KS, Xavier JS (2014) Nanotubular MnO_2/graphene oxide composites for the application of open air-breathing cathode microbial fuel cells. Biosensors Bioelectron 53:528–534. doi:10.1016/j.bios.2013.10.012
22. Mehdinia A, Ziaei E, Jabbari A (2014) Multi-walled carbon nanotube/SnO_2 nanocomposite: a novel anode material for microbial fuel cells. Electrochim Acta 130:512–518. doi:10.1016/j.electacta.2014.03.011
23. Rinaldi A, Mecheri B, Garavaglia V, Licoccia S, Di Nardo P, Traversa E (2008) Review—engineering materials and biology to boost performance of microbial fuel cells: a critical review. Energy Environ Sci 1:417–429. doi:10.1039/b806498a
24. Ghasemi M, Daud WRW, Hassan SHA, Oh S-E, Ismail M, Rahimnejad M, Jahim JM (2013) Review—nano-structured carbon as electrode material in microbial fuel cells: a comprehensive review. J Alloys Compd 580:245–255. doi:10.1016/j.jallcom.2013.05.094

25. Gadhamshetty V, Koratkar N (2012) Nano-engineered biocatalyst-electrode structures for next generation microbial fuel cells. Nano Energy 1:3–5. doi:10.1016/j.nanoen.2011.11.003
26. Baranton S, Coutanceau C, Léger J-M, Roux C, Capron P (2005) Alternative cathodes based on iron phthalocyanine catalysts for mini- or micro-DMFC working at room temperature. Electrochim Acta 51:517–525. doi:10.1016/j.electacta.2005.05.010
27. Garret RH, Grisham CM (1999) Electron transport and oxidative phosphorylation. In: Garret RH, Grisham CM (eds) Biochemistry, 2nd edn. Saunders College Publishing, Philadelphia, pp 673–689
28. Rismani-Yazdi H, Carver SM, Christy AD, Tuovinen OH (2008) Cathodic limitations in microbial fuel cells: an overview. J Power Sources 180:683–694. doi:10.1016/j.jpowsour.2008.02.074
29. Chae KJ, Choi M, Ajayi FF, Park W, Chang IS, Kim IS (2008) Mass transport through a proton exchange membrane (nafion) in microbial fuel cells. Energy Fuel 22:169–176. doi:10.1021/ef700308u
30. Zhang X, Cheng S, Wang X, Huang X, Logan BE (2009) Separator characteristics for increasing performance of microbial fuel cells. Environ Sci Technol 43:8456–8461. doi:10.1021/es901631p
31. Huggins T, Wang H, Kearns J, Jenkins P, Ren ZJ (2014) Biochar as a sustainable electrode material for electricity production in microbial fuel cells. Bioresour Technol 157:114–119. doi:10.1016/j.biortech.2014.01.058
32. Kumar GG, Sarathi VGS, Nahm KS (2013) Recent advances and challenges in the anode architecture and their modifications for the applications of microbial fuel cells. Biosensors Bioelectron 43:461–475. doi:10.1016/j.bios.2012.12.048
33. Peng X, Yu H, Lina A, Li N, Wang X (2013) Time behavior and capacitance analysis of nano-Fe_3O_4 added microbial fuel cells. Bioresour Technol 114:689–692. doi:10.1016/j.biortech.2013.07.037
34. Sevilla M, Sanchís C, Valdés-Solís T, Morallón E, Fuertes AB (2007) Synthesis of graphitic carbon nanostructures from sawdust and their application as electrocatalyst supports. J Phys Chem C 111:9749–9756. doi:10.1021/jp072246x
35. Vix-Guterl C, Couzi M, Dentzer J, Trinquecoste M, Delhaes P (2004) Surface characterizations of carbon multiwall nanotubes: comparison between surface active sites and raman spectroscopy. J Phys Chem B 108:19361–19367. doi:10.1021/jp047237s
36. Paulus UA, Schmidt TJ, Gasteiger HA, Behm RJ (2001) Oxygen reduction on a high-surface area Pt/Vulcan carbon catalyst: a thin-film rotating ring-disk electrode study. J Electroanal Chem 495:134–145. doi:10.1016/S0022-0728(00)00407-1
37. Kelly PT, He Z (2014) Nutrients removal and recovery in bioelectrochemical systems: a review. Bioresour Technol 153:351–360. doi:10.1016/j.biortech.2013.12.046
38. Xu F (2000) Applications of oxidoreductases: recent progress. Ind Biotechnol 1:38–50. doi:10.1089/ind.2005.1.38
39. Read ST, Dutta P, Bond PL, Keller J, Rabaey K (2010) Initial development and structure of biofilms on microbial fuel cell anodes. BMC Microbiol 10:98. doi:10.1186/1471-2180-10-98
40. Wu Y, Liu T, Li X, Li F (2014) Exogenous electron shuttle-mediated extracellular electron transfer of *Shewanella putrefaciens* 200: electrochemical parameters and thermodynamics. Environ Sci Technol 48:9306–9314. doi:10.1021/es5017312
41. Biffinger JC, Pietron J, Bretschger O, Nadeau LJ, Johnson GR, Williams CC, Nealson KH, Ringeisen BR (2008) The influence of acidity on microbial fuel cells containing *Shewanella oneidensis*. Biosensors Bioelectron 24:900–905. doi:10.1016/j.bios.2008.07.034
42. Pham TH, Boon N, De Maeyer K, Höfte M, Rabaey K, Verstraete W (2008) Use of *Pseudomonas* species producing phenazine-based metabolites in the anodes of microbial fuel cells to improve electricity generation. Appl Microbiol Biotechnol 80:985–993. doi:10.1007/s00253-008-1619-7
43. Jayapriya J, Ramamurthy V (2014) The role of electrode material in capturing power generated in *Pseudomonas* catalysed fuel cells. Can J Chem Eng 92:610–614. doi:10.1002/cjce.21895

44. Chalenko Y, Shumyantseva V, Ermolaeva S, Archakov A (2012) Electrochemistry of *Escherichia coli* JM109: direct electron transfer and antibiotic resistance. Biosensors Bioelectron 31:219–223. doi:10.1016/j.bios.2011.12.015
45. Park HS, Kim BH, Kim HS, Kim HJ, Kim GT, Kim M, Cahn IS, Park YK, Chang HI (2001) A novel electrochemically active and Fe(III)-reducing bacterium phylogenetically related to *Clostridium butyricum* isolated from a microbial fuel cell. Anaerobe 7:297–306. doi:10.1006/anae.2001.0399
46. Feng C, Li J, Qin D, Chen L, Zhao F, Chen S, Hongbo H, Yu C-P (2014) Characterization of exoelectrogenic bacteria *enterobacter* strains isolated from a microbial fuel cell exposed to copper shock load. PLoS One 9:e11379. doi:10.1371/journal.pone.0113379
47. Juang DF, Yang PC, Lee CH, Hsueh SC, Kuo TH (2011) Electrogenic capabilities of Gram negative and Gram positive bacteria in microbial fuel cell combined with biological wastewater treatment. Int J Environ Sci Technol 8:781–792. doi:10.1007/BF03326261
48. Yang Y, Xu M, Guo J, Sun G (2012) Bacterial extracellular electron transfer in bioelectrochemical systems. Process Biochem 47:1707–1714. doi:10.1016/j.procbio.2012.07.032
49. Du J, Catania C, Bazan GC (2014) Modification of abiotic–biotic interfaces with small molecules and nanomaterials for improved bioelectronics. Chem Mater 26:686–697. doi:10.1021/cm401912j
50. Devi UV, Puri P, Sharma NN, Ananthasubramanian M (2014) Electrokinetics of cells in dielectrophoretic separation: a biological perspective. BioNanoScience 4:276–287. doi:10.1007/s12668-014-0140-y
51. Kato S, Watanebe K (2010) Ecological and evolutionary interactions in syntrophic methanogenic consortia. Microbes Environ 25:145–151. doi:10.1264/jsme2.ME10122
52. Inoue S, Parra EA, Higa A, Jiang Y, Wang P, Buie CR, Coates JD, Lin L (2012) Structural optimization of contact electrodes in microbial fuel cells for current density enhancements. Sensor Actuat A Phys 177:30–36. doi:10.1016/j.sna.2011.09.023
53. Park IH, Christy M, Kim P, Nahm KS (2014) Enhanced electrical contact of microbes using Fe_3O_4/CNT nanocomposite anode in mediator-less microbial fuel cell. Biosensors Bioelectron 58:75–80. doi:10.1016/j.bios.2014.02.044
54. Xu S, Liu H, Fan Y, Schaller R, Jia J, Chaplen F (2012) Enhanced performance and mechanism study of microbial electrolysis cells using Fe nanoparticle-decorated anodes. Appl Microbiol Biotechnol 93:871–880. doi:10.1007/s00253-011-3643-2
55. Gai P-P, Zhao C-E, Wang Y, Abdel-Halim ES, Zhang J-R, Zhu J-J (2014) NADH dehydrogenase-like behaviour of nitrogen-doped graphene and its application in NAD^+-dependent dehydrogenase biosensing. Biosensors Bioelectron 62:170–176. doi:10.1016/j.bios.2014.06.043
56. Yehezkeli O, Tel-Vered R, Raichlin S, Willner I (2011) Nano-engineered flavin-dependent glucose dehydrogenase/gold nanoparticle-modified electrodes for glucose sensing and biofuel cell applications. ACS Nano 5:2385–2391. doi:10.1021/nn200313t
57. Wang G, He X, Zhou F, Li Z, Fang B, Zhang X, Wang L (2012) Application of gold nanoparticles/TiO_2 modified electrode for the electrooxidative determination of catechol in tea samples. Food Chem 135:446–451. doi:10.1016/j.foodchem.2012.04.139
58. Harvey D (2000) Appendix 3D: standard reduction potentials. In: Harvey D (ed) Modern analytical chemistry, 1st edn. McGraw-Hill Higher Education, New York, pp 743–747
59. Bagotsky VS (2006) Reactions at nonconsumable electrodes. In: Bagotsky VS (ed), Fundamentals of electrochemistry, 2 edn. Wiley, New York, pp 261–296
60. Leong JX, Daud WRW, Ghasemi M, Liew KB, Ismail M (2013) Ion exchange membranes as separators in microbial fuel cells for bioenergy conversion: a comprehensive review. Renew Sustain Energy Rev 28:575–587. doi:10.1016/j.rser.2013.08.052
61. Liew KB, Daud WRW, Ghasemi M, Leong JX, Lim SS, Ismail M (2014) Non-Pt catalyst as oxygen reduction reaction in microbial fuel cells: a review. Int J Hydrogen Energy 39:4870–4883. doi:10.1016/j.ijhydene.2014.01.062

62. Patnaik P (2003) Platinum. In: Patnaik P (ed) Handbook of inorganic chemicals. McGraw-Hill Higher Education, New York, pp 719–722
63. Tarasevich MR, Korchagin OV (2013) Electrocatalysis and pH (a review). Russ J Electrochem 49:600–618. doi:10.1134/S102319351307015X
64. Geniès L, Faure R, Durand R (1998) Electrochemical reduction of oxygen on platinum nanoparticles in alkaline media. Electrochim Acta 44:1317–1327. doi:10.1016/S0013-4686(98)00254-0
65. Zagal JH, Griveau S, Silva JF, Nyokong T, Bedioui F (2010) Metallophthalocyanine-based molecular materials as catalysts for electrochemical reactions. Coord Chem Rev 251:2755–2791. doi:10.1016/j.ccr.2010.05.001
66. Masa J, Zhao A, Xia W, Muhler M, Schuhmann W (2013) Metal-free catalysts for oxygen reduction in alkaline electrolytes: influence of the presence of Co, Fe, Mn and Ni inclusions. Electrochim Acta 128:271–278. doi:10.1016/j.electacta.2013.11.026
67. Kobayashi M, Niwa H, Saito M, Harada Y, Masaharu O, Ofuchi H, Terakura K, Ikeda T, Koshigoe Y, Ozaki J, Miyata S (2012) Indirect contribution of transition metal towards oxygen reduction reaction activity in iron phthalocyanine-based carbon catalysts for polymer electrolyte fuel cells. Electrochim Acta 74:254–259. doi:10.1016/j.electacta.2012.04.075
68. Qu L, Liu Y, Baek J-B, Dai L (2010) Nitrogen-doped graphene as efficient metal-free electrocatalyst for oxygen reduction in fuel cells. ACS Nano 4:1321–1326. doi:10.1021/nn901850u
69. Hu X, Xia D, Zhang L, Zhang J (2013) High crystallinity binuclear iron phthalocyanine catalyst with enhanced performance for oxygen reduction reaction. J Power Sources 231:91–96. doi:10.1016/j.jpowsour.2012.12.018
70. Chen Y, Zhu H, Rasmussen M, Scherson D (2010) Rational design of electrocatalytic interfaces: the multielectron reduction of nitrate in aqueous electrolytes. J Phys Chem Lett 1:1907–1911. doi:10.1021/jz1005253
71. Ghach W, Etienne M, Billard P, Jorand FPA, Walcarius A (2013) Electrochemically assisted bacteria encapsulation in thin hybrid sol-gel films. J Mater Chem B 1:1052–1059. doi:10.1039/C2TB00421F

Index

A

Acetaldehyde, 61, 385, 438–439, 443, 457–459, 461, 462, 466, 480, 483, 494
Acetylene, 294
Activation polarization, 359
Adatoms, 34–36, 50, 55, 71, 192–194, 204, 210, 262
Adsorbate, 103, 145, 170, 282, 283, 286, 288, 302, 303, 479
Adsorption, 2, 17, 21, 33, 36, 39, 40, 49, 66, 73, 74, 169, 170, 178, 189, 195, 203, 207, 310, 312, 314, 318, 324, 325, 333, 334, 338, 345–348, 352
Ag seed, 98, 107, 108, 120, 122, 123, 130, 148
Alcohol, 54, 195–202, 314, 318, 384–385, 422, 440–442, 466, 477–506, 518, 520, 533, 536, 539
Alkaline electrocatalysis, 32, 56, 73, 74, 171, 197, 201, 206, 283, 296, 314, 334, 478, 479, 482, 497, 500, 502, 503
Alkaline fuel cell (AFC), 293, 409, 410, 484, 492, 518
Alloy nanoparticles, 36, 51, 52, 55, 57, 65, 73–77, 96, 98, 101, 108, 114, 178, 201, 206, 259, 262, 283, 287, 293
Altervalent, 314, 315, 317, 319, 324, 341, 355, 356, 358, 360
Amphiphilic, 379
Anatase, 315, 316, 332, 333, 339, 344, 348, 350, 352–357, 360, 388
Anion exchange membrane (AEM), 478, 479, 487, 491, 502, 504, 506, 518
Anisotropic growth, 120, 136
Anode, 93, 110, 147, 152, 169, 171, 193, 196, 198, 203, 206, 214, 253, 367, 368, 372, 373, 384–385, 410, 423–428, 436–439, 461, 465, 466, 479, 484, 486, 488–492, 494, 500, 502–506, 518, 520, 533, 551–557, 560–562, 564, 565, 567, 570
Antibonding, 283, 302, 303, 328–330
Autoclave, 180, 297

B

Bacteria, 412, 551–553, 555, 557, 562, 570
Bending mode, 238
Bifunctional effect, 49–51, 101
Bimetallic nanocrystals, 51, 60, 169–214
Bimetallic tripods, 53
Binary alloys, 491
Binding energy, 69, 73, 196, 282, 286–291, 293, 294, 303, 332, 348, 353, 537
Binding sites, 369, 392, 393, 526, 528, 530
Bio-alcohols, 478, 506
Biocompatibility, 552, 553, 563, 564
Biofuel cell, 502, 504
Biomass, 195, 200, 436, 477–479, 505
Bio-refinery, 478, 506
Brust-Schiffrin reaction, 125
Butler-Volmer equation, 4, 6, 325

C

Cage-bell structure, 94, 104–116, 122, 148, 149
Capping agent, 32, 33, 37, 39–42, 44–47, 52, 54, 55, 58, 61, 64, 68–70, 75, 77, 126, 172, 176, 189
Carbon black, 13, 15, 21, 64, 72, 171, 173, 205, 297, 441, 497, 526, 527, 531, 533, 556, 563

K.I. Ozoemena, S. Chen (eds.), *Nanomaterials for Fuel Cell Catalysis*, Nanostructure Science and Technology, DOI 10.1007/978-3-319-29930-3

Carbon nanotubes (CNTs), 171, 173, 200, 206, 208, 296, 440, 526, 556, 563, 568
Catalyst, 1, 42, 93, 169, 243, 253, 281–303, 314, 367, 409, 436, 478, 518, 551
Catalyst loading, 255, 411, 464, 527, 558
Catalytic activity, 2, 42, 96, 169, 254, 281–303, 313, 382, 421, 437, 481, 526, 567
Cathode, 169, 171, 182, 195, 206, 207, 214
Cavitation, 414–419, 426, 429
C-C bond cleavage, 64, 439, 466, 478, 481, 489
Cellular electron acceptors, 557
Ceria (CeO_2), 313, 317, 319, 330, 334, 336, 339, 354, 360, 425, 428, 429, 482, 487, 506, 533
Cetyltrimethylammonium bromide (CTAB), 37, 42, 47, 59, 64, 117, 173, 175, 423
Characterization, 70, 94, 99, 107, 135, 137, 149, 154, 175, 178, 189, 339–342, 348
Charge separation, 385
Chemisorption, 55, 58, 145, 243, 286, 340, 347, 369, 397, 438, 458, 479
Chromatographic, 480
Citric acid (CA), 182, 186, 188, 441
CO adsorption, 17, 21, 37, 228, 233
Co-chemical reduction, 171–181, 213
Collection efficiency, 22, 207, 452, 453, 455
Composition, 51–53, 57, 58, 65, 75–77, 96, 103, 104, 114, 147, 171–173, 176, 178, 189, 192, 196, 197, 204, 205, 207–209, 214, 359
Concave, 45–48, 52, 56, 58–60, 62–64, 66–68, 180, 182, 183, 212, 495
Conducting polymer, 173, 200, 493, 531–533
Conjugation, 532
Convection, 2, 4, 10, 21
Coordinate solvent, 378
Core-sandwich-shell, 53
Corrosion, 170, 171, 207, 284, 368, 419, 490, 500, 521, 523, 524, 531, 533, 534, 536–538
CO tolerance, 54, 60, 195, 203, 241, 314, 347, 355, 373
Crystallographic orientation, 192, 203
Crystal Orbital Hamiltonian Population (COHP) formalism, 243
Cubic, 32, 34, 35, 37–40, 43, 45–49, 52–55, 57, 59, 61, 63, 66, 67, 70, 72, 73, 180, 182, 184–186, 189, 193, 203
Cubooctahedral, 45, 72, 255
Cytochrome, 557, 561, 562, 564

D

d-band center, 48, 49, 69, 74, 101, 133, 134, 147, 170, 207, 243, 283, 286, 293, 295, 303, 490
Dealloying, 264–267, 269, 272, 287
Decontamination, 33, 37, 39–42
Defect, 31, 107, 256, 262, 263, 266, 271, 297–300, 395, 526
Dehydrogenation, 197, 203, 396, 397, 399, 438, 439, 482, 483, 491, 503, 504
Dendritic, 94, 101, 103, 116–128, 130, 171, 189–192, 196, 211, 421, 422, 427
Density functional theory (DFT), 65, 70, 71, 101, 193, 230, 287, 290, 297, 330, 445, 491, 526
Desorption, 17, 33, 34, 39, 170, 203, 255, 284, 294, 310, 312, 318, 320, 321, 334, 345, 346, 348, 350, 352, 356, 395–397
Diffusion, 2, 3, 6, 8, 9, 12–16, 18, 21, 27, 67, 68, 174, 203, 207, 323, 340, 342
Diffusion-limited current, 447–449, 455
Direct formic acid fuel cells (DFAFCs), 44, 169, 202
Direct methanol fuel cell (DMFC), 105, 142, 145, 146, 169, 195, 202
Disproportionation, 310, 312, 314, 334, 335, 359
Dissociative adsorption, 201, 230, 232, 236, 240, 315–317, 320, 322, 324, 338, 358, 438, 480
Dissolution, 107, 117, 119, 140, 170, 171, 181, 182, 207, 256, 265, 287, 419, 429, 446, 462, 500, 523, 526
Doping, 388, 393
Dual path mechanism, 44
Dual-purpose supports, 538–539
Durability test, 77, 212, 213, 538

E

Electrocatalysis, 15–16, 31–77, 116, 169–214, 309–361
Electrochemical surface area (ESA), 48, 191, 207, 257, 284
Electrochemistry, 1–27, 54, 286, 334, 347, 355, 416, 418–420, 424
Electrode, 2, 32, 145, 172, 226, 258, 285, 309–361, 367, 410, 438, 480, 518, 551
Electrodeposition, 200, 419, 427, 428, 492, 497
Electrolysis, 9, 409, 410, 418, 461, 493, 518
Electron density, 145, 186, 294, 297, 299
Electronic conductivity, 179, 338, 536

Electron transfer, 1–4, 6, 8, 9, 57, 144, 171, 191, 201, 206, 295, 314, 338, 419, 449, 450, 455, 503, 504, 518, 553, 558–565, 567, 570
Electrooxidation, 32, 103, 179, 384–385, 437, 477–506
Electrostatic, 114–116, 443, 562
Elemental mapping, 95, 150, 176, 177, 180, 181
Energy density, 200, 409, 418, 436, 477, 481
Epitaxial growth, 143, 145, 146, 172, 184–189
Etchant, 107
E-TEK Pt catalyst, 52, 111, 532
Ethanol, 39, 93–154, 200, 225, 384, 435–466, 477, 517
Ethanol oxidation, 61–65, 201
Ethylene glycol, 39, 58, 173, 178, 227, 229, 421, 440–443, 477, 478, 484–486, 488, 491, 493, 503, 504, 520, 527
Evaluation, 1–27, 31, 33, 154, 332
Exoelectrogenesis, 561, 562
Extended Hückel molecular orbital (MO) theory, 243
Extended X-ray absorption fine structure (EXAFS), 174

F
Face-centered cubic (fcc), 31, 45, 107, 132, 143, 170
Fatty acid, 124
FeCo, 487, 491, 495, 502, 504, 527, 529
FeCu, 487, 493
Fermi level, 74, 134, 145, 170, 242–244, 283, 286, 330–331, 536
First-principles study, 291
Formaldehyde, 57, 68, 178, 179, 311, 318, 319, 321
Formic acid, 44–54, 63, 77, 172–174, 183, 193, 202–206, 321
Frank-vander Merwe mode, 184
Functionality, 371, 378, 394, 397–399

G
Galvanic replacement, 50, 58, 105, 171, 172, 181–184, 192–194, 210, 212, 295
Geometric and electronic effects, 101
Glycerol, 57, 173, 477, 478, 484–486, 489, 491, 493, 494, 503, 520, 527, 533
Grain size, 422, 428
Gram-negative bacteria, 560, 562
Gram-positive bacteria, 560, 562
Graphene, 13, 57, 60, 64, 171, 174, 179, 180, 191, 200, 201
Graphene oxide (GO), 15, 55, 57, 179, 180, 491, 533
Graphene quantum dots (GQDs), 294, 296–299

H
H_2PdCl_4, 56, 64, 174
H_2PtCl_4
Hammet constant, 294, 295
Heterodimer, 129, 130, 132, 134, 150, 151, 269
High-angle annular dark-field scanning TEM (HAADF-STEM), 95, 98, 142, 176, 177, 179, 181, 185, 190, 193, 194, 271, 341–343
High index facets, 45, 47, 50, 52, 59, 62–67, 180, 494–498
Hydrodynamic method, 21
Hydrogenation, 272, 333, 477, 567
Hydrogen bond, 126, 373, 374, 376, 378
Hydrogen evolution reaction (HER), 20–21, 225, 325, 326, 329, 330, 334, 358
Hydrogen peroxide, 12, 22, 25, 206, 444, 445, 466, 566, 567
Hydrogen production and storage, 368, 369, 397, 398
Hydrogen storage, 368, 390–397
Hydrophilicity, 375, 523, 526
Hydrophobicity, 500, 523
Hydrothermal, 57, 58, 68, 178–180, 297, 300
Hyper-d-metal, 313, 339
Hypo-d-oxide, 313, 314, 316–319, 321, 324, 327, 329, 330, 335, 336, 338–343, 347–350, 353–355, 357–360

I
Infrared (IR), 458, 483
In Situ Infrared Spectroscopy, 457–460
Interbonding, 313, 330–334, 342, 359
Interface, 2, 101, 116, 117, 132, 134, 144, 145, 176, 186, 238, 243, 283, 332, 342, 354, 425, 561
Interfacial engineering, 281–303
Interfacial vacancy defects, 96
Iron porphyrin, 383
Isosteric heat of adsorption, 394

J
J-coupling, 230

K

K_2PdCl_4, 172, 184, 186
K_2PtCl_4, 32, 58, 117, 122, 126, 172, 175, 178, 179, 188, 189, 191, 192
Kinetics, 4–9, 19, 27, 170, 178, 183, 192, 193, 207, 325, 346
Korringa constant, 231
Koutecky-Levich equation, 11

L

Latent storage, 309–361
Lattice mismatch, 48, 98, 101, 145, 269, 490
LiBH4, 395, 396
Ligand-mediated activity, 293–299

M

Macroporous, 522
Magneli phases, 315, 324, 335–337, 358
Marcus-Hush theory, 5
Mass activity, 51, 53, 68, 71, 73–77, 100, 192–194, 196, 201, 202, 208, 209, 211, 212
Mass transport, 2–4, 6, 8–10, 13, 16, 18, 27, 196, 272, 381, 383, 416, 418–420, 429, 451, 452, 500, 502, 553, 558, 564
Mechanism, 23, 50, 66, 70, 73, 96, 126, 152, 170, 172, 190, 195, 197, 201, 203, 214, 315, 323, 330, 347, 354
Mediator, 502, 553, 561–563, 565
Mesoporous, 60, 171, 196, 258, 369, 372, 398, 520, 522, 536
Metabolism, 557, 562
Metal-adsorbate bonding, 230, 232
Metal carbides, 531, 534–536
Metal-ligand interfacial bonding, 294
Metal nitride, 101, 536–538
Metal-organic framework (MOF), 367–399
Metal oxide, 105, 171, 284, 299–303, 317, 388, 425, 426, 441, 489, 499–502, 531, 533–534, 556
Methanol, 44, 93–154, 169, 225, 293, 377, 409, 436, 478, 517, 555
MgH_2, 395
Microbial fuel cell (MFC), 410, 551–571
Microbial secretion, 561
Microelectrode, 10, 18
Microemulsion, 61, 105, 142, 174, 440
Microkinetic model, 288
Microorganisms, 551
Microporous, 369, 372, 383, 384, 398, 522
Microwave heating synthesis, 196
Migration, 2, 3, 115, 130, 132, 134, 256, 314, 315, 330, 367, 373, 395
Mixed valence network, 315
Monolayer, 13, 51, 72, 101, 145, 171, 172, 176, 193–195, 209, 210, 213, 230, 269, 293, 316, 322, 323, 355, 359, 391
Monomer, 238
Multimetallic, 94, 95, 104, 270
Multiply twinned, 57, 58, 96, 98, 99, 107, 120, 122, 124

N

Nafion, 152, 233, 323, 346, 372–374, 397, 464, 532, 533, 558
Nanoaggregate, 421
Nanoclusters, 192, 338, 340, 342, 396, 526
Nanocomplexes, 426
Nanoconfinement, 395–397
Nanocrystal, 45, 117, 169–214, 257, 332, 397, 422, 495
Nanocubes (NCs), 40–42, 46–48, 51–54, 57–60, 63, 64, 66–72, 74, 75, 178–186, 198, 199, 212
Nanoelectrochemistry, 13
Nanofiber, 174, 175, 199, 200, 526
Nanoflowers, 56, 60, 62, 64
Nanomaterials, 1–27, 70, 73, 74, 77, 93–154, 176, 196, 200, 207
Nanoparticle, 12–15, 31–77, 94, 171, 225–244, 281–303, 323, 387, 411, 440, 479, 520, 555
Nanoporous, 269, 272, 273, 395, 490
Nanorod, 48, 49, 71, 135, 136, 138–140, 174, 182, 196, 212, 300
Nanowire, 102, 173, 181, 182, 186, 190, 193, 194, 196, 210–212
Ni/NiO, 489
Non platinum catalysts, 381
Nuclear magnetic resonance (NMR), 225–244, 505
Nuclear spin-lattice (T_1) and spin-spin (T_2) relaxation, 230

O

Octahedron, 31, 54, 117, 185, 198, 261, 335, 356
Oleic acid, 40, 60, 75, 95, 124, 140
Oleylamine (OAm), 40, 54, 59, 60, 75, 95, 96, 98, 103, 107, 121, 122, 124–130, 136, 140, 176, 190

Onset potential, 47, 55, 61, 65, 193, 196–198, 200, 202, 208, 211, 212, 235, 285, 294, 298, 383, 384, 438, 439, 466, 487, 493, 498
Organic linkers, 369, 371, 372, 390, 392, 393, 397, 398
Organic pollutant, 555
OrganoMetallic Fuel Cell (OMFC), 479, 486, 503–505
Overlayer, 51, 145, 490
Overpotential, 8, 13, 15, 16, 21, 44, 206, 253, 322, 323, 424, 426, 427, 439, 446, 447, 449, 453, 479, 489, 490, 493, 503, 553, 566, 567
Oxidation potential, 117, 196, 199, 203, 258, 264, 270, 385, 438, 567
Oxidative etching, 49
Oxide, 15, 34, 99, 171, 228, 284, 309–361, 371, 423–429, 436, 479, 518, 555
Oxidorectase, 560
Oxophilic, 54, 93, 114, 233, 260, 330, 438, 466, 489
Oxygen evolution reaction (OER), 310, 314, 318, 321, 324, 330, 335, 355, 358–360

P

Particle size, 16, 34, 53, 57, 70, 72–74, 76, 93, 99, 101, 108, 114, 119, 122, 191, 209, 227, 253–257, 259–264, 267, 270, 339, 340, 426, 427, 439, 440, 442, 443, 495, 498, 527, 536
Pathway, 15, 22, 44, 66, 133, 184, 198, 201, 203, 206, 213
Pd-Fe, 293
Perovskite oxide (ABO3), 300, 303
Peroxide, 70, 284, 286, 445, 446, 453, 455, 566
Photocatalyst, 368, 385–388
Pluronic, 101, 191, 192
Poisoning intermediate, 44, 195, 487, 499
Poly(amidoamine) dendrimer (G6-OH), 172, 208
Polyaniline (PANI), 174, 175, 200, 384, 493, 532
Polybenzimidazole (PBI), 372, 530
Polycrystalline, 45, 66, 198, 272, 310, 311, 337, 338, 348, 493, 499
Polyhedral, 31, 52, 63, 176, 196, 272, 371, 377, 392, 495
Polymer electrolyte membrane (PEM), 147, 281, 367, 368, 372–380, 397, 398
Polyol, 57, 64, 101, 173, 181, 491
Polyvalent, 315
Poly(vinyl pyrrolidone) (PVP), 39–42, 54, 55, 57, 58, 60, 61, 63, 178–180, 182, 186, 189, 192, 421, 422, 426, 427
Pore volume, 390–392, 399
Porphyrin, 383, 386, 567, 568
Potential cycling, 40, 49, 53, 66, 117, 209, 233, 255, 265, 266, 269, 270
Potential of zero charge, 237
Potentiodynamic, 309, 312, 314, 315, 318–321, 323, 333, 336–338, 340, 355, 356, 359, 360
Power density, 169, 372, 383, 384, 466, 490, 491, 493, 504, 520, 553, 555, 557, 558, 570
Precursor, 15, 21, 32, 57, 60, 171–174, 176, 182, 183, 186, 187, 189, 192
Primary oxide, 309–361, 482
Proton conduction, 372–376, 378–380, 397
Proton exchange membrane fuel cell (PEMFC), 169, 410, 436, 486, 518
Pt/C, 21, 23, 45, 46, 51, 52, 55, 57–60, 63–65, 68, 70, 72, 74–76, 99, 100, 117, 124, 144, 179, 182, 191, 198, 199, 201, 206, 208–213, 228, 253–256, 259, 260, 269–272, 285, 294, 312, 319, 321, 323, 328, 333, 335, 336, 338, 340, 348, 355, 356, 358, 447, 448, 450, 451, 453, 455, 464, 466, 487, 489, 491, 497, 537, 538
Pt_3Co, 60, 260, 265, 286
PtCu, 51, 59, 264, 287, 288
Pt_3Ni, 74, 75, 77, 261, 272, 274, 284–286
Pt–O binding, 255, 258
Pt-skin, 74, 77, 261, 267–269, 285, 286
PtSn, 438–442, 459, 460
Pulse voltammetry, 45, 46

Q

Quantum yield, 387

R

Reaction kinetics, 44, 117, 129, 173, 184, 187, 197, 206, 373, 395, 447
Reaction polarization, 310, 312–314, 318, 323, 334, 335, 359, 360
Reflux, 112, 440, 527
Renewable energy, 169, 368, 502, 517, 518
Rotating ring-disk electrode (RRDE), 10–12, 22, 24, 207, 294, 382, 444, 450–455
Ruthenium oxides, 115, 479
Rutile, 315, 335, 356

S

Sabatier principle, 286
Seed-mediated growth, 94, 117–120, 124, 130, 134, 135, 172, 184–192, 195, 203, 213
Segregation, 96, 192, 261, 267, 269, 271, 295, 442
SEIRAS, 226, 233, 235, 236, 242, 244
Selectivity, 77, 105, 111, 152, 178, 182, 272, 446, 457, 478, 479, 489, 492, 494, 505, 506, 533
Self-assembly, 135, 176, 369
Self-poisoning, 203
Semiconductor, 130, 134–140, 142, 144, 145, 148
Shape controlled nanoparticles, 59, 77
Single crystal, 17, 31–34, 36, 40, 42, 47, 49, 50, 54, 59, 61, 62, 66, 68, 70, 71, 73, 74, 186, 203
Sodium borohydride ($NaBH_4$), 42, 59, 60, 135, 172, 173, 227, 388, 389, 398, 440, 500, 501
Solid oxide fuel cells (SOFC), 195, 410, 423–429, 518
Solvothermal, 44, 47, 52, 75, 300, 388, 491, 495, 538
Sonochemistry, 414–416
Sonoelectrochemistry, 418–420, 428
Specific activity, 46, 64, 68–72, 74–76, 100, 132, 199, 201, 202, 207, 209, 211, 254–256, 260, 269, 272, 294, 295, 298
Spectroelectrochemistry, 249
Spherical, 12, 18, 32, 34, 35, 45, 47, 48, 52, 54, 60, 61, 64, 67, 71, 72, 122, 124, 142, 176, 340, 421, 422, 427
Spillover, 309–361, 395, 482
Square-wave potential, 63, 64, 496, 497
Stability, 34, 46, 47, 53–58, 60, 63–65, 69, 70, 77, 95, 101, 104, 107, 116, 133, 154, 171, 178, 179, 196–198, 200, 201, 204, 206, 207, 209, 211, 314, 324, 325, 328, 336, 338, 359, 360
Stellated heterogeneous nanostructure, 120
Stoichiometry, 300
Stripping, 36, 37, 53, 72, 76, 199, 233, 235, 236, 356
Strong metal-support interactions (SMSI), 313, 332, 487, 491
Structure-property relationship, 254, 261
Sulfonated polyether-ether ketones (SPEEK), 372
Sulfonation, 527
Surface cleaning, 37–42, 49, 77, 259, 263, 412, 416, 420
Surface disordering, 42–43
Surface-enhanced Raman scattering spectroscopy (SERS), 225
Surface functionalization, 284, 293, 294, 300, 526, 527
Surface geometry, 254
Surface site, 32–45, 49, 71, 203, 234, 240, 254, 258, 481, 482
Surface/volume ratio, 174, 178, 297, 447, 558
Surfactant, 21, 54–56, 59, 63, 65, 75, 94, 101, 112, 140, 192, 259, 263, 421–423, 426–429, 440, 441, 443
Synergy, 52, 96

T

T_1 value, 230
Tafel plot, 11, 100, 299, 319, 326, 335, 336, 450, 455
Tensile effect, 98, 101, 133
Tetrahedral-octahedral, 37, 43, 45
Tetrahexahedral (THH), 45, 50, 52, 53, 55, 56, 62–65
Thermal activation, 383
Thermal annealing, 267, 268, 348
Thermal decomposition, 126, 415, 441
Thermogravimetric analyses (TGA), 39, 55, 300
TiO_2, 21, 134, 313, 317, 319, 321, 322, 324, 330, 332–336, 338–342, 344, 345, 347–350, 352, 353, 355, 356, 358, 360
Tolerance, 54, 57, 58, 60, 64, 105, 152, 196–198, 209, 439, 532, 537, 539
Transmission electron microscopy (TEM), 95, 185, 525
Turnover number, 386

U

Ultrahigh vacuum (UHV), 262, 267, 348, 349, 352
Ultrasonication, 426–429
Ultrasound, 411–414, 416, 418–420, 422, 426, 428
Under potentially deposited (UPD), 34, 36, 42, 47, 48, 50, 70, 72, 76, 193, 255, 269, 270, 318, 320, 340, 356
UV/ozone (UVO), 42, 43, 69
UV photo-oxidation (UVPO), 39, 55

V

Volcano curve, 193, 286, 288, 325, 328, 329
Volmer-Weber mode, 124, 184
Voltammetry, 2, 6–15, 20, 21, 61, 62, 70, 99, 144, 208, 211, 285, 294, 295, 323, 355–357, 449, 458, 493, 498
Vulcan XC-72, 99, 128, 173, 335, 440, 441, 464, 482, 503, 525, 532, 533

W

WO_x, 340, 439
Wormlike micelle networks, 173

X

X-ray absorption near edge spectroscopy (XANES), 174, 439

Z

Zeolitic imidazolate framework (ZIF), 383, 384, 388

The manufacturer's authorised representative in the EU is Springer Nature Customer Service Centre GmbH, Europaplatz 3, 69115 Heidelberg, Germany. If you have any concerns regarding our products, please contact ProductSafety@springernature.com

Printed and bound by CPI Group (UK) Ltd, Croydon, CR0 4YY
06/10/2026
02228868-0005